STUDY GUIDE
FOR
CAMPBELL BIOLOGY

STUDY GUIDE

FOR

CAMPBELL BIOLOGY

LISA A. URRY • MICHAEL L. CAIN
STEVEN A. WASSERMAN • PETER V. MINORSKY
JANE B. REECE

MARTHA R. TAYLOR
Ithaca, New York

MICHAEL POLLOCK
Calgary, Alberta, Canada

ELEVENTH EDITION

 Pearson

330 Hudson Street, NY NY 10013

Courseware Portfolio Management, Director: Beth Wilbur
Courseware Portfolio Management Specialist: Josh Frost
Courseware Director, Content Development: Ginnie Simione Jutson
Director, Content Production and Digital Studio: Erin Gregg
Managing Producer, Science: Michael Early
Content Producer, Science: Lori Newman
Production Management and Composition: Progressive Publishing Services
Cover Design: Seventeenth Street Studios
Rights & Permissions Manager: Ben Ferrini
VP Product Marketing: Christy Lesko
Executive Product Marketing Manager: Christa Palaez
Field Marketing Manager: Kelly Galli
Manufacturing Buyer: Stacey Weinberger

Cover Photo Credit: Radius Images/Getty Images

2   17

ISBN-10:   0-134-44377-2
ISBN-13: 978-0-134-44377-5

Contents

Another name for the *Student Study Guide* for *Campbell Biology*, Eleventh Edition, could be "A Student Structuring Guide." This guide is designed to help you structure and organize your developing knowledge of biology and create your own personal understanding of the topics covered in the text. Take a few minutes to learn about some strategies of successful students, as well as some strategies for using this study guide.

Strategies of Successful Students

What does it take to be a successful student? How can you learn more efficiently and earn better grades? And how can you enjoy your biology course more?

Every successful student tackles learning in his or her own unique way. But there are commonalities that can be shared and results from educational research that can inform you as you develop your own approach. Here is a short list of suggestions that may help.

Make it interesting. Biology is fascinating and so much easier to learn when you love it. Try to bring a sense of wonder to your classes and your studying. Appreciate how much has been discovered and how much we have yet to learn about the living world.

Focus in lecture. What does it mean to focus? Obviously sleeping and texting are out, but so is taking down every word your professor says without thinking. If a lecture outline is provided, take notes on it. (There is evidence that many students learn better if they take notes by hand.) Try taking notes on one page of a notebook and then on the facing page rework, elaborate, or at least make legible what you wrote down in class. Doing this shortly after class helps you review the material while it is fresh and develop an organized set of notes to use in studying for exams.

Make reading active. Research has shown that simply rereading material does not improve test scores, but interacting with the reading assignment does. How do you interact with the printed word? Ask yourself questions and look for answers. Reword every concept heading as a question before you start reading. Make questions out of subheadings. If you are trying to answer a question, you are looking for meaning. Your brain is not just decoding words, it is working toward understanding those words.

Use your resources. What does your course provide—review sessions, office hours, tutors, study groups? Science is a social endeavor; interact with people as well as the material. Use the resources provided by your course and by your textbook: web sites, practice exams, tutorials. When possible, prepare for exams by practicing with questions from previous exams. Each professor has his or her own approach to writing questions. One way to improve your grade is to study how your professor asks questions.

Learn from your mistakes. Carefully review your exams—the questions you missed and even those you got right. If possible, go over the exam with an instructor. Review both *what* you missed and *why* you missed it so you can learn *how* to choose the correct answer on the next exam. Work to improve your test-taking skills. (See the section on tips for taking tests.)

Write your own questions. Making up your own questions forces you to do the same thing that your instructor does: determine what is most important and identify areas of common misunderstanding. Often your questions will closely resemble those that your instructor uses. If you have a study buddy, you can exchange questions and practice answering them. If your buddy's questions surprise you, discuss why he or she thought the material was important enough to be on an exam.

Try something new. If what you are doing isn't working, don't just keep trying to do it harder. Different study mechanisms work for different people. Experiment with new strategies. Research has shown that taking tests is an excellent way to learn. So answer all the questions in this study guide and go in search of more. Work with others and ask each other questions. Try studying in different locations and for differing lengths of time. Be creative and have fun learning.

Strategies for Using This Study Guide

This study guide is not a replacement for your textbook or biology class. However, it should help support and even streamline your learning process. Some students read the text chapters before lecture and then read the study guide chapters after class to reinforce and review. Others reverse that order, skimming the study guide before lecture and then carefully dissecting the text after the professor's presentation. Use this study guide to prepare for tests. With the textbook open so you can refer to essential diagrams, read the relevant study guide chapters for a quick review.

Each study guide chapter has the following seven sections:

- The listing of the **Key Concepts** from the text keeps you focused on the major themes of the chapter.
- The **Framework** identifies the overall picture; it provides a conceptual framework into which the chapter's information fits.
- The **Chapter Review** is a condensation of the textbook chapter, including a brief summary for each concept heading. All the bold terms in the text are also shown in boldface in this section. Interspersed in this review section are **Interactive Questions** that help you stop and actively review the material just covered.
- The **Word Roots** section presents the meanings of key biological prefixes, suffixes, and word roots. Examples of terms from the chapter are then defined. Breaking a complicated term into identifiable components will help you to recognize and learn many new biological terms.
- The **Structure Your Knowledge** section directs you to organize and relate the main concepts of the chapter. It helps you to piece together the key ideas into a bigger picture.
- In the **Test Your Knowledge** section, you are provided with objective questions to test your understanding. The multiple choice questions ask you to choose the best answer. Some answers may be partly correct; almost all choices have been written to test your ability to discriminate among alternatives or to point out common misconceptions.
- Suggested answers to the Interactive Questions, the Structure Your Knowledge section, and the Test Your Knowledge questions are provided in the **Answer Section** at the end of the book.

Using Concept Maps. What are the concept maps that appear throughout this study guide? A **concept map** is a diagram that organizes and relates ideas. The *structure* of a concept map is a hierarchically organized cluster of concepts, enclosed in boxes and connected with labeled lines that explicitly state the relationships among concepts. The *function* of a concept map is to help you structure your understanding of a topic and create meaning. The *value* of a concept map is in the thinking and organizing required to create it.

Developing a concept map requires you to evaluate the relative importance of a group of concepts (Which are most inclusive and important? Which are subordinate to other concepts?), arrange the concepts in meaningful clusters, and draw and label the connections between them.

This book uses concept maps in several ways. A map of a chapter may be presented in the Framework section to show the organization of the key concepts in that chapter. An Interactive Question may provide a skeleton map, with some concepts provided and some empty boxes for you to label. This technique is intended to help you become more familiar with concept maps and to illustrate one possible approach to organizing the concepts of a particular section.

You will also be asked to develop your own concept maps on certain subsets of ideas. In these cases, the Answer Section will present a suggested map. A concept map is an individual picture of your understanding at the time you make the map. As your understanding of an area develops, your map will evolve—sometimes becoming more complex and interrelated, sometimes becoming simplified and more streamlined. Do not look to the Answer Section for the "right" concept map. After you have organized your own thoughts, look at the answer map to make sure you have included the key concepts (although you may have added more), to check that the connections you have made are reasonable, and perhaps to see another way to organize the information.

Tips for Taking Multiple Choice Tests. When instructors write multiple choice questions, they ensure that at least one of the wrong answers sounds plausible. One way to avoid being caught by the plausible wrong answer is to read the stem of the question carefully without looking at the answers. Try to come up with a correct answer to the stem. Now when you read each answer carefully you are looking for one that matches what you already know is correct. If you aren't sure about a question, try rereading the question and each choice individually. Don't do all of your thinking in your head. Write in the margins and blank spaces of your test. Draw yourself diagrams and pictures. Write down what you do know, and it may jog your memory. If you still are not sure, mark the question and come back to it later. As you work through related questions, you may find information that helps you figure out that question. And remember, there is no substitute for good preparation, proper rest and nutrition, and a positive attitude.

Biology is a fascinating, broad, and exciting subject. *Campbell Biology*, Eleventh Edition, is filled with information organized in a manner that will help you build a conceptual framework of the major themes of biology. This *Student Study Guide* is intended to help you learn and recall information and, most importantly, to encourage and guide you as you develop your own understanding of and appreciation for biology.

Martha R. Taylor
Ithaca, New York

Michael Pollock
Calgary, Alberta, Canada

Chapter 1

Evolution, the Themes of Biology, and Scientific Inquiry

Key Concepts

1.1 **The study of life reveals unifying themes**

1.2 **The Core Theme: Evolution accounts for the unity and diversity of life**

1.3 **In studying nature, scientists make observations and form and test hypotheses**

1.4 **Science benefits from a cooperative approach and diverse viewpoints**

Framework

This chapter outlines the broad scope of biology, describes themes that unify the study of life, and examines the scientific construction of biological knowledge. A course in biology is neither a vocabulary course nor a classification exercise for the diverse forms of life. Biology is a collection of facts and concepts structured within theories and organizing principles. Recognizing the common themes within biology will help you structure your knowledge of the fascinating and challenging study of life.

Chapter Review

Biology is the scientific study of life with *evolution*, the process of change that has shaped life from its origin on Earth to today's diversity, as its organizing principle. The properties and processes of life include highly ordered structure, evolutionary adaptation, response to the environment, regulation, energy processing, reproduction, and growth and development.

1.1 The study of life reveals unifying themes

Theme: New Properties Emerge at Successive Levels of Biological Organization The scale of biology extends from the biosphere to molecules.

INTERACTIVE QUESTION 1.1

Write a brief description of each of the following levels of biological organization.

a. biosphere

b. ecosystems

c. communities

d. populations

e. organisms

f. organs and organ systems

g. tissues

h. cells

i. organelles

j. molecules

Biology today combines the powerful strategy of *reductionism*, which breaks down complex systems into simpler components, with **systems biology**, which studies the interactions of the parts of a system and models the system's dynamic behavior. The structural arrangement of and interactions among components at each level of biological organization lead to **emergent properties** at the next level. The form of a structure is usually well matched to its function. Structure and function are correlated at all levels of biological organization.

The cell is an organism's basic unit of structure and function—the lowest structural level capable of performing all the activities of life. Every cell is enclosed by a membrane. Cells of bacteria and archaea are **prokaryotic cells**, which lack a membrane-bound nucleus containing DNA and other membranous organelles. The **eukaryotic cell**—with a nucleus containing DNA and numerous organelles—is typical of all other living organisms.

Theme: Life's Processes Involve the Expression and Transmission of Genetic Information   The genetic information of a cell is encoded in **DNA (deoxyribonucleic acid)**. **Genes** are the units of inheritance that transmit information from parents to offspring. Genes are located on chromosomes, long DNA molecules that replicate before cell division and provide identical copies to daughter cells.

The biological instructions for the development and functioning of organisms are encoded in the arrangement of four kinds of nucleotides (A, T, C, G) on the two strands of a DNA double helix. Many genes contain information that directs the cell's production of proteins, and almost all cellular structures and actions involve one or more proteins.

Gene expression is the process by which a gene's information is converted into a cellular product. All forms of life use essentially the same genetic code of nucleotides. Information in DNA is transcribed to form an RNA intermediate that is translated to produce a protein. Genes also code for RNAs that serve other functions, such as being part of a cell's protein-manufacturing machinery and regulating gene expression.

INTERACTIVE QUESTION   1.2

Describe the pathway from DNA nucleotides to proteins.

All the genetic instructions an organism inherits make up its **genome**. Each of the two sets of chromosomes in a human cell contains about 3 billion nucleotide pairs.

The human genome and the genomes of many other organisms have been sequenced. Using a systems approach called **genomics**, scientists analyze whole sets of genes in one or more species. **Proteomes** are whole sets of proteins encoded by a genome; **proteomics** studies proteomes and their interactions.

Three research developments contribute to genomics and proteomics: "high-throughput" technology that can analyze biological materials rapidly; **bioinformatics**, which provides the computational tools to process and analyze the resulting data; and interdisciplinary research teams with specialists from many diverse scientific fields.

Theme: Life Requires the Transfer and Transformation of Energy and Matter   Living organisms require energy. Many **producers** use light energy to drive the storage of the chemical energy in sugars, which powers the cellular activities of plants. **Consumers** eat plants and other organisms, using the chemical energy in their foods to power their movement, growth, and other activities. In each use of energy to perform work, some energy is lost to the surroundings as heat.

INTERACTIVE QUESTION   1.3

Explain how the movement of energy in an ecosystem is similar to and different from that of chemical nutrients.

Theme: From Molecules to Ecosystems, Interactions Are Important in Biological Systems   All levels of the biological hierarchy are characterized by interactions among their components. Molecular interactions within organisms are crucial to proper functioning. Protein enzymes catalyze a cell's chemical reactions, which are often organized into chemical pathways. Many biological processes are controlled through **feedback regulation**, in which the product of a process regulates that process. In *negative feedback*, the response feeds back and reduces the original stimulus. In *positive feedback*, an end product speeds up the process. Regulatory mechanisms operate at all levels of the biological hierarchy.

Interactions between organisms may be mutually beneficial or may harm one or both participants. Both organisms and the environment are affected by interactions between them. These interactions result in the cycling of chemical nutrients between organisms and the environment.

Human interactions with the environment are responsible for various aspects of **climate change**. Global warming, a particularly significant feature of climate change, has already affected many organisms.

1.2   The Core Theme: Evolution accounts for the unity and diversity of life

Evolution explains how diverse organisms of the past and the present are related through descent from common ancestors.

Classifying the Diversity of Life   Of an estimated total of 10–100 million species, only about 1.8 million species have been identified. Each species is given a unique two-part name.

The classification of life is currently being reexamined, but all of life is now grouped into three domains. The prokaryotes are divided into domains **Archaea** and **Bacteria**. All eukaryotes are placed in domain **Eukarya**. Within the Eukarya, the mostly unicellular protists are being split into groups to better reflect evolutionary relationships.

Within this diversity, living forms share a universal genetic language of DNA and similarities in cell structure.

INTERACTIVE QUESTION   1.4

What is a commonly used criterion for placing plants, fungi, and animals into separate kingdoms?

Charles Darwin and the Theory of Natural Selection   In *The Origin of Species*, published in 1859, Charles Darwin presented his case for "descent with modification," the idea that present forms have diverged from a succession of ancestral forms. Darwin developed the theory of **natural selection** as the mechanism of evolution by drawing an inference from three observations: Individuals vary in many heritable traits, the overproduction of offspring sets up a competition for survival, and species are generally matched to their environments. From this, Darwin inferred that individuals with traits best suited to the environment leave more offspring than do less-fit individuals. This natural selection, or unequal reproductive success within a population, results in the gradual accumulation of favorable adaptations to the environment.

The Tree of Life   The underlying unity seen in the structures of related species, both living and in the fossil record, reflects the inheritance of those structures from a common ancestor. The diversity of species results from natural selection acting over millions of generations as populations adapted to different environments. The tree-like diagrams of evolutionary relationships reflect the branching genealogy extending from ancestral species. Similar species share a common ancestor at a more recent branch point on the tree of life. Distantly related species share a more ancient common ancestor.

INTERACTIVE QUESTION   1.5

Describe in your own words Darwin's theory of natural selection as the mechanism of evolutionary adaptation and the origin of new species.

1.3   In studying nature, scientists make observations and form and test hypotheses

Science is an approach to understanding the natural world that involves **inquiry**, the search for explanations of natural phenomena.

Explorations and Observations   Careful observations are the basis of scientific inquiry. Observations involve our senses and tools that extend our senses; **data**, both *quantitative* and *qualitative*, are recorded observations. Numerical data are analyzed through the mathematical techniques of *statistics* to ensure that they are not the result of chance fluctuations. Using **inductive reasoning**, generalizations can often be drawn from collections of observations.

Forming and Testing Hypotheses   Observations and inductions lead to the search for natural causes and explanations. A **hypothesis** is a tentative answer to a question or an explanation of observations.

A scientific hypothesis must be *testable*. An ideal approach is to frame two or more alternative hypotheses and design **experiments** to test each candidate explanation. A hypothesis cannot be *proven* to be true or incorrect. However, a hypothesis gains credibility when it is tested in various ways.

Deductive reasoning uses "if . . . then" logic to proceed from the general to the specific—from a general hypothesis to specific predictions of results if the general premise is correct.

Science seeks natural causes for natural phenomena; it does not address questions of the supernatural.

The Flexibility of the Scientific Method   The *scientific method*, as outlined by a structured series of steps, is rarely adhered to rigidly in scientific inquiry. Although testing ideas by gathering and interpreting data is central to the scientific process, three other circles of influence include initial exploration and discovery, analysis and feedback from the scientific community, and societal issues and benefits.

A Case Study in Scientific Inquiry: Investigating Coat Coloration in Mouse Populations   Beach and mainland mice are found in two distinct habitats in Florida and differ in coloration, although they are members of the same species. H. Hoekstra and her students tested the hypothesis that coloration patterns provide camouflage that protects mice from predation. To test the prediction that mice with coloration that did not match their habitat would be preyed on more than mice that were camouflaged by their coloration, they set out models of beach and mainland mice in both habitats. After recording signs of predation, they calculated the proportion of attacked mice in each habitat. In both cases, the mice whose coloration did not match their habitat had a higher predation rate. Thus, their experiment supports the camouflage hypothesis.

Experimental Variables and Controls   An experiment is a scientific test that involves measuring the effects of manipulating a factor within a system. Both the factor and the effects are types of experimental **variables**. The mouse coloration experimental design illustrates a **controlled experiment** in which subjects are divided into an *experimental group* and a *control group*. Both groups are alike except for the one variable that the experiment is trying to test. Another characteristic of science is that observations and experimental results must be repeatable.

INTERACTIVE QUESTION   1.6

a. Identify the control and experimental groups in the mouse camouflage experiment.

b. Identify the **independent variable** (the factor manipulated by the researchers) and the **dependent variable** (the factor that was measured) in the experiment.

c. Why were the results of this study presented as the proportion of attacks on camouflaged and non-camouflaged mice in each area rather than as the total number of attacks on non-camouflaged mice?

Theories in Science   A **theory** is broader in scope than a hypothesis, generates many specific hypotheses, and is supported by a large body of evidence. Still, a theory can be modified or even rejected when results and new evidence no longer support it.

1.4   Science benefits from a cooperative approach and diverse viewpoints

Building on the Work of Others   Most scientists work in teams and share their results with a broader research community in seminars, publications, and websites. Science is distinguished by adherence to the criteria of verifiable observations and hypotheses that are testable, and scientists often attempt to confirm the observations and experimental results of other colleagues. Scientists often share data on **model organisms**, which are easy to grow in the lab and are useful for answering particular questions that may have broad applications. Biological questions can be approached from different angles and levels of biological organization.

Science, Technology, and Society   Science and technology are interdependent: The information generated by science is applied by **technology** for specific purposes or products, and technological advances are used to extend scientific knowledge. The uses of scientific knowledge and technologies are influenced by and in turn influence politics, economics, and cultural values.

The Value of Diverse Viewpoints in Science   Women and many racial and ethnic groups have been underrepresented in scientific professions. A diversity of backgrounds and viewpoints is important to the progress of science.

INTERACTIVE QUESTION   1.7

a. Compare hypotheses and theories.

b. Compare science and technology.

Word Roots

bi- = life (*biology:* the scientific study of life; *bioinformatics:* the use of computers, software, and mathematical models to process and integrate biological information from large data sets)

eu- = true; **kary-** = nucleus (*eukaryotic cell:* a type of cell with a membrane-enclosed nucleus and organelles)

pro- = before (*prokaryotic cell:* a type of cell lacking a membrane-enclosed nucleus and organelles)

Structure Your Knowledge

1. Briefly describe in your own words each of the five unifying themes of biology presented in this chapter: organization, information, energy and matter, interactions, evolution.
   a. New properties emerge at successive levels of biological organization
   b. Life's processes involve the expression and transmission of genetic information
   c. Life requires the transfer and transformation of energy and matter
   d. From ecosystems to molecules, interactions are important in biological systems
   e. The core theme: evolution accounts for the unity and diversity of life

Test Your Knowledge

MULTIPLE CHOICE: *Choose the one best answer.*

1. The overarching theme of biology is
   a. scientific inquiry using inductive and deductive reasoning.
   b. the biological hierarchy.
   c. evolution.
   d. systems biology.

2. In an experiment similar to the camouflage experiment performed by Hoekstra and colleagues, a researcher found that more total predator attacks occurred on light model mice in beach habitats than on light models placed in inland habitats. From this the researcher concluded that
   a. the camouflage hypothesis is false.
   b. the experimental design was faulty because the results were not as predicted.
   c. the data that should be compared to draw a conclusion must include a control—a comparison with the total number of attacks on both models in each habitat.
   d. more data must be collected before a conclusion can be drawn.

3. Why can a hypothesis never be "proven" to be true?
   a. One can never collect enough data to be 100% sure.
   b. There may always be alternative untested hypotheses that might account for the results.
   c. Science is limited by our senses, and experimental error is always a factor.
   d. Science "evolves"; hypotheses and even theories are always changing.

4. Which of the following statements is an example of a positive feedback loop?
   a. Uterine contractions stimulate release of chemicals that stimulate more uterine contractions, leading to the birth of a baby.
   b. Although energy makes a one-way trip through an ecosystem, chemical nutrients are recycled.
   c. When cells have sufficient energy available, the pathways that break down sugars are turned off.
   d. A rise in CO_2 in the atmosphere correlates with increasing global temperature.

5. The smallest unit that shows the emergent properties that we associate with life is
   a. a DNA molecule.
   b. a cell.
   c. an organism.
   d. a population.

6. In a pond sample, you find a unicellular organism that has numerous chloroplasts and a whiplike flagellum. In which of the following groups do you think it should be classified?
   a. plant
   b. animal
   c. one of the proposed kingdoms of protists
   d. You cannot tell unless you see if it has a nucleus.

7. A biologist studying interactions among unicellular eukaryotes in a pond could *not* be working at which level in life's hierarchy?
   a. molecular level
   b. community level
   c. ecosystem level
   d. tissue level

8. With respect to information flow, transcription is _____ but translation is _____.
   a. DNA→RNA; RNA→protein
   b. RNA→DNA; DNA→protein
   c. RNA→protein; DNA→RNA
   d. DNA→protein; RNA→DNA

9. Which of the following sequences correctly lists life's hierarchical levels from lowest to highest?
   a. organ, tissue, organ system, organism, population
   b. organism, community, population, ecosystem, biosphere
   c. molecule, organelle, cell, tissue, organ, organism
   d. tissue, cell, organ, organism, community

10. Which of the following themes of biology is most related to the goals and practices of systems biology?
    a. Evolution accounts for the unity and diversity of life.
    b. Life's processes involve the expression and transmission of genetic information.
    c. Life requires energy transfer and transformation.
    d. New properties emerge at each level in the biological hierarchy.

11. The adaptation of an organism to its environment is best explained by
    a. Darwin's model of descent with modification.
    b. the transmission of acquired characteristics from parents to offspring.
    c. the fact that useful characteristics appear in response to environmental pressures.
    d. the structure of the DNA double helix.

12. Which of the following best accounts for the diversity of life?
    a. the uniform genetic code for translating DNA into proteins
    b. natural selection
    c. molecular interactions
    d. the transfer and transformation of energy and matter

UNIT 1 · The Chemistry of Life

Chapter 2

The Chemical Context of Life

Key Concepts

2.1 Matter consists of chemical elements in pure form and in combinations called compounds

2.2 An element's properties depend on the structure of its atoms

2.3 The formation and function of molecules depend on chemical bonding between atoms

2.4 Chemical reactions make and break chemical bonds

Framework

This chapter considers the basic principles of chemistry that explain the behavior of atoms and molecules. You will learn how subatomic particles—protons, neutrons, and electrons—are organized in atoms, how atoms are connected by covalent bonds, and how ions are attracted to each other in ionic bonds. Weak chemical bonds help to create the shapes and functions of molecules.

Chapter Review

2.1 Matter consists of chemical elements in pure form and in combinations called compounds

Matter is anything that takes up space and has mass. (Although sometimes the terms are used interchangeably, *mass* is the amount of matter in an object, whereas *weight* reflects gravity's pull on that mass.)

Elements and Compounds **Elements** are substances that cannot be chemically broken down to other types of matter. A **compound** is made up of two or more elements combined in a fixed ratio. The characteristics of a compound differ from those of its constituent elements, an example of emergent properties in higher levels of organization.

The Elements of Life **Essential elements** are those needed for an organism to live and reproduce. Carbon (C), oxygen (O), hydrogen (H), and nitrogen (N) make up 96% of living matter. The seven elements listed in Interactive Question 2.1 make up most of the remaining 4%. Some elements, such as iron (Fe) and iodine (I), may be required in very minute quantities and are called **trace elements**.

INTERACTIVE QUESTION 2.1

Fill in the names beside the symbols of the following elements commonly found in living matter.

Ca	Na
P	Cl
K	Mg
S	

Case Study: *Evolution of Tolerance to Toxic Elements* Serpentine soil contains toxic elements. Some plants exhibit evolutionary adaptations that enable them to grow in such soils.

2.2 An element's properties depend on the structure of its atoms

Each element has its own type of **atom**, the smallest unit of matter retaining the properties of that element.

Subatomic Particles   Three stable *subatomic particles* are important to your understanding of atoms. Uncharged **neutrons** and positively charged **protons** are packed tightly together to form the **atomic nucleus** of an atom. Negatively charged **electrons** form a large cloud around the positively charged nucleus.

Protons and neutrons have a similar mass of about 1.7×10^{-24} gram, or close to 1 dalton each. A **dalton** is the measurement unit for atomic mass. Relative to protons and neutrons, electrons have negligible mass.

Atomic Number and Atomic Mass   Each element has a characteristic **atomic number**, or number of protons in each of its atoms. Unless otherwise indicated, an atom has a neutral electrical charge, and thus the number of protons is equal to the number of electrons. A subscript to the left of the symbol for an element indicates its atomic number; a superscript indicates its **mass number**. The mass number is equal to the number of protons and neutrons in the nucleus and approximates the mass of an atom of that element in daltons. The term **atomic mass** refers to the total mass of an atom.

INTERACTIVE QUESTION   2.2

The difference between the mass number and the atomic number of an atom is equal to the number of _____. An atom of phosphorus, $^{31}_{15}P$, contains _____ protons, _____ electrons, and _____ neutrons. The atomic mass of phosphorus is approximately _____.

Isotopes   Although the number of protons is constant, the number of neutrons can vary among the atoms of an element, creating different **isotopes** that have slightly different masses but the same chemical behavior. Some isotopes are unstable; the nuclei of **radioactive isotopes** spontaneously decay, giving off particles and energy. Radioactive isotopes are important tools in biological research and medicine. Too great an exposure to radiation from decaying isotopes poses a significant health hazard.

Each radioactive isotope has a fixed rate of decay: its **half-life**, which is the number of years it takes for 50% of the "parent" isotope to decay into its "daughter" isotope. In a process called **radiometric dating**, scientists use the ratio of different isotopes to estimate how many half-lives (in years) have passed since a fossil or rock was formed.

The Energy Levels of Electrons   **Energy** is defined as the capacity to cause change—to do work. **Potential energy** is energy stored in matter as a consequence of its position or structure. The potential energy of electrons increases as their distance from the positively charged nucleus increases. Electrons can be located in different **electron shells**, each with a characteristic energy level and distance from the nucleus.

INTERACTIVE QUESTION   2.3

To move to a shell farther from the nucleus, an electron must _____ energy; an electron _____ energy when it moves to a closer shell.

Electron Distribution and Chemical Properties   The chemical behavior of an atom is a function of the distribution of its electrons—in particular, the number of **valence electrons** in its outermost electron shell, or **valence shell**. A valence shell of eight electrons is complete, resulting in an unreactive or *inert* atom. (The first shell, however, can hold only two electrons.) Atoms with incomplete valence shells are chemically reactive. The elements in each row, or *period*, of the *periodic table of the elements* have the same number of electron shells and are arranged in order of increasing number of electrons.

INTERACTIVE QUESTION   2.4

Draw an electron distribution diagram for the following atoms. *(Note that electrons do not pair up in the second and third shells until after 4 electrons are present.)*

a. $_1H$ **c.** $_8O$

b. $_6C$ **d.** $_{11}Na$

Electron Orbitals   An **orbital** is the three-dimensional space or volume within which an electron is most likely to be found. No more than two electrons can occupy the same orbital. The first electron shell can contain two electrons in a single spherical orbital, called the 1*s* orbital. The second electron shell can hold a maximum of eight electrons in its four orbitals, which are a 2*s* spherical orbital and three dumbbell-shaped *p* orbitals located along the *x-*, *y-*, and *z*-axes.

INTERACTIVE QUESTION 2.5

Fill in the blanks in the following concept map to help you review the atomic structure of atoms.

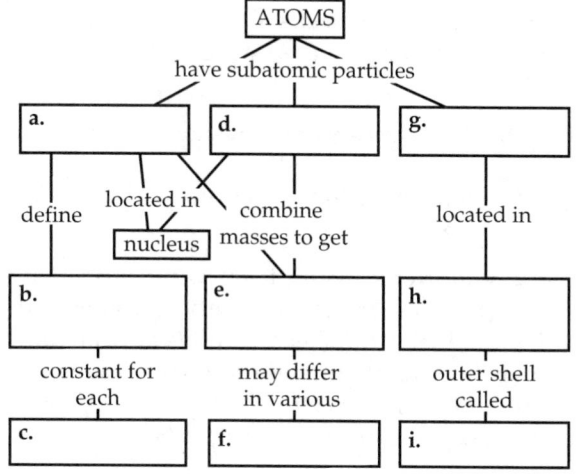

INTERACTIVE QUESTION 2.6

a. What are the valences of the four most common elements of living matter?

b. Draw the Lewis dot structures for the following molecules.

H_2 O_2 N_2 CH_4

Electronegativity is the attraction of a particular atom for shared electrons. If the atoms in a molecule have similar electronegativities, the electrons remain equally shared, and the bond is said to be a **nonpolar covalent bond**. If one element is more electronegative, it pulls the shared electrons closer to itself, creating a **polar covalent bond**. This unequal sharing of negatively charged electrons results in a "polarity" or separation of charges, with a slight negative charge ($\delta-$) associated with the more electronegative atom and a slight positive charge ($\delta+$) associated with the atom from which the electrons are pulled.

INTERACTIVE QUESTION 2.7

Explain whether the following molecules contain nonpolar or polar covalent bonds. (Hint: N and O both have high electronegativities. C and H have lower, and similar, electronegativities.)

a. nitrogen molecule N≡N   **c.** methane

$$H-\overset{\displaystyle H}{\underset{\displaystyle H}{C}}-H$$

b. ammonia

$$H\diagup\overset{\displaystyle N}{\underset{\displaystyle H}{|}}\diagdown H$$

d. formaldehyde

$$\overset{\displaystyle H}{\underset{\displaystyle H}{\diagdown}}C=O$$

2.3 The formation and function of molecules depend on chemical bonding between atoms

Atoms with incomplete valence shells can either share electrons with or completely transfer electrons to or from other atoms such that each atom is able to complete its valence shell. These interactions usually result in attractions called **chemical bonds**, which hold the atoms close together.

Covalent Bonds   The sharing of a pair of valence electrons forms a **covalent bond**. A **molecule** consists of two or more atoms held together by covalent bonds. A *molecular formula*, such as H_2, indicates only the kinds and numbers of atoms. Both an electron distribution diagram and a *Lewis dot structure* show the shared electrons in a molecule. A *structural formula*, such as H—H, uses a dash to indicate the shared electrons of a **single bond**. In an oxygen molecule (O_2), two pairs of valence electrons are shared between oxygen atoms, forming a **double bond** (O=O). The **valence**, or bonding capacity, of an atom usually equals the number of unpaired electrons in its valence shell.

Ionic Bonds   If two atoms are very different in their attraction for valence electrons, the more electronegative atom may completely strip an electron from the other atom, resulting in the formation of charged atoms called **ions**. The atom that lost the electron is a positively charged **cation**. The negatively charged atom that gained the electron is called an **anion**. An **ionic bond** may hold these ions together because of the attraction of their opposite charges.

Ionic compounds, or **salts**, often exist as three-dimensional crystalline lattice arrangements held together by electrical attractions. The number of ions present in a salt crystal is not fixed, but the atoms are present in specific ratios. Salts have strong ionic bonds when dry, but the crystal dissolves in water.

Molecules that are electrically charged are also referred to as *ions*.

INTERACTIVE QUESTION   2.8

Calcium ($_{20}$Ca) and chlorine ($_{17}$Cl) can combine to form the salt calcium chloride. Based on the number of electrons in their valence shells and their bonding capacities, what would the formula for this salt be? **a.** _____ Which atom becomes the cation? **b.** _____

Weak Chemical Interactions   Ionic bonds between ions dissociated in water and other weak bonds may form temporary interactions between molecules. Weak interactions within many large molecules help to create those molecules' three-dimensional functional shapes.

A hydrogen atom that is covalently bonded to an electronegative atom has a partial positive charge and can be attracted to a different nearby electronegative atom. This attraction is called a **hydrogen bond**.

All atoms and molecules are attracted to each other when in close contact by **van der Waals interactions**. Momentary uneven electron distributions produce changing positive and negative regions that create these weak attractions.

INTERACTIVE QUESTION   2.9

Draw the structural formula of a water molecule, showing in which direction the shared electrons are pulled. Indicate the areas with slight negative and positive charges that enable a water molecule to form hydrogen bonds with other polar molecules. Then draw a second water molecule and indicate a hydrogen bond between the two.

Molecular Shape and Function   A molecule's characteristic size and shape affect how it interacts with other molecules. When atoms form covalent bonds, their *s* and three *p* orbitals hybridize to form four teardrop-shaped orbitals in a tetrahedral arrangement. These hybrid orbitals dictate the specific shapes of different molecules.

INTERACTIVE QUESTION   2.10

Look at your sketch of a water molecule in Interactive Question 2.9. Why is it roughly V-shaped?

2.4   Chemical reactions make and break chemical bonds

Chemical reactions involve the making or breaking of chemical bonds. Matter is conserved in chemical reactions; the same number and kinds of atoms are present in both **reactants** and **products**, although the rearrangement of electrons and atoms causes the properties of these molecules to be different.

INTERACTIVE QUESTION   2.11

Fill in the missing coefficients for respiration—the conversion of glucose and oxygen to carbon dioxide and water, with the release of energy. Make sure that all atoms are conserved in the chemical reaction.

$$C_6H_{12}O_6 + \underline{}O_2 \rightarrow \underline{}CO_2 + \underline{}H_2O$$

Chemical reactions are theoretically reversible—the products of the forward reaction can become reactants in the reverse reaction. Increasing the concentrations of reactants can speed up the rate of a reaction. **Chemical equilibrium** is reached when the forward and reverse reactions proceed at the same rate, and the relative concentrations of reactants and products no longer change.

Word Roots

an- = not (*anion*: a negatively charged ion)
co- = together; **-valent** = strength (*covalent bond*: a type of strong chemical bond in which two atoms share one or more pairs of valence electrons)

electr- = electricity (*electronegativity:* the attraction of a given atom for the electrons of a covalent bond)

is- = equal (*isotope:* one of several forms of an element, each with the same number of protons but a different number of neutrons, thus differing in atomic mass)

neutr- = neither (*neutron:* a subatomic particle having no electrical charge, found in the nucleus of an atom)

pro- = before (*proton:* a subatomic particle with a single positive electrical charge, found in the nucleus of an atom)

Structure Your Knowledge

Take the time to write out or discuss your answers to the following questions. Then refer to the suggested answers at the end of the book.

1. Fill in the following chart concerning the major subatomic particles of an atom.

Particle	Charge	Mass	Location

2. Atoms can have various numbers associated with them.
   a. Define the following values and show where each of them is placed in relation to the symbol of an element such as C (use the most common isotope, C-12): atomic number, mass number.
   b. Define *valence*.
   c. Which of these three numbers is most related to the chemical behavior of an atom? Explain.

3. Explain how ionic bonds and the two types of covalent bonds can be arranged in an order that reflects the degree of electron sharing.

Test Your Knowledge

MULTIPLE CHOICE: *Choose the one best answer.*

1. Each element has its own characteristic atom in which
   a. the atomic mass is constant.
   b. the atomic number is constant.
   c. the mass number is constant.
   d. Two of the above are correct.

2. A subatomic particle goes into a restaurant and orders a sandwich. After it is finished, it asks the waiter: "How much do I owe you?" The waiter replies: "For you, sir, no charge!" What kind of subatomic particle went into the restaurant?
   a. a neutron
   b. an electron
   c. a proton

3. Which of the following is *not* a trace element in the human body?
   a. iodine
   b. zinc
   c. iron
   d. calcium

4. Congratulations! You have just discovered a new chemical element. In order to describe the properties of the new element, you determine both the number of protons and the number of neutrons in the nuclei of the atoms of the element. If you add these two numbers together, you will have
   a. the atomic number for your element.
   b. the mass number for your element.
   c. the valence for your element.
   d. the ionization potential.

5. Radioactive isotopes can be used in studies of metabolic pathways because
   a. their half-lives allow a researcher to time an experiment.
   b. they are more reactive.
   c. the cell does not recognize the extra protons in the nucleus, so isotopes are readily used in metabolism.
   d. their location or quantity can be experimentally determined because of their radioactivity.

6. Which of the following atomic numbers would describe the element that is least reactive?
   a. 8   c. 16
   b. 12   d. 18

Use this information to answer questions 7 through 12.

Six of the elements most common in living organisms are

$$^{12}_{6}C \quad ^{16}_{8}O \quad ^{1}_{1}H \quad ^{14}_{7}N \quad ^{32}_{16}S \quad ^{31}_{15}P$$

7. How many electrons does phosphorus have in its valence shell?
   a. 3   c. 7
   b. 5   d. 15

8. What is the approximate atomic mass of sulfur?
   a. 16   c. 48
   b. 32   d. 64

9. A radioactive isotope of carbon has the mass number 14. How many neutrons does this isotope have?
   a. 2   c. 8
   b. 6   d. 14

10. How many single covalent bonds is a nitrogen atom most likely to form?
    a. 1   c. 3
    b. 2   d. 4

11. Based on electron configuration, which of the following elements would have chemical behavior most like that of oxygen?
    a. S   c. C
    b. N   d. P

12. How many of the elements listed above are found next to each other (side by side) on the periodic table?
    a. one group of two
    b. two groups of two
    c. one group of two and one group of three
    d. one group of three

13. A carbon atom has 6 neutrons. How many protons are present in the nucleus of a carbon atom?
    a. 12
    b. 8
    c. 6
    d. Can't tell; more information is required.

14. Which of the following describes what happens as a chlorophyll pigment absorbs energy from sunlight?
    a. An electron moves to a higher electron shell and the electron's potential energy increases.
    b. An electron moves to a higher electron shell and its potential energy decreases.
    c. An electron drops to a lower electron shell and releases its energy as heat.
    d. An electron of sunlight is transferred to chlorophyll, producing a chlorophyll ion with higher potential energy.

15. An atom of argon has three electron shells, all of which are full. Its mass number is 40. How many neutrons does it have?
    a. 16   c. 22
    b. 20   d. 24

16. A covalent bond between two atoms is likely to be nonpolar if
    a. one of the atoms is much more electronegative than the other.
    b. the two atoms are about equally electronegative.
    c. the two atoms are of the same element.
    d. Both b and c are correct.

17. A triple covalent bond would
    a. not be possible.
    b. involve the bonding of six atoms.
    c. produce a triangularly shaped molecule.
    d. involve the sharing of six electrons.

18. A cation
    a. has gained an electron.
    b. is more likely to form in an atom with seven electrons in its valence shell.
    c. has a positive charge.
    d. Both b and c are correct.

For questions 19–21, choose from the following answers to identify the types of bonds in this diagram of a water molecule interacting with an ammonia molecule.
    a. nonpolar covalent bond
    b. polar covalent bond
    c. ionic bond
    d. hydrogen bond

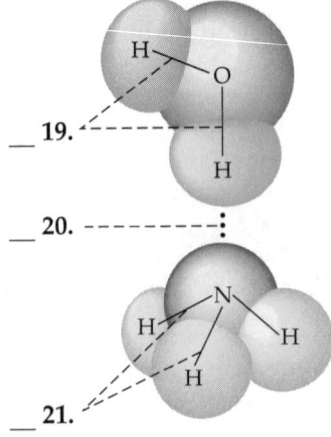

___ 19.

___ 20.

___ 21.

22. Which of the following explains how geckos can walk up walls?
    a. polar covalent bonds
    b. ionic bonds
    c. hydrogen bonds
    d. van der Waals interactions

23. What is the molecular shape of methane (CH_4)?
    a. planar or flat, with the four H extending out from the carbon
    b. tetrahedral, due to the hybridization of the *s* and three *p* orbitals of carbon
    c. circular, with the four H attached in a ring around the carbon
    d. linear, because all the bonds are nonpolar covalent bonds

24. A hydrogen bond
    a. is one in which hydrogen bonds to itself.
    b. is one in which hydrogen is covalently bonded to a different atom.
    c. is an ionic bond between two hydrogen ions (H^+).
    d. is relatively weak.

25. When an atom with one electron in the outer shell loses it, the atom becomes
    a. radioactive.
    b. a negatively charged ion.
    c. a positively charged ion.
    d. nonpolar.

26. What is the difference between a molecule and a compound?
    a. There is no difference; the terms are interchangeable.
    b. A molecule consists of two or more covalently bonded atoms; a compound contains two or more atoms held together by ionic bonds.
    c. A compound consists of two or more elements in a fixed ratio; a molecule has two or more co-valently bonded atoms of the same or different elements.
    d. Compounds always consist of molecules, but molecules are not always compounds.

27. Which of the following is true of a reaction in chemical equilibrium?
    a. The forward and reverse reactions are occur-ring at the same rate.
    b. The reactants and products are in equal concentration.
    c. The forward reaction has gone to completion.
    d. There are equal numbers of atoms on both sides of the equation.

28. What would be the probable effect of adding more product to a reaction that is in equilibrium?
    a. There would be no change because the reaction is in equilibrium.
    b. The forward reaction would stop because excess product is present.
    c. The forward reaction would increase and more product would be formed.
    d. The reverse reaction would increase and more reactants would be formed.

Chapter 3

Water and Life

Key Concepts

3.1 Polar covalent bonds in water molecules result in hydrogen bonding

3.2 Four emergent properties of water contribute to Earth's suitability for life

3.3 Acidic and basic conditions affect living organisms

Framework

Water makes up 70% to 95% of the cell content of living organisms and covers 75% of Earth's surface. Its unique properties make the planet's environment suitable for life and the internal environments of organisms suitable for the chemical and physical processes of life.

Hydrogen bonding between polar water molecules creates a cohesive liquid with a high specific heat and high heat of vaporization, both of which help to regulate environmental temperature. Ice floats and protects oceans and lakes from freezing. The polarity of water makes it a versatile solvent. The [H$^+$] in a solution is expressed as pH and determines if a solution is acidic, neutral, or basic. Buffers regulate an organism's pH.

Chapter Review

3.1 Polar covalent bonds in water molecules result in hydrogen bonding

A water molecule consists of two hydrogen atoms, each bonded to a more electronegative oxygen atom by a **polar covalent bond**. This **polar molecule** has a shape like a wide V with a partial positive charge on each hydrogen atom ($\delta+$) and a partial negative charge ($\delta-$) associated with the oxygen. Hydrogen bonds, electrical attractions between the hydrogen atom of one water molecule and the oxygen atom of a nearby water molecule, create a structural organization that leads to the emergent properties of water.

INTERACTIVE QUESTION 3.1

Draw the four water molecules that can hydrogen-bond to this water molecule. Label the bonds and the partial negative and positive charges that account for the formation of these hydrogen bonds.

3.2 Four emergent properties of water contribute to Earth's suitability for life

Cohesion of Water Molecules   Liquid water is unusually cohesive due to the constant forming and re-forming of hydrogen bonds that hold the molecules close together. This **cohesion** creates a more structurally organized liquid and helps water to be pulled upward in plants. The **adhesion** of water molecules to the walls of plant vessels also contributes to water transport. Hydrogen bonding between water molecules produces a high **surface tension** at the interface between water and air, making the surface unusually difficult to break.

Moderation of Temperature by Water **Thermal energy** is the **kinetic energy** (the energy of motion) associated with the random movement of atoms and molecules. **Temperature** measures the *average* kinetic energy of the molecules in a body of matter; thermal energy reflects the *total* kinetic energy in that matter, which relates to the volume of the body of matter. The thermal energy that transfers from a warmer to a cooler body of matter is defined as **heat**.

A **calorie (cal)** is the amount of heat it takes to raise 1 g of water 1°C. A **kilocalorie (kcal)** is 1,000 calories, the amount of heat required or released to change the temperature of 1 kg of water by 1°C. A **joule (J)** equals 0.239 cal; one calorie is 4.184 J.

Specific heat is the amount of heat absorbed or lost when 1 g of a substance changes its temperature by 1°C. Water's specific heat of 1 cal/g·°C is unusually high compared with other common substances. Heat must be absorbed to break hydrogen bonds before water molecules can move faster and the temperature can rise, and conversely, heat is released when hydrogen bonds form as water molecules slow down when the temperature of water drops. The high proportion of water in the environment and within organisms keeps temperature fluctuations within limits that permit life.

Vaporization or evaporation occurs when molecules of a liquid with sufficient kinetic energy overcome their attraction to other molecules and escape into the air as gas (vapor). The **heat of vaporization** is the quantity of heat that must be absorbed for 1 g of a liquid to be converted to a gas. Water has a high heat of vaporization (580 cal/g at 25°C) because a large amount of heat is needed to break the hydrogen bonds holding water molecules together. Water helps moderate Earth's climate as solar heat absorbed by tropical seas is dissipated during evaporation, and heat is released as moist tropical air moving poleward condenses to form rain.

As a liquid vaporizes, the surface left behind loses the kinetic energy of the escaping molecules and cools down. **Evaporative cooling** helps to protect terrestrial organisms from overheating and contributes to the stability of temperatures in lakes and ponds.

Floating of Ice on Liquid Water As water cools below 4°C, it expands because its molecules are moving too slowly to break hydrogen bonds. By 0°C, each water molecule is rigidly hydrogen-bonded to four other molecules in a crystalline lattice that spaces the molecules farther apart than is true for liquid water. Ice is thus less dense than liquid water and so it floats, insulating the water below and providing a solid substrate for some animals.

The following concept map is one way to show how the breaking and forming of hydrogen bonds are related to temperature moderation. Fill in the blanks and compare your choice of concepts to those given in the answer section. Or, even better, create your own map to help you understand how water stabilizes temperature.

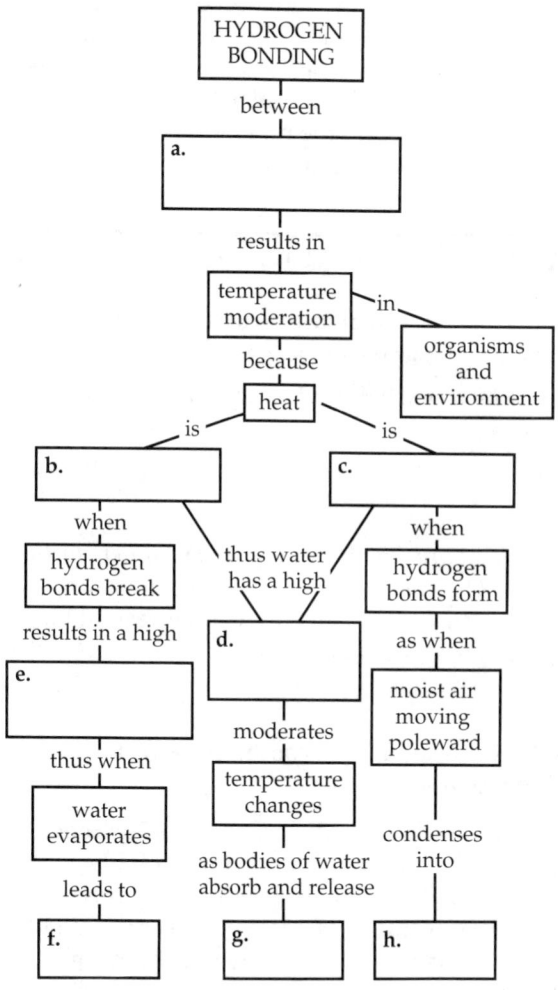

Water: The Solvent of Life A **solution** is a liquid homogeneous mixture of two or more substances; the dissolving agent is called the **solvent**, and the substance that is dissolved is the **solute**. Water is the solvent in an **aqueous solution**. The positive and negative regions of water molecules are attracted to oppositely charged ions or partially charged regions of polar molecules. Thus, solute molecules become surrounded by water molecules (a **hydration shell**) and dissolve into solution.

Ionic and polar substances are **hydrophilic**; they have an affinity for water due to electrical attractions and hydrogen bonding. Nonpolar and nonionic substances are **hydrophobic**; they will not easily mix with or dissolve in water.

Most of the chemical reactions of life take place in water. A **mole (mol)** is the amount of a substance that has a mass in grams numerically equivalent to its **molecular mass** (the sum of the mass of all atoms in the molecule). A mole of any substance has exactly the same number of molecules—6.02×10^{23}, called Avogadro's number. The **molarity** of a solution (abbreviated M) refers to the number of moles of a solute dissolved in 1 liter of solution.

INTERACTIVE QUESTION 3.3

a. How many grams of lactic acid ($C_3H_6O_3$) are in 1 liter of a 0.5 M solution of lactic acid? (^{12}C, ^{1}H, ^{16}O)

b. How many molecules of lactic acid are in the solution in *a*?

Possible Evolution of Life on Other Planets with Water The emergent properties of water support life on Earth. In the search for extraterrestrial life, *astrobiologists* look for evidence of water, as has been recently confirmed on Mars.

3.3 Acidic and basic conditions affect living organisms

A water molecule can dissociate into a **hydrogen ion**, H^+ (which binds to another water molecule to form a **hydronium ion**, H_3O^+), and a **hydroxide ion**, OH^-. (By convention 'H^+' is used to represent H_3O^+, although free hydrogen ions don't exist in aqueous solution.) In pure water at 25°C, the concentrations of H^+ and OH^- are the same; both are equal to 10^{-7} M.

Acids and Bases When acids or bases dissolve in water, the H^+ and OH^- balance shifts. An **acid** adds H^+ to a solution, whereas a **base** reduces H^+ in a solution by accepting hydrogen ions or by adding hydroxide ions (which then combine with H^+ and thus remove hydrogen ions). A strong acid or strong base dissociates completely when mixed with water. A weak acid or weak base reversibly dissociates, either releasing or binding H^+ until its equilibrium ratio is reached.

The pH Scale In an aqueous solution, the *product* of the $[H^+]$ and $[OH^-]$ is constant at 10^{-14}. Square

brackets indicate molar concentration. If $[H^+]$ is higher, then $[OH^-]$ is lower because the increased hydrogen ions combine with the hydroxide ions in solution and form water. Likewise, an increase in $[OH^-]$ causes an equivalent decrease in $[H^+]$.

The **pH** of a solution is defined as the negative log (base 10) of the $[H^+]$: pH $= -\log [H^+]$. For a neutral aqueous solution, $[H^+]$ is 10^{-7} M and the pH = 7. As the $[H^+]$ increases in an acidic solution, the pH value decreases. (This inverse relationship makes sense because the exponent becomes smaller: 10^{-4} is a larger number than 10^{-7}.) The difference between each unit of the pH scale represents a tenfold difference in the concentration of $[H^+]$ and $[OH^-]$.

INTERACTIVE QUESTION 3.4

Complete the following table to review your understanding of pH.

$[H^+]$	$[OH^-]$	pH	Acidic, Basic, or Neutral?
10^{-8}			
	10^{-7}		
		1	

Buffers Most cells have an internal pH close to 7. **Buffers** within the cell maintain a stable pH by accepting excess H^+ or donating H^+ when H^+ concentration decreases. Weak acid-base pairs that reversibly bind hydrogen ions are typical of most buffering systems.

INTERACTIVE QUESTION 3.5

The carbonic acid/bicarbonate system is an important biological buffer. Label the molecules and ions in this equation, and indicate which is the H^+ donor (acid) and which is the H^+ acceptor (base).

$$H_2CO_3 \rightleftharpoons HCO_3^- + H^+$$

In which direction will this reaction proceed

a. when the pH of a solution begins to fall?

b. when the pH rises above normal level?

Acidification: A Threat to Water Quality Fossil fuel combustion releases CO_2 to the atmosphere.

The oceans absorb about 25% of this CO_2, which lowers the pH of seawater. The resulting **ocean acidification** decreases the concentration of carbonate ions (CO_3^{2-}), an important ion needed for coral reef calcification.

INTERACTIVE QUESTION 3.6

a. Add to the formula in Interactive Question 3.5 to show how increasing $[CO_2]$ dissolving in water leads to a lower pH.

b. Use this formula to explain how a lower pH would affect the $[CO_3^{2-}]$ in the ocean.

$$HCO_3^- \rightleftharpoons CO_3^{2-} + H^+$$

c. Assuming a fairly constant $[Ca^{2+}]$ in the ocean, how would a change in $[CO_3^{2-}]$ affect the calcification rate—the production of calcium carbonate ($CaCO_3$)—by the coral in a reef ecosystem?

Word Roots

kilo- = a thousand (*kilocalorie:* a thousand calories; the amount of heat energy required to raise the temperature of 1 kg of water by 1°C)

hydr- = water; **phil-** = loving; **phob-** = fearing (*hydrophilic:* having an affinity for water; *hydrophobic:* having no affinity for water)

Structure Your Knowledge

1. Fill in the following table, which summarizes the emergent properties of water that contribute to the fitness of the environment for life.

2. To become proficient in the use of the concepts relating to pH, develop a concept map to organize your understanding of the following terms: pH, $[H^+]$, $[OH^-]$, acidic, basic, neutral, buffer, 0–14, acid-base pair. Remember to label connecting lines and add additional concepts as you need them. *A suggested concept map is given in the answer section, but remember that your concept map should represent your own understanding. The value of this exercise is in organizing these concepts for yourself.*

Test Your Knowledge

MULTIPLE CHOICE: *Choose the one best answer.*

1. Each water molecule is capable of forming
   a. three ionic bonds.
   b. four hydrogen bonds.
   c. two covalent bonds and two hydrogen bonds.
   d. four polar covalent bonds.

2. The properties of water make it an ideal and essential medium for life as we know it. These special properties are the result of
   a. the extreme polarity of water molecules.
   b. the ability of water to dissolve many biologically important molecules.
   c. the relatively high boiling point of water.
   d. the fact that water is colorless and odorless.

3. What accounts for the movement of water up the vessels of a tall tree?
   a. cohesion
   b. hydrogen bonding
   c. adhesion
   d. all of the above

Property of Water	Explanation of Property	Example of Benefit to Life
a.	Hydrogen bonds hold molecules together and adhere them to hydrophilic surface.	b.
High specific heat	c.	Temperature changes in environment and organisms are moderated.
d.	Hydrogen bonds must be broken for water to evaporate.	e.
Less dense as a solid	f.	g.
h.	i.	Most chemical reactions in life involve solutes dissolved in water.

4. Climates tend to be moderate near large bodies of water because
   a. a large amount of solar heat is absorbed because the temperature of water rises rapidly.
   b. water absorbs heat from the environment as it cools.
   c. the high specific heat of water helps to moderate air temperatures.
   d. a great deal of heat is absorbed as hydrogen bonds form and released as hydrogen bonds break.

5. You have three flasks containing 100 mL of different liquids. Each is warmed with 100 calories of heat. The temperature of the liquid in flask 1 rises 1°C; in flask 2 it rises 1.5°C; and in flask 3 it rises 2°C. Which of these liquids has the highest specific heat?
   a. the liquid in flask 1
   b. the liquid in flask 2
   c. the liquid in flask 3
   d. You cannot tell unless you know what liquid is in each flask.

6. A burn from steam at 100°C is more severe than a burn from boiling water because
   a. the steam is hotter than boiling water.
   b. steam releases a great deal of heat as it condenses on the skin.
   c. steam has a higher heat of vaporization than does water.
   d. steam stays on the skin longer than does boiling water.

7. Evaporative cooling is a result of
   a. the release of heat during the breaking of hydrogen bonds when water molecules escape.
   b. the absorption of heat as hydrogen bonds form.
   c. the reduction in the average kinetic energy of a liquid after energetic water molecules enter the gaseous state.
   d. both **a** and **c**.

8. Ice floats because
   a. air is trapped in the crystalline lattice.
   b. the formation of hydrogen bonds releases heat; warmer objects float.
   c. it insulates bodies of water so they do not freeze from the bottom up.
   d. hydrogen bonding spaces the molecules farther apart, creating a less dense structure.

9. Why is water such an excellent solvent?
   a. As a polar molecule, it can surround and dissolve ionic and polar molecules.
   b. It forms ionic bonds with ions, hydrogen bonds with polar molecules, and hydrophobic interactions with nonpolar molecules.
   c. It forms hydrogen bonds with itself.
   d. It is liquid and has a high surface tension.

10. What sort of molecules dissolve readily in water?
    a. nonpolar molecules
    b. polar molecules
    c. charged molecules
    d. both b and c

11. The molarity of a solution is equal to
    a. Avogadro's number of molecules in 1 liter of solvent.
    b. the number of moles of a solute in 1 liter of solution.
    c. the molecular mass of a solute in 1 liter of solution.
    d. the number of solute molecules in 1 liter of solvent.

12. The molecular mass of methane (CH_4) is 16 daltons. If you carefully measured out 1 mole of methane, how much would it weigh?
    a. 16 daltons
    b. 16 grams
    c. 6.02×10^{23} grams
    d. 4 grams

13. How many molecules of solute would be in 1 liter of a 0.1 M solution?
    a. 0.1
    b. 180
    c. 6×10^{22}
    d. 6×10^{23}

14. Consider two solutions labeled "A" and "B." Solution A has a pH of 6 and solution B has a pH of 9. Solution A has a hydrogen ion concentration that is
    a. 3 times higher than that of solution B.
    b. 3 times lower than that of solution B.
    c. 1,000 times higher than that of solution B.
    d. 1,000 times lower than that of solution B.

15. What does a buffer do?
    a. moderates pH changes
    b. maintains neutrality
    c. increases solubility
    d. contributes bicarbonate ions

Use the following pH values to answer questions 16–18: cola—2; orange juice—3; beer—4; coffee—5; human blood—7.4.

16. Which of these liquids has the *highest* molar concentration of H^+?

   a. cola   **c.** coffee

   b. orange juice   **d.** human blood

17. What is $[OH^-]$ for coffee?

   a. 1×10^{-5} **c.** 9

   b. 1×10^{5} **d.** 1×10^{-9}

18. How many hydronium ions (H_3O^+) are in 1 liter of beer?

   a. 1.0×10^{-4} **c.** 6.02×10^{19}

   b. 1.0×10^{-10} **d.** 6.02×10^{-19}

Carbon and the Molecular Diversity of Life

Key Concepts

4.1 **Organic chemistry is the study of carbon compounds**

4.2 **Carbon atoms can form diverse molecules by bonding to four other atoms**

4.3 **A few chemical groups are key to molecular function**

Framework

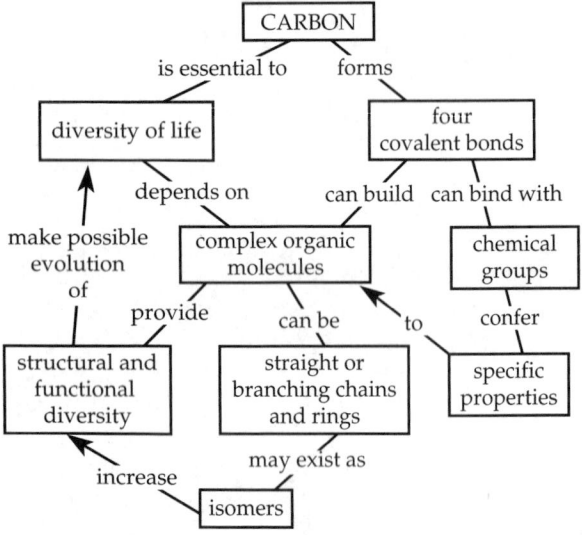

Chapter Review

4.1 Organic chemistry is the study of carbon compounds

Organic chemistry is the study of carbon-containing compounds. Early chemists could not make the complex molecules found in living organisms. Later these organic compounds were synthesized, supporting the view that the processes of life are governed by physical and chemical laws. Hydrogen, nitrogen, oxygen, phosphorus, and sulfur are also common in the major molecules of life.

Organic Molecules and the Origin of Life on Earth Stanley Miller's 1953 experiment showed that organic molecules could form under conditions believed to simulate those on early Earth.

INTERACTIVE QUESTION 4.1

How did Miller's classic experiment relate to the origin of life?

4.2 Carbon atoms can form diverse molecules by bonding to four other atoms

The Formation of Bonds with Carbon Carbon's **valence** of four is at the center of its ability to form large and complex molecules. When carbon forms four single covalent bonds, its hybrid orbitals create a tetrahedral shape. When two carbons are joined by a double bond, the other atoms bonded to the carbons are in the same plane, forming a flat molecule.

Molecular Diversity Arising from Variation in Carbon Skeletons Carbon skeletons can vary in length, branching, placement of double bonds, and the presence of ring structures. **Hydrocarbons** consist of only carbon and hydrogen. Hydrocarbon chains are hydrophobic due to their nonpolar C—H bonds, and they release energy when broken down.

Isomers are compounds with the same molecular formula but different structural arrangements and thus different properties. **Structural isomers** differ in the covalent arrangement of atoms and often in the location of double bonds. *Cis-trans* **isomers** (formerly called geometric isomers) have the same sequence of covalently bonded atoms overall but differ in structure due to the inflexibility of double bonds. A *cis isomer* has the same atoms attached to the carbons on the same side of the double bond; a *trans isomer* has these atoms on opposite sides of the double bond. **Enantiomers** are left- and right-handed versions of a molecule and can differ greatly in their biological activity because of their different shapes. An *asymmetric carbon* is one that is covalently bonded to four different kinds of atoms or groups of atoms. The atoms or groups of atoms around an asymmetric carbon can be arranged in two different, mirror-image enantiomeric forms.

INTERACTIVE QUESTION   4.2

Define structural isomers, *cis-trans* isomers, and enantiomers, and then identify the pairs of isomers among the following compounds. Which of the *cis-trans* isomers is the *cis* isomer?

4.3   A few chemical groups are key to molecular function

The Chemical Groups Most Important in the Processes of Life   The properties of organic molecules are largely determined by characteristic chemical groups attached to the carbon skeleton. Seven such functional groups are common in biological molecules. All except the sulfhydryl and methyl groups increase the solubility of organic compounds in water. The methyl group does not generally participate in chemical reactions, but it alters molecular shape and may serve as a signal on organic molecules.

The **hydroxyl group** consists of an oxygen and hydrogen (—OH). Organic molecules with hydroxyl groups are called **alcohols**, and their names usually end in *-ol*.

A **carbonyl group** consists of a carbon double-bonded to an oxygen ($\diagdown$CO). If the carbonyl group is at the end of the carbon skeleton, the compound is called an **aldehyde**. Otherwise, the compound is called a **ketone**.

A **carboxyl group** consists of a carbon double-bonded to an oxygen and also attached to a hydroxyl group (—COOH). Compounds with a carboxyl group are called **carboxylic acids** or **organic acids** because they tend to donate an H^+, becoming a carboxylate ion.

An **amino group** consists of a nitrogen atom bonded to two hydrogen atoms (—NH$_2$). Compounds with an amino group, called **amines**, can act as bases, picking up an H^+ as the amino group becomes —NH$_3^+$.

The **sulfhydryl group** consists of a sulfur atom bonded to a hydrogen (—SH). **Thiols** are compounds containing sulfhydryl groups.

A **phosphate group** is bonded to a carbon skeleton by an oxygen attached to a phosphorus atom that is bonded to three other oxygen atoms (—OPO$_3^{2-}$). This anion contributes negative charge to **organic phosphates**.

A **methyl group** consists of a carbon bonded to three H atoms (—CH$_3$). **Methylated compounds** may have their function modified due to the addition of the methyl group.

INTERACTIVE QUESTION   4.3

Practice recognizing the functional groups by circling and naming the groups you see in the following molecules.

ATP: An Important Source of Energy for Cellular Processes   **Adenosine triphosphate**, or **ATP**, consists of the organic molecule adenosine to which three phosphate groups are attached. When ATP reacts with water, the third phosphate is split off and energy is released.

The Chemical Elements of Life: A Review   Carbon, oxygen, hydrogen, nitrogen, and smaller quantities of sulfur and phosphorus, all capable of forming strong covalent bonds, are combined into the complex organic molecules of living matter.

Word Roots

carb- = coal (*carboxyl group:* a chemical group present in organic acids, consisting of a carbon atom double-bonded to an oxygen atom and also bonded to a hydroxyl group)

enanti- = opposite (*enantiomer:* one of two compounds that are mirror images of each other and that differ in shape due to the presence of an asymmetric carbon)

is- = equal (*isomer:* one of several organic compounds with the same molecular formula but different structures and therefore different properties)

sulf- = sulfur (*sulfhydryl group:* a chemical group that consists of a sulfur atom bonded to a hydrogen atom)

Structure Your Knowledge

1. Construct a concept map that illustrates your understanding of the characteristics and significance of the three types of isomers. *A suggested map is in the answer section. Comparing and discussing your map with that of a study partner would be most helpful.*

2. Fill in the following table to review the important chemical groups of organic compounds.

Chemical Group	Molecular Formula	Names and Characteristics of Compounds Containing Group
	—OH	
		Aldehyde or ketone; polar group
Carboxyl		
	—NH₂	
		Thiols; cross-links stabilize protein
Phosphate		
	—CH₃	

Test Your Knowledge

MULTIPLE CHOICE: *Choose the one best answer.*

1. Carbon's valence of four most directly results from
   a. its tetrahedral shape.
   b. the four valence electrons that can form four covalent bonds.
   c. its ability to form single, double, and triple bonds.
   d. its ability to form chains and rings of carbon atoms.

2. Protein structures may be stabilized by cross-links between
   a. hydroxyl groups.
   b. methyl groups.
   c. sulfhydryl groups.
   d. carboxyl groups.

3. Hydrocarbons are not soluble in water because
   a. they are hydrophilic.
   b. their C—H bonds are nonpolar.
   c. they do not ionize.
   d. they are lighter than water.

4. A carbonyl group can be found in a molecule that is an aldehyde or a(n)
   a. ketone.
   b. thiol.
   c. alcohol.
   d. organic phosphate.

5. Which of the following is *not* true of an asymmetric carbon atom?

   a. It is attached to four different atoms or groups.

   b. It results in right- and left-handed versions of a molecule.

   c. Its configuration is in the shape of a tetrahedron.

   d. It is found in *cis-trans* isomers.

6. Enantiomers are

   a. molecules that are mirror images of one another.

   b. organic acids with different numbers of carbons.

   c. structural isomers.

   d. molecules with a *cis* or *trans* configuration across a carbon-carbon double bond.

7. Which statement is *not* true about structural isomers?

   a. They have the same molecular formula but different chemical properties.

   b. Their atoms and bonds are arranged in different sequences.

   c. They are a result of restricted movement around a carbon double bond.

   d. Their possible numbers increase as carbon skeletons increase in size.

8. The fats stored in your body consist mostly of

   a. methyl groups.   c. carboxylic acids.

   b. alcohols.   d. hydrocarbons.

9. The orbitals of a carbon atom form a(n)

   a. pyramid.   c. icosahedron.

   b. tetrahedron.   d. cube.

10. The following ribose molecule contains how many asymmetric carbons?

   a. 1   b. 2   c. 3   d. 4

11. *Cis-trans* isomers require

   a. highly polar functional groups.

   b. carbon-carbon double bonds.

   c. branched carbon chains.

   d. ring-shaped carbon skeletons.

12. The chemical group that can cause an organic molecule to act as a base is

   a. —COOH.   c. —SH.

   b. —OH.   d. —NH₂.

13. The chemical group that confers acidic properties on organic molecules is

   a. —COOH.   c. —SH.

   b. —OH.   d. —NH₂.

MATCHING: *Match the formulas (a–f) to the terms at the right. Choices may be used more than once, and terms may have more than one correct answer.*

_____ 1. structural isomers

_____ 2. *cis-trans* isomers

_____ 3. can have enantiomers

_____ 4. carboxylic acid

_____ 5. can make cross-link in protein

_____ 6. hydrophilic

_____ 7. hydrocarbon

_____ 8. amino acid

_____ 9. organic phosphate

_____ 10. aldehyde

_____ 11. amine

_____ 12. ketone

Chapter 5

The Structure and Function of Large Biological Molecules

Key Concepts

5.1 Macromolecules are polymers, built from monomers

5.2 Carbohydrates serve as fuel and building materials

5.3 Lipids are a diverse group of hydrophobic molecules

5.4 Proteins include a diversity of structures, resulting in a wide range of functions

5.5 Nucleic acids store, transmit, and help express hereditary information

5.6 Genomics and proteomics have transformed biological inquiry and applications

Framework

A small number of monomers or subunits join to form a huge variety of large molecules with diverse structures and functions. The following table briefly summarizes the four classes of biological molecules.

Chapter Review

The most important biological molecules include lipids and the huge **macromolecules** of carbohydrates, proteins, and nucleic acids. These molecules have emergent properties that arise from their complex and unique structures.

5.1 Macromolecules are polymers, built from monomers

Polymers are chainlike molecules formed from the linking together of many similar or identical small molecules, called **monomers**. The macromolecules of carbohydrates, proteins, and nucleic acids are polymers.

The Synthesis and Breakdown of Polymers In the process of polymerization, monomers are joined by a **dehydration reaction**, in which one monomer provides a hydroxyl group (—OH) and the other contributes a hydrogen (—H) to release a water molecule. In **hydrolysis**, the bond between monomers is broken by the addition of water. The hydroxyl group of a water molecule is joined to one monomer while the hydrogen is bonded with the other. **Enzymes** catalyze both dehydration reactions and hydrolysis.

Class	Monomers	Functions
Carbohydrates	Monosaccharides	Energy source, raw materials, energy storage, structural compounds
Lipids		Energy storage (fats), membrane components (phospholipids), hormones (steroids)
Proteins	Amino acids	Enzymes, transport, movement, receptors, defense, structure, storage, hormones
Nucleic acids	Nucleotides	Heredity, various functions in gene expression

The Diversity of Polymers   Polymers are constructed from about 40 to 50 common monomers and a few rarer molecules. The seemingly endless variety of macromolecules arises from the essentially infinite number of possibilities in the sequencing of these basic building blocks.

INTERACTIVE QUESTION 5.1

Monomers are linked into polymers by _____ _____, which involve the _____ of a water molecule.

Polymers are broken down to monomers by _____, which involves the _____ of a water molecule.

5.2 Carbohydrates serve as fuel and building materials

Carbohydrates include sugars and their polymers.

Sugars   Most **monosaccharides** have the general formula of $(CH_2O)_n$. The number (n) of these units forming a sugar varies, with hexoses ($C_6H_{12}O_6$), trioses, and pentoses being most common. Sugar molecules also vary in the location of their carbonyl group and the spatial arrangement of parts around asymmetric carbons.

INTERACTIVE QUESTION 5.2

Fill in the blanks to review monosaccharides.

You can recognize a monosaccharide by its multiple (a) _____ groups and its one (b) _____ group, whose location determines whether the sugar is a(n) (c) _____ or a(n) (d) _____. In aqueous solutions, most five- and six-carbon sugars form (e) _____. The names for most sugars end in (f) _____.

Glucose is broken down to yield energy in cellular respiration. Monosaccharides also serve as the raw materials for the synthesis of other organic molecules. Two monosaccharides joined by a **glycosidic linkage** form a **disaccharide**.

INTERACTIVE QUESTION 5.3

Number the carbons in the following glucose and fructose molecules (each unlabeled corner of the ring represents a carbon. In glucose, carbon 1 is to the right of the O in the ring; in fructose, carbon 1 extends up from the ring on the left side.) Circle the atoms that will be removed by a dehydration reaction. Then draw the resulting sucrose molecule with its 1–2 glycosidic linkage.

Glucose   Fructose

Sucrose

Polysaccharides   **Polysaccharides** are storage or structural macromolecules. **Starch**, a storage molecule in plants, is a polymer made of glucose molecules joined by 1–4 glycosidic linkages that give starch a helical shape. Amylose is an unbranched polymer; amylopectin is a more complex branched form of starch with 1–6 linkages at branch points. Animals use **glycogen**, a highly branched polymer of glucose, as their energy storage molecule.

The structural polysaccharide **cellulose** is the major component of plant cell walls; it is the most abundant organic compound on Earth. It differs from starch by the configuration of the ring form of glucose (β instead of α) and the resulting geometry of the glycosidic bonds. In a plant cell wall, hydrogen bonds between hydroxyl groups hold the parallel, straight cellulose molecules together to form strong microfibrils.

Enzymes that digest the α linkages of starch are unable to hydrolyze the β linkages of cellulose. Only a few organisms (some prokaryotes, protists, and fungi) have enzymes that can digest cellulose.

Chitin is a branched structural polysaccharide formed from glucose monomers with a nitrogen-containing group. Chitin is found in the exoskeleton of arthropods and the cell walls of most fungi.

INTERACTIVE QUESTION   5.4

Fill in the following concept map that summarizes this section on carbohydrates.

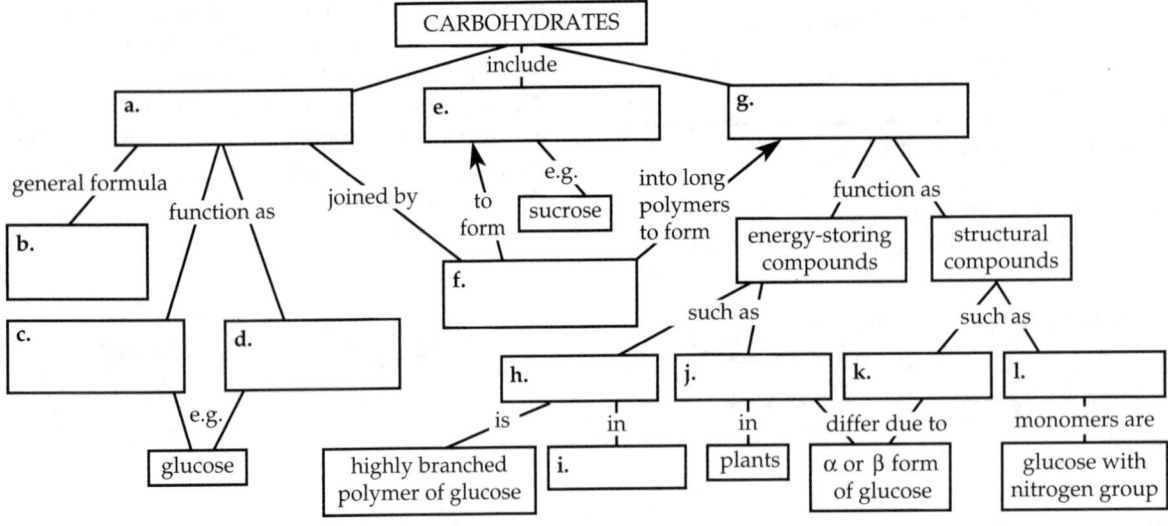

5.3 Lipids are a diverse group of hydrophobic molecules

Fats, phospholipids, and steroids are part of a diverse assemblage of biological molecules that are grouped together as **lipids** based on their hydrophobic behavior. Lipids do not form polymers and are not generally considered macromolecules.

Fats   **Fats** are composed of **fatty acids** attached to glycerol, a three-carbon alcohol. A fatty acid consists of a long hydrocarbon chain with a carboxyl group at one end. The nonpolar hydrocarbons make a fat hydrophobic.

A **triacylglycerol**, or fat (or *triglyceride*), consists of three fatty acids, each linked to glycerol by an ester linkage, a bond between a hydroxyl and a carboxyl group.

Fatty acids with double bonds in their carbon chains are called **unsaturated fatty acids**. The *cis* double bond creates a kink in the hydrocarbon chain and prevents fat molecules with unsaturated fatty acids from packing closely together and becoming solidified at room temperature. The fats of plants and fish are generally unsaturated and are called oils. **Saturated fatty acids** have no double bonds in their carbon chains. Most animal fats are saturated and solid at room temperature. Diets rich in saturated fats and in the **trans fats** made in the process of hydrogenating vegetable oils have been linked to cardiovascular disease.

Fats are excellent energy storage molecules, containing twice the energy reserves of carbohydrates such as starch on a per gram basis. Adipose tissue, made of fat storage cells, also cushions organs and insulates the body.

Phospholipids   **Phospholipids** consist of a glycerol linked to two fatty acids and a negatively charged phosphate group, to which other small molecules are attached. The phosphate head of this molecule is hydrophilic and water soluble, whereas the two fatty acid chains are hydrophobic. The unique structure of phospholipids makes them ideal constituents of cell membranes. The hydrophilic heads face the aqueous environment on either side of a membrane; the hydrophobic tails associate in the center of the phospholipid bilayer, shielded from water.

Steroids   **Steroids** are a class of lipids distinguished by four connected carbon rings with various chemical groups attached. **Cholesterol** is a common component of animal cell membranes and a precursor for other steroids, including many hormones.

INTERACTIVE QUESTION 5.5

Fill in this concept map to help you organize your understanding of lipids.

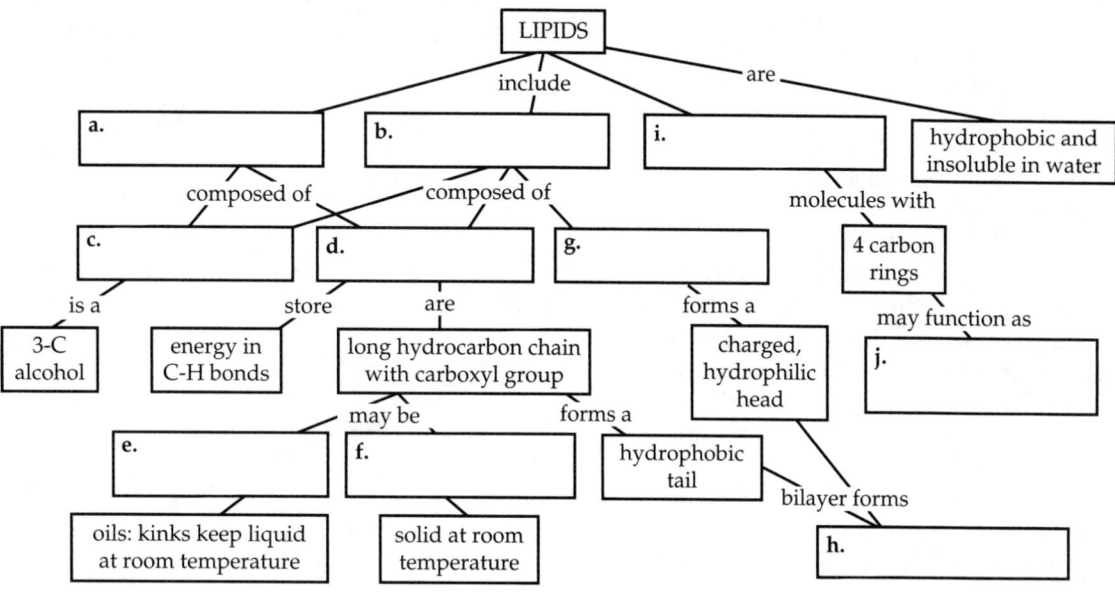

5.4 **Proteins include a diversity of structures, resulting in a wide range of functions**

Proteins are central to almost every function of life. Most enzymes, which function as **catalysts** that selectively speed up the chemical reactions of a cell, are proteins. A **polypeptide** is a polymer of amino acids, connected by peptide bonds. A **protein** is a functional molecule that consists of one or more polypeptides, each folded into a specific three-dimensional shape.

Amino Acid Monomers   **Amino acids** are composed of an asymmetric carbon (called the *alpha [α] carbon*) bonded to four partners: an amino group, a carboxyl group, a hydrogen atom, and a variable side chain called the R group. At the pH in a cell, the amino and carboxyl groups are usually ionized. The R group confers the unique physical and chemical properties of each amino acid. Side chains may be either nonpolar and thus hydrophobic, or polar or charged (acidic or basic) and thus hydrophilic.

Polypeptides (Amino Acid Polymers)   A **peptide bond** links the carboxyl group of one amino acid with the amino group of another. A string of amino acids making up a polypeptide has an amino end (N-terminus) and a carboxyl end (C-terminus). The series of amino acid residues, exclusive of the associated R groups, is called the *polypeptide backbone*. It is the kind and sequence of the R groups that determine the structure and function of the polypeptide.

INTERACTIVE QUESTION 5.6

a. Draw the amino acids alanine (R group: —CH₃) and serine (R group: —CH₂OH) and then show how a dehydration reaction will form a peptide bond between them.

b. Which of these amino acids has a polar R group? a nonpolar R group?

c. What does the following molecule segment represent? (Note the N—C—C—N—C—C sequence.) What would be attached at each of the three vertical bond lines?

$$
\begin{array}{c}
\text{H} \qquad\qquad \text{H} \qquad\qquad \text{H} \\
| \qquad\qquad\quad | \qquad\qquad\quad | \\
-\text{N}-\text{C}-\text{C}-\text{N}-\text{C}-\text{C}-\text{N}-\text{C}-\text{C}- \\
| \quad\; || \qquad\; | \quad\; || \qquad\; | \quad\; || \\
\text{H} \;\; \text{O} \qquad \text{H} \;\; \text{O} \qquad \text{H} \;\; \text{O}
\end{array}
$$

Protein Structure and Function   A protein has a unique three-dimensional shape, or structure, created by the twisting or folding of one or more polypeptide chains. Depending on the sequence of amino acids, various types of bonds form between parts of the chain as the protein is synthesized in the cell. *Globular proteins* are roughly spherical; *fibrous proteins* are long fibers. The unique structure of a protein enables it to recognize and bind to other molecules.

Primary structure is the genetically coded sequence of amino acids within a protein.

Secondary structure involves regions of coiling or folding of the polypeptide backbone, stabilized by hydrogen bonds between the oxygen (with a partial negative charge) of one peptide bond and the partially positive hydrogen attached to the nitrogen of another peptide bond. An **α helix** is a coil produced by hydrogen bonding between every fourth amino acid. A **β pleated sheet** is held by repeated hydrogen bonds along segments of the polypeptide backbone lying parallel to each other.

Tertiary structure, the overall three-dimensional shape of a protein, results from interactions between the various side chains (R groups) of the constituent amino acids. The following chemical interactions help produce the stable and unique shape of a protein: (1) **hydrophobic interactions** between nonpolar side groups clumped in the center of the molecule due to their repulsion by water; (2) van der Waals interactions among those non-polar side chains; (3) hydrogen bonds between polar side chains; (4) ionic bonds between negatively and positively charged side chains; and (5) strong covalent bonds, called **disulfide bridges**, may occur between the sulfhydryl side groups of cysteine monomers that have been brought close together by the folding of the polypeptide.

Quaternary structure occurs in proteins that are composed of more than one polypeptide. The individual polypeptide subunits are held together in a precise structural arrangement to form a functional protein.

INTERACTIVE QUESTION   5.7

In the following diagram of a portion of a protein, label the types of interactions that are shown. What level of structure are these interactions producing?

Polypeptide backbone

In the inherited blood disorder **sickle-cell disease**, a change in one amino acid affects the structure of a hemoglobin molecule, causing red blood cells to deform into a sickle shape that clogs tiny blood vessels.

If the bonds and interactions that maintain the three-dimensional structure are disrupted, the protein may undergo **denaturation**. This loss of shape and function may result from changes in pH or salt concentration or from exposure to elevated temperature or chemical agents that disrupt hydrogen bonds, ionic bonds, or disulfide bridges. Such denaturation also occurs if a protein is transferred to a nonpolar solvent; in that case, its hydrophilic regions cluster on the inside while its hydrophobic regions are on the outside interacting with the nonpolar solvent.

Some proteins can be misfolded, adopting a stable shape unlike that normally seen. Such misfolded proteins are associated with a variety of medical conditions.

Using the techniques of **X-ray crystallography**, nuclear magnetic resonance (NMR) spectroscopy, and bioinformatics, biochemists have identified the structure of more than 35,000 proteins. These structures can then be related to the specific functions of different regions of a protein. A significant number of proteins, called *intrinsically disordered proteins*, appear to have no distinct structure until they interact with some other molecule.

Now that you have gained experience with concept maps, create your own map to review what you have learned about proteins. Try to include the concepts of structure and function and look for cross-links on your map. Because proteins are complex, you may want to make several smaller maps to organize your knowledge. *One version of a large protein concept map is included in the answer section, but remember that the real value is in the thinking process you must go through to create your own.*

5.5   Nucleic acids store, transmit, and help express hereditary information

Genes are the units of inheritance that determine the primary structure of proteins. **Nucleic acids** are polymers made of nucleotide monomers.

The Roles of Nucleic Acids   **Deoxyribonucleic acid (DNA)** is the genetic material that is inherited from one generation to the next and is replicated whenever a cell divides so that all cells of an organism contain identical DNA. In the process known as **gene expression**, the instructions coded in DNA are transcribed to **ribonucleic acid (RNA)**, which then directs the synthesis of proteins, the ultimate enactors of the genetic program. In a eukaryotic cell, most DNA resides in the nucleus in the form of linear molecules complexed with protein. *Messenger RNA (mRNA)* carries the instructions for protein synthesis to ribosomes located in the cytoplasm.

The Components of Nucleic Acids   **Polynucleotides** are polymers of **nucleotides**—monomers that consist of a pentose (five-carbon sugar) covalently bonded to a phosphate group and a nitrogenous (nitrogen-containing) base. A nucleotide may contain more than one phosphate group; without the phosphate group it is called a *nucleoside*.

There are two families of nitrogenous bases. **Pyrimidines**, including cytosine (C), thymine (T), and uracil (U), are characterized by six-membered rings of carbon and nitrogen atoms. **Purines**, adenine (A) and guanine (G), contain two rings of carbon and nitrogen atoms. Adenine, guanine, and cytosine are present in DNA and RNA. Thymine is only in DNA; uracil is only in RNA. In DNA, the sugar is **deoxyribose**; in RNA, it is **ribose**. The nitrogenous base attaches to the 1′ carbon and a phosphate group attaches to the 5′ carbon of the sugar.

Nucleotide Polymers   Polynucleotides are formed through dehydration reactions in which nucleotides are joined by phosphodiester linkages, which join the sugar of one nucleotide with the phosphate of the next. The polymer has two distinct ends: a *5′ end* with a phosphate attached to the 5′ carbon of a sugar, and a *3′ end* with a hydroxyl group on the 3′ carbon of a sugar. The nitrogenous bases extend from this *sugar–phosphate backbone*. The unique sequence of bases in a gene codes for the specific amino acid sequence of a protein.

a. Label the three parts of this nucleotide. Indicate with an arrow where the phosphate group of the next nucleotide would attach to build a polynucleotide. Number the carbons of the pentose.

b. Is the base of this nucleotide a purine or a pyrimidine? How do you know?

c. Is this a DNA nucleotide or an RNA nucleotide? How do you know?

The Structures of DNA and RNA Molecules
DNA molecules consist of two polynucleotides (strands) spiralling in a **double helix**. The two sugar–phosphate backbones are **antiparallel**; that is, they run in opposite 5′ → 3′ directions. The nitrogenous bases pair and hydrogen-bond together in the inside of the molecule. Adenine pairs only with thymine; guanine always pairs with cytosine. Thus, the sequences of nitrogenous bases on the two strands of DNA are *complementary*. Because of this specific base-pairing property, DNA can replicate itself and precisely copy the genes of inheritance.

RNA molecules are usually single polynucleotides, although base-pairing within or between RNA molecules is common. For example, the functional shape of *transfer RNA (tRNA)*, an RNA involved in protein synthesis, involves several regions of complementary base-pairing.

INTERACTIVE QUESTION   5.10

Take the time to create a concept map that summarizes what you have just reviewed about nucleic acids. Compare your map with that of a study partner or explain it to a friend. Refer to Figures 5.24 and 5.25 in your textbook to help you visualize polynucleotides and the three-dimensional structures of DNA and RNA.

5.6 Genomics and proteomics have transformed biological inquiry and applications

The first techniques for *DNA sequencing* were developed in the 1970s. The Human Genome Project succeeded in sequencing the entire human *genome* in the early 2000s. Continuing increases in the speed of sequencing and decreases in its cost have spurred the development of *bioinformatics*, the use of computer software and computational tools to analyze large data sets.

Biologists can now analyze and compare whole genomes of different species (an approach called **genomics**) and large sets of proteins (**proteomics**). These new approaches have contributed to the fields of evolution, medical science, paleontology, community ecology, and conservation biology.

DNA and Proteins as Tape Measures of Evolution   Genes form the hereditary link between generations. Closely related members of the same species share many common DNA sequences and proteins. More closely related species have a larger proportion of their DNA and proteins in common. This "molecular genealogy" provides additional evidence in assessing evolutionary relationships.

Word Roots

di- = two; **sacchar-** = sugar (*disaccharide:* a double sugar, consisting of two monosaccharides joined by a glycosidic linkage formed by a dehydration reaction)

glyc- = sweet (*glycogen:* an extensively branched glucose-storage polysaccharide found in the liver and muscle of animals)

hydr- = water; **-lysis** = breaking (*hydrolysis:* a chemical reaction that breaks bonds between two molecules by the addition of water)

macro- = large (*macromolecule:* a giant molecule formed by the joining of smaller molecules, such as a polysaccharide, a protein, or a nucleic acid)

poly- = many; **mer-** = part (*polymer:* a long molecule consisting of many similar or identical monomers)

tri- = three (*triacylglycerol:* a lipid consisting of three fatty acids linked to one glycerol molecule; also called a fat or triglyceride)

Structure Your Knowledge

1. Describe the four structural levels in the functional shape of a protein.
2. Identify the type of monomer or group shown by the formulas a–g. Then match the chemical formulas with their descriptions. Answers may be used more than once.

a. _____

b. _____

c. _____

d. _____

e. _____

f. _____

g. _____

_____ 1. molecules that would combine to form a fat

_____ 2. molecule that would be attached to other monomers by a peptide bond

_____ 3. molecules or groups that would combine to form a nucleotide

_____ 4. molecules that are carbohydrates

_____ 5. molecule that is a purine

_____ 6. monomer of a protein

_____ 7. groups that would be joined by phosphodiester bonds

_____ 8. most nonpolar (hydrophobic) molecule

Test Your Knowledge

MATCHING: *Match the molecule with its class of molecule.*

_____ 1. glycogen

_____ 2. cholesterol

_____ 3. RNA

_____ 4. collagen

_____ 5. hemoglobin

_____ 6. a gene

_____ 7. triacylglycerol

_____ 8. enzyme

_____ 9. cellulose

_____ 10. chitin

A. carbohydrate

B. lipid

C. protein

D. nucleic acid

MULTIPLE CHOICE: *Choose the one best answer.*

1. Polymerization (the formation of polymers) is a process that
   a. creates bonds between R groups in the formation of a polypeptide.
   b. involves the removal of a water molecule in a dehydration reaction.
   c. links the nitrogenous base of one nucleotide with the phosphate of the next.
   d. may involve all of the above.

2. Which of the following statements is *not* true of a pentose?
   a. It can be found in nucleic acids.
   b. It can occur in a ring structure.
   c. It has the formula $C_5H_{10}O_5$.
   d. It has one carbonyl and five hydroxyl groups.

3. Where would you look for glycogen?
   a. in a steak
   b. in a potato
   c. in a milkshake
   d. in a cauliflower

4. Plants store most of their energy for later use as
   a. unsaturated fats.
   b. starch.
   c. sucrose.
   d. cellulose.

5. Maltose is made from joining two glucose molecules in a dehydration reaction. What is the molecular formula for this disaccharide?
   a. $C_6H_{12}O_6$
   b. $C_{10}H_{20}O_{10}$
   c. $C_{12}H_{22}O_{11}$
   d. $C_{12}H_{24}O_{12}$

6. Which of the following substances is the major component of the cell membrane of a fungus?
   a. cellulose
   b. chitin
   c. cholesterol
   d. phospholipids

7. A fatty acid that has the formula $C_{16}H_{32}O_2$ is
   a. saturated.
   b. unsaturated.
   c. branched.
   d. a trans fat.

8. Three molecules of the fatty acid in question 7 are joined to a molecule of glycerol ($C_3H_8O_3$). The resulting molecule has the formula
   a. $C_{48}H_{96}O_6$.
   b. $C_{48}H_{98}O_9$.
   c. $C_{51}H_{102}O_8$.
   d. $C_{51}H_{98}O_6$.

9. Which of the following is *not* a characteristic of fats and oils?
   a. They are polymers of fatty acids.
   b. They are hydrophobic.
   c. They contain fatty acid residues consisting mainly of carbon and hydrogen.
   d. They can contain C=C double bonds.

10. A steroid has
    a. four fused rings.
    b. three fused rings.
    c. two fused rings.
    d. either one or two rings, depending on what kind it is.

11. Which of the following molecules provides the most energy (kcal/g) when eaten and digested?
    a. glycogen
    b. starch
    c. fat
    d. protein

12. What maintains the secondary structure of a protein?
    a. peptide bonds
    b. disulfide bridges
    c. ionic bonds
    d. hydrogen bonds

13. What happens when a protein denatures?
    a. Its primary structure is disrupted.
    b. Its secondary and tertiary structures are disrupted.
    c. It flips inside out.
    d. Its hydrogen bonds, ionic bonds, hydrophobic interactions, disulfide bridges, and peptide bonds are disrupted.

14. The α helix of proteins is
    a. part of a protein's tertiary structure and is stabilized by disulfide bridges.
    b. a double helix.
    c. stabilized by hydrogen bonds and is commonly found in fibrous proteins.
    d. found in some regions of globular proteins and is stabilized by hydrophobic interactions.

15. What is the best description of the following molecule?

    a. chitin   c. tripeptide
    b. amino acid   d. protein

16. Which number(s) in the molecule in question 15 refer(s) to a peptide bond?
    a. 1   c. 3
    b. 2   d. 2 and 4

17. What *determines* the sequence of the amino acids in a particular protein?
    a. its primary structure
    b. the sequence of nucleotides in RNA, which was determined by the sequence of nucleotides in the gene for that protein
    c. the sequence of nucleotides in DNA, which was determined by the sequence of nucleotides in RNA
    d. the sequence of RNA nucleotides making up the ribosome

18. Which of the following is true of the subunits of DNA?
    a. They are called nucleotides.
    b. They contain four different kinds of phosphate groups.
    c. They contain a six-carbon sugar.
    d. They contain 20 different kinds of nitrogenous bases.

19. How are nucleotide monomers connected to form a polynucleotide?
    a. by covalent bonds between the sugar of one nucleotide and the phosphate of the next
    b. by hydrolysis, the removal of a water molecule and formation of a covalent bond between monomers
    c. by hydrogen bonds between complementary nitrogenous base pairs
    d. by ester linkages between the carboxyl group of one nucleotide and the hydroxyl group on the ribose of the next

20. If the nucleotide sequence of one strand of a DNA helix is 5'GCCTAA3', what would be the 3'–5' sequence on the complementary strand?
    a. GCCTAA   c. CGGATT
    b. CGGAUU   d. ATTCGG

21. Monkeys and humans share many of the same DNA sequences and have similar proteins, indicating that
    a. the two groups belong to the same species.
    b. the two groups share a relatively recent common ancestor.
    c. humans evolved from monkeys.
    d. the two groups evolved at about the same time.

22. Which of the following questions is *least* likely to be answered using genomics or proteomics?
    a. Are certain proteins more abundant and active in cancer cells than in other body cells?
    b. Did Neanderthals and *Homo sapiens* interbreed?
    c. How do polypeptides fold into their three-dimensional protein shapes?
    d. Are some products for sale in fish markets from endangered species whose harvest is illegal?

Chapter 6

A Tour of the Cell

Framework

The cell is the fundamental unit of life. The complexities in the processes of life are reflected in the complexities of the cell. It is easy to become overwhelmed by the number of new vocabulary terms for this array of cell structures. The diagram on the following page provides an organizational framework for the wealth of detail found in "a tour of the cell."

Chapter Review

The cell is the basic structural and functional unit of all organisms. In the hierarchy of biological organization, the capacity for life emerges from the structural order of the cell. All cells are related through common descent, although evolution has shaped diverse adaptations.

6.1 Biologists use microscopes and the tools of biochemistry to study cells

Microscopy   The development of microscopes enabled scientists to see and study cells beginning in the 1600s. The glass lenses of a **light microscope (LM)** refract (bend) the visible light passing through a specimen such that the projected image is magnified. *Magnification* is the ratio of the size of an image to the real size of the object. *Resolution* is a measure of the clarity of an image and is determined by the minimum distance two points must be separated to be seen as separate. The resolving power of the light microscope is limited, so that details finer than 0.2 μm (micrometer = 10^{-3} mm) cannot be distinguished. Staining specimens and using techniques such as phase-contrast, fluorescence, and confocal microscopy improve visibility by increasing the *contrast* between structures.

Many cellular structures, including most membrane-enclosed **organelles**, cannot be resolved by the light microscope. Their structures were first revealed when biologists began using the **electron microscope (EM)** in the 1950s. The electron microscope focuses a beam of electrons through a specimen or onto its surface. The short wavelength of the electron beam allows a resolution of about 2 nm (nanometer = 10^{-3} μm), about a hundred times greater than that of the light microscope.

In a **scanning electron microscope (SEM)**, an electron beam scans the surface of a specimen that is usually coated with a thin gold film. The beam excites electrons from the surface, which are detected and translated into an image on a video screen. This image appears three-dimensional.

In a **transmission electron microscope (TEM)**, a beam of electrons is passed through a thin section of a specimen stained with atoms of heavy metals. In both the SEM and TEM, electromagnets, acting as lenses, focus the image onto a monitor.

New techniques of light microscopy, such as *super-resolution microscopy,* are able to provide extremely high-resolution images. Modern cell biology integrates *cytology* with *biochemistry* to understand relationships between cellular structure and function.

INTERACTIVE QUESTION 6.1

a. Define *cytology*.

b. What do cell biologists use a TEM to study?

c. What does an SEM show best?

d. What advantages does light microscopy have over both TEM and SEM?

Cell Fractionation **Cell fractionation** is a technique that separates subcellular structures so that they can be identified and their functions studied. Cells are broken apart and the homogenate is separated into component fractions by centrifugation at increasing speeds.

6.2 **Eukaryotic cells have internal membranes that compartmentalize their functions**

Comparing Prokaryotic and Eukaryotic Cells All cells are bounded by a *plasma membrane*, which encloses a semifluid substance called **cytosol**. All cells contain *chromosomes* (with genes composed of DNA) and *ribosomes* (which make proteins).

Only members of the domains Bacteria and Archaea have **prokaryotic cells**, which are cells with no nucleus or membrane-enclosed organelles. The DNA of prokaryotic cells is concentrated in a region called the **nucleoid**. **Eukaryotic cells** (protists, fungi, animals, and plants) have a true *nucleus* enclosed in a double membrane and numerous membrane-bounded organelles suspended in cytosol. **Cytoplasm** refers to the region between the nucleus and the plasma membrane, and also to the interior of a prokaryotic cell.

Most bacterial cells range from 1 to 5 μm in diameter, whereas eukaryotic cells range from 10 to 100 μm. The small size of cells is influenced by geometry. Area is proportional to the square of linear dimension, whereas volume is proportional to its cube. The **plasma membrane** surrounding every cell must provide sufficient surface area for exchange of oxygen, nutrients, and wastes relative to the volume of the cell.

A Panoramic View of the Eukaryotic Cell Membranes compartmentalize the eukaryotic cell, providing local environments for specific metabolic functions and participating in metabolism through membrane-bound enzymes. All types of eukaryotic cells share many common structures, although there are also important differences between them.

INTERACTIVE QUESTION 6.2

a. Describe the molecular structure of the plasma membrane.

b. If a spherical eukaryotic cell has a linear dimension that is 10 times that of a spherical bacterial cell, how much more surface area, proportionally, would the eukaryotic cell have? Proportionally, how much more volume would it have?

c. What size, shapes, and surface adaptations help provide a sufficiently high ratio of surface area to volume to accommodate the exchange of materials across the plasma membrane?

6.3 **The eukaryotic cell's genetic instructions are housed in the nucleus and carried out by the ribosomes**

The Nucleus: Information Central The **nucleus** is surrounded by the **nuclear envelope**, a *double* membrane perforated by pores. A protein *pore complex* lining each pore regulates the movement of materials between the nucleus and the cytoplasm. The inner membrane is lined by the **nuclear lamina**, a layer of protein filaments that helps to maintain the shape of the nucleus. A *nuclear matrix* of fibers appears to extend throughout the nucleus.

Most of the cell's DNA is located in the nucleus, where it is organized into units called **chromosomes**; each chromosome consists of one DNA molecule complexed with protein to form a material called **chromatin**. Each eukaryotic species has a characteristic chromosome number. Individual chromosomes are visible under a microscope only when condensed in a dividing cell.

The **nucleolus**, a dense structure visible in the nondividing nucleus, synthesizes *ribosomal RNA (rRNA)* and combines it with protein (imported from the cytoplasm) to assemble ribosomal subunits, which then pass through nuclear pores to the cytoplasm.

Ribosomes: Protein Factories   **Ribosomes** are composed of protein and ribosomal RNA. Most of the proteins produced on *free ribosomes* are used within the cytosol. *Bound ribosomes*, attached to the endoplasmic reticulum or nuclear envelope, usually make proteins that will be included within membranes, packaged into organelles, or exported from the cell.

INTERACTIVE QUESTION   6.3

How does DNA in the nucleus direct protein synthesis in the cytoplasm?

6.4   The endomembrane system regulates protein traffic and performs metabolic functions in the cell

Many of the organelles in a cell are components of the **endomembrane system**—the nuclear envelope, endoplasmic reticulum, Golgi apparatus, lysosomes, vesicles, vacuoles, and the plasma membrane. Although not identical, all of these membranes are related either through direct contact or by the transfer of membrane segments by membrane-bound sacs called **vesicles**.

The Endoplasmic Reticulum: Biosynthetic Factory   The **endoplasmic reticulum (ER)** is a membranous system that is continuous with the nuclear envelope and encloses a network of interconnected tubules or sacs called cisternae. The internal compartment of the ER is called the *ER lumen* or cisternal space. Ribosomes are attached to the cytoplasmic surface of **rough ER**; **smooth ER** lacks ribosomes.

Smooth ER serves diverse functions in different cells: Its enzymes are involved in phospholipid and steroid (including sex hormone) synthesis, carbohydrate metabolism, and detoxification of drugs and poisons. Barbiturates, alcohol, and other drugs increase a liver cell's production of smooth ER, thus leading to an increased tolerance (and thus reduced effectiveness) for these and other drugs. Smooth ER also functions in storage and release of calcium ions during muscle contraction.

Proteins intended for secretion are manufactured by membrane-bound ribosomes and then threaded into the lumen of the rough ER. Many are covalently bonded to small carbohydrates to form **glycoproteins**. Secretory proteins are transported from the rough ER in membrane-bound **transport vesicles**.

Rough ER also manufactures membranes. Enzymes built into the membrane assemble phospholipids, and membrane proteins formed by bound ribosomes are inserted into the ER membrane. Transport vesicles transfer the ER membrane to other parts of the endomembrane system.

The Golgi Apparatus: Shipping and Receiving Center   The **Golgi apparatus** consists of a stack of flattened sacs. Vesicles that bud from the ER join to the *cis* face of a Golgi stack, adding to it their contents and membrane. According to the *cisternal maturation model*, Golgi products are processed and tagged as the cisternae themselves progress from the *cis* to the *trans* face. Glycoproteins often have their attached carbohydrates modified. The Golgi apparatus of plant cells manufactures some polysaccharides, such as pectins. Golgi products are sorted into vesicles, which pinch off from the *trans* face of the Golgi apparatus. These vesicles may have surface molecules that help direct them to the plasma membrane or to other organelles.

Lysosomes: Digestive Compartments   In many eukaryotic cells, **lysosomes** are membrane-enclosed sacs containing hydrolytic enzymes that digest macromolecules. Lysosomes provide an acidic pH for these enzymes.

In some unicellular eukaryotes, lysosomes fuse with *food vacuoles* to digest material ingested by **phagocytosis**. Macrophages, a type of white blood cell, use lysosomes to destroy ingested bacteria. Lysosomes also recycle a cell's own macromolecules by fusing with vesicles enclosing damaged organelles or small bits of cytosol, a process known as *autophagy*.

Vacuoles: Diverse Maintenance Compartments   **Vacuoles** are large vesicles. **Food vacuoles** are formed as a result of phagocytosis. **Contractile vacuoles** pump excess water out of freshwater unicellular eukaryotes. Vacuoles in plant cells may carry out hydrolysis, store organic compounds for the cell, or contain poisonous or unpalatable compounds that may protect the plant from predators. A large **central vacuole** is found in mature plant cells and encloses a solution called cell sap. A plant cell increases in size with a minimal addition of new cytoplasm as its vacuole absorbs water and expands.

The Endomembrane System: A Review   As membranes move from the ER to the Golgi apparatus and then to other organelles, their compositions, functions, and contents are modified.

Name the components of the endomembrane system shown in this diagram and list the functions of each of these membranes.

a.

b.

c.

d.

e.

f.

g.

free-living prokaryotic cells engulfed by an ancestral eukaryotic cell. The double membranes surrounding mitochondria and chloroplasts appear to have been part of the prokaryotic *endosymbiont.* These organelles grow and reproduce independently within the cell. They contain a small amount of DNA and ribosomes, and they synthesize some of their proteins.

Mitochondria: Chemical Energy Conversion Two membranes, each a phospholipid bilayer with unique embedded proteins, enclose a mitochondrion. A narrow intermembrane space exists between the smooth outer membrane and the convoluted inner membrane. The folds of the inner membrane, called **cristae**, create a large surface area and enclose the **mitochondrial matrix**. Many respiratory enzymes, mitochondrial DNA, and ribosomes are housed in this matrix. Other respiratory enzymes and proteins are built into the inner membrane. Although they appear static in TEM images, mitochondria form a dynamic network within the cell.

Chloroplasts: Capture of Light Energy Two membranes separated by a thin intermembrane space enclose a chloroplast. Inside the inner membrane is a membranous system of connected flattened sacs called **thylakoids**, inside of which is the thylakoid space. Photosynthetic enzymes are embedded in the thylakoids, which may be stacked together to form structures called **grana**. Chloroplast DNA, ribosomes, and many enzymes are found in the **stroma**, the fluid surrounding the thylakoids.

Plastids are a family of plant organelles that, along with chloroplasts, includes amyloplasts, which store starch, and chromoplasts, which contain pigments.

Sketch a mitochondrion and a chloroplast and label their membranes and compartments.

6.5 Mitochondria and chloroplasts change energy from one form to another

Cellular respiration, the metabolic processing of fuels to produce ATP, occurs within the **mitochondria** of eukaryotic cells. Photosynthesis occurs in the **chloroplasts** of plants and algae, which produce sugars from carbon dioxide and water by absorbing solar energy.

The Evolutionary Origins of Mitochondria and Chloroplasts According to the **endosymbiont theory**, both mitochondria and chloroplasts originated as

Peroxisomes: Oxidation   **Peroxisomes** are oxidative organelles filled with enzymes that function in a variety of metabolic pathways by removing hydrogen atoms and combining them with oxygen to produce hydrogen peroxide (H_2O_2). An enzyme that converts H_2O_2 to water is also packaged into peroxisomes. *Glyoxysomes*, found in plant seeds, contain enzymes that convert fatty acids to sugars for emerging seedlings.

INTERACTIVE QUESTION   6.6

Why are peroxisomes not considered part of the endomembrane system?

6.6   The cytoskeleton is a network of fibers that organizes structures and activities in the cell

The **cytoskeleton** of a eukaryotic cell is a dynamic network consisting of three types of protein structures: microtubules, microfilaments, and intermediate filaments.

Roles of the Cytoskeleton: Support and Motility   The cytoskeleton provides mechanical support to the cell, provides anchorage for organelles, and maintains cell shape. It also functions in cell motility (of both internal structures and the cell as a whole), usually interacting with special proteins called **motor proteins**.

Components of the Cytoskeleton   Let's review the three main types of fibers found in the cytoskeleton.

All eukaryotic cells have **microtubules**, which are hollow rods constructed of columns of globular proteins called tubulins. Microtubules change length through the addition or subtraction of tubulin *dimers*. In addition to providing the supporting framework of the cell, microtubules separate chromosomes during cell division and serve as tracks along which organelles move with the aid of motor proteins.

In animal cells, microtubules grow out from a region near the nucleus called a **centrosome**. A pair of **centrioles**, each composed of nine sets of triplet microtubules arranged in a ring, is located within the centrosome. Many other eukaryotic cells lack centrosomes with centrioles and evidently organize their microtubules by other means.

Cilia and **flagella** are locomotor extensions of some eukaryotic cells. Cilia are numerous and short; flagella occur one or two to a cell and are longer. Many unicellular eukaryotes use cilia or flagella to move through aqueous media. Cilia or flagella attached to the stationary cells of a tissue move fluid past the cell. A nonmotile, signal-receiving cilium, such as the *primary cilium* found on vertebrate animal cells, transmits environmental signals to a cell's interior.

Motile cilia and flagella are composed of two single microtubules surrounded by a ring of nine doublets of microtubules (a nearly universal "9 + 2" arrangement), all of which are enclosed in an extension of the plasma membrane. A **basal body**, with a "9 + 0" pattern of microtubule triplets, anchors a cilium or flagellum in the cell. ATP drives the sliding of the microtubule doublets past each other as the two "feet" of large motor proteins called **dyneins** alternately attach to adjacent doublets, pull down, release, and reattach. In conjunction with anchoring cross-linking proteins and radial spokes, this action causes the bending of the flagellum or cilium.

Microfilaments, probably present in all eukaryotic cells, are thin solid rods consisting of a twisted double chain of globular proteins called **actin**. Thus, microfilaments are also called actin filaments. *Cortical microfilaments* form a network just inside the plasma membrane that creates a semisolid or gel consistency in this cytoplasmic layer called the **cortex**. In some types of animal cells, microfilaments form a supportive core of small cytoplasmic extensions called microvilli.

In muscle cells, thousands of actin filaments interacting with thicker filaments made of the motor protein **myosin** produce muscle contraction. Actin and myosin interact in the formation of cellular extensions called **pseudopodia** during amoeboid movement. **Cytoplasmic streaming** in plant cells also involves actin-myosin interactions.

Intermediate filaments are intermediate in size between microtubules and microfilaments and are more diverse in their protein composition. Found in the cells of some animals, intermediate filaments are less dynamic than other cytoskeletal elements, and are important in maintaining cell shape. The nucleus is securely held in a web of intermediate filaments, and the nuclear lamina lining the inside of the nuclear envelope is composed of intermediate filaments.

INTERACTIVE QUESTION   6.7

Fill in the following table to organize what you have learned about the components of the cytoskeleton. You may wish to refer to the textbook for additional details.

Cytoskeleton Component	Structure and Monomers	Functions
Microtubules	a.	b.
Microfilaments (actin filaments)	c.	d.
Intermediate filaments	e.	f.

6.7   Extracellular components and connections between cells help coordinate cellular activities

Cell Walls of Plants   Plant **cell walls** are composed of microfibrils of cellulose embedded in a matrix of polysaccharides and protein.

The **primary cell wall** secreted by a young plant cell is relatively thin and flexible. Adjacent cells are glued together by the **middle lamella**, a thin layer of polysaccharides (called pectins). When they stop growing, some cells secrete a thicker and stronger **secondary cell wall** between the plasma membrane and the primary cell wall.

INTERACTIVE QUESTION   6.8

Sketch two adjacent plant cells and show the location of the primary and secondary cell walls and the middle lamella.

The Extracellular Matrix (ECM) of Animal Cells   Animal cells secrete an **extracellular matrix (ECM)** composed primarily of glycoproteins and other carbohydrate-containing molecules. **Collagen** forms strong fibers that are embedded in a network of proteoglycan complexes. These large complexes form when multiple **proteoglycans**, each consisting of a small core protein with many carbohydrate chains, attach to a long polysaccharide. Cells may be attached to the ECM by **fibronectins** and other glycoproteins

that bind to **integrins**, proteins that span the plasma membrane and bind to microfilaments of the cytoskeleton. Thus, information about changes outside and inside the cell can be communicated through a mechanical signaling pathway involving fibronectins, integrins, and the microfilaments of the cytoskeleton. Signals from the ECM appear to influence the activity of genes in the nucleus.

INTERACTIVE QUESTION   6.9

Label the indicated structures in this diagram of the ECM of an animal cell.

a. _____

b. _____

c. _____

d. _____

Plasma membrane ⸨

e. _____

EXTRACELLULAR FLUID

CYTOPLASM

Cell Junctions   **Plasmodesmata** are channels in plant cell walls that link most cells of a plant into a living continuum. Water, small solutes, and even some proteins and RNA molecules can move between adjacent cells through these membrane-lined channels.

Animal cells are connected by three types of junctions. At **tight junctions**, proteins hold adjacent cell membranes tightly together, creating an impermeable seal across a layer of epithelial cells. **Desmosomes** (also called *anchoring junctions*) are reinforced by intermediate filaments and rivet cells into strong sheets. **Gap junctions** (also called *communicating junctions*) are cytoplasmic connections that allow for the exchange of ions and small molecules between cells through protein-lined pores.

INTERACTIVE QUESTION   6.10

a. Return to your sketch of plant cells in Interactive Question 6.8 and draw in a plasmodesma.

b. List some of the functions of the extracellular matrix of animal cells.

6.8   A cell is greater than the sum of its parts

The compartmentalization and the many specialized organelles typical of cells exemplify the principle that structure correlates with function. Although the various cellular components have been considered separately, all elements of the well work together to produce a living entity.

Word Roots

centr- = the center; **-some** = a body (*centrosome:* a structure in the cytoplasm of animal cells that functions to organize microtubule assembly and is important during cell division; a centrosome has two centrioles)

chlor- = green (*chloroplast:* the site of photosynthesis in plants and other photosynthetic eukaryotes)

cili- = hair (*cilium:* a short appendage containing microtubules in eukaryotic cells; a motile cilium is specialized for locomotion or moving fluid past the cell)

cyt- = cell (*cytosol:* the semifluid portion of the cytoplasm)

-elle = small (*organelle:* any of several membrane-enclosed structures with specialized functions, suspended in the cytosol of eukaryotic cells)

endo- = inner (*endomembrane system:* the collection of membranes inside and surrounding a eukaryotic cell that includes the plasma membrane, nuclear envelope, smooth and rough endoplasmic reticulum, Golgi apparatus, lysosomes, vesicles, and vacuoles)

eu- = true (*eukaryotic cell:* a type of cell with a membrane-enclosed nucleus and organelles)

extra- = outside (*extracellular matrix:* the meshwork surrounding animal cells, consisting of glycoproteins, polysaccharides, and proteoglycans)

flagell- = whip (*flagellum:* a long cellular appendage specialized for locomotion)

glyc- = sweet (*glycoprotein:* a protein with one or more covalently attached carbohydrates)

lamin- = sheet/layer (*nuclear lamina:* a netlike array of protein filaments lining the inner surface of the nuclear envelope)

lys- = loosen (*lysosome:* a membrane-enclosed sac of hydrolytic enzymes in the cytoplasm of animal cells and some unicellular protists)

micro- = small; **-tubule** = a little pipe (*microtubule:* a hollow rod of tubulin proteins that makes up part of the cytoskeleton in all eukaryotic cells and is found in cilia and flagella)

nucle- = nucleus; **-oid** = like (*nucleoid:* a non-membrane-bounded region in a prokaryotic cell where the DNA is concentrated)

phag- = to eat (*phagocytosis:* a type of endocytosis in which large particulate substances or small organisms are taken up by a cell)

plasm- = molded; **-desma** = a band or bond (*plasmodesma:* an open channel through the cell wall that connects the cytoplasm of adjacent plant cells)

pro- = before; **kary-** = nucleus (*prokaryotic cell:* a type of cell lacking a membrane-enclosed nucleus and organelles)

pseudo- = false; **-pod** = foot (*pseudopodium:* a cellular extension of amoeboid cells used in moving and feeding)

thylak- = sac or pouch (*thylakoid:* a flattened, membranous sac inside a chloroplast)

vacu- = empty (*vacuole:* a relatively large vesicle whose specialized function varies in different kinds of cells)

Structure Your Knowledge

1. In the following table, write the organelles or structures that are associated with each of the listed functions in animal cells.

Functions	Associated Organelles and Structures
Cell division	a.
Information storage and transferal	b.
Energy conversions	c.
Manufacture of proteins, membranes, and other products	d.
Lipid synthesis, drug detoxification	e.
Digestion, recycling	f.
Oxidation, conversion of H_2O_2 to water	g.
Structural integrity	h.
Movement	i.
Exchange with the environment	j.
Cell-to-cell connections	k.

2. Write the functions of each of the following plant cell structures.

Plant Cell Structures	Functions
Cell wall	a.
Central vacuole	b.
Chloroplast	c.
Amyloplast	d.
Plasmodesmata	e.

3. Label the indicated structures in the following diagram of an animal cell.

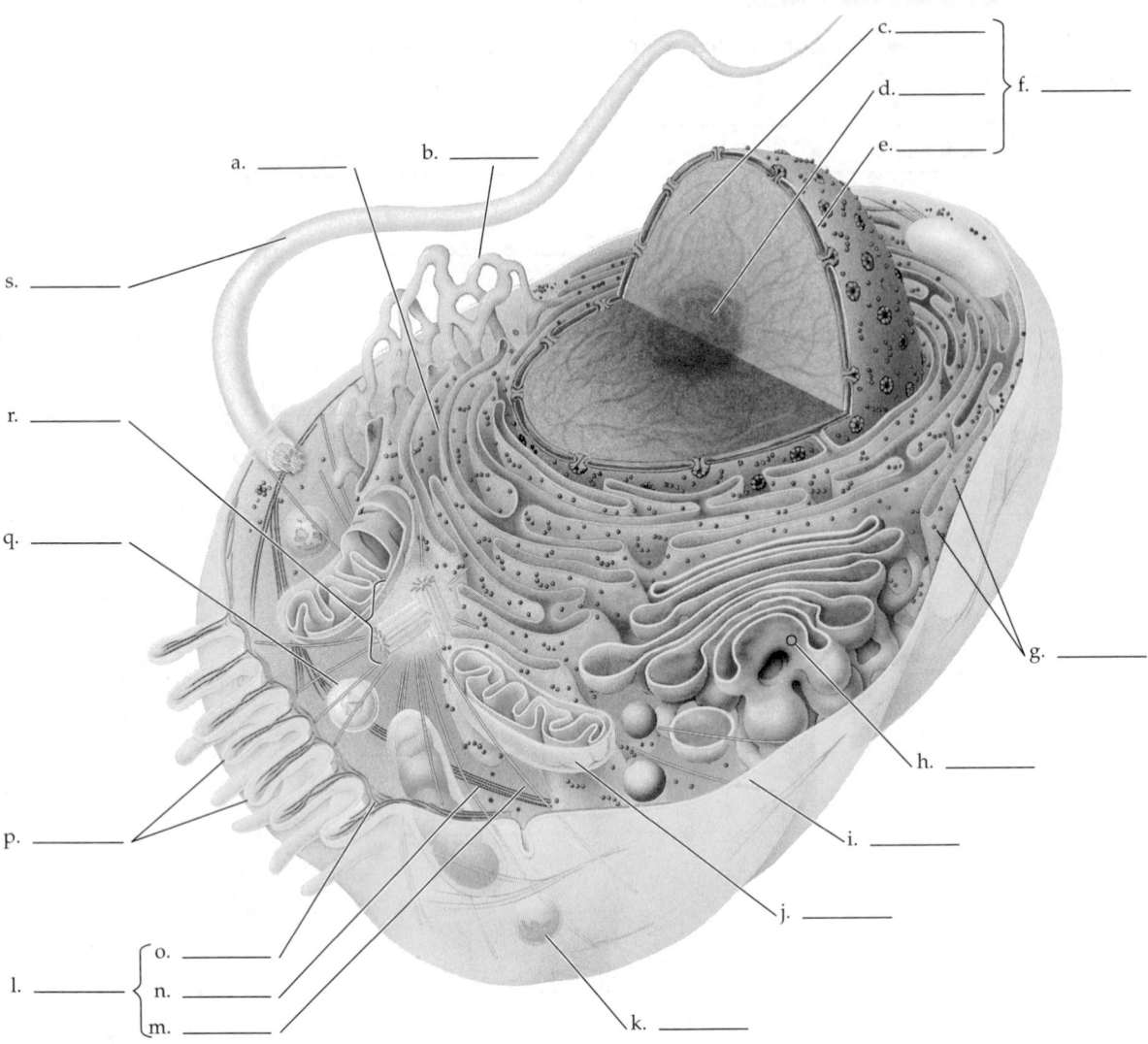

a. _____
b. _____
c. _____
d. _____
e. _____
f. _____
g. _____
h. _____
i. _____
j. _____
k. _____
l. _____
m. _____
n. _____
o. _____
p. _____
q. _____
r. _____
s. _____

4. Create a diagram or flowchart in the following space to trace the development of a secretory product (such as a digestive enzyme) from the DNA code to its export from the cell.

Test Your Knowledge

MULTIPLE CHOICE: *Choose the one best answer.*

1. The resolution of a microscope refers to the
   a. working distance between the lens and the specimen.
   b. sharpness or clarity of an image.
   c. amount of magnification of an image.
   d. wavelength of light.

2. The detailed structure of a chloroplast can be seen with the best resolution by using _____ microscopy.
   a. transmission electron
   b. scanning electron
   c. phase-contrast light
   d. deconvolution

3. In an animal cell fractionation procedure, one of the last pellets from the series of centrifugations would most likely contain
   a. ribosomes.   c. mitochondria.
   b. peroxisomes.   d. nuclei.

4. You are a linear DNA molecule complexed with protein. You reside within a nuclear envelope. You are a
   a. prokaryotic chromosome.
   b. eukaryotic chromosome.
   c. nucleolus.
   d. ribosome.

5. The cells of an ant and an elephant are, on average, the same size; an elephant just has more cells. What is the main advantage of small cell size?
   a. Small cells are easier to organize into tissues and organs.
   b. A small cell has a larger plasma membrane surface area than does a large cell, facilitating sufficient exchange of materials with its environment.
   c. The cytoskeleton of a large cell would have to be so large that cells would be too heavy.
   d. A small cell has a larger surface area relative to cytoplasmic volume, facilitating sufficient exchange of materials.

6. Plasma cells are cells of the immune system committed to production of large amounts of protein (antibodies) for export. These cells would be expected to contain
   a. many lysosomes.
   b. an extensive network of rough endoplasmic reticulum.
   c. an extensive network of smooth endoplasmic reticulum.
   d. a complex cytoskeletal network.

7. The pores in the nuclear envelope provide for the movement of
   a. proteins into the nucleus.
   b. ribosomal subunits out of the nucleus.
   c. mRNA out of the nucleus.
   d. all of the above

8. A fluorescent green tag is attached to a protein being synthesized and inserted into an area of rough ER membrane. Which of the following is a possible route for that protein to follow?
   a. rough ER → Golgi → lysosome → nuclear envelope → plasma membrane
   b. rough ER → transport vesicle → Golgi → vesicle → plasma membrane → food vacuole
   c. rough ER → transport vesicle → Golgi → vesicle → extracellular matrix
   d. rough ER → transport vesicle → Golgi → smooth ER → plasma membrane

9. Although lysosomal enzymes are necessary for breaking down food molecules, they would also be capable of damaging the cell. How is the cell protected from these enzymes?
   a. The enzymes are designed to work on foreign proteins, but they leave the cell's proteins intact.
   b. The phospholipid membrane that contains the enzymes provides an effective barrier between the interior of the lysosome and the rest of the cell.
   c. Lysosomal enzymes function best under the acid conditions in the lysosome, but they are less efficient in the cytoplasm, which is essentially neutral.
   d. Both **b** and **c** provide protection from lysosomal enzymes.

10. A plant cell grows primarily by
    a. increasing the number of vacuoles.
    b. synthesizing more cytoplasm.
    c. taking up water into its central vacuole.
    d. producing a secondary cell wall.

11. Which of the following is *not* a similarity among nuclei, chloroplasts, and mitochondria?
    a. They contain DNA.
    b. They are bounded by two (or more) phospholipid bilayer membranes.
    c. They can divide to reproduce themselves.
    d. They are associated with the endoplasmic reticulum system.

12. Which of the following is a major component of the plasma membrane of a plant cell?
    a. proteoglycans
    b. cellulose
    c. phospholipids
    d. pectins

13. Microtubules are components of or associated with all of the following *except*
    a. centrioles.
    b. amoeboid movement.
    c. tracks along which organelles can move using motor proteins.
    d. flagella and cilia.

14. Which of the following characteristics do the cristae of mitochondria, thylakoids, and microvilli share?
    a. have a large surface area of membrane that enhances their particular function
    b. are bounded by a double membrane
    c. have a shape reinforced by microfilaments
    d. are not derived from the endoplasmic reticulum system

15. The innermost portion of the cell wall of a plant cell specialized for support is the
    a. primary cell wall.
    b. secondary cell wall.
    c. middle lamella.
    d. plasma membrane.

16. Plasmodesmata in plant cells are similar in function to
    a. desmosomes.
    b. tight junctions.
    c. gap junctions.
    d. the extracellular matrix.

17. Glycoproteins and lipoproteins are proteins complexed with carbohydrates and lipids, respectively. Where in the cell are carbohydrates and lipids added to proteins?
    a. Golgi complex
    b. rough endoplasmic reticulum
    c. smooth endoplasmic reticulum
    d. ribosomes

18. What would a cell use a desmosome for?
    a. food
    b. attachment
    c. defense
    d. fuel

19. Which of the following is *incorrectly* paired with its function?
    a. peroxisome—contains enzymes that break down H_2O_2
    b. nucleolus—produces ribosomal RNA; assembles ribosome subunits
    c. Golgi apparatus—processes, tags, and ships cellular products
    d. lysosome—food sac formed by phagocytosis

Answer questions 20–23 using the following four choices.
    a. pancreatic cells that manufacture digestive enzymes
    b. macrophages that engulf bacteria
    c. epithelial cells lining the digestive tract
    d. ovarian cells that produce estrogen (a steroid hormone)

20. In which cells would you expect to find the most tight junctions?

21. In which cells would you expect to find the most lysosomes?

22. In which cells would you expect to find the most smooth endoplasmic reticulum?

23. In which cells would you expect to find the most bound ribosomes?

Membrane Structure and Function

Framework

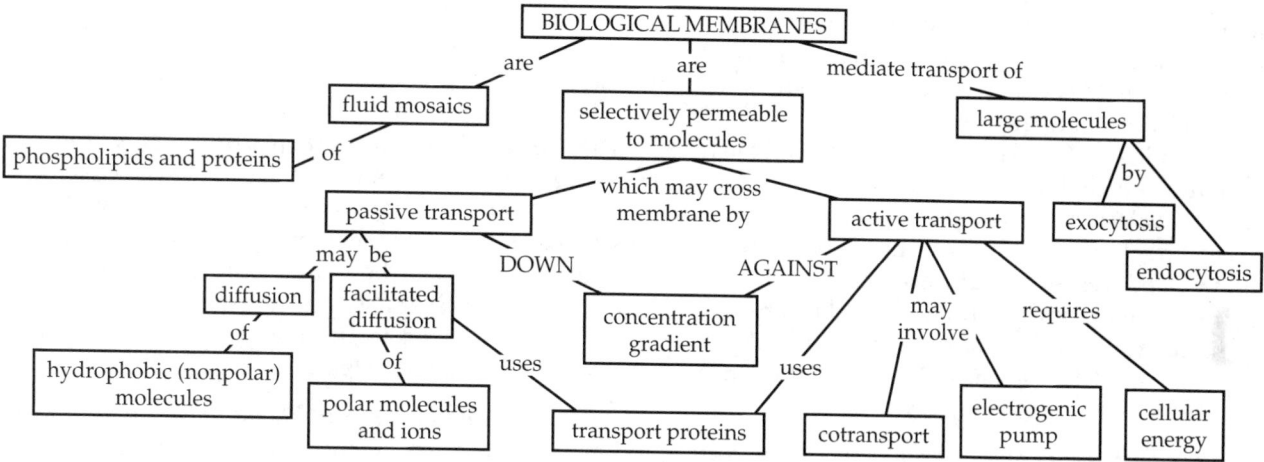

Chapter Review

The plasma membrane is the boundary of life. Like all biological membranes, it has **selective permeability**, allowing some materials to cross it more easily than others.

7.1 Cellular membranes are fluid mosaics of lipids and proteins

According to the **fluid mosaic model**, biological membranes consist of various proteins that are attached to or embedded in a bilayer of **amphipathic** phospholipids (having both hydrophilic and hydrophobic regions). Some evidence suggests the existence of stable regions of specialized lipids called *lipid rafts*.

INTERACTIVE QUESTION   7.1

Label the components in the following diagram of a small portion of a plasma membrane. Indicate whether regions are hydrophobic or hydrophilic.

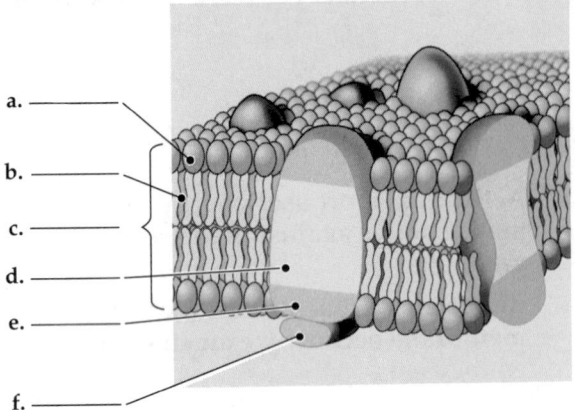

a.

b.

c.

d.

e.

f.

The Fluidity of Membranes   Membranes are held together primarily by weak hydrophobic interactions that allow the lipids and some of the proteins to drift laterally. Some membrane proteins seem to be held by attachments to the cytoskeleton or ECM; others appear to be directed in their movements.

   Phospholipids with unsaturated hydrocarbon tails maintain membrane fluidity at lower temperatures. The steroid cholesterol, common in the plasma membranes of animals, restricts movement of phospholipids and thus reduces fluidity at warmer temperatures; in plant cell membranes, where cholesterol is rare, related lipids serve a similar function. Cholesterol also prevents the close packing of lipids and thus enhances fluidity at lower temperatures.

Evolution of Differences in Membrane Lipid Composition   Variations in membrane lipid composition and the ability to change that composition in response to changing temperatures are evolutionary adaptations.

INTERACTIVE QUESTION   7.2

a. Cite some experimental evidence that indicates that membrane proteins drift.

b. How might the membrane composition of membrane lipids differ for two populations of fish if one lives in cold mountain lakes and the other in warm valley ponds?

Membrane Proteins and Their Functions   Each membrane has its own unique set of proteins, which determines most of the specific functions of that membrane. **Integral proteins** usually extend through the membrane (are *transmembrane* proteins), with a hydrophilic region associated with each side of the membrane. The hydrophobic midsection usually consists of one or more α-helical stretches of nonpolar amino acids. **Peripheral proteins** are attached to lipids and proteins on the surface of the membrane. Attachments of membrane proteins to the cytoskeleton and to fibers of the extracellular matrix provide support for the plasma membrane.

INTERACTIVE QUESTION   7.3

List the six major functions that membrane proteins may perform.

The Role of Membrane Carbohydrates in Cell-Cell Recognition   Recognition between cells is based on binding of membrane proteins to molecules on the surface of the plasma membrane. Often this recognition involves the carbohydrate portion of **glycolipids** and **glycoproteins**. These surface molecules vary from species to species, from individual to individual, and even among cell types.

Synthesis and Sidedness of Membranes   Membranes have distinct faces, or sides, due to the composition of the lipid layers and the directional orientation of their proteins. Carbohydrates attached to membrane proteins as they are synthesized in the ER are modified in the Golgi apparatus. Carbohydrates may also be attached to lipids in the Golgi apparatus. When transport vesicles fuse with the plasma membrane, these interior glycoproteins and glycolipids become located on the extracellular face of the membrane.

7.2 Membrane structure results in selective permeability

The plasma membrane permits a regular exchange of nutrients, waste products, respiratory gases, and inorganic ions.

The Permeability of the Lipid Bilayer   Hydrophobic, nonpolar molecules, such as hydrocarbons, CO_2, and O_2, can dissolve in and cross a membrane.

What types of molecules have difficulty crossing the plasma membrane? Why?

Transport Proteins Ions and polar molecules may move across the plasma membrane with the aid of **transport proteins**. Hydrophilic passageways through a membrane are provided for specific molecules or ions by *channel proteins*, such as **aquaporins**, which increase the rate of water passage. *Carrier proteins* may physically bind and transport a specific molecule across a membrane.

7.3 **Passive transport is diffusion of a substance across a membrane with no energy investment**

Diffusion is the movement of a substance down its **concentration gradient** due to random molecular motion (thermal energy). The diffusion of one solute is unaffected by the concentration gradients of other solutes. Remember that the cell does not expend energy when a substance diffuses across a membrane down its concentration gradient; therefore, the process is called **passive transport**.

Effects of Osmosis on Water Balance **Osmosis** is the diffusion of free water across a selectively permeable membrane. Water diffuses down its own concentration gradient, which is affected by the solute concentration. Clustering of water molecules around solute particles lowers the proportion of *free* water that is available to cross the membrane.

A solution of 1 *M* glucose is separated by a selectively permeable membrane from a solution of 0.2 *M* fructose and 0.7 *M* sucrose. The membrane is not permeable to the sugar molecules. Indicate which side initially has more free water molecules, and which side has fewer. Show the direction of osmosis.

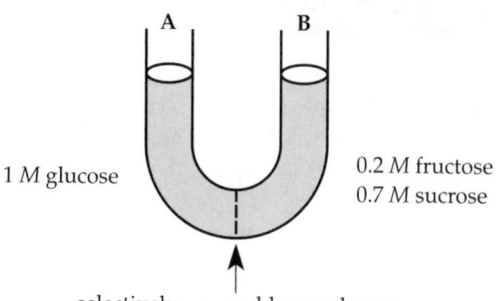

1 *M* glucose   0.2 *M* fructose   0.7 *M* sucrose

selectively permeable membrane

Tonicity, the tendency of a surrounding solution to cause a cell to gain or lose water, is affected by the relative concentrations of those solutes in the solution and in the cell that cannot cross the membrane. An animal cell will neither gain nor lose water in an **isotonic** environment. An animal cell placed in a **hypertonic** solution (which has more nonpenetrating solutes) will lose water and shrivel. If placed in a **hypotonic** solution, the cell will gain water, swell, and possibly lyse (burst). Cells without rigid cell walls must either live in an isotonic environment, or have adaptations for **osmoregulation**, the control of solute concentrations and water balance.

The cell walls of plants, fungi, prokaryotes, and some unicellular eukaryotes play a role in water balance in hypotonic environments. Water moving into the cell causes the cell to swell against its cell wall, creating *turgor pressure*. **Turgid** cells provide mechanical support for nonwoody plants. Plant cells in an isotonic surrounding are **flaccid**. In a hypertonic medium, a plant cell undergoes **plasmolysis**—that is, the plasma membrane pulls away from the cell wall as water exits and the cell shrivels.

a. What osmotic problems does the freshwater unicellular eukaryote *Paramecium* face, and what adaptations enable it to osmoregulate?

b. Compare the ideal osmotic environment for animal cells and plant cells.

Facilitated Diffusion: Passive Transport Aided by Proteins **Facilitated diffusion** involves the diffusion (down a concentration gradient) of polar molecules and ions across a membrane with the aid of transport proteins, either channel proteins or carrier proteins. Many **ion channels** are **gated channels**, which open or close in response to electrical or chemical stimuli. The binding of a solute to a carrier protein may cause a change in the protein's shape that translocates the binding site and the attached solute across the membrane.

INTERACTIVE QUESTION   7.7

Why is facilitated diffusion considered a form of passive transport?

7.4   Active transport uses energy to move solutes against their gradients

The Need for Energy in Active Transport   **Active transport**, which requires the expenditure of energy to transport a solute against its concentration gradient, is essential if a cell is to maintain internal concentrations of small molecules that differ from their concentrations outside the cell. The terminal phosphate group of ATP may be transferred to a carrier protein, inducing it to change its shape and translocate a bound solute across the membrane. The **sodium-potassium pump** works this way to exchange N^+ and K^+ across animal cell membranes, creating a higher concentration of potassium ions and a lower concentration of sodium ions within the cell.

How Ion Pumps Maintain Membrane Potential   Cells have a **membrane potential**, a voltage across the plasma membrane due to the unequal distribution of positive and negative ions on either side. This electrical potential energy results from the separation of opposite charges: The cytoplasm of a cell is negatively charged relative to the extracellular fluid. The membrane potential therefore favors the diffusion of cations into the cell and anions out of the cell. Both the membrane potential and the concentration gradient affect the diffusion of an ion; thus, an ion diffuses down its **electrochemical gradient**.

   Electrogenic pumps are membrane proteins that generate voltage across a membrane by the active transport of ions. A **proton pump** that transports H^+ out of the cell generates voltage across membranes in plants, fungi, and bacteria.

INTERACTIVE QUESTION   7.8

The sodium-potassium pump, the major electrogenic pump in animal cells, exchanges sodium ions for potassium ions, both of which are cations. How does this exchange generate a membrane potential?

Cotransport: Coupled Transport by a Membrane Protein   **Cotransport** is a mechanism through which the active transport of one substance is driven by an ATP-powered pump that transports another solute against its gradient. As that solute diffuses back down its concentration gradient through a cotransporter, the substance is carried across the membrane against its concentration gradient.

7.5   Bulk transport across the plasma membrane occurs by exocytosis and endocytosis

Bulk transport, like active transport, requires energy to transport larger biological molecules, packaged in vesicles, across the membrane.

Exocytosis   In **exocytosis**, the cell secretes large molecules by the fusion of Golgi-derived vesicles with the plasma membrane.

Endocytosis   In **endocytosis**, a region of the plasma membrane sinks inward and pinches off to form a vesicle containing material that had been outside the cell. **Phagocytosis** is a form of endocytosis in which pseudopodia wrap around a food particle or another large particle, creating a vacuole that then fuses with a lysosome containing hydrolytic enzymes. In **pinocytosis**, droplets of extracellular fluid are taken into the cell in small vesicles. **Receptor-mediated endocytosis** enables a cell to acquire specific substances from extracellular fluid.

INTERACTIVE QUESTION   7.9

a. How is cholesterol transported into human cells?

b. Explain why cholesterol accumulates in the blood of individuals with the disease familial hypercholesterolemia.

Word Roots

amphi- = dual (*amphipathic molecule:* a molecule that has both hydrophobic and hydrophilic regions)
aqua- = water; **por-** = a small opening (*aquaporin:* a channel protein in the plasma membrane that

specifically facilitates osmosis, the diffusion of free water across the membrane)

co- = together; **trans-** = across (*cotransport:* the coupling of the "downhill" diffusion of one substance to the "uphill" transport of another against its concentration gradient)

electr- = electricity; **-genic** = producing, produced by (*electrogenic pump:* an active transport protein that generates voltage across a membrane while pumping ions)

endo- = inner; **cyt-** = cell (*endocytosis:* cellular uptake of matter via formation of vesicles from the plasma membrane)

exo- = outer (*exocytosis:* cellular secretion of molecules by the fusion of vesicles containing them with the plasma membrane)

hyper- = exceeding; **-tonus** = tension (*hypertonic:* referring to a solution that, when surrounding a cell, will cause the cell to lose water)

hypo- = lower (*hypotonic:* referring to a solution that, when surrounding a cell, will cause the cell to take up water)

is- = same (*isotonic:* referring to a solution that, when surrounding a cell, causes no net movement of water into or out of the cell)

phag- = eat (*phagocytosis:* "cell eating"; endocytosis in which large particulate substances are taken up by a cell)

pin- = drink (*pinocytosis:* "cell drinking"; endocytosis in which the cell ingests extracellular fluid and its dissolved solutes)

plasm- = molded; **-lysis** = loosen (*plasmolysis:* a phenomenon in walled cells in which the cytoplasm shrivels and the plasma membrane pulls away from the cell wall; occurs when the cell loses water to a hypertonic environment)

Structure Your Knowledge

1. Create a concept map to illustrate your understanding of osmosis. This exercise will help you to practice using the words *hypotonic, isotonic,* and *hypertonic,* and to focus on the effect of these osmotic environments on plant and animal cells. Explain your map to a friend.

2. The following diagram illustrates passive and active transport across a plasma membrane. Use it to answer questions a–d.

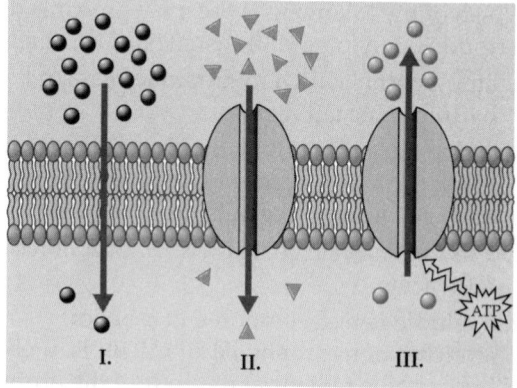

a. Which section shows facilitated diffusion? How can you tell? Explain whether the cell expends energy in this transport. What types of solutes are moved by this type of transport?

b. Which section shows active transport? List two ways that you can tell.

c. What types of solute molecules can diffuse through the membrane shown in section I?

d. Which of these sections are considered passive transport?

Test Your Knowledge

MULTIPLE CHOICE: *Choose the one best answer.*

1. If a single layer of phospholipids coats the water in a beaker, which parts of the molecules face the air?
   a. the phosphate groups
   b. the hydrocarbon tails
   c. both the heads and the tails because the molecules are amphipathic and lie sideways
   d. The molecules are arranged randomly, with no clear orientation.

2. Glycoproteins and glycolipids are important for
   a. facilitated diffusion.
   b. active transport.
   c. cell-cell recognition.
   d. exocytosis and endocytosis.

3. Which of the following is the most probable description of an integral transmembrane protein?

    a. amphipathic with a hydrophilic head and a hydrophobic tail region

    b. a globular protein with hydrophobic amino acids in the interior and hydrophilic amino acids arranged around the outside

    c. a fibrous protein coated with hydrophobic fatty acids

    d. a middle region composed of α helical stretches of hydrophobic amino acids, with hydrophilic regions at the ends of the protein facing the inside and outside of the cell

4. A cell is manufacturing receptor proteins for cholesterol. How would those proteins be oriented before they reach the plasma membrane?

    a. facing inside the ER lumen but outside the transport vesicle membrane

    b. facing inside the ER lumen and inside the transport vesicle

    c. attached outside the ER but facing inside the transport vesicle

    d. embedded in the hydrophobic center of both the ER and transport vesicle membranes

5. If you increased the amount of cholesterol in the membrane of a cell, you would make it

    a. more fluid at low temperatures.

    b. more fluid at high temperatures.

    c. less fluid at high temperatures.

    d. both a and c.

Use the following U-tube setup to answer questions 6 through 8.

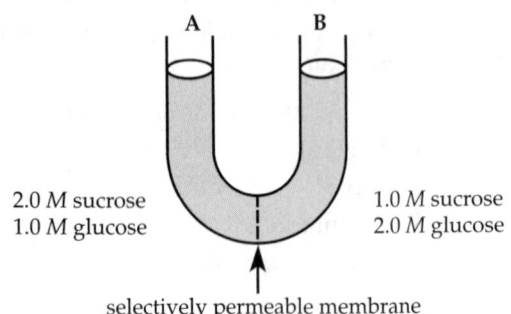

2.0 M sucrose
1.0 M glucose

1.0 M sucrose
2.0 M glucose

selectively permeable membrane

The solutions in the two arms of this U-tube are separated by a membrane that is permeable to water and glucose but not to sucrose. Side A is filled with a solution of 2.0 *M* sucrose and 1.0 *M* glucose. Side B is filled with 1.0 *M* sucrose and 2.0 *M* glucose.

6. *Initially*, the solution in side A, with respect to that in side B,

    a. has a lower solute concentration.

    b. has a higher solute concentration.

    c. has an equal solute concentration.

    d. is higher in the tube.

7. During the period *before* equilibrium is reached, which molecule(s) will show net movement through the membrane?

    a. water

    b. glucose

    c. water and sucrose

    d. water and glucose

8. *After* the system reaches equilibrium, what changes can be observed?

    a. The water level is higher in side A than in side B.

    b. The water level is higher in side B than in side A.

    c. The molarity of glucose is higher in side A than in side B.

    d. Both **a** and **c** have occurred.

9. How much work is expended in diffusion?

    a. an amount proportionate to the charge on the solute

    b. an amount proportionate to the concentration of the solvent

    c. an amount proportionate to the concentration of the solute

    d. No work is expended in diffusion.

10. Consider the following situation. Test tubes A and B each contain 10 mL of distilled water. Plant cells are added to tube A, and animal cells are added to tube B. After 10 minutes, the contents of each tube are examined microscopically. What is the expected result?

    a. Both types of cell have burst.

    b. The plant cells appear unchanged, while the animal cells have burst.

    c. The plant cells have burst, while the animal cells appear unchanged.

    d. The plant cells appear unchanged, while the animal cells have become shrunken and shriveled.

11. You observe plant cells under a microscope as they are placed in an unknown solution. First the cells plasmolyze; after a minute, the plasmolysis reverses and the cells appear normal. What would you conclude about the unknown solution?
    a. It is hypertonic to the plant cells, and its solute cannot cross the plant cell membranes.
    b. It is hypotonic to the plant cells, and its solute cannot cross the plant cell membranes.
    c. It is isotonic to the plant cells, but its solute can cross the plant cell membranes.
    d. It is hypertonic to the plant cells, but its solute can cross the plant cell membranes.

12. Water passes quickly through cell membranes because
    a. water movement is linked to ATP hydrolysis.
    b. it is a small, polar, charged molecule.
    c. aquaporins facilitate the movement of water across the membrane.
    d. it moves through hydrophobic channels.

13. Facilitated diffusion of ions across a cellular membrane requires _____, and the ions move _____.
    a. energy and channel proteins; against their electrochemical gradient
    b. cotransport proteins; against their electrochemical gradient
    c. cotransport proteins; down their electrochemical gradient
    d. channel proteins; down their electrochemical gradient

14. The membrane potential of a cell favors the movement of
    a. cations into the cell.
    b. anions into the cell.
    c. sodium out of the cell.
    d. both b and c.

15. Which of the following describes cotransport?
    a. active transport of two solutes through a cotransport protein
    b. passive transport of two solutes through a cotransport protein
    c. a pump such as the sodium-potassium pump that moves ions in two different directions
    d. transport of one solute against its concentration gradient in tandem with another that is diffusing down its concentration gradient

16. The proton pump in plant cells is the functional equivalent of an animal cell's
    a. cotransport mechanism.
    b. sodium-potassium pump.
    c. contractile vacuole for osmoregulation.
    d. ATP pump.

17. An animal cell moves potassium against its concentration gradient via
    a. active transport involving the hydrolysis of ATP.
    b. passive diffusion due to high cellular concentrations of potassium.
    c. cotransport using a concentration gradient of glucose to drive potassium transport.
    d. bulk transport via exocytosis.

18. Pinocytosis involves
    a. the fusion of a newly formed food vacuole with a lysosome.
    b. receptor-mediated endocytosis that involves binding of a ligand.
    c. the pinching in of the plasma membrane around small droplets of external fluid.
    d. the secretion of cell fluid.

19. Exocytosis may involve all of the following *except*
    a. ligands and coated pits.
    b. the fusion of a vesicle with the plasma membrane.
    c. a mechanism to rejuvenate the plasma membrane.
    d. a means of exporting large molecules.

20. In response to low blood pressure, the adrenal glands release aldosterone (a small steroid hormone). By which transport mechanism will aldosterone enter body cells?
    a. endocytosis
    b. simple diffusion
    c. facilitated diffusion
    d. active transport using a cotransport protein

An Introduction to Metabolism

Key Concepts

8.1 An organism's metabolism transforms matter and energy, subject to the laws of thermodynamics

8.2 The free-energy change of a reaction tells us whether or not the reaction occurs spontaneously

8.3 ATP powers cellular work by coupling exergonic reactions to endergonic reactions

8.4 Enzymes speed up metabolic reactions by lowering energy barriers

8.5 Regulation of enzyme activity helps control metabolism

Framework

CHANGES IN FREE ENERGY IN A CELL

RESPIRATION
Glucose → CO_2 + H_2O

BIOSYNTHESIS
Monomers → Macromolecules

Catabolic Reactions

Anabolic Reactions

Enzymes catalyze each step in catabolic and anabolic pathways

Decrease in FREE ENERGY
$-\Delta G$

Increase in FREE ENERGY
$+\Delta G$

EXERGONIC

ENDERGONIC

Energy Shuttle
ATP

Chapter Review

8.1 An organism's metabolism transforms matter and energy, subject to the laws of thermodynamics

Metabolism includes all of the chemical reactions taking place within an organism.

Organization of the Chemistry of Life into Metabolic Pathways These reactions are ordered into **metabolic pathways**, a sequence of steps, each controlled by an enzyme, that converts specific molecules to products.

Catabolic pathways release the energy stored in complex molecules through the breakdown of these molecules into simpler compounds. **Anabolic pathways**, sometimes called biosynthetic pathways, require energy to combine simpler molecules into more complicated ones. The energy released from catabolic pathways drives the anabolic pathways in a cell. The study of energy transformations in organisms, called **bioenergetics**, is central to understanding metabolism.

Forms of Energy **Energy** can be defined as the capacity to cause change. Some forms of energy can do work, such as moving matter against an opposing force. **Kinetic energy** is the energy of motion, of matter that is moving. This matter does its work by transferring its motion to other matter. The kinetic energy of randomly moving atoms or molecules is **thermal energy**. Thermal energy transferring from one body of matter to another is called **heat**.

Potential energy is the capacity of matter to cause change as a consequence of its location or arrangement. **Chemical energy** is a form of potential energy that is available for release in chemical reactions.

Energy can be converted from one form to another. Plants convert light energy to the chemical energy in sugar, and cells release this potential energy to drive cellular processes.

The Laws of Energy Transformation   **Thermodynamics** is the study of energy transformations in a collection of matter. In an *open system*, energy and matter may be exchanged between the *system* and its *surroundings*. (In an *isolated system*, such exchange does not occur.) Organisms are open systems.

The **first law of thermodynamics** states that energy can be neither created nor destroyed. According to this *principle of conservation of energy*, energy can be transferred and transformed from one kind to another, but the total energy of the universe is constant.

The **second law of thermodynamics** states that every energy transfer or transformation results in increasing disorder within the universe. **Entropy** is used as a measure of disorder or randomness. In every energy transfer or transformation, some of the energy is converted to thermal energy, a disordered form of energy, and released as heat.

A **spontaneous process** is one that is energetically favorable—that occurs without an input of energy. For a process to occur spontaneously, it must result in an increase in entropy. A nonspontaneous process, considered on its own, leads to a decrease in entropy and will occur only if energy is supplied.

A cell may become more ordered, but it does so with an attendant increase in the entropy of its surroundings. For example, an animal takes in and uses highly ordered organic molecules as a source of the matter and energy needed to create and maintain its own organized structure, but it returns heat and the simple molecules of carbon dioxide and water to the environment.

INTERACTIVE QUESTION   8.1

Complete the following concept map that summarizes some of the key ideas about energy.

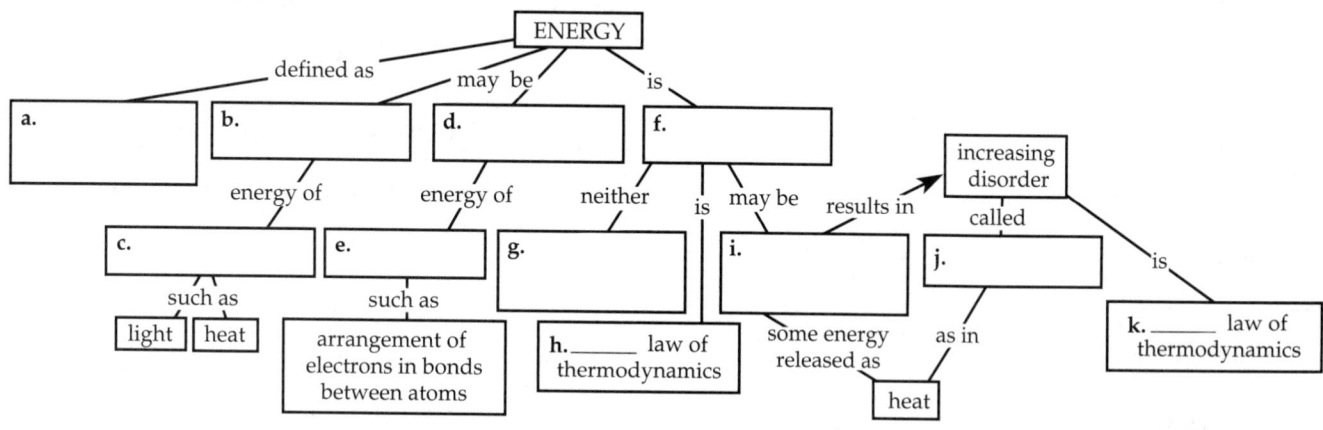

8.2 The free-energy change of a reaction tells us whether or not the reaction occurs spontaneously

Free-Energy Change, ΔG   The portion of a system's energy available to perform work when the system's temperature and pressure are uniform is defined as **free energy** (symbolized by *G*). The change in free energy during a chemical reaction is represented by $\Delta G = \Delta H - T\Delta S$. Enthalpy (*H*) is the total energy of a system; entropy (*S*) is a measure of its disorder; and *T* is absolute temperature (Kelvin units = °C + 273). The free energy of the system must decrease (−ΔG) for a reaction to be spontaneous: The system must lose energy (*H* must decrease), become more disordered (*TS* must increase), or both. Processes with a positive or zero ΔG are never spontaneous.

Free Energy, Stability, and Equilibrium   When ΔG is negative, the final state has less free energy than the initial state; thus, the final state is less likely to change and is more stable. A system rich in free energy has a tendency to change spontaneously to a more stable state. This change may be harnessed to perform work. A state of maximum stability is called *equilibrium*. At equilibrium in a chemical reaction, the forward and backward reactions occur at the same rate, and the relative concentrations of *products* and *reactants* stay the same. Moving toward equilibrium is spontaneous; the ΔG of the reaction is negative. Once at equilibrium, a system is at its minimum of free energy; it will not spontaneously change, and it can do no work.

INTERACTIVE QUESTION   8.2

Complete the following table to indicate how the free energy of a system (or a chemical reaction) relates to the system's stability, capacity to do work, tendency for spontaneous change, and equilibrium.

	Stability	Work Capacity	Spontaneous Change?	Equilibrium
System with high free energy				
System with low free energy				

Free Energy and Metabolism   An **exergonic reaction** ($-\Delta G$) proceeds with a net release of free energy and is spontaneous (energetically favorable). The magnitude of ΔG indicates the maximum amount of work an exergonic reaction can perform. **Endergonic reactions** ($+\Delta G$) are nonspontaneous; they must absorb free energy from the surroundings. The energy released by an exergonic reaction ($-\Delta G$) is equal to the energy required by the reverse reaction ($+\Delta G$).

Metabolic disequilibrium is essential to life. Metabolic reactions are reversible and could reach equilibrium if the cell did not maintain a steady supply of reactants and siphon off the products (as reactants for the next reaction or as waste products to be expelled).

INTERACTIVE QUESTION   8.3

Develop a concept map on free energy and ΔG. *The value in this exercise is for you to wrestle with and organize these concepts for yourself. Do not turn to the suggested concept map until you have worked on your own understanding. Remember that the concept map in the answer section is only one way of structuring these ideas—have confidence in your own organizational scheme.*

8.3   ATP powers cellular work by coupling exergonic reactions to endergonic reactions

Central to a cell's bioenergetics is **energy coupling**, the use of exergonic processes to drive endergonic ones. A cell usually uses ATP as the immediate source of energy for its *chemical, transport,* and *mechanical work.*

The Structure and Hydrolysis of ATP   **ATP (adenosine triphosphate)** consists of the nitrogenous base adenine bonded to the sugar ribose, which is connected to a chain of three phosphate groups (the triphosphate group). ATP can be hydrolyzed to ADP (adenosine diphosphate) and an inorganic phosphate molecule ($\textcircled{P}_i$), releasing 7.3 kcal (30.5 kJ) of energy per mole of ATP when measured under standard conditions. The ΔG of the reaction in the cell is estimated to be closer to -13 kcal/mol.

INTERACTIVE QUESTION   8.4

Label the three components (a through c) of the following ATP molecule.

d. Indicate which bond is likely to break. By what chemical mechanism is the bond broken?

e. Explain why this reaction releases so much energy.

How the Hydrolysis of ATP Performs Work ATP is used to couple exergonic and endergonic reactions in a cell. The free energy released from the hydrolysis of ATP is generally used to transfer the phosphate group to a reactant molecule, producing a **phosphorylated intermediate** that is more reactive (less stable). The hydrolysis of ATP also forms the basis for almost all transport and mechanical work in a cell, as phosphorylation and dephosphorylation change the shapes and binding affinities of proteins.

The Regeneration of ATP A cell regenerates ATP at a phenomenal rate. The formation of ATP from ADP and P_i is endergonic, with a ΔG of $+7.3$ kcal/mol (standard conditions). Cellular respiration (the catabolic processing of glucose and other organic molecules) provides the energy for the regeneration of ATP. Plants can also produce ATP using light energy.

INTERACTIVE QUESTION 8.5

Look at the following ATP cycle and explain why both the left and right sides of the cycle are examples of energy coupling.

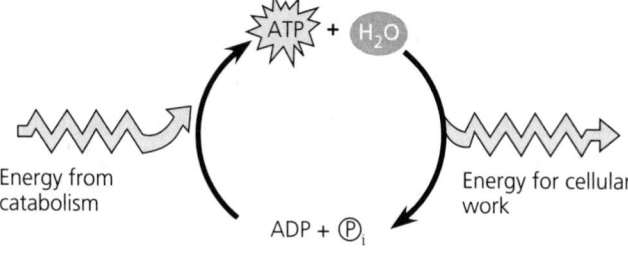

8.4 Enzymes speed up metabolic reactions by lowering energy barriers

Enzymes are biological **catalysts**—agents that speed the rate of a reaction but are unchanged by the reaction.

The Activation Energy Barrier Chemical reactions involve both the breaking and forming of chemical bonds. Energy must be absorbed to contort molecules to an unstable state in which bonds can break. Energy is released when new bonds form and molecules return to stable states with lower energy than the contorted state. **Activation energy**, or the *free energy of activation* (E_A), is the energy that must be absorbed by reactants to reach the unstable *transition state*, in which bonds are likely to break, and from which the reaction can proceed.

How Enzymes Speed Up Reactions The E_A barrier is essential to life because it prevents the energy-rich macromolecules of the cell from decomposing spontaneously. In the process called **catalysis**, enzymes are able to lower E_A so that specific reactions can proceed at cellular temperatures. Enzymes do not change the ΔG for a reaction.

INTERACTIVE QUESTION 8.6

In the following graph of an exergonic reaction with and without an enzyme catalyst, label parts a through e.

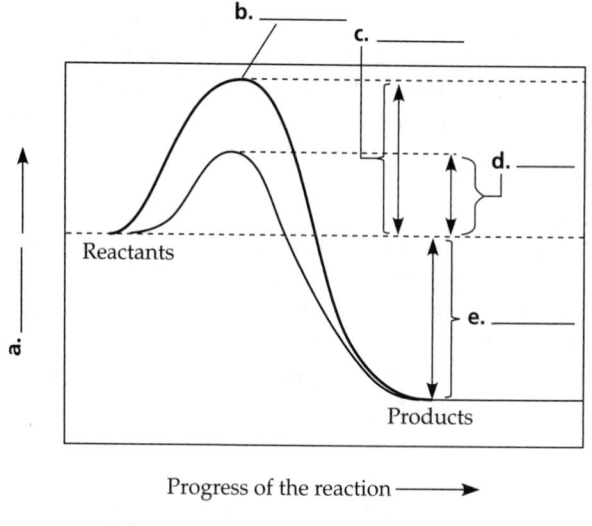

Progress of the reaction ⟶

Substrate Specificity of Enzymes Protein enzymes are macromolecules with characteristic three-dimensional shapes that result in their specificity for their particular **substrate**. The substrate attaches at the enzyme's **active site**, a pocket or groove found on the surface of the enzyme that has a shape compatible with that of the substrate. The substrate is temporarily bound to its enzyme, forming an **enzyme-substrate complex**. Interactions between substrate and active site cause the enzyme to change shape slightly, creating an **induced fit** that enhances the ability of the enzyme to catalyze the chemical reaction.

Catalysis in the Enzyme's Active Site The substrate is often held in the active site by hydrogen bonds or ionic bonds. The side chains (R groups) of some of the surrounding amino acids in the active site facilitate the conversion of substrate to product. The product then leaves the active site, and the catalytic cycle repeats, often at astonishing speed.

Whether an enzyme catalyzes the forward or backward reaction is influenced by the relative concentrations of reactants and products and the ΔG of the reactions. Enzymes catalyze reactions moving toward equilibrium.

There are several mechanisms by which enzymes can lower activation energy and speed reactions. Enzymes may catalyze reactions involving the joining of two reactants by properly orienting the substrates closely together. An induced fit can stretch critical bonds in the substrate molecule and make them easier to break. An active site may provide a microenvironment, such as a lower pH, that is necessary for a particular reaction. Side chains of amino acids in the active site may also actually participate in a reaction by forming brief covalent bonds with the substrate.

The rate of a reaction will increase with increasing substrate concentration up to the point at which all enzyme molecules are *saturated* with substrate molecules and working at full speed. At that point, only adding more enzyme molecules will increase the rate of the reaction.

INTERACTIVE QUESTION   8.7

In the following diagram of a catalytic cycle, sketch two appropriate substrate molecules and two products, show the substrate and product molecules in the active site, and label the key steps of the cycle.

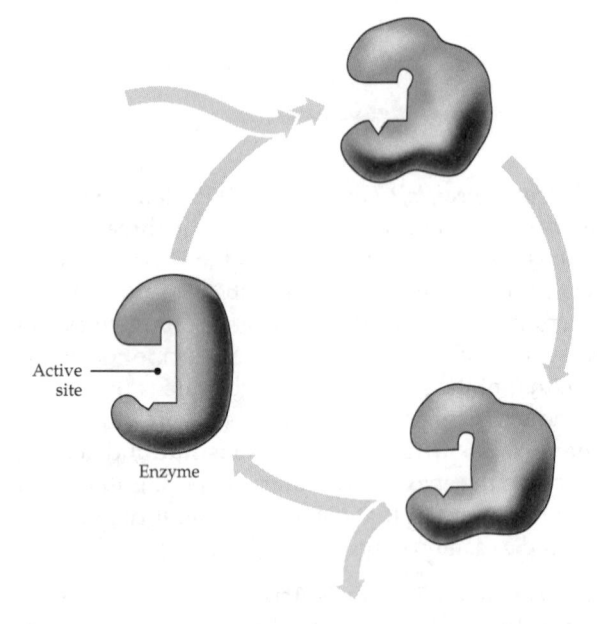

Active site

Enzyme

Effects of Local Conditions on Enzyme Activity   The speed of an enzyme-catalyzed reaction may increase with rising temperature up to the point at which increased thermal agitation begins to disrupt the weak bonds and interactions that stabilize protein shape.

Each enzyme has *optimal conditions* that include a temperature and pH that favor its most active shape.

Cofactors are small molecules that bind either permanently or reversibly with enzymes and are necessary for enzyme function. They may be inorganic, such as various metal ions, or organic molecules called **coenzymes**. Most vitamins are coenzymes or precursors of coenzymes.

Enzyme inhibitors disrupt the action of enzymes, either reversibly by binding to the enzyme with weak bonds, or irreversibly by attaching with covalent bonds. **Competitive inhibitors** compete with the substrate for the active site of the enzyme and typically resemble the substrate molecule. Increasing the concentration of substrate molecules may overcome this type of inhibition. **Noncompetitive inhibitors** bind to a part of the enzyme separate from the active site and impede enzyme action by changing the shape of the enzyme.

INTERACTIVE QUESTION   8.8

Return to the diagram in Interactive Question 8.7. Draw a competitive inhibitor and a noncompetitive inhibitor and indicate where each would bind to the enzyme molecule.

The Evolution of Enzymes   Mutations in genes may alter the amino acid sequence and thus change the substrate specificity or activity of an enzyme. If this novel function is beneficial, natural selection would be expected to preserve the mutated gene in the population. Researchers tested this hypothesis with the *E. coli* gene for β-galactosidase. After seven rounds of mutation and selection, the "evolved" enzyme digested a new substrate much better than did the original enzyme.

8.5   Regulation of enzyme activity helps control metabolism

Allosteric Regulation of Enzymes   In **allosteric regulation**, molecules may inhibit or activate enzyme activity when they bind to a site separate from the active site. Enzymes made of two or more polypeptides, each with its own active site, may have regulatory sites (sometimes called allosteric sites) located where subunits join. The entire unit may oscillate between two forms. The binding of an *activator* stabilizes the catalytically active shape, whereas an *inhibitor* stabilizes the inactive form of the enzyme. Allosteric enzymes may be critical regulators of both catabolic and anabolic pathways.

Through a phenomenon called **cooperativity**, the binding of a substrate molecule to one active site in a multisubunit enzyme changes the shape of all subunits such that their active sites are stabilized in the active form.

Metabolic pathways are commonly regulated by **feedback inhibition**, in which the end product acts as an allosteric inhibitor of an enzyme early in the pathway.

INTERACTIVE QUESTION 8.9

Both ATP and ADP serve as regulators of enzyme activity. In catabolic pathways, which of these two molecules would you predict acts as an inhibitor? Explain.

Predict which of these molecules would act as an activator.

Localization of Enzymes Within the Cell   Enzymes for several steps of a metabolic pathway may be associated in a multienzyme complex, facilitating the sequence of reactions. Specialized cellular compartments may contain high concentrations of the enzymes and substrates needed for a particular pathway. Enzymes are often incorporated into the membranes of cellular compartments. The complex internal structures of the cell facilitate metabolic order.

Word Roots

all- = different (*allosteric regulation:* the binding of a regulatory molecule at one site that affects the function of the protein at a different site)

ana- = up (*anabolic pathway:* a metabolic pathway that consumes energy to synthesize a complex molecule from simpler molecules)

bi- = life (*bioenergetics:* the overall flow and transformation of energy in an organism)

cata- = down (*catabolic pathway:* a metabolic pathway that releases energy by breaking down complex molecules into simpler molecules)

endo- = within (*endergonic reaction:* a nonspontaneous chemical reaction, in which free energy is absorbed from the surroundings)

ex- = out (*exergonic reaction:* a spontaneous reaction, in which there is a net release of free energy)

kinet- = movement (*kinetic energy:* the energy associated with the relative motion of objects)

therm- = heat (*thermodynamics:* the study of the energy transformations that occur in a collection of matter)

Structure Your Knowledge

This chapter introduced many new and complex concepts. See if you can step back from the details and answer the following general questions.

1. What is the relationship between the concept of free energy and metabolism?

2. What role do enzymes play in metabolism?

Test Your Knowledge

FILL IN THE BLANKS

_____ 1. the totality of an organism's chemical processes

_____ 2. pathways that use energy to synthesize complex molecules

_____ 3. the energy resulting from location or structure

_____ 4. the most random form of energy

_____ 5. term for the measure of disorder or randomness

_____ 6. the energy that must be absorbed by molecules to reach the transition state

_____ 7. inhibitors that decrease an enzyme's activity by binding to the active site

_____ 8. nonprotein organic molecules that bind to enzymes and are necessary for their functioning

_____ 9. regulatory device in which the product of a pathway binds to an enzyme early in the pathway

_____ 10. more reactive molecules created by the transfer of a phosphate group from ATP

MULTIPLE CHOICE: *Choose the one best answer.*

1. Catabolic and anabolic pathways are often coupled in a cell because
   a. the intermediates of a catabolic pathway are used in the anabolic pathway.
   b. both pathways use the same enzymes.
   c. the free energy released from one pathway is used to drive the other pathway.
   d. their enzymes are controlled by the same activators and inhibitors.

2. Which statement most closely reflects the first law of thermodynamics?
   a. The total amount of energy in the universe is always increasing.
   b. The total amount of energy in the universe is always decreasing.
   c. The total amount of energy in the universe is constant.
   d. The total amount of energy in the universe varies up and down.

3. When a cell breaks down glucose, only about 34% of the energy is captured in ATP molecules. The remaining 66% of the energy is
   a. used to increase the order necessary for life to exist.
   b. lost as heat, in accordance with the second law of thermodynamics.
   c. stored in starch or glycogen for later use by the cell.
   d. released when the ATP molecules are hydrolyzed.

4. When glucose and O_2 are converted to CO_2 and H_2O, changes in total energy, entropy, and free energy are correctly represented as
   a. $-\Delta H, -\Delta S, -\Delta G$.
   b. $-\Delta H, +\Delta S, -\Delta G$.
   c. $+\Delta H, +\Delta S, +\Delta G$.
   d. $+\Delta H, -\Delta S, +\Delta G$.

5. When amino acids join to form a protein, which of the following energy and entropy changes apply?
   a. $-\Delta H, +\Delta S, +\Delta G$.
   b. $-\Delta H, -\Delta S, +\Delta G$.
   c. $+\Delta H, +\Delta S, +\Delta G$.
   d. $+\Delta H, -\Delta S, +\Delta G$.

6. You have been told that a particular reaction is exergonic. What do you know about the reaction?
   a. It requires an input of energy to proceed.
   b. It violates the second law of thermodynamics.
   c. It will not occur in the absence of an enzyme.
   d. It releases energy as it proceeds.

7. One way in which a cell maintains metabolic disequilibrium is to
   a. siphon the products of a reaction off to the next step in a metabolic pathway.
   b. provide a constant supply of enzymes for critical reactions.
   c. use allosteric enzymes that can bind to activators or inhibitors.
   d. use the energy from anabolic pathways to drive catabolic pathways.

8. Which of the following statements would be predicted by the second law of thermodynamics?
   a. Energy conversions result in some energy becoming unavailable for useful work.
   b. Disorder cannot be created or destroyed.
   c. Energy is always increasing.
   d. All energy exists as chemical energy.

9. The ATP molecule is ideally suited to its role of energy carrier because
   a. its bonds contain more energy than those of most other substances.
   b. it is large and therefore contains vast quantities of energy.
   c. its phosphate bonds are easily broken to release energy.
   d. it exists in large quantities in living cells.

10. What is meant by an induced fit?
    a. The binding of the substrate is an energy-requiring process.
    b. A competitive inhibitor can outcompete the substrate for the active site.
    c. The binding of the substrate changes the shape of the active site, which can stress or bend substrate bonds.
    d. The binding of an activator to an allosteric site induces a more active form of the subunits of an enzyme.

11. In an experiment, changing the pH from 7 to 6 resulted in an increase in product formation. From this we could conclude that
    a. the enzyme's optimal pH is 6.
    b. this enzyme works best at a neutral pH.
    c. the temperature must have increased when the pH was changed to 6.
    d. the enzyme was in a more active shape at pH 6.

12. When substance A was added to an enzyme reaction, product formation decreased. The addition of more substrate did not increase product formation. From this we conclude that substance A could be
    a. product molecules.
    b. an allosteric enzyme.
    c. a competitive inhibitor.
    d. a noncompetitive inhibitor.

13. Penicillin binds to the active site of an enzyme that is involved in the synthesis of bacterial cell wall materials. The antibiotic action of penicillin is thus a result of
    a. competitive inhibition.
    b. allosteric inhibition.
    c. negative feedback.
    d. noncompetitive inhibition.

14. An enzyme raises which of the following parameters?
    a. ΔG
    b. ΔH
    c. the free energy of activation
    d. the speed of a reaction

15. Zinc, an essential trace element, may be found bound to the active site of some enzymes. Such zinc ions most likely function as a(n)
    a. coenzyme derived from a vitamin.
    b. cofactor necessary for catalysis.
    c. substrate of the enzyme.
    d. allosteric activator of the enzyme.

Use the following diagram to answer questions 16–18.

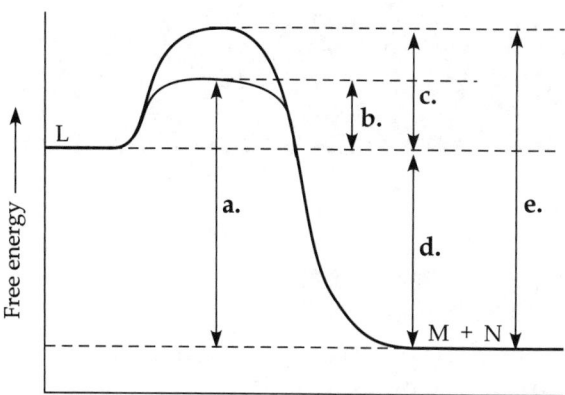

16. Which line in the diagram indicates the ΔG of the enzyme-catalyzed reaction L → M + N?

17. Which line in the diagram indicates the activation energy of the noncatalyzed reaction?

18. Which of the following terms best describes this reaction?
    a. nonspontaneous   c. endergonic
    b. $-\Delta G$ d. anabolic reaction

19. In cooperativity,
    a. a cellular organelle contains all the enzymes needed for a metabolic pathway.
    b. a product of a pathway serves as a competitive inhibitor of an enzyme early in the pathway.
    c. a molecule bound to the active site of one subunit of an enzyme affects the active site of other subunits.
    d. the product of one reaction serves as the substrate for the next reaction in intricately ordered metabolic pathways.

20. In the metabolic pathway A → B → C → D → E, what effect would molecule E likely have on the enzyme that catalyzes A → B?
    a. allosteric inhibitor
    b. allosteric activator
    c. competitive inhibitor
    d. feedback activator

Chapter 9

Cellular Respiration and Fermentation

Key Concepts

9.1 Catabolic pathways yield energy by oxidizing organic fuels

9.2 Glycolysis harvests chemical energy by oxidizing glucose to pyruvate

9.3 After pyruvate is oxidized, the citric acid cycle completes the energy-yielding oxidation of organic molecules

9.4 During oxidative phosphorylation, chemiosmosis couples electron transport to ATP synthesis

9.5 Fermentation and anaerobic respiration enable cells to produce ATP without the use of oxygen

9.6 Glycolysis and the citric acid cycle connect to many other metabolic pathways

Framework

The catabolic pathways of glycolysis and cellular respiration release the chemical energy in glucose and other fuels and use it to produce ATP. Glycolysis, occurring in the cytosol, produces ATP, pyruvate, and NADH; the latter two may then enter the mitochondria for respiration. A mitochondrion consists of a matrix in which the enzymes of the citric acid cycle are localized and a highly folded inner membrane in which electron transport chains are embedded. The redox reactions of the electron transport chain pump H^+ into the intermembrane space. ATP is produced by oxidative phosphorylation, using a chemiosmotic mechanism in which a proton-motive force drives protons through ATP synthases located in the membrane. Anaerobic respiration uses a final electron acceptor other than oxygen. Fermentation breaks down glucose to pyruvate, producing 2 ATP and recycling NADH to NAD^+.

Chapter Review

9.1 Catabolic pathways yield energy by oxidizing organic fuels

Catabolic Pathways and Production of ATP The breaking down of complex organic molecules in catabolic pathways releases energy that cells can use to do work. **Fermentation** occurs without oxygen and partially degrades sugars or other fuels to release energy. **Aerobic respiration** uses oxygen in the breakdown of glucose (or other energy-rich organic compounds) to yield carbon dioxide and water and release energy as ATP and heat. The *anaerobic respiration* of some prokaryotes does not use oxygen as a reactant but is a similar process. Aerobic respiration is often referred to as **cellular respiration**. This exergonic process has a free energy change of −686 kcal/mol of glucose.

INTERACTIVE QUESTION 9.1

Fill in the following summary equation for cellular respiration, starting with the sugar glucose.

$$\underline{\quad} + 6\ O_2 \rightarrow \underline{\quad} + 6\ H_2O + \underline{\quad}$$

Redox Reactions: Oxidation and Reduction Oxidation–reduction reactions or **redox reactions** involve the partial or complete transfer of one or more electrons from one reactant to another. **Oxidation** is the loss of electrons from one substance; **reduction** is the addition of electrons to another substance. The substance that loses electrons becomes oxidized and acts as a **reducing agent** (electron donor). By gaining electrons, a substance acts as an **oxidizing agent** (electron acceptor) and becomes reduced.

INTERACTIVE QUESTION 9.2

Fill in the appropriate terms in the following equation.

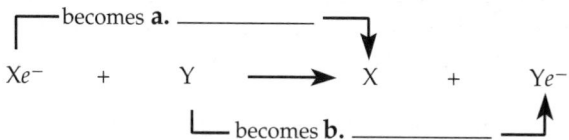

$$Xe^- + Y \longrightarrow X + Ye^-$$

with *becomes* **a.** _____ above and *becomes* **b.** _____ below

Xe^- is the reducing agent; it **c.** _____ electrons.

Y is the **d.** _____ ; it **e.** _____ electrons.

Oxygen strongly attracts electrons and is one of the most powerful oxidizing agents. As electrons shift toward a more electronegative atom, they give up potential energy. Thus, chemical energy is released in a redox reaction that relocates electrons closer to oxygen.

Organic molecules with an abundance of hydrogen are rich in "hilltop" electrons, which release their potential energy when they "fall" closer to oxygen.

INTERACTIVE QUESTION 9.3

a. In the conversion of glucose and O_2 to CO_2 and H_2O, which molecule becomes reduced?

b. Which molecule becomes oxidized?

c. What happens to the energy that is released in this redox reaction?

At certain steps in the oxidation of glucose, two hydrogen atoms are removed by enzymes called dehydrogenases, and the two electrons and one proton are passed to a coenzyme, **NAD$^+$** (nicotinamide adenine dinucleotide), reducing it to NADH.

Energy from respiration is slowly released in a series of redox reactions as electrons are passed from NADH down an **electron transport chain**, a group of carrier molecules located in the inner mitochondrial membrane (or in the plasma membrane of aerobic prokaryotes), to a stable location close to a highly electronegative oxygen atom, forming water.

INTERACTIVE QUESTION 9.4

a. NAD$^+$ is called a(n) _____ .

b. Its reduced form is _____ .

The Stages of Cellular Respiration: A Preview
Glycolysis, which occurs in the cytosol, breaks glucose into two molecules of pyruvate. Within the mitochondrial matrix or in the cytosol of prokaryotes, pyruvate is oxidized to acetyl CoA. The **citric acid cycle** then oxidizes acetyl CoA to CO_2. In some steps, electrons are transferred to NAD$^+$ (to produce NADH) or FAD (to produce FADH$_2$). NADH passes electrons to the electron transport chain, at the bottom of which they combine with H$^+$ and oxygen to form water. The energy released in this chain of redox reactions is used to synthesize ATP by **oxidative phosphorylation**, a process that includes electron transport and chemiosmosis.

Up to 32 molecules of ATP may be generated for each glucose molecule oxidized to CO_2. About 10% of this ATP is produced by **substrate-level phosphorylation**, in which an enzyme transfers a phosphate group from a substrate molecule to ADP.

INTERACTIVE QUESTION 9.5

Fill in the three stages of cellular respiration (**a–c**). What do the arrows labeled **d.** represent? Indicate whether ATP is produced by substrate-level or oxidative phosphorylation (**e–g**).

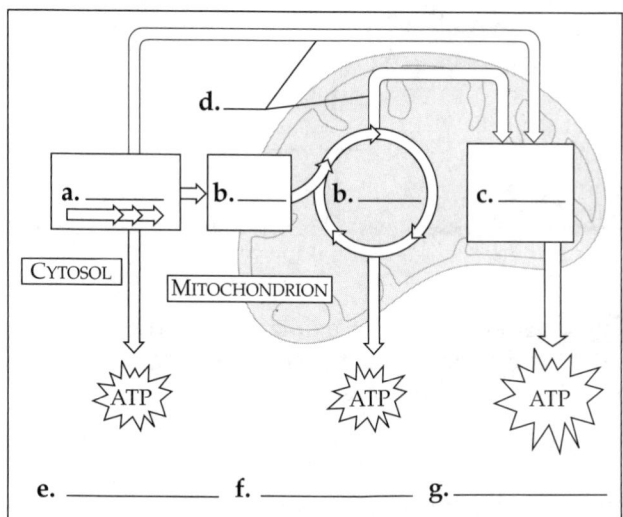

9.2   Glycolysis harvests chemical energy by oxidizing glucose to pyruvate

Glycolysis, a 10-step process occurring in the cytosol, has an energy-investment phase and an energy-payoff phase. Two molecules of ATP are consumed as glucose is split into two molecules of a three-carbon sugar (glyceraldehyde-3-phosphate). The oxidation of these molecules and the production of pyruvate yields 2 NADH and 4 ATP by substrate-level phosphorylation. For each molecule of glucose, glycolysis yields two molecules of pyruvate, 2 NADH, and a net gain of 2 ATP.

Enzymes catalyze each step in glycolysis. Kinases transfer phosphate groups; other enzymes cleave the six-carbon sugar and rearrange atoms in substrate molecules; and a dehydrogenase oxidizes glyceraldehyde-3-phosphate and reduces NAD^+.

INTERACTIVE QUESTION   9.6

Fill in the blanks in the following summary diagram of glycolysis.

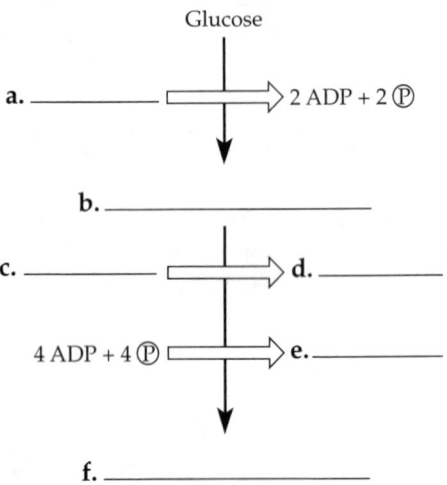

9.3   After pyruvate is oxidized, the citric acid cycle completes the energy-yielding oxidation of organic molecules

Oxidation of Pyruvate to Acetyl CoA   The pyruvate produced in glycolysis is actively transported into the mitochondrion. In a series of steps within a multienzyme complex, a carboxyl group is removed from the three-carbon pyruvate and released as CO_2; the remaining two-carbon group is oxidized to form acetate, with the accompanying reduction of NAD^+

to NADH; and coenzyme A is attached by its sulfur atom to the acetate, forming **acetyl CoA**.

The Citric Acid Cycle   In the citric acid cycle, the acetyl group of acetyl CoA is added to oxaloacetate to form citrate, which is progressively decomposed back to oxaloacetate. For each turn of the citric acid cycle, two carbons enter from acetyl CoA; two carbons exit completely oxidized as CO_2; three NADH and one $FADH_2$ are formed; and one ATP (GTP in most animal cells) is made by substrate-level phosphorylation. It takes two turns of the citric acid cycle to oxidize the two acetyl groups derived from a single glucose molecule.

INTERACTIVE QUESTION   9.7

Fill in the blanks in the following diagram of the citric acid cycle. Gray balls represent carbon atoms.

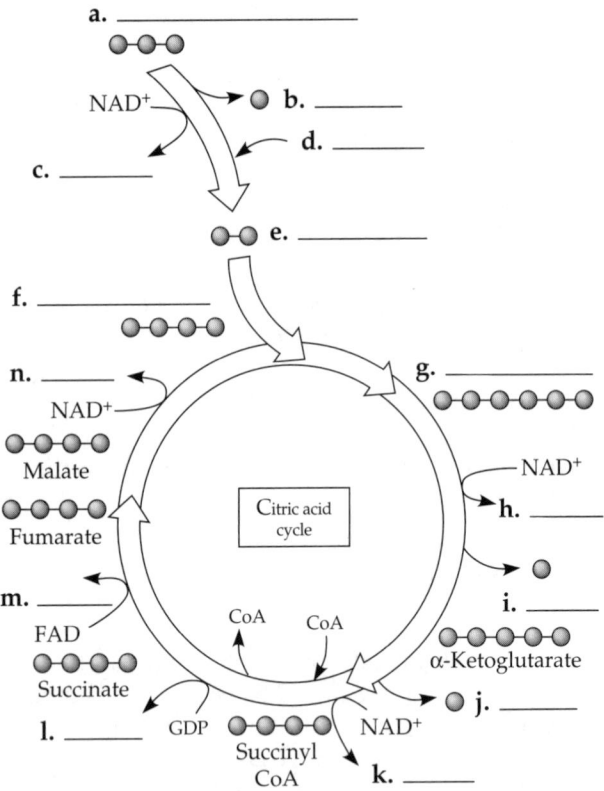

9.4   During oxidative phosphorylation, chemiosmosis couples electron transport to ATP synthesis

The first two stages of respiration have produced only 4 ATP molecules per glucose. A much more impressive yield of ATP is achieved through oxidative phosphorylation.

The Pathway of Electron Transport Thousands of electron transport chains are embedded in the cristae (infoldings) of the inner mitochondrial membrane (or in the plasma membrane of prokaryotes). The components of the electron transport chain are organized into four complexes, and most are proteins to which nonprotein *prosthetic groups* are tightly bound.

The electron carriers shift between reduced and oxidized states as they accept and donate electrons, with the free energy of electrons decreasing step by step. The transfer of electrons proceeds from NADH to a flavoprotein to an iron-sulfur protein (Complex I) to a mobile hydrophobic molecule called ubiquinone (CoQ or Coenzyme Q). Next, electrons are passed down a series of molecules called **cytochromes**, which are proteins with iron-containing heme groups. The last cytochrome, cyt a_3, passes two electrons to oxygen, which picks up two H^+, forming water.

$FADH_2$ adds its electrons to the chain at a lower energy level (within Complex II); thus, less energy is provided for ATP synthesis by $FADH_2$ as compared to NADH.

Chemiosmosis: The Energy-Coupling Mechanism **ATP synthase**, a protein complex embedded in the inner mitochondrial membrane or in the prokaryotic plasma membrane, uses the energy of a proton (H^+) gradient to make ATP, an example of the process called **chemiosmosis**.

The flow of H^+ down their gradient through the stator and rotor part of the ATP synthase complex causes the rotor and attached rod to rotate, activating catalytic sites in the knob portion, where ADP and inorganic phosphate join to make ATP.

Where does the proton gradient that powers ATP synthase come from? When some members of the electron transport chain pass electrons, they also accept and release protons, which are deposited in the intermembrane space. The potential energy of the proton gradient is referred to as the **proton-motive force**.

In mitochondria, exergonic redox reactions produce the proton gradient that drives the production of ATP. Chloroplasts use light energy to create the proton-motive force used to make ATP. Prokaryotes use proton gradients generated across the plasma membrane to transport molecules, make ATP, and rotate flagella. These are all examples of chemiosmosis—the use of an H^+ gradient across a membrane to drive cellular work.

INTERACTIVE QUESTION   9.8

Label the following diagram of oxidative phosphorylation in a mitochondrial membrane.

An Accounting of ATP Production by Cellular Respiration   About 30 to 32 ATPs may be produced per glucose molecule oxidized. These numbers are only estimates for three reasons: The ratio of NADH to ATP is not a whole number—experimental data indicate the production of 2.5 ATPs/NADH and of 1.5 ATPs/FADH$_2$; the electrons from NADH produced by glycolysis may be passed across the mitochondrial membrane to NAD$^+$ or FAD, depending on the type of shuttle used in the cell; and the proton-motive force generated by the electron transport chain is also used to power other work in the mitochondrion.

The efficiency of energy conversion in respiration is about 34%. The remaining 66% is lost as heat. In the brown fat cells of some mammals, an uncoupling protein allows protons to flow back across the inner mitochondrial membrane. Without the generation of ATP (which would inhibit cellular respiration), fats continue to be oxidized and heat is generated during hibernation.

INTERACTIVE QUESTION   9.9

Fill in the following tally for the maximum ATP yield from the oxidation of one molecule of glucose to six molecules of carbon dioxide.

Process	# ATP
Initial phosphorylation of glucose:	**a.** _____
Substrate-level phosphorylation: in glycolysis	**b.** _____
In **c.** _____	2
Oxidative phosphorylation:*	**d.** _____
Maximum Total	**e.** _____

*2.5 ATP for each of the **f.** _____ NADH from pyruvate → acetyl CoA and the **g.** _____ NADH from citric acid cycle; 1.5 ATP for each of the **h.** _____ FADH$_2$ from citric acid cycle; 2.5 or 1.5 ATP for each of the **i.** _____ NADH from glycolysis, depending on which shuttle passes electrons across the membrane.

9.5   **Fermentation and anaerobic respiration enable cells to produce ATP without the use of oxygen**

Organisms that generate ATP through anaerobic respiration have an electron transport chain that does not use oxygen as the final electron acceptor. Some bacteria use sulfate ions to accept electrons, generating H$_2$S instead of H$_2$O.

Is it possible for cells to generate ATP without using either oxygen or an electron transport chain? In fermentation, oxidation of glucose in glycolysis produces a net of 2 ATP by substrate-level phosphorylation, and NADH is recycled to NAD$^+$ by the transfer of electrons to pyruvate or derivatives of pyruvate.

Types of Fermentation   In **alcohol fermentation**, pyruvate is converted to acetaldehyde, and CO$_2$ is released. Acetaldehyde is then reduced by NADH to form ethanol (ethyl alcohol), and NAD$^+$ is regenerated. In **lactic acid fermentation**, pyruvate is reduced directly by NADH to lactate, recycling NAD$^+$. Muscle cells make ATP by lactic acid fermentation when energy demand is high and O$_2$ supply is low.

Comparing Fermentation with Anaerobic and Aerobic Respiration   Fermentation and aerobic and anaerobic respiration all use glycolysis, with NAD$^+$ as the oxidizing agent to convert glucose and other organic fuels to pyruvate. To oxidize NADH back to NAD$^+$, fermentation uses a molecule such as pyruvate or acetaldehyde as the final electron acceptor. Respiration uses oxygen or another electronegative molecule (in anaerobic respiration) as the final electron acceptor after electrons are passed down an electron transport chain.

Obligate anaerobes make ATP by fermentation or anaerobic respiration and are poisoned by oxygen. **Facultative anaerobes**, such as yeasts and many bacteria, can make ATP by fermentation or respiration, depending upon the availability of oxygen.

INTERACTIVE QUESTION   9.10

How much more ATP can be generated by aerobic respiration than by fermentation? Explain why.

The Evolutionary Significance of Glycolysis   Glycolysis is common to fermentation and respiration. This most widespread of all processes probably evolved in ancient prokaryotes. Its cytosolic location and independence of oxygen also provide evidence of its antiquity.

9.6   **Glycolysis and the citric acid cycle connect to many other metabolic pathways**

The Versatility of Catabolism   Fats, proteins, and carbohydrates can all be used in cellular respiration.

Proteins are digested into amino acids, which are then *deaminated* (the amino group is removed) and can enter glycolysis or the citric acid cycle at several points. The digestion of fats yields glycerol, which is fed into glycolysis, and fatty acids, which are broken down by **beta oxidation** to two-carbon fragments that enter the citric acid cycle as acetyl CoA.

Biosynthesis (Anabolic Pathways) The organic molecules of food also provide carbon skeletons for biosynthesis. Some monomers, such as amino acids, can be directly incorporated into the cell's macromolecules. Intermediate compounds of glycolysis and of the citric acid cycle serve as precursors for anabolic pathways. Carbohydrates, fats, and proteins can be interconverted to provide for a cell's needs.

Regulation of Cellular Respiration via Feedback Mechanisms Through feedback inhibition, the end product of an anabolic pathway inhibits an enzyme early in the pathway, thus preventing a cell from producing an excess of a particular substance.

The supply of ATP in the cell regulates respiration. The allosteric enzyme that catalyzes the third step of glycolysis, phosphofructokinase, is inhibited by ATP and activated by AMP (derived from ADP). Phosphofructokinase is also inhibited by citrate released from the mitochondria, thereby synchronizing the rates of glycolysis and the citric acid cycle. Other enzymes located at key intersections help to maintain metabolic balance.

Word Roots

aer- = air (*aerobic respiration:* a catabolic pathway for organic molecules using oxygen as the final electron acceptor in an electron transport chain and ultimately producing ATP)

an- = not (*facultative anaerobe:* an organism that makes ATP by aerobic respiration if oxygen is present but that switches to anaerobic respiration or fermentation if oxygen is not present)

chem- = chemical (*chemiosmosis:* an energy-coupling mechanism that uses energy stored in the form of an H^+ gradient across a membrane to drive cellular work, such as the synthesis of ATP)

glyc- = sweet; **-lysis** = split (*glycolysis:* a series of reactions that splits glucose into pyruvate; glycolysis occurs in almost all living cells, serving as the starting point for fermentation or cellular respiration)

Structure Your Knowledge

1. This chapter describes how catabolic pathways release chemical energy and store it in ATP. One of the best ways to learn the three main components of cellular respiration is to teach them to someone. Find two study partners and have each person learn and explain the important concepts of glycolysis, pyruvate oxidation and the citric acid cycle, or oxidative phosphorylation. Use diagrams to illustrate the process you are explaining.
2. Create a concept map to organize your understanding of oxidative phosphorylation.
3. Fill in the following table to summarize glycolysis, the citric acid cycle, oxidative phosphorylation, and fermentation. Base the main inputs and outputs of these processes on one glucose molecule.

Process	Brief Description	Main Inputs	Main Outputs
Glycolysis			
Pyruvate to acetyl CoA and citric acid cycle			
Oxidative phosphorylation			
Fermentation			

Test Your Knowledge

MULTIPLE CHOICE: *Choose the one best answer.*

1. When electrons move closer to a more electronegative atom,
   a. energy is released.
   b. chemical energy is stored.
   c. a proton gradient is established.
   d. water is produced.

2. In the reaction $C_6H_{12}O_6 + 6\,O_2 \rightarrow 6\,CO_2 + 6\,H_2O$,
   a. glucose becomes reduced.
   b. oxygen becomes reduced.
   c. water is a reducing agent.
   d. oxygen is a reducing agent.

3. In which of the following conversions is the first molecule becoming reduced to the second molecule?
   a. pyruvate $\rightarrow$ acetyl CoA
   b. glucose $\rightarrow$ pyruvate
   c. NADH + H^+ $\rightarrow$ NAD^+ + 2 H
   d. pyruvate $\rightarrow$ lactate

4. Some prokaryotes use anaerobic respiration, a process that
   a. does not involve an electron transport chain.
   b. produces ATP solely by substrate-level phosphorylation.
   c. uses a substance other than oxygen as the final electron acceptor.
   d. Both **a** and **b** are correct.

5. Which of the following reactions is *incorrectly* paired with its location?
   a. ATP synthesis—inner membrane of the mitochondrion, mitochondrial matrix, and cytosol
   b. fermentation—cell cytosol
   c. glycolysis—cell cytosol
   d. citric acid cycle—cristae of mitochondrion

6. In which organelle of a plant cell does the citric acid cycle occur?
   a. chloroplast
   b. mitochondrion
   c. endoplasmic reticulum
   d. nucleus

7. Which of the following compounds produces the most ATP when completely oxidized to CO_2?
   a. acetyl CoA   c. pyruvate
   b. glucose   d. fructose-1,6-bisphosphate

8. Aerobic eukaryotes produce CO_2 as a by-product during cellular respiration. Which of the following is a correct statement of the origin of CO_2 produced by the complete oxidation of 1 molecule of glucose?
   a. 1 CO_2 from glycolysis; 1 CO_2 from the oxidation of pyruvate to acetyl CoA; 4 CO_2 from the citric acid cycle
   b. 2 CO_2 from the oxidation of pyruvate to acetyl CoA; 4 CO_2 from the citric acid cycle
   c. 1 CO_2 from the oxidation of pyruvate to acetyl CoA; 2 CO_2 from the citric acid cycle
   d. 3 CO_2 from the citric acid cycle

9. Which of the following statements correctly describes the role of oxygen in cellular respiration?
   a. It is reduced in glycolysis as glucose is oxidized.
   b. It combines with H^+ diffusing through ATP synthase to produce H_2O.
   c. It is the final electron acceptor for the electron transport chain.
   d. It combines with the carbon removed during the citric acid cycle to form CO_2.

10. In the electron transport system, H^+ ions are pumped to
    a. the intermembrane space.
    b. the matrix of the mitochondrion.
    c. the citric acid cycle.
    d. the cytoplasm.

11. When glucose is oxidized to CO_2 and water, approximately 66% of its energy is transformed to
    a. heat.
    b. ATP.
    c. a proton-motive force.
    d. potential energy.

12. The energy required for the chemiosmotic synthesis of ATP is provided by
    a. a gradient of hydrogen ions.
    b. a gradient of sodium ions.
    c. ADP.
    d. water.

13. Which of the following statements correctly describes a metabolic effect of cyanide, a poison that blocks the passage of electrons along the electron transport chain?
    a. The pH of the intermembrane space becomes much lower than normal.
    b. Alcohol would build up in the mitochondria.
    c. NADH supplies would be exhausted, and ATP synthesis would cease.
    d. No proton gradient would be produced, and ATP synthesis would cease.

14. The citric acid cycle is best described as
    a. the principle site for ATP synthesis within the cell.
    b. the process by which CO2 is fixed into stable organic molecules in plants.
    c. the process in which oxygen is required during respiration.
    d. a process that stores chemical energy primarily in the form of intermediates such as FADH2 and NADH.

15. Fermentation produces less ATP than cellular respiration because
    a. NAD^+ is regenerated by alcohol or lactate production, without the electrons of NADH passing through the electron transport chain.
    b. pyruvate still contains most of the "hilltop" electrons that were present in glucose.
    c. its starting reactant is pyruvate and not glucose.
    d. Both a and b are correct.

16. If muscle cells do not receive enough oxygen from the blood, they use lactic acid fermentation to produce ATP. What do these cells gain from the reduction of pyruvate?
    a. ATP
    b. ATP and NAD^+
    c. CO_2 and NAD^+
    d. ATP and CO_2

17. Glucose made from six radioactively labeled carbon atoms is fed to yeast cells in the absence of oxygen. How many molecules of radioactive alcohol (C_2H_5OH) are formed from each molecule of glucose?
    a. 0   c. 2
    b. 1   d. 3

18. Glycolysis is considered one of the first metabolic pathways to have evolved because it
    a. relies on fermentation, which is characteristic of archaea and bacteria.
    b. is found only in prokaryotes, whereas eukaryotes use mitochondria to produce ATP.
    c. produces ATP only by substrate-level phosphorylation and does not involve redox reactions.
    d. is nearly universal, is located in the cytosol, and does not involve O_2.

19. Which of the following substances produces the *most* ATP per gram?
    a. glucose, because it is the starting place for glycolysis
    b. glycogen or starch, because they are polymers of glucose
    c. fats, because they are highly reduced compounds
    d. proteins, because of the energy stored in their tertiary structure

20. Fats and proteins can be used as fuel in the cell because they
    a. can be converted to glucose by enzymes.
    b. can be converted to intermediates of glycolysis or the citric acid cycle.
    c. can pass through the mitochondrial membrane to enter the citric acid cycle.
    d. contain more energy than glucose.

21. Which of the following statements is *false* concerning the enzyme phosphofructokinase?
    a. It is an allosteric enzyme.
    b. It is inhibited by citrate.
    c. It is inhibited by AMP.
    d. It is an early enzyme in the glycolytic pathway.

22. Brown fat, which is found in newborn infants and hibernating mammals, has uncoupler proteins that, when activated, make the inner mitochondrial membrane leaky to H^+. What is the function of brown fat?
    a. It produces more ATP than does regular fat and is also found in the flight muscles of ducks and geese, providing more energy for long-distance migrations.
    b. It lowers the pH of the intermembrane space, which results in the production of more ATP per gram than is produced by the oxidation of glucose or regular fat tissue.
    c. Because it dissipates the proton gradient, it generates heat through cellular respiration without producing ATP, thereby raising the body temperature of hibernating mammals or newborn infants.
    d. Its main function is insulation in the endothermic animals in which is it common.

Chapter 10

Photosynthesis

10.1 Photosynthesis converts light energy to the chemical energy of food

10.2 The light reactions convert solar energy to the chemical energy of ATP and NADPH

10.3 The Calvin cycle uses the chemical energy of ATP and NADPH to reduce CO_2 to sugar

10.4 Alternative mechanisms of carbon fixation have evolved in hot, arid climates

10.5 Life depends on photosynthesis

Framework

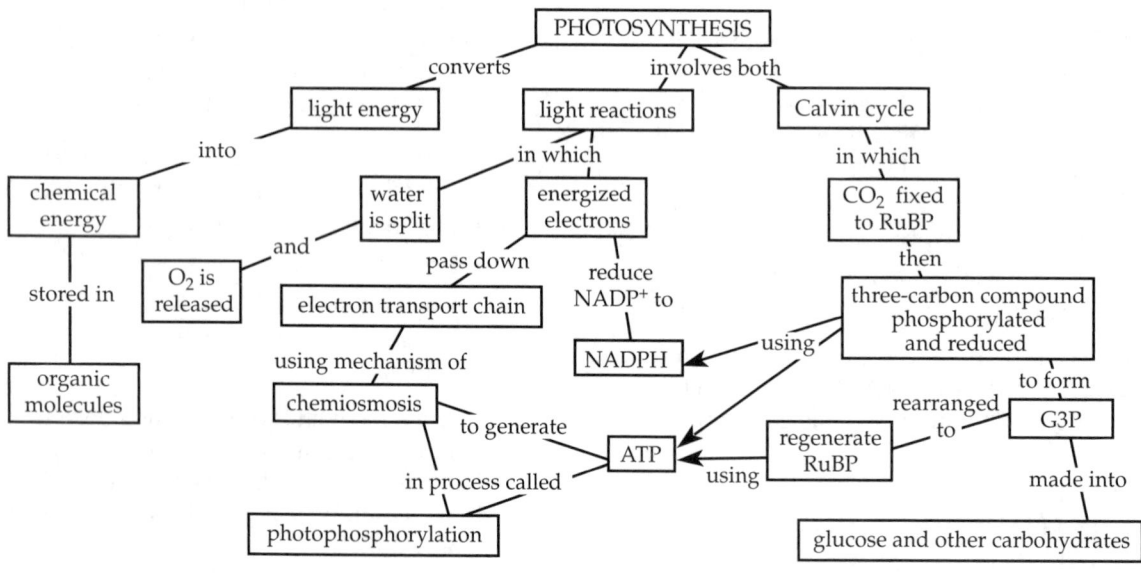

Chapter Review

In **photosynthesis,** the light energy of the sun is converted into chemical energy stored in organic molecules. In photosynthetic eukaryotes, photosynthesis takes place in organelles called **chloroplasts. Autotrophs** "feed themselves" in the sense that they make their own organic molecules from inorganic raw materials. Autotrophs are the *producers* of the biosphere. Plants, algae, some other unicellular eukaryotes, and some prokaryotes are photoautotrophs.

Heterotrophs are *consumers.* They may eat plants or animals or decompose organic litter, but almost all ultimately depend on photoautotrophs for food and O_2.

10.1 **Photosynthesis converts light energy to the chemical energy of food**

The first photosynthetic organisms were likely bacteria with clusters of photosynthetic enzymes and other molecules embedded in infoldings of the plasma membrane.

Chloroplasts: The Sites of Photosynthesis in Plants Chloroplasts are located mainly in the **mesophyll** tissue of the leaf. CO_2 enters and O_2 exits the leaf through **stomata**. Veins carry water from the roots to the leaves and distribute sugar to nonphotosynthetic tissue.

A chloroplast consists of a double membrane surrounding a dense fluid called the **stroma** and a membrane system of sacs called **thylakoids**, which encloses the *thylakoid space*. Thylakoid sacs may be stacked to form *grana*. **Chlorophyll**, the green pigment that absorbs the light energy that drives photosynthesis, is embedded in the thylakoid membrane.

INTERACTIVE QUESTION 10.1

Label the indicated parts in the following diagram of a chloroplast.

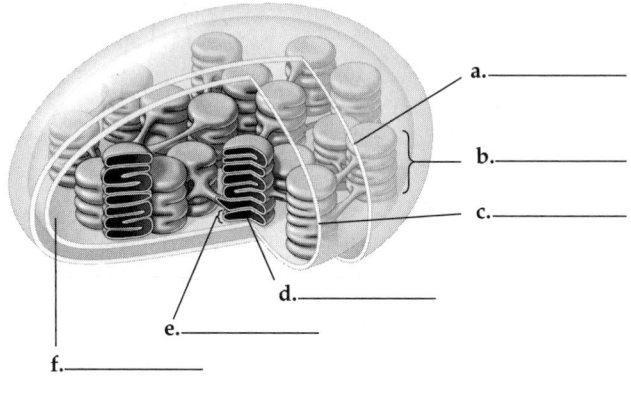

a._____

b._____

c._____

d._____

e._____

f._____

Tracking Atoms Through Photosynthesis: Scientific Inquiry Taking into account only the net consumption of water and considering glucose as the product (even though the direct product is a three-carbon sugar), the equation for photosynthesis is the reverse of respiration:

$$6\ CO_2 + 6\ H_2O + \text{Light energy} \rightarrow C_6H_{12}O_6 + 6\ O_2$$

The simplest equation is $CO_2 + H_2O \rightarrow [CH_2O] + O_2$.

Using evidence from bacteria that utilize hydrogen sulfide (H_2S) for photosynthesis, C. B. van Niel hypothesized that photosynthetic organisms need a hydrogen source and that plants use H_2O as their hydrogen source and release O_2. This hypothesis was later confirmed by experiments that used a heavy isotope of oxygen (^{18}O). Labeled O_2 was produced only when water, but not carbon dioxide, contained ^{18}O.

Photosynthesis, like respiration, is a redox process, but it differs in the direction of electron flow. The electrons increase their potential energy when they travel from water to reduce CO_2 into sugar, and light provides the energy for this endergonic process.

The Two Stages of Photosynthesis: A Preview Solar energy is converted into chemical energy in the **light reactions**. Light energy absorbed by chlorophyll drives the transfer of electrons and hydrogen ions from water to the electron acceptor **NADP$^+$**, which is reduced to **NADPH** and temporarily stores electrons. Oxygen is released when water is split. ATP is formed

---

INTERACTIVE QUESTION 10.2

Fill in the blanks in the following overview of photosynthesis in a chloroplast. Indicate the locations of the stages c and h.

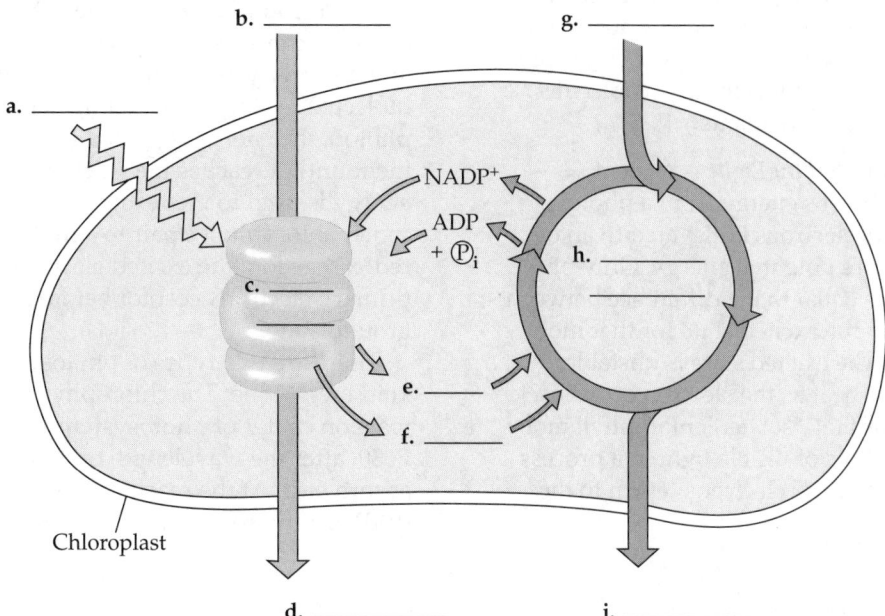

during the light reactions, using chemiosmosis in a process called **photophosphorylation**.

In the **Calvin cycle**, CO_2 is incorporated into existing organic compounds by **carbon fixation**, and these compounds are then reduced to form carbohydrate. NADPH and ATP from the light reactions supply the reducing power and chemical energy needed for the Calvin cycle.

10.2 The light reactions convert solar energy to the chemical energy of ATP and NADPH

The Nature of Sunlight   Electromagnetic energy, also called electromagnetic radiation, travels as rhythmic wave disturbances of electric and magnetic fields. The distance between the crests of electromagnetic waves, called their **wavelength**, ranges across the **electromagnetic spectrum**, from short gamma waves to long radio waves. The small band of radiation with wavelength of about 380 nm to 750 nm is called **visible light** and is the radiation that drives photosynthesis.

Light also behaves as if it consists of discrete particles called **photons**, which have a fixed quantity of energy. The amount of energy in a photon is inversely related to its wavelength.

Photosynthetic Pigments: The Light Receptors   *Pigments* are substances that absorb light. A **spectrophotometer** measures the amount of light of different wavelengths absorbed by a pigment. The **absorption spectrum** of **chlorophyll** *a*, the pigment that participates directly in the light reactions, shows that it absorbs violet-blue and red light best. Accessory pigments such as **chlorophyll** *b* and **carotenoids** absorb light of different wavelengths and broaden the spectrum of colors useful in photosynthesis. Some carotenoids function in *photoprotection* by absorbing excessive light energy that might damage chlorophyll or interact with oxygen to form reactive molecules. (These pigments act as antioxidants.)

Excitation of Chlorophyll by Light   When a pigment molecule absorbs energy from a photon, one of the molecule's electrons is elevated to an orbital where it has more potential energy. Only photons whose energy is equal to the difference between the ground state and the excited state for that molecule are absorbed. The excited state is unstable. Energy is released as heat as the electron drops back to its ground-state orbital. Isolated chlorophyll molecules also emit photons of visible light in a process called fluorescence, as their electrons return to the ground state.

An **action spectrum** shows the relative rates of photosynthesis under different wavelengths of light. The following graph includes both an absorption spectrum and an action spectrum. Label the left *y*-axis for the absorption spectrum and the right *y*-axis for the action spectrum. Then indicate the line that represents the absorption spectrum for chlorophyll *a* and the line for the action spectrum for photosynthesis. Why are these lines different?

Absorption and action spectra for photosynthesis

A Photosystem: A Reaction-Center Complex Associated with Light-Harvesting Complexes   Embedded in thylakoid membranes are numerous **photosystems**, each composed of several **light-harvesting complexes** and a **reaction-center complex**. A reaction-center complex contains two special chlorophyll *a* molecules and a **primary electron acceptor**. When a pigment molecule in a light-harvesting complex absorbs a photon, the energy is passed from pigment to pigment until it reaches the reaction center. The absorbed energy is used to move an electron from the reaction center chlorophyll *a* pair to a higher energy state. In a redox reaction, the excited electron is captured by the primary electron acceptor before it can return to the ground state.

There are two types of photosystems in the thylakoid membrane. The chlorophyll *a* molecules of the reaction center of **photosystem II (PSII)** are called P680, after the wavelength of light (680 nm) they absorb best. At the reaction center of **photosystem I (PSI)** are chlorophyll *a* molecules called P700.

INTERACTIVE QUESTION 10.4

Describe the components of a photosystem.

Linear Electron Flow Through a sequence called **linear electron flow**, electrons pass from water to $NADP^+$ through the two photosystems. A pigment molecule absorbs a photon of light, and the energy is relayed through other pigment molecules of the light-harvesting complex to the P680 pair of chlorophyll *a* molecules in the PS II reaction-center complex. An excited electron of P680 is trapped by the primary electron acceptor. The loss of this electron creates the charged form $P680^+$, which is a strong oxidizing agent. The electron hole in $P680^+$ is filled when an enzyme removes electrons from water, splitting it into two electrons, two H^+ (which are released into the thylakoid space), and an oxygen atom that immediately combines with another oxygen atom to form O_2.

The primary electron acceptor passes the photo-excited electron to an electron transport chain made up of plastoquinone (Pq), a cytochrome complex, and plastocyanin (Pc). The energy released as electrons "fall" through the electron transport chain is used to pump protons into the thylakoid space, contributing to the proton gradient used for the synthesis of ATP.

At the bottom of the electron transport chain, the electron passes to $P700^+$ in photosystem I. It replaces the photoexcited electron that was captured by its primary electron acceptor when PSI absorbed a photon. This primary electron acceptor passes the electron down a second electron transport chain through ferredoxin (Fd), from which the enzyme $NADP^+$ reductase transfers two electrons to reduce $NADP^+$ to NADPH using H^+ from the stroma.

INTERACTIVE QUESTION 10.5

Identify the components of linear electron flow in the following diagram. Circle the important products that will provide chemical energy and reducing power to the Calvin cycle.

Cyclic Electron Flow   In **cyclic electron flow**, electrons excited from P700 in PSI are passed from Fd to the cytochrome complex and back to P700. Several groups of photosynthetic bacteria have only a single photosystem and generate ATP by cyclic electron flow. Photosynthesis may have evolved in a form similar to cyclic electron flow. Cyclic electron flow may be photoprotective in eukaryotic photosynthesizers.

INTERACTIVE QUESTION   10.6

a. On the diagram in Interactive Question 10.5, sketch the path that electrons from P700 take during cyclic electron flow.

b. Why is neither O_2 nor NADPH generated by cyclic electron flow?

c. How, then, is ATP produced by cyclic electron flow?

A Comparison of Chemiosmosis in Chloroplasts and Mitochondria   Chemiosmosis in mitochondria and in chloroplasts is very similar. Electron transport chains built into a membrane pump protons across the membrane as electrons are passed down the chain in a series of redox reactions. In respiration, however, organic molecules provide the electrons, and chemical energy is transferred to ATP; in chloroplasts, by contrast, water provides the electrons, and light energy is transformed to the chemical energy of ATP.

In chloroplasts, the electron transport chain pumps protons from the stroma into the thylakoid space. As H^+ diffuses back through ATP synthase, ATP is formed on the stroma side, where it is available to the Calvin cycle.

INTERACTIVE QUESTION   10.7

a. In the light, the proton gradient across the thylakoid membrane is as great as 3 pH units. On which side is the pH lowest?

b. What three factors contribute to the formation of this large difference in H^+ concentration between the thylakoid space and the stroma?

10.3   **The Calvin cycle uses the chemical energy of ATP and NADPH to reduce CO_2 to sugar**

It takes three turns of the Calvin cycle to fix three molecules of CO_2 and produce one molecule of the three-carbon sugar **glyceraldehyde-3-phosphate (G3P)**. The cycle can be divided into three phases:

1. *Carbon fixation:* CO_2 is added to a five-carbon sugar, ribulose bisphosphate (RuBP), in a reaction catalyzed by the enzyme RuBP carboxylase-oxygenase (**rubisco**). The resulting unstable six-carbon intermediate splits into two molecules of 3-phosphoglycerate.

2. *Reduction:* Each molecule of 3-phosphoglycerate is then phosphorylated by ATP to form 1,3-bisphosphoglycerate. Two electrons from NADPH reduce this compound to G3P. The cycle must turn three times to create a net gain of one molecule of G3P.

3. *Regeneration of CO_2 acceptor (RuBP):* The rearrangement of five molecules of G3P into three molecules of RuBP requires three more ATP.

INTERACTIVE QUESTION 10.8

Label the three phases (a, b, and c) and key molecules in the following diagram of the Calvin cycle. How many ATP and NADPH are needed to synthesize one G3P molecule?

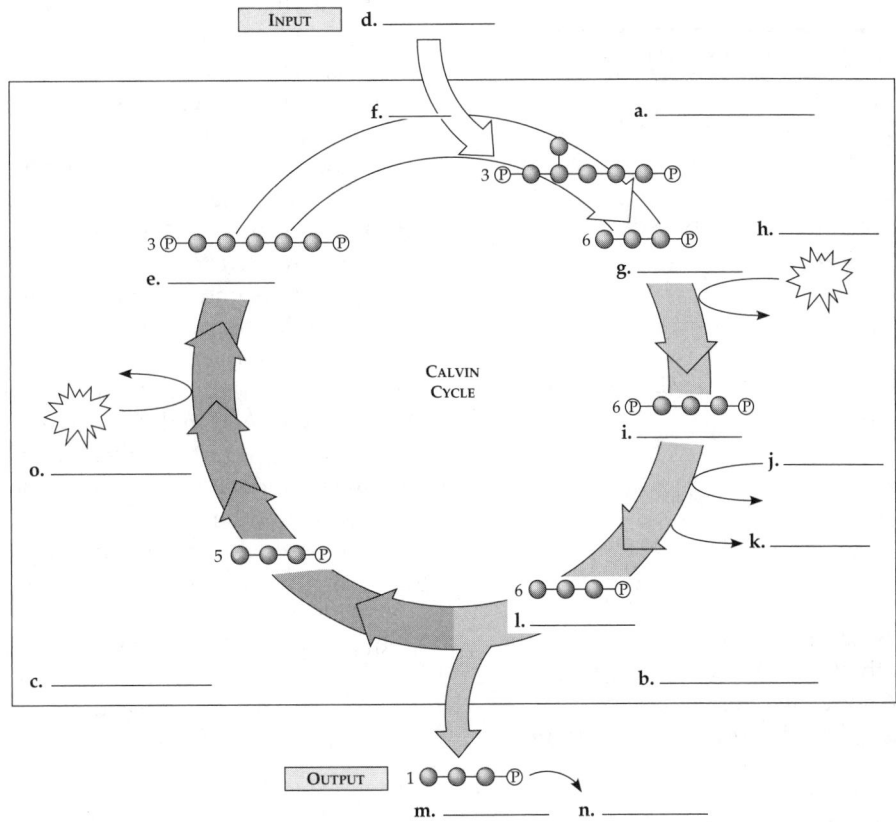

10.4 Alternative mechanisms of carbon fixation have evolved in hot, arid climates

Photorespiration: An Evolutionary Relic?   In most plants, the first product of carbon fixation is a three-carbon compound, 3-phosphoglycerate, formed in the Calvin cycle. When these **C₃ plants** close their stomata on hot, dry days to limit water loss, CO_2 concentration in the leaf air spaces falls, slowing the Calvin cycle. As more O_2 from the light reactions accumulates, rubisco adds O_2 to RuBP in place of CO_2. The product splits, and a two-carbon compound exits the chloroplast and is broken down to release CO_2. This seemingly wasteful process is called **photorespiration**.

INTERACTIVE QUESTION 10.9

What are two possible explanations for the existence of photorespiration, a process that can result in the loss of as much as 50% of the carbon fixed in the Calvin cycle?

C₄ Plants   In **C₄ plants**, CO_2 is first added to the three-carbon compound phosphoenolpyruvate (PEP), with the aid of an enzyme (**PEP carboxylase**) that has a high affinity for CO_2 and no affinity for O_2. The resulting four-carbon compound, formed in the mesophyll cells of the leaf, is transported to **bundle-sheath cells** tightly packed around the veins of the leaf. The compound is broken down to release CO_2 and pyruvate. The CO_2 is fixed in the Calvin cycle by rubisco, and the pyruvate is returned to the mesophyll cells where ATP is used to regenerate PEP.

Increasing atmospheric CO_2 concentrations and the resulting global climate change with an increase in temperature may affect the distribution of C_3 and C_4 plants and change the structure of various plant communities.

INTERACTIVE QUESTION   10.10

a. Where does the Calvin cycle take place in C_4 plants?

b. How can C_4 plants successfully utilize the Calvin cycle in hot, dry conditions when C_3 plants would be undergoing photorespiration?

c. Why does C_4 photosynthesis require more ATP than does C_3 photosynthesis?

CAM Plants   Many succulent plants close their stomata during the day, helping to prevent water loss. At night, they open their stomata and take up CO_2, incorporating it into a variety of organic acids in a mode of carbon fixation called **crassulacean acid metabolism (CAM)**. During daylight, **CAM plants** break these compounds down and release CO_2, allowing the Calvin cycle to proceed. Unlike the C_4 pathway, the CAM pathway does not structurally separate carbon fixation from the Calvin cycle; instead, the two processes are separated in time.

10.5   Life depends on photosynthesis

The Importance of Photosynthesis: A Review   Photosynthesis uses the energy of sunlight to drive the formation of ATP and NADPH, and then uses these molecules to produce G3P in the Calvin cycle. G3P can then be converted to other organic compounds. Some of this organic material is used as fuel in cellular respiration, just as in heterotrophs. Some G3P is converted to sucrose for transport within the plant; sugar beyond the immediate requirements of the plant is typically stored as starch. Much of the G3P is used in the form of glucose to produce the important structural molecule cellulose.

Directly or indirectly, photosynthesis provides food for most heterotrophs. Photosynthesis is also responsible for atmospheric oxygen as a product of the splitting of water in PSII.

Word Roots

auto- = self; **-troph** = nourish (*autotroph:* an organism that obtains organic food molecules without eating other organisms, but by using energy from the sun or from the oxidation of inorganic substances to make organic molecules from inorganic ones)

chlor- = green; **phyll-** = leaf (*chlorophyll:* a green, photosynthetic pigment located in membranes within the chloroplasts of plants and algae and in the membranes of certain prokaryotes)

electr- = electricity; **magnet-** = magnetic (*electromagnetic spectrum:* the entire spectrum of electromagnetic radiation, ranging in wavelength from less than a nanometer to more than a kilometer)

heter- = other (*heterotroph:* an organism that obtains organic food molecules by eating other organisms or substances derived from them)

meso- = middle (*mesophyll:* leaf cells specialized for photosynthesis, located in C_3 and CAM plants between the upper and lower epidermis; in C_4 plants, they are located between the bundle-sheath cells and the epidermis)

phot- = light (*photosystem:* a light-capturing unit located in the thylakoid membrane, consisting of a reaction-center complex surrounded by numerous light-harvesting complexes)

Structure Your Knowledge

1. You have already filled in the blanks in several diagrams of photosynthesis. To really understand this process, however, try to create your own representation. In a diagrammatic form, trace the flow of electrons and the production of ATP and NADPH in the light reactions. Then outline the three major stages in the production of G3P. Indicate where these reactions occur in the chloroplast. You can compare your diagram to the sketch in the answers section. Perhaps your study group can collaboratively create a clear, concise summary of this chapter.

2. Create a concept map to confirm your understanding of the chemiosmotic synthesis of ATP in photophosphorylation.

Test Your Knowledge

MULTIPLE CHOICE: *Choose the one best answer.*

1. Which of the following processes or structures is mismatched with its location?
   a. electron transport chain—thylakoid membrane
   b. Calvin cycle—stroma
   c. ATP synthase—double membrane surrounding chloroplast
   d. splitting of water—thylakoid space

2. A plant is growing under bright lights in a sealed container. What will happen to the atmosphere in the chamber?
   a. The amount of O_2 will increase and the amount of CO_2 will decrease.
   b. The amount of O_2 will decrease and the amount of CO_2 will increase.
   c. The amounts of both O_2 and CO_2 will increase.
   d. The amounts of both O_2 and CO_2 will decrease.

3. The immediate product of the Calvin cycle is glyceraldehyde-3-phosphate (G3P), a three-carbon molecule. What is *most* of this G3P used for?
   a. It provides the energy required to synthesize ATP.
   b. It is used as the starting point for the synthesis of six-carbon sugars, such as glucose.
   c. It is used to regenerate the RuBP that is required for the Calvin cycle.
   d. It is combined with CO_2 to produce four-carbon molecules such as oxaloacetate.

4. A spectrophotometer can be used to measure the
   a. absorption spectrum of a pigment.
   b. action spectrum of a reaction.
   c. amount of energy in a photon.
   d. wavelength of visible light.

5. Accessory pigments within chloroplasts are responsible for
   a. driving the splitting of water molecules.
   b. absorbing photons of different wavelengths of light and passing that energy to P680 or P700.
   c. providing electrons to the reaction-center chlorophyll *a* molecules.
   d. pumping H^+ across the thylakoid membrane to create a proton-motive force.

6. The following diagram is an absorption spectrum for an unknown pigment molecule. What color would this pigment appear to you?

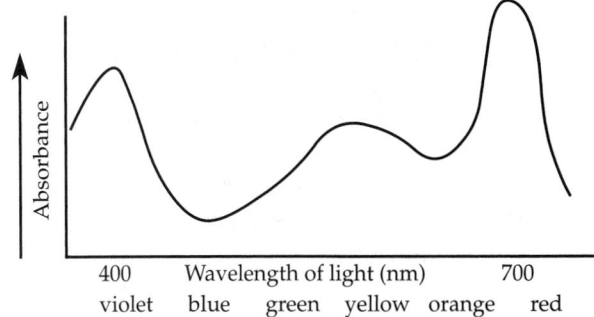

   a. violet   c. green
   b. blue   d. red

7. Linear electron flow along with chemiosmosis in the chloroplast results in the formation of
   a. ATP only.
   b. ATP and NADPH.
   c. ATP and O_2.
   d. ATP, NADPH, and O_2.

8. The chlorophyll known as P680$^+$ has its electron "holes" filled by electrons from
   a. photosystem I.
   b. photosystem II.
   c. water.
   d. accessory pigments.

9. The enzyme rubisco
   a. uses NADPH and ATP to make G3P.
   b. fixes CO_2 to RuBP.
   c. uses ATP to regenerate RuBP.
   d. removes electrons from water.

10. In photosynthesis, the splitting of water
    a. releases oxygen, which will be used in the Calvin cycle.
    b. provides electrons that reduce NADP$^+$ to NADPH.
    c. provides electrons that produce the electron gradient used in the chemiosmotic synthesis of ATP.
    d. occurs in photosystem II on the stroma side of the thylakoid membrane.

11. The reactions of the Calvin cycle occur in the
    a. stroma.
    b. cytoplasm.
    c. thylakoid space.
    d. intermembrane space.

12. A difference between electron transport in photosynthesis and respiration is that in photosynthesis,
    a. NADPH rather than NADH passes electrons to the electron transport chain.
    b. ATP synthase releases ATP into the stroma rather than into the cytosol.
    c. light provides the energy to push electrons to the top of the electron chain, rather than energy from the oxidation of food molecules.
    d. an H^+ concentration gradient rather than a proton-motive force drives the phosphorylation of ATP.

13. How does cyclic electron flow differ from linear electron flow?
    a. No NADPH is produced by cyclic electron flow.
    b. No O_2 is produced by cyclic electron flow.
    c. The cytochrome complex in the electron transport chain is not involved in cyclic electron flow.
    d. Both a and b are correct.

14. Chloroplasts could make carbohydrate in the dark if provided with
    a. ATP, NADPH, and CO_2.
    b. an artificially induced proton gradient.
    c. organic acids or four-carbon compounds.
    d. photons and CO_2.

15. The disaccharide sucrose is the form of carbohydrate transported throughout a plant body. How many turns of the Calvin cycle does it take to produce one molecule of sucrose?
    a. 2   c. 11
    b. 6   d. 12

16. Which of the following is common to both chloroplasts and mitochondria?
    a. the citric acid cycle
    b. chemiosmotic ATP synthesis
    c. carbohydrate synthesis
    d. splitting of water

17. RuBP carboxylase-oxygenase is the enzyme that
    a. can add O_2 to RuBP when CO_2 levels are low and O_2 levels are high.
    b. regenerates RuBP with the aid of ATP.
    c. adds CO_2 to RuBP in the carbon fixation stage.
    d. Both a and c are correct.

18. In C_4 plants,
    a. initial carbon fixation takes place in the mesophyll cells.
    b. photorespiration requires more energy than it does in C_3 plants.
    c. the Calvin cycle, which takes place in the bundle-sheath cells, uses PEP carboxylase instead of rubisco because of its greater affinity for CO_2.
    d. All of the above are correct.

19. CAM plants avoid photorespiration by
    a. keeping their stomata closed during the day.
    b. performing the Calvin cycle at night.
    c. fixing CO_2 into four-carbon compounds in the mesophyll, which then release CO_2 in the bundle-sheath cells.
    d. fixing CO_2 into organic acids during the night, which then provide CO_2 during the day.

20. In green plants, most of the ATP for synthesis of proteins, cytoplasmic streaming, and other cellular activities comes directly from
    a. photosystem II.
    b. cyclic electron flow.
    c. oxidative phosphorylation.
    d. photophosphorylation.

For each of the events listed in questions 21 through 26, indicate whether the event occurs during
    a. respiration
    b. photosynthesis
    c. both respiration and photosynthesis
    d. neither respiration nor photosynthesis

21. Chemiosmotic synthesis of ATP

22. Reduction of oxygen

23. Oxidative phosphorylation

24. Reduction of CO_2

25. Oxidation of NAD^+

26. Reduction of $NADP^+$

Chapter 11

Cell Communication

Key Concepts

11.1 External signals are converted to responses within the cell

11.2 Reception: A signaling molecule binds to a receptor protein, causing it to change shape

11.3 Transduction: Cascades of molecular interactions relay signals from receptors to target molecules in the cell

11.4 Response: Cell signaling leads to regulation of transcription or cytoplasmic activities

11.5 Apoptosis integrates multiple cell-signaling pathways

Framework

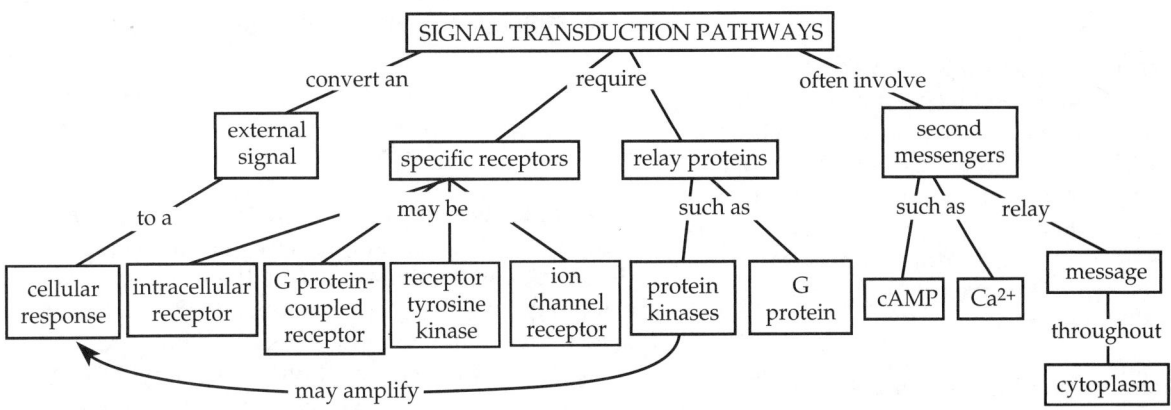

Chapter Review

Cell communication is critical to the development and functioning of multicellular organisms and also occurs between unicellular organisms. The similarity of mechanisms of cell signaling provides evidence for the evolutionary relatedness of all life.

11.1 External signals are converted to responses within the cell

Evolution of Cell Signaling   Each of the two mating types of yeast (**a** and **α**) secretes chemical factors that bind to receptors on the other mating type, initiating fusion (mating) of the cells. The series of steps involved in the conversion of a signal on a cell's surface to a cellular response is called a *signal transduction pathway.* Similarities among these pathways in prokaryotes, yeast, and multicellular organisms suggest an early evolution of cell-signaling mechanisms.

The concentration of signaling molecules allows some bacteria to sense their local density, a process called *quorum sensing.* Aggregations of bacterial cells called *biofilms* may form in response to signaling within a population. Quorum sensing may also allow pathogenic bacteria to produce toxins as part of an infection.

Local and Long-Distance Signaling   Chemical signals may be communicated by direct contact between cells through cytoplasmic connections (gap junctions or plasmodesmata) or through contact of membrane-bound surface molecules (cell–cell recognition in animal cells).

In *paracrine signaling* in animals, a signaling cell releases messenger molecules into the extracellular fluid, and these local regulators influence nearby cells. *Growth factors* are one class of local regulators. In another type of local signaling called *synaptic signaling*, a nerve cell releases neurotransmitter molecules, which diffuse across the narrow synapse to its target cell.

Hormones are chemical signals that travel to more distant cells. In hormonal or *endocrine signaling* in animals, the circulatory system transports hormones throughout the body to reach and bind to target cells that have appropriate receptors.

INTERACTIVE QUESTION 11.1

How do plant hormones (often called *plant growth regulators*) travel between secreting cells and target cells?

The Three Stages of Cell Signaling: A Preview
E. W. Sutherland's studies of epinephrine's effect on the hydrolysis of glycogen in liver cells established that cell signaling involves three stages: **reception** of a chemical signal by binding to a receptor protein either inside a target cell or on its surface; **transduction** of the signal, often by a **signal transduction pathway**, to a form that can elicit a cellular response; and the specific **response** of the cell.

11.2   Reception: A signaling molecule binds to a receptor protein, causing it to change shape

A signaling molecule acts as a **ligand**, which specifically binds to a receptor protein and usually activates the receptor by inducing a change in the receptor protein's shape or the aggregation of receptors.

Receptors in the Plasma Membrane   There are three major types of transmembrane receptors that bind with water-soluble signaling molecules and transmit information into the cell. Malfunctions of these receptors are associated with many human diseases.

The various receptors that work with the aid of a **G protein**, called **G protein-coupled receptors (GPCRs)**, are structurally similar, with seven helices of a single polypeptide spanning the plasma membrane. Binding of the appropriate extracellular signaling molecule to a GPCR activates the receptor, which then binds to and activates a specific G protein located on the cytoplasmic side of the membrane. This activation occurs when a GTP nucleotide replaces the GDP bound to the G protein. The G protein then usually activates a membrane-bound enzyme, after which it hydrolyzes its GTP and becomes inactive again. The activated enzyme triggers the next step in the pathway to the cell's response.

GPCR signaling systems are involved in the function of many hormones and neurotransmitters and in embryological development and sensory reception. Many bacteria produce toxins that interfere with G-protein function; up to 60% of all medicines influence G-protein pathways.

INTERACTIVE QUESTION 11.2

Explain why G protein-coupled receptor pathways shut down rapidly in the absence of a signal molecule.

Receptor tyrosine kinases (RTKs) are membrane receptors with enzymatic activity that can trigger several pathways at once. Part of the receptor protein extending into the cytoplasm functions as a tyrosine kinase, an enzyme that transfers phosphate groups from ATP to the amino acid tyrosine. Many RTKs exist as a single transmembrane monomer with a ligand binding site, a transmembrane α helix, and a cytoplasmic tail with a series of tyrosine amino acids. Ligand binding causes two receptor monomers to form a dimer, which causes the tyrosine kinases to phosphorylate the tyrosines on each other's cytoplasmic tails. Different relay proteins then bind to specific phosphorylated tyrosines and become activated, triggering many different transduction pathways in response to one type of signal.

INTERACTIVE QUESTION 11.3

Label the parts in the following diagram of an activated receptor tyrosine kinase dimer.

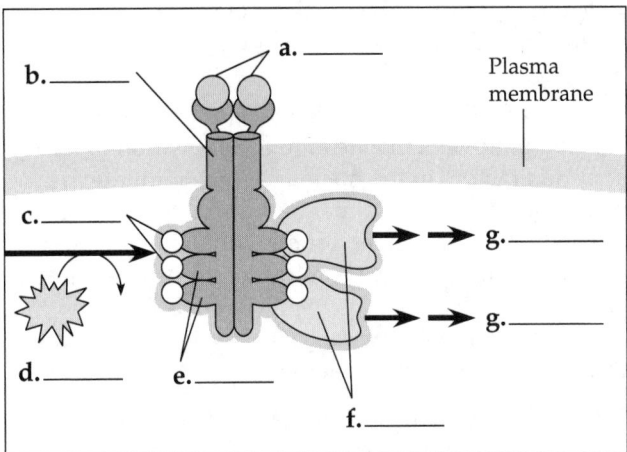

The binding of a signaling molecule to a **ligand-gated ion channel** opens or closes a "gate," thereby allowing or blocking the flow of specific ions through the receptor channel. The resulting change in ion concentration inside the cell triggers a cellular response. Neurotransmitters often bind to ligand-gated ion channels in the transmission of neural signals. Some voltage-gated ion channels respond to electrical signals.

A human G protein-coupled receptor has recently been crystallized and its structure determined. Understanding the structure and function of GPCRs and TRKs may facilitate the development of treatments for diseases such as asthma, heart disease, and cancers associated with malfunctioning receptors.

Intracellular Receptors   Hydrophobic chemical messengers and small signaling molecules such as the gas nitric oxide may cross a cell's plasma membrane and bind to receptors in the cytoplasm or nucleus of target cells. Steroid hormones activate receptors in target cells that function as *transcription factors* that regulate gene expression.

INTERACTIVE QUESTION 11.4

Describe the difference between signaling molecules that bind to cell membrane receptors or to intracellular receptors.

11.3 Transduction: Cascades of molecular interactions relay signals from receptors to target molecules in the cell

Multistep pathways enable a small number of extracellular signals to be amplified to produce a large cellular response and also provide opportunities for regulation and coordination.

Signal Transduction Pathways   The relay molecules in a signal transduction pathway are usually proteins, which interact as they pass the message from the extracellular signaling molecule to the protein that produces the cellular response.

Protein Phosphorylation and Dephosphorylation
Protein kinases are enzymes that transfer phosphate groups from ATP to proteins, often to the amino acids serine or threonine. Relay molecules in signal transduction pathways are often protein kinases, which are sequentially phosphorylated. Phosphorylation produces a shape change that usually activates each enzyme. Hundreds of different kinds of protein kinases regulate the activity of a cell's proteins.

INTERACTIVE QUESTION 11.5

a. What does a protein kinase do?

b. What is a **phosphorylation cascade**?

c. What does a **protein phosphatase** do?

Small Molecules and Ions as Second Messengers
Small, water-soluble molecules or ions often function as **second messengers**, which rapidly relay the signal from the receptor-bound ligand (the "first messenger") into a cell's interior.

Binding of an extracellular signal to a receptor such as a GPCR activates a G protein that may activate **adenylyl cyclase**, a membrane protein that converts ATP to cyclic adenosine monophosphate (**cyclic AMP** or **cAMP**). The cAMP often activates *protein kinase A*, which phosphorylates other proteins. Phosphodiesterase, a cytoplasmic enzyme, converts cAMP to inactive AMP, thereby removing the second messenger.

Some signal molecules may activate an *inhibitory* G protein that inhibits adenylyl cyclase.

The second messenger *cyclic GMP* (*cGMP*) is produced by smooth muscle cells in response to nitric oxide. Because cGMP causes relaxation of smooth muscle in blood vessels, blood flow in arteries is increased. The drug Viagra inhibits the conversion of cGMP to GMP, leading to increased blood flow to erectile tissue in the penis.

INTERACTIVE QUESTION   11.6

Label the components in the following diagram depicting the steps in a signal transduction pathway that uses cAMP as a second messenger.

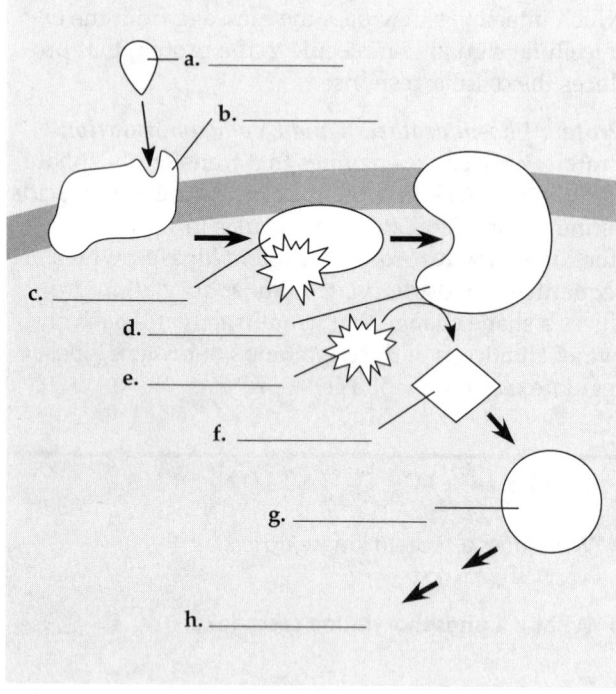

a. _____

b. _____

c. _____

d. _____

e. _____

f. _____

g. _____

h. _____

Calcium ions are widely used as a second messenger in many G protein and receptor tyrosine kinase pathways. Cytosolic concentration of Ca^{2+} is normally kept low by active transport of calcium out of the cell and into the endoplasmic reticulum. The release of Ca^{2+} from the ER involves another second messenger, **inositol trisphosphate (IP$_3$)**. In response to the reception of a signal, the enzyme phospholipase C cleaves a membrane phospholipid into two second messengers, IP_3 and **diacylglycerol (DAG)**. IP_3 binds to and opens IP_3-gated calcium channels in the ER. The calcium ions then activate cellular responses.

INTERACTIVE QUESTION   11.7

Fill in the blanks to review a G protein-coupled pathway that uses Ca^{2+} as a second messenger.

A **a.**_____ binds to a G protein-coupled receptor. An activated **b.**_____ activates the enzyme phospholipase C, which cleaves a **c.**_____ into DAG and

---

d._____, which binds to and opens a ligand-gated channel, releasing **e.**_____ from the **f.**_____.

11.4 Response: Cell signaling leads to regulation of transcription or cytoplasmic activities

Nuclear and Cytoplasmic Responses   Signal transduction pathways may lead to the activation of transcription factors in the nucleus, which regulate the expression of specific genes and ultimately to changes in production of specific proteins. Outside the nucleus, signaling pathways may also activate existing enzymes, open or close protein channels in membranes, or influence overall cell activity by orienting the growth of the cytoskeleton.

Regulation of the Response   As one aspect of response regulation, a signal transduction pathway amplifies a signal in an enzyme cascade. Because each successive enzyme in the pathway remains active long enough to process multiple molecules, which then activate the next step, a cell's response to a signal is amplified.

As a result of their particular set of receptor proteins, relay proteins, and response proteins, different cells can respond to different signals or can exhibit different responses to the same molecular signal. Pathways may branch to produce multiple responses, or two pathways may interact ("cross-talk") to mediate a single response.

Scaffolding proteins are large relay proteins to which other relay proteins attach, increasing the efficiency of signal transduction in a pathway. Scaffolding proteins that permanently attach networks of signaling-pathway proteins at synapses have been identified in brain cells.

INTERACTIVE QUESTION   11.8

How does each of the following *inactivation* mechanisms discontinue a cell's response to a signal and maintain the cell's ability to respond to fresh signals?

a. reversible binding of signaling molecules

b. GTPase activity of G protein

c. phosphodiesterase

d. protein phosphatases

11.5 Apoptosis integrates multiple cell-signaling pathways

In the best understood type of "programmed cell death," called **apoptosis**, cellular components are chopped up and packaged into vesicles that are then shed and engulfed by scavenger cells. Signals that trigger apoptosis can be external or internal.

Apoptosis in the Soil Worm **Caenorhabditis elegans**   Apoptosis occurs numerous times during normal development in *C. elegans.* The *ced-9* gene produces a protein that inhibits the activity of Ced-3 and Ced-4, protein products of genes *ced-3* and *ced-4.* When a cell receives a death signal on its membrane receptor, Ced-9 protein becomes inactivated, and the apoptosis pathway activates proteases and nucleases that hydrolyze the cell's proteins and DNA. Ced-3 is the main *caspase* (protease) of apoptosis in this nematode.

Apoptotic Pathways and the Signals That Trigger Them   One type of apoptotic pathway in mammals involves proteins, such as cytochrome *c*, that are released from mitochondria when proteins in the outer membrane are triggered to form pores. In other cases, binding of a death-signaling ligand to a cell-surface receptor leads to the activation of caspases that carry out apoptosis. Other signals can come from the nucleus when the DNA has suffered irreparable damage or from the endoplasmic reticulum in response to extensive protein misfolding.

Programmed apoptosis is part of normal development. Faulty cell suicide programs have been implicated in some neurological diseases and in cancer.

INTERACTIVE QUESTION   11.9

a. What can one conclude from the fact that the proteins involved in mitochondrial apoptosis in mammals are similar to the Ced-3, Ced-4, and Ced-9 proteins of nematodes?

b. Give some examples of programmed cell death in humans.

Word Roots

trans- = across (*signal transduction pathway*: a series of steps linking a mechanical, chemical, or electrical stimulus to a specific cellular response)

-yl = substance or matter (*adenylyl cyclase*: an enzyme that converts ATP to cAMP in response to an extracellular signal)

Structure Your Knowledge

1. Why is cell signaling such an important aspect of a cell's life?
2. Briefly describe the three stages of cell signaling.
3. Some signaling pathways alter a protein's activity; others result in the production of new proteins. Explain the mechanisms for these two different responses.
4. How does a phosphorylation cascade produce an amplified response to a signal molecule?

Test Your Knowledge

MULTIPLE CHOICE: *Choose the one best answer.*

1. What is a key difference between a local regulator and a hormone?
   a. Local regulators are small, hydrophobic molecules; hormones are either larger polypeptides or steroids.
   b. Local regulators diffuse to neighboring cells; hormones usually travel throughout the plant or animal body to distant target cells.
   c. Local regulators initiate short-term responses; hormones trigger longer-lasting responses to environmental stimuli.
   d. Local regulators often open ligand-gated channels and affect ion concentrations in a cell; hormones bind with intracellular receptors and affect gene expression.

2. Steroid hormones
   a. bind to cell-surface receptors.
   b. do not require specific receptor molecules.
   c. remain outside the cells that they influence.
   d. influence gene expression.

3. A signaling molecule that binds to a plasma-membrane protein receptor functions as a
   a. ligand.
   b. second messenger.
   c. protein kinase.
   d. receptor protein.

4. Which of the following best describes a G protein?
   a. a type of membrane-receptor protein with seven transmembrane α helices
   b. a protein on the cytoplasmic side of a membrane that becomes activated by a transmembrane receptor protein
   c. a membrane-bound enzyme that converts ATP to cAMP
   d. a membrane-bound protein that cleaves phospholipids to produce second messengers

5. Which of the following is a key difference between GPCRs and RTKs?
   a. GPCRs use cGMP, but RTKs use cAMP.
   b. GPCRs are membrane bound, but RTKs are found in the cytoplasm.
   c. GPCRs are common in animal cells, but RTKs are common in plant cells.
   d. GPCRs generally activate a single pathway, but RTKs may trigger many signal transduction pathways.

6. Many human diseases (including bacterial infections) and the medicines used to treat them produce their effects by influencing which of the following?
   a. cAMP concentrations in the cell
   b. Ca^{2+} concentrations in the cell
   c. G-protein pathways
   d. ligand-gated ion channels

7. A G protein becomes activated when
   a. GDP is replaced by GTP.
   b. GTP is hydrolyzed.
   c. it binds Ca^{2+} ions.
   d. a phosphate group is transferred to it by a protein kinase.

8. If a protein is activated by the addition of a phosphate group, it will most likely be inactivated by the action of
   a. a protein kinase.
   b. a ligand.
   c. a protein phosphatase.
   d. cAMP.

9. One function of the second messenger IP_3 is to
   a. bind to and activate protein kinase A.
   b. activate transcription factors.
   c. convert ATP to cAMP.
   d. bind to and open ligand-gated calcium channels on the ER.

10. Signal amplification is most often achieved by
    a. a phosphorylation cascade involving multiple relay proteins.
    b. the binding of multiple signaling molecules.
    c. branching pathways that produce multiple cellular responses.
    d. the activation of transcription factors that increase gene expression.

11. From studying the effects of epinephrine on liver cells, Sutherland concluded that
    a. there is a 1-to-10,000 correlation between the number of epinephrine molecules bound to receptors and the number of glucose molecules released from glycogen.
    b. epinephrine enters liver cells and binds to receptors that function as transcription factors to turn on the gene for glycogen phosphorylase.
    c. there is a "second messenger" that transmits the signal of epinephrine binding on the plasma membrane to the enzymes involved in glycogen breakdown inside the cell.
    d. epinephrine functions as a ligand to open ion channels in the plasma membrane that allow Ca^{2+} to enter and initiate a response.

12. You are a Ca^{2+} ion in a typical animal cell. Where are you most likely located?
    a. in the nucleus
    b. in the cytoplasm
    c. in a calciosome
    d. in the endoplasmic reticulum

13. Which of the following molecules is *incorrectly* matched with its description?
    a. scaffolding protein—large relay protein that may bind with several other relay proteins to increase the efficiency of a signaling pathway
    b. protein phosphatase—enzyme that transfers a phosphate group from ATP to a protein, causing a shape change that usually activates that protein
    c. adenylyl cyclase—enzyme attached to plasma membrane that converts ATP to cAMP in response to an extracellular signal
    d. phospholipase C—enzyme that may be activated by a G protein or receptor tyrosine kinase and cleaves a plasma-membrane phospholipid into the second messengers IP_3 and DAG

14. When epinephrine binds to cardiac (heart) muscle cells, it speeds their contraction. When it binds to muscle cells of the small intestine, it inhibits their contraction. How can the same hormone have different effects on muscle cells?

    a. Cardiac cells have more receptors for epinephrine than do intestinal muscle cells.

    b. Epinephrine circulates to the heart first and is in higher concentration around cardiac cells.

    c. The two types of muscle cells have different signal transduction pathways for epinephrine and thus have different cellular responses.

    d. Epinephrine binds to G protein-coupled receptors in cardiac cells, which increase the response to a signal. Epinephrine binds to receptor tyrosine kinases in intestinal muscle cells, which decrease the response to a signal.

15. The gene *ced-9* codes for a protein that inhibits the Ced-3 and Ced-4 caspase proteins in *C elegans*. For development to proceed normally, *ced-9*

    a. should be expressed in all cells, but its protein product must remain inactive.

    b. should be expressed in all cells, but its product should be inactivated when cells programmed to die receive the proper signal.

    c. should not be expressed except when a cell receives a signal to die.

    d. should not be expressed in those cells that are not programmed to die during development.

Chapter 12

The Cell Cycle

Key Concepts

12.1 Most cell division results in genetically identical daughter cells

12.2 The mitotic phase alternates with interphase in the cell cycle

12.3 The eukaryotic cell cycle is regulated by a molecular control system

Framework

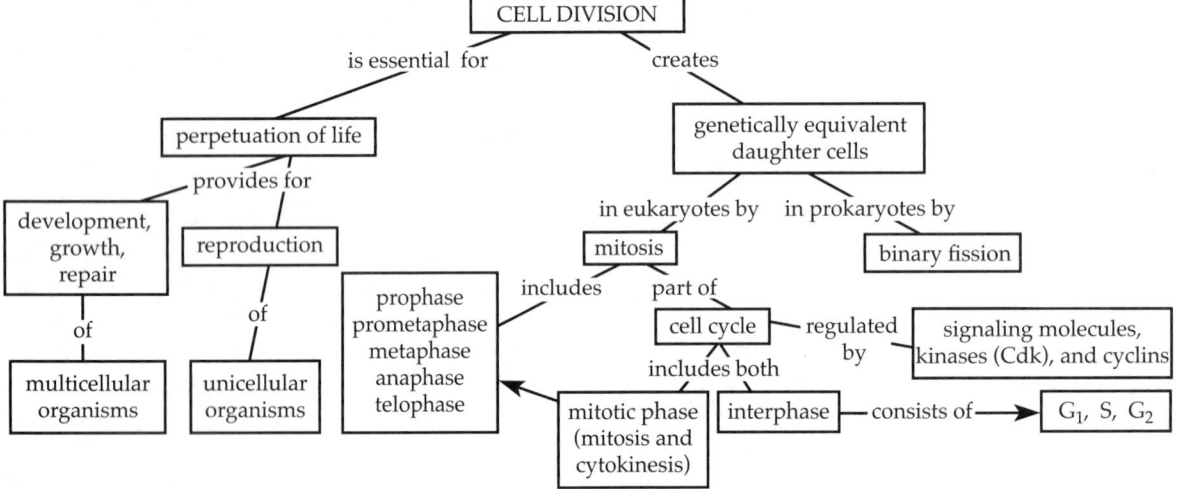

Chapter Review

Cell division, the reproduction of cells, creates duplicate offspring in unicellular organisms and provides for growth, development, and repair in multicellular organisms. The **cell cycle** extends from the formation of a new cell by the division of its parent cell to its own division into two daughter cells.

12.1 Most cell division results in genetically identical daughter cells

The process of re-creating a structure as intricate as a cell necessitates the exact replication and equal division of the DNA containing the cell's genetic program.

Cellular Organization of the Genetic Material

A cell's complement of DNA is called its **genome**, which is organized into **chromosomes**. Each eukaryotic chromosome is a very long, linear DNA molecule with associated proteins that help structure the chromosome and control the activity of genes, the units of inheritance. The entire complex of DNA and protein is called **chromatin**.

Every diploid eukaryotic species has a characteristic number of chromosomes in each **somatic cell**; reproductive cells, or **gametes** (egg and sperm), have half that number of chromosomes.

Distribution of Chromosomes During Eukaryotic Cell Division
Prior to cell division, a cell replicates its DNA. Duplicated chromosomes consist of two

identical **sister chromatids** attached along their length by proteins called *cohesins*. This attachment is called *sister chromatid cohesion*. Each sister chromatid has a **centromere**, a region where proteins bind to specific centromeric DNA sequences and hold the chromatids closely together. Regions on either side of the centromere are called the chromatid *arms*.

The two sister chromatids separate during **mitosis** (the division of the genetic material in the nucleus), and then the cytoplasm divides during **cytokinesis**, producing two separate, genetically equivalent cells. By definition, when sister chromatids separate each is considered a chromosome in its own right.

A type of cell division called *meiosis* produces daughter cells that have half the number of chromosomes as the parent cell. When a sperm fertilizes an egg (both of which are formed by meiosis in animals), the chromosome number is restored, and the somatic cells of the new offspring again have two sets of chromosomes.

INTERACTIVE QUESTION 12.1

a. How many chromosomes are in your somatic cells?

b. How many chromosomes are in your gametes?

c. How many chromatids are in one of your somatic cells that has duplicated its chromosomes prior to mitosis?

12.2 The mitotic phase alternates with interphase in the cell cycle

Phases of the Cell Cycle The cell cycle consists of the **mitotic (M) phase**, which includes mitosis and cytokinesis, and **interphase**, during which the cell grows and duplicates its chromosomes. Interphase, usually lasting 90% of the cell cycle, includes the G_1 **phase**, the **S phase**, and the G_2 **phase**. Mitosis is conventionally divided into five stages: **prophase, prometaphase, metaphase, anaphase,** and **telophase.**

INTERACTIVE QUESTION 12.2

a. What event is common to all three subphases of interphase?

b. What key event happens during the S phase?

The Mitotic Spindle: A Closer Look The **mitotic spindle** consists of fibers made of microtubules and associated proteins, and its assembly, in animal cells, begins in the **centrosome**, a type of *microtubule-organizing center*. A pair of centrioles is centered in each centrosome, but centrioles are not required for cell division. The single centrosome duplicates during interphase.

During prophase, the nucleoli disappear and the chromatin fibers coil and fold into visible chromosomes, consisting of sister chromatids joined at their centromeres and by sister chromatid cohesion. As spindle microtubules grow out from them, the two centrosomes move to opposite poles of the cell. Radial arrays of shorter microtubules, called *asters*, extend from the centrosomes. During prometaphase, some of the spindle microtubules attach to each chromatid's **kinetochore**, a structure of protein associated with DNA located at the centromere region. Alternate movement of the chromosome on opposite kinetochore microtubules results in each chromosome locating at the midline of the cell. At metaphase, the centromeres of all the chromosomes are aligned at the **metaphase plate**, an imaginary plane across the midline of the spindle. Nonkinetochore microtubules extend out from each centrosome and overlap at the midline. Aster microtubules contact the plasma membrane.

The cohesins joining sister chromatids are cleaved by the enzyme *separase* to begin anaphase, and the now separate chromosomes move toward the poles. Motor proteins appear to "walk" a chromosome along the kinetochore microtubules by depolymerizing them at their kinetochore ends. In some types of cells, motor proteins at the spindle poles appear to "reel in" the chromosomes while the microtubules depolymerize at the poles. The extension of the spindle poles away from each other as an animal cell elongates is due to the overlapping *nonkinetochore microtubules* "walking" past each other, also using motor proteins.

In telophase, equivalent sets of chromosomes are at the two poles of the cell. Nuclear envelopes form, nucleoli reappear, cytokinesis begins, and the spindle microtubules depolymerize.

Cytokinesis: A Closer Look **Cleavage** is the process that separates the two daughter cells in animals. A **cleavage furrow** forms, as a ring of actin microfilaments interacting with myosin proteins begins to contract on the cytoplasmic side of the membrane. The cleavage furrow deepens until the dividing cell is pinched in two.

In plant cells, a **cell plate** forms from the fusion of Golgi-derived membrane vesicles containing cell wall materials. The membrane of the enlarging cell plate joins with the plasma membrane, separating the two daughter cells. A new cell wall develops between the cells from the contents of the cell plate.

INTERACTIVE QUESTION   12.3

The following diagrams will depict the five stages of mitosis in an animal cell after you draw in the missing chromosomes and kinetochore microtubules in stages c, d, and e. For simplicity, this cell has only four chromosomes, as you can see in stages b and f. Identify the stages and label the indicated structures.

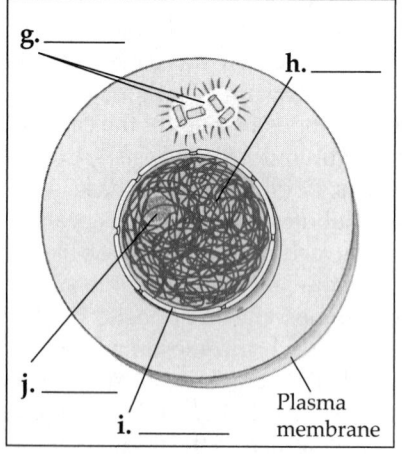

g. _____
h. _____
j. _____
i. _____
Plasma membrane

a. _____

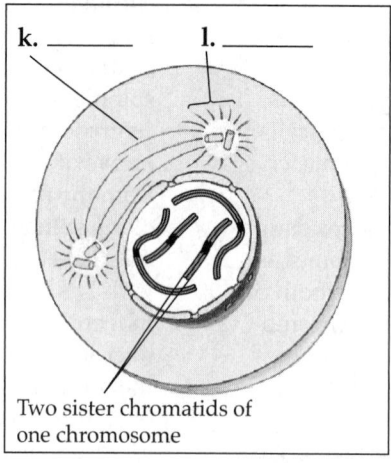

k. _____   l. _____
Two sister chromatids of one chromosome

b. _____

m. _____

c. _____

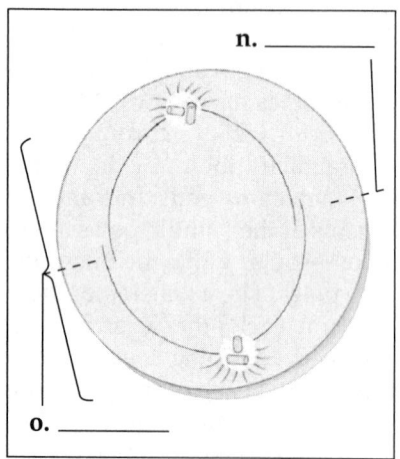

n. _____
o. _____

d. _____

e. _____

p. _____
q. _____

f. _____

Binary Fission in Bacteria   Single-celled eukaryotes reproduce asexually by a process of "dividing in half" known as **binary fission**, which involves mitosis. The binary fission of prokaryotes does not involve mitosis. The bacterial chromosome is a single circular DNA molecule with associated proteins. It begins to replicate at the **origin of replication**, and one of these duplicated origins moves to the opposite pole of the cell. Replication is completed as the cell doubles in size and the plasma membrane pinches inward to divide the two identical daughter cells. The mechanisms of chromosome movement and daughter cell separation are not fully understood but involve some proteins that resemble eukaryotic actin and tubulin.

The Evolution of Mitosis   Evidence for the evolution of mitosis from prokaryotic cell division includes (1) the relatedness of several proteins (such as actin and tubulin) involved in both types of division and (2) some possible intermediate stages seen in some unicellular eukaryotes, in which chromosomal separation takes place within an intact nuclear envelope.

12.3   The eukaryotic cell cycle is regulated by a molecular control system

Normal cell growth, development, and maintenance depend on proper control of the timing and rate of cell division.

The Cell Cycle Control System   Experiments that fuse two cells at different phases of the cell cycle indicate that cytoplasmic signaling molecules drive the cell cycle.

A **cell cycle control system**, consisting of a set of molecules that function cyclically, coordinates the events of the cell cycle. Important internal and external signals are monitored to determine whether the cell cycle will proceed past the three main **checkpoints** in the G_1, G_2, and M phases.

Protein kinases are enzymes that activate or inactivate other proteins by phosphorylating them. The changing concentrations of **cyclins**, regulatory proteins that attach to these kinases, affect the activity of **cyclin-dependent kinases**, or **Cdks**.

A cyclin–Cdk complex called **MPF**, for "maturation-promoting factor" or "M-phase-promoting factor," triggers passage past the G_2 checkpoint into M phase. In addition to phosphorylating proteins and other kinases that initiate mitotic events, MPF activates a protein breakdown process that destroys its cyclin component and thus MPF activity during anaphase. The Cdk portion of the complex remains to associate with new cyclin synthesized during the S and G_2 phases of the next cycle.

INTERACTIVE QUESTION   12.4

a. What is MPF?

b. Describe the relative concentrations of MPF and its constituent molecules throughout the cell cycle:

  MPF

  Cdk

  cyclin

Other Cdk proteins and cyclins appear to control the movement of a cell past the G_1 checkpoint. If a mammalian cell does not receive a go-ahead signal at the G_1 checkpoint, the cell will usually exit the cell cycle to a nondividing state called the G_0 **phase**.

An internal signal is required to move past the M phase checkpoint into anaphase. Only after all chromosomes are attached at their kinetochores to spindle microtubules is the enzyme separase activated and the cohesins holding sister chromatids together are cleaved. Additional checkpoints have been identified within S phase and between anaphase and telophase.

Growing animal cells in cell culture has allowed researchers to identify chemical and physical factors that affect cell division. Certain nutrients and regulatory proteins called **growth factors** have been found to be essential for cells to divide in culture. Mammalian fibroblasts have receptors on their plasma membranes for *platelet-derived growth factor (PDGF)*, which is released from blood platelets at the site of an injury. Binding of PDGF initiates a signal-transduction pathway that enables cells to pass the G_1 checkpoint.

Density-dependent inhibition of cell division appears to involve the binding of cell-surface proteins of adjacent cells, which sends a cell-division-inhibiting signal to both cells. Most animal cells also show **anchorage dependence** and must be attached to a substratum in order to divide.

Loss of Cell Cycle Controls in Cancer Cells   When grown in cell culture, cancer cells do not exhibit density-dependent inhibition or anchorage dependence, do not depend on growth factors to divide, do not undergo apoptosis when appropriate, and may continue to divide indefinitely instead of stopping after the typical 20 to 50 divisions of normal mammalian cells in culture. Cells in culture that can divide indefinitely are said to have undergone **transformation**.

When a normal cell undergoes changes that begin its conversion to a cancer cell, the body's immune system may destroy it. If it proliferates, a mass of abnormal cells develops within a tissue. A **benign tumor** remains at its original site and can be removed by surgery. The cells of a **malignant tumor** are considered to be *transformed* cells. They have significant genetic changes that enable them to invade and disrupt the functions of one or more organs. An individual with a malignant tumor is said to have cancer. Malignant tumor cells may have abnormal metabolism and unusual numbers of chromosomes. They may stimulate the growth of blood vessels, providing the tumor with a blood supply. They lose their attachments to other cells and may enter the blood and lymph systems and spread to other sites, a process called **metastasis**. Radiation treatment preferentially affects cancer cells because they are unable to repair the resulting DNA damage. Chemotherapy involves drugs that affect stages in the cell cycle and therefore target actively dividing cells.

Defects in cell-signaling pathways that affect the cell cycle are involved in the development of cancer. Advances in molecular techniques to analyze DNA and the levels of specific cancer-associated proteins increasingly enable physicians to match chemotherapy to a specific tumor in a particular patient.

Word Roots

ana- = up, throughout, again (*anaphase:* the fourth stage of mitosis, in which the chromatids of each chromosome have separated and the daughter chromosomes are moving to the poles of the cell)

bi- = two (*binary fission:* a method of asexual reproduction by "division in half." In prokaryotes, binary fission does not involve mitosis)

centr- = the center; **-mere** = a part (*centromere:* in a duplicated chromosome, the region on each sister chromatid where it is most closely attached to the other chromatid by proteins that bind to the centromeric DNA)

chrom- = colored (*chromatin:* the complex of DNA and proteins that makes up eukaryotic chromosomes)

cycl- = a circle (*cyclin:* a protein that occurs in a cyclically fluctuating concentration and plays an important role in regulating the cell cycle)

cyt- = cell; **-kinesis** = movement (*cytokinesis:* division of the cytoplasm to form two separate daughter cells)

gamet- = a wife or husband (*gamete:* a haploid reproductive cell, such as an egg or sperm)

inter- = between (*interphase:* the period in the cell cycle when the cell is not dividing)

mal- = bad or evil (*malignant tumor:* a cancerous tumor containing cells that have significant genetic and cellular changes and are capable of invading and surviving in new sites)

meta- = between (*metaphase:* the third stage of mitosis, in which the chromosomes, attached to microtubules at their kinetochores, are all aligned at the metaphase plate)

mit- = a warp thread (*mitosis:* a process of nuclear division that allocates replicated chromosomes equally to each of the daughter nuclei)

pro- = before (*prophase:* the first stage of mitosis, in which the chromatin condenses into discrete chromosomes and the mitotic spindle begins to form)

-some = body (*centrosome:* a structure in animal cells that functions as a microtubule-organizing center and is important during cell division)

tel- = an end (*telophase:* the fifth and final stage of mitosis, in which daughter nuclei are forming and cytokinesis has typically begun)

Structure Your Knowledge

1. In the following photomicrograph of onion root tip cells, identify the cell cycle stage for the indicated cells.

a. ⎯⎯⎯⎯⎯⎯⎯⎯⎯

b. ⎯⎯⎯⎯⎯⎯⎯⎯⎯

c. ⎯⎯⎯⎯⎯⎯⎯⎯⎯

d. ⎯⎯⎯⎯⎯⎯⎯⎯⎯

2. Describe one chromosome as it proceeds through an entire cell cycle, starting with G_1 of interphase and ending with telophase of mitosis.

3. Draw a sketch of one half of a mitotic spindle. Identify and list the functions of the components.

Test Your Knowledge

FILL IN THE BLANK: *Identify the appropriate phase of the cell cycle.*

_____ 1. most cells that will no longer divide are in this phase

_____ 2. sister chromatids separate and chromosomes move apart

_____ 3. mitotic spindle begins to form

_____ 4. cell plate forms or cleavage furrow pinches cells apart

_____ 5. replication of DNA

_____ 6. chromosomes line up at central plane

_____ 7. nuclear membranes form around separated chromosomes

_____ 8. chromosomes become visible

_____ 9. kinetochore–microtubule interactions move chromosomes to midline

_____ 10. restriction point occurs in this phase

MULTIPLE CHOICE: *Choose the one best answer.*

1. One of the differences between prokaryotic binary fission and eukaryotic mitosis is that
   a. cytokinesis does not occur in prokaryotic cells.
   b. mitosis involves protein such as tubulin and actin, but no such molecules are involved in prokaryotic binary fission.
   c. mitosis takes place within the nucleus but binary fission does not.
   d. the duplicated single chromosome does not separate along a mitotic spindle in prokaryotic cells.

2. A plant cell has 12 chromosomes at the end of mitosis. How many *chromosomes* would it have in the G_2 phase of its next cell cycle?
   a. 6   b. 12   c. 24   d. 48

3. How many *chromatids* would this plant cell have in the G_2 phase of its cell cycle?
   a. 6   b. 12   c. 24   d. 48

4. The longest part of the cell cycle is
   a. prophase.   c. mitosis.
   b. G_1 phase.   d. interphase.

5. In animal cells, movement of the chromosomes is accomplished using _____, but division of the cytoplasm requires the action of _____.
   a. microtubules; intermediate filaments
   b. microtubules; actin filaments
   c. actin filaments; microtubules
   d. cohesins; cyclins

6. In eukaryotic organisms, chromosomes are duplicated
   a. before mitosis.
   b. after mitosis.
   c. during mitosis.
   d. before or after mitosis—it depends on what kind of organism you're talking about.

7. When do chromatids become chromosomes?
   a. during the S phase of interphase
   b. during prophase, when chromosome condensation occurs
   c. when cohesins are degraded by separase
   d. when two daughter nuclei are formed in telophase

8. A cell in which of the following phases has the *least* amount of DNA?
   a. G_0 c. prophase
   b. G_2 d. metaphase

9. Which of the following is *not* true of a cell plate?
   a. It begins to form during telophase at the site of the metaphase plate.
   b. It results from the fusion of microtubules in a dividing plant cell.
   c. When it fuses with the plasma membrane, it separates daughter plant cells.
   d. A cell wall is laid down between its membranes.

10. A cell that passes the restriction point in G_1 will most likely
    a. undergo chromosome duplication.
    b. continue to divide only if it is a cancer cell.
    c. show a drop in MPF concentration.
    d. move into the G_0 phase.

11. During cell division, cyclin combines with cyclin-dependent kinase (Cdk) to form a complex called MPF. This complex
    a. moves the cell past the G_2 checkpoint into mitosis.
    b. ensures that DNA has been repaired prior to S phase of mitosis.
    c. promotes the degradation of cyclin, but leaves Cdk intact.
    d. both **a** and **c**

12. Which of the following would *not* be exhibited by cancer cells?
    a. changing levels of MPF concentration
    b. passage through all checkpoints
    c. loss of density-dependent inhibition and anchorage dependence
    d. apoptosis

13. What initiates the separation of sister chromatids?
    a. a rapid rise in Cdk concentration
    b. movement past the G_2 checkpoint
    c. a signal pathway initiated by the binding of a growth factor
    d. activation of a regulatory protein complex and the enzyme separase after all kinetochores are attached to microtubules

14. Knowledge of the cell cycle control system will be most beneficial to understanding
    a. growth of crop plants.
    b. human growth and development.
    c. cancer.
    d. cardiovascular disease.

15. Vinblastine is a chemotherapeutic drug used to treat certain cancers. It acts by interfering with microtubule assembly. The effectiveness of vinblastine must be related to
    a. prevention of cleavage furrow formation.
    b. disruption of mitotic spindle formation.
    c. inhibition of DNA synthesis.
    d. inhibition of kinetochore motor proteins.

Chapter 13

Meiosis and Sexual Life Cycles

Key Concepts

13.1 Offspring acquire genes from parents by inheriting chromosomes

13.2 Fertilization and meiosis alternate in sexual life cycles

13.3 Meiosis reduces the number of chromosome sets from diploid to haploid

13.4 Genetic variation produced in sexual life cycles contributes to evolution

Framework

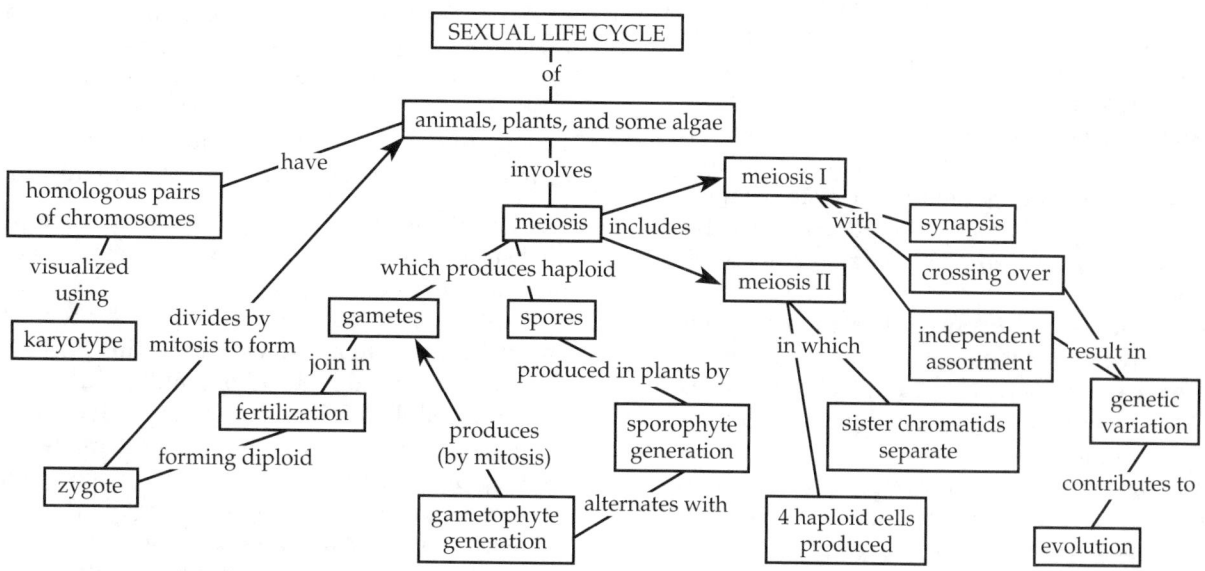

Chapter Review

Genetics is the scientific study of the transmission of traits from parents to offspring (**heredity**) and the **variation** between and within generations.

13.1 Offspring acquire genes from parents by inheriting chromosomes

Inheritance of Genes　The appearance of parental traits in offspring involves the inheritance of discrete units of information coded in segments of DNA known as **genes**. Most genes contain instructions for synthesizing enzymes and other proteins that then guide the development of inherited traits.

When **gametes** (sperm and eggs) fuse in fertilization, genes from both parents are transmitted to offspring. The DNA of a eukaryotic cell is packaged into a species-specific number of chromosomes present in all **somatic cells**. A gene's **locus** is its location on a chromosome.

Comparison of Asexual and Sexual Reproduction
In **asexual reproduction**, a single parent passes copies of all its genes to its offspring. A **clone** is a group of genetically identical offspring. In **sexual reproduction**, an individual receives a unique combination of genes inherited from two parents.

13.2   Fertilization and meiosis alternate in sexual life cycles

An organism's **life cycle** is the sequence of stages from conception to production of its own offspring.

Sets of Chromosomes in Human Cells   Chromosomes differ in the genes that they carry and in their physical characteristics (arm length, centromere location, banding pattern, etc.). Two chromosomes of similar genetic and physical characteristics are known as **homologous chromosomes** or **homologs**; a homologous pair of each type of chromosome is present in every somatic cell. A gene controlling a particular character is found at the same locus on each chromosome of a homologous pair.

A **karyotype** is an ordered display of an individual's condensed chromosomes. Isolated somatic cells are stimulated to undergo mitosis, arrested in metaphase, and stained. Digital photographs are used to arrange chromosomes into homologous pairs by size and shape.

Sex chromosomes determine the sex of a person: Females have two homologous X chromosomes; males have nonhomologous X and Y chromosomes. Chromosomes other than the sex chromosomes are called **autosomes**.

Chromosomes exist in homologous pairs because you received one set of chromosomes (that is, one chromosome of each type) from each parent. The somatic cells of your body contain two chromosome sets and are called **diploid** cells. The diploid complement of chromosomes is represented as $2n$, where n is the number of chromosomes in a set. Gametes are **haploid cells** and contain a single set of chromosomes. The haploid number (n) of chromosomes for humans is 23, and a haploid set is made up of 22 autosomes and one sex chromosome (either X or Y).

INTERACTIVE QUESTION   13.1

a. How many chromosomes are there in the somatic cells of an animal in which $2n = 14$? _____ How many chromosomes are in its gametes? _____

b. If $n = 14$, how many chromosomes are there in diploid somatic cells? _____ How many sets of homologous chromosomes are in the gametes? _____

c. If $2n = 28$, how many chromatids are there in a cell after chromosome duplication has occurred prior to cell division? _____ What is the difference between sister chromatids and nonsister chromatids?

Behavior of Chromosome Sets in the Human Life Cycle   **Fertilization**, or fusion of sperm and egg, produces a **zygote** containing both a paternal and a maternal set of chromosomes. The diploid zygote then divides by mitosis to produce the somatic cells of the body, all of which contain the diploid number ($2n$) of chromosomes.

Meiosis is a special type of cell division that halves the chromosome number and provides a haploid set of chromosomes to each gamete. Gametes are produced by meiosis from specialized *germ cells* in the gonads (ovaries and testes). An alternation between diploid and haploid numbers of chromosomes, involving the processes of fertilization and meiosis, is characteristic of sexually reproducing organisms.

The Variety of Sexual Life Cycles   In most animals, meiosis occurs in the formation of gametes, which are the only haploid cells in the life cycle.

Plants and some species of algae have a type of life cycle called **alternation of generations**, which includes both diploid and haploid multicellular stages. The multicellular diploid *sporophyte* produces haploid *spores* by meiosis. These spores undergo mitosis and develop into a multicellular haploid organism, the *gametophyte*, which produces gametes by mitosis. Gametes fuse to form a diploid zygote, which develops into the next sporophyte generation.

In most fungi and some protists, the only diploid stage is the zygote. Meiosis occurs after the gametes fuse, producing haploid cells that divide by mitosis to create a unicellular or multicellular haploid organism. Gametes are produced by mitosis in these organisms. Note that mitosis can occur in haploid or diploid cells, but meiosis occurs only in diploid cells.

INTERACTIVE QUESTION 13.2

Complete the following diagrams of sexual life cycles by filling in the type of cell division, event, or cells. Indicate whether the cell labeled l. in the bottom cycle is *n* or *2n*.

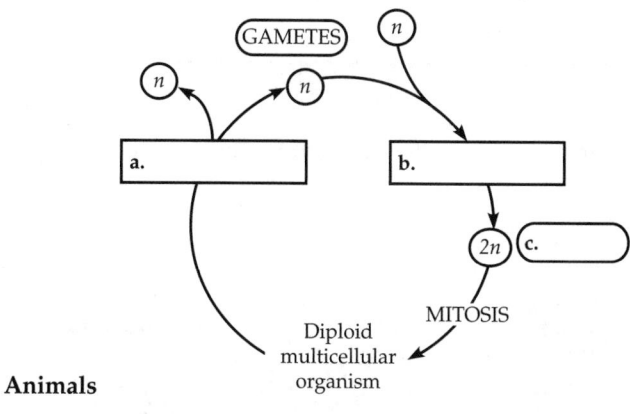

Animals

- -

Plants and some algae

- -

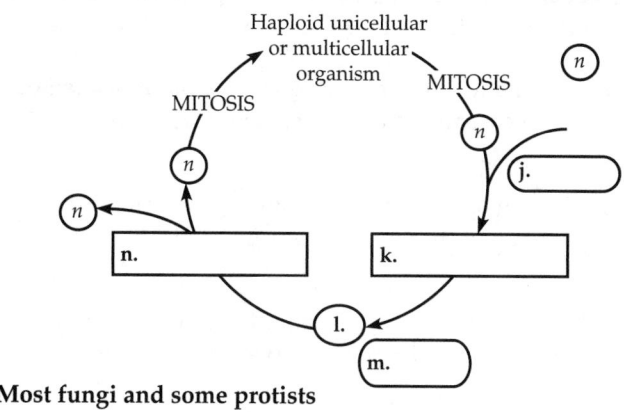

Most fungi and some protists

13.3 Meiosis reduces the number of chromosome sets from diploid to haploid

In meiosis, chromosome duplication is followed by two consecutive cell divisions—**meiosis I** and **meiosis II**—producing four haploid daughter cells, each with one set of chromosomes.

The Stages of Meiosis In interphase, each duplicated chromosome consists of two genetically identical sister chromatids that remain attached at the centromere and along their length by *sister chromatid cohesion*. During early prophase I, homologous chromosomes pair up and genetic material is exchanged by **crossing over** between nonsister chromatids. Crossovers remain visible as X-shaped regions called **chiasmata**.

In metaphase I, homologous pairs line up at the metaphase plate with their kinetochores attached to spindle fibers from opposite poles. Each pair separates in anaphase I, with one homolog moving toward each pole. In telophase I, a haploid set of chromosomes, each composed of two sister chromatids (usually containing some regions of nonsister chromatid DNA), reaches each pole. Cytokinesis usually occurs during telophase I. There is no replication of genetic material prior to the second division of meiosis.

Meiosis II looks like a regular mitotic division, in which chromosomes line up individually on the metaphase plate. Sister chromatid cohesion at the centromere breaks down, and sister chromatids separate and move apart in anaphase II. These sister chromatids, however, are not genetically identical due to crossing over in prophase I. (Homologs may have had different *alleles*.) At the end of telophase II, there are four genetically distinct haploid daughter cells.

Crossing Over and Synapsis During Prophase I Early in prophase I, homologs align precisely along their length and the DNA of nonsister chromatids is broken at corresponding points. A **synaptonemal complex** forms and attaches the homologs together. During this **synapsis**, or association, the DNA breaks are repaired, joining segments of nonsister chromatids together. The synaptonemal complex disassembles, but homologs remain attached where proteins called *cohesins* continue to hold the original (but now crossed-over) sister chromatids together.

A Comparison of Mitosis and Meiosis Mitosis produces daughter cells that are genetically identical to the parent cell. Meiosis produces haploid cells that differ genetically from their parent cell and from each other.

Three unique events are responsible for this result. In prophase I, homologous chromosomes pair and are held together (synapsis), and genetic material is exchanged by crossing over between nonsister chromatids. In metaphase I, chromosomes line up in homologous pairs, not as individual chromosomes, at the metaphase plate. And during anaphase I, the homologous pairs separate and one homolog (with sister chromatids still attached at the centromere) goes to each pole.

Homologs are held together by sister chromatid cohesion at regions of crossing over. In anaphase I, enzymes cleave the cohesin proteins along the arms of sister chromatids, and the homologs can separate. The cohesins at the *centromere* hold the sister chromatids together until anaphase II.

INTERACTIVE QUESTION   13.3

The following diagrams represent some of the stages of meiosis (*not presented in the right order*). Label these stages. What is the diploid number for this cell? _____

a. _____ b. _____ c. _____ d. _____ e. _____ f. _____

Now place these stages in the correct sequence.

_____ _____ _____ _____ _____ _____

13.4 Genetic variation produced in sexual life cycles contributes to evolution

Origins of Genetic Variation Among Offspring
Mutations that result in different alleles of a gene are the original source of genetic variation. In sexual reproduction, however, three mechanisms produce the genetic variation evident in each generation: independent assortment of chromosomes, crossing over, and random fertilization.

Each homologous pair lines up independently at the metaphase plate; the orientation of the original maternal and paternal chromosomes is random. Due to this *independent assortment* of chromosomes, the number of possible combinations of maternal and paternal chromosomes in gametes is 2^n, where n is the haploid number for that species.

In prophase I, homologous segments of nonsister chromatids are exchanged by crossing over, forming **recombinant chromosomes** with new genetic combinations of maternal and paternal alleles on the same

chromosome. These no-longer-equivalent sister chromatids assort independently during meiosis II.

The random nature of fertilization adds to the genetic variability established in meiosis.

INTERACTIVE QUESTION   13.4

a. How many different assortments of maternal and paternal chromosomes are possible in a human gamete?

b. In a human zygote, how many diploid combinations of parental chromosomes are possible?

c. Why does the answer to question b still underestimate the possible genetic variations in a zygote?

The Evolutionary Significance of Genetic Variation Within Populations In Darwin's theory of evolution by natural selection, the genetic variation present in a population results in adaptation as the individuals with the variations best suited to the environment produce the most offspring. The processes of sexual reproduction and mutation are the sources of this variation. Although sexual reproduction is more energetically expensive than asexual reproduction and may disrupt successful combinations of alleles in stable environments, it is widespread among organisms and almost universal among animals. The example of the bdelloid rotifers indicates that genetic variation is important for long-term evolutionary success. The asexual bdelloids achieve genetic variation through other means. Genetic variation appears to be evolutionarily advantageous.

Word Roots

a- = not or without (*asexual reproduction:* generation of offspring from a single parent, occurring without the fusion of gametes)

-apsis = juncture (*synapsis:* the pairing and physical connection of duplicated homologous chromosomes during prophase I of meiosis)

auto- = self (*autosome:* a chromosome that is not directly involved in determining sex)

chiasm- = marked crosswise (*chiasma:* the X-shaped microscopically visible region where crossing over has occurred between homologous nonsister chromatids)

di- = two (*diploid cell:* a cell containing two sets of chromosomes [2*n*], one set inherited from each parent)

fertil- = fruitful (*fertilization:* the union of haploid gametes to produce a diploid zygote)

hapl- = single (*haploid cell:* a cell containing only one set of chromosomes [*n*])

homo- = like (*homologous chromosomes:* a pair of like chromosomes that possess genes for the same characters at corresponding loci)

kary- = nucleus (*karyotype:* a display of the chromosome pairs of a cell)

mei- = less (*meiosis:* a modified type of cell division that results in cells with half the number of chromosome sets as the original cell)

som- = body (*somatic cell:* any cell in a multicellular organism except a gamete or its precursors)

Structure Your Knowledge

1. Label the following diagram to review the terms that describe duplicated chromosomes in a diploid cell.

2. Describe the key events of the following stages of meiosis.
   a. interphase
   b. prophase I
   c. metaphase I
   d. anaphase I
   e. metaphase II
   f. anaphase II

3. Create a concept map to help you organize your understanding of the similarities and differences between mitosis and meiosis. *Compare your map with those of some classmates to see different ways of organizing the material.*

Test Your Knowledge

MULTIPLE CHOICE: *Choose the one best answer.*

1. What does meiosis do?
   a. It produces haploid cells from diploid cells.
   b. It produces diploid cells from haploid cells.
   c. It reproduces a diploid set of chromosomes.
   d. It fuses an egg cell with a sperm cell.

2. Asexual reproduction of a diploid organism
   a. is impossible because the chromosome number would double each generation.
   b. usually involves spores produced by meiosis.
   c. produces variation among sibling offspring.
   d. produces offspring identical to each other and the parent.

3. Homologous chromosomes
   a. have identical DNA sequences in their genes.
   b. have genes for the same characters at the same loci.
   c. are found in gametes containing X chromosomes but not in those containing a Y chromosome.
   d. separate in meiosis II.

4. Which of the following represents a clone?
   a. a group of seedlings produced by seeds from the same tree
   b. you and your brother
   c. the cuttings from your spider plant that you gave to six of your friends
   d. the litter of new kittens that you're trying to give away.

5. What is a karyotype?
   a. a genotype of an individual
   b. a pictorial display of an individual's chromosomes
   c. a unique combination of chromosomes found in a gamete
   d. a species-specific diploid number of chromosomes

6. A reduction in the number of chromosomes per cell occurs
   a. during meiosis I.
   b. during meiosis II.
   c. during mitosis.
   d. during the S phase of interphase.

7. In most fungi and some protists,
   a. gametes are formed by meiosis.
   b. the organism is haploid and may be multicellular or unicellular.
   c. the gametophyte generation produces gametes by mitosis.
   d. reproduction is exclusively asexual.

8. In the alternation of generations found in plants,
   a. the sporophyte generation produces spores by mitosis.
   b. the gametophyte generation produces gametes by mitosis.
   c. the zygote develops into a sporophyte generation by meiosis.
   d. spores develop into the haploid sporophyte generation.

9. A synaptonemal complex would be found during
   a. early prophase of mitosis.
   b. the fusion of gametes in fertilization.
   c. metaphase I of meiosis.
   d. early prophase of meiosis I.

10. During meiosis I,
    a. homologous chromosomes separate, reducing the chromosome number by half.
    b. crossing over between nonsister chromatids occurs.
    c. paternal and maternal chromosomes assort randomly.
    d. all of the above occur.

11. Compared with one of the four cells produced by meiosis, a cell in G_2 before meiosis has
    a. twice as much DNA and twice as many chromosomes.
    b. twice as much DNA but the same number of chromosomes.
    c. four times as much DNA and twice as many chromosomes.
    d. four times as much DNA and four times as many chromosomes.

12. Meiosis II is similar to mitosis because
    a. sister chromatids separate.
    b. homologous chromosomes line up independently on the metaphase plate.
    c. DNA replication precedes the division.
    d. nonsister chromatids separate when cohesins at the centromeres are cleaved.

13. The Texan nine-banded armadillo (*Dasypus novemcinctus*) has a diploid chromosome number of 64 (i.e., $2n = 64$). How many *chromatids* would be found in an armadillo cell in prophase I of meiosis?
    a. 32
    b. 64
    c. 128
    d. You can't tell based on this information.

14. Which of the following is a key difference between mitosis and meiosis?

    a. Only in mitosis are daughter cells genetically identical to the parent cell.

    b. Sister chromatids separate during mitosis but not meiosis.

    c. DNA synthesis does not occur prior to meiosis

    d. Only mitosis has a metaphase.

15. Which of the following is *not* true of homologous chromosomes?

    a. They behave independently in mitosis.

    b. They are joined in synapsis during the S phase of meiosis.

    c. They are held together during synapsis by a synaptonemal complex.

    d. Crossing over between nonsister chromatids of homologous chromosomes is indicated by the presence of chiasmata.

16. Which of the following statements is *not* true?

    a. Sister chromatid cohesion at areas where crossing over has occurred holds homologous chromosomes together until anaphase I.

    b. Recombinant chromosomes are produced when both maternal and paternal chromosomes independently assort into the same gamete.

    c. In mitosis, separation of sister chromatids in anaphase results in two identical cells.

    d. Separation of sister chromatids in anaphase II does not usually result in two identical cells.

17. In a species with a diploid number of 6, how many different combinations of maternal and paternal chromosomes would be possible in *gametes*?

    a. 6   c. 12

    b. 8   d. 64

18. In a sexually reproducing species with a diploid number of 8, how many different combinations of chromosomes would be possible in the diploid *offspring*?

    a. 16   c. 256

    b. 64   d. 512

19. The calculation of offspring in question 18 includes only variation resulting from

    a. crossing over.

    b. random fertilization.

    c. independent assortment of chromosomes.

    d. both **b** and **c**.

20. Which of the following is *least* likely to be a source of genetic variation in sexually reproducing organisms?

    a. crossing over

    b. replication of DNA during S phase before meiosis I

    c. independent assortment of chromosomes

    d. random fertilization of gametes

21. Which of the following statements describes why or how recombinant chromosomes add to genetic variability?

    a. They are formed as a result of random fertilization when two sets of chromosomes combine in a zygote.

    b. They randomly orient during metaphase II, and the nonequivalent sister chromatids separate in anaphase II.

    c. Genetic material from two parents is combined on the same chromosome.

    d. Both **b** and **c** are true.

22. What is the evolutionary significance of bdelloid rotifers, which have not engaged in sex in 40 million years?

    a. They show that asexual reproduction is more advantageous because it ensures that successful combinations of alleles are passed to offspring.

    b. Although their 400 species originated through sexual reproduction, each species has successfully cloned itself since it arose.

    c. They exchange genetic material through horizontal gene transfer while in a state of suspended animation during dry periods, supporting the idea that genetic variation is evolutionarily advantageous.

    d. Their long history shows that sexual reproduction and genetic variation are not required for evolutionary success.

Chapter 14

Mendel and the Gene Idea

Key Concepts

14.1 **Mendel used the scientific approach to identify two laws of inheritance**

14.2 **Probability laws govern Mendelian inheritance**

14.3 **Inheritance patterns are often more complex than predicted by simple Mendelian genetics**

14.4 **Many human traits follow Mendelian patterns of inheritance**

Framework

Through his work with garden peas in the 1860s, Gregor Mendel developed the fundamental principles of inheritance. This chapter describes the basic monohybrid and dihybrid crosses that Mendel performed to establish that inheritance involves genes that segregate independently in the formation of gametes and recombine to form offspring. The rules of probability can be applied to predict the outcome of genetic crosses.

The phenotypic expression of genotype may be affected by such factors as incomplete dominance, codominance, multiple alleles, pleiotropy, epistasis, and polygenic inheritance, as well as the environment. Genetic counseling for inherited disorders uses Mendelian principles to analyze human pedigrees and genetic testing.

Chapter Review

Gregor Mendel's experiments with garden peas provided evidence to replace the "blending" hypothesis of inheritance with the "particulate" hypothesis, in which discrete heritable units—genes—are passed from parents to offspring.

14.1 Mendel used the scientific approach to identify two laws of inheritance

Mendel's Experimental, Quantitative Approach
Mendel's garden peas were an excellent choice for studying the mechanisms of heredity. They are available in many varieties, their fertilization is easily controlled, and the characteristics of their many offspring can be quantified.

Mendel studied seven **characters**, or heritable features, that occurred in alternative forms called **traits**. He used **true-breeding** varieties of pea plants, which means that self-fertilizing parents always produce offspring with the parental form of the character. To follow the transmission of these well-defined traits, Mendel performed **hybridizations** in which he mated (*crossed* by cross-pollinating) contrasting true-breeding varieties, and then allowed the next generation to self-pollinate. The true-breeding parental plants are the **P generation** (parental); the offspring of the first cross are the F_1 **generation** (first filial); and the next generation, from the self- or cross-pollination of F_1 plants, is known as the F_2 **generation**.

The Law of Segregation Mendel found that the F_1 offspring did not show a blending of the parental traits. Instead, only one of the parental traits of the character was found in the hybrid offspring. In the F_2 generation, however, the missing parental trait reappeared in the ratio of 3:1—three offspring with the *dominant* trait shown by the F_1 to one offspring with the reappearing *recessive* trait.

Mendel's explanation for this inheritance pattern contains four parts: (1) alternate forms of genes, called **alleles**, account for variations in characters; (2) an organism has two copies of a gene for each trait, one inherited from each parent; (3) when two different alleles occur together, one of them, called the **dominant allele**, determines the organism's appearance and the other, the **recessive allele**, has no observable effect on appearance; and (4) allele pairs separate (segregate) during the formation of gametes (Mendel's **law of segregation**), so an egg or sperm carries only one allele for each inherited character. This explanation is consistent with the behavior of chromosomes during meiosis.

A **Punnett square** can be used to predict the results of simple genetic crosses. Dominant alleles are often symbolized by a capital letter, recessive alleles by a small letter.

Fill in the following diagram of a cross of round-seeded and wrinkled-seeded pea plants. The round allele (*R*) is dominant; the wrinkled allele (*r*) is recessive.

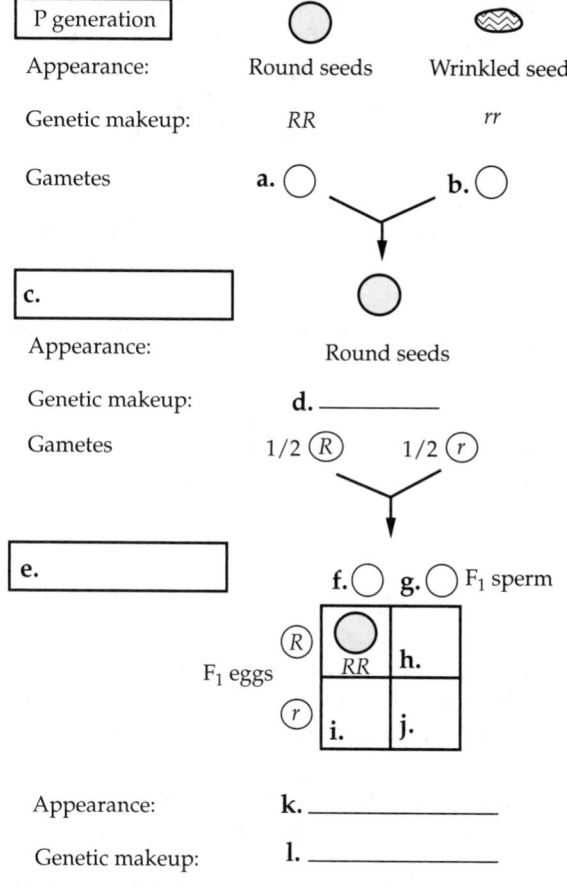

homozygote to determine the genotype of the phenotypically dominant organism.

A tall pea plant is crossed with a recessive dwarf pea plant. What will the phenotypic and genotypic ratio of offspring be if

a. the tall plant was *TT*?

b. the tall plant was *Tt*?

The Law of Independent Assortment   In a **monohybrid cross**, the inheritance of a single character is followed through the crossing of **monohybrids**, F_1 offspring that are heterozygous for that character. Mendel used **dihybrid crosses** between F_1 **dihybrids**, which are heterozygous for two characters, to determine whether the two characters from the parent plants were transmitted together or independently.

If the two pairs of alleles segregate independently, gametes from an F_1 hybrid generation (*AaBb*) should contain four combinations of alleles in equal quantities (*AB, Ab, aB, ab*). The random fertilization of these four types of gametes would result in 16 (4 × 4) gamete combinations that produce four phenotypic categories in a ratio of 9:3:3:1 (nine offspring showing both dominant traits, three showing one dominant trait and one recessive, three showing the opposite dominant and recessive traits, and one showing both recessive traits). Mendel obtained these ratios in the F_2 progeny of dihybrid crosses, providing evidence for the **law of independent assortment**. This principle states that two or more genes assort independently—each gene's pair of alleles segregates independently from the alleles of other genes in the formation of gametes. This law applies to genes located on different chromosomes or sufficiently far apart on the same chromosome.

An organism that has a pair of identical alleles is said to be **homozygous** for that gene and is called a **homozygote**; Mendel's true-breeding parental stocks were homozygous. If the organism has two different alleles, it is called a **heterozygote** and is said to be **heterozygous**. Homozygotes are true-breeding; heterozygotes are not, as they produce gametes with one or the other allele that can combine to produce offspring that are dominant homozygotes, heterozygotes, and recessive homozygotes. **Phenotype** is an organism's observable traits; **genotype** is its genetic makeup.

In a **testcross**, an organism expressing the dominant phenotype is crossed with a recessive

A true-breeding tall, purple-flowered pea plant (*TTPP*) is crossed with a true-breeding dwarf, white-flowered plant (*ttpp*). Assume that the genes are on different chromosomes.

a. What is the phenotype of the F_1 generation?

b. What is the genotype of the F_1 generation?

c. What four types of gametes are formed by F_1 plants?

_____   _____   _____   _____

d. Fill in the following Punnett square to show the offspring of the F$_2$ generation. Shade each phenotype a different color so you can see the ratio of offspring.

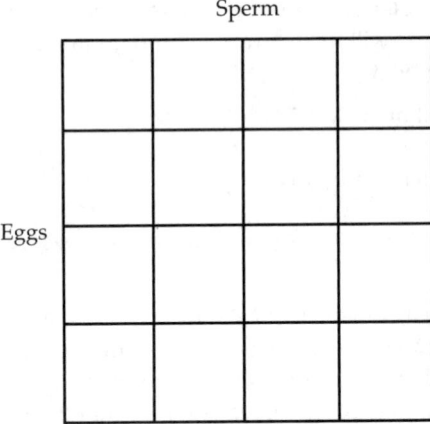

Sperm

Eggs

e. List the phenotypes and ratios found in the F$_2$ generation.

_____   _____

_____   _____

f. What is the ratio of tall plants to dwarf plants? _____

Of purple-flowered plants to white-flowered plants? _____

(Note that the alleles for each individual character segregate as in a monohybrid cross.)

14.2 Probability laws govern Mendelian inheritance

The probability scale goes from 0 to 1; the probabilities of all possible outcomes must add up to 1. The probability of an event occurring is the number of times that event could occur over all the possible events; for example, the probability of drawing an ace of any suit from a deck of cards is $\frac{4}{52}$. The outcome of independent events is not affected by previous or simultaneous but separate trials.

The Multiplication and Addition Rules Applied to Monohybrid Crosses   The **multiplication rule** states that the probability that a certain combination of independent events will occur together is equal to the product of the separate probabilities of the independent events. The probability of a particular genotype being formed by fertilization is equal to the product of the probabilities of forming each type of gamete needed to produce that genotype.

If a genotype can be formed in more than one way, then, according to the **addition rule**, its probability is equal to the sum of the separate probabilities of the different, mutually exclusive ways it can occur. For example, a heterozygote offspring can occur if the egg contains the dominant allele and the sperm the recessive ($\frac{1}{2} \times \frac{1}{2} = \frac{1}{4}$ probability), or vice versa ($\frac{1}{4}$). Therefore, a heterozygote offspring would be the predicted result from a monohybrid cross half of the time ($\frac{1}{4} + \frac{1}{4} = \frac{1}{2}$).

Solving Complex Genetics Problems with the Rules of Probability   The probability of a particular genotype arising from a cross can be determined by considering each gene involved as a separate monohybrid cross and then multiplying the probabilities of all the independent events involved in the final genotype. When more than one outcome is involved, the addition rule is also used.

The larger the sample size, the more closely the results will conform to statistical predictions.

INTERACTIVE QUESTION   14.4

a. In the following cross, what is the probability of obtaining offspring that show all three dominant traits? Assume that the genes are on different chromosomes.

$$AaBbcc \times AabbCC$$

probability of offspring that are $A\_B\_C\_$ = _____

(The _indicates that the second allele can be either dominant or recessive without affecting the phenotype determined by the first dominant allele.)

b. What is the probability that the offspring of this cross will show at least two dominant traits? _____

c. What is the probability of obtaining offspring that show only one dominant trait? _____

14.3 Inheritance patterns are often more complex than predicted by simple Mendelian genetics

Extending Mendelian Genetics for a Single Gene   In the case of an allele showing **complete dominance**, the phenotype of the heterozygote is indistinguishable from that of the dominant homozygote. Intermediate phenotypes are characteristic of alleles showing **incomplete dominance**. The F$_1$ hybrids have a phenotype intermediate between that of the parents. The F_2 offspring show a 1:2:1 phenotypic and genotypic ratio.

Alleles that exhibit **codominance** will each affect the phenotype in separate, distinguishable ways, as in the case of the MN blood group.

The Relationship Between Dominance and Phenotype
It is important to recognize that relationships between alleles (dominant/recessive, incompletely dominant, codominant) are related to the expression of the allelic forms. There are also multiple levels of phenotype, depending on how the results of gene expression are examined. **Tay-Sachs disease** is a lethal disorder in which brain cells are unable to metabolize certain lipids that then accumulate and damage the brain. The Tay-Sachs allele is recessive at the *organismal* level—a heterozygote does not have the disease. At the *biochemical* level, the enzyme activity level is intermediate between that of both homozygotes (incomplete dominance), and at the *molecular* level, the alleles are codominant in that each produces its enzyme product, either functional or dysfunctional.

Whether or not an allele is dominant or recessive has no relation to how common it is in a population.

Most genes exist in more than two allelic forms. The gene that determines human blood groups has three alleles. The alleles I^A and I^B are codominant with each other; each codes for an enzyme that attaches a carbohydrate to the surface of red blood cells. The allele *i* results in no attached carbohydrate.

INTERACTIVE QUESTION   14.5

List all the possible genotypes for the following blood groups.

a. A

b. B

c. AB

d. O

Pleiotropy is the ability of a single gene to produce multiple phenotypic effects. Certain hereditary diseases with complex sets of symptoms are caused by pleiotropic alleles.

Extending Mendelian Genetics for Two or More Genes   In **epistasis**, the expression of a gene at one locus may affect the expression of another gene. F_2 ratios that are variations on the typical 9:3:3:1 often indicate epistasis.

INTERACTIVE QUESTION   14.6

Consider an example in which the *M/m* gene (for pigment production) is epistatic to the *B/b* gene (for deposition of pigment). Suppose that a dominant allele *M* is necessary for the production of the black pigment melanin; mm individuals are white. A dominant allele *B* results in the deposition of a lot of pigment in an animal's hair, producing a black color. The genotype *bb* produces

Phenotype	Genotype	Ratio
Black		
	M_bb	

---

gray hair, assuming the presence of at least one *M allele*. Two black animals heterozygous for both genes are bred. Fill in the following table for the offspring of this *MmBb* × *MmBb* cross.

Quantitative characters, such as height or skin color, vary along a continuum in a population. Such phenotypic gradations are usually due to **polygenic inheritance**, in which two or more genes have an additive effect on one character. A recent study identified more than 180 genes that influence height in humans. Each dominant allele contributes one "unit" to the phenotype. A polygenic character may result in a normal distribution (forming a bell-shaped curve) of the character within a population.

INTERACTIVE QUESTION   14.7

The height of spike weed is a result of polygenic inheritance involving three genes. The dominant allele of each of these genes contributes an additional 5 cm to the base height of the plant, which is 10 cm. The tallest plant (*AABBCC*) can reach a height of 40 cm.

a. If a tall plant (*AABBCC*) is crossed with a base-height plant (*aabbcc*), what is the height of the F_1 plants?

b. How many different phenotypes will there be in the F_2?

Nature and Nurture: The Environmental Impact on Phenotype   The phenotype of an individual is the result of complex interactions between its genotype and the environment. Genotypes have a phenotypic range within which the environment influences phenotypic expression. Polygenic characters are often **multifactorial**, meaning that a combination of genetic and environmental factors affects phenotype.

A Mendelian View of Heredity and Variation   The phenotypic expression of most genes is influenced by other genes and by the environment. An organism's genotype can refer to its entire genome and its phenotype can include its appearance, anatomy, physiology, and behavior. Mendel's principles of segregation and independent assortment, however, still form the basis for modern genetics.

14.4   Many human traits follow Mendelian patterns of inheritance

Pedigree Analysis   A family **pedigree** is a family tree with the history of a particular trait shown across the generations. By convention, circles represent females, squares are used for males, and solid symbols indicate individuals that express the trait in question. Parents are joined by a horizontal line, and offspring are listed below parents from left to right in order of birth. The genotype of individuals in the pedigree can often be deduced by following the patterns of inheritance. Knowing the genotypes of individuals within a pedigree may allow predictions regarding the likely genotypes of their offspring.

Recessively Inherited Disorders   Only homozygous recessive individuals express the phenotype for the thousands of genetic disorders that are inherited as simple recessive traits. **Carriers** of the disorder are heterozygotes and are usually phenotypically normal, but may transmit the recessive allele to their offspring.

INTERACTIVE QUESTION   14.8

Consider the following pedigree for the trait albinism (lack of skin pigmentation) in three generations of a family. (Solid symbols represent individuals with albinism.) From your knowledge of Mendelian inheritance, answer the questions that follow.

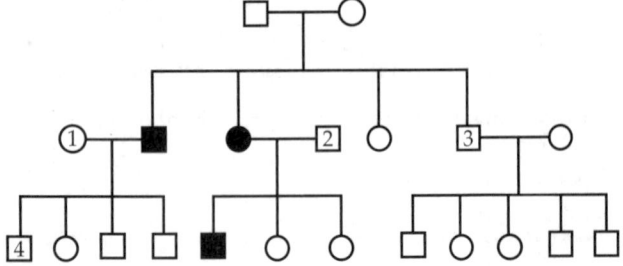

a. Is this trait caused by a dominant or recessive allele? How can you tell?

b. Determine the genotypes of the parents in the first generation. (Let *A* and *a* represent the alleles.) Genotype of father _____; genotype of mother _____. How can you tell?

c. Determine the probable genotypes of the mates of the offspring with albinism in the second generation and grandson #4 in the third generation. Genotypes: mate #1 _____; mate #2 _____; grandson #4 _____.

d. Can you determine the genotype of son #3 in the second generation? Why or why not?

   The likelihood of two individuals carrying the same rare deleterious allele increases when they have common ancestors. Consanguineous matings, between siblings or close relatives, are indicated on pedigrees by double lines. Because human populations have historically been geographically and culturally isolated, certain genetic disorders are more common in specific groups.

   Cystic fibrosis, the most common lethal genetic disease in the United States, occurs more frequently in people of European descent than in other groups. The recessive allele results in defective chloride channels in certain cell membranes. Accumulating extracellular chloride leads to the buildup of thickened mucus in various organs and a predisposition to bacterial infections.

   Sickle-cell disease is the most common inherited disease among African Americans. Due to a single amino acid substitution in the hemoglobin protein, red blood cells deform into a sickle shape when blood-oxygen concentration is low, triggering blood clumping and other pleiotropic effects. Heterozygous individuals are said to have *sickle-cell trait* but are usually healthy.

INTERACTIVE QUESTION   14.9

a. What is the probability that a mating between two carriers will produce an offspring with a recessively inherited disorder? _____

b. What is the probability that a phenotypically normal child produced by a mating of two heterozygotes will be a carrier? _____

Dominantly Inherited Disorders A few human disorders are due to dominant genes. In *achondroplasia*, dwarfism results from a single copy of a mutant allele.

Dominant lethal alleles are rarer than recessive lethals because the harmful allele cannot be masked in the heterozygote. Lethal dominant alleles typically lead to the death of affected individuals before they can reproduce. A late-acting lethal dominant allele can be passed on if the symptoms do not develop until after reproductive age. Molecular geneticists have developed a method to detect the lethal gene for **Huntington's disease**, a degenerative disease of the nervous system that does not develop until later in life.

Multifactorial Disorders Many diseases have genetic (usually polygenic) and environmental components. These multifactorial disorders include heart disease, diabetes, cancer, and certain mental illnesses.

Genetic Testing and Counseling Genetic tests to identify carriers have been developed for an increasing number of heritable disorders. Insurance companies and employers are now legally prohibited from making decisions based on results of genetic testing.

INTERACTIVE QUESTION 14.10

If two prospective parents both have siblings who have a recessive genetic disorder, what is the chance that they would have a child who inherits the disorder? If this couple has two children without the disorder, what is the chance that a third child would inherit the disorder? Explain both answers.

Amniocentesis is a procedure that extracts a small amount of amniotic fluid from the sac surrounding the fetus. The fluid is analyzed biochemically, and fetal cells are cultured for several weeks and then tested for certain genetic disorders and karyotyped to check for chromosomal defects.

Chorionic villus sampling (CVS) is a technique in which a small amount of fetal tissue is suctioned from the placenta. These rapidly growing cells can be karyotyped immediately, and this procedure can be performed earlier in pregnancy. A new technique isolates fetal cells from maternal blood, which can then be cultured and tested.

Ultrasound is a simple, noninvasive procedure that can reveal major abnormalities.

Some genetic disorders can be detected at birth. Routine screening tests are used to detect phenylketonuria (PKU) in newborns. If detected, dietary adjustments can lead to normal development.

Word Roots

-centesis = a puncture (*amniocentesis:* a technique of prenatal diagnosis in which amniotic fluid is obtained by aspiration from a needle inserted into the uterus)

co- = together (*codominance:* situation in which the phenotypes of both alleles are exhibited in the heterozygote)

di- = two (*dihybrid cross:* a cross between two organisms that are each heterozygous for both of the characters being followed)

epi- = beside; **-stasis** = standing (*epistasis:* a type of gene interaction in which the phenotypic expression of one gene alters that of another independently inherited gene)

heter- = different (*heterozygous:* having two different alleles for a given gene)

homo- = alike (*homozygous:* having two identical alleles for a given gene)

mono- = one (*monohybrid cross:* a cross between two organisms that are heterozygous for the character being followed [or the self-pollination of a heterozygous plant])

ped- = a child (*pedigree:* a diagram of a family tree showing the occurrence of heritable characters in parents and offspring over multiple generations)

phen- = appear (*phenotype:* the observed physical and physiological traits of an organism, determined by its genetic makeup)

plei- = more (*pleiotropy:* the ability of a single gene to have multiple effects)

poly- = many (*polygenic inheritance:* an additive effect of two or more genes on a single phenotypic character)

Structure Your Knowledge

1. Relate Mendel's two laws of inheritance to the behavior of chromosomes in meiosis that you studied in Chapter 13.

2. How many different types of gametes can be formed by the following genotypes? What generalized formula describes this number?

$Aa =$ $AaBb =$

$AaBbCc =$ $AABbCc =$

3. How can you tell whether alleles are completely dominant/recessive, incompletely dominant, or codominant?

Test Your Knowledge

MATCHING: *Match each of the following terms with its description.*

Term

_____ **1.** gene

_____ **2.** allele

_____ **3.** character

_____ **4.** trait

_____ **5.** dominant allele

_____ **6.** recessive allele

_____ **7.** genotype

_____ **8.** phenotype

_____ **9.** homozygous

_____ **10.** heterozygous

_____ **11.** testcross

_____ **12.** monohybrid cross

Description

a. has no effect on phenotype in a heterozygote

b. a variant form of a character

c. an individual having two identical alleles for a gene

d. a cross between individuals heterozygous for a single character

e. an alternative version of a gene

f. an individual having two different alleles for a gene

g. a heritable feature that varies among individuals

h. an organism's appearance or observable traits

i. a cross between an individual with an unknown genotype and a homozygous recessive individual

j. determines phenotype in a heterozygote

k. the genetic makeup of an individual

l. a heritable unit that determines a character; can exist in different forms

MULTIPLE CHOICE: *Choose the one best answer.*

1. According to Mendel's law of segregation,

   a. there is a 50% probability that a gamete will get a dominant allele.

   b. allele pairs segregate independently of other pairs of alleles in gamete formation.

   c. allele pairs separate in gamete formation.

   d. there is a 3:1 ratio in the F_2 generation.

2. The F_2 generation

   a. has a phenotypic ratio of 3:1.

   b. is the result of the self-fertilization or crossing of F_1 individuals.

   c. can be used to determine the genotype of individuals with the dominant phenotype.

   d. has 16 different genotypic possibilities.

3. A 1:1 phenotypic ratio in a testcross indicates that the

   a. alleles are dominant.

   b. parent with a dominant phenotype was a heterozygote.

   c. alleles segregated independently.

   d. alleles are codominant.

4. Which phase of meiosis is most directly related to the law of independent assortment?

   a. prophase I

   b. metaphase I

   c. metaphase II

   d. anaphase II

5. After obtaining two heads from two tosses of a coin, the probability of obtaining a head on the next toss is

   a. ½.   **c.** ⅛.

   b. ¼.   **d.** ⅜.

6. The probability of tossing three coins simultaneously and obtaining three heads is

   a. ½.   **c.** ⅛.

   b. ¼.   **d.** ⅜.

7. The probability of tossing three coins simultaneously and obtaining two heads and one tail is

   a. ½.   **c.** ⅛.

   b. ¼.   **d.** ⅜.

8. In the F_2 of a dihybrid cross involving two independently assorting genes, what proportion of the offspring will be true-breeding?
   a. ¹⁄₁₆
   b. ⅛
   c. ³⁄₁₆
   d. ¼

9. In guinea pigs, the brown coat color allele (*B*) is dominant over red (*b*), and the solid color allele (*S*) is dominant over spotted (*s*). The F_1 offspring of a cross between true-breeding brown, solid-colored guinea pigs and red, spotted pigs are crossed. What proportion of their offspring (F_2) would be expected to be red and solid colored?
   a. ⅑
   b. ¹⁄₁₆
   c. ³⁄₁₆
   d. ⁹⁄₁₆

10. A true-breeding dwarf corn plant with red ears is mated to a true-breeding tall corn plant with yellow ears. The F_1 generation is self-pollinated, and in the F_2 generation the following offspring appear: 10 tall red-eared plants, 30 dwarf red-eared plants, 90 dwarf yellow-eared plants, 30 tall yellow-eared plants. Which of the following statements is true?
    a. Tall and red are both dominant.
    b. Tall and yellow are both dominant.
    c. Dwarf and red are both dominant.
    d. Dwarf and yellow are both dominant.

11. You think that two alleles for coat color in mice show incomplete dominance. What is the best and simplest cross to perform in order to test your hypothesis?
    a. a testcross of a homozygous recessive mouse with a mouse of unknown genotype
    b. a cross of F_1 mice to look for a 1:2:1 ratio in the offspring
    c. a reciprocal cross in which the sex of the mice of each coat color is reversed
    d. a cross of two true-breeding mice of different colors to look for an intermediate phenotype in the F_1

12. Consider a couple with three children. The father's blood type is A, the mother's blood type is B, and the children's blood types are A, B, and O. If this couple has another child, what is the likelihood that it will have type AB blood?
    a. 0
    b. ¼
    c. ½
    d. 1

13. A dominant allele *P* causes the production of purple pigment; *pp* individuals are white. A dominant allele *C* is also required for color production; *cc* individuals are white. What proportion of offspring will be purple from a *ppCc* × *PpCc* cross?
    a. ⅛
    b. ⅜
    c. ½
    d. ¾

14. In humans, earwax can be wet or dry. The form of earwax is determined by a single gene, and the "wet" allele (*W*) is dominant to the "dry" allele (*w*). A woman and her mother both have wet earwax, but her father has dry earwax. Her partner is a man whose parents both had dry earwax. What are the genotypes of the woman and the man, respectively?
    a. *Ww, ww*
    b. *ww, Ww*
    c. *Ww, Ww*
    d. *ww, ww*

15. You have blood type B, your mother has blood type A, and your father has blood type B. You have two siblings whose blood types are A and AB. What is your genotype?
    a. $I^A I^B$
    b. $I^B i$
    c. $I^B I^B$
    d. ii

16. A family has three children who have been typed for the MN blood group. The phenotypes of the children are M, MN, and N. What are the genotypes of the parents?
    a. *MM, NN*
    b. *MN, NN*
    c. *MN, MN*
    d. It is impossible to tell from this information.

Genetics Problems

One of the best ways to learn genetics is to work problems. You can't memorize genetic knowledge; you have to practice using it. Work through the following problems, methodically setting down the information you are given and what you are to determine. Write the symbols used for the alleles and genotypes, and the phenotypes resulting from those genotypes. Avoid using Punnett squares; although useful for learning basic concepts, they are much too laborious. Instead, break complex crosses into their monohybrid components and rely on the rules of multiplication and addition. Always look at the answers you get to see if they make logical sense. And remember that study groups are great for going over your problems and helping each other "see the light."

1. Summer squash are either white or yellow. To get white squash, at least one of the parental plants must be white. The allele for which color is dominant?

2. For the following crosses, determine the probability of obtaining an offspring with the indicated genotype.

Cross	Offspring	Probability
AAbb × *AaBb*	*AAbb*	a.
AaBB × *AaBb*	*aaBB*	b.
AABbcc × *aabbCC*	*AaBbCc*	c.
AaBbCc × *AaBbcc*	*aabbcc*	d.

3. True-breeding tall red-flowered plants are crossed with dwarf white-flowered plants. The resulting F_1 generation consists of all tall pink-flowered plants. Assuming that height and flower color are each determined by a single gene on different chromosomes, predict the results of an F_1 cross of dihybrid plants. Choose appropriate symbols for the alleles of the genes. List the phenotypes and predicted ratios for the F_2 generation.

4. Blood typing has been used as evidence in paternity cases, when the blood types of the mother and child may indicate that a man alleged to be the father could not possibly have fathered the child. For the following combinations of mother and child, indicate the blood group(s) of men who would be exonerated (could *not* be the father of that child).

Blood Group of Mother	Blood Group of Child	Man Exonerated If He Belongs to Blood Group(s)
AB	A	a.
B	A	b.
A	AB	c.
O	O	d.

5. In rabbits, *CC* is normal, *Cc* results in rabbits with deformed legs, and *cc* is lethal. For a gene for coat color, the genotype *BB* produces black, *Bb* brown, and *bb* white. Give the phenotypic ratio of offspring from a cross of a deformed-leg, brown rabbit with a deformed-leg, white rabbit.

6. Polydactyly (extra fingers and toes) results from a dominant allele. A father is polydactyl, the mother has the normal phenotype, and they have had one normal child. What is the genotype of the father? Of the mother? What is the probability that a second child will have the normal number of digits?

7. In dogs, black (*B*) is dominant to chestnut (*b*), and solid color (*S*) is dominant to spotted (*s*). What are the genotypes of the parents in a mating that produced ⅜ black solid, ⅜ black spotted, ⅛ chestnut solid, and ⅛ chestnut spotted puppies? (*Hint:* First determine what genotypes the offspring must have before you deal with the fractions.)

8. When hairless hamsters are mated with normal-haired hamsters, about one-half the offspring are hairless and one-half are normal. When hairless hamsters are crossed with each other, the ratio of normal-haired to hairless is 1:2. How do you account for the results of the first cross? How would you explain the unusual ratio obtained in the second cross?

9. Two pigs whose tails are exactly 25 cm in length are bred over 10 years and they produce 96 piglets with the following tail lengths: 6 piglets at 15 cm, 25 at 20 cm, 37 at 25 cm, 23 at 30 cm, and 5 at 35 cm.
   a. How many pairs of genes are regulating the tail length character? *Hint:* Count the number of phenotypic classes, or determine the sum of the ratios of the classes. In a monohybrid cross, the F_2 ratios add up to 4 (3:1 or 1:2:1). In a dihybrid cross, the F_2 ratios add up to 16 (9:3:3:1 or some variation if the genes are epistatic or quantitative).
   b. What offspring phenotypes would you expect from a mating between a pig with a 15-cm-long tail and a pig with a 30-cm-long tail?

10. Fur color in rabbits is determined by a single gene, for which there are four alleles. Four phenotypes are possible: black, Chinchilla (gray color caused by white hairs with black tips), Himalayan (white with black patches on extremities), and white. The black allele (C) is dominant over all other alleles, the Chinchilla allele (C^{ch}) is dominant over Himalayan (C^h), and the white allele (c) is recessive to all others.

    a. A black rabbit is crossed with a Himalayan, and the F_1 consists of a ratio of 2 black to 2 Chinchilla. Explain what you can and cannot determine about the genotypes of the parents.

    b. A second cross was done between a black rabbit and a Chinchilla. The F_1 contained a ratio of 2 black to 1 Chinchilla to 1 Himalayan. Explain what you can and cannot determine about the genotypes of the parents.

11. Suppose that in hamsters, a dominant allele B determines black coat color and bb produces brown. A separate gene, however, shows epistasis over the B/b gene. The dominant E allele prevents the deposition of pigment in the hair, resulting in a golden coat color. An ee genotype is necessary for a black or brown hamster. A pet owner wants to know the genotypes of her three hamsters, so she breeds them and makes note of the offspring of the litters. Determine the genotypes of the three hamsters.

    a. golden female (hamster 1) × golden male (hamster 2) offspring: 7 golden, 1 black, 1 brown

    b. black female (hamster 3) × golden male (hamster 2) offspring: 8 golden, 5 black, 2 brown

12. The ability to taste phenylthiocarbamide (PTC) is controlled in humans by a single dominant allele (T). A nontasting woman married a tasting man, and they had three children: two tasting boys and a nontasting girl. All the grandparents could taste. Create a pedigree for this family for this trait. Solid symbols should signify nontasters (tt). Whenever possible, indicate whether tasters are TT or Tt.

13. Two true-breeding varieties of garden peas are crossed. One parent had red axial flowers, and the other had white terminal flowers. All F_1 individuals had red terminal flowers. If 100 F_2 offspring were counted, how many of them would you expect to have red axial flowers?

14. You cross true-breeding red-flowered plants with true-breeding white-flowered plants, and the F_1s are all red-flowered plants. The F_2s, however, occur in a ratio of 9 red:6 pale purple:1 white. How many genes are involved in the inheritance of this color character? Explain why the F_1s are all red and how the 9:6:1 ratio of phenotypes in the F_2 occurred.

15. A black guinea pig crossed with a guinea pig with albinism produces 12 black offspring. When a second black guinea pig is crossed with this same one, with albinism 7 black and 5 ones with albinism are obtained. What is the best explanation for this genetic outcome? Write genotypes for the parents and offspring of both crosses.

16. In sesame plants, the one-pod condition (P) is dominant to the three-pod condition (p), and normal leaf (L) is dominant to wrinkled leaf (l). Pod type and leaf type are inherited independently. Determine the genotypes for the two parents for all possible matings producing the following offspring:

    a. 318 one-pod, normal leaf and 98 one-pod, wrinkled leaf

    b. 323 three-pod, normal leaf and 106 three-pod, wrinkled leaf

    c. 401 one-pod, normal leaf

    d. 150 one-pod, normal leaf; 147 one-pod, wrinkled leaf; 51 three-pod, normal leaf; and 48 three-pod, wrinkled leaf

    e. 223 one-pod, normal leaf; 72 one-pod, wrinkled leaf; 76 three-pod, normal leaf; and 27 three-pod, wrinkled leaf

17. Imagine that a newly discovered, recessively inherited disease is expressed only in individuals with type O blood, although the disease and blood group are independently inherited. A normal man with type A blood and a normal woman with type B blood have already had one child with the disease. The woman is now pregnant for a second time. What is the probability that the second child will also have the disease?

18. In mice, black fur (B) is dominant to white (b). At a different locus, a dominant allele (A) produces a band of yellow just below the tip of each hair in mice with black fur. This gives a frosted appearance known as agouti. Expression of the recessive allele (a) results in a solid coat color. If mice that are heterozygous at both loci are crossed, what is the expected phenotypic ratio of their offspring?

Chapter 15

The Chromosomal Basis of Inheritance

Key Concepts

15.1 Morgan showed that Mendelian inheritance has its physical basis in the behavior of chromosomes: *scientific inquiry*

15.2 Sex-linked genes exhibit unique patterns of inheritance

15.3 Linked genes tend to be inherited together because they are located near each other on the same chromosome

15.4 Alterations of chromosome number or structure cause some genetic disorders

15.5 Some inheritance patterns are exceptions to standard Mendelian inheritance

Framework

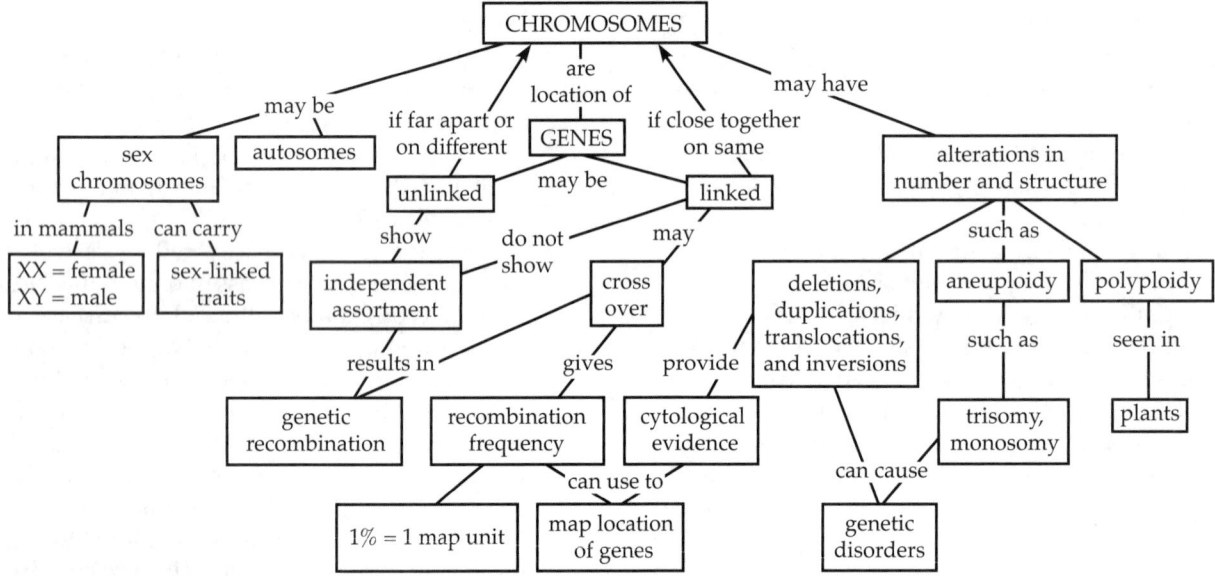

Chapter Review

Mendel's laws, combined with cytological evidence of the process of meiosis, led to the **chromosome theory of inheritance**: Genes occupy specific positions (loci) on chromosomes, and it is the random alignment of pairs of homologous chromosomes at metaphase I and the separation of homologs in anaphase I that result in the independent assortment and segregation of alleles in gamete formation.

15.1 Morgan showed that Mendelian inheritance has its physical basis in the behavior of chromosomes: *scientific inquiry*

Morgan's Choice of Experimental Organism
T. H. Morgan worked with fruit flies, *Drosophila melanogaster*, which are prolific and rapid breeders. Fruit flies have only four pairs of chromosomes; the sex chromosomes occur as XX in female flies and as XY in male flies.

The normal phenotype found most commonly in nature for a character is called the **wild type**, whereas alternative traits, assumed to have arisen as mutations, are called *mutant phenotypes*. For fruit flies, a mutant allele is designated by an italicized lowercase letter and the wild-type allele is designated by the italicized lowercase letter with a superscript +.

Correlating Behavior of a Gene's Alleles with Behavior of a Chromosome Pair

Morgan discovered a mutant white-eyed male fly, which he then mated with a wild-type red-eyed female. The F_1 offspring were all red-eyed. In the F_2, however, all female flies were red-eyed, whereas half of the males were red-eyed and half were white-eyed. Morgan deduced that the gene for eye color was located on the X chromosome. Males have only one X, so their phenotype is determined by the eye-color allele they inherit from their mother. This association of a specific gene with a chromosome provided evidence for the chromosome theory of inheritance.

INTERACTIVE QUESTION 15.1

Complete the following summary of Morgan's crosses involving the mutant white-eyed fly by filling in the Punnett square and indicating the genotypes and phenotypes of the F_2 generation. (X^w stands for the mutant recessive white allele; X^{w+} for the wild-type red allele.) The Y reminds you that this eye color gene is not present on the Y chromosome.

P Generation		
Phenotype	red-eyed (wild-type) female	× white-eyed male
Genotype	$X^{w+}X^{w+}$	X^wY

F_1 Generation		
Phenotype	red-eyed female	× red-eyed male
Genotype	$X^{w+}X^w$	$X^{w+}Y$

F_2 Generation

Eggs / Sperm (Punnett square)

Genotypes

_____ _____ _____ _____

Phenotypes

_____ _____ _____ _____

15.2 Sex-linked genes exhibit unique patterns of inheritance

The Chromosomal Basis of Sex

Sex is a phenotypic character usually determined by sex chromosomes. In humans and other mammals, females, who are XX, produce eggs that each contain an X chromosome. Males, who are XY, produce two kinds of sperm, each with either an X or a Y chromosome.

Other sex-determination systems include X-0 (in grasshoppers and some other insects), Z-W (in birds and some fishes and insects), and haplo-diploid (in bees and ants).

Whether the gonads of an embryo develop into testes or ovaries depends on the presence or absence of the gene *SRY*, found on the Y chromosome, whose protein product regulates many other genes. The gene *WNT4* is specifically associated with ovarian development. Genes located on either sex chromosome are called **sex-linked genes**. The 78 genes (coding for about 25 proteins) located on the Y chromosome are called *Y-linked genes*. About half of these genes are expressed only in the testis. The term **X-linked genes** refers to the approximately 1,100 genes located on the X chromosome.

Sex is now recognized as more complex than the simple male/female binary distinction would imply. An individual's sense of their gender identification is part of the sex phenotype.

Inheritance of X-Linked Genes

X-linked genes code for many characters that are not related to sex. Males inherit X-linked alleles from their mothers; daughters inherit X-linked alleles from both parents. Recessive X-linked traits are seen more often in males, because they are *hemizygous* for X-linked genes.

Duchenne muscular dystrophy is an X-linked disorder resulting from the lack of a key muscle protein. **Hemophilia** is an X-linked trait characterized by excessive bleeding due to the absence of one or more blood-clotting proteins.

X Inactivation in Female Mammals

Only one of the X chromosomes is fully active in most mammalian female somatic cells. The other X chromosome is condensed to form a **Barr body** located adjacent to the nuclear envelope. As a result of X inactivation, both males and females have an equal dosage of most X-linked gene products. The selection of which X chromosome is inactivated is a random event occurring independently in embryonic cells. A gene called *XIST* becomes active on the X chromosome that forms the Barr body. Its RNA products appear to trigger DNA methylation and X inactivation.

INTERACTIVE QUESTION   15.2

Two normal color-sighted individuals have two children and seven grandchildren. Fill in the probable genotype of each of the numbered individuals in the following pedigree. *Squares are males, circles are females, and solid symbols represent color blindness. Use the superscript N for the normal allele and n for the recessive allele for color blindness.*

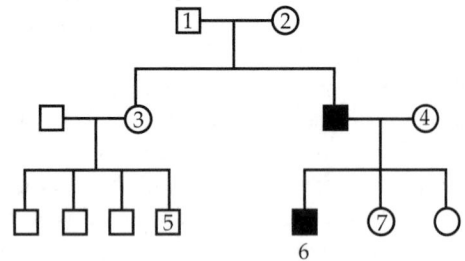

Genotypes:

1. _____   4. _____   7. _____

2. _____   5. _____

3. _____   6. _____

15.3 **Linked genes tend to be inherited together because they are located near each other on the same chromosome**

How Linkage Affects Inheritance   Morgan performed a testcross of F_1 dihybrid wild-type flies with flies that were homozygous recessive for black bodies and vestigial wings. He found that the offspring were not in the predicted 1:1:1:1 phenotypic ratio. Rather, most of the offspring were the same phenotypes as the P generation parents—either wild type (gray, normal wings) or double mutant (black, vestigial). Morgan deduced that these traits were inherited together because their genes were located on the same chromosome.

Genetic recombination results in offspring with combinations of traits that differ from those of either P generation parent.

Genetic Recombination and Linkage   Let's first look at the recombination of genes that are not linked. Consider a testcross between a dihybrid heterozygote and a doubly recessive homozygote. The dihybrid parent will produce four types of gametes; the homozygous recessive parent only one type. Of the four types of offspring that will be produced,

one-half will be **parental types** and have phenotypes like one or the other of the P generation parents that produced the F_1 dihybrid. The other half of the offspring, called **recombinant types (recombinants)**, will have combinations of the two traits that are unlike those of the P generation parents. This 50% frequency of recombination is observed when two genes are located on different chromosomes (or are sufficiently far apart on the same chromosome), and it results from the random alignment of pairs of homologous chromosomes at metaphase I and the resulting independent assortment of the two unlinked genes.

INTERACTIVE QUESTION   15.3

In a testcross between a heterozygote tall, purple-flowered pea plant and a dwarf, white-flowered plant,

a. what are the phenotypes of offspring that are parental types?

b. what are the phenotypes of offspring that are recombinants?

Linked genes, on the other hand, do not assort independently, and one would not expect to see recombination of parental traits in the offspring. Recombination of linked genes does occur, however, due to **crossing over**, the reciprocal trade between nonsister (a maternal and a paternal) chromatids of paired homologous chromosomes during prophase of meiosis I.

Recombinant chromosomes produced by crossing over, independent assortment, and random fertilization all contribute to the generation of new assortments of alleles, providing an abundance of genetic variation on which natural selection can work.

INTERACTIVE QUESTION   15.4

With unlinked genes, an equal number of parental and recombinant offspring are produced. With linked genes, are more or fewer parentals than recombinants produced? Explain your answer.

Mapping the Distance Between Genes Using Recombination Data: Scientific Inquiry   A **genetic map** is an ordered list of genes on a chromosome. The percentage of recombinant offspring produced in a genetic cross is called the *recombination frequency*. A. H. Sturtevant suggested that recombination frequencies reflect the relative distance between genes; for genes that are farther apart, the probability that crossing over will occur between them is greater than for genes that are closer together. Sturtevant used recombination data to create a **linkage map**, in which one **map unit** is defined as equal to a 1% recombination frequency.

The sequence of genes on a chromosome can be determined by finding the recombination frequency between different pairs of genes. Linkage cannot be determined if genes are so far apart that crossovers between them are almost certain. They would then have the 50% recombination frequency typical of unlinked genes. Such genes are *physically connected* but *genetically unlinked*. Distant genes on the same chromosome may be mapped by adding the recombination frequencies between them and intermediate genes.

Sturtevant and his colleagues found that the genes for the various known mutations of *Drosophila* clustered into four groups of linked genes (*linkage groups*), providing additional evidence that genes are located on chromosomes—in this case, on the four *Drosophila* chromosomes.

The frequency of crossing over may vary along the length of a chromosome, and a linkage map provides the sequence but not the exact location of genes on chromosomes. *Cytogenetic maps* locate gene loci in reference to visible chromosomal features. With recent advances in DNA sequencing, physical maps are produced that provide the number of DNA nucleotides between genes.

INTERACTIVE QUESTION 15.5

The following recombination frequencies have been determined for several gene pairs. Create a linkage map for these genes, showing the map unit distance between loci.

j, k: 12%   *j, m:* 9%   *k, l:* 6%   *l, m:* 15%

15.4 Alterations of chromosome number or structure cause some genetic disorders

Abnormal Chromosome Number   **Nondisjunction** occurs when a pair of homologous chromosomes does not separate properly in meiosis I or when sister chromatids do not separate in meiosis II. As a result, a gamete may receive either two copies or no copies

of that chromosome. A zygote formed with one of these aberrant gametes has a chromosomal alteration known as **aneuploidy**, a nontypical number of a particular chromosome. The zygote will be either **trisomic** for that chromosome (chromosome number is $2n + 1$) or **monosomic** ($2n - 1$). Aneuploid organisms that survive usually have a set of symptoms caused by the abnormal dosage of genes. A mitotic nondisjunction early in embryonic development is also likely to be harmful.

Polyploidy is a chromosomal alteration in which an organism has more than two complete chromosomal sets, as in *triploidy* ($3n$) or *tetraploidy* ($4n$). Polyploidy is common in the plant kingdom and has played an important role in the evolution of plants.

INTERACTIVE QUESTION 15.6

a. What is the difference between an organism with a trisomy and a triploid organism?

b. Which of these two organisms is likely to exhibit the more deleterious effects as a result of its chromosomal anomaly?

Alterations of Chromosome Structure   Chromosome breakage can result in chromosome fragments that are lost, called **deletion**; that join to a sister chromatid (or a nonsister chromatid), called **duplication**; that rejoin the original chromosome in the reverse orientation, called **inversion**; or that join a nonhomologous chromosome, called **translocation**. An unequal crossover can result in a deletion and duplication in nonsister chromatids, caused by unequal exchange between chromatids.

INTERACTIVE QUESTION 15.7

Two nonhomologous chromosomes have gene orders, respectively, of *A-B-C-D-E-F-G-H-I-J* and *M-N-O-P-Q-R-S-T*. What types of chromosome alterations would have occurred if daughter cells were found to have a gene sequence of *A-B-C-O-P-Q-G-J-I-H* on the first chromosome?

Human Disorders Due to Chromosomal Alterations The frequency of aneuploid zygotes may be fairly high in humans, but development is usually so disrupted that embryos spontaneously abort. Some genetic disorders, expressed as *syndromes* of characteristic traits, are the result of aneuploidy.

Down syndrome, also called *trisomy 21*, results in characteristic facial features, short stature, correctable heart defects, and developmental delays. The incidence of Down syndrome increases for older mothers, and prenatal screening for trisomies is now offered to all pregnant women.

XXY males exhibit *Klinefelter syndrome*, a condition in which the individual has abnormally small testes, is sterile, and may have subnormal intelligence. Males with an extra Y chromosome do not exhibit any well-defined syndrome.

Trisomy X results in females who are healthy and distinguishable only by karyotype, although they are slightly taller than average and at risk of learning disabilities. Monosomy X individuals (X0) exhibit *Turner syndrome* and are phenotypically female, sterile individuals with short stature and usually normal intelligence.

Structural alterations of chromosomes, such as deletions or translocations, may be associated with specific human disorders, such as *cri du chat* syndrome and *chronic myelogenous leukemia* (*CML*). The reciprocal translocation involved in CML occurs during mitosis and produces a short, recognizable *Philadelphia chromosome*. The translocation results in a fused gene associated with unregulated cell division.

INTERACTIVE QUESTION   15.8

Why do most sex chromosome aneuploidies have fewer deleterious effects than do autosomal aneuploidies?

15.5   Some inheritance patterns are exceptions to standard Mendelian inheritance

Genomic Imprinting   Over 200 traits in mammals depend on which parent supplied the alleles for the trait. **Genomic imprinting**, which occurs during gamete formation, determines whether or not an allele will be expressed in an offspring depending on which parent supplied the allele. When the next generation makes gametes, old maternal or paternal imprints are removed, and alleles are again imprinted according to the sex of the parent.

Most of the mammalian genes subject to imprinting identified so far are involved in embryonic development. The addition of methyl groups to cytosine nucleotides may inactivate (or activate in some cases) the imprinted gene, assuring that the developing embryo has only one active copy.

Inheritance of Organelle Genes   Exceptions to Mendelian inheritance occur for *extranuclear* (or *cytoplasmic*) *genes*. Such genes are located on small circles of DNA in mitochondria and plant plastids and are transmitted to offspring in the cytoplasm of the egg. Some rare human disorders are caused by mitochondrial mutations, and maternally inherited mitochondrial defects may contribute to some cases of diabetes, heart disease, and Alzheimer's disease. New techniques for creating hybrid eggs with nuclear genes from one female and mitochondria from another may allow the transmission of mitochondrial defects to be avoided.

Word Roots

an- = not, no; **eu-** = good, true (*aneuploidy:* a chromosomal aberration in which one or more chromosomes are present in extra copies or are deficient in number)

hem- = blood (*hemophilia:* a human genetic disease caused by a sex-linked recessive allele and characterized by excessive bleeding following injury)

mono- = one; **som-** = body (*monosomic:* referring to a cell that has only one copy of a particular chromosome instead of the normal two)

poly- = many (*polyploidy:* a chromosomal alteration in which the organism possesses more than two complete chromosome sets)

re- = again; **com-** = together; **bin-** = two at a time (*recombinant:* an offspring whose phenotype differs from that of the true-breeding P generation parents; also refers to the phenotype itself)

trans- = across (*translocation:* an aberration in chromosome structure resulting from attachment of a chromosomal fragment to a nonhomologous chromosome)

tri- = three (*trisomic:* referring to a diploid cell that has three copies of a particular chromosome)

Structure Your Knowledge

1. Mendel's law of independent assortment applies to genes that are on different chromosomes. However, at least two of the genes Mendel studied were actually located on the same chromosome. Explain why genes located more than 50 map units apart behave as though they are not linked. How can one determine whether these genes are linked, and what the relative distance is between them?

2. You have found a new mutant phenotype in fruit flies that you suspect is recessive and X-linked. What is the single, best cross you could make to confirm your predictions?

3. Women born with an extra X chromosome (XXX) are generally healthy and indistinguishable in appearance from normal XX women. What is a likely explanation for this finding? How could you test this explanation?

4. Various human disorders or syndromes are related to chromosomal abnormalities. What explanation can you give for the adverse phenotypic effects associated with these chromosomal alterations?

Test Your Knowledge

MULTIPLE CHOICE: *Choose the one best answer.*

1. Thomas Hunt Morgan firmly established the location of genes on chromosomes. What was his evidence?
   a. Cells have two copies of each gene; cells have two copies of each chromosome.
   b. A gene for eye color in *Drosophila* is localized to the X chromosome.
   c. Genes segregate independently during meiosis; chromosomes segregate independently during meiosis.
   d. Pairs of alleles segregate during meiosis; homologous chromosomes separate during meiosis.

2. A wild type is
   a. the phenotype found most commonly in nature.
   b. the dominant allele.
   c. designated by a small letter if it is recessive or a capital letter if it is dominant.
   d. a trait found on a sex chromosome.

3. Sex-linked traits
   a. are coded for by genes located on a sex chromosome.
   b. are found in only one sex, depending on the sex-determination system of the species.
   c. are always inherited from the mother in mammals and fruit flies.
   d. depend on whether the gene was inherited from the mother or the father.

4. Which of the following sex chromosome compositions would result in an individual that was physiologically female?
   a. X0   b. XY   c. XXY   d. XYY

5. A cross of a wild-type red-eyed female *Drosophila* with a violet-eyed male produces all red-eyed offspring. If the gene is X-linked, which of the following should the reciprocal cross (violet-eyed female × red-eyed male) produce? (Assume that the red allele is dominant to the violet allele.)
   a. all red-eyed flies
   b. 3 red-eyed flies to 1 violet-eyed fly
   c. a 1:1 ratio of red and violet eyes in both males and females
   d. red-eyed females and violet-eyed males

6. In chickens, males have two identical sex chromosomes called Z. Females have two different sex chromosomes, one Z and one W. Which of the following statements is true?
   a. The genetic makeup of the sperm determines the sex of the chick.
   b. A hen inherits her single W from her father.
   c. A rooster inherits both his Z chromosomes from his father.
   d. None of the above is true.

7. A 1:1:1:1 ratio of offspring from a dihybrid testcross indicates that the
   a. genes are linked.
   b. dominant organism was homozygous.
   c. genes are not linked or are more than 50 map units apart.
   d. genes are 25 map units apart.

8. Genes *A* and *B* are linked and 12 map units apart. A heterozygous individual, whose parents were *AAbb* and *aaBB*, would be expected to produce gametes in which of the following frequencies?
   a. 44% *AB*   6% *Ab*   6% *aB*   44% *ab*
   b. 6% *AB*   44% *Ab*   44% *aB*   6% *ab*
   c. 12% *AB*   38% *Ab*   38% *aB*   12% *ab*
   d. 38% *AB*   12% *Ab*   12% *aB*   38% *ab*

9. Two true-breeding *Drosophila* are crossed: a normal-winged, red-eyed female and a miniature-winged, vermilion-eyed male. The F$_1$s all have normal wings and red eyes. When F$_1$ offspring were crossed with miniature-winged, vermilion-eyed flies, the following offspring resulted:

   233 normal wing, red eye

   247 miniature wing, vermilion eye

   7 normal wing, vermilion eye

   13 miniature wing, red eye

   From these results, you could conclude that the alleles for miniature wings and vermilion eyes are
   a. both X-linked and recessive.

b. located on autosomes and dominant.

c. recessive, and that these genes are located 4 map units apart.

d. recessive, and that these genes are located 20 map units apart.

10. A female tortoiseshell cat is heterozygous for the gene that determines black or orange coat color, which is located on the X chromosome. A male tortoiseshell cat

    a. cannot occur.

    b. is hemizygous at this locus.

    c. must have resulted from a nondisjunction and has a Barr body in each of his cells.

    d. must have two alleles for coat color in each of his cells, one from his father and one from his mother.

11. Consider three genes on the X chromosome: A, B, and C. The recombination frequency between A and C is 20%, the recombination frequency between B and C is 30%, and the recombination frequency between A and B is 50%. What conclusions can be drawn from these data?

    a. A, B, and C will all segregate independently.

    b. A and C will appear linked, but A and B will segregate independently.

    c. The distance along the chromosome between A and B is greater than the distance between A and C.

    d. Both **b** and **c** are valid conclusions.

12. Which of the following chromosomal alterations does not alter genic balance but may affect gene expression and thus phenotype?

    a. deletion   **c.** duplication

    b. inversion   **d.** nondisjunction

13. Genomic imprinting

    a. explains cases in which the phenotypic effect of an allele depends on the sex of the parent from whom that allele is inherited.

    b. may involve the silencing of an allele by methylation such that offspring inherit only one active copy of a gene.

    c. occurs more often in females because of the larger maternal contribution of cytoplasm.

    d. involves both **a** and **b**.

14. There is a type of night-blindness that is produced by a recessive X-linked allele. Is it possible for a female to suffer from night-blindness?

    a. Yes, because all daughters of a night-blind mother will be night-blind.

    b. Yes, if her father is night-blind and her mother is a heterozygote with normal night vision, she will be night-blind.

    c. Yes, because all daughters of a night-blind man will be night-blind.

    d. No, because females can't express X-linked genes.

15. Tests of social behavior given to Turner syndrome volunteers (who are X0) found a correlation between scores on "behavioral inhibition" tasks and the source of the lone X chromosome. These test results appear to be an example of

    a. genomic imprinting of a gene on the X chromosome.

    b. Y-linked inheritance.

    c. meiotic nondisjunction.

    d. chromosomal reciprocal translocation.

16. Suppose that alleles for an X-linked character for wing shape in flies show incomplete dominance. The X^+ allele codes for pointed wings, the X^r for round wings, and X^+X^r individuals have oval wings. In a cross between an oval-winged female and a round-winged male, the following offspring were observed: oval-winged females, round-winged females, pointed-winged males, and round-winged males. A rare pointed-winged female was noted. Cytological study revealed that she had two X chromosomes. Which of the following events could account for this unusual offspring?

    a. a crossover between the two X chromosomes

    b. a nondisjunction in meiosis II between two X^+ chromatids

    c. a nondisjunction between the X and Y chromosomes, producing some sperm with no sex chromosome

    d. Both **b** and **c** together could produce an X^+X^+ female when an XX egg was fertilized by a sperm in which there was no sex chromosome.

17. Some girls who fail to undergo puberty are found to have Swyer syndrome, a condition in which they are externally female but have an XY genotype. Which of the following statements may explain the origin of this syndrome?

    a. A mutation in the *XIST* gene, which codes for RNA molecules that coat the X chromosome and initiate X-inactivation, must have occurred.

    b. A nondisjunction in the egg from the mother resulted in both sex chromosomes coming from the father. Genomic imprinting of the father's

X chromosome then caused the development of a female phenotype.

c. These individuals are actually XXY; the second X is not seen because it is condensed into a Barr body. They have small testes and are sterile but otherwise appear female.

d. A mutation or deletion of the *SRY* gene on the Y chromosome prevented development of testes and production of the male sex hormones required for a male phenotype.

18. A mutation in a mitochondrial gene has been linked to a rare muscle-wasting disease in humans. This disease is

   a. found more often in males than in females.

   b. found more often in females than in males.

   c. inherited in a simple Mendelian fashion.

   d. inherited from the mother and is equally likely to occur in male offspring as in female offspring.

19. The genetic event that results in Turner syndrome (X0) is probably

   a. nondisjunction.   c. genomic imprinting.

   b. deletion.   d. monoploidy.

20. A cross between a wild-type mouse and a dwarf mouse homozygous for a recessive mutation in the *Igf2* gene produces heterozygous offspring that are normal size if the dwarf parent was the mother but dwarf if the dwarf parent was the father. Which of the following genetic events explains these results?

   a. sex-linked inheritance

   b. monosomy

   c. genomic imprinting

   d. inheritance of mitochondrial genes

Genetics Problems

Again, one of the best ways to learn genetics is to do problems.

1. The following pedigree traces the inheritance of a genetic trait.

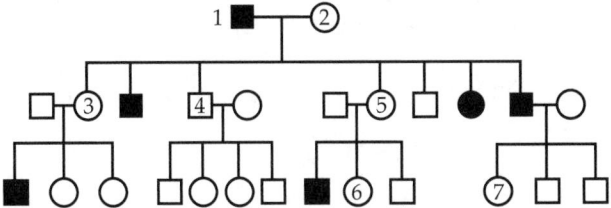

a. What type of inheritance does this trait show?

b. Choose an appropriate allele labeling system, and give the predicted genotype for the following individuals:

   1. _____   5. _____
   2. _____   6. _____
   3. _____   7. _____
   4. _____

c. What is the probability that a child of individual #6 and a phenotypically normal male will have this trait?

2. Use the following recombination frequencies to determine the order of the genes on the chromosome. a, c: 10%; a, d: 30%; b, c: 24%; b, d: 16%

3. In guinea pigs, black (*B*) is dominant to brown (*b*), and solid color (*S*) is dominant to spotted (*s*). A heterozygous black, solid-colored pig is mated with a brown, spotted pig. The offspring from several litters are as follows: black solid: 16; black spotted: 5; brown solid: 5; and brown spotted: 14. Are these genes linked or unlinked? If they are linked, how many map units are they apart?

4. A woman is a carrier for an X-linked lethal allele that causes an embryo with the allele to spontaneously abort. She has nine children. How many of these children do you expect to be boys?

5. A dominant sex-linked allele *B* produces white bars on black chickens, as seen in the Barred Plymouth Rock breed. A clutch of chicks has equal numbers of black and barred chicks. (Remember that sex is determined by the Z-W system in birds: ZZ are males, ZW are females.)

   a. If only the females are found to be black, what were the genotypes of the parents?

   b. If males and females are evenly represented in the black and barred chicks, what were the genotypes of the parents?

6. Red-green color blindness is caused by a sex-linked recessive allele. A color-blind man marries a woman with normal vision whose father was color-blind. What is the probability that they will have a color-blind daughter? What is the probability that their first son will be color-blind? (Note the different wording in the two questions.)

Chapter 16

The Molecular Basis of Inheritance

Key Concepts

16.1 **DNA is the genetic material**

16.2 **Many proteins work together in DNA replication and repair**

16.3 **A chromosome consists of a DNA molecule packed together with proteins**

Framework

This chapter outlines the key evidence that was gathered to establish DNA as the molecule of inheritance. Watson and Crick's double helix, with its rungs of specifically paired nitrogenous bases and twisting phosphate-sugar side ropes, provided the three-dimensional model that explained DNA's ability to encode a great variety of information and produce exact copies of itself through semiconservative replication. The replication of DNA is an extremely fast and accurate process involving many enzymes and proteins. Eukaryotic chromosomes are complexes of DNA and protein that exhibit varying levels of packing during the cell cycle.

Chapter Review

Deoxyribonucleic acid, or DNA, is the genetic material that is transmitted from one generation to the next and encodes the blueprints that direct and control the biochemical, anatomical, physiological, and behavioral traits of organisms. DNA is precisely copied in the process of **DNA replication**.

16.1 DNA is the genetic material

The Search for the Genetic Material: Scientific Inquiry Chromosomes were shown to carry hereditary information and to consist of proteins and DNA. Until the 1940s, proteins seemed the most likely candidate to be the genetic material. The role of DNA was established through work with microorganisms—bacteria and viruses.

In 1928, F. Griffith was working with two strains of *Streptococcus pneumoniae*. When he mixed the remains of heat-killed pathogenic bacteria with harmless bacteria, some bacteria were changed into disease-causing bacteria. These bacteria incorporated external genetic material in a process called **transformation**, which results in a change in genotype and phenotype. Scientists later determined that DNA was the molecule that transformed bacteria.

Viruses consist of DNA (or sometimes RNA) contained in a protein coat. They reproduce by infecting a cell and commandeering that cell's metabolic machinery. **Bacteriophages**, or **phages**, are viruses that infect bacteria. In 1952, A. Hershey and M. Chase showed that DNA was the genetic material of a phage known as T2 that infects the bacterium *Escherichia coli*.

INTERACTIVE QUESTION 16.1

Hershey and Chase devised an experiment using radioactive isotopes to determine whether it was a phage's DNA or protein that entered the bacteria and served as the genetic material of T2 phage.

a. How did they label phage protein?

b. How did they label phage DNA?

After infecting separate samples of *E. coli* with the differently labeled T2 cells, they blended and centrifuged the samples to isolate the bacterial cells from the lighter viral particles.

c. Where was the radioactivity found in the samples with labeled phage protein?

d. Where was the radioactivity found in the samples with labeled phage DNA?

e. What did Hershey and Chase conclude from these results?

In 1950, E. Chargaff noted that the percentages of the four nitrogenous bases in DNA were species specific. Chargaff also determined that the number of adenines and thymines was approximately equal, and the number of guanines and cytosines was also equal. The variation in base composition among species and the A=T and G=C properties of DNA became known as *Chargaff's rules.*

Building a Structural Model of DNA: Scientific Inquiry   By the early 1950s, the arrangement of covalent bonds in a nucleic acid polymer was established, but the three-dimensional structure of DNA was yet to be determined.

In X-ray crystallography, an X-ray beam passed through a substance produces an X-ray diffraction photo with a pattern of spots that a crystallographer interprets as information about three-dimensional molecular structure. J. Watson saw an X-ray photo produced by R. Franklin that indicated the helical shape of DNA and that it consisted of two strands—thus the term **double helix**.

Watson and F. Crick constructed wire models of a double helix that had the paired nitrogenous bases on the inside of the helix and two sugar-phosphate chains running in opposite directions (**antiparallel**) on the outside. The helix makes one full turn every 3.4 nm; 10 layers of nucleotide pairs, stacked 0.34 nm apart, are present in each turn of the helix.

To produce the molecule's uniform 2-nm width, a purine base must pair with a pyrimidine base. The side groups of the bases permit two hydrogen bonds to form between adenine and thymine, and three hydrogen bonds between guanine and cytosine. This complementary pairing explains Chargaff's rules. Van der Waals attractions between the closely stacked bases help hold the molecule together.

In 1953, Watson and Crick published a paper in *Nature* reporting the double helix as the molecular model for DNA.

INTERACTIVE QUESTION   16.2

Review the structure of DNA by labeling the following diagrams.

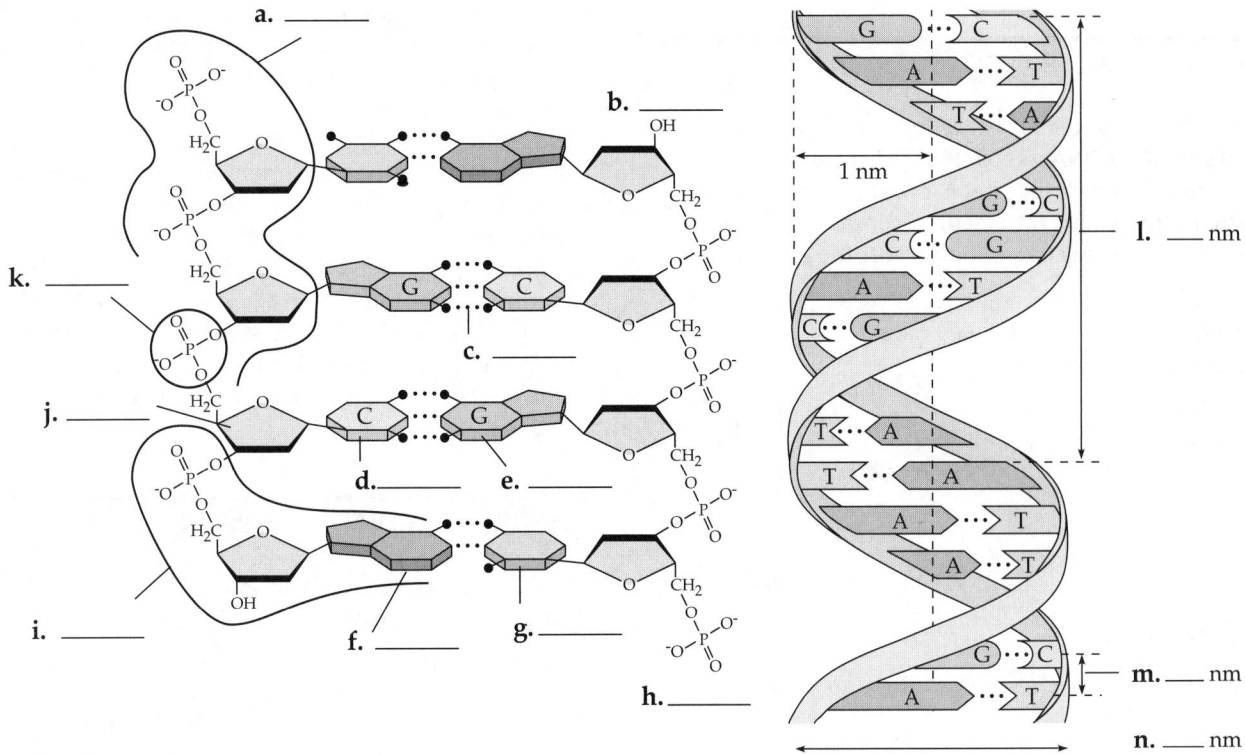

16.2 Many proteins work together in DNA replication and repair

The Basic Principle: Base Pairing to a Template Strand Watson and Crick noted that the base-pairing rule of DNA sets up a mechanism for its replication. Each side of the double helix is an exact complement to the other. When the two sides separate, each strand serves as a template for rebuilding a double-stranded molecule identical to the "parental" molecule.

The **semiconservative model** of DNA replication predicts that the two daughter DNA molecules each have one parental strand and one newly formed strand. In contrast, a conservative model predicts that the parental double helix reforms and the duplicated molecule is totally new, whereas a dispersive model predicts that all four strands of the two DNA molecules are a mixture of parental and new DNA.

M. Meselson and F. Stahl tested these models by growing *E. coli* in a medium with ^{15}N, a heavy isotope that the bacteria incorporated into their nitrogenous bases. Cells with labeled DNA were transferred to a medium with a lighter isotope, ^{14}N. Samples were removed after one and two generations of bacterial growth, and DNA was extracted and centrifuged. The locations of the density bands in the centrifuge tubes confirmed the semiconservative model of DNA replication.

INTERACTIVE QUESTION   16.3

Using different colors for heavy (parental) and light (new) strands of DNA, sketch the DNA molecules formed in two replication cycles after *E. coli* were moved from medium containing ^{15}N to medium containing ^{14}N. Show the resulting density bands in the centrifuge tubes.

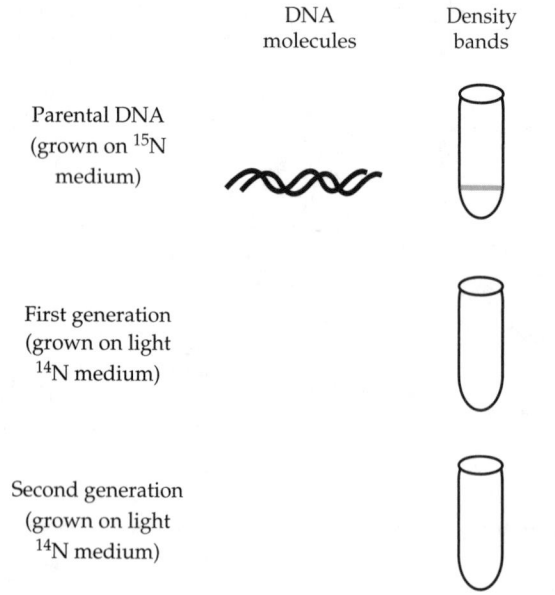

DNA Replication: A Closer Look Replication of most bacterial chromosomes begins at a single **origin of replication**, where proteins that initiate replication bind to a specific sequence of nucleotides and separate the two strands to form a replication "bubble." Replication proceeds in both directions in the two Y-shaped **replication forks**. Eukaryotic chromosomes have many origins of replication.

An enzyme called **helicase** unwinds the helix and separates the parental strands at each replication fork. **Single-strand binding proteins** keep the separated strands apart while they serve as templates. **Topoisomerase** helps relieve the strain from the tighter twisting of DNA strands in front of helicase. **DNA polymerases** can extend a DNA nucleotide strand, but they cannot initiate DNA synthesis. When the two DNA strands have been separated, 5 to 10 RNA nucleotides base-pair with the exposed DNA nucleotides. An enzyme called **primase** connects these ribonucleotides to form an RNA **primer**.

DNA polymerases now extend the RNA primer, connecting DNA nucleotides to the growing end of the new DNA strand. DNA polymerase III and I are involved in replication in *E. coli*; at least 11 different DNA polymerases have been discovered so far in eukaryotes. A nucleotide base-pairs with its complementary base on the template strand; it loses two phosphate groups, and the hydrolysis of this pyrophosphate to two inorganic phosphates ($\circledP_i$) provides the energy for the dehydration reaction that results in polymerization.

Remember that the two strands of a DNA molecule are antiparallel—they are oriented in opposite directions. The sugar of each nucleotide is connected to its own phosphate group at its 5′ carbon and connects to the phosphate group of the adjacent nucleotide by its 3′ carbon. Thus, the polarity of a strand of DNA runs from the 5′ end where the first nucleotide's phosphate group is exposed, to the 3′ end where the nucleotide at the other end has a hydroxyl group attached to its 3′ carbon.

INTERACTIVE QUESTION   16.4

Look back to Interactive Question 16.2 and label the 5′ and 3′ ends of the left strand of the DNA molecule.

DNA polymerases add nucleotides only to the free 3′ end of a primer or growing DNA strand; the DNA strand is elongated in a 5′ → 3′ direction. The **leading strand** is the new continuous strand being formed along one template strand as DNA polymerase III (DNA pol III) remains in the progressing replication fork.

The **lagging strand** is created as a series of short segments, called **Okazaki fragments**, which are elongated in the 5′ → 3′ direction away from the replication fork. Each fragment on the lagging strand requires a primer. A continuous strand of DNA is produced after DNA polymerase I (DNA pol I) replaces the RNA primer with DNA nucleotides and an enzyme called **DNA ligase** joins the sugar-phosphate backbones of the fragments.

Working out the details of DNA replication is an area of active research. The various proteins that function in replication form a large DNA replication complex, but it is not clear whether the complex moves along the DNA or the DNA actively moves through the complex. In eukaryotic cells, many such complexes may anchor to the nuclear matrix.

INTERACTIVE QUESTION 16.5

In this diagram of bacterial DNA replication, label the following items: leading and lagging strands, Okazaki fragment, DNA pol III, DNA pol I, DNA ligase, helicase, single-strand binding proteins, primase, RNA primer, and 5′ and 3′ ends of parental DNA.

Proofreading and Repairing DNA   Initial pairing errors in nucleotide placement may occur as often as 1 per 100,000 base pairs. The amazing accuracy of DNA replication (one error in 10 billion nucleotides) is achieved because DNA polymerases check each newly added nucleotide against its template and remove incorrect nucleotides. Other enzymes also fix incorrectly paired nucleotides, a process called **mismatch repair**.

A large number of different DNA repair enzymes monitor and repair damaged DNA. In **nucleotide excision repair**, the damaged strand is cut out by a **nuclease** and the gap is correctly filled through the action of a DNA polymerase and ligase. In skin cells, nucleotide excision repair frequently corrects *thymine dimers* caused by ultraviolet rays in sunlight. The excision repair mechanism is defective in individuals with the disorder xeroderma pigmentosum. Because thymine dimers cannot be corrected, the resulting mutations frequently cause skin cancers.

Evolutionary Significance of Altered DNA Nucleotides   A mutation results when DNA containing uncorrected mismatched nucleotides is replicated and the change is passed on to daughter cells (or to offspring, if the mutation is in a gamete). Although usually harmful, these permanent genetic changes provide the variation on which natural selection operates.

Replicating the Ends of DNA Molecules   Because DNA polymerase cannot attach nucleotides to the 5′ end of a growing DNA strand, repeated replications cause a progressive shortening of the 5′ end of newly synthesized linear DNA strands. Multiple repetitions of short nucleotide sequences at the ends of chromosomes, called **telomeres**, protect an organism's genes from being eroded during successive DNA replications. Telomeres also prevent staggered ends from being recognized as damaged DNA by cellular monitoring systems.

The unavoidable shortening of telomeres may limit cell division in somatic cells. In eukaryotic germ cells, however, the enzyme *telomerase* lengthens telomeres. Some somatic cancer cells and "immortal" strains of cultured cells produce telomerase and are thus capable of unregulated cell division.

INTERACTIVE QUESTION   16.6

Draw the last Okazaki fragment being formed on the lagging strand of a linear DNA molecule. Indicate how this results in a shortening of the end of the DNA molecule.

16.3 A chromosome consists of a DNA molecule packed together with proteins

A bacterial cell's double-stranded, circular DNA molecule (associated with a small amount of protein) is supercoiled and tightly packed into a region of the cell called the nucleoid.

In eukaryotes, each chromosome consists of a single, extremely long DNA double helix precisely associated with a large amount of protein, forming a complex called **chromatin**. This DNA fits into the nucleus through a multilevel system of coiling and folding.

Histones are small, positively charged proteins that bind tightly to the negatively charged DNA. Unfolded chromatin appears as a string of beads, each bead a **nucleosome** consisting of the DNA helix wound around a protein core of four pairs of the main histone types. The "string" between beads is called *linker* DNA. The amino end (*histone tail*) of each of the eight histones extends outward. Also called the *10-nm fiber* because of its diameter, nucleosomes are the basic unit of DNA packing. A fifth histone attaches to the DNA near the bead.

The *30-nm fiber* is a tightly coiled fiber of nucleosomes, histone tails, and linker DNA. The *looped domain* consists of loops of the 30-nm chromatin fiber attached to a protein scaffold, forming a *300-nm fiber*. In a metaphase chromosome, looped domains coil and fold, further compacting the chromatin into chromatids that are 700 nm wide and visible with a light microscope.

Some DNA packing is evident in interphase chromosomes, and a chromosome's looped domains appear to be attached to specific locations of the nuclear lamina inside the nuclear envelope. A technique for "painting" individual chromosomes different colors has shown that each chromosome occupies a discrete area in the interphase nucleus.

Certain regions of chromatin, called **heterochromatin**, are in a highly condensed state during interphase. The more open form of interphase chromatin, called **euchromatin**, is available for the transcription of genes.

INTERACTIVE QUESTION   16.7

List the successive levels of packing in a metaphase chromosome.

Word Roots

helic- = a spiral (*helicase*: an enzyme that untwists the double helix of DNA at replication forks, separating the two strands)

-phage = that which eats (*bacteriophage*: a virus that infects bacteria)

tel- = an end (*telomere*: the repetitive DNA at the end of a eukaryotic chromosome that protects the organism's genes from being eroded during successive rounds of replication)

trans- = across (*transformation*: a change in genotype and phenotype due to the assimilation of external DNA by a cell)

Structure Your Knowledge

1. Summarize the evidence and techniques Watson and Crick used to deduce the double-helix structure of DNA.
2. Review your understanding of DNA replication by describing the key enzymes and proteins (in the order of their functioning) that direct replication.

Test Your Knowledge

MULTIPLE CHOICE: *Choose the one best answer.*

1. One of the reasons most scientists thought proteins were the carriers of genetic information was that
   a. the protein content of duplicating cells always doubles prior to division.
   b. proteins are much more complex and heterogeneous than nucleic acids.
   c. early experimental evidence pointed to proteins as the hereditary material.
   d. proteins are found in DNA.

2. Transformation involves
   a. the uptake of external genetic material.
   b. the creation of a strand of RNA from a DNA molecule.
   c. the infection of bacterial cells by phage.
   d. the type of semiconservative replication shown by DNA.

3. The DNA of an organism has thymine as 20% of its bases. What percentage of its bases would be guanine?
   a. 20%   b. 30%   c. 40%   d. 60%

4. Which of the following most closely represents Chargaff's rule?
   a. A pairs with T and C pairs with G
   b. %A + %T = %C + %G
   c. %A + %G = %C + %T
   d. Strands in a DNA double helix are antiparallel.

5. T2 phage is grown in *E. coli* with radioactive phosphorus and then allowed to infect other unlabeled *E. coli*. The culture is blended to separate the viral coats from the bacterial cells and then centrifuged. Which of the following statements best describes the expected results of such an experiment?
   a. Both viral and bacterial DNA molecules are labeled; radioactivity is found in the liquid above the pellet.
   b. Viral DNA is labeled; radioactivity is found in the pellet.
   c. Viral proteins are labeled; radioactivity is found in the liquid but not in the pellet.
   d. Both viral and bacterial proteins are labeled; radioactivity is present in both the liquid and the pellet.

6. Watson and Crick concluded that each base could not pair with itself because
   a. there would not be room for the helix to make a full turn every 3.4 nm.
   b. the uniform width of 2 nm would not permit two purines or two pyrimidines to pair together.
   c. identical bases could not hydrogen-bond together.
   d. they would be on antiparallel strands.

7. In their classic experiment, Meselson and Stahl
   a. provided evidence for the semiconservative model of DNA replication.
   b. were able to separate phage protein coats from *E. coli* by using a blender.
   c. found that DNA labeled with ^{15}N was of intermediate density.
   d. found that DNA composition was species specific.

8. The joining of nucleotides in the polymerization of DNA requires energy from
   a. the hydrolysis of the terminal phosphate group of ATP.
   b. RNA nucleotides.
   c. the phosphate groups of the sugar-phosphate backbone.
   d. the hydrolysis of the pyrophosphates removed from the joining nucleotides.

9. DNA polymerase is not able to begin copying a DNA strand on its own. How is the replication of DNA initiated?
   a. A coenzyme joins to the polymerase to allow it to begin synthesizing DNA.
   b. Short single-stranded DNA molecules called primers exist in the nucleus, and they bind to the DNA to allow synthesis to begin.
   c. An enzyme called primase creates a short RNA primer that can be extended by DNA polymerase.
   d. Telomerase extends the DNA strand to create a starting point.

10. During DNA replication, replacement of RNA primers and connection of the Okazaki fragments are carried out by
    a. primase and telomerase, respectively.
    b. ligase and telomerase, respectively.
    c. DNA polymerase I and ligase, respectively.
    d. helicase and ligase, respectively.

11. In a DNA double helix, the nucleotides in a single strand are held together by ____ bonds, and the two strands are held together by ____ bonds.
    a. hydrogen; covalent
    b. hydrogen; ionic
    c. ionic; hydrogen
    d. covalent; hydrogen

12. How does DNA synthesis along the lagging strand differ from that on the leading strand?
    a. Nucleotides are added to the 5′ end instead of the 3′ end.
    b. Ligase is the enzyme that polymerizes DNA on the lagging strand.
    c. An RNA primer is needed on the lagging strand but not on the leading strand.
    d. Okazaki fragments, which each grow 5′ → 3′, must be joined along the lagging strand.

13. Which of the following is least related to the others in the list, in terms of function?

  a. primase

  b. DNA polymerase

  c. ligase

  d. RNA polymerase

Use the following diagram to answer questions 14 through 17.

a. O^-
$$O{=}P{-}O{-}CH_2$$
$$O^-$$
                   A

    O

$$O{=}P{-}O{-}CH_2$$
$$O^-$$
  b.   T

    O

        c.
$$O{=}P{-}O{-}CH_2$$
$$O^-$$
              e. G

  d. HO

14. Which letter indicates the 5′ end of this single DNA strand?

  a.   **b.**   **c.**   **d.**   **e.**

15. At which letter would the next nucleotide be added?

  a.   **b.**   **c.**   **d.**   **e.**

16. Which letter indicates a phosphodiester bond formed by DNA polymerase?

  a.   **b.**   **c.**   **d.**   **e.**

17. The base sequence of the DNA strand made from this template would be (from bottom to top)

  a. A T G   **c.** T A C.

  b. C A T.   **d.** U A C.

18. Which of the following statements about telomeres is correct?

  a. They are ever-shortening tips of chromosomes that may signal cells to stop dividing at maturity.

  b. They are highly repetitive sequences at the tips of chromosomes that protect the leading strand during replication.

  c. They are enzymes in germ cells that allow these cells to undergo repeated divisions.

  d. All of the above are correct.

19. You are trying to test your hypothesis that DNA replication is *conservative*— that is, that the parental strands separate, newly made complementary strands join together to make a new DNA molecule, and the parental strands then rejoin. You take a sample of *E. coli* grown in a medium containing only heavy nitrogen (^{15}N) and transfer it to a medium containing light nitrogen (^{14}N). After allowing time for only one DNA replication, you centrifuge a sample and compare the density band(s) formed to the bands formed from bacteria grown on either normal ^{14}N or ^{15}N medium. Which band location would support your hypothesis of *conservative* DNA replication?

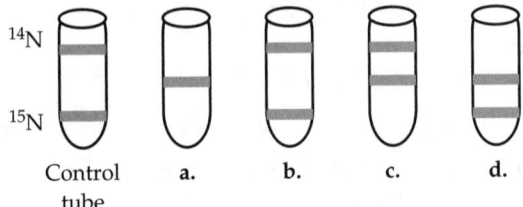

    Control   **a.**   **b.**   **c.**   **d.**
    tube

Bands after one replication on ^{14}N

20. Given the experimental procedure explained in question 19, which centrifuge tube (obtained after one DNA replication) would represent the band distribution indicating that DNA replication is *semiconservative*?

21. Which of the following structures is smallest?

  a. looped domain

  b. histone

  c. nucleosome

  d. 30-nm fiber

22. Biologists have learned from the technique of "painting chromosomes" with different-colored molecular tags that

  a. the two homologs of a pair of chromosomes differ enough that they stain different colors.

  b. chromosome packing occurs only as the cell prepares for mitosis or meiosis.

  c. heterochromatin is concentrated at the tips and centromeres of chromosomes.

  d. in the interphase nucleus, each chromosome appears to occupy a specific area.

Chapter 17

Gene Expression: From Gene to Protein

Key Concepts

17.1 Genes specify proteins via transcription and translation

17.2 Transcription is the DNA-directed synthesis of RNA: *a closer look*

17.3 Eukaryotic cells modify RNA after transcription

17.4 Translation is the RNA-directed synthesis of a polypeptide: *a closer look*

17.5 Mutations of one or a few nucleotides can affect protein structure and function

Framework

This chapter examines the pathway from DNA to RNA to proteins. The instructions of DNA are transcribed to a sequence of codons in mRNA. In eukaryotes, mRNA is processed before it leaves the nucleus. Complexed with ribosomes, mRNA is translated as a sequence of amino acids in a polypeptide as tRNAs match their anticodons to the mRNA codons. Mutations, which alter the nucleotide pairs in DNA, may alter the protein product. A gene may be defined as a DNA sequence whose final product is either a polypeptide or an RNA molecule.

Chapter Review

The DNA-directed synthesis of a functional product is called **gene expression**. Generally, this product is a protein, but in some cases an untranslated RNA molecule may be the functional gene product.

17.1 Genes specify proteins via transcription and translation

Evidence from Studying Metabolic Defects   In 1902, A. Garrod first suggested that genes determine phenotype through the action of enzymes, reasoning that inherited diseases were caused by an inability to make certain enzymes.

G. Beadle and E. Tatum used X-rays to produce "nutritional mutants" of the haploid bread mold *Neurospora crassa*. Mutants that failed to grow on minimal growth medium were transferred to minimal medium supplemented with specific nutrients, and growth was assessed. In this way the specific metabolic deficit for each mutant was identified. By supplementing different precursors of the arginine pathway, A. Srb and N. Horowitz identified three classes of *Neurospora* mutants. They reasoned that the arginine pathway was blocked at a different enzymatic step in each class of mutants. Such research supported Beadle and Tatum's *one gene–one enzyme hypothesis*.

Biologists revised this hypothesis to one gene–one protein. Because many proteins consist of more than one polypeptide chain, the axiom changed to the *one gene–one polypeptide hypothesis*. (Exceptions include genes that code for more than one protein due to alternative splicing and genes that code for untranslated RNA molecules.)

Basic Principles of Transcription and Translation   RNA is the link between a gene and the protein for which it codes.

Transcription is the synthesis of **messenger RNA (mRNA)** or another type of RNA using information stored in DNA. **Translation** is the RNA-directed synthesis of a polypeptide, changing from the language of nucleotides to that of amino acids. **Ribosomes** are the sites of translation—the sites of the synthesis of a polypeptide.

In prokaryotes, which lack a nucleus, transcription and translation can occur simultaneously. In eukaryotes, the mRNA, called *pre-mRNA*, is processed before it exits the nucleus and enters the cytoplasm, where translation occurs. The initial RNA transcript of any gene is called the **primary transcript**.

INTERACTIVE QUESTION   17.1

a. In what three ways does RNA differ from DNA?

b. Fill in the following sequence in the flow of genetic information, often called the *central dogma*. Above each arrow, write the name of the process involved.

_____ → _____ → _____

The Genetic Code   A sequence of three nucleotides provides 4^3 (64) possible unique sequences of nucleotides, more than enough to code for the 20 common amino acids. The translation of nucleotides into amino acids uses a **triplet code** to specify each amino acid.

The nucleotide base triplets along the **template strand** of a gene are transcribed into complementary mRNA **codons**. One strand of a DNA molecule can serve as the template strand for one gene and the nontemplate strand for another. The RNA molecule is also antiparallel to the template DNA strand. The mRNA is complementary to the DNA template because its nucleotides follow the same base-pairing rules, with the exception that uracil substitutes for thymine in RNA. The term *codon* can also refer to the DNA nucleotide triplets on the *nontemplate* strand, which may be called the **coding strand**.

During translation, the sequence of codons, read in the 5′ → 3′ direction, determines the sequence of amino acids in the polypeptide.

In the early 1960s, M. Nirenberg added artificial "poly U" mRNA to a test tube containing all the biochemical ingredients necessary for protein synthesis and obtained a polypeptide containing the single amino acid phenylalanine. By the mid-1960s, molecular biologists had deciphered all 64 codons. Three codons function as stop signals, or termination codons. The codon AUG both codes for methionine and functions as an initiation codon, a start signal for translation.

The code is often redundant, meaning that more than one codon may specify a single amino acid. The code is never ambiguous: No codon specifies two different amino acids.

The nucleotide sequence on mRNA is read in the correct **reading frame**, starting at a start codon and reading each triplet sequentially.

INTERACTIVE QUESTION   17.2

Practice using the codon table in your textbook. Determine the amino acid sequence for a polypeptide coded for by the following mRNA transcript:

5′-AUGCCUGACUUUAAGUAG-3′

The genetic code of codons and their corresponding amino acids is almost universal. A bacterial cell can translate the genetic messages of human cells. The near universality of a common genetic language provides compelling evidence of the antiquity of the code and the evolutionary connection of all living organisms.

17.2   Transcription is the DNA-directed synthesis of RNA: *a closer look*

Molecular Components of Transcription   The **promoter** is the DNA sequence where **RNA polymerase** attaches to the template strand and initiates transcription. RNA polymerase joins RNA nucleotides that are complementary to the DNA template strand in a 5′ → 3′ direction and does not require a primer. In bacteria, the **terminator** is the sequence that signals the end of transcription. A **transcription unit** is the sequence of DNA that is transcribed into one RNA molecule.

Bacteria have one type of RNA polymerase. Eukaryotes have three types; the one that synthesizes mRNA is called RNA polymerase II.

Synthesis of an RNA Transcript   The specific binding of RNA polymerase to the promoter determines where transcription starts and which DNA strand is used as the template. The promoter includes the transcription **start point** and recognition sequences, such as the **TATA box** common in eukaryotes, upstream from the start point. In eukaryotes, **transcription factors** must first recognize and bind to the promoter before RNA polymerase II can attach, at which point the assembly is called the **transcription initiation complex**.

RNA polymerase unwinds the double helix, exposing DNA nucleotides for base pairing with RNA nucleotides, and joins the nucleotides to the 3′ end of the growing polymer. The new RNA peels away from the DNA template, and the DNA double helix re-forms. Several molecules of RNA polymerase may be transcribing simultaneously along a gene, enabling a cell to produce large quantities of mRNA.

In prokaryotes, transcription ends after RNA polymerase transcribes the terminator sequence. In eukaryotes, polymerase continues past a polyadenylation signal sequence (coding for 5'-AAUAAA-3'), and proteins cut loose the pre-mRNA.

INTERACTIVE QUESTION 17.3

Describe the key steps of transcription as they occur in eukaryotes:

a.

b.

c.

17.3 Eukaryotic cells modify RNA after transcription

In eukaryotes, pre-mRNA is modified by **RNA processing** before it leaves the nucleus.

Alteration of mRNA Ends   A modified guanine nucleotide (**5' cap**) is attached to the 5' end of a pre-mRNA, and a string of adenine nucleotides, called a **poly-A tail**, is added to the 3' end. The 5' cap and poly-A tail may facilitate transport of mRNA from the nucleus. They also protect the ends of the mRNA from hydrolytic enzymes and aid in the recruitment of ribosomes in the cytoplasm. The cap and tail are attached to the untranslated regions (UTRs) at the 5' and 3' ends.

Split Genes and RNA Splicing   Long segments of non-coding nucleotide sequences, known as **introns** or *in*tervening sequences, occur within the boundaries of eukaryotic genes. The remaining coding regions are called **exons** because they are *ex*pressed in protein synthesis (with the exception of the 5' and 3' UTRs, which are not translated). After a primary transcript is made of a gene, introns are removed and exons joined together before the mRNA leaves the nucleus—a process called **RNA splicing**.

The signals for RNA splicing are sets of a few nucleotides at either end of each intron. Small RNAs within a large protein complex called a **spliceosome** snip an intron out of the RNA transcript and connect the adjoining exons. Such RNA molecules that act as enzymes are called **ribozymes**.

In some cases of RNA splicing, intron RNA catalyzes its own removal. RNA has several properties that allow it to function as an enzyme. RNA is single-stranded and can base-pair with itself, forming specific three-dimensional structures. In addition, some of the bases in RNA contain functional groups that can participate in catalysis. Finally, RNA can bond to other nucleic acid molecules, allowing it to precisely locate splicing regions.

Some introns are involved in regulating gene activity. **Alternative RNA splicing** allows different polypeptides to be produced from a single gene. Exons may code for polypeptide **domains**, functional segments of a protein, such as binding sites and active sites. Introns may facilitate recombination of exons between different alleles or even between different genes. Such *exon shuffling* can result in novel proteins.

INTERACTIVE QUESTION 17.4

How does the mRNA that leaves the nucleus differ from pre-mRNA?

17.4 Translation is the RNA-directed synthesis of a polypeptide: *a closer look*

Molecular Components of Translation   **Transfer RNA (tRNA)** molecules carry amino acids to ribosomes, where they are added to a growing polypeptide. Each tRNA carries a specific amino acid and has a nucleotide triplet, called an **anticodon**, that base-pairs with a complementary codon on mRNA. Due to hydrogen bonding between complementary nucleotide base sequences, these single-stranded, short RNA molecules fold into a three-dimensional, roughly L-shaped structure. The anticodon is at one end of the L; the 3' end is the attachment site for its specific amino acid.

Transfer RNA is transcribed in the nucleus of a eukaryote and moves into the cytoplasm. In both bacterial and eukaryotic cells, tRNA molecules are used repeatedly. Each amino acid has a specific **aminoacyl-tRNA synthetase** that attaches it to its appropriate tRNA molecules to create an aminoacyl tRNA, or charged tRNA. The hydrolysis of ATP drives this process.

Sixty-one codons for amino acids can be read from mRNA, but there are only about 45 different tRNA molecules. A phenomenon known as **wobble** enables the third nucleotide of some tRNA anticodons to pair with more than one kind of nucleotide in the codon. Thus, a tRNA may recognize more than one mRNA codon, all of which code for the same amino acid carried by that tRNA.

INTERACTIVE QUESTION   17.5

a. Explain what is meant by saying that tRNA is a translator.

b. Using some of the codons and the amino acids you identified in Interactive Question 17.2, fill in the following table.

DNA Triplet 3′ → 5′	mRNA Codon 5′ → 3′	Anticodon 3′ → 5′	Amino Acid
			methionine
		GCA	
TTC			
	UAG		

Ribosomes facilitate the specific pairing of tRNA anticodons with mRNA codons during protein synthesis. Ribosomes consist of a large and a small subunit, each composed of proteins and RNA molecules called **ribosomal RNA (rRNA)**. Subunits are constructed in the nucleolus in eukaryotes. Prokaryotic ribosomes are smaller and differ enough in molecular composition that some antibiotics can inhibit them without affecting eukaryotic ribosomes.

A large and a small subunit join to form a ribosome when they attach to an mRNA molecule. Ribosomes have a binding site for mRNA and three tRNA binding sites: a **P site** (peptidyl tRNA-binding site) that holds the tRNA carrying the growing polypeptide chain, an **A site** (aminoacyl tRNA-binding site) that holds the tRNA carrying the amino acid to be added next, and an **E site** (exit site) from which discharged tRNAs leave the ribosome. A ribosome also has an *exit tunnel* through which the growing polypeptide passes out of the ribosome. Evidence indicates that rRNA is primarily responsible for both the structure and function of a ribosome.

Building a Polypeptide   Each of the three stages of protein synthesis—initiation, elongation, and termination—requires the aid of protein "factors." The first two stages also require energy provided by the hydrolysis of GTP (guanosine triphosphate).

The initiation stage begins as the small ribosomal subunit binds to an mRNA and an initiator tRNA carrying methionine, which attaches to the start codon AUG on the mRNA. With the aid of proteins called *initiation factors* and the hydrolysis of GTP, the large subunit of the ribosome attaches to the small one, forming a *translation initiation complex*. The initiator tRNA fits into the P site.

The addition of amino acids in the elongation stage involves several proteins called *elongation factors* and occurs in a three-step cycle.

In codon recognition, an aminoacyl-tRNA base-pairs with the mRNA codon in the A binding site. This step requires energy from the hydrolysis of GTP.

In the peptide bond formation step, an RNA molecule of the large subunit catalyzes the formation of a peptide bond between the carboxyl end of the polypeptide held in the P site and the amino group of the new amino acid in the A site. The polypeptide is now held by the tRNA in the A site.

In translocation, the tRNA carrying the growing polypeptide is translocated to the P site, a process requiring energy from the hydrolysis of GTP. The empty tRNA from the P site is moved to the E site and released. The next mRNA codon moves into the A site as the mRNA moves through the ribosome.

Termination occurs when a stop codon—UAA, UAG, or UGA—reaches the A site of the ribosome. A *release factor* binds to the stop codon and hydrolyzes the bond between the polypeptide and the tRNA in the P site. The completed polypeptide leaves through the exit tunnel of the large subunit. With the hydrolysis of two GTP and the aid of other protein factors, the two ribosomal subunits and other components dissociate. Review translation by completing Interactive Question 17.6.

In the following diagrams of polypeptide synthesis, name the stages (1–4), identify the components (a–l), and then briefly describe what happens in each stage. (This diagram does not include the initiation stage.)

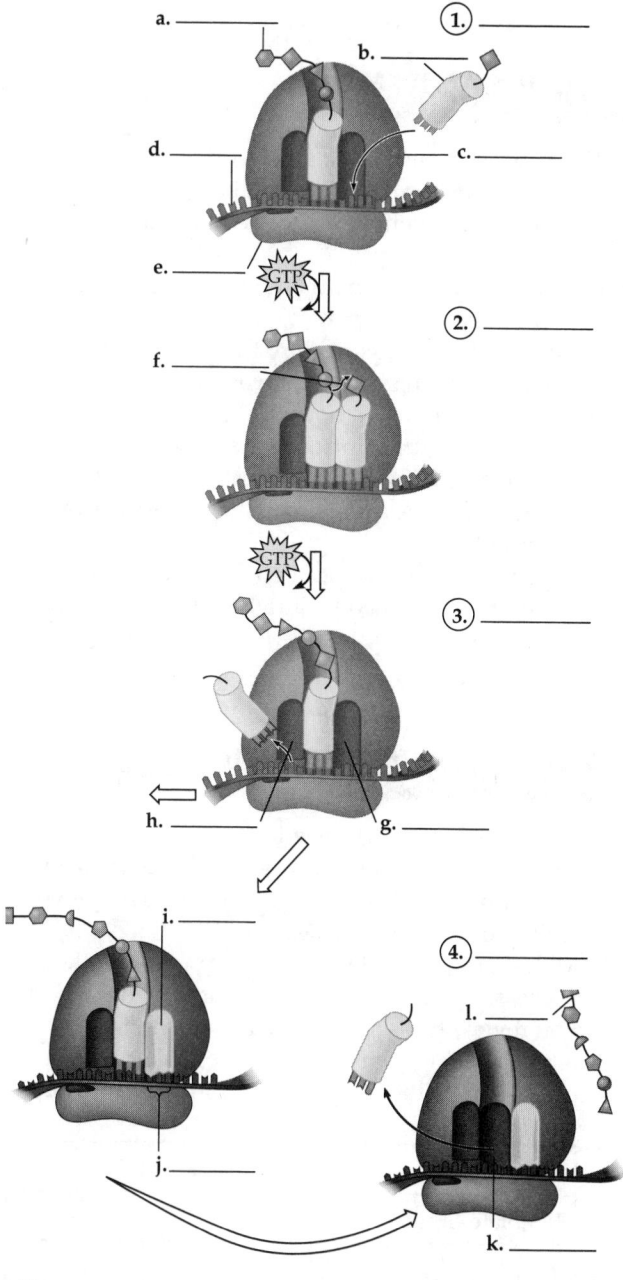

Completing and Targeting the Functional Protein
During and following translation, a polypeptide folds spontaneously into its secondary and tertiary structures.

The protein may need to undergo *post-translational modifications*: Amino acids may be chemically modified; one or more amino acids at the beginning of the chain may be removed; segments of the polypeptide may be cut into two or more pieces; or several polypeptides may associate into a protein with a quaternary structure.

All ribosomes are identical, whether they are free ribosomes that synthesize cytosolic proteins or bound ribosomes attached to the endoplasmic reticulum (ER) that make membrane and secretory proteins. Polypeptide synthesis begins in the cytoplasm. If a protein is destined for the endomembrane system or for secretion, its polypeptide chain will begin with a **signal peptide**. This short sequence of amino acids is recognized by a protein–RNA complex called a **signal-recognition particle (SRP)**, which attaches the ribosome to a receptor protein that is part of a translocation complex on the ER membrane. As the growing polypeptide threads into the ER, the signal peptide is usually removed.

Other signal peptides direct some proteins made in the cytosol to specific sites such as mitochondria, chloroplasts, or the interior of the nucleus. In such cases, protein synthesis is completed before transport of the polypeptide.

What determines if a ribosome becomes bound to the ER?

Making Multiple Polypeptides in Bacteria and Eukaryotes An mRNA may be translated simultaneously by several ribosomes in strings called **polyribosomes** (or **polysomes**). In bacterial cells, transcription and translation can occur simultaneously.

17.5 Mutations of one or a few nucleotides can affect protein structure and function

Mutations, changes in the genetic information of a cell (or virus), may be either large scale (involving long segments of a chromosome) or small scale (such as **point mutations**, which affect a single nucleotide pair). If the mutation occurs in a cell that gives rise to a gamete, it may be passed on to offspring.

Types of Small-Scale Mutations The first type of small-scale mutation is a **nucleotide-pair substitution**, in which one nucleotide and its complementary partner is replaced with another pair of nucleotides. Due to the redundancy of the genetic code, some substitutions do not change the amino acid translation and are called **silent mutations**. A **missense mutation**, which results in the insertion of a different amino acid, may not alter the character of the protein if the new amino acid has similar properties or is not located in a region crucial to that protein's function.

A nucleotide-pair substitution that results in a different amino acid in a critical portion of a protein, such as the active site of an enzyme, may significantly impair protein function. Occasionally such a change proves beneficial.

Nonsense mutations occur when a point mutation changes an amino acid codon into a stop codon, prematurely halting the translation of the polypeptide chain and usually creating a nonfunctional protein.

A second type of small-scale mutations involves **insertions** or **deletions** of nucleotide pairs. If these are not in multiples of three, they will alter the reading frame. All nucleotides downstream from the mutation will be improperly grouped into codons, creating extensive missense mutations and usually ending in nonsense. Such **frameshift mutations** almost always produce nonfunctional proteins.

New Mutations and Mutagens *Spontaneous mutations* include nucleotide-pair substitutions, insertions, deletions, and longer mutations that occur during DNA replication or recombination.

Physical agents (such as X-rays and UV light) and a variety of chemical agents can cause mutations and are called **mutagens**. Chemical mutagens include nucleotide analogs that substitute for normal nucleotides and then pair incorrectly in DNA synthesis and agents that chemically alter nucleotides and change their pairing. Tests can measure the mutagenic effects of chemicals and thus their potential carcinogenic risk.

INTERACTIVE QUESTION 17.8

Define the following terms and explain what type of small-scale mutation could cause each of these types of mutations.

a. silent mutation

b. missense mutation

c. nonsense mutation

d. frameshift mutation

What Is a Gene? Revisiting the Question Our definition of a gene has evolved from Mendel's heritable factors, to Morgan's loci along chromosomes, to one gene–one polypeptide. Research continually refines our understanding of the structural and functional aspects of genes, which now include introns, promoters, and other regulatory regions. Currently, the best working definition of a gene is that it is a region of DNA whose final functional product is either a polypeptide or an RNA molecule.

Word Roots

anti- = opposite (*anticodon:* a nucleotide triplet at one end of a tRNA molecule that base-pairs with a particular complementary codon on an mRNA molecule)

exo- = out, outside, without (*exon:* a sequence within a primary transcript that remains in the RNA after RNA processing; also the region of DNA from which this sequence was transcribed)

intra- = within (*intron:* a noncoding, intervening sequence within a primary transcript that is removed during RNA processing; also the region of DNA from which this sequence was transcribed)

poly- = many (*polyribosome:* a group of several ribosomes attached to, and translating, the same mRNA molecule)

Structure Your Knowledge

1. You have been introduced to several types of RNA in this chapter. List four of these types and their functions. What explains the functional versatility of RNA molecules?

2. Make sure you understand and can explain the processes of transcription and translation. To help you review these processes, fill in the following table describing various aspects of eukaryotic gene expression (either by yourself or in a study group).

	Transcription	Translation
Template		
Location		
Molecules involved		
Enzymes involved		
Control—start and stop		
Product		
Product processing		

3. What is the genetic code? Explain redundancy and the wobble phenomenon. What is the significance of the fact that the genetic code is nearly universal?

4. Prepare a concept map showing the types and consequences of small-scale mutations.

Test Your Knowledge

MULTIPLE CHOICE: *Choose the one best answer.*

1. A series of studies on mutants of *Neurospora* identified three classes of mutants that needed arginine added to minimal media in order to grow. The production of arginine includes the following steps: precursor → ornithine → citrulline → arginine. What nutrient(s) have to be supplied to the mutants that had a defective enzyme for the ornithine → citrulline step in order for them to grow?
   a. ornithine
   b. citrulline
   c. either ornithine or citrulline
   d. the precursor, ornithine, and citrulline

2. Transcription involves the transfer of information from
   a. DNA to RNA.
   b. mRNA to an amino acid sequence.
   c. DNA to an amino acid sequence.
   d. the nucleus to the cytoplasm.

3. Some individuals suffer from *developmental verbal dyspraxia* (DVS), a genetic disorder that prevents them from properly using verbs in speech. A triplet on the template strand of the gene responsible for DVS has the sequence 3'-CGA-5'. When this gene is transcribed and the resulting message is translated, what is the base sequence of the anticodon on the tRNA appropriate for this triplet?
   a. 3'–CGA–5'
   b. 5'–CGA–3'
   c. 5'–CGU–3'
   d. 3'–GCT–5'

4. A bacterial gene 600 nucleotides long can code at most for a polypeptide of how many amino acids?
   a. 200   c. 600
   b. 300   d. 1,800

5. Which of the following is a statement of the *central dogma* of biology?
   a. Survival of the fittest.
   b. A = T; C ≡ G.
   c. DNA → RNA → protein.
   d. Your karma ran over my dogma.

6. How is the template strand for a particular gene determined?
   a. It is the DNA strand that runs from the 3' → 5' direction.
   b. It is established by the location and orientation of the promoter.
   c. It doesn't matter which strand is the template because they are complementary and will produce the same mRNA.
   d. It is signaled by a polyadenylation signal sequence.

7. Which enzyme synthesizes tRNA?
   a. DNA polymerase
   b. RNA polymerase
   c. aminoacyl-tRNA synthetase
   d. ribosomal RNA

8. Which of the following is *not* involved in the formation of a eukaryotic transcription initiation complex?
   a. TATA box in a promoter
   b. transcription factors
   c. small RNA molecules
   d. RNA polymerase II

9. Which of the following is true of RNA processing?
   a. Exons are excised before the mRNA is translated.
   b. Signal peptides are added to the 5' end of the transcript.
   c. Assemblies of protein and small RNAs, called spliceosomes, may catalyze splicing.
   d. Large quantities of rRNA are assembled into ribosomes.

10. Prokaryotes, but not eukaryotes, are capable of
   a. simultaneous transcription and translation.
   b. alternative splicing of introns.
   c. transcription using RNA polymerase.
   d. 5' capping of mRNA.

11. Some mutations have little effect on protein function. Since the gene sequence controls protein structure, how is this possible?

    a. Some mutations change a triplet coding for an amino acid into another triplet coding for the same amino acid.

    b. Some mutations substitute a chemically similar amino acid so that the function of the protein is preserved.

    c. Some substituted amino acids are in nonessential regions of the protein.

    d. All of a, b, and c would minimize the effect of mutations on protein function.

12. A ribozyme is an

    a. exception to the one gene–one RNA molecule axiom.

    b. enzyme that adds the 5′ cap and poly-A tail to mRNA.

    c. RNA molecule that functions as an enzyme.

    d. enzyme that produces both small and large ribosomal subunits.

13. Which of the following would not be found in a bacterial cell?

    a. tRNA

    b. rRNA

    c. signal peptide

    d. polyribosome

14. Which of the following is transcribed and then translated to form a protein product?

    a. an intron

    b. a gene for a transcription factor

    c. 5′ and 3′ UTRs

    d. a gene for rRNA

15. What type of bonding is responsible for maintaining the shape of the tRNA molecule?

    a. peptide bonding between amino acids

    b. covalent bonding between sulfur atoms

    c. hydrogen bonding between base pairs

    d. covalent bonding between base pairs

16. Place the following events in the synthesis of a polypeptide in the proper order.

    1. A peptide bond forms.

    2. An aminoacyl tRNA matches its anticodon to the codon in the A site.

    3. A tRNA translocates from the A site to the P site, and an unattached tRNA exits from the E site.

    4. The large subunit attaches to the small subunit, with the initiator tRNA in the P site.

    5. A small subunit binds to an mRNA and an initiator tRNA.

       a. 4-5-3-2-1

       b. 4-5-2-1-3

       c. 5-4-1-2-3

       d. 5-4-2-1-3

17. Translocation in the process of translation involves

    a. the hydrolysis of GTP.

    b. movement of the tRNA in the A site to the P site.

    c. movement along the mRNA a distance of one triplet.

    d. all of the above.

18. Which of the following type of molecule catalyzes the formation of a peptide bond?

    a. RNA polymerase

    b. rRNA

    c. mRNA

    d. aminoacyl-tRNA synthetase

19. Which of the following is not true of an anticodon?

    a. It lines up in the 5′ → 3′ direction along the 5′ → 3′ mRNA strand.

    b. It consists of three nucleotides extending from one loop of a tRNA molecule.

    c. It may pair with more than one codon.

    d. Its base uracil base-pairs with adenine.

20. Changes in a polypeptide following translation may involve the

    a. addition of sugars or lipids to certain amino acids.

    b. enzymatic addition of amino acids at the beginning of the chain.

    c. removal of poly-A from the end of the chain.

    d. addition of a 5′ cap of a modified guanosine.

21. A signal peptide
    a. is most likely to be found on cytosolic proteins produced by bacterial cells.
    b. directs an mRNA molecule into the lumen of the ER.
    c. is a sign to bind the small ribosomal unit at the initiation codon.
    d. would be the first 20 or so amino acids of a protein destined for a membrane location or for secretion from the cell.

22. Consider the following sequence: **the big red dog ate the fat cat**. Which of the following would represent the effect of a frameshift mutation on this sequence?
    a. the big red dog ate the fit cat
    b. the igr edd oga tet hef atc at
    c. the big xre ddo gat eth efi tca t
    d. Either **b** or **c** would represent a frameshift mutation.

23. The type of mutation responsible for sickle cell anemia is
    a. a silent mutation.
    b. a point mutation.
    c. a nucleotide-pair substitution.
    d. Both **b** and **c** describe the type of mutation.

Use the following diagram of coupled transcription and translation in bacteria to answer questions 24–28.

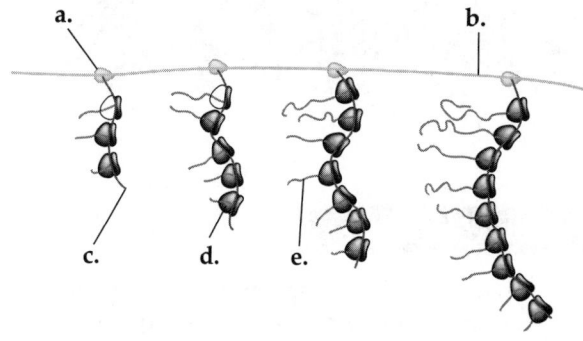

24. Which letter refers to an mRNA molecule?

25. Which letter refers to a forming polypeptide?

26. Which letter refers to RNA polymerase?

27. Which letter refers to a ribosome?

28. Which letters indicate structures or molecules containing nucleotides?
    a. a and b
    b. b, c, and d
    c. b, d, and e
    d. c and d

Chapter 18

Regulation of Gene Expression

Key Concepts

18.1 Bacteria often respond to environmental change by regulating transcription

18.2 Eukaryotic gene expression is regulated at many stages

18.3 Noncoding RNAs play multiple roles in controlling gene expression

18.4 A program of differential gene expression leads to the different cell types in a multicellular organism

18.5 Cancer results from genetic changes that affect cell cycle control

Framework

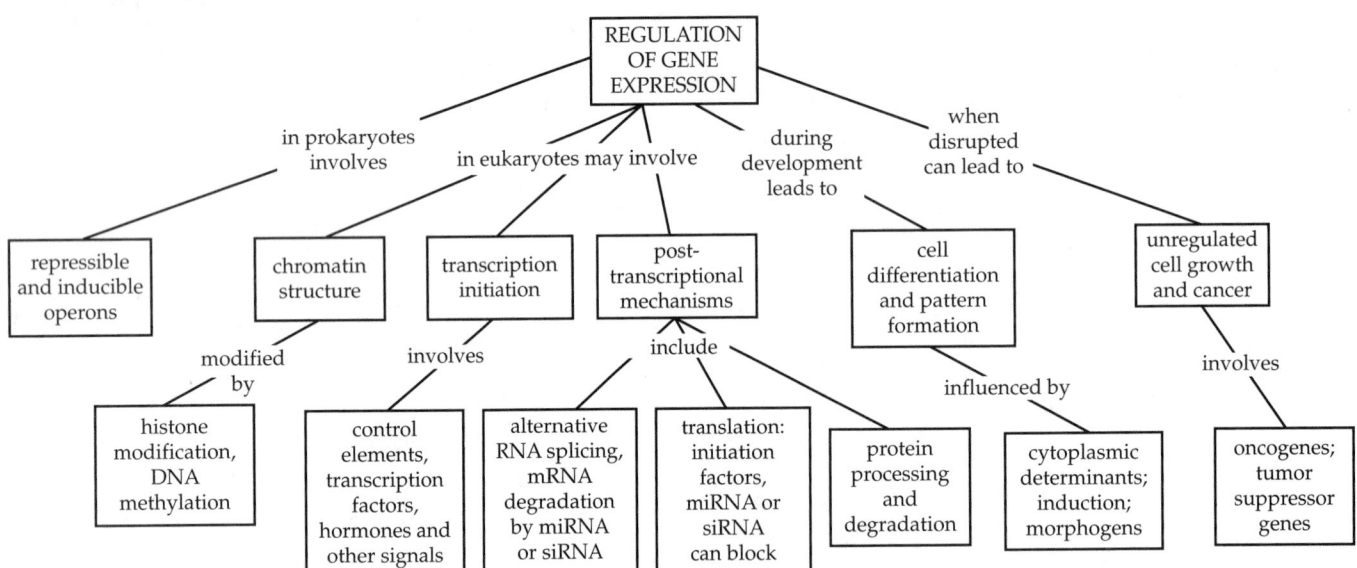

Chapter Review

18.1 Bacteria often respond to environmental change by regulating transcription

Bacteria can respond to short-term environmental fluctuations by regulating enzyme activity through *feedback inhibition*, usually of the first enzyme in a metabolic pathway. A second type of response involves regulation of enzyme production through the control of gene transcription in longer-term response to changing metabolic needs. F. Jacob and J. Monod described the *operon model* for gene regulation in 1961.

Operons: The Basic Concept   In bacteria, genes for the different enzymes of a single metabolic pathway may be grouped together into one transcription unit and served by a single promoter. Thus, these genes are *coordinately controlled*. An **operator** is a segment of DNA within or near the promoter that controls the access of RNA polymerase to the genes. An **operon** is the DNA segment that includes the clustered genes, the promoter, and the operator.

A **repressor** is a protein that binds to a specific operator, blocking attachment of RNA polymerase and thus turning an operon "off." **Regulatory genes** code for repressor proteins. These allosteric proteins, which may assume active or inactive shapes, are usually produced at a slow but continuous rate. The activity of the repressor protein may be determined by the presence or absence of a **corepressor**.

In the *trp* operon, the repressor, coded for by the regulatory gene *trpR*, is synthesized in its inactive form. Tryptophan is the corepressor that binds to the *trp* repressor, changing it into its active shape, which has a high affinity for the *trp* operator and switches off the *trp* operon. When the tryptophan concentration of the cell falls, tryptophan is no longer bound to repressor proteins and the operator is no longer repressed. RNA polymerase then attaches to the promoter, and mRNA for the enzymes needed for tryptophan synthesis is produced.

Repressible and Inducible Operons: Two Types of Negative Gene Regulation   The transcription of a *repressible operon,* such as the *trp* operon, is inhibited when a specific small molecule binds to and activates a repressor. The transcription of an *inducible operon* is stimulated when a specific small molecule binds to and inactivates a repressor.

The *lac* operon, controlling lactose metabolism in *E. coli,* is an inducible operon that contains three genes. The *lac* repressor protein, coded for by the regulatory gene *lacI,* is innately active, binding to the *lac* operator and switching off the operon. Allolactose, an isomer of lactose, acts as an **inducer**, a small molecule that binds to and *inactivates* the repressor protein, so that the operon can be transcribed.

INTERACTIVE QUESTION 18.1

In the following diagram of the *lac* operon, an inducible operon, identify components a through i.

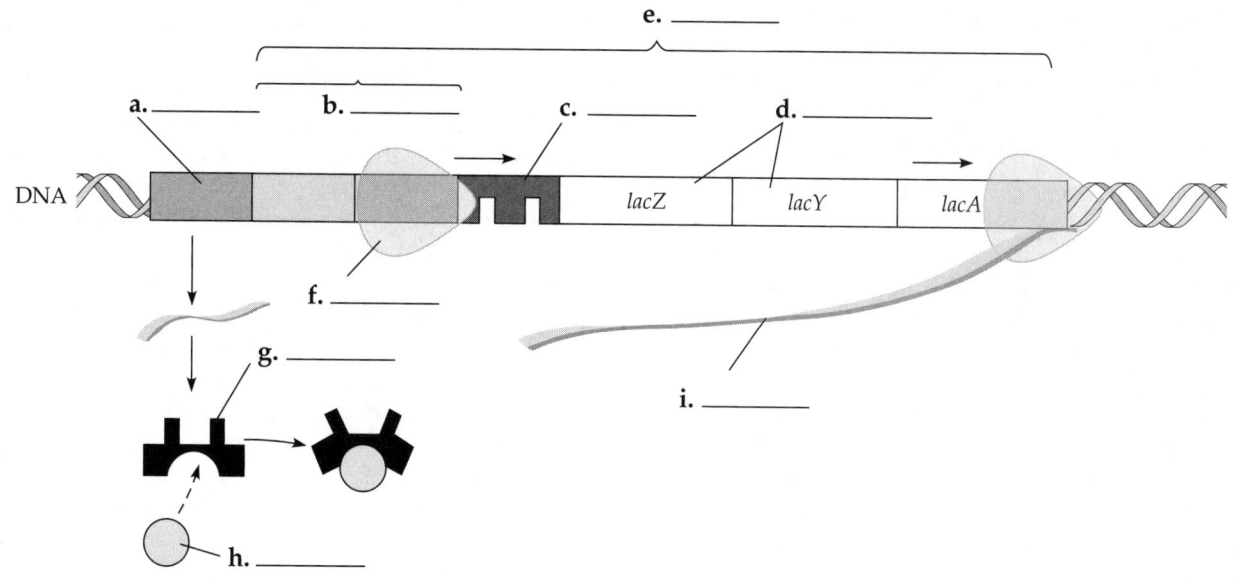

INTERACTIVE QUESTION   18.2

a. *Repressible enzymes* usually function in _____ pathways. The pathway's product serves as a(n) _____ to activate the repressor and turn off enzyme synthesis, thus preventing overproduction of the product of the pathway. Genes for repressible enzymes are usually switched _____ and the repressor is synthesized in a(n) _____ form.

b. *Inducible enzymes* usually function in _____ pathways. Nutrient molecules serve as _____ to stimulate production of the enzymes necessary for their breakdown. Genes for inducible enzymes are usually switched _____ and the repressor is synthesized in a(n) _____ form.

Positive Gene Regulation   *E. coli* cells preferentially use glucose. Should glucose levels fall, transcription of operons for other catabolic pathways can be increased through the action of the allosteric regulatory protein *cAMP receptor protein (CRP)*, which is an **activator**. When glucose is scarce, **cyclic AMP (cAMP)** accumulates in the cell and binds with CRP, changing it to its active shape. The active CRP attaches at the upstream end of the promoter region and stimulates transcription by facilitating the binding of RNA polymerase.

The regulation of the *lac* operon includes both negative control by the *lac* repressor protein and positive control by CRP when it is complexed with cAMP.

18.2   Eukaryotic gene expression is regulated at many stages

All cells control the expression of their genes in response to their external and internal environments. Eukaryotic cells in multicellular organisms also regulate gene expression to produce specialized cells.

Differential Gene Expression   Differences between cells with the same genome are the result of **differential gene expression**. Gene expression is most commonly controlled at transcription, but in eukaryotes, it can be regulated at various stages.

Regulation of Chromatin Structure   DNA packing and the location of a gene's promoter (relative to nucleosomes and to the attachment to the chromosome scaffold) may help control which genes are available for transcription. Genes within highly condensed heterochromatin are usually not expressed.

Histone acetylation, the attachment of acetyl groups (—COCH₃) to the histone tails extending from a nucleosome, opens up chromatin structure, giving transcription proteins greater access to genes. The addition of methyl groups to histone tails leads to condensation of the chromatin and reduced transcription.

In a process called **DNA methylation**, methyl groups are added to DNA bases (usually cytosine). Genes are generally more heavily methylated in cells in which they are not expressed. Methylation patterns are usually passed on during subsequent DNA replications. Such methylation patterns account for *genomic imprinting* in mammals, in which either a maternal or a paternal allele of certain genes is permanently regulated.

Epigenetic inheritance is the transmission of traits by mechanisms that do not involve changes in DNA sequences, such as chromatin modifications that affect gene expression in future generations of cells.

INTERACTIVE QUESTION   18.3

a. Give an example of highly methylated and inactive DNA that is common in female mammalian cells.

b. Explain whether the removal of acetyl groups from histone tails in a nucleosome would increase or decrease the transcription of a gene.

Regulation of Transcription Initiation   A typical eukaryotic gene consists of a promoter sequence, to which a *transcription initiation complex* that includes RNA polymerase II attaches, and a sequence of introns interspersed among the coding exons. After transcription, RNA processing of the pre-mRNA removes the introns and adds a 5′ cap and a poly-A tail at the 3′ end. **Control elements** are noncoding sequences that help regulate transcription by serving as binding sites for transcription factors.

General transcription factors bind to RNA polymerase and to each other to initiate the transcription of all protein-coding genes. A few bind to DNA sequences such as the TATA box in the promoter. The binding of *specific transcription factors* to control elements, however, greatly increases or decreases transcription rates of particular genes.

Proximal control elements are located close to the promoter. **Enhancers** are groups of *distal control elements* located far upstream or downstream from a gene or within an intron. According to a current model, a protein-mediated bend in the DNA brings activators bound to enhancers into contact with *mediator proteins* and general transcription factors at the promoter. These protein–protein interactions assemble the initiation complex. Hundreds of transcription activators have been identified. Most have a *DNA-binding domain* and one or more *activation domains* that bind other regulatory proteins.

Some specific transcription factors act as repressors in various ways, such as by blocking the binding of activators or by binding to specific control elements in an enhancer. Some activators and repressors influence chromatin structure by recruiting proteins that either acetylate histones or remove acetyl groups from histones. Most eukaryotic gene repression may occur by this *silencing* at the level of chromatin modification.

Only a dozen or so unique nucleotide sequences have been found in control elements, but each enhancer contains about 10 different control elements. The *combination* of these elements ensures that a specific collection of activator proteins must be present to activate transcription. Thus, the specific activators and repressors made in a cell determine which genes are expressed.

Most eukaryotic genes coding for enzymes in the same metabolic pathway are scattered on the chromosomes. The coordinate control of these genes typically involves a specific collection of control elements that are associated with each related gene and require the same activators.

For example, steroid hormones may activate multiple genes when they bind to specific receptor proteins within a cell that function as transcription activators for several genes. The same chemical signal will activate genes with the same control elements and thus coordinate gene expression.

Although each chromosome occupies a distinct area in the interphase nucleus, loops of chromatin from various chromosomes may extend into RNA polymerase-rich areas called *transcription factories*.

INTERACTIVE QUESTION 18.4

Label the components of the following diagram that illustrates how enhancers, activators, mediator proteins, and transcription factors facilitate the formation of a transcription initiation complex.

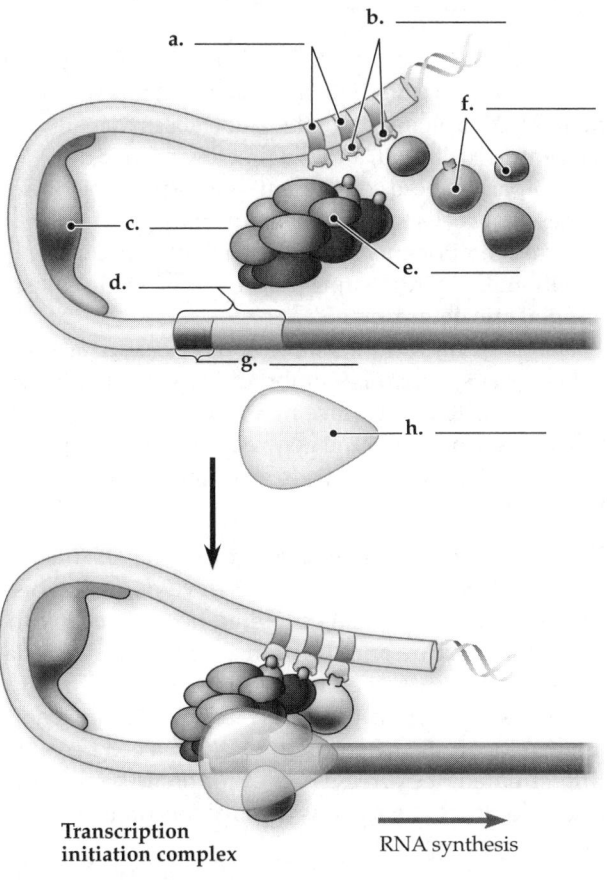

Transcription initiation complex

RNA synthesis

Mechanisms of Post-Transcriptional Regulation Gene expression, measured by the amount of functional protein that is made, can be regulated at any post-transcriptional step. At the level of RNA processing, **alternative RNA splicing** may produce different mRNA molecules when regulatory proteins control which segments of the primary transcript are chosen as introns and which are chosen as exons. The majority of human genes that have multiple exons probably undergo alternative splicing.

The translation of eukaryotic mRNA can be delayed by regulatory proteins that bind to the untranslated region (UTR) at the 5' or 3' end and prevent ribosome attachment. A great deal of mRNA is synthesized and stored in egg cells; translation may begin when translation initiation factors are activated following fertilization.

In contrast to prokaryotic mRNA, which is degraded after a few minutes, eukaryotic mRNA can last hours or even weeks. The UTR at the 3' end of mRNA may contain nucleotide sequences that affect how long an mRNA lasts.

Following translation, polypeptides are often cleaved or chemical groups added to yield an active protein. Some proteins must be transported to target locations.

Selective degradation of proteins may serve as a control mechanism in a cell. Molecules of the protein ubiquitin are added to mark proteins for destruction. Protein complexes called proteasomes recognize and degrade the marked proteins.

INTERACTIVE QUESTION 18.5

a. How is the process of alternate RNA splicing related to the control of gene expression?

b. The untranslated regions (UTR) at both the 5' and 3' ends of an mRNA may contribute to regulation of gene expression. Describe their effects.

18.3 Noncoding RNAs play multiple roles in controlling gene expression

Only 1.5% of the human genome is protein-coding DNA; an additional small percentage codes for genes for ribosomal and transfer RNA. Mounting evidence indicates that a large amount of the genome may be transcribed into non-protein-coding RNAs (*noncoding*

RNAs or *ncRNAs*). A large population of RNA molecules may be involved in regulating gene expression.

Effects on mRNAs by MicroRNAs and Small Interfering RNAs   Small RNA molecules called **microRNAs (miRNAs)** are processed from longer RNA precursors. These short (about 22 nucleotides) single-stranded RNA molecules associate with a complex of proteins and then base-pair with any mRNA that has 7 or 8 nucleotides of a complementary sequence. This miRNA–protein complex either blocks translation or degrades the mRNA. An estimated one-half of all human genes may be regulated by miRNAs.

Small interfering RNAs (siRNAs) appear to inhibit the expression of genes by the same mechanism as miRNAs. They differ from miRNAs in the structure of their double-stranded RNA precursor. The blocking of gene expression by siRNAs, called **RNA interference (RNAi)**, is used to study gene function. The RNAi pathway may have evolved as a defense against RNA viruses. Many species apparently produce double-stranded RNA precursors that are cut into small RNAs such as siRNAs.

Chromatin Remodeling by ncRNAs   In some yeasts, siRNAs are involved in the formation of heterochromatin. Evidence indicates that an RNA transcript produced from DNA in the centromeric region is copied into double-stranded RNA and associated with proteins. This complex then targets the centromere region, where it initiates chromatin remodeling into heterochromatin.

Newly discovered *piwi-associated RNAs (piRNAs)* also induce heterochromatin formation, thus blocking transcription of some transposons. They also appear to help reestablish proper methylation patterns in the genome during gamete formation.

In addition to these small ncRNAs, researchers have identified a variety of **long noncoding RNAs (lncRNAs)** with specific functions. One such lncRNA is transcribed from the XIST gene and is responsible for inactivating one X chromosome in the cells of female mammals.

The Evolutionary Significance of Small ncRNAs
Small ncRNAs are also found in prokaryotic cells. The CRISPR-Cas9 system, which protects prokaryotes from viruses, involves small ncRNAs and has become an important research tool.

The evolution of morphological complexity in eukaryotes may be linked to the extra levels of gene regulation that are provided by small ncRNAs. Evidence indicates that siRNAs evolved first, followed by miRNAs, and then piRNAs, which are found only in animals. The various classes of small ncRNAs are particularly important in embryonic development.

INTERACTIVE QUESTION 18.6

a. Describe how miRNAs regulate gene expression.

b. Do siRNAs regulate gene expression by affecting translation or transcription? Explain.

18.4   A program of differential gene expression leads to the different cell types in a multicellular organism

A Genetic Program for Embryonic Development   The three key processes of embryonic development are (1) cell division, the production of large numbers of cells; (2) cell **differentiation**, the formation of cells specialized in structure and function; and (3) **morphogenesis**, the physical processes that produce body structures and shape. All three processes are based on differential gene expression resulting from differences in gene regulation in cells.

Cytoplasmic Determinants and Inductive Signals   The cytoplasm of an unfertilized egg cell contains maternal mRNA, proteins, and other substances that are unevenly distributed, and the first few mitotic divisions separate these components and expose the nuclei in these new cells to different environments. These maternal components of the egg that influence early development by regulating gene expression are called **cytoplasmic determinants**.

The other important source of developmental control is signals from other embryonic cells in the form of contact with cell-surface molecules or secreted molecules. Change in the gene expression of target cells resulting from signals from other cells is called **induction**.

Sequential Regulation of Gene Expression During Cellular Differentiation   A cell's developmental history leads to its eventual differentiation as a cell with a specific structure and function. The term **determination** is used to describe the condition in which a cell is irreversibly committed to its fate, even if it has not yet developed its final structure. When a cell becomes differentiated, it expresses genes for *tissue-specific proteins*, and that expression is usually controlled at the level of transcription.

Research has shown how muscle cells become determined and differentiated. The embryonic precursor cells from which muscle cells arise have the potential to develop into a number of different cell types. Once they become committed to becoming muscle cells, they are called *myoblasts*. Researchers have identified several "master regulatory genes" that cause myoblast determination. One of these is *myoD*, which codes for MyoD protein, a transcription factor that binds to specific control elements and initiates transcription of other muscle-specific transcription factors. These secondary transcription factors then coordinately activate muscle-protein genes. MyoD also turns on genes that block the cell cycle and stop cell division, and it activates its own transcription, thereby maintaining the cell's differentiated state.

INTERACTIVE QUESTION 18.7

a. What is the difference between determination and differentiation?

b. The MyoD protein is able to transform some, but not all, differentiated cells into muscle cells. Why doesn't it work on all kinds of cells?

Pattern Formation: Setting Up the Body Plan **Pattern formation** is the spatial ordering of cells and tissues into their characteristic structures and locations. An animal's three major body axes are laid out early in development. Cytoplasmic determinants and inductive signals provide the molecular cues, called **positional information**, that tell a cell where it is located relative to the body axes and neighboring cells.

Anatomical, genetic, and biochemical studies of *Drosophila* development have led to the discovery of some common developmental principles. A fruit fly's body consists of a series of segments grouped into the head, thorax, and abdomen. The anterior-posterior and dorsal-ventral axes are determined by positional information provided by cytoplasmic determinants localized in the unfertilized egg. Nurse cells and follicle cells surrounding each egg cell supply mRNAs and nutrients needed for development. A fertilized egg develops into a segmented larva. Metamorphosis of a pupa produces an adult fly.

In the 1940s, E. B. Lewis studied developmental mutants and was able to map certain mutations that control pattern formation to specific genes, called **homeotic genes**. In the 1970s, C. Nüsslein-Volhard and E. Wieschaus undertook a search for the genes that control segment formation. They studied mutations that were **embryonic lethals**, which prevented the development of viable larvae. They exposed flies to a chemical mutagen and then performed many thousands of crosses to detect recessive mutations that caused the death of embryos or resulted in larvae with abnormal segmentation. Of the 1,200 genes they found that were essential for pattern formation, they identified 120 genes necessary for normal segmentation.

Maternal effect genes are genes of the mother that, when mutant, result in mutant offspring, regardless of the offspring's genotype. They code for proteins or mRNA that are deposited in the unfertilized egg. The specific maternal effect genes that are responsible for axis determination in *Drosophila* are called *egg-polarity genes*.

One egg-polarity gene is **bicoid**. The product of the *bicoid* gene is concentrated at one end of the embryo and is responsible for determining its anterior end. Offspring of a mother with two mutant alleles for this gene have two tail regions and lack the front half of the body. Researchers located *bicoid* mRNA concentrated in the most anterior end of egg cells. Following fertilization, the mRNA is translated into Bicoid protein, which diffuses posteriorly, forming a gradient in the early embryo. Gradients of such substances, which are called **morphogens**, establish an embryo's axes or other features—an example of the *morphogen gradient hypothesis*. Gradients of proteins translated from maternal mRNAs that determine the posterior end and establish the dorsal-ventral axis have also been identified. Positional information encoded by the embryo's genes then establishes the proper number of segments and finally triggers the formation of each segment's characteristic structures.

INTERACTIVE QUESTION 18.8

What type of evidence established that Bicoid protein is a morphogen that determines the anterior end of a fruit fly?

18.5 **Cancer results from genetic changes that affect cell cycle control**

Types of Genes Associated with Cancer Chemical carcinogens, X-rays, or certain viruses often cause the changes in the genes that regulate cell growth and division that lead to cancer. **Oncogenes**, or cancer-causing genes, were first found in certain types of viruses. Similar genes were later recognized in the genomes of humans and other animals. Cellular **proto-oncogenes**, which code for proteins that stimulate cell growth and division, may become oncogenes by several mechanisms, resulting in an overproduction or increased activity of growth-stimulating proteins.

Mutations in **tumor-suppressor genes** can contribute to the onset of cancer when they result in a decrease in the activity of proteins that prevent uncontrolled cell growth.

INTERACTIVE QUESTION   18.9

a. List three genetic changes that can convert a proto-oncogene into an oncogene.

b. List three possible functions of tumor-suppressor proteins.

Interference with Normal Cell-Signaling Pathways   In about 30% of human cancers, the *ras* proto-oncogene is mutated. The ***ras* gene** codes for a G protein that connects a growth-factor receptor on the plasma membrane to a cascade of protein kinases that leads to the production of a cell-cycle-stimulating protein. A mutation may create a hyperactive version of the Ras protein that relays a signal without the binding of a growth factor.

The ***p53* gene** is mutated in about 50% of human cancers. It codes for a tumor-suppressor protein that is a specific transcription factor for several genes. The p53 protein often activates the *p21* gene, whose product binds to cyclin-dependent kinases, halting the cell cycle and allowing time for the cell to repair damaged DNA. It also activates several miRNAS that inhibit the cell cycle. The p53 protein can also activate genes involved in DNA repair. Should DNA damage be irreparable, p53 activates "suicide genes" that initiate apoptosis.

The Multistep Model of Cancer Development   More than one mutation appears to be needed to produce a cancerous cell. Mutation of a single proto-oncogene can stimulate cell division, but usually both alleles for several tumor-suppressor genes must be defective to allow uncontrolled cell growth.

Genomics research has identified four major subtypes of breast cancer. They differ in the expression of three signal receptors: intracellular receptors for estrogen (ERa) and progesterone (PR) and a cell-surface receptor (HER2) for a growth factor. The subtypes differ in prognosis and treatment, so the identification of a patient's cancer subtype has profound clinical significance.

Inherited Predisposition and Environmental Factors Contributing to Cancer   A genetic predisposition to certain cancers may involve the inheritance of an oncogene or a recessive mutant allele for a tumor-suppressor gene. Approximately 15% of colorectal cancers involve inherited mutations, often in the tumor-suppressor gene *APC*, which regulates cell migration and adhesion.

About 5–10% of breast cancer cases are linked to an inherited mutant allele for either *BRCA1* or *BRCA2*, both of which appear to be tumor-suppressor genes involved in a cell's DNA damage repair pathway.

The ultraviolet radiation in sunlight and the chemicals in cigarette smoke may contribute to cancer through their DNA-damaging effects.

The Role of Viruses in Cancer Tumor viruses appear to be involved in about 15% of human cancers. Viruses can interfere with gene regulation when the insertion of their genetic material into a cell's DNA introduces an oncogene or affects a proto-oncogene or tumor-suppressor gene. Viral proteins may also inactivate p53 or other tumor-suppressor proteins.

Word Roots

epi-   = beside (*epigenetic inheritance:* inheritance of traits transmitted by mechanisms not directly involving the nucleotide sequence of a genome)

morph- = form; **-gen** = produce (*morphogen:* a substance that provides positional information in the form of a concentration gradient along an embryonic axis)

proto-   = first, original; **onc-** = tumor (*proto-oncogene:* a normal cellular gene that has the potential to become an oncogene, which is involved in triggering molecular events that lead to cancer)

Structure Your Knowledge

1. Complete the following concept map to help you review the mechanisms by which bacteria regulate gene expression in response to varying metabolic needs.

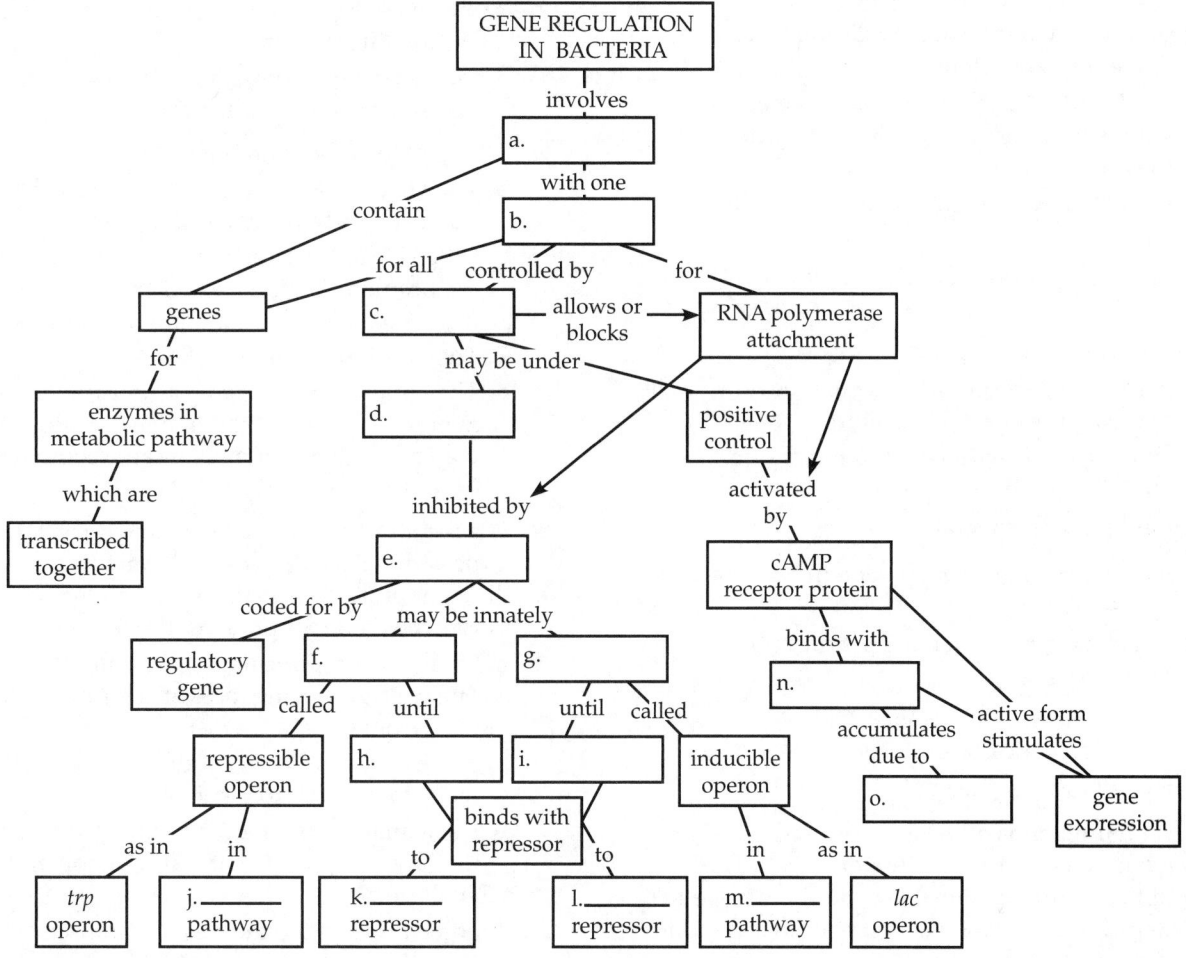

2. Fill in the following table to help you organize the major mechanisms that can regulate the expression of eukaryotic genes.

Level of Control	Examples
Chromatin modification	**a.**
Transcriptional regulation	**b.**
RNA processing	**c.**
RNA degradation	**d.**
Translational regulation	**e.**
Protein processing; degradation	**f.**

3. How might the mechanism of transcriptional regulation differ for cytoplasmic determinants and for the cell-cell signaling involved in induction?

Test Your Knowledge

MULTIPLE CHOICE: *Choose the one best answer.*

1. Inducible enzymes are
   a. usually involved in anabolic pathways.
   b. produced when a small molecule inactivates the repressor protein.
   c. produced when an activator molecule enhances the attachment of RNA polymerase to the operator.
   d. regulated by inherently inactive repressor molecules.

2. In *E. coli,* tryptophan switches off the *trp* operon by
   a. inactivating the repressor protein.
   b. inactivating the gene for the first enzyme in the pathway (feedback inhibition).
   c. binding to the repressor and increasing the latter's affinity for the operator.
   d. binding to the operator.

3. In the control of gene expression in bacteria, a regulatory gene
   a. has its own promoter and is transcribed continuously.
   b. is not contained in the operon it controls.
   c. codes for repressor proteins.
   d. is or does all of the above.

4. Certain mutations of the *lac* operon in *E. coli* involve a change to the operator region. This modified operator is unable to bind the repressor protein, whether or not the repressor is bound to allolactose. In such cells
   a. enzymes for the utilization of lactose would always be produced, even in the absence of lactose.
   b. enzymes for the utilization of lactose would never be produced, even in the presence of lactose.
   c. RNA polymerase would be unable to bind to the promoter region to initiate RNA synthesis.
   d. RNA polymerase would be unable to bind to the operator region to initiate RNA synthesis.

5. Proinsulin is a polypeptide created directly by translating an mRNA molecule. To become insulin, it is cleaved and the pieces joined covalently. This is an example of
   a. post-transcriptional regulation
   b. heterochromatin regulation
   c. alternative splicing
   d. post-translational regulation

6. DNA methylation of cytosine bases
   a. initiates the acetylation of histones.
   b. may be a mechanism of epigenetic inheritance when methylation patterns are repeated in daughter cells.
   c. increases the rate of transcription by loosening chromatin structure.
   d. increases the binding of inhibitors to distal control elements in an enhancer.

7. Which of the following is *not* true of enhancers?
   a. They may be located thousands of nucleotides upstream from the genes they affect.
   b. When bound with activators, they interact with the promoter region and other transcription factors to produce an initiation complex.
   c. They typically contain a TATA box.
   d. They may coordinate the transcription of enzymes involved in the same metabolic pathway when they contain the same combination of control elements.

8. The results of the Human Genome Project have indicated that a eukaryotic cell can produce more different proteins than it has genes. Given that most genes code for proteins, how is this possible?
   a. Some transcripts can be spliced in more than one way, producing messages for multiple proteins from the same gene.
   b. Some proteins can be produced in the cell without information from genes.
   c. Darwinian theory requires extra proteins as the raw material for natural selection and evolution.
   d. Some proteins are the result of transcription from the genes of viruses that have infected the cell.

9. A eukaryotic gene typically has all of the following associated with it *except*
   a. a promoter.
   b. an operator.
   c. enhancers containing distal control elements.
   d. introns and exons.

10. Which of the following would you expect to find as part of a receptor protein that binds with a steroid hormone?
   a. a domain that binds to DNA and protein-binding domains
   b. an activated promoter region that allows attachment of RNA polymerase
   c. a binding domain that enables it to bend the DNA to bring bound activators closer to the promoter
   d. transmembrane domains that facilitate the protein's localization in the nuclear membrane

11. Proteasomes are
    a. complexes of proteins that excise introns.
    b. single-stranded RNA molecules complexed with proteins that block translation of or degrade mRNA.
    c. small, positively charged proteins that form the core of nucleosomes.
    d. enormous protein complexes that degrade proteins marked with ubiquitin.

12. Which of the following statements explains why a larger portion of the DNA in a human cell is transcribed than would be predicted by the 1.5% of the genome that is protein-coding DNA?
    a. Multiple enhancer regions are being transcribed to amplify the transcription of protein-coding genes.
    b. Much of this non-protein-coding RNA functions to regulate the translation or degradation of mRNAs.
    c. Many of these transcriptions produce miRNAs and siRNAs that regulate the transcription of other genes.
    d. The additional DNA that is transcribed represents introns that are excised from the primary transcript in the production of mRNA.

13. Which of the following is *not* descriptive of ncRNAs?
    a. They are thought to have facilitated the evolution of morphological complexity.
    b. The inhibition of gene expression by RNA can be observed experimentally and is called RNA interference (RNAi).
    c. The newly discovered piRNAs induce heterochromatin formation and appear to help reestablish methylation patterns during gamete formation.
    d. They are often found in "transcription factories" within an interphase nucleus where they loosen loops of chromatin and enhance transcription of genes on multiple chromosomes.

14. Which of the following descriptions is *not* part of the process by which miRNA regulates gene expression?
    a. A longer RNA precursor is processed into a single-stranded RNA of about 22 nucleotides.
    b. The miRNA forms a complex with proteins.
    c. If the miRNA and an mRNA molecule are complementary all along their length, the miRNA is degraded and translation proceeds.
    d. If the match between the miRNA and an mRNA is less complete, then translation is blocked.

15. In snails, shell coiling can be either right-handed or left-handed. The pattern of coiling in a snail is related to the genotype of its mother and not to the genotype of the individual. This curious observation suggests that shell coiling is controlled by
    a. a maternal effect gene.
    b. an oncogene.
    c. a piRNA.
    d. a morphogen.

16. Pattern formation in animals is based on
    a. positional information a cell receives from gradients of morphogens.
    b. the induction of cells by the nurse cells in the mother's ovary.
    c. the packing of chromatin in the nucleus.
    d. the differentiation of cells that then migrate together to form tissues and organs.

17. What would be the fate of a *Drosophila* larva that inherits two copies of a mutant *bicoid* gene (one mutant allele from each heterozygous parent)?
    a. It develops two heads, one at each end of the larva.
    b. It develops two tails, one at each end of the larva.
    c. It develops normally but, if female, produces mutant larvae that have two tail regions.
    d. It develops into an adult with legs growing out of its head.

18. How is the coordinated transcription of eukaryotic genes involved in the same pathway regulated?
    a. The genes are transcribed in one transcription unit, although each gene has its own promoter.
    b. The genes are located in the same region of the chromosome, and enzymes acetylate the entire region so that transcription may begin.
    c. All the genes respond to the same general transcription factors, although they may respond to different specific transcription factors.
    d. The genes have the same combination of control elements in their enhancers.

19. In the following hypothetical embryo, a high concentration of a morphogen called morpho is needed to activate gene *P*; gene *Q* is active at or above medium concentrations of morpho; and gene *R* is expressed so long as any quantity of morpho is present. A different morphogen, called phogen, activates gene *S* and inactivates gene *Q* when at medium to high concentrations. If morpho and phogen are diffusing from their sites of production at the opposite ends of the embryo, which genes will be expressed in region 2 of this embryo? (Assume a gradient of morphogen concentrations in the three regions, from high at the source, to medium in the middle, and to low at the opposite end.)

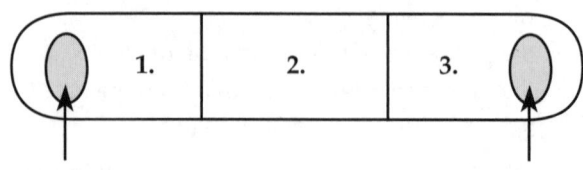

Morpho   Phogen
produced here   produced here

    a. genes *P*, *Q*, *R*, and *S*
    b. genes *P*, *Q*, and *R*
    c. genes *Q* and *R*
    d. genes *R* and *S*

20. Which of the following might a proto-oncogene code for?
    a. DNA polymerase
    b. RNA polymerase
    c. receptor protein for growth factors
    d. an enhancer

21. The most commonly mutated gene in human tumors is
    a. *p53*.
    b. *ras*.
    c. *APC*.
    d. *BRCA2*.

22. A tumor-suppressor gene could cause the onset of cancer if
    a. both alleles have mutations that decrease the activity of the gene product.
    b. it is inherited from a parent in mutated form.
    c. a proto-oncogene has also become an oncogene.
    d. both **a** and **c** have occurred.

23. Apoptosis is
    a. a cell suicide program that may be initiated by p53 protein in response to DNA damage.
    b. metastasis, or the spread of cancer cells to a new location in the body.
    c. the mutation of a G protein into a hyperactive form.
    d. the transformation of a proto-oncogene to an oncogene by a point mutation.

24. Which of the following would most likely account for a family history of colorectal cancer?
    a. a diet that is high in fats and fiber
    b. inheritance of one mutated *APC* allele that regulates cell adhesion and migration
    c. inheritance of the *ras* oncogene, which locks the G protein in an active configuration
    d. inheritance of a proto-oncogene

Key Concepts

19.1 A virus consists of a nucleic acid surrounded by a protein coat

19.2 Viruses replicate only in host cells

19.3 Viruses and prions are formidable pathogens in animals and plants

Framework

A **virus** is an infectious particle consisting of a genome of single-stranded or double-stranded DNA or RNA enclosed in a protein capsid and sometimes within a membrane envelope derived from the host. Viruses replicate using the metabolic machinery of their bacterial, animal, or plant host. Viral infections may destroy the host cell and cause diseases within the host organism. Viruses may have evolved from plasmids or transposons.

Chapter Review

19.1 A virus consists of a nucleic acid surrounded by a protein coat

The Discovery of Viruses: Scientific Inquiry The search for the cause of tobacco mosaic disease led to the discovery of viruses. The infectious agent could neither be filtered from infected sap nor cultivated on nutrient media, but it was able to replicate within plants. In 1935, W. Stanley crystallized the infectious particle, now known as tobacco mosaic virus (TMV). Since that time, many viruses have been seen with the electron microscope.

Structure of Viruses A virus is an infectious particle consisting of genes inside a protein coat and, for some viruses, a membranous envelope. Viral genomes may be single-stranded or double-stranded DNA or RNA. Viral genes are usually contained on a single linear or circular nucleic acid molecule.

The **capsid**, or protein shell, is built from a large number of often identical protein subunits (*capsomeres*) and may be rod-shaped (*helical viruses*), polyhedral (*icosahedral viruses*), or more complex in shape. **Viral envelopes**, which are derived from membranes of the host cell but also include viral proteins and glycoproteins, may cloak the capsids of viruses that infect animals. Some viruses also contain a few viral enzymes.

Complex capsids are found among **bacteriophages**, or **phages**, viruses that infect bacteria. Of the phages that infect the bacterium *E. coli*, T2, T4, and T6 have similar capsid structures consisting of an icosahedral head and a protein tail piece with tail fibers for attaching to a bacterium.

19.2 Viruses replicate only in host cells

Viruses are obligate intracellular parasites that lack metabolic enzymes and other equipment needed to replicate. Each virus type has a limited **host range** due to proteins on the outside of the virus that recognize only specific receptor molecules on the host cell surface.

General Features of Viral Replicative Cycles Once the viral genome enters the host cell, the cell's enzymes, nucleotides, amino acids, ribosomes, ATP, and other resources are used to replicate the viral genome and produce capsid proteins. Many DNA viruses use host DNA polymerases to copy their genome, whereas RNA viruses use virus-encoded RNA polymerases for replicating their RNA genome.

After replication, viral nucleic acid and capsid proteins spontaneously self-assemble to form new viruses within the host cell. Hundreds or thousands of newly formed virus particles are released, often destroying the host cell in the process.

Replicative Cycles of Phages   A **lytic cycle** culminates in lysis of the host cell and release of newly produced phages. **Virulent phages** replicate only by a lytic cycle.

The T4 phage uses its tail fibers to stick to a receptor site on the surface of an *E. coli* cell. The sheath of the tail contracts and thrusts its viral DNA into the cell, leaving the empty capsid behind. The *E. coli* cell begins to transcribe and translate phage genes, one of which codes for an enzyme that chops up host cell DNA. Nucleotides from the degraded bacterial DNA are used to produce viral DNA. Capsid proteins are assembled into phage tails, tail fibers, and heads. The viral components assemble into phage particles. These are released after the manufacture of an enzyme that damages the bacterial cell wall, causing the cell to swell and burst.

In a **lysogenic cycle**, a virus replicates its genome without killing its host. **Temperate phages** can replicate by lytic and lysogenic cycles.

When the phage lambda (λ) injects its DNA into an *E. coli* cell, it can begin a lytic cycle, or its DNA may be incorporated into the host cell's chromosome and begin a lysogenic cycle as a **prophage**. Most of the genes of the inserted phage genome are repressed by a protein coded for by a prophage gene. Reproduction of the host cell replicates the phage DNA along with the bacterial DNA. The prophage may exit the bacterial chromosome, usually in response to environmental stimuli, and start a lytic cycle.

Several disease-causing bacteria would be harmless except for the expression of prophage genes that code for toxins.

Mutations that alter bacterial receptor sites provide some protection against infection by phages. Phage DNA may also be degraded by **restriction enzymes** that cut foreign DNA; host DNA is protected from these enzymes by specific DNA methylation patterns.

Many bacteria and archaea possess the *CRISPR-Cas* defense system, which involves specific repetitive DNA sequences. The DNA of an infecting phage is integrated into the host genome between two CRISPR repeat sequences. A subsequent attack by a similar phage causes transcription of the CRISPR region, and the resulting RNA molecules are cut into pieces. These RNA molecules, complexed with Cas proteins, bind to complementary sequences in the phage genome, leading to degradation of the phage DNA.

INTERACTIVE QUESTION   19.1

In the following diagram of lytic and lysogenic cycles, describe steps numbered 1–8 and label structures a–e.

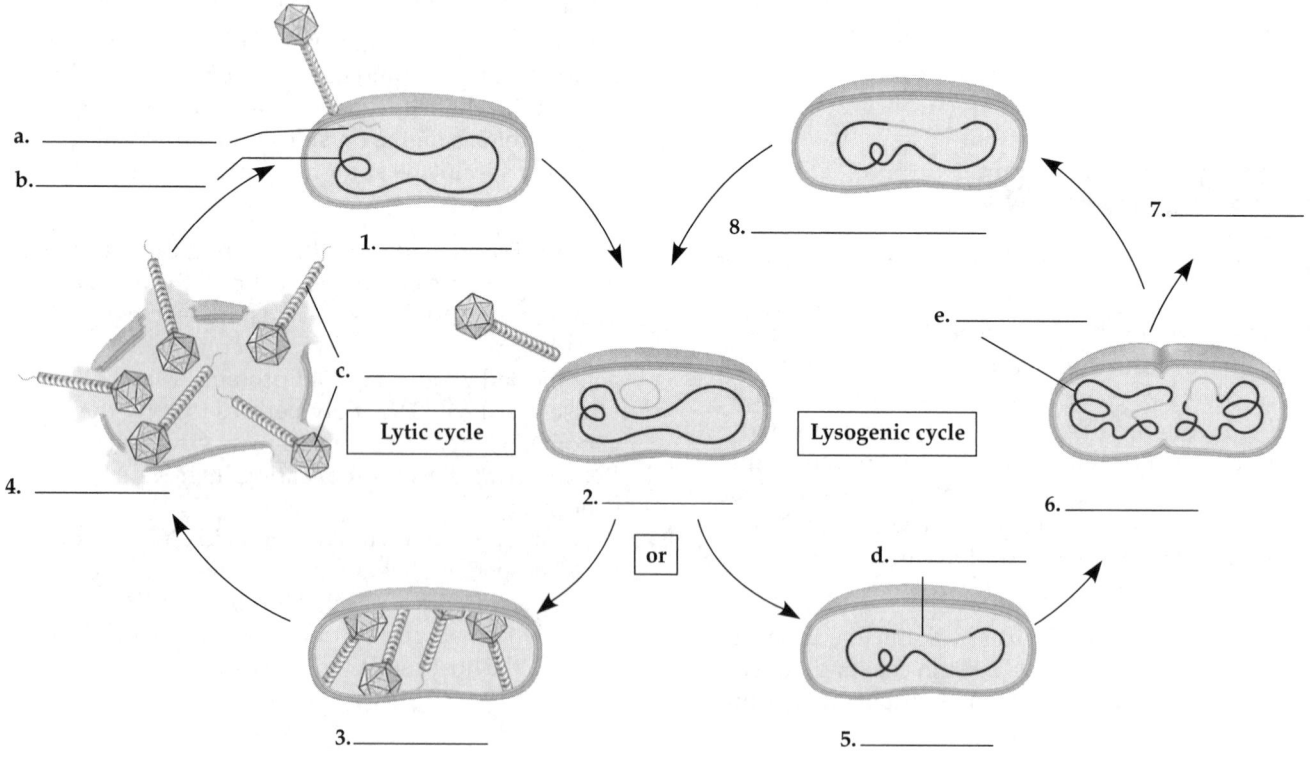

Replicative Cycles of Animal Viruses Animal viruses are classified based on their type of genome: double- or single-stranded DNA or RNA. A membranous envelope surrounds the capsid of almost all animal viruses that have RNA genomes.

What is the replicative cycle of an enveloped RNA virus? Glycoproteins extending from the viral membrane attach to receptor sites on a host cell plasma membrane. The capsid enters the cell, and the viral genome replicates using a viral RNA polymerase. Viral RNA then directs the synthesis of viral proteins. Viral glycoproteins are produced and embedded in the endoplasmic reticulum (ER) membrane, and then transported to the plasma membrane. New viruses bud off within an envelope that is derived from the host's plasma membrane and bears viral glycoproteins.

Herpesviruses replicate within the host cell nucleus and are temporarily cloaked in a membrane derived from the host cell nuclear envelope. This membrane is later shed and replaced by one derived from the Golgi apparatus. The double-stranded DNA of the herpesvirus can remain latent as a mini-chromosome in the nucleus of certain nerve cells until it initiates herpes infections in times of stress.

The single-stranded RNA of animal class IV viruses can serve directly as mRNA. The RNA genome of class V viruses must first be transcribed into a strand of complementary RNA (using a viral enzyme packaged inside the capsid) that then serves as mRNA and as a template for making genome RNA.

In the complicated replicative cycle of **retroviruses** (class VI), the viral RNA genome is transcribed into double-stranded DNA by a viral enzyme, **reverse transcriptase**. This viral DNA is then integrated into a chromosome, where it is transcribed by the host cell into viral RNA, which acts both as new viral genome and as mRNA for viral proteins. **HIV (human immunodeficiency virus)** is a retrovirus that causes **AIDS (acquired immunodeficiency syndrome)**. The integrated viral DNA remains as a **provirus** within the host cell DNA. New viruses, assembled with two copies of the RNA genome and two molecules of reverse transcriptase within a capsid, bud off, covered in host cell plasma membrane studded with viral glycoproteins.

INTERACTIVE QUESTION 19.2

Summarize the flow of genetic information during replication of a retrovirus. Indicate the enzymes that catalyze this flow.

_____ → _____ → _____

Enzymes:

Evolution of Viruses Viruses may have evolved from fragments of cellular nucleic acids that moved from one cell to another. Sources of viral genomes may have been *plasmids*, self-replicating circles of DNA found in bacteria and yeast, and *transposons*, segments of DNA that can change locations within a cell's genome. Thus, viruses, plasmids, and transposons are all *mobile genetic elements*.

The genomes of viruses are often more similar to those of their host cells than to the genomes of viruses infecting other hosts. Recent sequencing of viral genomes, however, has revealed genetic similarities among viruses that otherwise seem distantly related. Scientists debate whether the recently discovered mimivirus, which has genes for cellular machinery, evolved before or after the first cells.

19.3 **Viruses and prions are formidable pathogens in animals and plants**

Viral Diseases in Animals The symptoms of a viral infection may be caused by viral-programmed toxins produced by infected cells, cells killed or damaged by the virus, or the body's defense mechanisms fighting the infection.

Vaccines are harmless variants or derivatives of pathogens that induce the immune system to react against the actual disease agent. Vaccinations have greatly reduced the incidence of many viral diseases.

Unlike bacteria, viruses use the host's cellular machinery to replicate, and few drugs have been found to treat or cure viral infections. Some antiviral drugs resemble nucleosides and interfere with viral nucleic acid synthesis.

Emerging Viruses Examples of *emerging viruses* include HIV, Ebola virus, West Nile virus, chikungunya virus, and Zika virus. A general outbreak of a disease is called an **epidemic**. The flu epidemic of 2009 spread rapidly, becoming a global epidemic or **pandemic**.

The sudden emergence of viral diseases may be linked to the mutation of an existing virus (more common in RNA viruses, which have higher mutation rates), to the dissemination of an existing virus to a more widespread population (as in HIV), or to the spread from one host species to another (as in the 2009 flu pandemic, which likely passed to humans from pigs). Many new human diseases are thought to originate by the third mechanism.

Influenza types B and C infect only humans; type A infects other animals as well as humans. If different strains of virus undergo genetic recombination within an animal's cells and accumulate mutations that allow the virus to infect human cells, the recombinant virus may be highly pathogenic. The H1N1 virus (named for the form of the viral surface proteins hemagglutinin and neuraminidase) caused both the 1918 and 2009 flu pandemics. The H5N1 virus is carried by wild and domestic birds but can be transmitted to humans in some cases. The human mortality rate for "avian flu" infections is greater than 50%, but thus far the virus is not easily transmitted from person to person. Global climate change may allow virus vectors such as mosquitoes to expand their ranges, thereby increasing the spread of viral diseases.

Viral Diseases in Plants   Most plant viruses are RNA viruses. Plant viral diseases may spread through *vertical transmission* from a parent plant via infected seeds or cuttings or through *horizontal transmission* from an external source. Plant injuries increase susceptibility to viral infections, and insects can act as carriers of viruses.

INTERACTIVE QUESTION   19.3

How does a virus spread throughout a plant? Are there cures for viral plant diseases?

Prions: Proteins as Infectious Agents   **Prions** are protein infectious agents that may be linked to several degenerative brain diseases, such as scrapie and mad cow disease in animals and Creutzfeldt-Jacob disease and kuru in humans. Prions are misfolded forms of a protein and apparently cause disease by converting normal cellular proteins into the misfolded prion version.

Word Roots

caps-   = a box (*capsid:* the protein shell that encloses a viral genome)

-phage = that which eats (*bacteriophage:* a virus that infects bacteria)

pro-   = before (*provirus:* a viral genome that is permanently inserted into a host genome)

retro- = backward (*retrovirus:* an RNA virus that replicates by transcribing its RNA into DNA and then inserting the DNA into a cellular chromosome)

Structure Your Knowledge

1. Create a concept map that describes the lytic and lysogenic cycles of a phage.

2. Complete the following concept map to help organize your understanding of viruses.

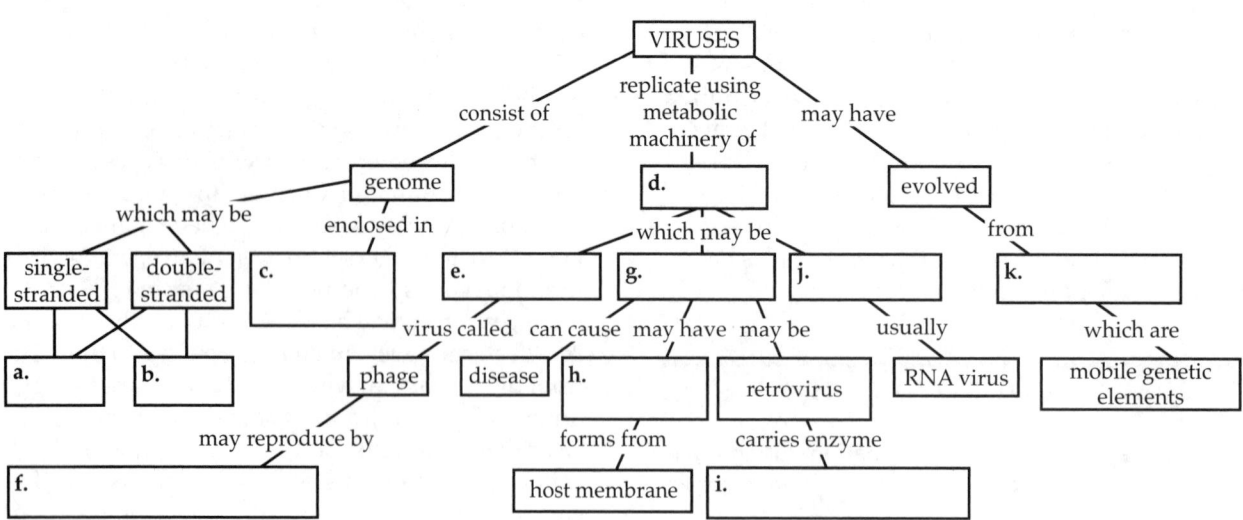

Test Your Knowledge

MULTIPLE CHOICE: *Choose the one best answer.*

1. The study of viruses has provided information on all of the following topics *except*
   a. the molecular biology of all organisms.
   b. the sexual replicative cycles of viruses.
   c. new techniques for manipulating genes.
   d. the causes, prevention, and treatment of diseases.

2. Beijerinck concluded that the cause of tobacco mosaic disease was not a filterable toxin because
   a. the infectious agent could not be cultivated on nutrient media.
   b. a plant sprayed with filtered sap would develop the disease.
   c. the infectious agent replicated and could be passed on from a plant infected with filtered sap.
   d. the filtered sap was infectious even though microbes could not be found in it.

3. Viral genomes may be any of the following *except*
   a. single- or double-stranded DNA.
   b. double-stranded RNA.
   c. misfolded infectious proteins.
   d. a linear single-stranded RNA molecule.

4. The reverse transcriptase carried by retroviruses
   a. uses viral RNA as a template for making complementary RNA strands.
   b. protects viral DNA from degradation by restriction enzymes.
   c. translates RNA into proteins.
   d. uses viral RNA as a template for DNA synthesis.

5. T2 and T4 are two types of bacteriophage that infect *E. coli*. Suppose a bacteriophage is artificially constructed from T2 phage capsomeres and T4 phage DNA. If this phage is used to infect *E. coli*, the newly produced phages will have
   a. T4 protein and T4 DNA.
   b. T4 protein and T2 DNA.
   c. T2 protein and T2 DNA.
   d. T2 protein and T4 DNA.

6. A virus has a base ratio of $(A + G)/(U + C) = 1$. What type of virus is this?
   a. a retrovirus
   b. a single-stranded RNA virus
   c. a double-stranded DNA virus
   d. a double-stranded RNA virus

7. Plant viruses can spread throughout a plant
   a. in water carried up from the roots.
   b. through gap junctions.
   c. through plasmodesmata.
   d. using viral transport proteins.

8. Bacterial restriction enzymes
   a. cut up phage DNA.
   b. replicate phage DNA.
   c. produce a strand of DNA using RNA as a template.
   d. insert phage DNA between CRISPR sequences.

9. Drugs that are effective in treating viral infections
   a. inhibit the action of viral ribosomes.
   b. interfere with the synthesis of viral nucleic acid.
   c. change the cell-recognition sites on the host cell.
   d. are vaccines that stimulate the immune system to develop immunity.

10. Which of the following is true of prions?
    a. They are emerging viruses.
    b. They are fast-acting infectious agents.
    c. They probably evolved from transposons.
    d. They are infectious proteins that may convert brain proteins into misfolded forms.

11. An RNA viral genome may be replicated by
    a. DNA polymerase from the host.
    b. RNA polymerase coded by viral genes and carried in the viral capsid.
    c. reverse transcriptase that synthesizes RNA.
    d. RNA polymerase from the host.

12. The CRISPR-Cas system allows a bacterium to
    a. degrade DNA from viruses that it has previously been exposed to.
    b. degrade DNA from viruses that it has never been exposed to.
    c. prevent phages from entering the cell.
    d. prevent prophages from exiting the bacterial chromosome.

Chapter 20

DNA Tools and Biotechnology

Key Concepts

20.1 DNA sequencing and DNA cloning are valuable tools for genetic engineering and biological inquiry

20.2 Biologists use DNA technology to study gene expression and function

20.3 Cloned organisms and stem cells are useful for basic research and other applications

20.4 The practical applications of DNA-based biotechnology affect our lives in many ways

Framework

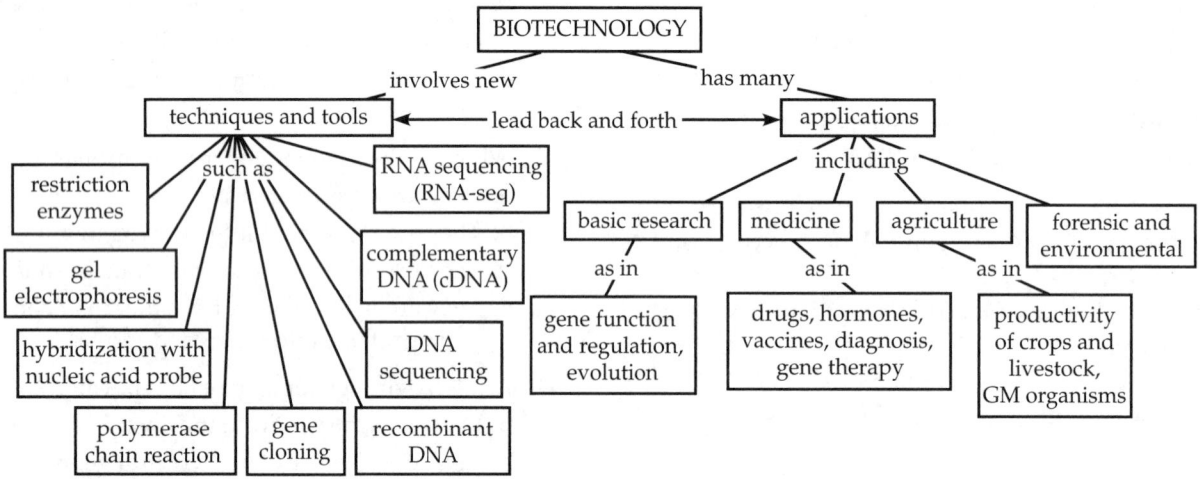

Chapter Review

Advances in **DNA technology** contribute to **biotechnology,** the use of organisms or their components to manufacture desirable products.

20.1 DNA sequencing and DNA cloning are valuable tools for genetic engineering and biological inquiry

The complementary nature of the two strands of a DNA molecule is the basis for **nucleic acid hybridization**. This base pairing of one strand of nucleic acid to complementary sequences of another strand is central to most techniques used in **genetic engineering,** the direct manipulation of genetic material for practical purposes.

DNA Sequencing   *Dideoxyribonucleotide* (or *dideoxy*) *chain termination sequencing* was the earliest automated **DNA sequencing** technique. Although it is still used for simple small sequencing tasks, it has largely been supplanted by more rapid and efficient sequencing techniques.

"Next generation sequencing" techniques use DNA fragments and allow *sequencing by synthesis* of complementary strands, one nucleotide at a time. Individual genomic DNA fragments are isolated and amplified by the polymerase chain reaction (PCR). The 5′ ends of about 10^6 copied single strands are "captured" by a bead. Two million beads, each holding a different fragment, are placed on a multiwell plate. These fragments are sequenced "in parallel" as nucleotides are added one at a time and washed off. A flash of light is recorded whenever a nucleotide is joined to a growing strand. Using "high-throughput" DNA technology, computer software analyzes each fragment and "stitches" them together into a whole genomic sequence.

"Third-generation sequencing" continues to speed up and lessen the expense of DNA sequencing. In these methods, a single, very long DNA molecule can be sequenced on its own. One model involves moving a single strand of DNA through a *nanopore* in a membrane and measuring each base's interruption of an electrical current.

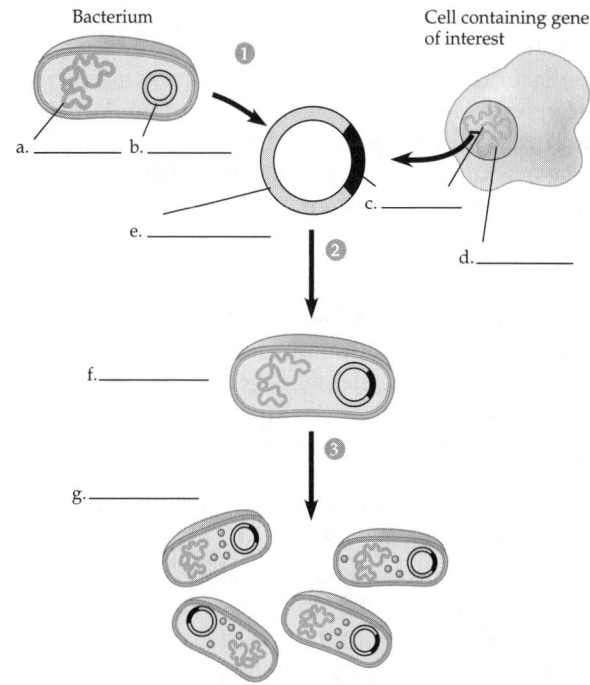

INTERACTIVE QUESTION 20.1

In what ways would third-generation sequencing be different from sequencing by synthesis?

Making Multiple Copies of a Gene or Other DNA Segment

Various methods have been developed for **DNA cloning**—making multiple copies of a specific DNA segment. One of the approaches makes use of the **plasmids** of bacterial cells. **Recombinant DNA**, which contains nucleotide sequences from different organisms or species, may be made by inserting foreign DNA into plasmids. These plasmids are put back into bacterial cells, where they will replicate as the *recombinant bacteria* reproduce to form clones of identical cells. Such **gene cloning** *amplifies* or provides multiple copies of the gene and may also be used to produce protein coded for by the foreign DNA. Bacterial plasmids are widely used as **cloning vectors**, DNA molecules that can move foreign DNA into a cell and replicate there.

INTERACTIVE QUESTION 20.2

The following schematic diagram depicts an overview of gene cloning. Identify components a–g. Briefly describe what is happening in the three numbered steps of the process. For what two purposes might the bacteria produced in step 3 be used?

Using Restriction Enzymes to Make a Recombinant DNA Plasmid

Restriction endonucleases, or **restriction enzymes**, protect bacteria from phages or other organisms by cutting up foreign DNA. Most restriction enzymes recognize short nucleotide sequences, called **restriction sites**, and cut both strands of DNA at specific points within them. The bacterial cell protects its own DNA from restriction by methylating nucleotide bases within its own restriction sequences.

A restriction site is usually symmetrical; the same sequence, often of four to eight nucleotides, runs in opposite directions on the two strands. The most useful restriction enzymes cut the sugar-phosphate backbone in a staggered way, leaving **sticky ends** of short single-stranded sequences on both sides of the resulting **restriction fragment**.

DNA from different sources can be combined in the laboratory when the various DNA molecules are cut by the same restriction enzyme, and the complementary bases on the resulting sticky ends of the restriction fragments form hydrogen-bonded base pairs. DNA ligase is used to seal the strands together.

INTERACTIVE QUESTION 20.3

Which of the following DNA sequences would most likely function as a restriction site for a restriction enzyme? Why?

··CAGCAG··	··GTGCTG··	··GAATTC··
··GTCGTC··	··CACGAC··	··CTTAAG··

Gel electrophoresis is a technique that separates a mixture of nucleic acids (or proteins) on the basis of their size and electrical charge. DNA molecules migrate through the electric field produced in a thin slab of gel toward the positive electrode. Linear molecules of DNA move at a rate inversely proportional to their length, producing in the gel band patterns containing fragments of decreasing size.

Amplifying DNA: The Polymerase Chain Reaction (PCR) and Its Use in Cloning   The **polymerase chain reaction (PCR)** can produce billions of copies of a section of DNA in only a few hours. The desired DNA segment need not be purified from the starting material, and very small samples can be used. DNA containing the region of interest is incubated with the four nucleotides, a special heat-stable type of DNA polymerase, and specially synthesized primers that bind upstream (at the 5′ end) from the target sequence on each of the DNA strands. The solution is heated to denature (separate) the DNA strands, then cooled so the primers can anneal (hydrogen-bond) to complementary sequences. DNA polymerase then adds nucleotides to the 3′ ends of the primers. The solution is heated again and the process repeated. The heating process would normally denature the DNA polymerase so that it would need to be added in each cycle. To prevent this problem, heat-stable DNA polymerases from the thermophilic prokaryotes *Thermus aquaticus* (*Taq* polymerase) and *Pyrococcus furiosus* (*Pfu* polymerase) are used.

INTERACTIVE QUESTION   20.4

a. When PCR is used to prepare a DNA fragment for cloning, what sequence is often included in the primers?

b. What type of gene is often included in the plasmids for gene cloning that helps identify the cells that have taken up a plasmid?

c. Why even bother cloning genes in cells, given that PCR produces so many copies so fast?

Expressing Cloned Eukaryotic Genes   Differences between prokaryotic and eukaryotic mechanisms for gene expression can be overcome by using an **expression vector**, a cloning vector that has an active prokaryotic promoter just upstream from the eukaryotic gene insertion site. Using a *complementary DNA* (*cDNA*) version of mRNA removes the problem of long introns,

especially given that bacteria lack RNA-splicing machinery.

Yeast cells are eukaryotic hosts used to clone and express eukaryotic genes; they are easy to grow and have plasmids that serve as vectors. Plant and animal cells in culture can serve as hosts and may be necessary when a protein must be modified following translation.

More efficient means for introducing DNA into cells include **electroporation**, in which an electric pulse briefly opens holes in the plasma membrane through which DNA can enter; injection into cells using microscopically thin needles; and the use of *Agrobacterium tumefaciens* to introduce DNA into plant cells.

The ability of species to express genes transferred from very different species illustrates the shared evolutionary history of all life. For example, the vertebrate *Pax-6* gene, which triggers development of the single-lens vertebrate eye, will lead to the formation of a compound eye when introduced into a fly embryo. Similarly, the fly *Pax-6* gene, when transferred into a frog embryo, will produce a vertebrate eye.

20.2   Biologists use DNA technology to study gene expression and function

Analyzing Gene Expression   Using a **nucleic acid probe** (either RNA or DNA) complementary to the mRNA transcribed from a specific gene allows researchers to determine in which cells that particular gene is being expressed. The technique of *in situ* **hybridization** uses labeled probes to detect the locations of specific mRNAs within an intact organism.

Sample sets of mRNA can be studied using the technique called **reverse transcriptase-polymerase chain reaction (RT-PCR)**. The enzyme reverse transcriptase is used to make a single-stranded DNA *reverse transcript* of mRNA. DNA polymerase synthesizes the second strand, producing **complementary DNA (cDNA)**. The cDNA is amplified by PCR using primers specific for that gene, and gel electrophoresis produces bands of amplified DNA only in samples in which the gene has been transcribed to mRNA. An advanced technique that avoids the gel electrophoresis step is *quantitative RT-PCR* (*qT-PCR*). In this process the product is identified and quantified by measuring fluorescence of a dye bound to a double-stranded PCR product.

In order to study the expression of interacting groups of genes, researchers isolate the mRNA made in different tissues or at different developmental stages. Using cDNA made from this mRNA in **DNA microarray assays**, scientists can test all the genes expressed in a tissue for hybridization with short, single-stranded DNA fragments from thousands of genes arrayed on a grid (called a *DNA chip*). The intensity of fluorescence in each well measures the relative expression of that gene.

As DNA sequencing has become more rapid and inexpensive, cDNA samples made from mRNA isolated from different sources can now be sequenced to identify the genes being expressed (a method called **RNA sequencing** or **RNA-seq**). This technique does not require existing labeled probes and provides quantitative information regarding gene expression over a wider range than is possible with microarray assays. RNA-seq also provides additional data on gene expression, including levels of alternatively spliced mRNAs.

Determining Gene Function The function of unknown genes can be studied using *in vitro* **mutagenesis**, in which changes are made to a cloned gene, the gene is returned to the cell, and it "knocks out" the normal gene. Changes in physiology or developmental patterns that result from the missing gene product are monitored.

A new and very powerful gene editing system, called the **CRISPR-Cas9 system**, is becoming widely used in genetic engineering. Cas9 is a bacterial nuclease that cuts double-stranded DNA complementary to a guide RNA. The attack on the targeted DNA typically renders the associated gene nonfunctional. Once the gene is specifically knocked out, the effects of the gene loss can be investigated. The CRISPR-Cas9 system can also be modified to repair specific mutations in DNA and so may be useful for gene therapy. Engineering a modified gene to enhance its selective advantage creates a **gene drive** that causes the modified form to spread rapidly in a population.

Another way to stop the expression of selected genes in cells is called **RNA interference (RNAi)**. Synthetic double-stranded RNA molecules that match a gene sequence trigger the breakdown or block the translation of that gene's mRNA.

In **genome-wide association studies**, researchers test for *genetic markers*—sequence variations or polymorphisms—that may be associated with certain conditions or diseases. **Single nucleotide polymorphisms (SNPs)** are single base-pair variations that occur in at least 1% of the human population. These genetic markers can now be detected by microarray analysis or PCR. When a SNP is identified in individuals with a given disease, it is assumed to be close to or within a gene that contributes to the disease, and the region of DNA can be sequenced and studied.

INTERACTIVE QUESTION   20.5

a. What are some of the benefits of determining the nucleotide sequence of a gene of unknown function?

b. Describe a research method you could use to identify the tissues in which a particular gene is expressed.

c. Describe a research method that enables you to compare which genes are expressed and what their relative rates of expression are in several tissues.

d. In one study with early nematode embryos, RNAi was used to prevent the expression, one gene at a time, of 86% of nematode genes. What did this allow researchers to do?

20.3 Cloned organisms and stem cells are useful for basic research and other applications

Organismal cloning involves producing genetically identical individuals (clones) from a single cell of a multicellular organism. One focus of organismal cloning is the generation of **stem cells**, relatively unspecialized cells that continue to reproduce themselves and can, under proper conditions, differentiate into one or more types of cells.

Cloning Plants: Single Cell Cultures F. C. Steward was the first to clone plants by growing new carrot plants from differentiated root cells. Such **totipotent** cells retain the ability to give rise to all the specialized cell types of an organism.

Cloning Animals: Nuclear Transplantation Early evidence of totipotency in animal cells was provided by the work of Briggs, King, and Gurdon, who transplanted nuclei from embryonic and tadpole cells into *enucleated* frog egg cells, a method called *nuclear transplantation* (or *somatic cell nuclear transfer*). The ability of the transplanted nucleus to direct normal development was inversely related to its developmental age.

In 1997, Scottish researchers reported cloning an adult sheep by transplanting a nucleus from a fully differentiated mammary cell into an unfertilized enucleated egg cell, and then implanting the resulting early embryo into a surrogate mother. The mammary cell was induced to dedifferentiate by culturing it in a nutrient-poor medium.

The *reproductive cloning* of numerous mammals has shown that cloned animals do not always look and behave identically. Environmental influences and random events play a role in development.

Although numerous mammals have been cloned successfully, most cloned embryos fail to develop normally, and many cloned animals have various defects. What is a likely cause of these developmental failures?

Stem Cells of Animals   Embryonic stem (ES) cells taken from early embryos can be cultured indefinitely and can differentiate into a wide variety of cell types. Thus, ES cells are **pluripotent**. *Adult stem cells* have been isolated from various tissues and grown in culture. Such cells are capable of producing multiple (but not all) types of cells.

Stem cell research has the potential to provide cells to repair organs that are damaged or diseased. *Therapeutic cloning* of embryonic stem cells, although different from reproductive cloning of humans, still raises ethical issues. The transformation of adult stem cells into *induced pluripotent stem (iPS)* cells may provide sources of model cells for studying diseases and potential treatments and may someday provide a patient's own iPS cells for regenerative treatments.

20.4   The practical applications of DNA-based biotechnology affect our lives in many ways

Medical Applications   DNA technology is identifying genes responsible for genetic diseases; it is hoped this will lead to new ways to diagnose, treat, and even prevent those disorders. DNA microarray assays allow comparisons of gene expression in healthy and diseased tissues.

PCR, labeled nucleic acid probes, and RT-PCR are being used to identify pathogens and to diagnose infectious diseases and genetic disorders. An allele for various diseases or increased risk for heart disease, Alzheimer's, and various cancers may be identified when closely linked with a known SNP.

Treatments for diseases have improved as it becomes increasingly possible to correlate an individual's genetic risk factors with treatment options, perhaps leading to a future of personalized medicine. Such individualized treatments require information about the patient in the form of a *genetic profile*. At present a genetic profile consists of a set of genetic markers, but advances in sequencing technology may soon make sequencing individual genomes economically feasible.

Gene therapy may provide the means for correcting genetic disorders in individuals by replacing or supplementing defective genes. New genes would be introduced into somatic cells of the affected tissue. For the correction to be permanent, the cells must be types that actively reproduce within the body, such as bone marrow cells. A few trials have used retroviral vectors to carry a normal allele into bone marrow cells to treat severe combined immunodeficiency (SCID), but results have been mixed.

The CRISPR-Cas9 system allows targeted gene editing and does not require a viral vector. It has already been possible to correct mutations in mice, but early work on human cells suggests that there are significant difficulties to overcome before the CRISPR-Cas9 system can be used therapeutically.

What are some of the practical and ethical considerations in human gene therapy?

As the molecular bases for cancers and other diseases become known, researchers are developing drugs that can block the function of proteins responsible for such diseases. Many pharmaceutical proteins are produced using biotechnology. Engineering host cells so that they secrete a protein as it is made simplifies its purification.

Transgenic animals are produced by injecting a foreign *transgene* into egg cells fertilized *in vitro*. The eggs are then transplanted into surrogate mothers. "Pharm" animals have been engineered to produce large quantities of a pharmaceutical protein, often by secretion in the animal's milk.

Forensic Evidence and Genetic Profiles   In criminal cases, the **genetic profile** of a suspect can now be compared with that of crime-related DNA samples. Variations in the number of **short tandem repeats (STRs)** found at various loci are now commonly used in DNA analysis. Even if only very small samples are available, PCR can amplify specific STRs using differently colored fluorescent tags, which are then separated and their length (corresponding to the number of repeats) determined by electrophoresis. Forensic tests are able to provide a high statistical probability that matching genetic profiles come from the same individual.

A bloody crime has occurred, and police have collected several blood samples. The following table shows the number of repeats for three STR markers from those samples. (Normally, 13 STR markers would be compared.)

Source of sample	STR marker 1	STR marker 2	STR marker 3
Blood at scene	12, 15, 18, 19	9, 14, 17	10, 15, 16, 20
Victim	15, 19	9, 17	15, 20
Suspect 1	12, 21	15, 17	10, 18
Suspect 2	12, 18	14, 17	10, 16

Briefly explain how these data were obtained. Which suspect would you charge with the crime?

Environmental Cleanup Genetically engineered microorganisms that are able to extract heavy metals (such as copper, lead, and nickel) may become important in mining and in cleaning up mining waste. Engineering organisms to degrade chlorinated hydrocarbons and other toxic compounds is an area of active research.

Agricultural Applications The production of transgenic animals is aimed at increasing productivity or beneficial traits.

Many crops have been engineered with a bacterial gene making the plants resistant to herbicides. Crop plants are being engineered to be resistant to infectious pathogens and insects and to grow in soils having high salinity.

Safety and Ethical Questions Raised by DNA Technology Safety regulations have focused on potential hazards of engineered microbes. Most public concern, however, now centers on agriculture—on **genetically modified organisms (GMOs)**, which contain artificially acquired genes from either a different variety or a different species. One environmental risk of transgenic plants is that their herbicide- or insect-resistant genes might pass to wild plants, creating "super weeds." Some fear that GM foodstuffs may be hazardous to human health. An international Biosafety Protocol requires exporters to label GM organisms in bulk food shipments so that importing countries can decide on potential environmental or health risks. In the United States, several federal agencies evaluate and regulate new products and procedures. Use of GMOs is more controversial in Europe, and individual European Union countries may ban GMOs outright or require strict labeling of GM crop products.

Ethical questions about human genetic information include who should have access to information about a person's genome and how that information should be used. Potential ethical, environmental, and health issues must be considered in the development of these powerful genetic techniques and remarkable products of biotechnology.

Word Roots

electro- = electricity (*electroporation:* a technique to introduce recombinant DNA into cells by applying a brief electrical pulse, which creates temporary holes in the cells' plasma membranes, through which DNA can enter.)

-genesis = origin, birth (in vitro *mutagenesis:* a technique used to discover the function of a gene by cloning it, introducing specific changes into the cloned gene's sequence, reinserting the mutated gene into a cell, and studying the phenotype of the mutant)

poly- = many; **morph-** = form (*single nucleotide polymorphism:* a single base-pair site in a genome where nucleotide variation is found in at least 1% of the population)

Technique or Tool	Brief Description	Some Uses in DNA Technology
DNA sequencing	a.	
Restriction enzymes	b.	
Gel electrophoresis	c.	
PCR	d.	
Nucleic acid probe	e.	
RT-PCR	f.	
RNA-seq	g.	
In vitro mutagenesis	h.	
RNA interference (RNAi)	i.	

Structure Your Knowledge

1. Describe several examples of the many possible applications for DNA technology in agriculture and in medicine.
2. Fill in the table on the previous page on the basic tools used in DNA technology.
3. Describe how the complementary nature of DNA strands forms the basis for many of the techniques of genetic engineering.

Test Your Knowledge

MULTIPLE CHOICE: *Choose the one best answer.*

1. The role of restriction enzymes in DNA technology is to
   a. provide a vector for the transfer of recombinant DNA.
   b. produce cDNA from mRNA.
   c. produce a cut (usually staggered) at specific restriction sites on DNA.
   d. reseal "sticky ends" after base pairing of complementary bases.

2. Yeast has become important in genetic engineering because it
   a. has plasmids that can be genetically engineered.
   b. facilitates the study of eukaryotic gene regulation and expression.
   c. grows readily and rapidly in the laboratory.
   d. does all of the above.

3. Gel electrophoresis is a means of separating fragments of DNA based on
   a. charge.   c. genetic information.
   b. nucleotide sequence.   d. size.

4. A plasmid has two antibiotic resistance genes, one for ampicillin and one for tetracycline. It is treated with a restriction enzyme that cuts in the middle of the ampicillin gene. DNA fragments containing a human globin gene were cut with the same enzyme. The plasmids and fragments are mixed, treated with ligase, and used to transform bacterial cells. Clones that have taken up the recombinant DNA
   a. can grow on plates with ampicillin but not with tetracycline.
   b. can grow on plates with tetracycline but not with ampicillin.
   c. can grow on plates with tetracycline and ampicillin.
   d. cannot grow with any antibiotics.

5. Two DNA fragments created by cutting both with the same restriction enzyme can be recombined
   a. if both DNA fragments come from the same species.
   b. if both DNA fragments come from animals.
   c. if both DNA fragments come from eukaryotes.
   d. regardless of the sources of the DNA.

6. The following segment of DNA has restriction sites I and II, which create restriction fragments a, b, and c. Which of the gels produced by electrophoresis would represent the separation and identity of these fragments?

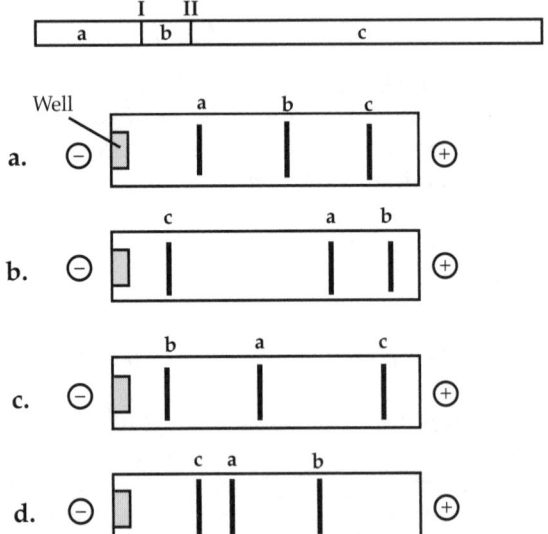

7. Because Neanderthal individuals died out relatively recently, their bones have not fossilized (turned to stone). It has been possible to isolate bits of DNA from Neanderthal bones, although there isn't much and what is available is in poor condition. What will be your first step in preparing this DNA for analysis?
   a. reverse transcription
   b. electrophoresis
   c. DNA fingerprinting
   d. polymerase chain reaction

8. You wish to express a eukaryotic gene product in a prokaryotic expression system. What is the advantage of using cDNA as your starting point?
   a. cDNA does not contain introns.
   b. cDNA is entirely artificial.
   c. cDNA and prokaryotic DNA use the same genetic code.
   d. cDNA is easier to prepare than genomic DNA.

Use the following choices to answer questions 9–12.

- a. restriction enzyme
- b. reverse transcriptase
- c. DNA ligase
- d. DNA polymerase

9. Which is the first enzyme used in the production of a recombinant plasmid?

10. Which is the last enzyme involved in making recombinant plasmids?

11. Which is the first enzyme used in the production of cDNA?

12. Which enzyme is used in the polymerase chain reaction?

13. You are attempting to introduce a gene that imparts resistance to larval moths in bean plants. Which of the following vectors are you most likely to use?
- **a.** phage DNA
- **b.** *E. coli* plasmid
- **c.** Ti plasmid
- **d.** yeast plasmid

14. STRs (short tandem repeats) are a valuable tool for
- **a.** DNA microarray assays.
- **b.** acting as probes following gel electrophoresis.
- **c.** genetic profiling.
- **d.** RNAi (RNA interference).

15. You have affixed the chromosomes from a cell onto a microscope slide. Which of the following would *not* make a good radioactively labeled probe to help map a particular gene to one of those chromosomes? (Assume that the DNA is single stranded.)
- **a.** cDNA made from the mRNA transcribed from the gene
- **b.** a portion of the amino acid sequence of that protein
- **c.** mRNA transcribed from the gene
- **d.** a sequence of nucleotides determined from a known sequence of amino acids in the protein product of the gene

16. Which of the following has the greatest potential for generating different cell types?
- **a.** ES cells
- **b.** iPS cells
- **c.** adult stem cells
- **d.** reprogrammed differentiated body cells

17. A "pharm" animal is a(n)
- **a.** transgenic animal that produces large quantities of a pharmaceutical product.
- **b.** animal used by the pharmaceutical industry to test new medical treatments.
- **c.** cloned animal that was produced from an adult cell nucleus inserted into an egg.
- **d.** genetically modified organism (GMO) whose production is permitted in the United States but not in the European Union.

18. Petroleum-lysing bacteria are being engineered for the treatment of oil spills. Which of the following is the most realistic danger of these bacteria to the environment?
- **a.** mutations leading to the production of a strain pathogenic to humans
- **b.** extinction of natural microbes due to the competitive advantage of the "petro-bacterium"
- **c.** destruction of natural oil deposits
- **d.** contamination of the water

19. Which of the following would be useful in signaling the presence of a disease-causing allele even if the gene has not yet been identified?
- **a.** RNA interference
- **b.** *in situ* hybridization
- **c.** single nucleotide polymorphisms
- **d.** RT-PCR

20. The following restriction fragment contains a gene whose recessive allele is lethal. The normal allele has restriction sites for the restriction enzyme PSTI at sites I and II. The recessive allele lacks restriction site I. An individual whose sister had the lethal trait is being tested to determine if he is a carrier of that allele. Which of the following band patterns would be produced on a gel if he is a carrier (heterozygous for the gene)?

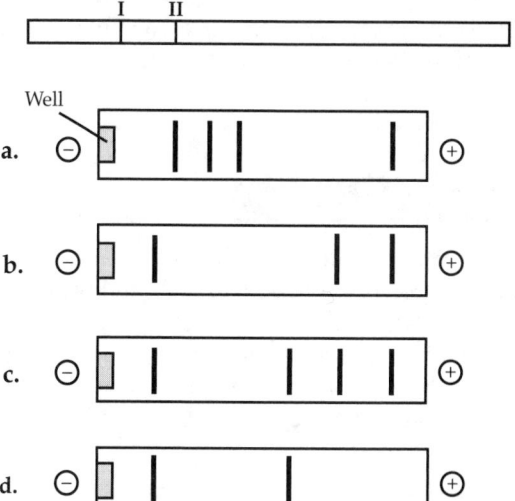

Chapter 21

Genomes and Their Evolution

Key Concepts

21.1 The Human Genome Project fostered development of faster, less expensive sequencing techniques

21.2 Scientists use bioinformatics to analyze genomes and their functions

21.3 Genomes vary in size, number of genes, and gene density

21.4 Multicellular eukaryotes have a lot of noncoding DNA and many multigene families

21.5 Duplication, rearrangement, and mutation of DNA contribute to genome evolution

21.6 Comparing genome sequences provides clues to evolution and development

Framework

This chapter introduces genomics and bioinformatics, new approaches to analyzing and comparing the genomes of life's diverse organisms. Researchers in these fields address questions about genome organization, gene expression, growth and development, and evolution. The various types of non-protein-coding DNA in the human genome are described. The chapter also covers the processes that contribute to genome evolution.

Chapter Review

The complete genome sequences of humans, chimpanzees, and numerous other eukaryotes and prokaryotes are enabling the study of whole sets of genes and their interactions, called **genomics**. The new field of **bioinformatics** is applying computational methods to the analysis of the ever-growing volume of biological data.

21.1 The Human Genome Project fostered development of faster, less expensive sequencing techniques

The international effort to sequence the human genome, called the **Human Genome Project**, was begun in 1990 and declared "virtually completed" in 2006. Technological advances during this project have tremendously accelerated DNA sequencing. Today's sequencing machines are "high-throughput" devices, able to analyze biological materials rapidly and produce enormous volumes of data.

J. C. Venter, founder of the company Celera Genomics, developed a **whole-genome shotgun approach** that relies on powerful computer programs to order the large number of sequenced, overlapping short fragments cut and cloned from a chromosome. This approach is now widely used. Newer "next-generation" sequencing techniques have both increased the speed and decreased the cost of sequencing genomes.

Metagenomics is an approach in which an environmental sample that includes DNA from many species is sequenced, and then computers sort the sequences into specific genomes.

INTERACTIVE QUESTION 21.1

What types of organisms have been best studied using metagenomics?

21.2 Scientists use bioinformatics to analyze genomes and their functions

Centralized Resources for Analyzing Genome Sequences The National Center for Biotechnology Information (NCBI) and other genome centers maintain websites with databases of DNA sequences as well as software for analyzing and comparing sequence data. The ever-expanding NCBI database of sequences, called GenBank, makes the resources of bioinformatics available to researchers worldwide. A Protein Data Bank contains all the three-dimensional protein structures determined thus far.

INTERACTIVE QUESTION 21.2

The software program BLAST, available on the NCBI website, allows a researcher to compare a DNA sequence to every sequence in GenBank to identify similar regions. What are some uses of this function?

Identifying Protein-Coding Genes and Understanding Their Functions Now that genes can be studied directly, geneticists seek to determine phenotype from genotype. Through the process called **gene annotation**, sequences are analyzed to identify protein-coding genes and their functions. Computer software can scan DNA sequences for signs of genes, such as start and stop signals; RNA-splicing sites; and *expressed sequence tags*, or *ESTs* (short sequences from known cDNAs stored in databases).

The sequences of newly identified genes are compared with sequences of known genes of other species to look for similarities that might indicate the gene's function. A combination of biochemical techniques (identifying the amino acid sequence, three-dimensional structure and binding sites of the protein coded for by the sequence) and functional studies (*knocking out* the gene with RNAi or the CRISPR-Cas9 system to determine the effect on phenotype) helps to reveal protein function.

Understanding Genes and Gene Expression at the Systems Level An ongoing research project called ENCODE (Encyclopedia of DNA Elements) analyzes the human genome to identify protein-coding and noncoding RNA genes, regulatory sequences, and epigenetic chromatin modifications. One surprising finding of this collaborative project is that about 75% of the genome is transcribed at some point in the cell types studied, even though less than 2% codes for proteins. At least 80% of the genome contains one or more functional elements. Study of these elements has been extended to the genomes of the nematode *C. elegans* and the fruit fly *D. melangaster*.

The aim of another large-scale project, the Roadmap Epigenomics Project, is the characterization of the *epigenome* in human cells, including stem cells,

normal tissues, and tissues from individuals with specific diseases.

Proteomics is the identification and study of entire protein sets (**proteomes**) coded for by a genome. With compiled lists of DNA sequences and proteins now available, researchers are studying the functional integration of these components in biological systems. This **systems biology** approach seeks to define gene circuits and protein interaction networks, using computers, mathematics, and new software to process and integrate huge amounts of data. To map the protein interaction network in the yeast *S. cerevisiae*, for example, researchers created doubly mutant cells, knocking out one pair of genes at a time, in order to test for gene interactions.

The Cancer Genome Atlas is taking a systems biology approach to analyzing changes in genes and patterns of gene expression in cancer cells. The project catalogues the mutations found in several common cancers and identifies genes of known and unknown functions that may provide new targets for therapies. Microarrays and RNA-seq are used to analyze gene expression in patients with various diseases. Such approaches may help to match treatments with a person's unique genetic makeup.

21.3 Genomes vary in size, number of genes, and gene density

Genome Size Of the thousands of genomes that have been sequenced, most are of bacteria, whose genomes have between 1 and 5 million base pairs (Mb). Archaeal genomes appear to be of a similar size, whereas most animals and plants have genomes of at least 100 Mb. Among eukaryotes, there does not appear to be a correlation between genome size and an organism's phenotype.

Number of Genes Bacteria and archaea have from 1,500 to 7,500 genes; eukaryotes range from about 5,000 genes for unicellular fungi to 40,000 for some multicellular organisms. The human genome contains fewer than 21,000 genes. Alternative splicing of the 10 or so exons in most human genes can yield many different proteins for each gene. Small RNAs (such as miRNAs) that regulate gene expression may contribute to greater organismal complexity.

Gene Density and Noncoding DNA Eukaryotes have fewer genes per million base pairs than bacteria or archaea. Mammals appear to have the lowest gene density. Most of the DNA of eukaryotic genomes is noncoding DNA that is located within and between genes and includes introns, complex regulatory sequences, and other non-protein-coding DNA.

INTERACTIVE QUESTION   21.3

Refer to the organisms listed in Table 21.1 in your text to answer the following questions.

a. Which organism has the highest gene density? _____ the lowest gene density? _____

b. Which organism has the largest number of genes? _____ the smallest number? _____

c. Which organism has the largest haploid genome size? _____ the smallest genome size? _____

d. What is the estimated number of genes in the human genome? _____ Explain the fact that there are many more different polypeptides than genes.

21.4   Multicellular eukaryotes have a lot of noncoding DNA and many multigene families

About 1.5% of the human genome consists of exons that code for proteins, rRNA, or tRNA. The rest includes gene-related regulatory sequences (5%), introns (20%), gene fragments and nonfunctional former genes called **pseudogenes**, and sequences present in many copies, called **repetitive DNA**. About 15% of the genome is unique noncoding DNA. A small percentage of the genome consists of genes for small noncoding RNAs; these genes are included in the percentage values for introns and unique noncoding DNAs.

Noncoding DNA, which was previously referred to as "junk DNA," may turn out to have important functions. Almost 500 identical regions of noncoding DNA have been identified in humans, rats, and mice, a higher level of sequence conservation than for protein-coding regions in these species. The ENCODE project has demonstrated key roles for much unique noncoding DNA.

Transposable Elements and Related Sequences
Much of the repetitive DNA (44% of the human genome) is either made up of or related to *transposable genetic elements* or **transposable elements**. These are stretches of DNA that can move about within a genome through a process called *transposition*.

Transposons move about a genome as a DNA intermediate, either by a "cut-and-paste" mechanism or a "copy-and-paste" mechanism. The enzyme *transposase*, generally encoded by the transposon, is required for

both mechanisms. **Retrotransposons**, which make up the majority of transposable elements, are first transcribed into an RNA intermediate. This RNA transcript is converted back to DNA by reverse transcriptase, which is coded for by the retrotransposon itself.

Transposable elements may be represented as multiple (although not identical) copies of transposons or as related sequences that have lost the ability to move. In humans, about 10% of the genome is made up of *Alu elements*. Many of these 300-nucleotide-long sequences are transcribed into RNA. Some of these RNAs may help regulate gene expression.

About 17% of the human genome consists of *LINE-1*, or *L1*, retrotransposons. These long sequences normally have a low rate of transposition, although they have been found to be more active in developing brain cells in rats.

INTERACTIVE QUESTION   21.4

Explain why retrotransposons always move by the "copy-and-paste" mechanism.

Other Repetitive DNA, Including Simple Sequence DNA   About 14% of the human genome is repetitive DNA that appears to have arisen from mistakes in DNA replication or recombination. Scattered large-segment duplications account for 5–6% of the human genome. **Simple-sequence DNA**, by contrast, makes up 3% of the human genome and consists of multiple copies of tandemly repeated sequences. When the repeat consists of two to five nucleotides, the unit is called a **short tandem repeat**, or **STR**. The variation in repeat numbers between individuals is the basis of genetic profiles. Much of a genome's simple sequence DNA is located at centromeres, where it functions in cell division and chromatin organization, and at telomeres, which protect the tips of chromosomes.

Short repetitive sequences complicate genome sequencing using the whole-genome shotgun approach.

Genes and Multigene Families   Sequences coding for proteins, tRNAs, and rRNAs make up 1.5% of the human genome. Including introns and regulatory sequences, the amount of gene-related DNA is 25%. More than half of the coding DNA occurs in **multigene families**, collections of similar or identical genes.

With the exception of the genes for histone proteins, *identical* multigene families code for RNA products. The genes coding for the three largest rRNA molecules are arranged in a single transcription unit repeated in huge tandem arrays, enabling cells to produce the millions of ribosomes needed for protein synthesis.

Examples of multigene families of *nonidentical* genes are the two families of genes that code for globins, including the α and β polypeptide subunits of hemoglobin. Different versions of each globin subunit are clustered together on two different chromosomes and are expressed at the appropriate times during development. The families also include several pseudogenes.

INTERACTIVE QUESTION   21.5

For each of the following types of DNA sequences found in the human genome, write the letter of the correct description and the percentage of the genome (listed beneath the descriptions) in the blanks provided.

Types of DNA	Description	%
1. Exons or rRNA-/tRNA-coding	_____	_____
2. Introns	_____	_____
3. Regulatory sequences	_____	_____
4. Transposable elements and related sequences	_____	_____
5. *Alu* elements	_____	_____
6. *L1* sequences	_____	_____
7. Unique noncoding DNA	_____	_____
8. Large-segment duplications	_____	_____
9. Simple sequence DNA	_____	_____

Descriptions

A. DNA in centromeres and telomeres, also STRs

B. multiple copies of mostly movable sequences

C. gene fragments and pseudogenes

D. protein- and RNA-coding sequences

E. family of short sequences related to transposable elements

F. multiple copies of large sequences

G. longer retrotransposons with low rate of transposition

H. enhancers, promoters, and other such sequences

I. noncoding sequences within genes

Choices of percentages: 1.5, 3, 5, 5–6, 10, 15, 17, 20, and 44 (These percentages do not add up to 100 because some of these types of DNA are subsets of other categories, and some types are not listed.)

21.5   Duplication, rearrangement, and mutation of DNA contribute to genome evolution

Duplication of Entire Chromosome Sets   Extra sets of chromosomes may arise by accidents in meiosis. The resulting extra genes might diverge through mutation, leading to genes with novel functions. Although rare in animals, polyploidy is common in plants.

Alterations of Chromosome Structure   Using genomic sequence information, researchers can compare the locations of DNA sequences on chromosomes among different species and reconstruct the evolutionary history of chromosomal rearrangements. Chromosomal rearrangements such as duplications and inversions of chromosomes are thought to contribute to speciation because matings between individuals from populations with differing chromosomal rearrangements would be less successful.

Duplication and Divergence of Gene-Sized Regions of DNA   Errors such as unequal crossing over during meiosis (as may occur between copies of a transposable element on misaligned nonsister chromatids) and slippage of template strands during DNA replication might lead to the duplication of genes.

The α-globin and β-globin gene families appear to have evolved from a common ancestral globin gene, which was duplicated and then diverged. The two divergent copies were then translocated to different chromosomes where further duplication and divergence occurred. The multiple duplications and mutations within each family have created the current α- and β-globin families of genes with related functions along with several intervening pseudogenes.

In other cases, mutation of a duplicated gene may lead to a protein product with a new function.

INTERACTIVE QUESTION   21.6

Lysozyme and α-lactalbumin have similar amino acid sequences but different functions. The genes for both proteins are found in mammals, but birds have only the gene for lysozyme. What does this observation suggest about the evolution of these genes?

Rearrangements of Parts of Genes: Exon Duplication and Exon Shuffling   Unequal crossing over can lead to a gene with a duplicated exon. Exons often code for domains of a protein, and their duplication could provide a protein with enhanced structure and function. Errors in meiotic recombination could also lead to *exon shuffling* of different exons within a gene or between nonallelic genes.

How Transposable Elements Contribute to Genome Evolution   Recombination events can take place between homologous transposable element sequences that are scattered throughout the genome, causing chromosomal mutations that may occasionally be beneficial to the organism. Transposable elements that insert within a gene may disrupt its functioning; those that insert within regulatory sequences may increase or decrease gene expression. A transposable element can also move a copy of a gene or an exon to a new location. The increased genetic diversity provided by these mechanisms provides raw material for natural selection.

INTERACTIVE QUESTION   21.7

a. Explain two ways in which exon shuffling could occur.

b. What is a potential benefit of exon duplication and shuffling?

21.6   Comparing genome sequences provides clues to evolution and development

Comparing Genomes   Comparisons of *highly conserved* genes illuminate the evolutionary relationships among species that are distantly related. Such analyses support the theory that bacteria, archaea, and eukaryotes represent the three domains of life, and also demonstrate the advantages of using model organisms to study both basic biological processes and human biology.

By comparing the genomes of two closely related species, researchers can identify the small differences that translate into the phenotypic divergence of the species. The human and chimpanzee genomes differ in single nucleotide substitutions by only 1.2%. Insertions or deletions of larger regions in the genome result in an additional 2.7% difference; a third of the human duplications are not in the chimpanzee genome. In addition, there are more *Alu* elements in the human genome.

Comparisons of genetic changes since species diverged show that some genes are changing faster in humans than in the chimpanzee or mouse. Many of these more-quickly-evolving genes encode transcription factors; one example is the *FOXP2* gene, which appears to function in vocalization in vertebrates and in speech and language in humans. The recently sequenced Neanderthal genome revealed that the Neanderthal and human *FOXP2* genes encode identical proteins, suggesting that Neanderthals may have been capable of speech.

Comparisons of human genomes have revealed several million single nucleotide polymorphisms (SNPs)—single base-pair sites that vary in at least 1% of the population. Comparisons have also discovered inversions, deletions, duplications, and a surprisingly high number of *copy-number variants* (*CNVs*) in which some individuals have one or multiple copies of a gene or genetic region rather than the normal two. CNVs likely affect phenotype and may play a role in diseases and disorders. Genetic markers such as CNVs, SNPs, and STRs will contribute to the study of human evolution.

Widespread Conservation of Developmental Genes Among Animals   Biologists in the field of evolutionary developmental biology (**evo-devo**) compare developmental processes to understand how they have evolved and how minor changes in gene sequence or regulation may lead to diverse forms of life.

A sequence of 180 nucleotides called a **homeobox**, which codes for a *homeodomain*, has been found in *Drosophila* homeotic genes. The same or very similar homeobox nucleotide sequences have been identified in homeotic genes of many animals. Homeotic genes in the fruit fly and mouse are found in the same linear sequence on chromosomes. Related sequences are found in regulatory genes of yeast and plants. These similarities indicate that the homeobox sequence must have arisen early and been conserved through evolution as part of the genes involved in the regulation of gene expression and development.

Homeotic genes are often called *Hox* genes in animals. Proteins with homeodomains probably coordinate the transcription of groups of developmental genes.

Many other genes involved in development, such as those coding for components of signaling pathways, are highly conserved. The differing patterns of expression of these genes in different body areas may explain the development of animals with different body plans.

If all *Hox* genes contain the same or very similar homeobox sequences, how can they control different developmental sequences?

Word Roots

pseudo- = false (*pseudogene:* a DNA segment very similar to a real gene but that does not yield a functional product)

retro- = backward (*retrotransposon:* a transposable element that moves within a genome by means of an RNA intermediate, a transcript of the retrotransposon DNA)

Structure Your Knowledge

1. About 25% of the human genome relates to the production of proteins or RNA products (exons, introns, or regulatory sequences). Is the remaining 75% just "junk"? Describe the following types of noncoding DNA, including some of their possible functions.
   a. transposable elements
   b. *Alu* elements
   c. *L1* sequences
   d. simple sequence DNA
   e. pseudogenes

2. Describe some of the processes that contribute to genome evolution.

Test Your Knowledge

MULTIPLE CHOICE: *Choose the one best answer.*

1. Which of the following has decreased the time and expense required to sequence genomes?
   a. whole-genome shotgun approach
   b. "next-generation" sequencing techniques
   c. powerful computer programs that order the sequenced fragments
   d. All of the above contribute to faster and less expensive genome sequencing.

2. You have been given a soil sample from the Amazon rain forest. What approach would you use to identify the prokaryotic species present in the sample?
   a. culture the prokaryotes on a minimal medium
   b. extract and sequence DNA from the sample and compare the results to known DNA sequences
   c. metagenomic analysis
   d. identify known proteins in the sample

3. In the process called gene annotation, computer software can identify protein-coding genes in DNA sequences. For which of the following components is the software probably *not* scanning?
   a. RNA-splicing sites
   b. SNPs (single nucleotide polymorphisms)
   c. ESTs (expressed sequence tags) collected from cDNA sequences
   d. start or stop signals for transcription

4. Why is proteomics important in the systems biology approach?
   a. The interactions of networks of proteins are central to the functioning of cells and organisms.
   b. Determining the proteins expressed in a cell identifies the genes more accurately than can be done through genomics.
   c. The three-dimensional structure of a protein can be used to predict its function.
   d. Comparing the proteins produced by a normal allele and the allele associated with a disease can facilitate improved treatments.

5. Bacterial genes have an average length of 1,000 base pairs; human genes average about 27,000 base pairs. Which of the following statements is the best explanation for that difference?
   a. Prokaryotes have smaller, but many more, individual genes.
   b. Prokaryotes are more ancient organisms; longer genes arose later in evolution.
   c. Prokaryotes are unicellular; humans have many types of differentiated cells.
   d. Prokaryotic genes do not have introns; human genes have multiple introns.

6. Which of the following statements best explains the discovery that a complex human has roughly the same number of genes as the simple nematode *C. elegans*?
   a. The unusually long introns in human genes are involved in regulation of gene expression.
   b. More than one polypeptide can be produced from a human gene by alternative splicing.
   c. The human genome has a high proportion of noncoding DNA.
   d. The large number of copy-number variants (CNVs) in the human genome provides a great deal of genetic variability.

7. What is a pseudogene?
   a. a gene that has been rendered nonfunctional by mutation
   b. a transposon that codes for a transposase
   c. a functional but slightly altered copy of an existing gene
   d. a unique noncoding DNA

8. Which of the following techniques can be used to determine the function of a newly identified gene?
   a. comparisons with genes of known functions that have similar sequences
   b. blockage of gene function with RNAi to see the effect on the phenotype
   c. searches for similar domains of known function in other proteins
   d. all of the above

9. Which of the following is common to both retrotransposons and retroviruses?
   a. reverse transcription of DNA from RNA
   b. a membranous envelope
   c. the ability to infect other cells
   d. repeated short DNA sequences

10. The protein tissue plasminogen activator (TPA) has three types of domains. Which of the following statements best explains why one of each of these types of domains is found in three different proteins (epidermal growth factor, fibronectin, and plasminogen)?
    a. The genes for all four proteins are members of a multigene family.
    b. The gene for TPA was the first gene to evolve; the other three genes each lost two of the domains present in the ancestral TPA gene.
    c. The gene for TPA arose by exon shuffling involving the other three genes.
    d. Several exon duplications led to the evolution of the TPA gene.

11. Genes that are highly conserved are useful for
    a. identifying genes that may have been involved in speciation events.
    b. determining the function of newly identified genes.
    c. establishing the sequence of divergence of closely related species.
    d. tracing the relationships of groups that diverged early in the evolution of life.

12. An identical or very similar nucleotide sequence that has been found in developmental regulatory genes in many diverse organisms is called a
    a. homeodomain.
    b. homeobox.
    c. retrotransposon.
    d. homeotic gene.

13. Which of the following events has probably *not* contributed to genome evolution?
    a. duplications and inversions of chromosome sections
    b. highly conserved sequences
    c. recombination between transposable elements on different chromosomes
    d. exon duplication and shuffling

14. Which of the following approaches would be most useful in tracing human evolution?
    a. evo-devo and the comparison of developmental genes in plants and animals
    b. metagenomics and proteomics
    c. systems biology and the use of "knock-out" experiments
    d. analysis of single nucleotide polymorphisms and copy-number variants across individuals from the same and different populations

15. Compared to genes in mice and chimpanzees, the most rapidly evolving human genes code for
    a. α- and β-globin.
    b. lysozyme.
    c. transcription factors.
    d. ribosomal RNAs.

4 Mechanisms of Evolution

Chapter 22

Descent with Modification:
A Darwinian View of Life

Key Concepts

22.1 The Darwinian revolution challenged traditional views of a young Earth inhabited by unchanging species

22.2 Descent with modification by natural selection explains the adaptations of organisms and the unity and diversity of life

22.3 Evolution is supported by an overwhelming amount of scientific evidence

Framework

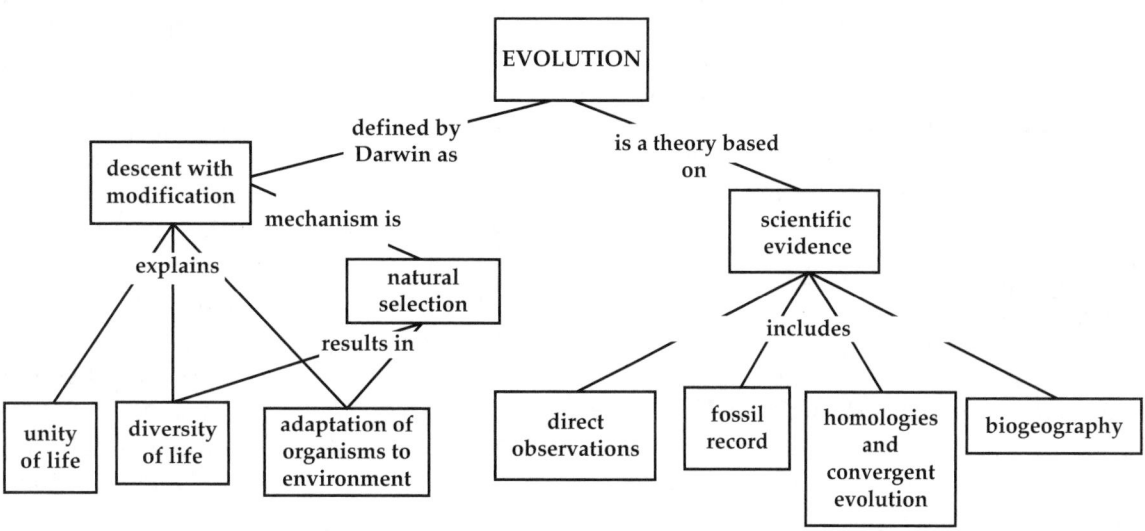

Chapter Review

Charles Darwin presented his scientific explanation for the adaptations of organisms to their environments, the diversity of life, and the unity of life in his book *On the Origin of Species*. **Evolution** may be defined in Darwin's terms as *descent with modification,* or more narrowly as the changes in a population's genetic composition over time.

Evolution can be viewed as both the *pattern* of evolutionary change observable in the natural world and the *process* or mechanisms underlying those changes.

22.1 The Darwinian revolution challenged traditional views of a young Earth inhabited by unchanging species

Scala Naturae *and Classification of Species* The Greek philosopher Aristotle proposed that all forms of life were permanent and perfect and could be arranged on a "scale of nature" of increasing complexity. The Old Testament account of creation also asserts that species are perfect and fixed.

In the 1700s, Linnaeus developed both a *binomial* system for naming organisms (genus and species) and a hierarchy of classifications grouping his named species.

Ideas About Change over Time **Fossils** are remnants or traces of past organisms, usually found in sedimentary rocks formed through the compression of layers of sand and mud into superimposed layers called **strata**. **Paleontology**, the study of fossils, originated with the work of Cuvier. He maintained that the extinctions and differences in the fossils found in different strata were the result of local sudden catastrophic events and were not indicative of evolution.

Hutton proposed that immense changes in Earth's geology are the cumulative result of slow but continuous processes. Lyell, a contemporary of Darwin, proposed that geologic processes operate at a uniform rate and continue in the present.

Darwin took two ideas from the observations of Hutton and Lyell: Earth must be very old, and very slow processes could produce substantial changes in species.

Lamarck's Hypothesis of Evolution Lamarck's hypothesis, published in 1809, explained the mechanism of evolution using two principles: the *use and disuse* of body parts, leading to their development or deterioration, and the *inheritance of acquired characteristics*. Although current genetic knowledge rejects his mechanism, Lamarck proposed the key idea that evolution is the best explanation for both the fossil record and the adaptation of organisms to the environment.

INTERACTIVE QUESTION   22.1

a. Write the capital letter representing the theory or philosophy and the lowercase letter(s) representing its proponent(s) in the blanks preceding the following seven descriptions.

A. extinctions result of catastrophic events   a. Aristotle

B. inheritance of acquired characteristics   b. Cuvier

C. gradual geologic changes   c. Darwin

D. descent with modification   d. Hutton

E. classification system   e. Lamarck

F. scale of nature   f. Linnaeus

G. uniform rate of geologic processes   g. Lyell

Theory   Proponent(s)

1. _____ _____ Describes the diversity of God's creations by naming and classifying species

2. _____ _____ The history of Earth is marked by sudden floods or droughts that resulted in extinctions

3. _____ _____ Proposed mechanism of evolution in which modifications due to use or disuse are passed on to offspring

4. _____ _____ Profound change is the product of slow but continuous processes

5. _____ _____ Earth contains fixed species on a continuum from simple to complex

6. _____ _____ All of life is related; present-day species differ from ancestral species

7. _____ _____ Geologic processes have constant rates throughout time

b. Now place the people listed in a through g in chronological order according to the dates of their ideas regarding change over time.

_____ _____ _____ _____ _____ _____ _____

22.2 Descent with modification by natural selection explains the adaptations of organisms and the unity and diversity of life

Darwin's Research Darwin was 22 years old when he sailed from Great Britain on the HMS *Beagle*. He spent the voyage collecting thousands of specimens of the fauna and flora of South America, observing the various adaptations of organisms living in very diverse habitats and making special note of the geographic distribution of the distinctly South American species. Darwin also read and was influenced by Lyell's *Principles of Geology*.

Adaptations are inherited characteristics that contribute to an organism's survival and reproduction in a specific environment. Darwin proposed that populations change through **natural selection**, a process in which individuals with beneficial characteristics produce more offspring than others because of those characteristics. In 1844 he wrote an essay on the origin of species and natural selection but did not publish it. In 1858 Darwin received Wallace's manuscript describing a nearly identical theory. Darwin then published *On the Origin of Species by Means of Natural Selection* in 1859.

***Ideas from* The Origin of Species** Darwin's concept of descent with modification explains that all organisms are related through descent from a common ancestor and develop increasing modifications as they adapt to various habitats. The history of life is analogous to a tree, with a common ancestor at the fork of each new branch and with present-day species at the tips of the youngest twigs. Most branches (probably 99% of all species that have ever lived) end in extinction, accounting for the morphological gaps sometimes present between related groups.

Darwin used the example of **artificial selection** in the breeding of domesticated plants and animals as evidence that selection among the variations present in a population can lead to substantial changes. Malthus's essay on human population growth supported Darwin's idea of the overproduction of offspring.

INTERACTIVE QUESTION 22.2

List the two observations from which Darwin drew the two inferences that explain natural selection.

Observation 1:

Observation 2:

Inference 1: Individuals whose inherited traits give them a better chance of surviving and reproducing in a given environment tend to leave more offspring.

Inference 2: This unequal survival and reproduction leads to an accumulation of favorable traits in a population over generations.

Key Features of Natural Selection Remember these three points about evolution by natural selection: (1) Natural selection results in the evolution of populations, not individuals; (2) natural selection affects only those traits that are heritable and that differ among individuals in a population; and (3) natural selection depends on the specific environmental factors present in a region at a given time. If the environment changes, different adaptations may be favored.

22.3 Evolution is supported by an overwhelming amount of scientific evidence

Direct Observations of Evolutionary Change One of the four types of data that document the pattern and process of evolution involves direct studies of evolution occurring on a short time scale. As an example, the beak length of soapberry bugs has been shown to correlate with the size of the fruits that house the seeds on which they feed. Beak lengths have changed as populations of soapberry bugs have become adapted to introduced plant species, often in relatively short amounts of time.

The development of new antibiotics is usually followed rapidly by the evolution of resistance to them. Antibiotic resistance has developed in a number of pathogenic bacteria, making treatment of some infections very difficult. Current strains of MRSA—methicillin-resistant *Staphylococcus aureus*—are resistant to multiple antibiotics and can cause potentially lethal infections.

The evolution of drug-resistant bacteria and changes in soapberry bug beak length illustrate three facets of natural selection: It is an editing, not a creative, mechanism that selects for variations already present in a population. It is regionally and temporally specific, selecting for traits that are beneficial in the local environment and at the current time. And it can result in rapid changes in populations, particularly in organisms that reproduce rapidly.

INTERACTIVE QUESTION   22.3

a. Explain how the rapid evolution of drug resistance in bacteria is an example of natural selection.

b. How might multidrug-resistant strains of MRSA have evolved?

Homology   **Homology** is similarity resulting from common ancestry. The forelimbs of all mammals are **homologous structures**, containing the same skeletal elements regardless of function or external shape. Comparative anatomy and embryology illustrate that evolution is a remodeling process in which ancestral structures become modified for new functions.

Vestigial structures, which may be of little value to the organism, are historical remnants of ancestral structures.

Homologies can also be seen on a molecular level. DNA and an essentially universal genetic code, which have been passed along through all branches of evolution, are important evidence that all forms of life descended from common ancestors and are thus related. Homologous genes are found across the wide diversity of living organisms.

Homology is evident on different hierarchical levels, reflecting evolutionary history in a nested pattern of descent. An **evolutionary tree** is a diagram of evolutionary relationships. Each branch point represents the common ancestor of all the species beyond that point. Evolutionary trees are hypotheses based on the best available data.

Distantly related organisms may appear similar as a result of **convergent evolution**, the independent evolution of similar characteristics. These **analogous** structures arise as a result of evolutionary adaptation to similar environments, rather than through descent from a recent common ancestor.

The Fossil Record   The fossil record demonstrates that present and past organisms differ, and that many species have become extinct. Fossils also allow us to trace the evolution of new groups, as in the origin of whales from land mammals.

Biogeography   The study of the geographic distributions of species, or **biogeography**, provides a fourth type of evidence of evolution. The distribution of organisms has been influenced by *continental drift*, the slow movements of Earth's continents. The single landmass called **Pangaea** formed 250 million years ago and began to break apart 200 million years ago. The gradual separation of continents helps to explain and predict where fossils of different groups and their descendants are found.

Islands often have **endemic** species found nowhere else and usually closely related to species on the nearest island or mainland. Widely separated areas having similar environments are not likely to be populated by closely related species. Rather, each area is more likely to have species that are taxonomically related to those of their region, regardless of environment.

INTERACTIVE QUESTION   22.4

Complete the following concept map that summarizes the main sources of evidence for evolution.

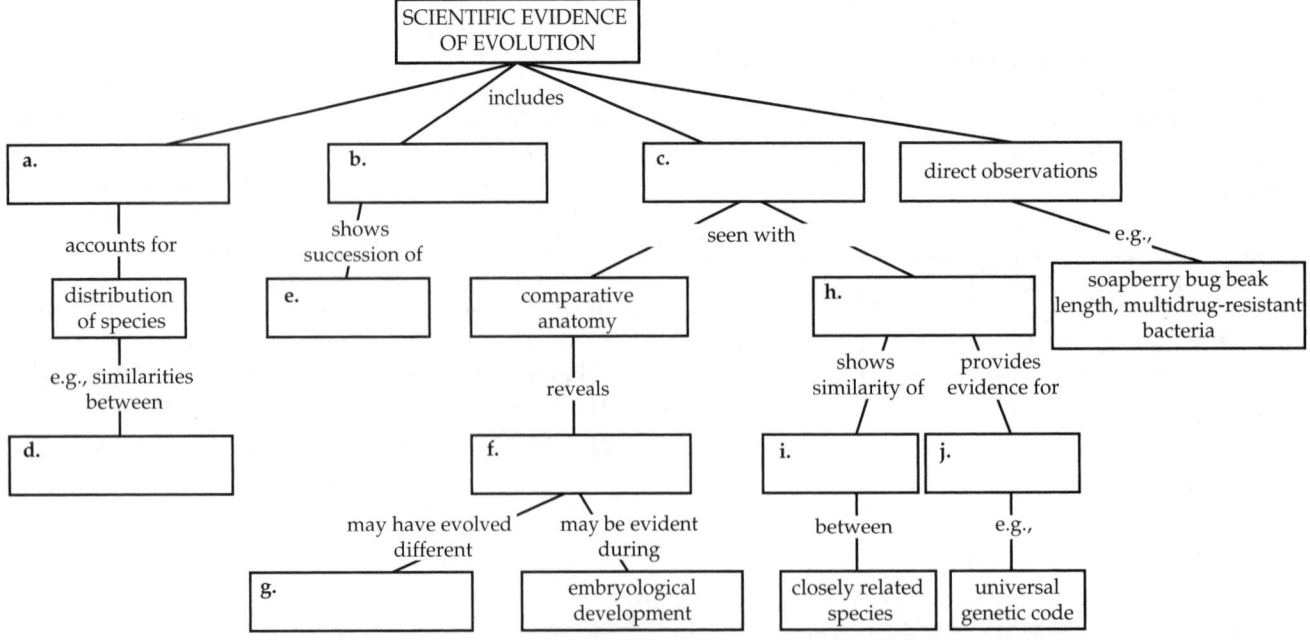

What Is Theoretical About the Darwinian View of Life? A scientific *theory* is a unifying concept with broad explanatory power and with predictions that have been and continue to be tested by experiments and observations. The *pattern* of evolution has been established using a great deal of evidence, ranging from fossils to embryology to DNA sequences. The *process* of natural selection makes sense of many otherwise unexplained observations.

Word Roots

bi- = life; **ge-** = the Earth (*biogeography*: the scientific study of the past and present distribution of species)

end- = within (*endemic*: referring to a species that is confined to a specific geographic area)

hom- = like, resembling (*homology*: similarity in characteristics resulting from a shared ancestry)

pale- = ancient (*paleontology*: the scientific study of fossils)

Structure Your Knowledge

1. Explain in your own words the main components of Darwin's theory of evolution.
2. Briefly explain the following statement: In Darwin's model of descent with modification, individuals are selected, but populations evolve.

Test Your Knowledge

MULTIPLE CHOICE: *Choose the one best answer.*

1. Which of the following slogans best describes the process of natural selection?
   a. Survival of the fittest.
   b. Do unto others before they do unto you.
   c. At the end of the game whoever has the most offspring wins.
   d. Nature, red in tooth and claw.

2. Which of the following scientists proposed that geological and biological changes are the result of catastrophic events?
   a. Cuvier
   b. Hutton
   c. Lyell
   d. Darwin

3. Darwin proposed that new species evolve from ancestral forms by the
   a. gradual accumulation of adaptations to changing or different environments.
   b. struggle for limited resources.
   c. accumulation of mutations.
   d. excessive production of offspring.

4. Each of the following influenced Darwin as he synthesized the theory of evolution by natural selection *except*
   a. the biogeographic distribution of species such as the mockingbirds on the Galápagos Islands.
   b. Lyell's book, *Principles of Geology*, on the gradualness of geologic changes.
   c. examples of artificial selection that produce rapid changes in domesticated species.
   d. Mendel's paper in which he described his "laws of inheritance."

5. Darwin's publication of *On the Origin of Species* was accelerated by similar ideas developed by
   a. Lamarck
   b. Fitzroy
   c. Lyell
   d. Russel

6. The smallest unit that can evolve is a(n)
   a. individual.
   b. mating pair.
   c. species.
   d. population.

7. Which of the following statements is *not* considered part of the process of natural selection?
   a. More offspring are produced than are able to survive and reproduce.
   b. Individuals with traits best adapted to the environment are likely to leave more offspring.
   c. Many adaptive traits may be acquired during an individual's lifetime, contributing to that individual's reproductive success.
   d. Unequal reproductive success leads to gradual change in a population.

8. Darwin observed that organisms are capable of producing more offspring than can survive and reproduce. This idea was influenced by Darwin's reading of
   a. Hutton.
   b. Linnaeus.
   c. Malthus.
   d. Lamarck.

9. What might you conclude from the observation that the bones in your arm and hand are similar to the bones that make up a bat's wing?
   a. The bones in the bat's wing are vestigial structures, no longer useful as "arm" bones.
   b. The bones in a bat's wing are homologous to your arm and hand bones.
   c. Bats lost their opposable digits during the course of evolution.
   d. Our ancestors could fly.

10. The remnants of pelvic and leg bones in a snake
    a. are vestigial structures.
    b. show that lizards evolved from snakes.
    c. are homologous structures.
    d. provide evidence for loss of acquired characteristics.

11. The hypothesis that whales evolved from land-dwelling ancestors is supported by
    a. evidence from the biogeographic distribution of whales.
    b. molecular comparisons of whales, fish, and reptiles.
    c. the ability of captive whales to be trained to walk.
    d. fossils of extinct whales that had increasingly reduced hind limbs.

12. Which of the following sources of evidence provides the best support for Darwin's claim that all of life is descended from a common ancestor?
    a. the fossil record
    b. comparative embryology
    c. the classification system of taxonomy
    d. molecular biology

13. When cytochrome *c* molecules are compared, yeasts and molds are found to differ by approximately 46 amino acids per 100 residues (amino acids in the protein); insects and vertebrates are found to differ by 29 amino acids per 100 residues. What can one conclude from these data?
    a. Very little, unless the DNA sequences for the cytochrome *c* genes are compared.
    b. Insects and vertebrates diverged from a common ancestor more recently than did yeasts and molds.
    c. Yeasts and molds diverged from a common ancestor more recently than did insects and vertebrates.
    d. The evolution of cytochrome *c* occurred more rapidly in yeasts and molds than in insects and vertebrates.

14. Which of the following is an example of convergent evolution?
    a. the evolution of multiple-drug resistance in MRSA
    b. similarities between the marsupial Tasmanian wolf and the eutherian North American wolf
    c. two very different plants that are found in different habitats but evolved from a fairly recent common ancestor
    d. the many different bill sizes and shapes of finches on the Galápagos Islands

Use the following evolutionary tree representing the relationships among a group of vertebrates to answer questions 15 and 16. The letters at each branch point indicate the common ancestor for groups beyond that point.

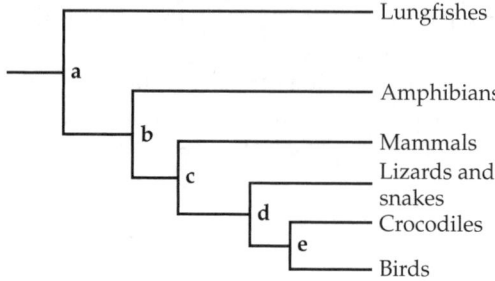

15. Which was the last common ancestor of crocodiles and the lineage of lizards and snakes?

16. Which of the following can be concluded from this evolutionary tree?
    a. Mammals are more closely related to amphibians than they are to birds.
    b. Mammals are more closely related to lizards and snakes than they are to birds.
    c. Birds are more closely related to lizards and snakes than they are to mammals.
    d. Birds and mammals are more closely related to each other than birds are to lizards and snakes.
    e. Lungfishes are not related to any of these groups.

17. Evolution by natural selection would be impossible in a population
    a. that has been removed from its natural environment.
    b. in which all individuals have the same number of surviving and fertile offspring.
    c. of microorganisms that are haploid and have a very short life span.
    d. that is not increasing because it has reached the environmental limits to its growth.

The Evolution of Populations

Framework

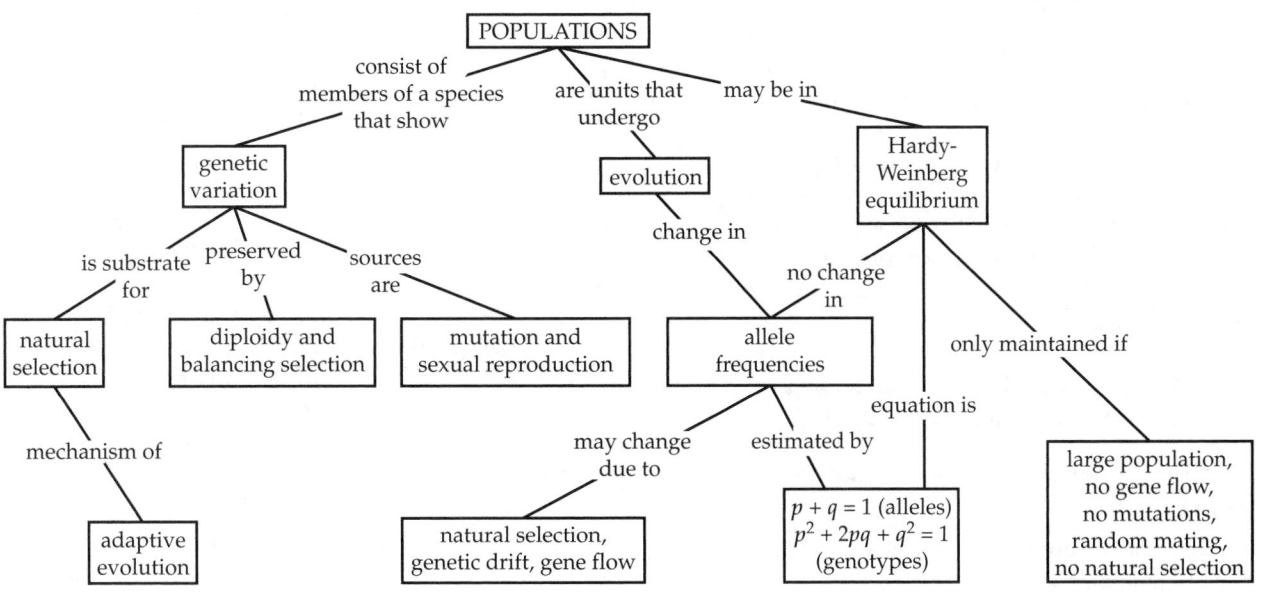

Chapter Review

Although individuals are selected for or against by natural selection, it is *populations* that evolve. **Microevolution** is defined as changes in the allele frequencies of a population from generation to generation. Although other mechanisms can result in microevolution, natural selection is the principal engine of evolutionary adaptation.

23.1 Genetic variation makes evolution possible

Genetic Variation   Genetic variation—differences in genes or other DNA sequences among individuals—is the raw material for natural selection. Some characters vary as distinct, either-or phenotypes and are usually determined by a single gene locus. Other phenotypic differences vary along a continuum and are usually affected by two or more gene loci.

Measures of genetic variation in a population include *gene variability*, the average percent of loci that are heterozygous, and *nucleotide variability*, differences in nucleotide base-pair sites between individuals of a population.

Sources of Genetic Variation   New alleles can originate by *mutation*, a change in the sequence of nucleotides in DNA. Mutations that occur in somatic cells in animals cannot be passed on to the next generation. Point mutations that occur in noncoding DNA or do not change the amino acid sequence of a protein usually result in **neutral variation**, genetic variation that does not affect selection. Mutations that do alter the phenotype are usually harmful. Rarely, however, a mutant allele may actually enhance an individual's reproductive success.

Chromosomal mutations are generally deleterious. Duplication of small DNA segments, introduced by transposable elements or errors in cell division, may provide extra loci that could eventually take on new functions by mutation.

Mutation rates in animals and plants average about one in every 100,000 genes per generation. Mutation produces genetic variation very rapidly in prokaryotes and viruses due to their short generation spans.

In a sexually reproducing population, most of the genetic variation comes from the reshuffling of alleles into new combinations in each individual offspring. The processes of crossing over and independent assortment in meiosis and random combination of gametes in fertilization create unique genetic combinations in offspring.

INTERACTIVE QUESTION   23.1

a. What is a major source of genetic variation for prokaryotes and viruses?

b. What is the major source of genetic variation for plants and animals?

c. Explain why your answers to a and b are different.

23.2 **The Hardy-Weinberg equation can be used to test whether a population is evolving**

Gene Pools and Allele Frequencies   A **population** is a localized group of members of a species capable of interbreeding to create fertile, viable offspring. The

gene pool is the term for all the alleles at all the loci present in a population. If all individuals in a population are homozygous for the same allele, the allele is said to be *fixed*. More often, two or more alleles are present in the gene pool in some relative proportion or frequency. In a case with two alleles at a particular gene locus, the letters p and q represent the frequencies of the two alleles within the population, and their combined frequencies must equal 1:
$p + q = 1$.

INTERACTIVE QUESTION   23.2

In a population of 200 mice, 98 are homozygous dominant for brown fur (*BB*), 84 are heterozygous (*Bb*), and 18 are homozygous recessive for white fur (*bb*).

a. The genotype frequencies of this population are

_____*BB* _____*Bb* _____*bb*.

b. The allele frequencies of this population are

_____*B* allele _____*b* allele.

The Hardy-Weinberg Equation   If only Mendelian segregation and recombination of alleles in sexual reproduction are involved, the frequencies of alleles and genotypes in a population will remain constant from one generation to the next. Such a nonevolving gene pool is said to be in **Hardy-Weinberg equilibrium**.

The allele frequency within a population determines the proportion of gametes that will contain that allele. The random combination of gametes will yield offspring with genotypes that reflect and reconstitute the allele frequencies of the previous generation.

With the equation for Hardy-Weinberg equilibrium, the frequencies of genotypes in the next generation can be calculated from the probability of each combination of alleles. According to the rule of multiplication, the probability that two gametes containing the same allele will come together is equal to $(p \times p)$ or p^2, or to $(q \times q)$ or q^2. A p allele and a q allele can combine in two different ways, depending on which parent contributes which allele; therefore, the frequency of a heterozygous offspring is equal to $2pq$. The sum of the frequencies of all possible genotypes in the population adds up to 1:
$p^2 + 2pq + q^2 = 1$.

INTERACTIVE QUESTION 23.3

Use the allele frequencies you determined in Interactive Question 23.2 to predict the genotype frequencies of the next generation.

Frequencies of

B (p) = _____ b (q) = _____

BB = p^2 = _____ Bb = 2pq = _____ bb = q^2 = _____

Hardy-Weinberg equilibrium is maintained only if all of the following five conditions are met: (1) no mutations, (2) random mating (because nonrandom mating changes genotype frequencies), (3) no natural selection (no unequal survival and reproductive success), (4) an extremely large population to offset chance fluctuations (called genetic drift), and (5) no gene flow or movement of alleles into or out of the population.

If the frequency of homozygous recessive individuals is known (q^2), then the frequency of q may be estimated as the square root of q^2 (assuming the population is in Hardy-Weinberg equilibrium for that gene).

INTERACTIVE QUESTION 23.4

Practice using the Hardy-Weinberg equation so that you can easily determine genotype frequencies from allele frequencies, and vice versa.

a. The allele frequencies in a population are A = 0.6 and a = 0.4. Predict the genotype frequencies for the next generation.

AA _____ Aa _____ aa _____

b. What would the allele frequencies be for the generation you predicted in part a?

A _____ a _____

c. Suppose you are able to determine the actual genotype frequencies in the population and find that these frequencies differ significantly from what you predicted in part a. What would such results indicate?

23.3 Natural selection, genetic drift, and gene flow can alter allele frequencies in a population

Natural Selection Individuals with traits that are better suited to their environment tend to be more successful in producing offspring, and they pass their alleles for these adaptive traits to the next generation in disproportionate numbers, resulting in **adaptive evolution**.

Genetic Drift Chance deviations from expected results are more likely to occur in a small sample. Fluctuation in a population's allele frequencies from one generation to the next as a result of chance events is called **genetic drift**.

Genetic drift that occurs when only a few individuals colonize a new area is known as the **founder effect**. Allele frequencies in the small sample are unlikely to be representative of the parent population.

The **bottleneck effect** occurs when some disaster or other factor reduces population size dramatically, and the few surviving individuals are unlikely to represent the genetic makeup of the original population. Genetic drift will remain a factor until the population grows large enough for chance events to have less of an impact.

In summary, genetic drift has a larger effect on small populations, causes gene frequencies to randomly fluctuate over time, can lead to the loss of genetic variation within populations, and can cause fixation of harmful alleles.

Gene Flow **Gene flow**, the migration of individuals or the transfer of gametes between populations, may change allele frequencies. Differences in allele frequencies between populations tend to be reduced by gene flow. Gene flow may transfer alleles that either improve or reduce the ability of populations to adapt to local conditions.

INTERACTIVE QUESTION   23.5

Fill in the following concept map that summarizes three causes of microevolution. Even better, create your own concept map to help you review microevolution.

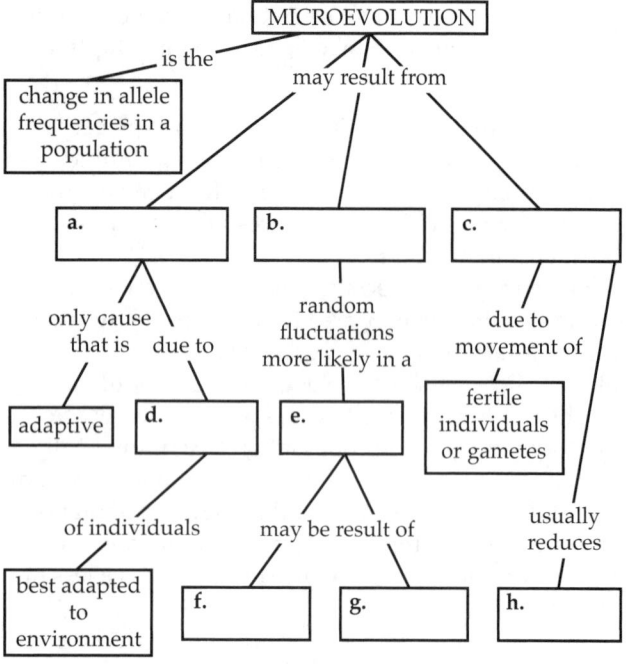

23.4 Natural selection is the only mechanism that consistently causes adaptive evolution

Natural Selection: A Closer Look   **Relative fitness** is a measure of an individual's contribution to the gene pool of the next generation, relative to the contributions of others.

The frequency distribution of a trait may be affected by three modes of selection. **Directional selection** occurs most frequently during periods of environmental change, when individuals on one end of a phenotypic range may be favored. **Disruptive selection** occurs when the environment favors individuals on both extremes of a phenotypic range. **Stabilizing selection** acts against extreme phenotypes and favors more intermediate forms, tending to reduce phenotypic variation.

The Key Role of Natural Selection in Adaptive Evolution   Because natural selection favors certain genotypes, it is a nonrandom process. As natural selection increases the frequencies of alleles that enhance survival and reproduction, the match between a species and its environment increases.

Sexual Selection   **Sexual selection** is the selection for characteristics that enhance an individual's chances of obtaining mates. Sexual selection can lead to **sexual dimorphism**, the distinction between males and females on the basis of secondary sexual characteristics. Sexual selection may involve **intrasexual selection**, in which individuals of the same sex compete for mates, or **intersexual selection**, in which individuals of one sex (usually female) discriminate in choosing a mate. Also called *mate choice*, intersexual selection may be based on showy traits that reflect the general health of the male, and thus the quality of his genes.

Balancing Selection   How is it that genetic variation is not eliminated as natural selection continually selects for the most favorable alleles? The diploidy of most eukaryotes maintains genetic variation by hiding recessive alleles in heterozygotes, enabling those alleles to persist in the population.

Two types of **balancing selection** can maintain stable frequencies of two or more phenotypes in a population. In **frequency-dependent selection**, a phenotype's reproductive success declines if it becomes too common in the population. **Heterozygote advantage** tends to maintain two or more alleles at a locus when heterozygotes have survival and reproductive advantages.

INTERACTIVE QUESTION   23.6

a. Why hasn't the highly deleterious sickle-cell allele been selected against and eliminated from the gene pool of the U.S. population?

b. Why is this allele at such a relatively high frequency in the gene pool of some African populations?

Why Natural Selection Cannot Fashion Perfect Organisms   Natural selection can act only on variations that are available; new alleles do not arise as they are needed. Each species has evolved from a long line of ancestral forms, many of whose structures have been co-opted for new situations. Adaptations are often compromises between the need to do different things, such as swim and walk. And finally, chance events affect a population's evolutionary history.

Word Roots

inter- = between (*intersexual selection:* a form of natural selection in which individuals of one sex (usually the females) are choosy in selecting their mates from the other sex; also called mate choice)

intra- = within (*intrasexual selection:* a form of natural selection in which there is direct competition among individuals of one sex for mates of the opposite sex)

micro- = small (*microevolution:* a change in the allele frequencies of a population over time)

Structure Your Knowledge

1. **a.** What is Hardy-Weinberg equilibrium?
   b. Define the variables of the equation for Hardy-Weinberg equilibrium. Make sure you can use this equation to determine allele frequencies and predict genotype frequencies.

2. It seems that natural selection would work toward genetic unity; individuals with the genotypes that are most fit produce the most offspring, increasing the frequency of adaptive alleles and eliminating less beneficial alleles from the population. Yet there remains a great deal of variability within populations of a species. Describe some of the factors that contribute to this genetic variability.

Test Your Knowledge

MULTIPLE CHOICE: *Choose the one best answer.*

1. Mutations are rarely a direct source for microevolution in eukaryotes because they
   a. are most often harmful and do not get passed on.
   b. do not directly produce most of the genetic variation present in a diploid population.
   c. occur very rarely and are passed on only when they occur in gametes.
   d. All of the above are true.

2. A change in allele frequencies is also known as
   a. microevolution.
   b. contingency.
   c. neutral variation.
   d. natural selection.

3. A scientist observes that the height of a certain species of asters decreases as the altitude on a mountainside increases. She gathers seeds from samples at various altitudes, plants them in a uniform environment, and measures the height of the new plants. All of her experimental asters grow to approximately the same height. From this she concludes that
   a. height is not a quantitative trait.
   b. the trend in height that she observed was due to genetic variations.
   c. the differences in the parent plants' heights were due to directional selection.
   d. the height variation she initially observed was an example of environmental influence.

4. Mice have an estimated 1,000 olfactory receptor genes. This is most likely a result of
   a. gene flow.
   b. gene duplication.
   c. neutral variation.
   d. directional selection.

5. Which of the following provides most of the genetic variation in plant and animal populations?
   a. mutations
   b. sexual reproduction
   c. sexual selection
   d. recessive masking in heterozygotes

6. In the Hardy-Weinberg equation, what does the term "$2pq$" represent?
   a. the frequency of the heterozygote genotype
   b. the frequency of the dominant phenotype
   c. the frequency of the recessive homozygote genotype
   d. the frequency of the dominant allele

7. If a population has the following genotype frequencies—$AA = 0.42$, $Aa = 0.46$, and $aa = 0.12$—what are the allele frequencies?
   a. $A = 0.42; a = 0.12$
   b. $A = 0.6; a = 0.4$
   c. $A = 0.65; a = 0.35$
   d. $A = 0.84; a = 0.24$

8. In a population with two alleles, B and b, the allele frequency of b is 0.4. What would be the frequency of heterozygotes if the population is in Hardy-Weinberg equilibrium?
   a. 0.16
   b. 0.24
   c. 0.48
   d. You cannot tell from this information.

9. In a population that is in Hardy-Weinberg equilibrium for two alleles, C and c, 16% of the population shows a recessive trait. Assuming C is dominant to c, what percent shows the dominant trait?
   a. 36%
   b. 48%
   c. 60%
   d. 84%

10. In a study of a population of field mice, you find that 48% of the mice have a coat color that indicates that they are heterozygous for a particular gene. What would be the frequency of the dominant allele in this population?
    a. 0.24
    b. 0.48
    c. 0.60
    d. You cannot estimate allele frequency from this information.

11. In a random sample of a population of shorthorn cattle, 73 animals were red ($C^R C^R$); 63 were roan, a mixture of red and white ($C^R C^r$); and 13 were white ($C^r C^r$). Estimate the allele frequencies of C^R and C^r, and explain whether or not the population is in Hardy-Weinberg equilibrium.
    a. $C^R = 0.64$, $C^r = 0.36$; because the population is large and a random sample was chosen, the population is in equilibrium.
    b. $C^R = 0.7$, $C^r = 0.3$; the genotype ratio is not what would be predicted from these frequencies, and the population is not in equilibrium.
    c. $C^R = 0.7$, $C^r = 0.3$; the genotype ratio is close to what would be predicted from these frequencies, and the population is in equilibrium.
    d. $C^R = 1.04$, $C^r = 0.44$; the allele frequencies add up to greater than 1, and the population is not in equilibrium.

12. Genetic drift is likely to be seen in a population
    a. that has a low mutation rate.
    b. in which natural selection is occurring.
    c. that is very small.
    d. for which environmental conditions are changing.

13. Porphyria variegata is a genetic disease characterized by red urine, light sensitivity, and liver damage. This is normally a rare disorder, but it is quite common among Afrikaners of South Africa. All Afrikaners are descended from about 30 families that settled in South Africa in the seventeenth century. The high frequency of porphyria variegata is a result of
    a. the founder effect.
    b. natural selection.
    c. new mutations.
    d. heterozygote advantage.

14. Cystic fibrosis is a very serious genetic disorder that occurs in homozygotes for a recessive allele. Although the disease is extremely serious, this recessive allele appears with surprisingly high frequency in Caucasian populations. It has been suggested that siblings of individuals with cystic fibrosis may have some protection from certain bacterial infections of the gut. If this is true, then the recessive allele is maintained at high frequency by
    a. the founder effect.
    b. new mutations.
    c. heterozygote advantage.
    d. frequency-dependent selection.

15. Which of the following mechanisms would lead to the presence of two distinct phenotypic classes within a population?
    a. disruptive selection
    b. heterozygote advantage
    c. directional selection
    d. frequency-dependent selection

16. Sexual selection
    a. selects for traits that enhance an individual's chance of mating.
    b. increases the size of individuals.
    c. results in individuals better adapted to the environment.
    d. selects for traits that increase fertility.

17. Careful studies have shown that robin eggs of intermediate size are more likely to produce viable offspring than smaller or larger eggs. This is an example of
    a. directional selection.
    b. disruptive selection.
    c. stabilizing selection.
    d. genetic drift.

18. Mice that are homozygous for a lethal recessive allele die shortly after birth. In a large breeding colony of mice, you find that a surprising 5% of all newborns die from this trait. In checking lab records, you discover that the same proportion of offspring have been dying from this trait in this colony for the past three years. (Mice breed several times a year and have large litters.) How might you explain the persistence of this lethal allele at such a high frequency?
    a. Homozygous recessive mice have a reproductive advantage.
    b. A large mutation rate keeps producing this lethal allele.
    c. There is some sort of heterozygote advantage and perhaps selection against the homozygous dominant trait.
    d. Because this is a diploid species, the recessive allele cannot be selected against when it is in the heterozygote.

19. If an allele is recessive and lethal in homozygotes shortly after birth, then
    a. the allele is present in the population at a frequency of 0.001.
    b. the relative fitness of the homozygous recessive genotype is 0.
    c. the allele will most likely remain in the population at a low frequency because it cannot be selected against when in a heterozygote.
    d. Both **b** and **c** are correct.

20. Which of these types of selection is *mismatched* with its example?
    a. disruptive—a population of black-bellied seed-crackers consists of birds with either small bills (more effective at eating soft seeds) or large bills (able to crack hard seeds)
    b. intrasexual—elephant seal males are more than four times larger than females; males fight over areas of beach where females congregate during breeding season
    c. stabilizing—the frequencies of A, B, AB, and O blood groups remain constant in a population
    d. frequency-dependent—as fish of one coloration become more numerous, predators form a "search image" for that coloration and preferentially feed on them

21. In an area of erratic rainfall, a biologist found that, in a certain wildflower population, homozygous plants with alleles for curled leaves reproduced better in dry years, whereas homozygous plants with alleles for flat leaves reproduced better in wet years. This situation would tend to
    a. result in gene flow in the wildflower population.
    b. lead to balancing selection as a result of the heterozygote advantage of wildflowers with both alleles.
    c. preserve genetic variation in this population.
    d. lead to disruptive selection in years of high rainfall.

The Origin of Species

Framework

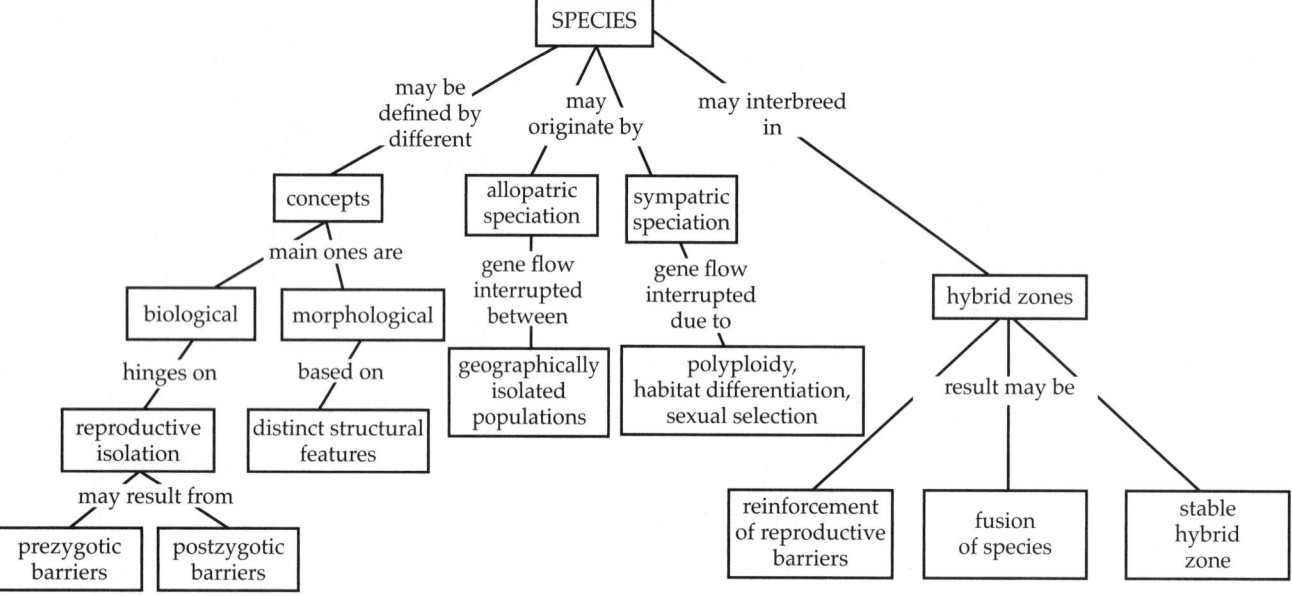

Chapter Review

Speciation, the splitting of one species into two or more species, creates the diversity of life while maintaining the relatedness or unity of life. **Microevolution** involves evolutionary changes in allele frequencies in a population. **Macroevolution** considers changes above the species level, such as the origin of new taxonomic groups.

24.1 The biological species concept emphasizes reproductive isolation

Different kinds of organisms are most often characterized by their physical form or morphology, although differences in physiology, biochemistry, and DNA sequences also support the existence of distinct species.

The Biological Species Concept According to the **biological species concept**, a **species** is a group of populations whose members have the potential to interbreed in nature and produce viable, fertile offspring, but do not successfully interbreed with members of other species. *Gene flow* between populations tends to hold a species together genetically.

The **reproductive isolation** that is necessary for the formation of a new species results from barriers that prevent individuals of different species from producing viable, fertile offspring, or **hybrids**. **Prezygotic barriers** function before the formation of a zygote by preventing mating between members of different species or successful fertilization should gametes meet. Should a hybrid zygote form, **postzygotic barriers** may prevent it from developing into a viable, fertile adult.

INTERACTIVE QUESTION 24.1

For the following examples, indicate whether it is a prezygotic or a postzygotic barrier and name the type of reproductive isolation.

Pre- or Post-	Type of Isolation	Example
a.		Two species of frogs mate in a laboratory setup and produce viable but sterile offspring.
b.		Two species of sea urchins release gametes at the same time, but the sperm fail to fuse with eggs of a different species.
c.		The genital openings of two species of land snails cannot line up because their shells spiral in opposite directions.
d.		Two species of short-lived mayflies emerge during different weeks in spring.
e.		Two species of salamanders mate and produce offspring, but the hybrid's offspring are sterile.
f.		Two similar species of birds have different mating rituals.
g.		Embryos of two species of mice bred in the lab usually abort.
h.		Peepers breed in woodland ponds; leopard frogs breed in swamps.

Although it is very useful, the biological species concept has some important limitations. It does not work for species that are asexual, and reproductive isolation cannot be determined for extinct species. Also, gene flow does occur between some species that remain morphologically and ecologically different.

Other Definitions of Species   Most species have been identified on the basis of physical characteristics, an approach called the **morphological species concept**. The **ecological species concept** defines species on the basis of their ecological niche, their interactions with the biotic and abiotic factors in the specific environments in which they are found.

INTERACTIVE QUESTION 24.2

Fill in the following table to review three of the approaches that biologists have proposed for conceptualizing a species.

Concept	Emphasis
biological	a.
b.	anatomical differences, most commonly used
c.	unique roles in specific environments

24.2 Speciation can take place with or without geographic separation

Allopatric ("Other Country") Speciation   **Allopatric speciation** may occur when geographic isolation interrupts gene flow between two subpopulations. Geographic separation alone is not a reproductive barrier in the biological sense. Intrinsic reproductive barriers may arise coincidentally as allopatric populations go down separate evolutionary paths due to mutation, genetic drift, and natural selection. Evidence of allopatric speciation has been found in both field and laboratory studies.

Sympatric ("Same Country") Speciation   **Sympatric speciation** occurs when reproductive barriers prevent gene flow between populations that share the same area.

Mistakes during cell division may lead to **polyploidy**, the presence of extra sets of chromosomes. An **autopolyploid** has more than two sets of chromosomes that have all come from the same species. Failures in cell division can produce tetraploids ($4n$), which can fertilize themselves or other tetraploids but cannot reproduce with diploids from the parent population, resulting in reproductive isolation in just one generation.

Polyploid species may also arise when two different species interbreed. The resulting hybrids may propagate asexually but are usually sterile due to difficulties in the meiotic production of gametes. Future cell division mistakes, however, can result in the production of a fertile **allopolyploid**.

Polyploid speciation has been frequent and important in plant evolution. Researchers now hybridize plants by inducing meiotic and mitotic errors to create new species.

INTERACTIVE QUESTION   24.3

a. A new plant species B forms by autopolyploidy from species A, which has a chromosome number of $2n = 10$. How many chromosomes would species B have?

b. If species A were to hybridize with species C ($2n = 14$) and produce a new allopolyploid species D, how many chromosomes would species D have?

Sexual selection may lead to sympatric speciation. In a laboratory study of two sympatric species of cichlids, mate choice based on coloration was shown to be the reproductive barrier that normally separates the two species.

Sympatric speciation in animals may involve isolation within the geographic range of the parent population based on different resource usage. Natural selection for traits that maximize usage of these different resources may lead to the evolution of other reproductive barriers.

Allopatric and Sympatric Speciation: A Review
New species form by allopatric speciation (involving geographic isolation) or by sympatric speciation (involving reproductive isolation within a population).

INTERACTIVE QUESTION   24.4

a. Differentiate between allopatric and sympatric speciation.

b. How might reproductive barriers arise in each type of speciation?

24.3   Hybrid zones reveal factors that cause reproductive isolation

Patterns Within Hybrid Zones   Areas in which members of different species come into contact and successfully interbreed are called **hybrid zones**. In the hybrid zone found between the yellow-bellied toad and the fire-bellied toad, the frequencies of alleles specific to one species range from 100% near that species' edge of the hybrid zone to 0% at the edge near the other species. What keeps these hybrids from introducing alleles of each other's species into the parent populations? Their poor survival and reproduction mean that these hybrids rarely produce viable offspring with the parent species.

Hybrid Zones and Environmental Change   Changes to the environment have the potential to create or alter hybrid zones. Hybrid zones may also be a means of transferring alleles from one species to another.

Hybrid Zones over Time   In a process called **reinforcement**, natural selection may strengthen prezygotic barriers to reproduction when hybrids are less fit than members of the parent species. In other cases, weak reproductive barriers may allow sufficient gene flow that barriers break down even more, resulting in the *fusion* of the two hybridizing species into a single species. Many hybrid zones exhibit stability, with hybrids continuing to be formed but with the two parent gene pools remaining separate.

INTERACTIVE QUESTION   24.5

Use the following diagrams to explain the three possible outcomes for a hybrid zone over time.

a.

b.

c.

24.4 Speciation can occur rapidly or slowly and can result from changes in few or many genes

The Time Course of Speciation   In the fossil record, new forms often appear rather suddenly, persist unchanged for a long time, and then disappear. **Punctuated equilibria** is the term used to describe these long periods of stasis punctuated by episodes of relatively rapid speciation and change. For other species, the fossil record indicates a much more gradual divergence.

Various studies suggest that once divergence begins, the process of speciation may be relatively rapid. For example, the chromosomes of experimentally produced hybrid sunflowers came to resemble those of the wild sunflower hybrid species in only five generations of breeding in the lab.

INTERACTIVE QUESTION 24.6

If the process of speciation appears to occur relatively rapidly, why don't we see new species evolving all the time? And how does your answer relate to the concern about the current high rate of extinction of Earth's species?

Studying the Genetics of Speciation   Researchers have been able to identify the genes that play a key role in speciation in some organisms. For example, reproductive isolation in two species of monkey flowers has been linked to a few genes, including a flower-color gene that affects pollinator preference.

From Speciation to Macroevolution   As species diverge and diversify, some differences may accumulate that lead to the formation of whole new groups of organisms, becoming part of the large-scale evolutionary changes of macroevolution.

Word Roots

all- = other; **patr-** = father (*allopatric speciation:* the formation of new species in populations that are geographically isolated from one another)

auto- = self; **poly-** = many (*autopolyploid:* an individual that has more than two chromosome sets that are all derived from a single species)

macro- = large (*macroevolution:* evolutionary change above the species level)

post- = after (*postzygotic barrier:* a reproductive barrier that prevents hybrid zygotes produced by two different species from developing into viable, fertile adults)

sym- = together (*sympatric speciation:* the formation of new species in populations that live in the same geographic area)

Structure Your Knowledge

1. How are speciation and microevolution different?
2. In which type of speciation, sympatric or allopatric, would you expect reproductive barriers to arise more easily? Explain your answer.
3. What does the term *punctuated equilibria* describe?

Test Your Knowledge

MULTIPLE CHOICE: *Choose the one best answer.*

1. Which of the following types of isolation is *not* a type of *intrinsic* reproductive isolation?
   a. behavioral
   c. gametic
   b. geographic
   d. temporal

2. Which concept of species applies to the work of both Linnaeus, who first named organisms according to genus and species, and Darwin?
   a. biological
   c. molecular
   b. ecological
   d. morphological

3. For which of the following is the biological species concept *least* appropriate?
   a. plants
   c. prokaryotes
   b. animals
   d. fungi

4. Which of the following is *not* a mechanism associated with sympatric speciation?
   a. geographic isolation
   b. autopolyploidy
   c. habitat differentiation
   d. sexual selection

5. A horse ($2n = 64$) and a donkey ($2n = 62$) can mate and produce a mule. How many chromosomes would there be in a mule's cells?
   a. 62
   c. 64
   b. 63
   d. 126

6. Which mechanism leads to the origin of new species after a geographical barrier splits a population into two isolated populations?
   a. allopatric speciation
   b. sympatric speciation
   c. gene flow
   d. polyploidy

7. A tetraploid plant species (with four identical sets of chromosomes) is probably the result of
   a. allopolyploidy.
   b. autopolyploidy.
   c. hybridization and nondisjunction.
   d. both a and c.

8. Prezygotic barriers to reproduction between related species are often strengthened within hybrid zones. This phenomenon is called
   a. reinforcement.
   b. sexual selection.
   c. resource partitioning.
   d. habitat differentiation.

9. A botanist identifies a new species of plant that has 32 chromosomes. It grows in the same habitat with three similar species: species A ($2n = 14$), species B ($2n = 16$), and species C ($2n = 18$). Which of the following is a possible speciation mechanism for the new species?
   a. change in a key developmental gene that causes the plants to flower at different times
   b. autopolyploidy, perhaps due to a nondisjunction in the formation of gametes of species B
   c. allopolyploidy, a hybrid of species A and C
   d. Either answer b or c could account for the formation of this new plant species.

10. Which of the following would *not* contribute to allopatric speciation?
    a. genetic drift
    b. gene flow
    c. different selection pressures
    d. founder effect

11. Morphological and genetic comparisons group 30 species of snapping shrimp (genus *Alpheus*) into 15 pairs of closely related species. What is the best explanation for the fact that one member of each pair lives on the Atlantic side of the Isthmus of Panama, whereas the other member of each pair lives on the Pacific side?
    a. Different predator pressures in the Atlantic and Pacific selected for differences that resulted in the reproductive isolation of these species.
    b. The pairs of species arose by allopatric speciation when the Isthmus of Panama separated their ancestral species.
    c. The 30 species of snapping shrimp evolved by sympatric speciation before the Isthmus of Panama formed.
    d. These 15 pairs of species illustrate the pattern of punctuated equilibrium, in which most speciation events take place in a short period of time.

12. When hybrids in a hybrid zone can breed with each other and with both parent species, and also have equal fitness to the parent species, one would predict that
    a. the hybrid zone would be stable.
    b. reinforcement of reproductive barriers would keep the parent species separate.
    c. reproductive barriers would lessen and the two parent species would fuse.
    d. a new hybrid species would form by sympatric speciation.

13. This chapter introduced several research studies that illustrate various aspects of speciation. Which of the following is *not* an accurate description of a conclusion based on one of these studies?
    a. Two closely related species of monkey flowers have different pollinators (bumblebees and hummingbirds). Transferring an allele for flower color between the two species resulted in both types of pollinators visiting flowers with the transferred flower allele. Thus, this reproductive barrier based on pollinator choice may have been influenced by a change in a single gene locus.
    b. The apparent stability of the hybrid zone between two species of *Bombina* toads may relate to the narrow width of the zone. Even though hybrid toads have increased rates of embryonic mortality and a variety of morphological abnormalities, members of both parent species continue to migrate into the zone, leading to the production of hybrids.
    c. Populations of mosquitofish inhabit ponds in the Bahamas. In ponds that contain predatory fishes, natural selection has favored a body shape that facilitates rapid bursts of speed. In mate choice experiments, female mosquitofish choose males with body shapes similar to their own. The reproductive isolation between populations is forming as a by-product of natural selection for predator avoidance.
    d. In the many species of cichlids found in Lake Victoria, the reproductive barriers between species as a result of female mate choice based on male coloration appear to be breaking down because the introduced Nile perch is a strong predator of cichlids.

Chapter 25

The History of Life on Earth

Key Concepts

25.1 Conditions on early Earth made the origin of life possible

25.2 The fossil record documents the history of life

25.3 Key events in life's history include the origins of unicellular and multicellular organisms and the colonization of land

25.4 The rise and fall of groups of organisms reflect differences in speciation and extinction rates

25.5 Major changes in body form can result from changes in the sequences and regulation of developmental genes

25.6 Evolution is not goal oriented

Framework

Starting with hypotheses on how life may have originated, this chapter describes some of the major events in life's history, such as the rise of photosynthetic prokaryotes, the origin of eukaryotic cells and multicellularity, and the colonization of land. The fossil record has been used to establish a geologic record, in which divisions between eras and periods are marked by major biological transitions. These transitions have been influenced by plate tectonics, mass extinctions, and adaptive radiations. The mechanisms underlying the major changes seen in the fossil record involve changes in and control of developmental genes.

Chapter Review

The history of life on Earth as chronicled in the fossil record illustrates **macroevolution,** evolutionary changes above the species level.

25.1 Conditions on early Earth made the origin of life possible

Scientists have proposed a four-stage hypothesis for the origin of life on early Earth: (1) the abiotic synthesis of small organic molecules; (2) the joining of these small molecules into macromolecules; (3) the packaging of these molecules into **protocells**; and (4) the origin of self-replicating molecules.

Synthesis of Organic Compounds on Early Earth   Earth formed about 4.6 billion years ago and was bombarded by huge rocks and ice until about 4 billion years ago. As Earth cooled, water vapor condensed into oceans, and the atmosphere probably contained nitrogen and its oxides, carbon dioxide, methane, ammonia, and hydrogen.

In the 1920s, A. Oparin and J. Haldane independently hypothesized that conditions on early Earth—in particular the reducing atmosphere, lightning, and UV radiation—favored the synthesis of organic compounds from simpler molecules.

In 1953, S. Miller experimentally supported the Oparin-Haldane hypothesis using an apparatus that simulated the hypothetical conditions of early Earth. Numerous laboratory replications using various combinations of atmospheric gases have also produced organic compounds.

The openings of volcanoes may have provided the necessary reducing environments for the synthesis of organic molecules. Another hypothesis is that organic compounds were produced in deep-sea **hydrothermal vents**. Deep-sea **alkaline vents** surrounded by the acidic early ocean waters may have provided the energy for the synthesis of organic compounds in the form of a pH gradient.

Meteorites have been found to contain carbon compounds, including many amino acids, lipids, sugars, and nitrogenous bases. Thus, meteorites may have been another source of organic compounds.

Abiotic Synthesis of Macromolecules   Researchers have created polymers by dripping solutions of amino acids or RNA nucleotides onto hot sand, clay, or rock. Amino acid polymers may have served as weak catalysts on early Earth.

Protocells   Life requires accurate replication of genetic information and a metabolism to provide building blocks and carry out this replication. Laboratory experiments have shown that membrane-bound *vesicles* can form spontaneously from mixtures of organic ingredients. Vesicle formation is speeded by the presence of *montmorillonite* clay, thought to have been common on early Earth. Such collections of abiotically produced organic molecules surrounded by a membrane can maintain a unique internal chemistry, undergo growth and reproduction, and perform some metabolic reactions.

Self-Replicating RNA   RNA probably functioned as the first hereditary material. **Ribozymes** are enzyme-like RNA catalysts. Some laboratory-produced ribozymes can make complementary copies of their sequences. In the so-called "RNA world," vesicles with self-replicating, catalytic RNA could grow, split, and pass some of their RNA to daughter protocells. As the more successful offspring continued to reproduce and be acted on by natural selection, metabolic and hereditary improvements may have accumulated. DNA, a more stable genetic molecule, eventually replaced RNA as the carrier of genetic information.

INTERACTIVE QUESTION   25.1

Why do we say that, for life to have begun, the ability to carry out reproduction and metabolism must have evolved?

25.2   The fossil record documents the history of life

The Fossil Record   The fossil record, based on the sequence of fossils found in the *strata* of sedimentary rocks, provides evidence of great changes in the organisms that have lived on Earth. The record is incomplete, however, because (1) large numbers of species that lived probably left no fossils, (2) geologic processes destroy many fossils, and (3) only a fraction of existing fossils have been found.

How Rocks and Fossils Are Dated   The order in which fossils appear in the strata of sedimentary rocks indicates their relative age.

Radiometric dating is used to determine the actual ages of rocks and fossils. Each radioactive isotope has a fixed rate of decay—its **half-life**, which is the number of years it takes for 50% of the "parent" isotope to

decay to its "daughter" isotope. During an organism's lifetime, it accumulates isotopes in proportions equal to their relative abundance in the environment. After the organism dies, its radioactive isotopes decay at a fixed rate. Carbon-14, with its half-life of 5,730 years, can date fossils up to about 75,000 years old. Isotopes that decay more slowly can be used to date older fossils. The age of volcanic rock can be used to infer the age of fossils associated with such layers of rock.

INTERACTIVE QUESTION   25.2

A fossil has ⅛ of the atmospheric ratio of carbon-14 to carbon-12. Estimate the age of this fossil.

The Origin of New Groups of Organisms   The fossil record can shed light on the origin of new groups of organisms. Following changes in skull bones and teeth, the gradual evolution of mammals can be traced from a group of tetrapods called synapsids through a sequence of ancestors in the group cynodonts.

25.3   Key events in life's history include the origins of unicellular and multicellular organisms and the colonization of land

Geologists have established a **geologic record** of Earth's history. The Hadean, Archaean, and Proterozoic eons encompass the first 4 billion years. The Phanerozoic eon covers the last half billion years and is divided into three eras: the Paleozoic, Mesozoic, and Cenozoic. These eras are delineated by major extinction events.

The First Single-Celled Organisms   The oldest known fossils are 3.5-billion-year-old fossil **stromatolites**, which are rocklike layers of prokaryotes and sediment.

Banded iron formations in marine sediments and the rusting of iron in terrestrial rocks that began about 2.7 billion years ago provide evidence of oxygen accumulation from the photosynthesis of ancient cyanobacteria. The relatively rapid accumulation of atmospheric oxygen after 2.4 billion years ago (the "oxygen revolution") caused the extinction of many prokaryotic groups, relegated others to anaerobic habitats, and was associated with the evolution of adaptations such as cellular respiration.

Fossils that are generally accepted as eukaryotic date from 1.8 billion years ago. The cytoskeleton of eukaryotic cells allows them to change shape and engulf other cells. It is widely accepted that eukaryotes arose through **endosymbiosis**, when an archaean prokaryote cell engulfed a heterotrophic bacterium. The engulfed cell was not digested but rather became a stable *endosymbiont* within the *host cell* and evolved into the mitochondria. A later step in the **serial endosymbiosis** involved a photosynthetic bacterium becoming an endosymbiont that gave rise to the plastids.

INTERACTIVE QUESTION   25.3

List some of the evidence supporting an endosymbiotic origin of mitochondria and plastids.

The Origin of Multicellularity   A great diversity of single-celled eukaryotes evolved from the structurally complex first eukaryotic cells. The first multicellular eukaryotes gave rise to algae, plants, fungi, and animals.

The oldest identified multicellular fossils that can be assigned to a recognized taxonomic group are that of small red algae that date to 1.2 billion years ago. Larger algae and animals known as the Ediacaran biota are represented by fossils that date to 635 to 541 million years ago (mya).

Fossils of sponges, cnidarians, and molluscs date from the late Proterozoic. The appearance of many of the major phyla of animals between 535 and 525 million years ago is referred to as the **Cambrian explosion**. Molecular (DNA) estimates indicate that many animal phyla began to diverge as early as 700 million years ago.

INTERACTIVE QUESTION   25.4

What lifestyle change may be related to the diversity of forms that evolved in the 10 million years known as the Cambrian explosion?

The Colonization of Land   Cyanobacteria coated damp terrestrial surfaces more than a billion years ago, but plants, fungi, and animals began to move onto land only about 500 million years ago. This colonization was associated with adaptations that helped prevent dehydration and permitted reproduction on land. Plants and fungi appear to have colonized the land together in symbiotic associations known as *mycorrhizae*. Arthropods colonized land about 450 million years ago; the first tetrapod fossils appeared about 365 million years ago.

25.4 The rise and fall of groups of organisms reflect differences in speciation and extinction rates

Large-scale processes such as plate tectonics, mass extinctions, and adaptive radiations have influenced rates of speciation and extinction and thus the rise and fall of major groups.

Plate Tectonics   The theory of **plate tectonics** describes how the continents are part of great plates of Earth's crust that float on the molten mantle below. Earthquakes occur and mountains are formed in regions where these shifting plates slide past or collide with each other. In the past billion years, *continental drift* has brought most of the landmasses together to form a supercontinent three separate times.

The formation of the supercontinent **Pangaea** about 250 million years ago destroyed and altered habitats, causing many species to become extinct. The shifting of landmasses north or south greatly changes their climate. The separation and drifting apart of continents creates huge geographic isolation events, promoting allopatric speciation. Continental drift helps explain the past and current distributions of organisms.

INTERACTIVE QUESTION   25.5

Fossil evidence indicates that marsupials evolved in what is now Asia, yet their greatest diversity is found in Australia. How can you account for this biogeographic distribution?

Mass Extinctions   A species may become extinct due to a change in its physical or biological environment. **Mass extinctions** have occurred during periods of major environmental change. There have been five mass extinctions over the past 500 million years.

The Permian mass extinction, occurring at the boundary between the Paleozoic and Mesozoic eras 252 mya, claimed about 96% of marine animal species and many terrestrial organisms. These extinctions may be related to massive volcanic eruptions in Siberia, which probably released enough CO_2 to warm the global climate and acidify the oceans.

Nutrients released by the volcanoes would have stimulated the growth of microorganisms, whose decomposition by aerobic bacteria would have lowered oxygen concentrations.

The Cretaceous mass extinction occurred 66 mya and claimed more than one-half of the marine species and many families of land plants and animals, including all dinosaurs (except birds). At the time of the Cretaceous mass extinction, a thin layer of clay rich in iridium (an element common in meteorites and other extraterrestrial objects) was deposited. This layer may represent the fallout from a huge cloud of dust created when an asteroid or comet collided with Earth. This cloud would have blocked sunlight and severely affected global climate.

The current rate of extinction is estimated to be 100 to 1,000 times the typical rate seen in the fossil record. A sixth mass extinction caused by human alteration of the global environment may occur in the next few centuries or millennia.

The fossil record indicates it takes 5 to 10 million years for the diversity of life to recover following a mass extinction. Once an evolutionary lineage becomes extinct, it cannot reappear. Mass extinctions may alter the types of organisms making up an ecological community.

Adaptive Radiations   **Adaptive radiations** are periods in which many new species evolve and fill various ecological niches. Adaptive radiations have occurred following mass extinctions, the evolution of major adaptations that allowed organisms to exploit new ecological roles, the diversification of other organisms that provided new sources of food, and the colonization of new regions.

INTERACTIVE QUESTION   25.6

a. Mammals originated 180 mya but did not change much until after 66 mya. Explain this observation.

b. Give examples of adaptive radiations following evolutionary innovations that opened new ecological roles.

c. What factors contributed to the radiation of the thousands of endemic species of the Hawaiian Islands?

25.5 **Major changes in body form can result from changes in the sequences and regulation of developmental genes**

Effects of Developmental Genes   The combination of evolutionary and developmental biology, called "evo-devo," explores how slight changes in developmental genes can result in major morphological differences between species.

Heterochrony is an evolutionary change in the rate or timing of development. A minor genetic alteration that affects the relative growth rates of body parts can produce a very differently proportioned adult form. **Paedomorphosis** is the retention in the adult of juvenile traits of an ancestral species and can result from changes in the timing of reproductive development.

Homeotic genes control the spatial arrangement of body parts. Changes in *Hox* genes (whose products provide positional information in animal embryos), or in how they are expressed, can drastically alter body form.

The Evolution of Development   Changes in the number and sequences of developmental genes as well as changes in gene regulation may produce new body forms.

INTERACTIVE QUESTION   25.7

a. Researchers identified the key difference in the *Hox* gene *Ubx* of *Artemia* (brine shrimp, a crustacean) and *Drosophila* (fruit fly, an insect) that is responsible for the suppression of legs in *Drosophila*. What was the significance of this study?

b. Marine populations of three-spine stickleback fish have protective spines on their ventral surface; some lake populations do not. How did researchers determine that the loss or reduction of these spines was caused by a difference in the regulation of a developmental gene *(Pitx1)* and not in the gene itself?

25.6 **Evolution is not goal oriented**

Evolutionary Novelties   Often very complex organs, such as the eyes of vertebrates and some molluscs, have evolved gradually from simpler structures that served similar needs in ancestral species.

Evolutionary novelties may also evolve by the gradual modification of existing structures for new functions. *Exaptation* is the term for structures that evolved and functioned in one setting and were then co-opted for a new function.

Evolutionary Trends   The fossil record documents that *Equus*, the modern horse, descended from its much smaller, browsing, multitoed ancestor, *Hyracotherium*, through a series of speciation episodes that produced many different species and diverging trends.

INTERACTIVE QUESTION   25.8

a. According to the model of *species selection*, "differential speciation success" plays a role in macroevolution similar to the role of differential reproductive success in microevolution. Explain.

b. Do evolutionary trends indicate that evolution is goal directed? Explain.

Structure Your Knowledge

1. List the sequence of four stages that could have led to the origin of the first living cells.
2. Label the eons, eras, and key events shown on the following clock analogy of Earth's history.

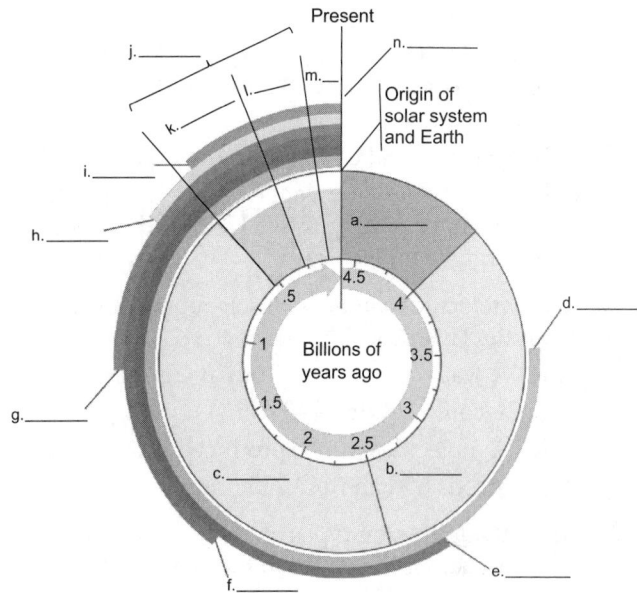

3. Describe three major processes that have influenced the changing diversity of life on Earth.
4. What genetic changes would most likely lead to the evolution of new morphological forms?

Word Roots

heter- = different (*heterochrony:* evolutionary changes in the timing or rate of development)

macro- = large (*macroevolution:* evolutionary change above the species level, such as the emergence of a new group of organisms through a series of speciation events)

paed- = child (*paedomorphosis:* the retention in the adult organism of the juvenile features of its evolutionary ancestors)

proto- = first (*protocell:* an abiotic precursor of a living cell that had a membrane-like structure and maintained an internal chemistry different from its surroundings)

stromat- = something spread out; **-lite** = a stone (*stromatolite:* layered rock that results from the activities of prokaryotes that bind thin films of sediment together)

Test Your Knowledge

MULTIPLE CHOICE: *Choose the one best answer.*

1. The primitive atmosphere of Earth may have favored the abiotic synthesis of organic molecules because it
   a. was highly oxidative.
   b. was reducing (near openings of volcanoes) and had energy sources in the form of lightning and UV radiation.
   c. had a great deal of methane and organic fuels.
   d. had plenty of water vapor, carbon, oxygen, and nitrogen, providing the C, H, O, and N needed for organic molecules.

2. RNA molecules can be both self-replicating and catalytic. This probably means that
   a. RNA was the first hereditary material.
   b. RNA could produce ATP.
   c. RNA molecules can be produced abiotically.
   d. protocells had an RNA membrane.

3. All of the following types of evidence support the hypothesis that life on Earth formed spontaneously *except* for
   a. laboratory studies involving the formation and reproduction of vesicles.
   b. the fossil record.
   c. the abiotic synthesis of polymers from monomers dripped onto hot rocks.
   d. the production of organic compounds in the Miller-Urey experimental apparatus.

4. The half-life of carbon-14 is 5,730 years. A 22,920-year-old fossil would have what proportion of the normal ratio of carbon-14 to carbon-12?
   a. ¼
   b. ⅙
   c. ⅛
   d. 1/16

5. How many half-lives should have elapsed if a sample contains 12.5% of the normal quantity of a parent isotope?
   a. 2
   b. 3
   c. 4
   d. 5

6. Which of the following characteristics would increase the likelihood that organisms would be well represented in the fossil record?
   a. abundant, widespread, and exist over a long time period
   b. have hard shells or skeletons
   c. live in regions where geology favors fossil formation
   d. All of the above contribute.

7. Stromatolites are
   a. layered communities of prokaryotes, fossils of which represent the oldest known prokaryotes.
   b. early prokaryotic fossils found in sediments around hydrothermal vents.
   c. the earliest protocells that formed when lipids assembled into a bilayer surrounding organic molecules.
   d. the ancestral photosynthetic eukaryote that originated through serial endosymbiosis.

8. Look back at the Geologic Record in Table 25.1. Which eon was the longest? In which eon did the oldest fossils of prokaryotes appear?
   a. Hadean; Archaean
   b. Archaean; Archaean
   c. Proterozoic; Archaean
   d. Proterozic; Proterozoic

9. Early cyanobacteria evolved a form of photosynthesis that permanently changed the Earth's atmosphere by releasing which of the following gases?
   a. carbon dioxide
   b. nitrogen
   c. oxygen
   d. hydrogen sulfide

10. According to the endosymbiont theory,
    a. the first primitive cells may have originated when a lipid bilayer spontaneously formed around abiotically produced polymers.
    b. plants were able to colonize land with fungi as symbionts within their roots.
    c. the mitochondria and plastids of eukaryotic cells were originally prokaryotic endosymbionts that became permanent parts of the host cell.
    d. the infoldings and specializations of the plasma membrane led to the evolution of the endomembrane system.

11. Which of the following facts supports the idea that mitochondria were once free-living bacteria?

    a. Mitochondria generate ATP.

    b. All eukaryotes contain mitochondria or mitochondrial remnants.

    c. Mitochondria use atmospheric oxygen in the production of ATP.

    d. Mitochondria have their own DNA that replicates independently of the host cell's DNA.

12. Two eukaryotic groups that appear to have colonized the land together are

    a. plants and fungi.

    b. animals and fungi.

    c. animals and plants.

    d. plants and algae.

13. The Permian mass extinction between the Paleozoic and Mesozoic eras

    a. may have been caused by extreme volcanism, which produced multiple environmental effects.

    b. coincided with the breaking apart of Pangaea.

    c. made way for the adaptive radiation of mammals, birds, and pollinating insects.

    d. appears to have been caused by a large asteroid striking Earth.

14. The more than 500 species of fruit fly on the various Hawaiian Islands, all apparently descended from a single ancestral species, are an excellent example of

    a. biogeography.

    b. adaptive radiation.

    c. exaptation.

    d. an evolutionary trend.

15. The evolution of major changes in animal body form is most likely to involve

    a. duplication of *Hox* genes.

    b. changes in regulatory sequences that determine where developmental genes are expressed.

    c. genetic changes that affect the growth rates of different body parts during development.

    d. All of the above can contribute to major morphological differences between species.

16. The evolution of the middle ear bones of mammals from bones of the jaw hinge of cynodonts is an example of

    a. heterochrony.

    b. exaptation.

    c. the gradual refinement of a structure with the same function.

    d. an evolutionary trend.

17. Heterochrony refers to

    a. a change in the rate or timing of developmental events.

    b. the condition of a cell living inside a host cell.

    c. rapid speciation to fill a variety of environmental niches.

    d. the production of living cells from nonliving materials.

18. What is meant by the concept of species selection?

    a. Characteristics that increase the probability that a species will separate into two or more species will accumulate in the most successful species.

    b. Natural selection can act on the species level as well as on the population level.

    c. The species that last the longest and speciate the most often influence the direction of evolutionary trends.

    d. The colonization of new and diverse habitats can lead to the adaptive radiation of numerous species.

5 The Evolutionary History of Biological Diversity

Phylogeny and the Tree of Life

Key Concepts

26.1 Phylogenies show evolutionary relationships

26.2 Phylogenies are inferred from morphological and molecular data

26.3 Shared characters are used to construct phylogenetic trees

26.4 An organism's evolutionary history is documented in its genome

26.5 Molecular clocks help track evolutionary time

26.6 Our understanding of the tree of life continues to change based on new data

Framework

A goal of systematics is to reconstruct the phylogenetic history of species. Cladistics uses the distribution of shared derived characters to identify monophyletic taxa, or clades. The best phylogenetic hypotheses are parsimonious trees based on morphological and molecular homologies as well as on fossil evidence. Comparisons of genomes from the three domains indicate that horizontal gene transfers were common in the early history of life and suggest that the base of the tree of life may have been a tangled web.

Chapter Review

Phylogeny is the evolutionary history of a species or group. **Systematics** focuses on classifying organisms and determining evolutionary relationships.

26.1 Phylogenies show evolutionary relationships

Binomial Nomenclature   **Taxonomy** is the scientific discipline that names and classifies species. As instituted by Linnaeus, each species is assigned a two-part Latinized name—a **binomial**—consisting of the name of the **genus** and the specific epithet, which designates one species in that genus. By convention, genus names are capitalized, and both genus and species names are italicized.

Hierarchical Classification   In the Linnaean taxonomic system, genera are grouped into **families**, which are then placed into the increasingly broader categories of **orders**, **classes**, **phyla**, and **kingdoms**, and more recently into **domains**. A taxonomic unit at any level is called a **taxon**.

Linking Classification and Phylogeny   A **phylogenetic tree** represents a hypothesis about evolutionary relationships among groups. Its branching patterns often (but not always) correspond to hierarchical Linnaean classification.

Each dichotomous **branch point** on a phylogenetic tree represents the divergence of two taxa from a common ancestor. **Sister taxa**, sharing an immediate common ancestor, are each other's closest relatives.

Trees are **rooted**, with the common ancestor to all the taxa in the tree often on the left. A **basal taxon** is one that diverged early in the evolutionary history of a group.

What We Can and Cannot Learn from Phylogenetic Trees   Phylogenetic trees show patterns of descent, which do not always correspond to phenotypic similarity. Branching sequences do not indicate when species evolved or the amount of genetic change in each lineage. Also, species located next to each other on the tree shared a common ancestor; one did not evolve from the other.

Applying Phylogenies   The information contained in phylogenetic trees has practical applications in such areas as molecular forensics and crop improvement.

INTERACTIVE QUESTION   26.1

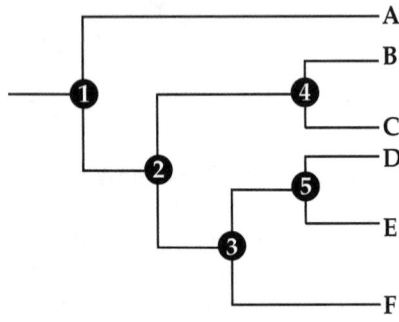

a. In this hypothetical phylogenetic tree, which number represents the common ancestor of all the taxa?

b. Which letter represents the basal taxon?

c. Which of the terminal taxa (A–F) are sister taxa?

26.2   **Phylogenies are inferred from morphological and molecular data**

Morphological and Molecular Homologies   **Homologies** are similarities due to shared ancestry. The more similar organisms' morphologies or DNA sequences

are, the more likely it is those organisms are closely related.

Sorting Homology from Analogy   **Analogy** is similarity due to convergent evolution, in which unrelated species develop similar features because natural selection has led to similar adaptations. In general, when two complex structures share many similar features, the more likely it is that those structures were inherited from a common ancestor. And when genes from different organisms share many nucleotide sequences, it is more likely that the genes are homologous.

INTERACTIVE QUESTION   26.2

What two complications may make it difficult to determine phylogenetic relationships based on morphological similarities between species? Give examples.

Evaluating Molecular Homologies   Molecular comparisons are often complicated by insertion or deletion mutations that change the lengths of homologous regions of DNA. Computer programs can identify and align homologous DNA segments properly for nucleotide comparisons. Statistical tools help distinguish "distant" homologies in divergent sequences from coincidental molecular homoplasies.

26.3   **Shared characters are used to construct phylogenetic trees**

Cladistics   **Cladistics** bases classification on common ancestry and nests clades within clades. A **clade** is a **monophyletic** group that consists of an ancestral species and all of its descendant species. A **paraphyletic** group excludes some species that share a common ancestor with other species in the group, and a **polyphyletic** group includes several distantly related species but not their most recent common ancestor.

Shared ancestral characters are found in a particular clade but originated in an ancestor that is not part of that clade. **Shared derived characters** are unique to a particular clade.

To determine the branching sequence of a group of related species, the group is compared to an **outgroup**, a species or group of species that diverged before the group being studied (the **ingroup**). A comparison of the characters that are present in each taxon of the ingroup indicates the sequence in which shared derived characters evolved and determines the branch points used to produce a phylogenetic tree.

INTERACTIVE QUESTION 26.3

Place the taxa (outgroup, A, B, C, and D) on the following phylogenetic tree based on the presence or absence of the characters 1–4 as shown in the table. Indicate before each branch point the number for the shared derived character that evolved in the ancestor of the clade.

	Outgroup O	A	Taxa B	C	D
1	0	1	1	1	1
Characters 2	0	0	1	0	1
3	0	1	1	0	1
4	0	0	1	0	0

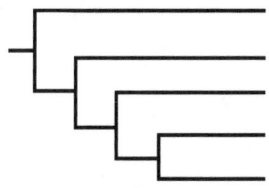

Phylogenetic Trees with Proportional Branch Lengths The branching pattern of most phylogenetic trees is relative rather than absolute, indicating only the order in which members of each clade last shared a common ancestor. The branch length of some trees can be scaled to reflect rates of evolutionary change (for instance, number of changes in DNA sequences) or time (using the dates of branch points as indicated in the fossil record).

Maximum Parsimony and Maximum Likelihood Systematists use morphological characters or molecular comparisons to choose among many possible phylogenetic trees using the principle of **maximum parsimony**—the approach that the smallest number of evolutionary changes is the simplest explanation and thus the best hypothesis to consider first. According to **maximum likelihood**, a tree should be the most likely one based on probabilities about rates of DNA changes. Computer programs search for the most parsimonious and most likely trees.

Phylogenetic Trees as Hypotheses A phylogenetic tree represents the best hypothesis of the relationships among a group of species; the more data that can be compared, the more reliable the tree becomes.

Phylogenetic hypotheses can be used to make and test predictions. *Phylogenetic bracketing* predicts that features shared by two closely related organisms will be present in their common ancestor and all its descendants. Fossil discoveries, for example, support the prediction that dinosaurs, as descendants of the common ancestor of birds and crocodiles, built nests and brooded their eggs.

INTERACTIVE QUESTION 26.4

According to the principle of parsimony, the evolution of the four-chambered heart should place birds and mammals in the same clade. Why does the most accepted evolutionary tree show them as separate branches from the reptilian line?

26.4 **An organism's evolutionary history is documented in its genome**

Molecular systematics using nucleic acid or other molecular comparisons make it possible both to compare genetic divergence among groups within a species and to reconstruct phylogeny among very distantly related species. The ability to span this length of time depends on genes that evolve either very slowly or very rapidly.

Gene Duplications and Gene Families Gene duplications are important evolutionary mutations, increasing the number of genes in the genome. Repeated duplications can result in groups of related genes, or *gene families*.

Orthologous genes are homologous genes found in different species; such genes may diverge after the two species separated. **Paralogous genes** result from gene duplication in the same genome, and as they diverge, they provide new opportunities for evolutionary change within a species.

Genome Evolution The existence of widespread orthologous genes indicates that all organisms share many biochemical and developmental pathways. The number of genes a species has does not necessarily correlate with phenotypic complexity.

---

a. Give an example of genes that would be compared to uncover phylogenetic relationships among the earliest branches on the tree of life.

b. Give an example of genes that evolve very rapidly and are used to discriminate among closely related species or even among populations of the same species.

26.5   Molecular clocks help track evolutionary time

Molecular Clocks   Some regions of DNA appear to evolve at constant rates, and comparisons of the number of nucleotide substitutions in orthologous genes can serve as **molecular clocks** to estimate the time since two species branched from their common ancestor. The number of differences in paralogous genes is proportional to the time since the ancestral gene was duplicated.

Some genes appear to have a reliable average rate of evolution. Graphs that plot nucleotide or amino acid differences against the times for known evolutionary branch points can be used to estimate phylogenetic branchings that are not evident from the fossil record.

Assuming that harmful mutations are removed quickly from the gene pool, but neutral mutations that have little effect on fitness should occur at a constant rate, explain why different genes might have a different clock speed.

Natural selection, which favors some DNA changes over others, may disrupt the smooth running of the molecular clock. Molecular clocks are less reliable when used to date evolutionary divergences that occurred billions of years ago. When molecular clocks are calibrated using many genes, fluctuations due to natural selection may average out. Molecular and fossil-based estimates of divergence times in vertebrate evolution agree closely.

Applying a Molecular Clock: Dating the Origin of HIV   By comparing nucleotide sequences of HIV from samples taken at various times during the epidemic, researchers have estimated that the HIV-1 M strain first infected humans around 1910.

26.6   Our understanding of the tree of life continues to change based on new data

From Two Kingdoms to Three Domains   Historically, taxonomists divided the diversity of life into two kingdoms—plants and animals. In the five-kingdom system, the prokaryotes were set apart from the eukaryotes and placed in kingdom Monera. The Protista contained mostly unicellular eukaryotes, whereas kingdoms Plantae, Fungi, and Animalia consisted of multicellular eukaryotes.

The current three-domain system creates a taxon above the kingdom level. The domains Bacteria and Archaea, although both consisting of single-celled prokaryotes, differ in many characteristics. The domain Eukarya includes all the eukaryotes: plants, fungi, animals, and many groups of mostly single-celled organisms.

The Important Role of Horizontal Gene Transfer   Comparisons of rRNA genes indicate that the first major split in the tree of life was the divergence of the bacteria from the other two domains. Analysis of other genes, however, suggests that eukaryotes share a more recent common ancestor with bacteria than with archaea.

Genome comparisons from the three domains indicate that **horizontal gene transfer**, perhaps through exchange of transposable elements and plasmids, viral infection, or even through fusions of different organisms, occurred during the early history of life. Over 200 instances of horizontal gene transfer of transposons have been reported in eukaryotes. It appears that horizontal gene transfer was common in the early history of life, leading some scientists to suggest that the tree of life is best represented as a tangled network of branches.

Chapter 26: Phylogeny and the Tree of Life 193

INTERACTIVE QUESTION 26.7

This tangled web of life represents a current hypothesis on the origin of the three domains. Identify the domains a, b, and c, and the two major episodes of horizontal gene transfer labeled d and e.

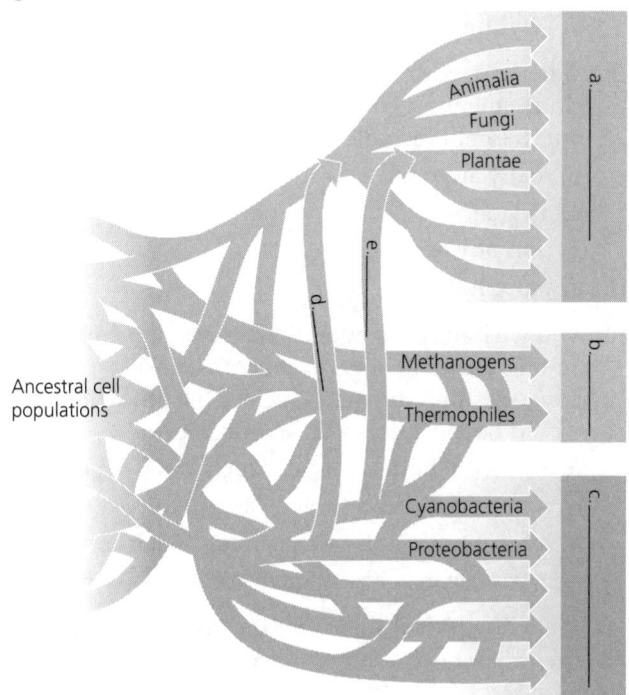

Word Roots

bi- = two; **nom-** = name (*binomial:* the two-part Latinized format for naming a species)

clad- = branch (*clade:* a group of species that includes an ancestral species and all its descendants)

hom- = like, resembling (*homoplasy:* a similar [analogous] structure or molecular sequence that has evolved independently in two species)

mono- = one (*monophyletic:* pertaining to a group of taxa that consists of a common ancestor and all its descendants)

pars- = to spare or save (*maximum parsimony:* a principle that states that one should first investigate the simplest explanation that is consistent with the facts)

phyl- = tribe; **-geny** = origin (*phylogeny:* the evolutionary history of a species or group of related species)

Structure Your Knowledge

1. Draw a phylogenetic tree that best represents the relationships among taxa A–E described as follows: Taxon A is the basal taxon. B and C are sister taxa. They share a more recent common ancestor with D than with E.
2. Describe some of the tools that systematists use for constructing phylogenetic trees.
3. What is a molecular clock, and how can it be used?

Test Your Knowledge

MULTIPLE CHOICE: *Choose the one best answer.*

1. Related families are grouped into the next-highest taxon, which is called a(n)
   a. class.
   b. phylum.
   c. order.
   d. kingdom.

2. Which binomial designation for our species is properly formatted?
   a. Homo sapiens
   b. *Homo sapiens*
   c. *Homo Sapiens*
   d. homo sapiens

3. Three of the following trees describe the same phylogenetic relationships among taxa A, B, C, D, E, and F. Which tree shows a different phylogeny?

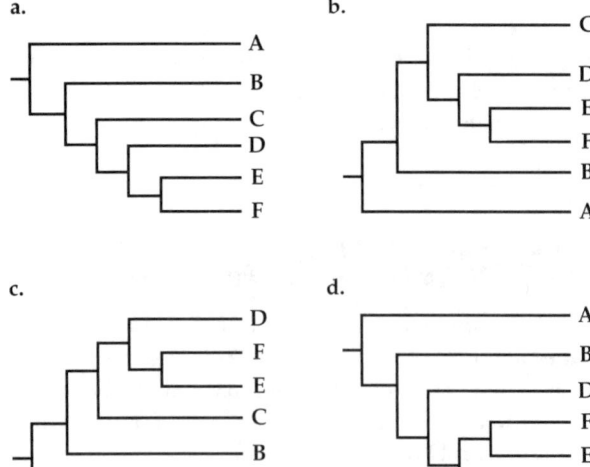

4. In a phylogenetic tree, a two-way branch point is called a(n)

   a. homology.   c. dichotomy.

   b. analogy.   d. paralogy.

5. Convergent evolution may occur

   a. when homologous structures are adapted for different functions.

   b. as a result of adaptive radiation.

   c. when species are widely separated geographically.

   d. when species have similar ecological roles.

6. A taxon is considered to be a clade if it

   a. is monophyletic.

   b. contains more than one living species.

   c. corresponds to a Linnean category.

   d. is based on homologous DNA sequences.

7. A common feature of all living birds is the presence of feathers. Are feathers a shared derived character?

   a. yes, because feathers distinguish birds from other groups

   b. no, because there is evidence that some dinosaurs had feathers

   c. it depends on whether you are comparing birds to other vertebrates or comparing one bird group to another

   d. no, because there are other features that distinguish birds from other vertebrates

8. The greatest number of shared derived characters should be found in two organisms that were traditionally placed in the same

   a. order.

   b. species.

   c. class.

   d. phylum.

9. A taxon such as the class Reptilia—which includes the most common ancestor of all reptiles but does not include the birds—is

   a. really an order.

   b. a clade.

   c. polyphyletic.

   d. paraphyletic.

10. You wish to study evolutionary relationships among the great apes: orangutans, gorillas, chimpanzees, and humans. Which of the following would be most useful as an outgroup?

    a. an invertebrate chordate, such as a lancelet

    b. a jawless vertebrate, such as a lamprey

    c. a mammal that is not a primate, such as a cat

    d. a primate that is not a great ape, such as a baboon

The following phylogenetic tree includes the dates of divergence for taxa A through E as determined from the fossil record. Use this tree to answer questions 11–15.

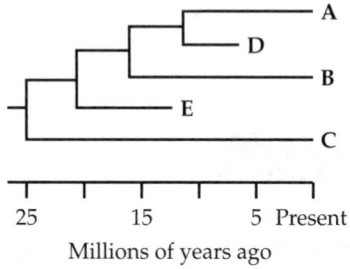

Millions of years ago

11. Which letter represents the basal taxon?

12. Which letter or letters refer to extinct groups?

    a. C

    b. D

    c. D and E

    d. B and D

13. Which two extant (living) taxa are most closely related?

    a. A and B

    b. A and C

    c. A and D

    d. B and C

14. How many million years ago did taxa A and E last share a common ancestor?
    a. 25
    b. 20
    c. 15
    d. 10

15. The technique that enables biologists to make predictions about certain characteristics of taxon D by comparing shared derived characters of taxa A and B is called
    a. cladistics.
    b. systematics.
    c. phylogenetic bracketing.
    d. maximum parsimony.

16. Which of the following approaches would allow a biologist studying the evolution of four similar species of birds to choose the best phylogenetic tree from all possible phylogenies?
    a. Draw the simplest tree and choose that one.
    b. From a comparison of DNA sequences, determine the number of evolutionary events required for each tree and then choose the most parsimonious tree.
    c. Compare the entire genomes of each species; the two most similar genomes are the two species that are most closely related.
    d. Determine which species can interbreed; those that can interbreed evolved from a common ancestor most recently.

17. A comparative study of which of the following would provide the best data on the early evolution of fungi and plants?
    a. mitochondrial DNA
    b. DNA from a chloroplast
    c. the amino acid sequence of chlorophyll
    d. DNA for ribosomal RNA

18. A comparative study of which of the following would provide the best data on the ancestry of people from Germany, Italy, and Spain?
    a. mitochondrial DNA
    b. ribosomal RNA
    c. paralogous genes
    d. the fossil record

19. Which of the following segments of DNA would likely have the fastest molecular clock rate?
    a. noncoding DNA that has a regulatory function
    b. a pseudogene (gene that has lost sequences needed for expression)
    c. a gene for an essential enzyme
    d. a gene for rRNA

20. Which of the following is the best description of our current hypothesis of the tree of life?
    a. The tree of life consists of three domains: Bacteria, Archaea, and Eukarya.
    b. The domain Archaea is known to be the first branch; domains Bacteria and Eukarya are more closely related to each other.
    c. There was substantial horizontal gene transfer between different organisms during the early history of life.
    d. Both a and c represent our current hypothesis.

21. Which of the following makes establishing the phylogeny of the three domains of life most challenging?
    a. the lack of orthologous genes among the three domains
    b. the unreliability of the molecular clock for such early events
    c. horizontal gene transfer
    d. the scarcity of fossils from that early time period

Chapter 27

Bacteria and Archaea

Key Concepts

27.1 Structural and functional adaptations contribute to prokaryotic success

27.2 Rapid reproduction, mutation, and genetic recombination promote genetic diversity in prokaryotes

27.3 Diverse nutritional and metabolic adaptations have evolved in prokaryotes

27.4 Prokaryotes have radiated into a diverse set of lineages

27.5 Prokaryotes play crucial roles in the biosphere

27.6 Prokaryotes have both beneficial and harmful impacts on humans

Framework

This chapter presents the morphology, reproduction, phylogeny, and ecology of prokaryotes. These generally single-celled organisms greatly outnumber and outweigh all eukaryotes combined and flourish in all habitats, including those that are too harsh for any other forms of life. The collective impact of these microscopic organisms is huge.

Chapter Review

27.1 Structural and functional adaptations contribute to prokaryotic success

Prokaryotic cells are usually 0.5–5 µm in diameter. The three most common cell shapes are spheres (cocci), rods (bacilli), and spirals.

Cell-Surface Structures Prokaryotic cell walls maintain cell shape and protect the cell from bursting in a hypotonic surrounding. Most prokaryotes plasmolyze in a hypertonic environment. The cell walls of bacteria (but not those of archaea) contain **peptidoglycan**, a polymer composed of modified sugars cross-linked by short polypeptides.

The **Gram stain** is an important tool for identifying bacteria as **gram-positive** (bacteria whose walls containing a thick layer of peptidoglycan retain the purple stain) or **gram-negative** (bacteria with a thin peptidoglycan layer that does not retain the purple stain and a more complex wall including an outer lipopolysaccharide membrane). Gram-negative bacteria are often more toxic and more resistant to antibiotics. Certain antibiotics, such as penicillin, inhibit the cross-linking of peptidoglycan in the bacterial cell wall.

Many prokaryotes secrete a dense, sticky **capsule** or a thinner *slime layer* outside the cell wall that may serve as protection from a host's immune system and as glue for adhering to a substrate or other cells. Some bacteria produce **endospores**, which are tough-walled, resistant, dormant cells formed in response to a lack of nutrients.

Some prokaryotes may also attach by means of surface hair-like appendages called **fimbriae**. Fimbriae are shorter and more numerous than **pili**, sometimes called *sex pili*, which are specialized for pulling cells together prior to the transfer of DNA between cells.

Motility Many prokaryotes exhibit **taxis**, an oriented movement in response to chemical, light, or other stimuli. Many prokaryotes are equipped with flagella, either scattered over the cell surface or concentrated at one or both ends of the cell. These flagella differ from eukaryotic flagella in their lack of a plasma membrane covering, and in their size, structure, and function. Although similar in size and rotational mechanism, the flagella of bacteria and archaea are composed of different and unrelated proteins, indicating an independent origin.

ATP-driven pumps and the diffusion of H^+ back into the cell power the motor (the rings embedded in the cell wall and plasma membrane), which turns a hook and rotates the filament of the flagellum. Of the 21 proteins shown to be universally required in bacterial flagella, 19 are homologous to cellular proteins with other functions. Thus, flagella may have evolved as a series of steps from simpler structures, as proteins took on new functions (*exaptation*).

Internal Organization and DNA Extensive internal compartmentalization is not found in prokaryotic cells, although some may have membranes, usually infoldings of the plasma membrane that function in respiration or photosynthesis.

The circular DNA chromosome, which has relatively little protein associated with it, is concentrated in a region called the **nucleoid**. Smaller rings of independently replicating DNA, called **plasmids**, may carry a few genes.

The ribosomes of prokaryotes are smaller than eukaryotic ribosomes and differ in their protein and RNA content. Some antibiotics work by binding to prokaryotic ribosomes and blocking protein synthesis.

Reproduction Prokaryotes reproduce rapidly by *binary fission* and have a huge reproductive potential. Their population growth, however, is regulated by the supply of nutrients, accumulation of toxic wastes, competition, or predation.

INTERACTIVE QUESTION 27.1

Fill in the following table with brief descriptions of the characteristics of prokaryotic cells.

Property	Description
Cell shape	a.
Cell size	b.
Cell surface	c.
Motility	d.
Internal membranes	e.
Genome	f.
Reproduction and growth	g.

27.2 **Rapid reproduction, mutation, and genetic recombination promote genetic diversity in prokaryotes**

Rapid Reproduction and Mutation Mutations, although statistically rare, produce a great deal of genetic diversity because prokaryotes have such short generation times and large population sizes. In *E. coli* populations, scientists have documented rapid adaptation—occurring in 20,000 generations—to a challenging new, low-glucose environment.

Genetic Recombination The mechanisms of *genetic recombination* in prokaryotes are different from the eukaryotic mechanisms of meiosis and fertilization. Three mechanisms can transfer DNA within species of prokaryotes and between different species (called *horizontal gene transfer*).

Prokaryotes can take up DNA from the environment in a process called **transformation**. The foreign DNA is integrated into the chromosome by an exchange of homologous DNA segments. Many bacteria have surface proteins specialized for uptake of DNA from closely related species.

Bacteriophages can transfer genes from one prokaryote to another by a process called **transduction**. A random piece of DNA may be accidentally packaged within a phage and introduced into another prokaryotic cell. Through crossing over, the newly introduced DNA (from the donor cell) can be incorporated into the recipient cell's chromosome, creating a recombinant cell.

In **conjugation**, DNA is transferred while two prokaryotic cells are temporarily joined. In *E. coli*, a donor cell attaches to another cell by a pilus and transfers DNA to the recipient, possibly through a "mating bridge" structure. The ability to form pili and donate DNA usually results from the presence of an **F factor**, a DNA segment that is either part of the chromosome or a plasmid.

Bacterial cells containing the F factor as the **F plasmid** are called F$^+$ cells. During conjugation, a single strand of the plasmid DNA is transferred through a mating bridge to the recipient cell. The single parental strands in both cells replicate, and the recipient cell is now a recombinant F$^+$ cell.

Cells in which the F factor is inserted into the bacterial chromosome are called *Hfr cells,* for "high frequency of recombination." When an Hfr cell undergoes conjugation, one parental strand of part of the F factor transfers an attached strand of the bacterial chromosome to the recipient cell. Movements disrupt the mating, usually resulting in a partial transfer of genes and F factor. Replication in both donor and recipient cells produces double-stranded DNA. Crossovers between homologous regions of the transferred DNA and the recipient cell's chromosome produce new genetic combinations in this recombinant F$^-$ cell.

R plasmids carry genes that code for various mechanisms of resistance to antibiotics. Many R plasmids also have genes coding for pili and are transferred to nonresistant cells during conjugation, contributing to the medical problem of antibiotic-resistant pathogens.

INTERACTIVE QUESTION 27.2

Complete the following concept map that summarizes the genetic characteristics of prokaryotes.

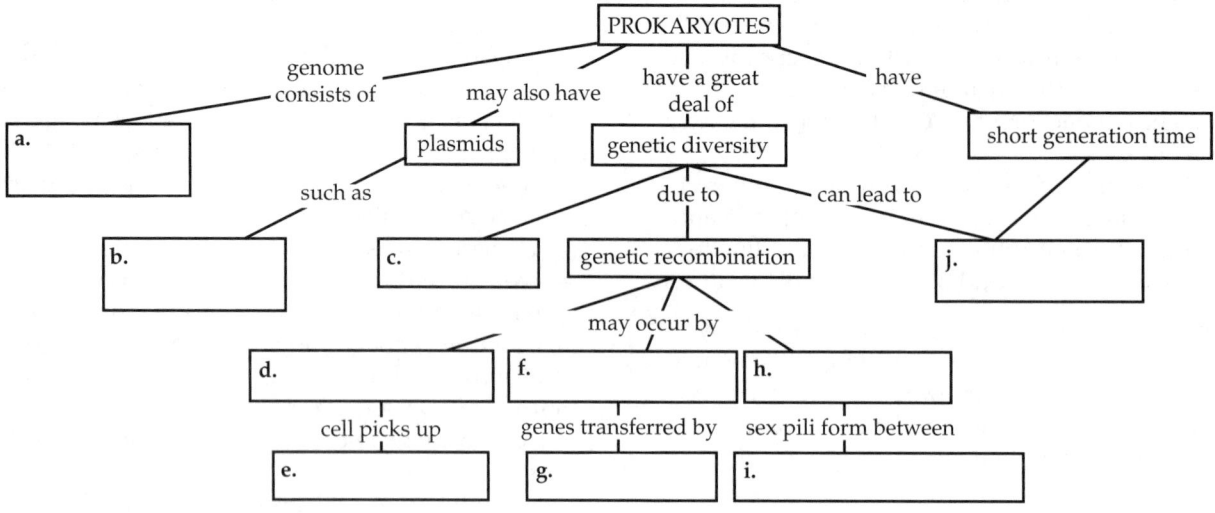

27.3 Diverse nutritional and metabolic adaptations have evolved in prokaryotes

Nutrition refers to how an organism obtains both the energy (*photo-* or *chemotroph*) and the carbon it needs for synthesizing organic compounds (*auto-* or *heterotroph*).

Thus, there are four major nutritional categories: (1) photoautotrophs, which use light energy and CO_2 (or HCO_3^-) to synthesize organic compounds; (2) chemoautotrophs, which obtain energy by oxidizing inorganic substances (such as H_2S, NH_3, or Fe^{2+}) and need only an inorganic carbon source; (3) photoheterotrophs, which use light energy but must obtain carbon in organic form; and (4) chemoheterotrophs, which use organic molecules as both an energy and a carbon source.

The Role of Oxygen in Metabolism   **Obligate aerobes** need O_2 for cellular respiration; **obligate anaerobes** are poisoned by O_2. Some obligate anaerobes use fermentation, others use **anaerobic respiration** to break down nutrients, with substances other than O_2 serving as the final electron acceptor. **Facultative anaerobes** can use O_2 but also can grow in anaerobic conditions using fermentation or anaerobic respiration.

Nitrogen Metabolism   Nitrogen is an essential component of proteins and nucleic acids. Some cyanobacteria and some methanogens obtain nitrogen through **nitrogen fixation**, the conversion of atmospheric N_2 to ammonia (NH_3).

Metabolic Cooperation   The cyanobacterium *Anabaena* forms filaments in which most cells carry out photosynthesis, while a few cells called **heterocysts** perform nitrogen fixation. Members of the filament share nutrients through intercellular connections. **Biofilms** are surface-coating colonies of one or more species. Biofilms are characterized by intercellular signaling, polysaccharides and proteins that adhere cells to each other and to the substrate, and channels in the colony for movement of nutrients and wastes.

INTERACTIVE QUESTION 27.3

This is a micrograph of the cyanobacterium *Anabaena*. What is its mode of nutrition? What is the name and function of the spherical cell indicated by the arrow?

27.4 Prokaryotes have radiated into a diverse set of lineages

An Overview of Prokaryotic Diversity Molecular comparisons of the small-subunit ribosomal RNA indicated that prokaryotes should be divided into domains Bacteria and Archaea. Continuing analysis of more than 1,700 entire genomes is rearranging phylogenetic hypotheses. Such analyses have expanded with the field of *metagenomics*, which enables researchers to sample prokaryotic genes directly from the environment. This ongoing research has shown that the genetic diversity of prokaryotes is huge and that horizontal gene transfer has been extensive in the evolution of prokaryotes.

Bacteria Bacteria show great diversity in their modes of nutrition and metabolism. Five major groups are described in the text.

Proteobacteria are a nutritionally diverse clade of aerobic and anaerobic gram-negative bacteria. The five subgroups of proteobacteria are alpha proteobacteria, which include the nitrogen-fixing *Rhizobium* species and the *Agrobacterium* species used in the genetic engineering of plants; beta proteobacteria, which include the nitrogen-recycling soil bacterium *Nitrosomonas*; gamma proteobacteria, which include autotrophic sulfur bacteria, intestinal inhabitants such as *Escherichia coli* and some serious pathogens; delta proteobacteria, including the colony-forming myxobacteria; and epsilon proteobacteria, the mostly pathogenic group that includes *Helicobacter pylori*, which causes stomach ulcers.

The second major group of bacteria is the chlamydias, which are obligate intracellular parasites of animal cells. One of these gram-negative species is the most common cause of blindness and nongonococcal urethritis, a sexually transmitted disease.

Spirochetes, the third group of bacteria, are helical gram-negative heterotrophs that move in a corkscrew fashion. Spirochetes cause syphilis and Lyme disease.

Cyanobacteria perform plantlike, oxygen-generating photosynthesis and, as *phytoplankton*, they are important producers in freshwater and marine ecosystems. Some filamentous cyanobacteria have cells specialized for nitrogen fixation.

The gram-positive bacteria make up the diverse, fifth major group. The colony-forming subgroup actinomycetes includes the species that cause tuberculosis and leprosy. Most actinomycetes are colonial soil bacteria, some of which are cultured to produce antibiotics. There are also diverse solitary species, including those responsible for anthrax and botulism, and the species of *Staphylococcus* and *Streptococcus*. Mycoplasmas, the smallest of all cells, are the only bacteria that lack cell walls.

Archaea Archaea have some characteristics in common with bacteria, some in common with eukaryotes, and some unique to themselves. Many archaea are **extremophiles**, species that live in extreme habitats. **Extreme halophiles** (saltlovers) live in extremely saline waters. **Extreme thermophiles** may be found in hot sulfur springs and near deep-sea *hydrothermal vents*.

Methanogens have a unique type of energy metabolism in which CO_2 is used to oxidize H_2, producing the waste product methane (CH_4). These strict anaerobes often live in swamps and marshes, are important decomposers in sewage treatment, and are gut inhabitants that contribute to the nutrition of cattle and other herbivores.

The methanogens and many extreme halophiles are placed in the clade Euryarchaeota. Most thermophiles fit into the Crenarchaeota. Not all archaea are extremophiles; numerous archaea have been found in more moderate habitats.

Archaeal phylogeny continues to change as new groups are discovered. Metagenomic studies have resulted in the identification of the TACK supergroup, consisting of Crenarchaeota and three new groups of archaea: Thaumarchaeota, Aigarchaeota, and Korarchaeota. Another new group, the lokiarchaeotes, may shed light on the origin of eukaryotic cells.

INTERACTIVE QUESTION 27.4

a. Evidence indicates that mitochondria evolved from an aerobic alpha proteobacterium through endosymbiosis. From which bacterial group did chloroplasts likely evolve?

b. What is metagenomics? How does it contribute to prokaryotic phylogeny?

27.5 Prokaryotes play crucial roles in the biosphere

Chemical Recycling As **decomposers**, prokaryotes return carbon, nitrogen, and other elements to the environment for assimilation into new living forms. Through photosynthesis and nitrogen fixation, prokaryotes convert inorganic compounds into forms that other organisms can use.

Ecological Interactions   **Symbiosis** is an ecological relationship involving close contact between organisms of different species. The larger organism is called the **host**, and the smaller is known as the **symbiont**. In **mutualism**, both species benefit. In **commensalism**, one organism benefits while the other is neither harmed nor helped. In **parasitism**, a **parasite** eats portions of its host. **Pathogens** are parasites that cause disease.

INTERACTIVE QUESTION   27.5

Explain the important role of chemoautotrophic bacteria in hydrothermal vent communities.

27.6   Prokaryotes have both beneficial and harmful impacts on humans

Mutualistic Bacteria   Many of the 500–1,000 species of bacteria living in the human intestines are mutualists that aid in food digestion and synthesize important nutrients that are used by the host.

Pathogenic Prokaryotes   About one-half of all human diseases are caused by pathogenic bacteria. Pathogens most commonly cause disease by producing toxins. **Exotoxins** are secreted proteins that cause such diseases as botulism and cholera. **Endotoxins**, which are lipopolysaccharides released from the outer membrane of gram-negative bacteria that have died, cause such diseases as typhoid fever and *Salmonella* food poisoning.

Improved sanitation and the development of antibiotics have decreased the incidence of bacterial disease in developed countries. The evolution of antibiotic-resistant strains of pathogenic bacteria, however, poses a serious health threat. Horizontal gene transfer also spreads genes connected with virulence, as in the emergence of the dangerous *E. coli* strain O157:H7.

Prokaryotes in Research and Technology   Prokaryotes are involved in the production of a wide variety of common foods. In addition, bacteria have been used in gene cloning and production of transgenic organisms. Particularly significant is the CRISPR-Cas system, which allows genes in almost any organism to be modified. Soil bacteria have been a rich source of antibiotics. **Bioremediation** is the use of organisms to remove environmental pollutants. Prokaryotes are used to treat sewage and clean up oil spills. They can be used to produce biodegradable plastics.

Prokaryotes have been engineered to make vitamins, antibiotics, ethanol, and other products.

Word Roots

an- = without, not; **aer-** = the air (*anaerobic respiration:* a catabolic pathway in which inorganic molecules other than oxygen accept electrons at the end of electron transport chains)

endo- = inner, within (*endotoxin:* a toxic component of the outer membrane of certain gram-negative bacteria that is released only when the bacteria die)

exo- = outside (*exotoxin:* a toxic protein secreted by a prokaryote or other pathogen and that produces specific symptoms, even if the pathogen is no longer present)

-gen = that which produces (*methanogen:* an organism that produces methane as a waste product of the way it obtains energy; all known methanogens are in domain Archaea)

hal- = salt; **-phile** = loving (*extreme halophile:* an organism that lives in a highly saline environment)

mutu- = reciprocal (*mutualism:* a symbiotic relationship in which both participants benefit)

-oid = like, form (*nucleoid:* a non-membrane-bounded region in a prokaryotic cell where its chromosome is located)

sym- = with, together; **bi-** = life (*symbiosis:* an ecological relationship between organisms of two different species that live together in direct contact)

taxis = to arrange (*taxis:* an oriented movement toward or away from a stimulus)

therm- = temperature (*extreme thermophile:* an organism that thrives in hot environments (often 60–80°C or hotter)

Structure Your Knowledge

1. One might think of prokaryotes as primitive, simple single cells. This chapter, however, provides many examples of how diverse, highly adaptable, metabolically versatile, cooperative, and communal they can be. List some examples of these prokaryotic characteristics.

2. Describe four positive ways in which prokaryotes have an impact on our lives and on the world around us.

Test Your Knowledge

FILL IN THE BLANKS

1. _____the name for spherical prokaryotes

2. _____region in which the prokaryotic chromosome is found

3. _____common staining technique that distinguishes two groups of bacteria

4. _____surface appendages of prokaryotes used for adherence to substrate

5. _____an oriented movement in response to light or chemical stimuli

6. _____resistant cell that can survive harsh conditions

7. _____surface-coating cooperative colonies of prokaryotes

8. _____proteins that are secreted by pathogens and are potent poisons

9. _____use of organisms to remove environmental pollutants

10. _____type of respiration that uses molecules other than O_2 as final electron acceptors

MULTIPLE CHOICE: *Choose the one best answer.*

1. The Gram stain distinguishes among bacteria based upon differences in
   a. flagellar structure.
   b. cell wall chemistry.
   c. photosynthetic pigments.
   d. ribosomal RNA.

2. It is thought that feathers first arose in dinosaurs as a means of maintaining body heat. Although they were not originally selected for flight, they have properties that make them useful for flight. This change in function is an example of
   a. exaptation.
   b. convergent evolution.
   c. artificial selection.
   d. genetic drift.

3. Some prokaryotes have specialized internal membranes that
   a. have attached ribosomes and function in protein synthesis.
   b. arose from endosymbiosis of smaller prokaryotes, evolving into mitochondria and chloroplasts.
   c. form from infoldings of the plasma membrane and may function in cellular respiration or photosynthesis.
   d. separate plasmids from the single prokaryotic chromosome located in the nucleoid.

4. Which of the following statements is *not* descriptive of conjugation between an Hfr bacterium and an F⁻ bacterium?
   a. The Hfr cell forms sex pili and transfers one strand of its chromosome into an F⁻ cell.
   b. Random movements often break the mating bridge before the entire single strand of its chromosome and F factor is transferred.
   c. Hfr conjugation is the major mechanism that transfers antibiotic resistance genes between bacteria.
   d. Crossing over between homologous genes creates a recombinant recipient cell.

5. Which of the following processes involves viruses known as bacteriophages?
   a. transduction
   b. transformation
   c. exaptation
   d. conjugation

6. Which of the following are obligate anaerobes?
   a. methanogens
   b. cyanobacteria
   c. *Rhizobium* species
   d. actinomycetes

7. Based on the way you obtain nutrients and energy, how would you classify yourself?
   a. photoautotroph
   b. photoheterotroph
   c. chemoautotroph
   d. chemoheterotroph

8. Which of the following groups is not part of the TACK supergroup?
   a. thaumarchaeota
   b. aigarchaeota
   c. cyanobacteria
   d. korarchaeota

9. The placement of prokaryotes into major clades within the domains Archaea and Bacteria has been based on
   a. fossil records of prokaryotes that have been recently discovered.
   b. molecular comparisons.
   c. shape, motility, and Gram-stain results.
   d. nutritional modes and ecology.

10. Which of the following groups is *incorrectly* described?

    a. Proteobacteria—diverse gram-negative bacteria, including pathogens such as *Salmonella* and *H. pylori* and beneficial species such as *Rhizobium*

    b. Chlamydias—intracellular parasites, including a species that causes blindness and nongonococcal urethritis

    c. Spirochetes—helical heterotrophs, including pathogens that cause syphilis and Lyme disease

    d. Cyanobacteria—photosynthetic filaments of cells; many species use the pigment bacteriorhodopsin and inhabit very saline waters

11. Which of the following statements accurately describes a role of prokaryotes in chemical recycling?

    a. Chemoheterotrophic prokaryotes may function as decomposers to return chemical elements to an ecosystem.

    b. Photoheterotrophs convert atmospheric CO_2 into organic compounds.

    c. Chemoautotrophs such as the soil bacterium *Burkholderia glathei* increase the availability of nitrogen for growing seedlings.

    d. All of the above are correct.

12. Some cyanobacteria are capable of nitrogen fixation, a process that

    a. oxidizes nitrogen-containing compounds to produce ATP.

    b. allows these bacteria to live in anaerobic environments.

    c. removes soil nitrogen and returns N_2 gas to the atmosphere.

    d. converts N_2 to ammonia, making nitrogen available for incorporation into proteins and nucleic acids.

13. O157:H7, a pathogenic strain of *E. coli*, has over 1,000 genes that are not present in the harmless strain K-12. Many of these genes are associated with phage DNA in the bacterial chromosome. These observations suggest that

    a. this virulent strain has many more R plasmids than does the harmless strain and is resistant to almost all antibiotics.

    b. at least some of these genes were acquired through phage-mediated horizontal gene transfer.

    c. virulence genes appear to have been acquired through conjugation with K-12.

    d. all of the above are correct.

Chapter 28

Protists

Key Concepts

28.1 Most eukaryotes are single-celled organisms

28.2 Excavates include protists with modified mitochondria and protists with unique flagella

28.3 SAR is a highly diverse group of protists defined by DNA similarities

28.4 Red algae and green algae are the closest relatives of land plants

28.5 Unikonts include protists that are closely related to fungi and animals

28.6 Protists play key roles in ecological communities

Framework

This chapter surveys the incredible diversity of protists—eukaryotes that are not fungi, plants, or animals. The tree to the right represents a phylogenetic hypothesis of all extant eukaryote groups, organized into four larger "supergroups." Dotted lines indicate as-yet-uncertain evolutionary relationships.

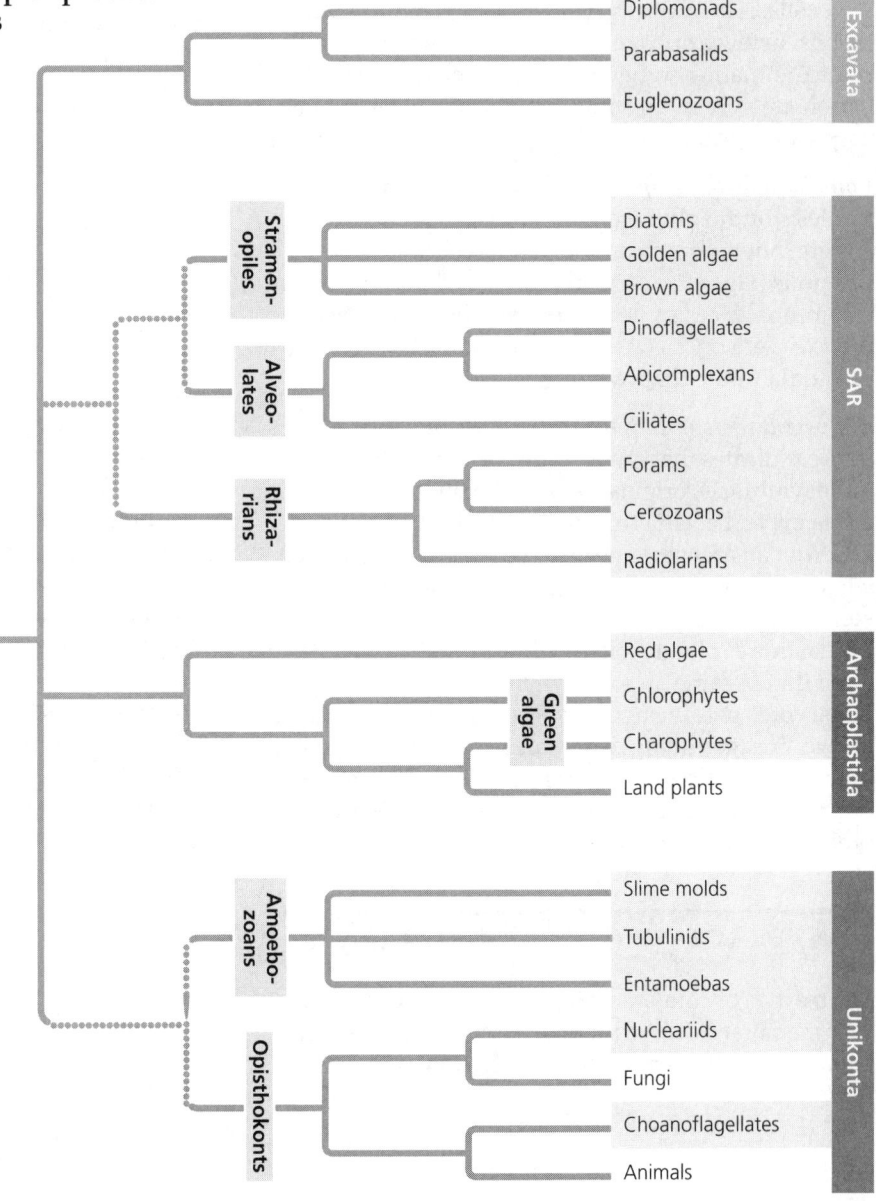

Chapter Review

The traditional kingdom Protista has been abandoned; many lineages of this former group are recognized as their own kingdoms. The informal term **protist** is still used for this diverse assembly of mostly unicellular groups of eukaryotes.

28.1 Most eukaryotes are single-celled organisms

Eukaryotic cells are structurally more complex than prokaryotic cells, with membrane-enclosed organelles and a cytoskeleton, the structural support of which enables changes in cell shape and movement.

Structural and Functional Diversity in Protists
Only three eukaryotic lineages are not protists. Most protists are unicellular, many with structurally complex cells. Nutritionally, protists can be photoautotrophs, heterotrophs, or **mixotrophs**, which are both photosynthetic and heterotrophic. These modes of nutrition are spread throughout the various protist lineages. Reproduction and life cycles are highly diverse.

Four Supergroups of Eukaryotes   A current hypothesis of the phylogeny of protists includes four supergroups, shown on the phylogenetic tree on the previous page. The sequence of their divergence from a common ancestor is not known. The *amitochondriate protists* were once considered the oldest lineage, but new data refute that hypothesis.

Endosymbiosis in Eukaryotic Evolution   Much evidence indicates that mitochondria evolved through **endosymbiosis**, originating from an endosymbiotic alpha proteobacteria that had been engulfed by a host cell from an archaeal lineage. The host may have been a lokiarchaeote, or a complex cell in which a cytoskeleton had evolved.

Plastids evolved from a photosynthetic cyanobacterium that became an endosymbiont of a heterotrophic eukaryote. This lineage of cells eventually gave rise to two lineages of **algae**: red algae and green algae. In several instances of **secondary endosymbiosis**, a red or green alga was engulfed by a heterotrophic eukaryote, leading to new protist lineages.

INTERACTIVE QUESTION   28.1

The plastids of chlorarachniophytes are surrounded by four membranes and enclose a vestigial nucleus called a *nucleomorph*. What do these facts suggest?

28.2 Excavates include protists with modified mitochondria and protists with unique flagella

The supergroup **Excavata** was originally based on morphological studies of the cytoskeleton and an "excavated" feeding groove found in some members. Molecular evidence indicates that each of the three groups of excavates—the diplomonads, parabasalids, and euglenozoans—is monophyletic, and recent genomic studies support the monophyly of the supergroup.

Diplomonads   *Mitosomes*, the modified mitochondria of **diplomonads**, lack functional electron transport chains. As a result, most diplomonads inhabit anaerobic environments. Diplomonads lack plastids, but they possess two equal-sized nuclei and multiple flagella. Many diplomonads are parasitic.

Parabasalids   Like diplomonads, **parabasalids** lack plastids. Their modified mitochondria, known as *hydrogenosomes*, function anaerobically and release H_2.

Euglenozoans   The diverse clade of **euglenozoans** includes predatory heterotrophs, photoautotrophs, mixotrophs, and parasites, all of which have a spiral or crystalline rod inside their flagella.

A single large mitochondrion containing a mass of DNA called a *kinetoplast* is characteristic of the group called **kinetoplastids**. These protists include free-living heterotrophs as well as a number of parasites of plants, animals, and other protists.

The **euglenids** are characterized by one or two flagella that emerge from one end of the cell. Photosynthetic *Euglena* switch to heterotrophy in the absence of sunlight. Many heterotrophic euglenids engulf their prey by phagocytosis.

INTERACTIVE QUESTION   28.2

Sleeping sickness is caused by the kinetoplastid *Trypanosoma* and transmitted by the African tsetse fly. Why is it so difficult for the human immune system to attack this parasite successfully?

28.3 SAR is a highly diverse group of protists defined by DNA similarities

The large, diverse supergroup known as **SAR** is proposed based on recent whole-genome analyses and includes the stramenopiles, alveolates, and rhizarians. Some evidence suggests that the stramenopiles and alveolates originated more than a billion years ago when their common ancestor engulfed a red alga (secondary endosymbiosis). Some members of these groups, however, completely lack plastids.

Stramenopiles   The **stramenopiles** include several important photosynthetic groups, all characterized by hairlike projections on a flagellum. In most cases, a shorter "smooth" flagellum is paired with the "hairy" one.

Diatoms are a major component of marine and freshwater phytoplankton; unique boxlike silica walls protect these unicellular protists. Massive quantities of fossilized diatom walls make up *diatomaceous earth*. Diatom blooms are part of a biological carbon "pump" that sequesters carbon in ocean deposits.

The color of **golden algae** results from yellow and brown carotenoids. Most golden algae are unicellular, biflagellated members of marine and freshwater *plankton*, although some species are colonial. They are all photosynthetic; some species are mixotrophic.

The mostly marine, multicellular **brown algae** are the largest and most complex algae. Some brown algal "seaweeds" have tissues and organs that are analogous to those found in plants. They may have a rootlike **holdfast** and a stemlike **stipe**, which supports leaflike **blades**. Some brown algae have floats, which keep the photosynthesizing blades near the surface. The giant, fast-growing kelps live in deeper waters.

Some multicellular algae have an **alternation of generations**. The multicellular, haploid *gametophyte* produces gametes. After fertilization, the diploid zygote grows into the multicellular *sporophyte*, which then produces motile zoospores by meiosis. The gametophyte and sporophyte may be similar in appearance (**isomorphic**) or distinct (**heteromorphic**).

Alveolates   Membrane-bounded alveoli under the plasma membrane are characteristic of the **alveolates**.

Dinoflagellates include both heterotrophs and photoautotrophs. Photosynthetic dinoflagellates make up a proportion of marine and freshwater *phytoplankton*. The beating of two flagella in grooves on the cell surface between the cellulose plates of the cell produces a characteristic spinning movement. Blooms, or population explosions, of dinoflagellates are responsible for the often-harmful "red tides." Many dinoflagellates are heterotrophic.

Nearly all **apicomplexans** are animal parasites, and most have complex life cycles that often include sexual and asexual stages and several host species. The parasites spread through their hosts as tiny infectious cells called *sporozoites*. These have an apical complex of organelles specialized for invading host cells and a modified plastid called an apicoplast.

INTERACTIVE QUESTION   28.3

a. What are some of the factors that complicate the control of malaria, caused by the apicomplexan *Plasmodium*?

b. Why might the apicoplast of *Plasmodium* provide an effective target for therapeutic drugs?

Ciliates are characterized by the cilia they use to move and feed. Ciliates have two types of nuclei: Large macronuclei control everyday functions of the cell and contain multiple copies of the genome. Small micronuclei are exchanged in a process called **conjugation**, which results in genetic variation. Ciliates reproduce asexually by binary fission.

INTERACTIVE QUESTION   28.4

Label the indicated structures in the following diagram of a *Paramecium*. To which clade, subgroup, and supergroup does this organism belong?

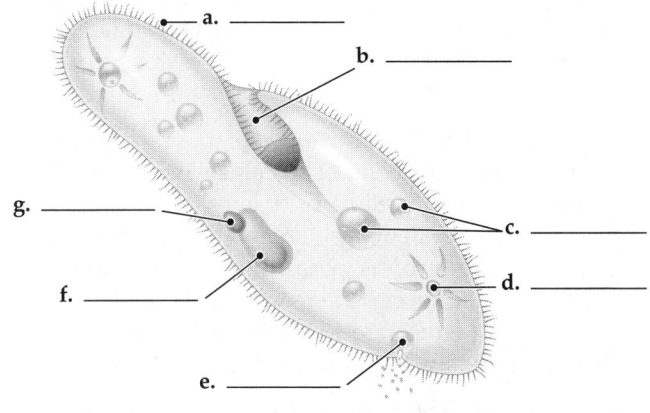

Rhizarians   The third subgroup of the SAR clade is the **rhizarians**. **Amoebas**, which move and feed with their cellular extensions called **pseudopodia**, are spread across many eukaryotic taxa. The rhizarian amoebas have threadlike pseudopodia.

Slender pseudopodia containing bundles of microtubules help **radiolarians** phagocytize microscopic food organisms. These primarily marine organisms have delicate internal silica skeletons.

Foraminiferans, or **forams**, are known for their porous shells, called **tests**, made of organic material and calcium carbonate. Pseudopodia extending through the pores function in swimming, test formation, and feeding. Many forams obtain nourishment from symbiotic algae. Foram species are well represented in the fossil record.

Cercozoans are a large group of amoeboid and flagellated protists that feed with threadlike pseudopodia. Most cercozoans are parasites or predators, although one group is mixotrophic and at least one cercozoan is autotrophic.

INTERACTIVE QUESTION 28.5

a. How are foram fossils used?

b. The cercozoan *Paulinella chromatophora* has a unique photosynthetic structure called a chromatophore. What does genetic and morphological evidence suggest about the origin of the chromatophore?

28.4 Red algae and green algae are the closest relatives of land plants

Descendants of the heterotrophic protist that acquired a cyanobacterium endosymbiont evolved into red algae and green algae. Land plants arose from the green algal lineage. The supergroup **Archaeplastida** is a monophyletic group that includes red algae, green algae, and land plants.

Red Algae The photosynthetic pigment phycoerythrin produces the color of **red algae** and allows them to absorb those wavelengths of light that penetrate into deep water. Red algae are mostly photosynthetic, but some heterotrophic parasitic forms are known. Most red algae species are marine and multicellular, the largest of which are called seaweeds. Red algae reproduce sexually, but have no flagellated gametes.

Green Algae The chloroplasts of **green algae** resemble those of land plants, and molecular and cytological evidence indicates that green algae and plants are closely related. Some systematists favor including green algae in an extended plant kingdom, Viridiplantae. The two main green algal groups are chlorophytes and charophytes. The latter group is very closely related to land plants.

Most chlorophytes live in fresh water. Unicellular forms may be phytoplankton, inhabitants of damp soil, or symbionts in other eukaryotes. Some chlorophytes are colonial, and some have developed true multicellularity.

The life cycles of most chlorophytes include sexual stages (with biflagellated gametes) and asexual stages. Some multicellular green algae have an alternation of generations.

INTERACTIVE QUESTION 28.6

Describe three different mechanisms by which increased size and complexity have evolved in chlorophytes.

28.5 Unikonts include protists that are closely related to fungi and animals

The fourth supergroup, **Unikonta**, includes animals, fungi, and some protists. Molecular systematics supports two major clades: the amoebozoans and the opisthokonts (animals, fungi, and closely related protists). Some studies support the close relationship of these two clades; other studies do not.

Research has yet to establish the root of the eukaryotic tree. Two researchers have proposed that the unikonts were the first eukaryotes to diverge. They base this hypothesis on the DHFR and TS genes, which in bacteria and the unikont groups remain separate (the ancestral character), and which are fused (the derived character) in members of the Excavata, SAR, and Archaeplastida.

Amoebozoans Slime molds, tubulinids, and entamoebas are grouped as **amoebozoans**, a clade that includes many species of amoeba with lobe- or tube-shaped pseudopodia.

The resemblance of the slime molds to fungi is the result of convergent evolution. Molecular evidence indicates that they descended from unicellular ancestors. Plasmodial slime molds engulf food particles by phagocytosis as they grow through leaf litter or rotting logs. This multinucleate mass, called a plasmodium, is the feeding stage. Under harsh conditions, fruiting bodies form that function in sexual reproduction.

During the feeding stage of the life cycle, cellular slime molds consist of haploid solitary amoeboid cells. As food is depleted, the individual cells congregate into a sluglike, motile mass. Some cells form a stalk that supports asexual fruiting bodies, which produce resistant spores.

The cellular slime mold *Dictyostelium discoideum* is an experimental model for studying the evolution of multicellularity. Research indicates that a recognition system involving cell surface proteins allows "noncheater" cells to aggregate differentially from "cheater" cells (mutant cells that always become spore cells).

Tubulinids include a diverse group of unicellular amoebas found free-living in freshwater, marine, or soil habitats, where they consume bacteria and other protists.

Amoebas of the genus *Entamoeba* are parasites of vertebrates and some invertebrates. *E. histolytica* causes amebic dysentery in humans.

Of the protists in the clade **opisthokonts**, nucleariids are more closely related to fungi, and choanoflagellates are more closely related to animals.

INTERACTIVE QUESTION 28.7

Fill in the names of the two major clades and the groups of protists on this phylogenetic tree of the supergroup Unikonta.

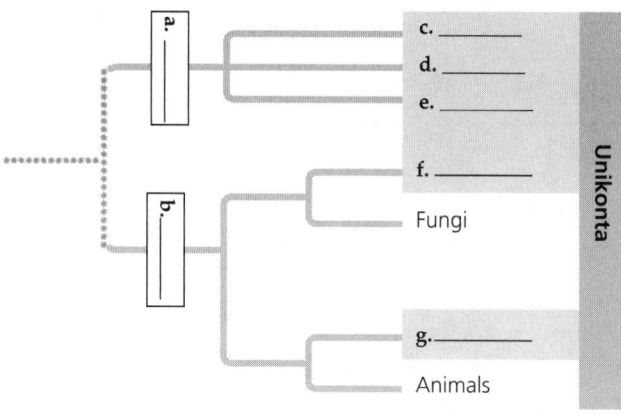

28.6 Protists play key roles in ecological communities

Symbiotic Protists Dinoflagellates are important photosynthetic symbionts in coral. Many termites harbor wood-digesting protists in their guts. Other protists are destructive parasites of plants and animals, some causing serious economic or health concerns for humans.

Photosynthetic Protists Many protists are **producers**. Most aquatic food webs are based on photosynthetic protists and prokaryotes. Aquatic protists contribute about 30% of the world's photosynthesis.

INTERACTIVE QUESTION 28.8

Satellite data have shown a negative correlation, particularly in tropical and midlatitude oceans, between increasing sea surface temperatures and the biomass of photosynthetic protists and prokaryotes. What may account for this decrease in the growth of producers? What effects might these changes have on marine ecosystems and the global carbon cycle?

Word Roots

archae- = ancient (*Archaeplastida:* one of four supergroups of eukaryotes. This monophyletic group, which includes red algae, green algae, and land plants, descended from an ancient protist ancestor that engulfed a cyanobacterium)

con- = with, together (*conjugation:* in ciliates, a sexual process in which two cells exchange haploid micronuclei but do not reproduce)

heter- = different; **morph-** = form (*heteromorphic:* a condition in the life cycle of plants and certain algae in which the sporophyte and gametophyte generations differ in morphology)

is- = same (*isomorphic:* alternating generations in which the sporophytes and gametophytes look alike, although they differ in chromosome number)

opisth- = posterior; **-kont** = flagellum (*opisthokonts:* an extremely diverse clade of eukaryotes that includes fungi, animals, and certain protists)

pseudo- = false; **-podium** = foot (*pseudopodium:* a cellular extension of amoeboid cells used in moving and feeding)

stramen- = straw; **pil-** = hair (*stramenopiles:* one of the three major subgroups for which the "SAR" eukaryotic supergroup is named. This clade arose by secondary endosymbiosis and includes diatoms and brown algae)

uni- = one (*Unikonta:* one of four supergroups of eukaryotes)

Structure Your Knowledge

1. Evidence indicates that all plastids (except for one recent example) evolved from a cyanobacterium that was engulfed by an ancestral heterotrophic eukaryote (primary endosymbiosis). Fill in the blanks in the following diagram, which depicts the diversification of this ancestral eukaryote into red algae and green algae and then to various protist groups.

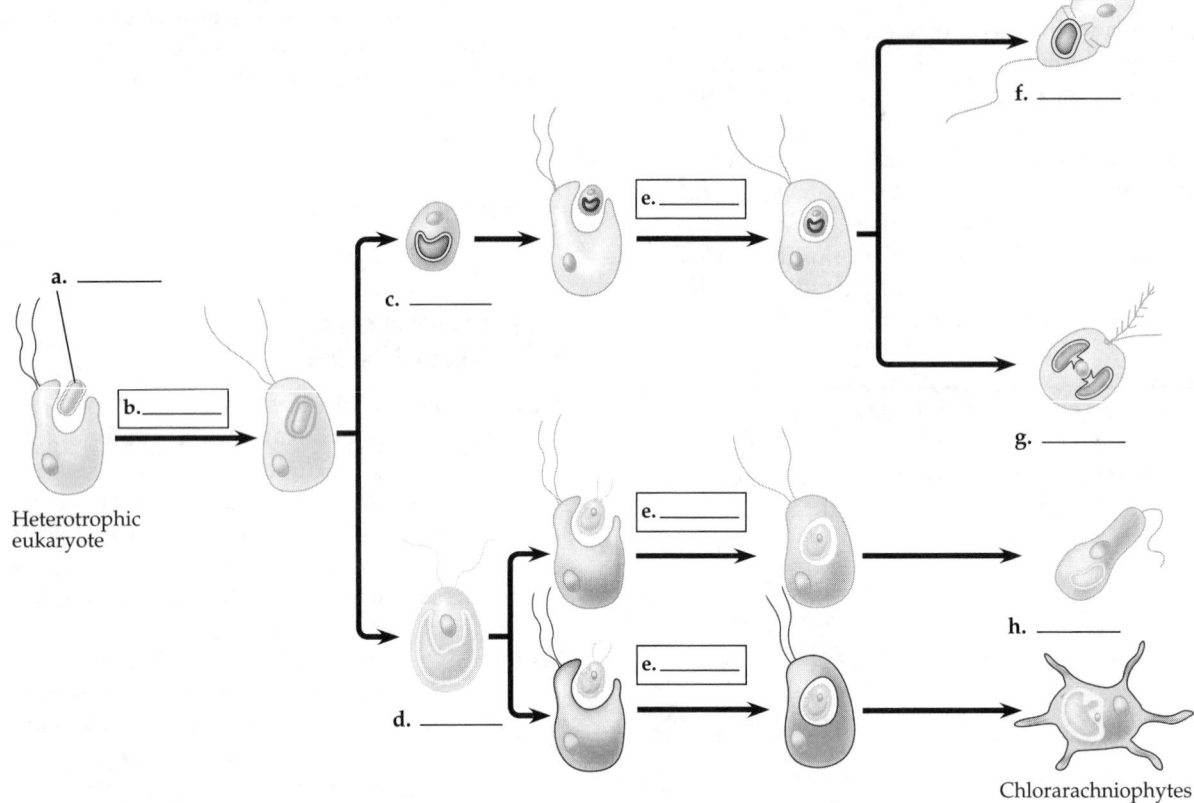

Heterotrophic eukaryote

Chlorarachniophytes

Test Your Knowledge

MATCHING: *Match the protist clades (a–h) with their characteristics and examples (1–8).*

_____   **1.** modified mitochondria; *Giardia* and *Trichomonas*

_____   **2.** plant-type chloroplasts; *Chlamydomonas* and *Ulva*

_____   **3.** lobe-shaped pseudopodia; slime molds, tubulinids, entamoebas

_____   **4.** phycoerythrin, no flagellated stage; *Porphyra*

_____   **5.** hairy and smooth flagella; diatoms, golden and brown algae

_____   **6.** amoebas with threadlike pseudopodia

_____   **7.** subsurface sacs; dinoflagellates, apicomplexans, ciliates

_____   **8.** flagella with spiral or crystalline rod; *Trypanosoma* and *Euglena*

A. alveolates
B. amoebozoans
C. diplomonads and parabasalids
D. euglenozoans
E. rhizarians
F. green algae
G. red algae
H. stramenopiles

MULTIPLE CHOICE: *Choose the one best answer.*

1. Mixotrophs
   a. have unicellular and multicellular parts of their life cycles that differ in morphology.
   b. have both bacterial and archaeal features as a result of endosymbiosis.
   c. can be both heterotrophic and autotrophic.
   d. can be both parasitic and free-living.

2. According to the theory of secondary endosymbiosis,
   a. multicellularity evolved when primitive cells incorporated prokaryotic cells that then took on specialized functions.
   b. the symbiotic associations found in coral result from the incorporation of dinoflagellates into coral animals.
   c. eukaryotic cells descended from cells that had obtained their plastids through endosymbiosis were engulfed and themselves became plastids in heterotrophic eukaryotic cells.
   d. mitochondria originated first from a heterotrophic prokaryotic endosymbiont; plastids originated from a secondary endosymbiosis of a photosynthetic prokaryote.

3. Evidence for secondary endosymbiosis is provided by a structure called a nucleomorph. This structure is characteristic of
   a. chlorachniophytes.   c. golden algae.
   b. nucleariids.   d. diplomonads.

4. The diplomonads and parabasalids are unique among eukaryotes in that they
   a. lack double membranes around their nuclei even though diplomonads have two nuclei.
   b. lack plastids, and all other protist lineages have at least some members that are autotrophic.
   c. have modified mitochondria and obtain energy anaerobically.
   d. were an early branch from bacteria rather than from archaea.

5. Massive fish kills have resulted from the protist *Pfiestra shumwayae*. This organism is
   a. symbiotic.   c. a dinoflagellate.
   b. parasitic.   d. a, b, and c

6. Genetic variation is generated in the ciliate *Paramecium* when
   a. a micronucleus replicates its genome many times and becomes a macronucleus.
   b. micronuclei are exchanged in conjugation.
   c. mutations occur in the many copies of the genome contained in the macronucleus and are passed on to offspring.
   d. plasmids are exchanged in conjugation.

7. Which of the following is *not* true of seaweeds?
   a. Most are classified as charophytes, the closest living relatives of land plants.
   b. They may be a source of food and commercial products.
   c. They often have rootlike structures that anchor them tightly to withstand the turbulence of waves.
   d. They are multicellular photoautotrophs found in red, green, and brown algal groups.

8. Among the unikonts, animals are most closely related to
   a. plants.   c. slime molds.
   b. fungi.   d. choanoflagellates.

9. Green algae are believed to share a common ancestor with plants because
   a. they have flagellated gametes.
   b. they are the only protist lineage with plastids that evolved from cyanobacteria.
   c. their chloroplasts are similar in structure and pigment composition to those of plants.
   d. they are the only algae that exhibit alternation of generations.

10. You feed using pseudopodia, and you have an internal skeleton made of silica. You are a(n)
    a. radiolarian.   c. diatom.
    b. entamoeba.   d. dinoflagellate.

11. Which supergroup includes the land plants?
    a. Unikonta   c. Excavata
    b. Archaeplastida   d. SAR

12. Phytoplankton
    a. form the basis of most marine food chains.
    b. are mutualistic symbionts that provide food for coral reef communities.
    c. are unicellular heterotrophs that float near the ocean surface.
    d. Both a and c are correct.

13. The supergroup known as SAR is so named because it contains
    a. slime molds, archaeoplastids, and red algae.
    b. stramenopiles, alveolates, and rhizarians.
    c. seaweeds, apicomplexans, and radiolarians.
    d. stramenopiles, animals, and rhizarians.

14. How do diatoms affect global CO_2 levels?
    a. As surface seawater warms, the photosynthetic output of diatoms increases, removing more CO_2.
    b. Blooms of diatoms kill fish and other consumers, increasing decomposition and raising CO_2 levels.
    c. CO_2 absorbed by diatoms is "pumped" to the ocean floor when they die, lowering CO_2 levels.
    d. Because their glasslike walls are made of silica and not carbon, diatoms do not affect CO_2 levels.

Chapter 29

Plant Diversity I: How Plants Colonized Land

Key Concepts

29.1 **Plants evolved from green algae**

29.2 **Mosses and other nonvascular plants have life cycles dominated by gametophytes**

29.3 **Ferns and other seedless vascular plants were the first plants to grow tall**

Framework

The evolution of land plants involved adaptations to terrestrial habitats. Plants are also responsible for providing oxygen that aerobic heterotrophs require, for providing food for terrestrial heterotrophs, and for creating habitats for a variety of organisms. This chapter describes the evolution, life cycles, and ecological importance of bryophytes and seedless vascular plants.

Chapter Review

For most of Earth's history, life was confined to aquatic environments. Cyanobacteria and protists had colonized the land by 1.2 billion years ago. Small plants began to move onto land about 500 million years ago, and tall plants appeared about 385 million years ago. Plants enabled other forms of life to survive on land. Plants have diversified into 290,000 known species.

29.1 Plants evolved from green algae

Morphological and Molecular Evidence Plants are multicellular, photosynthetic eukaryotes, as are certain groups of algae. Plants have chloroplasts containing chlorophylls *a* and *b* and cell walls of cellulose, but these characteristics are also found in several algal groups. Only the green algal charophytes and land plants share the following distinctive traits: rings of cellulose-synthesizing proteins, which produce the cellulose microfibrils of cell walls; similarities in sperm cells (in those land plants that have flagellated sperm); and the formation of a phragmoplast, a group of microtubules in the middle of which the cell plate forms in mitosis.

Results from the analysis of nuclear and chloroplast genes confirm that charophytes, including *Zygnema* and *Coleochaete*, are the closest living relatives of land plants.

Adaptations Enabling the Move to Land The tough polymer **sporopollenin** protects charophyte zygotes during fluctuations in water levels. Ancient charophytes living along the edges of ponds and lakes may have given rise to the first plants to colonize land. Sporopollenin walls also encase plant spores.

INTERACTIVE QUESTION   29.1

List some of the benefits and challenges of a terrestrial habitat for the first land plants.

Derived Traits of Plants The following derived traits distinguish land plants: alternation of generations, multicellular dependent embryos, walled spores produced in sporangia, multicellular gametangia, and apical meristems.

Alternation of generations is found in all land plants. In this life cycle, the multicellular haploid **gametophyte** alternates with the multicellular diploid **sporophyte**. The sporophyte produces haploid **spores**, which are reproductive cells that develop directly into new organisms.

INTERACTIVE QUESTION 29.2

The gametophyte produces a. _____ by b. _____.
Following fertilization, the c. _____ divides by
d. _____ to develop into the e. _____. The
sporophyte produces f. _____ by g. _____.
Spores germinate and develop into the h. _____.

Plant embryos are protected and nourished within the female gametophyte plant, which is the basis for referring to plants as **embryophytes**. The embryo has *placental transfer cells* that facilitate the transfer of nutrients from parental tissues.

Spores are produced and protected within multicellular **sporangia** on the sporophyte plant. Diploid cells called **sporocytes**, or spore mother cells, undergo meiosis and produce haploid spores in sporangia. The tough polymer sporopollenin protects spores as they disperse on land.

In the gametophyte, gametes are produced within multicellular **gametangia**. Sperm are produced in **antheridia**. In many groups, sperm are flagellated and swim to the eggs. A single egg cell is produced and fertilized within an **archegonium**, where the zygote develops into an embryo. Gametangia have been lost in many seed plants.

Regions of cell division at the tips of roots and shoots, called **apical meristems**, produce linear growth and cells that differentiate into a protective epidermis and internal plant tissues. The elongation of roots and shoots provides increasing access to environmental resources.

Additional adaptive traits evolved in many species. The leaves and stems of many land plants are coated by a waxy **cuticle**, which protects from water loss and microbial attack. Most plants also have **stomata**, pores through which gases are exchanged and water evaporates.

Early plant fossils show symbiotic associations between plants and fungi. Such *mycorrhizae* facilitate nutrient absorption and may have helped early plants, which lacked roots, colonize land.

The Origin and Diversification of Plants Fossils of spores with chemical and structural similarities to plant spores appeared 470 million years ago, and spores embedded in plant tissue have been found that date back 450 million years.

The 10 extant plant phyla may be grouped as nonvascular plants (bryophytes), seedless vascular plants, and seed plants.

Vascular plants have **vascular tissue**, a transport system composed of cells joined into tubes. Liverworts, mosses, and hornworts do not have an extensive transport system and are referred to as "nonvascular" plants and informally called **bryophytes**. Molecular and morphological evidence indicates that the bryophytes do not constitute a clade.

Vascular plants—about 93% of extant plant species—do form a clade and are divided into smaller clades. The **lycophytes** (club mosses and their relatives) and the **monilophytes** (ferns and their relatives) are informally called **seedless vascular plants**. The seedless vascular plants, like the bryophytes, may be referred to as a *grade*, a group of organisms that share a key adaptation but do not form a monophyletic clade.

A **seed** is an embryo enclosed with a supply of nutrients within a protective coat. The majority of living plants are seed plants, a clade that contains two groups. **Gymnosperms** are called "naked seed" plants because their seeds are not enclosed. **Angiosperms** are flowering plants, in which seeds develop inside chambers in a flower.

INTERACTIVE QUESTION 29.3

Why are the seedless vascular plants considered a grade and not a clade?

29.2 **Mosses and other nonvascular plants have life cycles dominated by gametophytes**

Today there are three phyla of nonvascular plants: **liverworts** (phylum Hepatophyta), **mosses** (phylum Bryophyta), and **hornworts** (phylum Anthocerophyta).

Bryophyte Gametophytes In the bryophytes, the gametophyte is the dominant generation; sporophytes are present only part of the time. When moss spores germinate in a moist habitat, they grow into a mass of green filaments called a **protonema**. Meristem-containing "buds" grow into gamete-producing upright structures called **gametophores**. Some mosses have conducting tissues in their "stems." These tissues apparently evolved independently from the vascular tissue of vascular plants.

Bryophyte gametophytes are generally only a few cells thick and low growing. They are anchored by long, tubular cells or filaments of cells called **rhizoids**, which do not play a major role in water and mineral absorption.

Most mosses have separate male and female gametophytes. Gametangia are enclosed in protective tissue. Sperm are produced in antheridia, from which they swim to the single egg retained in each archegonium. Zygotes develop into embryos, which rely on nutrients transported by placental transfer cells to develop into sporophytes. Many bryophyte species also reproduce asexually, as in the *brood bodies* produced by some mosses.

Bryophyte Sporophytes   Although they are usually photosynthetic when young, bryophyte sporophytes remain attached to and dependent on the parental gametophytes. A sporophyte has a **foot** that obtains nutrients from the gametophyte, a **seta** (stalk) that conducts materials and may elongate for spore dispersal, and a sporangium or **capsule** in which millions of spores are produced by meiosis. The specialized "toothed" **peristome** gradually releases spores to be dispersed by wind currents.

The sporophytes of mosses and hornworts are often larger than those of liverworts and have stomata.

INTERACTIVE QUESTION   29.4

Review the life cycle of a typical moss plant by filling in the following blanks.

The dominant generation is the **a.** _____.

Female gametophytes produce eggs in **b.** _____.

Male gametophytes produce sperm in **c.** _____.

Sperm **d.** _____ through a film of water to fertilize the egg.

The zygote remains in the archegonium and develops into the **e.** _____, still attached to the female gametophyte.

Spores are formed by the process of **f.** _____ in the **g.** _____.

When shed, spores develop into the **h.** _____.

Ecological and Economic Importance of Mosses   Mosses are widely distributed and common in moist habitats. Some are adapted to very cold or dry habitats due to their ability to dehydrate and rehydrate without dying. Research has shown that mosses reduce the rate at which nitrogen is lost by water erosion in a sandy ecosystem. Some species harbor nitrogen-fixing cyanobacteria.

Sphagnum, or "peat moss," is an abundant wetland moss that forms undecayed organic deposits known as **peat**, which is used as a source of fuel. The decay-resistant compounds in peat bogs can preserve organisms for thousands of years. Peat is harvested for use as a soil conditioner and a packing material for plants. Peatlands serve as huge stores of organic carbon and help to stabilize atmospheric CO_2 concentrations. Current overharvesting of *Sphagnum*, as well as its accelerated decomposition due to the drying of some peatlands resulting from global climate change, can increase the release of stored CO_2 and contribute to global warming.

29.3 **Ferns and other seedless vascular plants were the first plants to grow tall**

Origins and Traits of Vascular Plants   The earliest fossils of vascular plants date back 425 million years. These tiny plants had branched, independent sporophytes. The branching made possible more complex bodies and multiple sporangia. They lacked roots and some other adaptations of vascular plants. Seedless vascular plants remain limited to damp habitats because of their swimming sperm.

In extant vascular plants, the diploid sporophyte is the larger and more complex plant.

Vascular plants are characterized by the presence of two systems of vessels. Water and minerals are conducted up from roots in **xylem**. The xylem of most vascular plants includes tube-shaped cells called **tracheids** with cell walls that are *lignified*—strengthened by the polymer **lignin**. **Phloem** transports sugars and other organic nutrients through cells arranged into tubes.

Unlike the rhizoids of bryophytes, **roots** both anchor vascular plants and absorb water and nutrients. Roots may have evolved from subterranean stems, either once in a common ancestor or independently in different lineages of vascular plants.

INTERACTIVE QUESTION   29.5

Describe the traits of vascular plants that enable them to grow tall. What are the adaptive advantages of increased height?

Leaves increase a plant's photosynthetic surface area. Lycophytes, the oldest lineage of extant vascular plants, have single-veined leaves known as **microphylls**. The branching vascular systems of **megaphylls**, typical of most other vascular plants, support larger leaves with greater photosynthetic capacity. Megaphylls appear in the fossil record 370 million years ago, 40 million years after the appearance of microphylls.

Sporophylls are modified leaves bearing sporangia. In ferns, clusters of sporangia called **sori** are produced on the undersides of sporophylls. Cone-like structures called **strobili** are formed from groups of sporophylls in many lycophytes and most gymnosperms.

Most seedless vascular plants are **homosporous**, producing only one kind of spore, which develops into bisexual gametophytes. **Heterosporous** species produce two kinds of spores: **megaspores** that develop into female gametophytes and **microspores** that develop into male gametophytes. All seed plants and some seedless vascular plants are heterosporous.

INTERACTIVE QUESTION   29.6

In the following diagram of the life cycle of a fern, label the processes (in the boxes) and structures (on the lines). Indicate which portion of the life cycle is haploid and which is diploid.

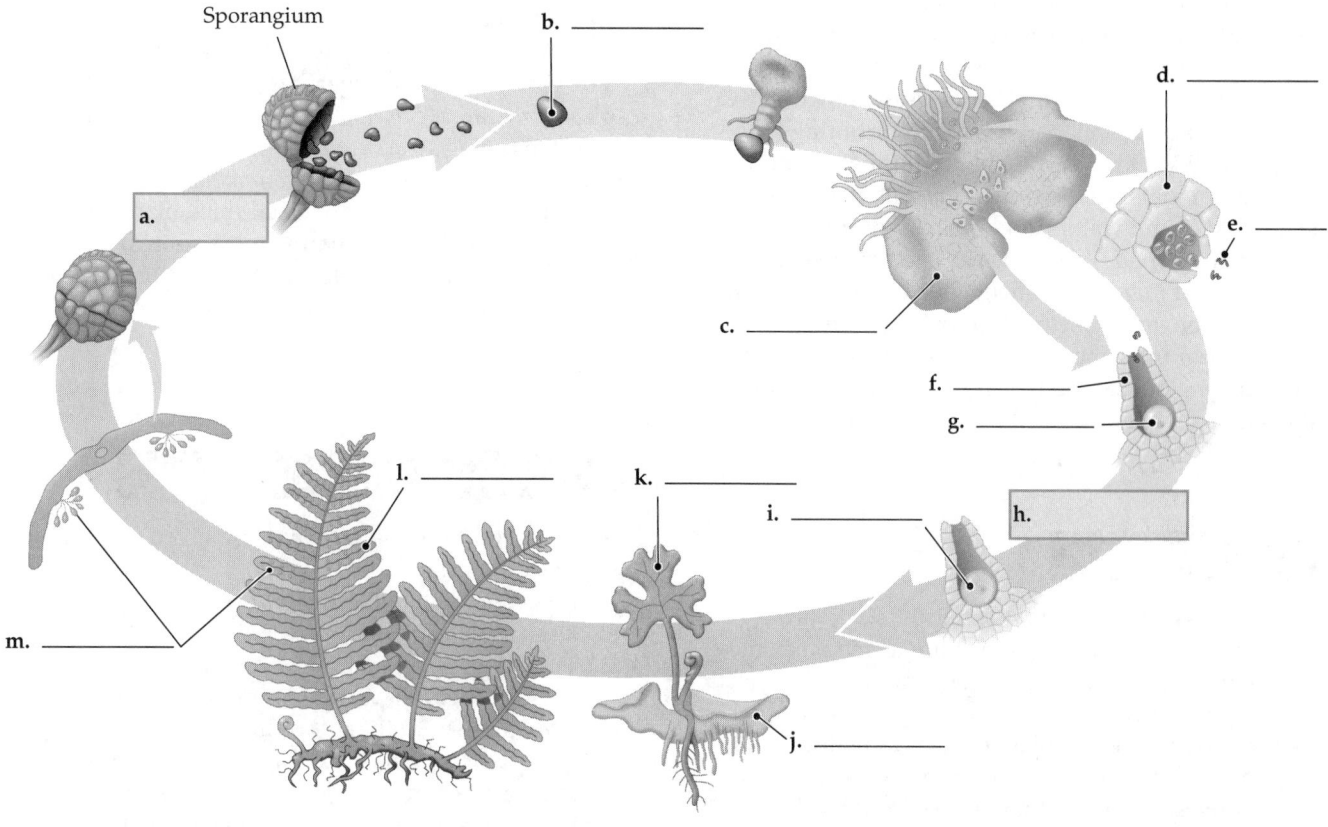

Sporangium

b. _____

d. _____

a.

e. _____

c. _____

f. _____

g. _____

h.

i. _____

k. _____

l. _____

m. _____

j. _____

Classification of Seedless Vascular Plants   The two clades of extant seedless vascular plant are phylum Lycophyta and phylum Monilophyta.

Lycophytes were a major part of the landscape during the Carboniferous period. One evolutionary line, the giant lycophyte trees, became extinct when the climate became drier. The herbaceous lycophytes survived and are represented today by the club mosses, spike mosses, and quillworts. Many tropical species grow on trees as *epiphytes*—plants that anchor to other organisms but are not parasites. Sporophytes grow upright with many small leaves; horizontal stems grow along the soil surface and produce dichotomously branching roots.

Monilophytes, which include ferns, horsetails, and whisk ferns, differ from lycophytes in having megaphylls and roots that branch at various points. Ferns were found in the great forests of the Carboniferous period and are the most numerous seedless vascular plants in the modern flora. Large fern leaves, or fronds, often grow from horizontal stems. Most fern species are homosporous; their gametophytes are small.

Horsetails grew as tall plants during the Carboniferous period. A few species of the genus *Equisetum* are the only representatives found today. These small, upright plants grow in damp locations and have green, jointed stems.

In spite of their dichotomous branching and lack of true roots, molecular evidence indicates that *Psilotum*, the whisk fern, along with the related genus, *Tmesipteris*, form a clade closely related to ferns.

The Significance of Seedless Vascular Plants   The seedless vascular plants formed early forests during the Devonian and early Carboniferous. Plant growth enhanced the release of chemicals from rocks. Calcium and magnesium reacted with CO_2 in rain water and washed into the oceans, where carbon became stored in marine rocks. Such processes decreased CO_2 levels in the atmosphere, perhaps contributing to the global cooling at the end of the Carboniferous period. Dead plant material did not completely decay in the stagnant swamp waters, and great accumulations of peat developed. When the sea later covered the swamps and marine sediments piled on top, heat and pressure converted the peat to coal, again removing carbon from the atmosphere for long periods of time. The burning of coal today is returning this carbon to the atmosphere and contributing to global warming.

Word Roots

angi- = vessel (*gametangium:* multicellular plant structure in which gametes are formed. Female gametangia are called archegonia; male gametangia are called antheridia)

bry- = moss; **-phyte** = plant (*bryophyte:* an informal name for a moss, liverwort, or hornwort; a nonvascular plant that lives on land but lacks some of the terrestrial adaptations of vascular plants)

gymn- = naked; **-sperm** = seed (*gymnosperm:* a vascular plant that bears naked seeds—seeds not enclosed in specialized chambers)

hetero- = different; **spor-** = a seed (*heterosporous:* referring to a plant species that has two kinds of spores: microspores, which develop into male gametophytes, and megaspores, which develop into female gametophytes)

homo- = like (*homosporous:* referring to a plant species that has a single kind of spore, which develops into a bisexual gametophyte)

mega- = large (*megaspores:* a spore from a heterosporous plant that develops into a female gametophyte)

micro- = small; **-phyll** = leaf (*microphyll:* in lycophytes. a small leaf with a single unbranched vein)

peri- = around; **-stoma** = mouth (*peristome:* a ring of interlocking, toothlike structures on the upper part of the moss capsule, often specialized for gradual spore discharge)

-phore = bearer (*gametophore:* the mature gamete-producing structure of a moss gametophyte)

proto- = first; **-nema** = thread (*protonema:* a mass of green, branched, one-cell-thick filaments produced by germinating moss spores)

rhiz- = root; **-oid** = like, form (*rhizoid:* a long tubular single cell or filament of cells that anchors bryophytes to the ground)

Structure Your Knowledge

1. The evolution of land plants involved adaptations to a terrestrial habitat. List some of these adaptations that evolved in the bryophytes and the seedless vascular plants.
2. The diagram on the following page presents a widely held view of plant phylogeny. Fill in blanks a–e with the informal names for the major groups, and blanks f–k with the common names for the modern plant groups (list the three groups included in clade i). To what events do the numbers 1, 2, and 3 refer?

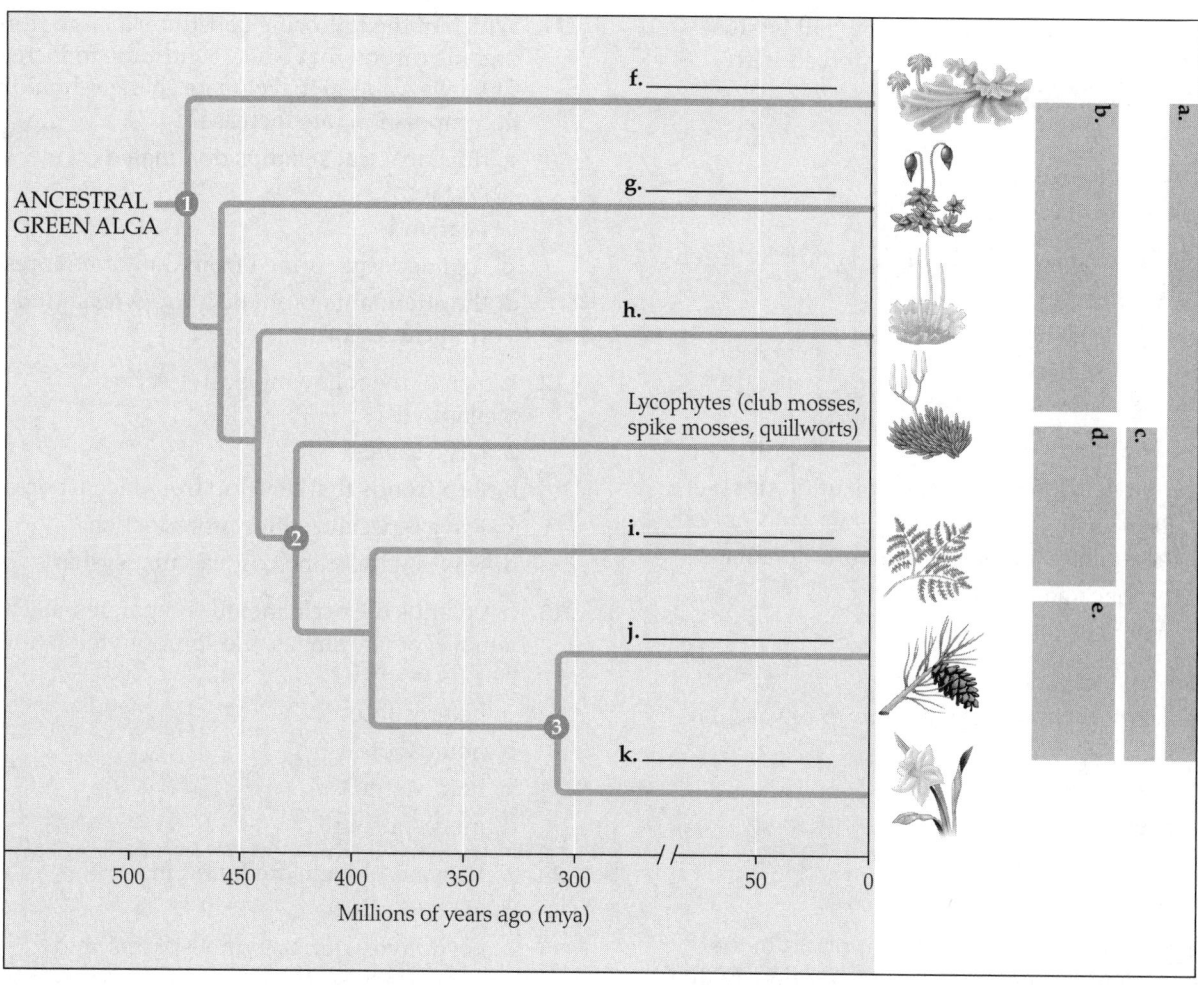

ANCESTRAL GREEN ALGA

f. _____

g. _____

h. _____

Lycophytes (club mosses, spike mosses, quillworts)

i. _____

j. _____

k. _____

500   450   400   350   300   50   0

Millions of years ago (mya)

Test Your Knowledge

MULTIPLE CHOICE: *Choose the one best answer.*

1. Which of the following is a derived trait of plants?
   a. specific structure for flagellated sperm
   b. rings of cellulose-synthesizing proteins
   c. formation of phragmoplasts
   d. apical meristems

2. Plants are thought to be most closely related to charophytes based on
   a. rosette cellulose-synthesizing proteins in their plasma membranes.
   b. phragmoplast formation of cell plates during mitosis.
   c. similarities in sperm cells (in those land plants with flagellated sperm).
   d. all of the above characteristics.

3. Early land plants lacked true roots but were able to gain nutrients through a symbiotic relationship with
   a. fungi.
   b. algae.
   c. nitrogen-fixing bacteria.
   d. insects.

4. Which of the following groups is most likely the closest relative of seed plants?
   a. monilophytes
   b. hornworts
   c. lycophytes
   d. mosses

5. Vascular plants are able to grow tall because their vascular systems are reinforced with
   a. sporopollenin.
   b. lignin.
   c. calcium salts.
   d. microtubules.

6. Where would you look for the sporophyte generation of a bryophyte?
   a. on the ground
   b. under the ground
   c. on a gametophyte
   d. in an antheridium

7. Bryophytes differ from other plant groups because
   a. their gametophyte generation is dominant.
   b. they lack gametangia.
   c. they have flagellated sperm.
   d. they are not embryophytes.

8. Which of the following is a group of vascular plants?
   a. lycophytes
   b. liverworts
   c. mosses
   d. hornworts

9. Bryophytes were the dominant plants in the first 100 million years of plant evolution. By the Carboniferous period, seedless vascular plants formed giant forests. Why could these plants outcompete bryophytes?
   a. Their protected embryos were better able to withstand dry conditions, providing a selective advantage in dominating terrestrial habitats.
   b. They did not require water for their sperm to swim to fertilize the eggs, allowing them to colonize dry habitats.
   c. Their dominant generation was diploid, so they could grow faster than haploid bryophytes.
   d. Their vascular tissue enabled them to grow tall, competing for light and more widely dispersing their spores.

10. Which of the following plant groups is *incorrectly* paired with its gametophyte generation?
    a. lycophyte—either green or nonphotosynthetic tiny plants
    b. moss—green matlike plant
    c. fern—frond growing from horizontal stem
    d. liverwort—flattened, low-growing green plant

11. Which of the following constitute a large store of organic carbon that could contribute to increasing CO_2 levels and global climate change should its decomposition rate increase?
    a. huge tropical swamps dominated by mosses and ferns
    b. peatlands
    c. coal deposits formed from Cambrian forests
    d. the abundant epiphytic lycophytes found in tropical regions

12. Which of the following best describes megaphylls?
    a. large leaves
    b. fern fronds that develop from megaspores
    c. leaves specialized for reproduction
    d. leaves with branching vascular systems

13. If a plant's life cycle includes separate male and female gametophytes, the sporophyte plant must be
    a. heterosporous.
    b. homosporous.
    c. homologous.
    d. megasporous.

14. Xylem and phloem are found in
    a. all land plants.
    b. bryophytes, ferns, gymnosperms, and angiosperms.
    c. the vascular plants, which include lycophytes, monilophytes, gymnosperms, and angiosperms.
    d. only the vascular plants with seeds.

15. If you could take a time machine back to the Carboniferous period, which of the following would you most likely encounter?
    a. creeping mats of low-growing bryophytes
    b. fields of tall grasses swaying in the wind
    c. swampy forests dominated by tree lycophytes, horsetails, and ferns
    d. huge forests of naked-seed trees filling the air with pollen

Chapter 30

Plant Diversity II: The Evolution of Seed Plants

Key Concepts

30.1 **Seeds and pollen grains are key adaptations for life on land**

30.2 **Gymnosperms bear "naked" seeds, typically on cones**

30.3 **The reproductive adaptations of angiosperms include flowers and fruits**

30.4 **Human welfare depends greatly on seed plants**

Framework

The reduction and protection of the gametophyte within the parent sporophyte, the protection and dispersal of embryos in seeds, and the use of pollen for the transfer of sperm are reproductive adaptations of seed plants that enhance their success on land.

Gymnosperms bear their seeds "naked" on modified sporophylls. Angiosperms are the most diverse and widespread of plants, owing much of their success to their efficient reproductive apparatus—the flower—and their seed-dispersal mechanism—the fruit. Agriculture is based almost entirely on angiosperms.

Chapter Review

Seed plants originated about 360 million years ago. One of the key adaptations that enabled seed plants to become the dominant members of terrestrial ecosystems is the **seed**—an embryo packaged with food within a protective coat.

30.1 Seeds and pollen grains are key adaptations for life on land

Advantages of Reduced Gametophytes The extremely reduced gametophytes of seed plants develop within the sporangia of the sporophyte plant, where they are nourished and also protected from drying out and UV radiation.

Heterospory: The Rule Among Seed Plants Unlike most seedless plants, which are *homosporous*, all seed plants are *heterosporous*. Each megaspongium produces a single functional *megaspore*, which gives rise to the female gametophyte; a microsporangium produces many *microspores*, which give rise to male gametophytes.

Ovules and Production of Eggs The megasporangium is retained in the parent sporophyte, surrounded by protective **integuments** derived from sporophyte tissue. An **ovule** consists of the integument(s), megasporangium, and megaspore. The female gametophyte develops from the megaspore and produces one or more eggs.

Pollen and Production of Sperm A microspore develops into a male gametophyte contained in a sporopollenin-protected **pollen grain**. The transfer of pollen to the part of a plant containing ovules is called **pollination**. When pollen grains germinate, they release sperm through a pollen tube to the female gametophyte.

INTERACTIVE QUESTION 30.1

Why is the evolution of pollen an important terrestrial adaptation?

The Evolutionary Advantage of Seeds   Unicellular spores are the only protected stage in bryophytes and seedless vascular plants. An ovule develops into a seed consisting of a sporophyte embryo surrounded by a food supply and enclosed within a protective coat. Seeds not only provide food to an embryo, but also can withstand harsh environmental conditions during dormancy and disperse offspring.

Label the parts in the following generalized diagrams of an unfertilized ovule and a seed of a gymnosperm. Indicate whether structures are diploid or haploid tissues.

Ovule   Seed

30.2   Gymnosperms bear "naked" seeds, typically on cones

Seed plants include two sister clades: Gymnosperms have "naked" seeds that are exposed on sporophylls that usually form cones. The most widespread gymnosperm group is the cone-bearing **conifers**. Angiosperms are plants in which seeds develop inside chambers.

The Life Cycle of a Pine   The pine tree is a heterosporous sporophyte. In the microsporangia on pollen cones, microsporocytes undergo meiosis and give rise to microspores, each of which develops into a pollen grain enclosing a male gametophyte. Scales of an ovulate cone hold two ovules, each of which contains a megasporangium. A megasporocyte undergoes meiosis, and one of the resulting megaspores undergoes repeated divisions to produce a female gametophyte in which a few archegonia develop. Most pines have both types of cone on the same tree.

Following pollination, a pollen tube grows and digests its way through the megasporangium. Fertilization may occur more than a year after pollination. The zygote develops into a sporophyte embryo, which is nourished by the remaining female gametophyte tissue and is enclosed within a seed coat derived from the integument of the parent sporophyte.

a. Describe a pollen cone and the formation of a male gametophyte.

b. Describe an ovulate cone and the formation of a female gametophyte.

Early Seed Plants and the Rise of Gymnosperms   Transitional species of heterosporous, wood-producing, but seedless vascular plants date from 380 million years ago. The first seed plants to appear in the fossil record date to about 360 mya. The earliest gymnosperm fossils from a species in an extant lineage are about 305 million years old.

During the Carboniferous period, early gymnosperms lived in ecosystems dominated by seedless vascular plants. Drier conditions during the Permian gave gymnosperms a selective advantage with their seeds, pollen, and small leaves covered in thick cuticles. Gymnosperms supported the giant herbaceous dinosaurs of the Mesozoic era, which extended from 252 to 66 mya.

Gymnosperm Diversity   There are four gymnosperm phyla. Cycadophyta includes cycads, which have large cones, motile sperm, and palmlike leaves. Cycads thrived during the Mesozoic era. Today, 75% of cycad species are endangered. Ginkgophyta has only one extant species: the deciduous, ornamental ginkgo tree (*Gingko biloba*). Gnetophyta includes three genera: the bizarre *Welwitschia*, the tropical *Gnetum*, and the desert shrub *Ephedra*. The largest gymnosperm phylum is Coniferophyta. Most conifers are evergreens with needle-shaped or scale-like leaves. Coniferous trees are some of the largest and oldest living organisms.

30.3   The reproductive adaptations of angiosperms include flowers and fruits

The diverse and widespread angiosperms (phylum Anthophyta) include more than 90% of all plants—over 250,000 species.

Characteristics of Angiosperms   The **flower**, the unique reproductive structure of an angiosperm, has up to four types of modified leaves called floral organs. The outer **sepals** are usually green, whereas **petals** are brightly colored in most flowers that are pollinated by insects and birds. **Stamens** produce microspores, which develop into pollen grains containing male gametophytes. A stamen has a stalk, called a **filament**, and a terminal **anther**, in which pollen is produced. **Carpels** produce megaspores, which develop into female gametophytes. The carpel has a sticky **stigma**, which receives pollen, and a **style**, which leads to the **ovary**. The ovary contains ovules, which develop into seeds.

Flowers may have multiple carpels, and in many cases these are fused. A single carpel is called a simple **pistil**, and a compound pistil is formed from multiple fused carpels. Many aspects of flowers can differ, including symmetry (radial or bilateral), color, size, and fragrance.

A **fruit** is a mature ovary that functions in the protection and dispersal of seeds. As seeds develop, the ovary enlarges and forms the thickened wall (pericarp) of the fruit. Fruits may be fleshy or dry, and many show modifications that enhance their dispersal. The dry fruits of grasses are major foods for humans.

In the angiosperm life cycle, the two highly reduced gametophytes are produced within the flower. Pollen grains contain the male gametophytes, consisting of a *generative cell* (which divides to form two sperm) and a *tube cell*. Ovules contain the female gametophyte, called an **embryo sac**, which consists of a few cells.

Most flowers have some mechanism to ensure **cross-pollination**, which increases genetic variation. A pollen grain germinates on the stigma and extends a pollen tube down the style, through the **micropyle**, and into the ovule, where it releases two sperm cells into the embryo sac. In a process called **double fertilization**, one sperm unites with the egg to form the zygote, and the other sperm fuses with the two polar nuclei in the large central cell of the female gametophyte.

INTERACTIVE QUESTION 30.4

In the following diagram of the life cycle of an angiosperm, label the indicated structures and processes. Which structures represent the male and female gametophyte generations?

The zygote divides and forms the sporophyte embryo, consisting of a rudimentary root and one or two seed leaves, called **cotyledons**. The **endosperm**, which develops from the triploid central cell, serves as a food reserve for the embryo. Double fertilization may be an adaptation that ensures that food reserves are only provided when an egg has been fertilized.

Angiosperm Evolution   Although details of their evolutionary origin remain unclear, angiosperms are thought to have originated about 140 million years ago. By 100 mya, angiosperms were becoming more dominant and increased in diversity and importance by the end of the Cretaceous (66 mya).

Although fossils of 125-million-year-old *Archaefructus* and *Leefructus* indicate that this group was herbaceous and aquatic, all fossil seed plants proposed to be close relatives of angiosperms were woody, suggesting that the angiosperm common ancestor was woody. Evidence indicates that extant gymnosperms diverged from the ancestors of angiosperms 305 mya. The extinct group Bennettitales had flowerlike structures and may be more closely related to angiosperms than to gymnosperms.

The woody shrub *Amborella trichopoda*, water lilies, and star anise are living members of early diverging angiosperm lineages. These basal angiosperms lacked *vessel elements*. Early angiosperms may have been shrubs with small flowers and simple water-conducting cells.

Animals have influenced the evolution of plants and vice versa. Both plant defenses against herbivory and herbivores that can overcome plant defenses would be favored by natural selection. Beneficial relationships with pollinating animals would also affect evolutionary adaptations.

Pollinators can enter bilateral flowers from only one direction and thus more specifically pick up and disperse pollen to flowers of the same species. A study comparing the number of species in clades with bilaterally symmetrical flowers with the number in sister clades with radially symmetrical flowers supports the hypothesis that bilateral flower shape promotes speciation, perhaps by reducing gene flow between diverging populations.

Angiosperm Diversity   The angiosperms were traditionally divided into two main groups, **monocots** and **dicots**. Recent DNA comparisons reveal that although the monocots are a monophyletic group, the dicots are not. The clade **eudicots** includes most previously classified dicots, but four other former dicot lineages have been identified. The three early-diverging lineages are called **basal angiosperms**. They include *Amborella trichopoda*, the water lilies, and the clade of star anise and its relatives. A lineage known as the **magnoliids** is more closely related to monocots and eudicots than to the basal angiosperms.

About one-quarter of angiosperm species are monocots, such as orchids, palms, and grasses (which include many important crops). More than two-thirds of angiosperms are eudicots, such as legumes, the rose family, and flowering trees.

INTERACTIVE QUESTION   30.5

Fill in the following table, which compares characteristics of monocots and eudicots.

Characteristic	Monocot	Eudicot
Cotyledons		
Leaf venation		
Stems		
Roots		
Pollen		
Floral organs		

30.4   Human welfare depends greatly on seed plants

Products from Seed Plants   Almost all of our food crops are angiosperms, and angiosperms are also used to feed livestock. The dramatic evolution of domesticated plants is a result of artificial selection in plant breeding. Seed plants also supply spices, wood, and medicines.

Threats to Plant Diversity   The growing human population has increasing demands for space, food, and natural resources. Tropical rain forests, where plant diversity is greatest, are rapidly being destroyed, probably contributing to global climate change and resulting in the loss of potential new food crops and medicines.

Word Roots

anth- = a flower (*anther:* the terminal pollen sac of a stamen, where pollen grains containing sperm-producing male gametophytes form)

carp- = a fruit (*carpel:* the ovule-producing reproductive organ of a flower, consisting of the stigma, style, and ovary) endo- = inner (*endosperm:* a nutrient-rich tissue, formed by the union of a sperm cell with two polar nuclei during double fertilization, that provides nourishment to the developing embryo in angiosperm seeds)

Structure Your Knowledge

1. Describe the characteristics of seed plants that are evolutionary adaptations to a terrestrial habitat.
2. What adaptations helped angiosperms become the most successful and widespread land plants?
3. Current morphological and molecular evidence suggests six angiosperm lineages. Label those lineages on the following proposed phylogenetic tree.

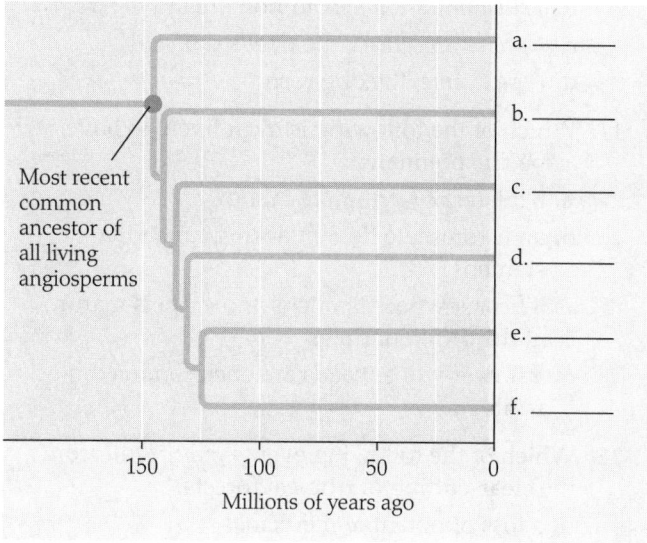

Most recent common ancestor of all living angiosperms

a. _____
b. _____
c. _____
d. _____
e. _____
f. _____

150   100   50   0
Millions of years ago

Test Your Knowledge

TRUE OR FALSE: *Indicate T or F, and then correct the false statements.*

_____ 1. All photoautotrophic, multicellular eukaryotes are plants.

_____ 2. Heterosporous plants produce male and female spores.

_____ 3. The gametophyte generation is most reduced in the gymnosperms.

_____ 4. *Ginkgo*, cycads, and conifers are naked-seed plants.

_____ 5. A sporangium produces spores, no matter what group it is found in.

_____ 6. A fruit consists of an embryo, nutritive material, and a protective coat.

_____ 7. A stamen consists of a filament and an anther that produces microspores, which give rise to pollen grains.

_____ 8. The female gametophyte in angiosperms consists of haploid cells in which a few archegonia develop.

MULTIPLE CHOICE: *Choose the one best answer.*

1. In which of the following groups must sperm no longer swim to reach the female gametophyte?
   a. bryophytes
   b. ferns
   c. most gymnosperms and angiosperms
   d. both **b** and **c**

2. Which of the following is *not* a characteristic of seed plants?
   a. dominant gametophyte generation
   b. dominant sporophyte generation
   c. homospory
   d. flagellated sperm

3. For a pine seed, which of the following choices states the number of generations and correctly describes those generations?
   a. two: seed coat and food supply from female gametophyte and the sporophyte embryo
   b. two: seed coat from integument of the parent sporophyte and new sporophyte embryo
   c. three: seed coat from the parent sporophyte, food supply from the gametophyte, and the sporophyte embryo
   d. three: seed coat from the female gametophyte, food supply from the parent sporophyte, and the sporophyte embryo

4. Gymnosperms rose to dominance during the
   a. Devonian period, when they successfully competed with the short-statured bryophytes.
   b. Carboniferous period, when they formed important components of the great "coal forests."
   c. Permian period, when they replaced the seedless vascular plants as the climate became drier.
   d. Mesozoic era, when they replaced angiosperms as the largest and most widespread group of plants.

5. In a pine tree, where would you find a microsporangium?
   a. within the embryo sac in an ovule
   b. in the pollen sacs in an anther
   c. on a scale in a pollen cone
   d. on a scale in an ovulate cone

6. Which of the following is a way in which angiosperms differ from gymnosperms?
   a. pollination by insects instead of by wind
   b. nutritive tissue packaged into the seed
   c. fruits, which develop from ovules
   d. none of the above

7. Which of the following correctly describes the path a pollen tube takes to reach the female gametophyte in an angiosperm?
   a. stigma, style, ovary, ovule, embryo sac
   b. anther, stigma, filament, ovule, ovum
   c. stigma, filament, carpel, ovary, ovule
   d. anther, stigma, ovary, ovule, embryo sac

8. Which of the following pairs is mismatched?
   a. endosperm—triploid
   b. gametophyte—diploid
   c. tube nucleus—haploid
   d. zygote—diploid

9. Which of the following sentences describes a likely advantage of double fertilization in angiosperms?
   a. Two embryos are produced in a seed, thus increasing the reproductive output of a plant.
   b. This is a remnant of the two sperm produced by the ancestral angiosperm; there does not appear to be an adaptive advantage to this trait.
   c. The endosperm does not form unless fertilization of the egg occurs, so plants do not waste nutrients on infertile ovules.
   d. The triploid endosperm resulting from the fusion of two sperm with the central cell has a higher nutrient content than the haploid food supply surrounding gymnosperm embryos.

10. Which of the following angiosperm species would you expect to be wind-pollinated?
    a. *Vanilla plainfolia*, a vanilla orchid
    b. *Triticum aestivum*, common bread wheat
    c. *Narcissus poeticus*, the poet's daffodil
    d. *Rosa canina*, the dog rose

11. Which of the following is most likely to have a specific pollinator?
    a. a bilaterally symmetrical flower
    b. an incomplete flower lacking functional stamens
    c. a flower whose stamens and carpels mature at the different times
    d. a flower whose seeds are enclosed in edible fruits

12. Which of the following events may result from the clear-cutting of tropical forests?
    a. a loss of potential medicines
    b. an increase in atmospheric CO_2 levels, adding to global climate change
    c. extinction of many plant and animal species
    d. all of the above

Chapter 31

Fungi

Key Concepts

31.1 Fungi are heterotrophs that feed by absorption

31.2 Fungi produce spores through sexual or asexual life cycles

31.3 The ancestor of fungi was an aquatic, single-celled, flagellated protist

31.4 Fungi have radiated into a diverse set of lineages

31.5 Fungi play key roles in nutrient cycling, ecological interactions, and human welfare

Framework

This chapter describes the morphology, life cycles, evolutionary history, diversity, and economic and ecological importance of the kingdom Fungi. Fungi play an essential ecological role, both as decomposers and in their mycorrhizal association with plant roots. A flagellated protist may have been the common ancestor of fungi and animals.

Chapter Review

About 100,000 fungal species have been described; mycologists estimate that as many as 1.5 million fungal species may exist. Fungi are found in nearly every habitat on Earth.

31.1 Fungi are heterotrophs that feed by absorption

Nutrition and Ecology   Fungi are heterotrophs that obtain their nutrients by absorption; many secrete digestive enzymes into the surrounding food and absorb the resulting small organic molecules. Fungi may be decomposers that break down nonliving organic material, parasites that absorb nutrients from living cells, or mutualists that feed on, but also benefit, their hosts.

Body Structure   Most fungi are composed of multicellular filaments. Some are single-celled **yeasts**, which may inhabit liquid or moist habitats. Some species can grow as both filaments and yeasts.

A typical multicellular fungal body consists of a network of filamentous **hyphae**, which form a mass called a **mycelium**. This body form provides an extensive surface area for absorption of nutrients. The tubular cell walls are strengthened by **chitin**, which offsets the osmotic flow of water that follows the absorption of nutrients. Although fungi are nonmotile, the rapid growth of their hyphae enables them to enter new food territory.

The hyphae of most fungi are divided into cells by cross-walls called **septa**, which usually have pores through which nutrients and cell organelles can pass. **Coenocytic fungi** lack septa and consist of a continuous cytoplasmic mass containing many nuclei.

Specialized Hyphae in Mycorrhizal Fungi   Special hyphae allow some fungi to trap and feed on animals, while other fungi extract nutrients from plants using modified hyphae called *haustoria*. Of particular importance are the fungi that participate in mutually beneficial relationships (called **mycorrhizae**) with plants using branching hyphae called **arbuscules**. **Ectomycorrhizal fungi** form hyphal sheaths around a root and also grow into spaces in the root cortex. **Arbuscular mycorrhizal fungi** extend their hyphae through root cell walls, pushing in the plant cell membrane to form haustoria within root cells.

Mycorrhizal fungi disperse to new areas by producing haploid **spores**.

INTERACTIVE QUESTION   31.1

In what way(s) do mycorrhizal fungi benefit plants? In what way(s) do plants benefit the fungi?

31.2 Fungi produce spores through sexual or asexual life cycles

Fungi release huge quantities of spores, produced either sexually or asexually, which aids in dispersal.

Sexual Reproduction   Nuclei of hyphae and spores of most fungal species are haploid. Sexual reproduction involves the release of sexual-signaling **pheromones**, which may cause the hypha of suitable mating types to grow toward and fuse with each other. Such cytoplasmic fusion, called **plasmogamy**, forms a mycelium with genetically different nuclei, or a **heterokaryon**. Nuclei from the two parents may pair up in cells and divide in tandem, forming **dikaryotic** cells. After a period of time, the nuclei fuse (**karyogamy**), forming the only diploid stage in most fungi. Meiosis produces genetically diverse haploid spores.

Asexual Reproduction   Many fungi reproduce asexually by spores; some species reproduce only asexually.

Molds are rapidly growing mycelia that reproduce asexually by spores. Many molds can reproduce sexually if they encounter other mating types. Asexual reproduction in single-celled yeasts occurs by cell division or budding.

Molds and yeasts for which no sexual stage is known are traditionally called **deuteromycetes**. When mycologists discover a sexual stage in one of these fungi, they classify the species accordingly. Genetic analysis is also used to classify fungi.

INTERACTIVE QUESTION   31.2

Briefly define each of the following terms that relate to the structure and reproduction of fungi.

a. mycelium

b. septa

c. coenocytic

d. plasmogamy

e. heterokaryon

f. dikaryon

g. karyogamy

31.3 The ancestor of fungi was an aquatic, single-celled, flagellated protist

Fungi and Animalia are more closely related to each other than to plants or to other eukaryotes.

The Origin of Fungi   Although most fungi lack flagella, phylogenetic systematics indicates that they evolved from a flagellated ancestor. The fungi, animals, and their protistan relatives form a clade called **opisthokonts**.

DNA data indicate that fungi are most closely related to **nucleariids**, a group of single-celled protists, suggesting that the ancestor of fungi was unicellular. Evidence indicates that fungi and animals evolved from different single-celled ancestors, and thus evolved multicellularity independently. Molecular clock estimates suggest that the ancestors of animals and fungi diverged as early as 1.5 billion years ago. Although fungi most likely originated in water, the oldest accepted fossils of fungi are of terrestrial species that date to about 460 million years ago.

Basal Fungal Groups   Genomic studies have identified chytrids in the genus *Rozella* as an early-diverging lineage and placed them in a clade of unicellular fungi with flagellated spores called "cryptomycota." The cryptomycota entirely lack chitin, which suggests that the chitinous cell wall evolved after they diverged from other fungi.

The Move to Land   Cyanobacteria, algae, and fungi probably colonized land before the arrival of plants (about 470 million years ago). Early fossils of land plants (from 405 mya) include evidence of mycorrhizae. Both a fungus and its plant partner must express certain genes to establish a symbiosis. Plant genes required for the formation of mycorrhizae are present in all major plant lineages, suggesting that these *sym* genes were present in the common ancestor of land plants.

INTERACTIVE QUESTION   31.3

Fill in the blanks in the following phylogenetic tree of the opisthokonts.

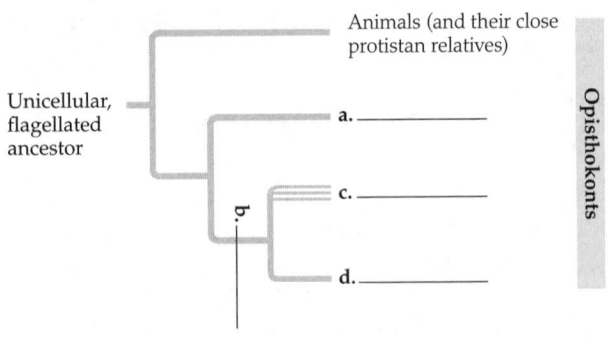

31.4 Fungi have radiated into a diverse set of lineages

Recent metagenomic studies indicate that there may be as many as 1.5 million species. Although fungi have typically been classified in five major groups, recent discoveries like the cryptomycota and previously unknown unicellular fungi suggest that fungal diversity may be much higher than has been supposed.

Chytrids   The paraphyletic **chytrids** are unique in having flagellated **zoospores** and may include some of the earliest fungal groups to diverge. Metagenomic studies have found new clades of this ubiquitous group in hydrothermal vent and other marine communities. Chytrids include aquatic decomposers and parasites. Some are mutualists in the digestive tracts of sheep and cattle.

Zygomycetes   **Zygomycetes** are a paraphyletic group that includes decomposing molds, parasites, and commensal symbionts.

Rhizopus stolonifer, the black bread mold, is a common zygomycete that spreads horizontal coenocytic hyphae across its food substrate and erects hyphae with bulbous black sporangia containing hundreds of asexual haploid spores. Sexual reproduction involving plasmogamy between mycelia of different mating types produces a heterokaryotic **zygosporangium** with a tough protective coat that remains dormant until conditions are favorable. Karyogamy is followed by meiosis to produce genetically diverse haploid spores.

Glomeromycetes   Analysis of fungal genomes has identified the **glomeromycetes** as a separate clade. Almost all glomeromycetes are arbuscular mycorrhizal fungi. Hyphae that invade plant root cells branch into tiny treelike arbuscles. More than 80% of all plants form mycorrhizae with glomeromycetes.

Ascomycetes   The **ascomycetes**, or *sac fungi*, are found in a wide variety of habitats. They produce sexual ascospores in saclike **asci**, usually contained in fruiting bodies called **ascocarps**.

Ascomycetes range in complexity from yeasts to elaborate cup fungi and morels. Some are serious plant pathogens; others are important decomposers. Many ascomycete species form symbiotic associations called lichens with algae or cyanobacteria, and some form mycorrhizae with plants.

Asexual reproduction involves spores called **conidia**, which are produced in chains or clusters at the ends of hyphae called conidiophores. Plasmogamy between different mating types produces dikaryotic hyphae. Within asci formed at the tips of these hyphae, karyogamy is followed by meiosis, yielding genetically diverse ascospores.

Genome comparisons show that the genome of the bread mold *Neurospora* has three-fourths as many genes as *Drosophila* and one-half as many genes as humans, but an unusually high gene density, as it has little noncoding DNA.

Basidiomycetes   The **basidiomycetes**, which include mushrooms, puffballs, and shelf fungi, produce a characteristic club-shaped cell called a **basidium**. These *club fungi* include important decomposers, mycorrhizae-forming mutualists, and the plant parasites rusts and smuts.

In response to environmental stimuli, an elaborate fruiting body called a **basidiocarp** is formed from the long-lived dikaryotic mycelium. Karyogamy and meiosis occur in numerous basidia, producing huge numbers of haploid, genetically diverse basidiospores.

INTERACTIVE QUESTION   31.4

Indicate whether the following diagrams (1, 2, 3) are from a zygomycete, an ascomycete, or a basidiomycete life cycle. Identify the labeled structures and whether spores are produced asexually or sexually.

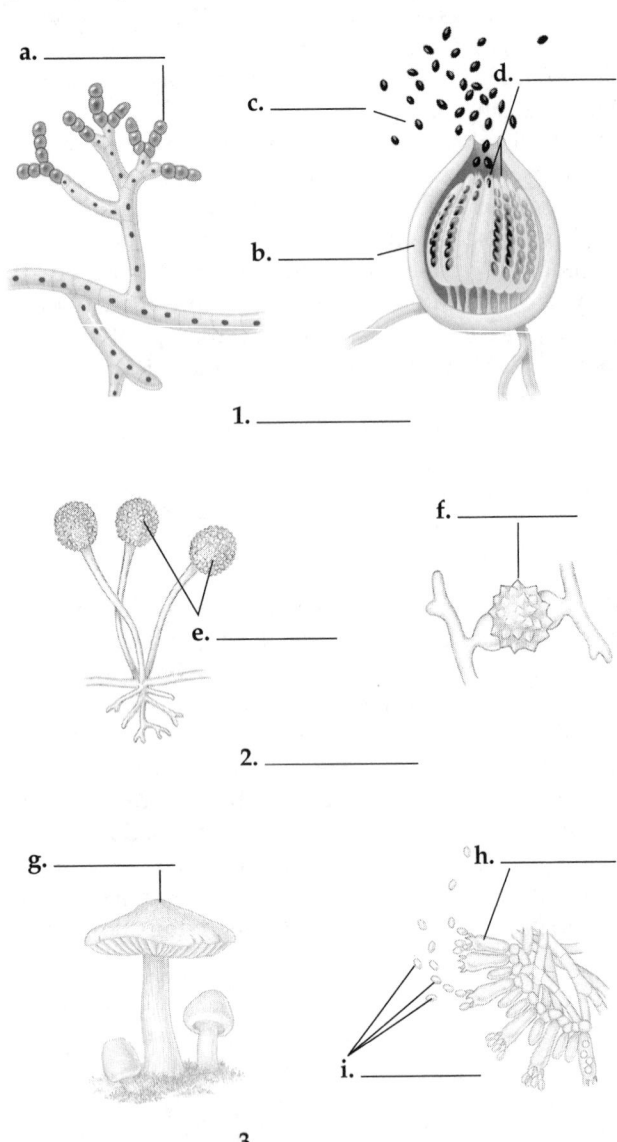

a. _____

c. _____

d. _____

b. _____

1. _____

f. _____

e. _____

2. _____

g. _____

h. _____

i. _____

3. _____

The leaves or other parts of all plant species studied thus far have symbiotic **endophytes,** which benefit plants by producing toxins that deter herbivores, or increasing the plant's tolerance for heat, drought, or heavy metals. They may also play a role in defending against pathogens.

Fungal inhabitants of the guts of grazing mammals help digest plant material. Farmer leaf-cutter ants raise crops of fungi by gathering leaves and feeding them to the fungi.

Lichens are symbiotic associations of millions of photosynthetic organisms in a lattice of fungal hyphae. The fungus is usually an ascomycete, and the partner is a unicellular or filamentous green alga or a cyanobacterium. The alga provides the fungus with food and, in the case of cyanobacteria, nitrogen. The fungus creates most of the mass of the lichen, provides protection for the alga, and absorbs water and minerals.

Lichens reproduce asexually, either as fragments or as tiny clusters called **soredia**. In addition, the fungal component may reproduce sexually.

Lichens are important colonizers of bare rock and soil and may have helped facilitate the colonization of land by plants.

Fungi as Parasites   Fungal parasites of plants are common, killing chestnut and pine trees and spoiling grain and fruit crops. An ascomycete forms ergots on rye that can cause serious symptoms when accidentally milled into flour.

A parasitic chytrid has been implicated in the recent worldwide decline in frog populations. Fungal infections in animals are called **mycoses**. Parasitic fungi cause ringworm, athlete's foot, vaginal yeast infections, and lung infections. Systemic mycoses are very serious, and many mycoses are opportunistic.

Practical Uses of Fungi   Mushrooms are eaten, and fungi are used to ripen some cheeses. *Saccharomyces cerevisiae* is the yeast used in baking, brewing, and molecular research and biotechnology. Some fungi are sources of medical compounds, including antibiotics.

31.5   **Fungi play key roles in nutrient cycling, ecological interactions, and human welfare**

Fungi as Decomposers   Fungi are decomposers of organic matter, facilitating the essential recycling of chemical elements for plant growth.

Fungi as Mutualists   Mycorrhizae are very common and important in natural ecosystems and agriculture.

Word Roots

coen- = common; **cyt-** = cell (*coenocytic fungus:* a fungus that lacks septa and hence whose body is made up of a continuous cytoplasmic mass that may contain hundreds or thousands of nuclei)

di- = two; **kary-** = nucleus (*dikaryotic:* referring to a fungal mycelium with two haploid nuclei per cell, one from each parent)

hetero- = different (*heterokaryon*: a fungal mycelium that contains two or more haploid nuclei per cell)

myc- = fungus; **rhiz-** = root (*mycorrhiza*: a mutualistic association of plant roots and a fungus)

-osis = a condition of (*mycosis*: general term for a fungal infection)

plasm- = plasm; **-gamy** = marriage (*plasmogamy*: in fungi, the fusion of the cytoplasm of cells from two individuals; occurs as one stage of sexual reproduction, followed later by karyogamy)

Structure Your Knowledge

1. Fill in the following table to summarize the five major groups of fungi.

Group	Examples and Characteristics
Chytrids	**a.**
Zygomycetes	**b.**
Glomeromycetes	**c.**
Ascomycetes	**d.**
Basidiomycetes	**e.**

2. The fungi include members that are decomposers, parasites, and mutualists. How does each of these lifestyles relate to the ecological and economic importance of fungi?

Test Your Knowledge

FILL IN THE BLANKS

_____ 1. group of protists most closely related to fungi

_____ 2. clade that includes fungi, animals, and related flagellated protists

_____ 3. component of cell walls in most fungi

_____ 4. produced by fungi for their absorptive nutritional mode

_____ 5. fungus living inside a plant leaf that benefits the plant

_____ 6. mutualistic associations between plant roots and fungi

_____ 7. fungal group that diverged earliest

_____ 8. distinctive feature of glomeromycetes

MULTIPLE CHOICE: *Choose the one best answer.*

1. The major difference between fungi and plants is that fungi
   a. have an absorptive form of nutrition.
   b. do not have cell walls.
   c. are multinucleate but not multicellular.
   d. Both **a** and **c** are correct.

2. A fungus that is both a parasite and a decomposer
   a. digests the nonliving portions of its host's body.
   b. lives off the sap within its host's body.
   c. first lives as a parasite and then consumes the host after it dies.
   d. causes athlete's foot and vaginal infections.

3. Which of the following is the correct sequence of events during sexual reproduction in a typical fungus?
   a. karyogamy, plasmogamy, germination, meiosis
   b. plasmogamy, karyogamy, meiosis, germination
   c. germination, plasmogamy, meiosis, karyogamy
   d. meiosis, karyogamy, germination, plasmogamy

4. A fungus whose sexual stage is absent or unknown is
   a. a zygomycete.
   b. a basidiomycete.
   c. a chytrid.
   d. a deuteromycete.

5. Which of the following groups is characterized by the absence of chitin?
   a. lichens
   b. yeasts
   c. cryptomycota
   d. molds

6. Fungi and animals appear to be more closely related to each other than either is to plants
   a. because neither of them are photosynthetic.
   b. based on molecular analyses.
   c. because they appear in the fossil record at the same time and moved onto land together.
   d. based on the homologous ultrastructure of their flagella.

7. How can fungi be classified as opisthokonts, which means "posterior flagella," when most fungi lack flagella?
   a. Most fungal lineages appear to have lost their flagella during their evolution.
   b. Members of the most primitive lineage of fungi, the chytrids, have flagellated zoospores.
   c. Fungi are closely related to animals, and because animals, which do have flagella, are opisthokonts, fungi must be opisthokonts too.
   d. Both a and b are the best reasons given here for their classification as opisthokonts.

8. The names given to three of the fungal phyla are based on the
   a. structures in which karyogamy occurs during sexual reproduction.
   b. locations of plasmogamy during sexual reproduction.
   c. locations of the heterokaryotic stage in the life cycle.
   d. structures that produce asexual spores.

9. Which of the following statements accurately describes ascomycetes?
   a. Sexual reproduction occurs by conjugation.
   b. Spores often line up in a sac in the order in which they were formed by meiosis.
   c. Asexual spores form in sporangia on erect hyphae.
   d. All hyphae are dikaryotic.

10. A stable association between a fungus and a green alga or a cyanobacterium is a
    a. mold.
    b. yeast.
    c. lichen.
    d. chytrid.

From the following phylogenetic tree showing the proposed relationships among fungi, choose the letter that is associated with each of the lineages named in questions 11 through 15.

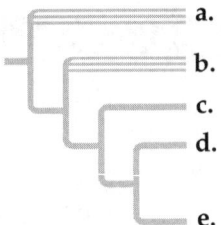

11. Ascomycetes _____

12. Basidiomycetes _____

13. Chytrids _____

14. Glomeromycetes _____

15. Zygomycetes _____

An Overview of Animal Diversity

Framework

Animals are multicellular, eukaryotic heterotrophs that ingest their food. The last common ancestor of animals, probably a flagellated protist, lived around 770 million years ago. Most animal phyla diversified during the 20 million years of the Cambrian explosion. Animal body plans differ in symmetry, tissue layers, and presence of a coelom. Genomic analyses contribute to the ongoing evaluation of animal phylogeny.

Chapter Review

32.1 Animals are multicellular, heterotrophic eukaryotes with tissues that develop from embryonic layers

Nutritional Mode   Most animals ingest their food, eating either other organisms or nonliving organic matter.

Cell Structure and Specialization   Animals are multicellular eukaryotes, and their cells lack walls. Extracellular proteins, most commonly collagen, provide support and hold adjacent cells together. Most animals have **tissues**, groups of specialized cells with a common structure and function. Muscle and nervous tissues are unique to animals.

Reproduction and Development   The diploid stage is usually dominant, and most animals reproduce sexually, with a flagellated sperm fertilizing a larger, nonmotile egg. The zygote undergoes a series of mitotic divisions, called **cleavage**, usually passing through a **blastula** stage during embryonic development. The infolding process of **gastrulation** produces layers of embryonic tissues, resulting in the **gastrula** stage. Some animals begin life as a **larva** and later undergo a dramatic change called **metamorphosis** to assume an adult body type.

All animals have developmental genes, many of which contain DNA sequences called *homeoboxes*. A unique homeobox-containing family of genes, called *Hox* genes, is involved in regulating gene expression in the embryonic development of most animals.

INTERACTIVE QUESTION   32.1

a. What is a larva?

b. How does a larva differ from a juvenile form of an animal?

c. By what process does a larva transform into a juvenile?

32.2 The history of animals spans more than half a billion years

Chemical evidence of particular steroids that are produced by a group of modern sponges suggests that animals had arisen by 710 million years ago (mya). Molecular clock studies have estimated that sponges originated 700 mya. Molecular analyses indicate that the common ancestor of all extant animals lived about 770 mya.

Steps in the Origin of Multicellular Animals   The common ancestor of living animals probably resembled choanoflagellates—suspension-feeding, flagellated protists. Evidence that choanoflagellates are closely related to animals includes molecular data and morphological similarities to the collar cells found in sponges and some other animals. Genomic comparisons have shown that animal cadherin proteins involved in cell attachment contain seven domains also found in the cadherin-like protein of choanoflagellates, as well as a highly conserved, novel domain.

Neoproterozoic Era (1 Billion–541 Million Years Ago)   The earliest generally accepted fossils of animals, known as the **Ediacaran biota**, date from about 560 mya. Some are thought to be sponges; others may be related to cnidarians and molluscs. Some animals from the Ediacaran period (635–541 mya) had protective shells and some fossils show evidence of predation.

Paleozoic Era (541–252 Million Years Ago)   Fossils from the **Cambrian explosion** (a burst of evolutionary change about 535–525 mya) represent about half of all extant phyla and include the first large animals with hard skeletons. Most of the Cambrian explosion fossils are of **bilaterians**—animals with a complete digestive tract and two-sided or bilateral symmetry. The Cambrian explosion saw the origin of large predators with novel prey-capturing adaptations and defensive adaptations in their prey.

INTERACTIVE QUESTION   32.2

Describe the three hypotheses for the rapid diversification of animal phyla during the Cambrian period.

During the Ordovician, Silurian, and Devonian periods, animal diversity continued to increase. Fishes became the top predators of the seas. Arthropods appeared on land by 450 million years ago. Vertebrates began to adapt to land around 365 million years ago. Amphibians and amniotes (reptiles and mammals) are the two terrestrial groups that survive today.

Mesozoic Era (252–66 Million Years Ago)   The first coral reefs formed, some reptiles returned to water, and flight appeared in pterosaurs and birds. Large predatory and herbivorous dinosaurs emerged, as did the first mammals.

Cenozoic Era (65.5 Million Years Ago to the Present)   Mass extinctions marked the beginning of this period and included the disappearance of the nonflying dinosaurs and marine reptiles. Mammals began to exploit the vacated niches.

32.3   Animals can be characterized by "body plans"

An animal's set of morphological and developmental traits is often referred to as a **body plan**. Some features, such as the steps involved in gastrulation, have remained unchanged for more than 500 million years. Other aspects of body plans may have changed several times, become lost, or evolved independently in different lineages. Body plans are significant for *evo-devo*, the intersection of evolution and development.

Symmetry   Many sponges lack body symmetry. An animal with **radial symmetry** has a round or barrel shape, and any slice through its center divides the animal into mirror images. Animals with **bilateral symmetry** have **dorsal** (top) and **ventral** (bottom) sides, **anterior** (front) and **posterior** (back) ends, and left and right sides. Bilateral symmetry is associated with the concentration of sensory organs and a central nervous system in the head end, which is an adaptation facilitating unidirectional movement.

Tissues   True tissues are functional groups of specialized cells separated from other tissues by membranous layers. Sponges lack true tissues. During gastrulation, an embryo develops concentric layers of cells called *germ layers*: **Ectoderm** develops into the outer body covering and, in some phyla, into the central nervous system; **endoderm** lines the developing digestive tube, or archenteron, and gives rise to the lining of the digestive tract and associated organs. A few animal groups, including cnidarians, have only these two germ layers and are said to be **diploblastic**. The bilaterally symmetrical animals are **triploblastic**; they have a third, middle layer, the **mesoderm**, from which arise muscles and most other organs.

Body Cavities   A fluid- or air-filled **body cavity** between the digestive tract and the outer body wall is called a **coelom**. A "true" coelom is completely lined by mesodermally derived tissue. Animals with a true coelom are called **coelomates**. Animals known as **pseudo-coelomates** have a body cavity, called a pseudocoelom, formed from mesoderm and endoderm. Triploblastic animals that have solid bodies are called **acoelomates**.

A fluid-filled body cavity cushions internal organs, allows organs to grow and move independently of the outer body wall, and also functions as a skeleton in soft-bodied animals.

A *grade* refers to a group of animals that share a similar body plan. During animal evolution, coeloms and pseudocoeloms have evolved and been lost many times. Thus the grade of coelomate or pseudocoelomate cannot be used to distinguish *clades* (which include an ancestral species and all of its descendants).

INTERACTIVE QUESTION 32.3

Compare the body plans (symmetry, tissues, and body cavity or coelom) of sponges, cnidarians, and bilateral animals.

Protostome and Deuterostome Development In general, **protostome development** differs from **deuterostome development** in three features: cleavage, coelom formation, and fate of the blastopore.

Protostome development, typical of molluscs and annelids, often involves **spiral cleavage**, in which the planes of cell division are diagonal, and newly formed cells fit in the grooves between cells of adjacent tiers. **Determinate cleavage** sets the developmental fate of each embryonic cell very early. Deuterostome development, typical of echinoderms and chordates, involves **radial cleavage**, in which parallel and perpendicular cleavage planes result in aligned tiers of cells. In **indeterminate cleavage**, typical of most animals with deuterostome development, cells from early cleavage divisions retain the capacity to develop into complete embryos.

The **archenteron**, which forms during gastrulation, develops into the digestive tract. In protostome development, the coelom forms from splits within solid masses of mesoderm. In deuterostome development, the coelom forms from mesodermal outpocketings of the archenteron.

The **blastopore** forms during gastrulation and is the opening leading to the developing archenteron. In protostome ("first mouth") development, the blastopore develops into the mouth, and a second opening at the end of the archenteron becomes the anus. In deuterostome development, the blastopore forms

the anus, and the second opening develops into the mouth.

INTERACTIVE QUESTION 32.4

Fill in the following table to review some of the differences in protostome and deuterostome development.

	Protostome	Deuterostome
Cleavage		
Coelom formation		
Blastopore fate		

32.4 Views of animal phylogeny continue to be shaped by new molecular and morphological data

The Diversification of Animals The relationships among the recognized animal phyla are determined using genomics, morphological traits, and comparisons of the following types of genes: ribosomal RNA, *Hox*, protein-coding, and mitochondrial genes.

There are five major points reflected in the proposed phylogenetic tree of living animals: (1) All animals share a common ancestor and thus represent a clade called Metazoa. (2) Sponges are the sister group to all other animals, branching from the base of the metazoan tree. (3) Eumetazoa is a clade of animals with true tissues. The **eumetazoans** include all animals except sponges and a few other groups. Basal members of the clade include the diploblastic, radially symmetrical phyla Cnidaria and Ctenophora (comb jellies). (4) Most animal phyla belong to the clade Bilateria. The *bilaterians* are defined by the shared derived characters of bilateral symmetry and three germ layers. (5) There are three major clades of bilaterian animals: Deuterostomia, Lophotrochozoa, and Ecdysozoa. Chordata, in the clade Deuterostomia, is the one phylum that includes **vertebrates**, animals with a backbone. All other phyla consist entirely of **invertebrates**.

The clade **Deuterostomia** includes hemichordates (acorn worms), echinoderms, and chordates. The clade **Ecdysozoa** includes the arthropods and nematodes. Its name refers to the shedding of an exoskeleton, called *ecdysis*. The clade **Lophotrochozoa** is named based on a feeding apparatus called a

lophophore (found in animals such as ectoprocts) and a larval stage called the **trochophore larva** (which is shared by molluscs, annelids, and some other members of the lophotrochozoans).

Future Directions in Animal Systematics   Ongoing research on animal phylogeny is addressing three questions: (1) Are sponges monophyletic? The sequencing and comparison of the genomes of various sponges may support the current hypothesis that they are. (2) Are ctenophores basal metazoans? Although some researchers suggest that ctenophores are basal metazoans, most recent evidence confirms sponges as the sister group to other animals. (3) Are acoelomate flatworms basal bilaterians? Although a recent analysis has placed acoelomate flatworms within Deuterostomia, some molecular studies have identified them as basal bilaterians.

includes organisms that have lophophores or trochophore larvae)

meso- = middle (*mesoderm:* the middle primary germ layer in a triploblastic animal embryo)

meta- = boundary, turning point; **-morph** = form (*metamorphosis:* a developmental transformation that turns an animal larva into either an adult or an adult-like stage that is not yet sexually mature)

proto- = first; **-stoma** = mouth (*protostome development:* a developmental mode distinguished by development of the mouth from the blastopore, spiral cleavage, and formation of the body cavity when solid masses of mesoderm split)

pseudo- = false (*pseudocoelomate:* an animal whose body cavity is lined by tissue derived from mesoderm and endoderm)

tri- = three (*triploblastic:* possessing three germ layers)

Word Roots

a- = without; **coel-** = hollow (*acoelomate:* a solid-bodied animal lacking a cavity between the gut and outer body wall)

arch- = ancient, beginning (*archenteron:* the endoderm-lined cavity, formed during gastrulation, that develops into the digestive tract of an animal)

bi- = two (*bilaterian:* member of a clade of animals with bilateral symmetry and three germ layers)

blast- = bud, sprout; **-pore** = a passage (*blastopore:* in a gastrula, the opening of the archenteron that typically develops into the mouth in protostomes and the anus in deuterostomes)

deuter- = second (*deuterostome development:* a developmental mode distinguished by development of the anus from the blastopore, radial cleavage, and body cavity forming as outpockets of mesoderm)

di- = two (*diploblastic:* having two germ layers)

ecdys- = putting off (*Ecdysozoa:* one of three main lineages of bilaterian animals; many are molting animals)

ecto- = outside; **-derm** = skin (*ectoderm:* the outermost primary germ layer in animal embryos)

endo- = within (*endoderm:* the innermost primary germ layer in animal embryos)

gastr- = stomach, belly (*gastrulation:* the inward folding of the blastula-stage embryo, producing a three-layered gastrula)

loph- = a crest, tuft; **troch-** = a wheel (*Lophotrochozoa:* one of three main lineages of bilaterian animals;

Structure Your Knowledge

1. Fill in the blanks in the following simplified tree that represents current hypotheses of animal phylogeny. In the table on the following page, list some of the characteristics of clades **a** through **d** and name some representative animals in clades **e**, **f**, and **g**.

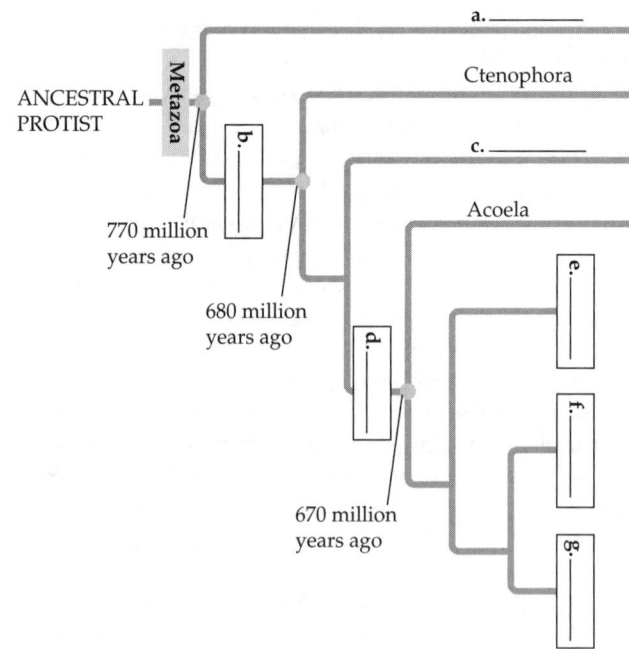

Clade	Characteristics/examples
a.	
b.	
c.	
d.	
e.	
f.	
g.	

Test Your Knowledge

MULTIPLE CHOICE: *Choose the one best answer.*

1. Which of the following characteristics is found only in animals?
   a. homeobox-containing genes
   b. flagellated sperm
   c. heterotrophic nutrition
   d. *Hox* genes

2. The process during which the germ layers are established is
   a. gastrulation.
   b. blastulation.
   c. metamorphosis.
   d. cleavage.

3. Which of the following statements supports the hypothesis that choanoflagellates are closely related to animals?
   a. Choanoflagellate cells and the collar cells of sponges are morphologically similar.
   b. Choanoflagellates are the only protists with true tissues.
   c. Choanoflagellates are bilaterally symmetrical.
   d. Choanoflagellates show spiral cleavage.

4. Sponges differ from other animals in that
   a. they are completely sessile.
   b. they have radial symmetry and are suspension feeders.
   c. their simple body structure has no true tissues, and they have no symmetry.
   d. they have no flagellated cells.

5. Which of the following lists provides the best description of cnidarians?
   a. no real symmetry, two germ layers, tentacles that capture prey
   b. radial symmetry, two germ layers
   c. no real symmetry, without true tissues, choanocytes for trapping food particles
   d. bilateral symmetry, three germ layers, simple muscles and a nerve net

6. A true coelom
   a. is characteristic of sponges.
   b. is fully lined by endoderm.
   c. is lined by endoderm and mesoderm.
   d. is fully lined by mesoderm.

7. Which of the following characteristics are descriptive of protostome development?
   a. spiral and determinate cleavage, blastopore becomes mouth, coelom from a split in mass of mesoderm
   b. radial and determinate cleavage, blastopore becomes mouth, pseudocoelom surrounded by endoderm and mesoderm
   c. spiral and indeterminate cleavage, coelom from a split in a solid mass of mesoderm
   d. radial and determinate cleavage, coelom from mesoderm outpockets, blastopore becomes anus

8. Powered flight evolved in pterosaurs and birds during
   a. the Cenozoic era.
   b. the Mesozoic era.
   c. the Proterozoic era.
   d. the Neoproterozoic era.

9. Which of the following is *not* suggested as a cause of the Cambrian explosion?
   a. availability of new habitats and resources as animals moved onto land
   b. changes in predator–prey relationships that served as selective forces triggering evolutionary adaptations
   c. origin of *Hox* genes and other regulatory sequences that increased developmental possibilities
   d. increase in atmospheric O_2 that supported more active lifestyles

Chapter 33

An Introduction to Invertebrates

Key Concepts

33.1 Sponges are basal animals that lack tissues

33.2 Cnidarians are an ancient phylum of eumetazoans

33.3 Lophotrochozoans, a clade identified by molecular data, have the widest range of animal body forms

33.4 Ecdysozoans are the most species-rich animal group

33.5 Echinoderms and chordates are deuterostomes

Framework

This chapter surveys the amazing diversity of invertebrate animals. Characteristics and representatives of the major animal phyla are presented. The groups covered are the sponges, cnidarians, clade Lophotrochozoa (flatworms, syndermatids, two lophophorate phyla, molluscs, and annelids), clade Ecdysozoa (nematodes and arthropods), and clade Deuterostomia (echinoderms and chordates).

Chapter Summary

More than 95% of known animal species are **invertebrates**, animals that lack a backbone.

33.1 Sponges are basal animals that lack tissues

Sponges (phylum Porifera) are sessile animals found mainly in marine waters. Water is drawn through pores in the body wall of these saclike animals into a central cavity, the **spongocoel**, and flows out through the **osculum**. Sponges are **filter feeders**, collecting food particles by the action of collared, flagellated **choanocytes** lining the spongocoel.

Sponges are *basal animals*, originating near the base of the animal phylogenetic tree. The resemblance between choanocytes and the cells of choanoflagellates supports the molecular evidence that animals originated from a choanoflagellate-like ancestor.

Sponges lack true tissues. In the **mesohyl**, or gelatinous matrix between the two body cell layers, are **amoebocytes**. These cells take up food, digest it, and carry nutrients to other cells. Amoebocytes also form skeletal fibers, which may be sharp spicules or flexible fibers. They are *totipotent*, and this ability to change into other cell types enables a sponge to reshape its body in response to environmental changes.

Most sponges are **hermaphrodites**, which produce both eggs and sperm, although they typically function sequentially as one sex or the other. Sperm, released into the water, fertilize eggs retained in neighboring sponges. Flagellated larvae disperse to a suitable substratum and develop into sessile adults. Sponges produce defensive compounds that may have pharmaceutical uses.

INTERACTIVE QUESTION 33.1

Give the locations and functions of each of the following cells:

a. choanocytes

b. amoebocytes

33.2 Cnidarians are an ancient phylum of eumetazoans

One of the oldest lineages of *eumetazoans*, animals with true tissues, is the phylum Cnidaria. DNA evidence indicates it originated about 680 million years ago. The cnidarians include hydras, jellies, and corals. Their simple anatomy consists of a two-cell-layered sac with a central **gastrovascular cavity** and a single opening serving as both mouth and anus. This radially symmetrical body plan has two forms: **polyp**, which is a sessile, cylindrical form with mouth and tentacles extending upward, and **medusa**, which is a flattened, mouth-down polyp that moves by passive drifting and weak body contractions. Both body forms occur in the life histories of some cnidarians.

Cnidarians use their ring of tentacles, armed with **cnidocytes**, to capture prey. Cnidocytes contain *cnidae*, which are capsules that can evert and discharge long threads. Specialized stinging cnidae are called **nematocysts**. A gelatinous mesoglea is sandwiched between the epidermis and the gastrodermis. Cells of these two layers have bundles of microfilaments arranged into contractile fibers. A *nerve net* is associated with simple sensory receptors and coordinates the contraction of these cells against the hydrostatic skeleton of the gastrovascular cavity, producing movement.

Evidence indicates that early in its history, phylum Cnidaria diverged into two major clades.

Medusozoans   Cnidarians that produce a medusa belong to clade Medusozoa. Most *hydrozoans* alternate between an asexually reproducing polyp and a sexually reproducing medusa form. The common freshwater hydras exist only in polyp form.

The medusa stage is more prevalent in the *scyphozoans*. The sessile polyp stage often does not occur in the jellies of the open ocean.

Cubozoans have a box-shaped medusa stage. Many species, such as the sea wasp, have highly toxic cnidocytes.

Anthozoans   Sea anemones and corals occur only as polyps and belong to the clade Anthozoa. Many corals secrete a calcified **exoskeleton**, and the accumulation of such skeletons produces coral.

INTERACTIVE QUESTION   33.2

In the following diagrams, name the two cnidarian body plans and identify the indicated structures.

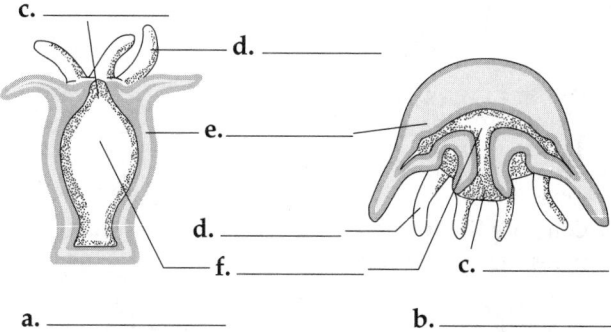

33.3 **Lophotrochozoans, a clade identified by molecular data, have the widest range of animal body forms**

Clade Bilateria consists of bilaterally symmetrical animals with triploblastic development. Most have

a complete digestive tract and a coelom. DNA evidence indicates that the bilaterian common ancestor lived about 670 million years ago.

Molecular data suggest that there are three major clades of bilaterians: Lophotrochozoa, Ecdysozoa, and Deuterostomia. The clade Lophotrochozoa is named for the *lophophore*, a feeding structure found in some members, and the *trochophore larva* found in others. This morphologically diverse group includes 18 phyla.

Flatworms   Free-living flatworms live in marine, freshwater, and damp terrestrial habitats, and many are parasitic. The flatworms (phylum Platyhelminthes) are triploblastic *acoelomates*. Most flatworms have a branching gastrovascular cavity, which functions in both digestion and distribution of food. Gas exchange and diffusion of nitrogenous wastes occur across the body surface. **Protonephridia** are networks of tubules with ciliated *flame bulbs* that pull fluid out of the body and function in osmoregulation.

Early flatworms diverged into two lineages. The Catenulida is a small clade of mostly freshwater species that typically reproduce by posterior asexual buds. The Rhabditophora is a larger, diverse clade of flatworms.

Rhabditophorans include freshwater **planarians** and other, mostly marine, free-living flatworms. Planarians move using cilia to glide on secreted mucus. Eyespots on the head detect light, and lateral head flaps detect chemicals. Their nervous system consists of pairs of anterior ganglia and ventral nerve cords. Planarians reproduce asexually by regeneration or sexually by copulation between hermaphroditic worms.

Most rhabditophorans are parasites, living in or on other animals. Many have a tough outer covering and suckers for attaching to their host. One group, the trematodes, has complex life cycles, usually including asexual and sexual stages and intermediate hosts in which larvae develop. Blood flukes are trematodes that use snails as their intermediate hosts and cause schistosomiasis in humans.

Members of a second group of parasitic rhabditophorans, the tapeworms, have bodies that consist of a scolex with suckers and hooks for attaching to the host's intestinal lining, and a ribbon of proglottids packed with reproductive organs. The life cycles of tapeworms may also include intermediate hosts.

INTERACTIVE QUESTION   33.3

a. Describe the digestive system of a planarian.

b. Explain how tapeworms, which lack a digestive system, obtain nutrients.

Rotifers and Acanthocephalans   Recent analysis supports placing rotifers and acanthocephalans (previously separate phyla) in the new phylum Syndermata.

Rotifers are smaller than many protists but have an **alimentary canal**, with separate mouth and anus, and other organs. Fluid in the *pseudocoelom* functions as a hydrostatic skeleton and distributes nutrients. A crown of cilia draws microscopic food particles into the mouth and then to its grinding jaws.

Some rotifer species reproduce totally asexually by **parthenogenesis**, in which female offspring develop from unfertilized eggs. Other parthenogenetic species may undergo sexual reproduction under certain conditions. Despite the disadvantages of asexual reproduction (such as faster accumulation of harmful mutations), bdelloids are an asexual clade of rotifers that appear to have been reproducing without males for over 50 million years.

The 1100 species of acanthocephalans are parasites of vertebrates, and recent studies indicate that they originated within the rotifers. These sexually reproducing organisms require two or more hosts to complete their complex life cycles.

Lophophorates: Ectoprocts and Brachiopods   These two phyla, known as lophophorates, have a *lophophore*, a crown of ciliated tentacles around the mouth that function in feeding. They have a true coelom.

Ectoprocts, commonly called bryozoans, are tiny, mostly marine animals that live in colonies and are often encased in a hard exoskeleton that has pores through which their lophophores extend. Some species are important reef builders. **Brachiopods**, or lamp shells, attach to the seafloor by a stalk and open their hinged shell to allow water to flow through the lophophore.

Molluscs   Molluscs are soft-bodied, mostly marine animals, most of which are protected by a shell. The molluscan body plan has three main parts: a muscular **foot** used for movement, a **visceral mass** containing the internal organs, and a **mantle** that covers the visceral mass and may secrete a shell. In many molluscs, a **mantle cavity**—a water-filled chamber formed by the extension of the mantle—encloses the gills, anus, and excretory pores. A rasping **radula** is used for feeding by many molluscs. The nervous system includes a nerve ring and nerve cords. Most molluscs have an open circulatory system. Metanephridia remove wastes from the hemolymph.

Most molluscs have separate sexes. Many marine molluscs have a life cycle that includes a ciliated larva called a trochophore. Four of the eight molluscan clades are discussed in the text.

Chitons are oval marine animals with shells that are divided into eight dorsal plates. Chitons cling to and creep slowly over rocks, feeding on algae.

Most gastropods are marine, although there are many freshwater species, and some snails and slugs are terrestrial. Most gastropods have single coiled shells. Many gastropods have distinct heads with eyes at the tips of tentacles. Moving by the rippling of the foot, most gastropods graze on plant material. Land snails lack gills; the lining of the mantle cavity functions in gas exchange.

Bivalves, such as clams, oysters, mussels, and scallops, have the two halves of their shells hinged at the mid-dorsal line. Most bivalves are suspension feeders; water flows into and out of the mantle cavity through siphons, and food particles are trapped in the mucus that coats the gills and then are swept to the mouth by cilia.

Cephalopods are rapid-moving marine predators. The mouth has beak-like biting jaws and is surrounded by tentacles. The shell is reduced and internal in most living species, absent in some octopuses, and external only in the chambered nautilus. The foot has been modified to form parts of the tentacles and the muscular excurrent siphon, which animals may use to jet-propel themselves when water from the mantle cavity is forcibly expelled.

Cephalopods are the only molluscs with a *closed circulatory system*. They have a well-developed nervous system, sense organs, and a complex brain—important features for active predators. Shelled **ammonites** were the dominant invertebrate predators until their extinction at the end of the Cretaceous period.

Many freshwater bivalves and terrestrial gastropods are at risk of extinction due to habitat loss, pollution, introduction of non-native species, and overharvesting.

INTERACTIVE QUESTION   33.4

a. Describe the three parts of the molluscan body plan.

1.

2.

3.

b. Compare the feeding behaviors and activity levels of snails, clams, and squid.

1. Snails:

2. Clams:

3. Squid:

Annelids Annelids are segmented worms found in marine, freshwater, and damp soil habitats. Current molecular evidence indicates that this coelomate phylum can be divided into two main clades: Errantia and Sedentaria.

The large and diverse clade Errantia includes mostly marine, mobile species. Many errantians have parapodia on each segment that function in locomotion and often in gas exchange. Each parapodium has numerous chaetae or bristles made of chitin. These predators or grazers tend to have well-developed jaws and sensory organs.

Clade Sedentaria includes burrowing or tube-dwelling species that are generally less mobile than those in Errantia. Many groups formerly classified as polychaetes and oligochaetes, such as earthworms and the leeches, are now classified in clade Sedentaria.

Most leeches inhabit freshwater environments. Many feed on small invertebrates, whereas others are parasites that temporarily attach to animals, slit a hole through the skin, and suck the blood of the host.

Earthworms eat through the soil, and their castings improve soil texture. The earthworm has a closed circulatory system, and respiration occurs across the highly vascularized skin. Septa partition the coelom into segments, in each of which is found a pair of excretory metanephridia. Earthworm movement results from the coordinated contraction of longitudinal and circular muscles surrounding each segment. The nervous system consists of cerebral ganglia and fused ventral nerve cord. Earthworms are hermaphrodites; sperm are exchanged between worms during mating.

INTERACTIVE QUESTION 33.5

Identify the structures shown in the following illustration of a body segment of an earthworm.

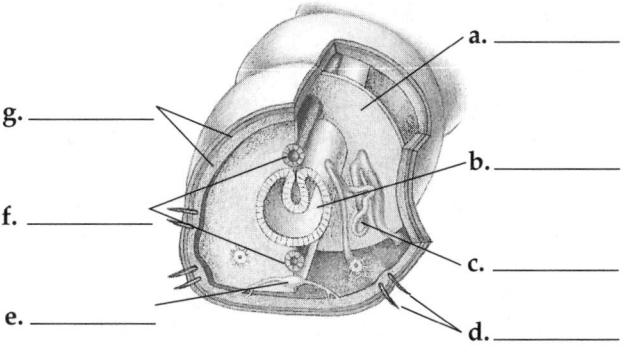

g. _____
f. _____
e. _____
a. _____
b. _____
c. _____
d. _____

33.4 Ecdysozoans are the most species-rich animal group

The clade Ecdysozoa includes animals that **molt**, shedding their external **cuticle** as they grow. Among the eight ecdysozoan phyla are two of the most abundant and successful of animal groups, the nematodes and the arthropods.

Nematodes Nematodes, or roundworms, inhabit water, soil, and the bodies of plants and animals. These cylindrical worms have a tough cuticle, which is periodically shed as they grow. Fluid in the pseudocoelom circulates nutrients absorbed from the alimentary canal.

Numerous nematode species are ecologically important decomposers. Other nematodes are serious agricultural pests and animal parasites. Parasitic nematodes can enlist the cells of their host either to supply nutrients or to expand to house them.

INTERACTIVE QUESTION 33.6

Compare the locomotion of an earthworm and that of a nematode.

Arthropods In terms of species diversity, distribution, and numbers, **arthropods** are the most successful group of animals. Characteristics of arthropods include segmentation, which allows for regional specialization; a hard exoskeleton; and jointed appendages.

The oldest fossil arthropods are from the Cambrian explosion. The *lobopods*, with identical body segments, may have been the group from which arthropods evolved. Trilobites were early arthropods with fairly uniform appendages. Studies show that onychophorans, a closely related invertebrate group, share all the arthropod *Hox* genes, including the unusual *Ubx* and *abd-A* genes. Thus, changes in the sequence or regulation of these genes (and not the origin of new *Hox* genes) may underlie the increased diversity in arthropod body segments.

A layered cuticle of chitin and protein completely covers the body as an exoskeleton, providing protection and points of attachment for muscles. To grow, an arthropod must molt, shedding its exoskeleton and secreting a larger one. The exoskeleton probably first evolved as protection and anchorage for muscles in marine organisms.

INTERACTIVE QUESTION   33.7

a. Arthropods were among the first animals to colonize land (about 450 mya). How did their exoskeletons facilitate that move?

b. List some of the functions which the jointed appendages of arthropods have become modified to perform.

c. What types of sensory organs have evolved in arthropods?

In arthropods, a heart pumps *hemolymph* through an **open circulatory system** consisting of short arteries and a network of sinuses known as the *hemocoel*. The embryonic coelom becomes reduced during development. Gas exchange in most aquatic species occurs through gills, whereas most insects have tracheal systems of branching internal ducts.

Molecular and morphological data suggest an early divergence into three evolutionary lineages: **chelicerates**, **myriapods**, and **pancrustaceans**.

Chelicerates (sea spiders, horseshoe crabs, scorpions, spiders, mites, and ticks) have clawlike **chelicerae**, which serve as pincers or fangs used for feeding; an anterior cephalothorax; and an abdomen. The earliest chelicerates were large, predatory **eurypterids**. Horseshoe crabs and sea spiders are the few marine chelicerates that survive today.

The majority of modern chelicerates are **arachnids**. Most ticks are bloodsucking parasites on reptiles or mammals. Many mites are parasites of plants and animals. The six pairs of appendages in arachnids include chelicerae, sensing or feeding appendages called *pedipalps*, and four pairs of walking legs. In most spiders, **book lungs**, consisting of stacked internal plates, function in gas exchange. Many spiders spin characteristic webs of silk from special abdominal glands.

INTERACTIVE QUESTION   33.8

a. What are chelicerae?

b. How do spiders trap, kill, and eat their prey?

Millipedes and centipedes are members of the clade Myriapoda. These terrestrial groups have a pair of antennae and three pairs of mouthparts, including jaw-like mandibles. Millipedes are vegetarians with two pairs of walking legs per segment. Centipedes are carnivores with one pair of legs per segment and poison claws on their front segment.

Recent phylogenomic studies suggest that insects and crustaceans form a clade named Pancrustacea.

The mostly aquatic crustaceans have two pairs of antennae; three or more pairs of mouthpart appendages, including the hard mandibles; walking legs on the thorax; and appendages on the abdomen. Larger crustaceans have gills. Nitrogenous wastes pass by diffusion through thin areas of the cuticle, and a pair of glands regulates the salt balance of the hemolymph. Sexes usually are separate. One or more swimming larval stages occur in most aquatic crustaceans.

Isopods are a large group of mostly small aquatic crustaceans but also include terrestrial pill bugs. Lobsters, crayfish, crabs, and shrimp are large crustaceans called *decapods*. The decapod's cuticle is hardened by calcium carbonate, and a carapace covers the dorsal side of the cephalothorax. The small, very numerous *copepods* are important members of marine and freshwater plankton communities. Shrimplike krill are a major food source for baleen whales. Barnacles are sessile, shelled crustaceans that strain food from the water with their appendages.

Insects and their relatives (clade Hexapoda) have more known species than all other eukaryotic groups combined. An insect body consists of a head, a thorax, and an abdomen. The head has one pair of antennae, eyes, and several pairs of mouthparts modified for various types of ingestion. The nervous system consists of a cerebral ganglion (brain) and paired ventral nerve cords with segmental ganglia. Malpighian tubules, outpocketings of the digestive tract, function in excretion. Trachael tubes constitute the respiratory system.

The oldest insect fossils are from about 415 mya, but a major diversification occurred in the Carboniferous and Permian periods with the evolution of flight and the modification of mouthparts for specialized feeding on plants. A diversification of insects appears to have accompanied the radiation of flowering plants about 100 mya.

In **incomplete metamorphosis**, the young (nymphs) are smaller versions of the adult and pass through several molts before developing wings and becoming sexually mature. In **complete metamorphosis**, the larva looks entirely different from the adult. Larvae eat and grow; adults primarily reproduce and disperse. Metamorphosis occurs in a pupal stage.

Insect reproduction is usually sexual; fertilization is usually internal. Sperm may be stored in a spermatheca in the female.

Insects are classified into more than 30 orders. Common groups of flying insects are beetles (Coleoptera); flies and mosquitoes (Diptera); ants, bees, and wasps (Hymenoptera); butterflies and moths (Lepidoptera); bugs (Hemiptera); and grasshoppers and crickets (Orthoptera). Insects affect humans as pollinators of crops, as food, as vectors of disease, and as competitors for food.

INTERACTIVE QUESTION 33.9

Compare and contrast the following features of crustaceans and insects.

Feature	Crustaceans	Insects
Habitat	a.	
Locomotion	b.	
Respiration	c.	
Excretion	d.	
Appendages	e.	

33.5 Echinoderms and chordates are deuterostomes

DNA evidence supports Deuterostomia as a clade that includes echinoderms and chordates. Some animals with deuterostome developmental features are not members of that clade.

Echinoderms   Most **echinoderms** are sessile or slow-moving marine animals. They have a thin skin covering an endoskeleton of calcareous plates. A **water vascular system** with a network of hydraulic canals controls extensions, called **tube feet**, that function in locomotion and feeding. Sexual reproduction usually involves separate sexes and external fertilization.

Bilateral larvae metamorphose into adults that typically have five parts radiating from the center.

Sea stars (clade Asteroidea) have multiple arms extending from a central disk. Nerve cords radiate from a central nerve ring into the arms. Tube feet lining the undersurfaces of these arms are used to creep slowly and to grasp and open prey. The gripping of tube feet is the result of adhesive chemicals. The starfish everts its stomach through its mouth and slips it into a slightly opened bivalve shell to eat. Sea stars have remarkable abilities to regenerate lost body parts. A small group in clade Asteroidea, the *sea daisies* have a minute, disk-shaped body with a five-fold symmetry.

Brittle stars (clade Ophiuroidea) have distinct central disks and move by lashing their long, flexible arms. They may be predators, scavengers, or suspension feeders.

Sea urchins and sand dollars, members of clade Echinoidea, have no arms but are able to move slowly using their five rows of tube feet. Long spines also aid a sea urchin's movement. Its mouth is ringed by complex jawlike structures used to eat seaweed.

Members of clade Crinoidea, sea lilies live attached to the substrate by stalks; feather stars crawl about. Their long, flexible arms extend upward from around the mouth and are used in suspension feeding.

Sea cucumbers (clade Holothuroidea) are elongated animals that bear little resemblance to other echinoderms other than having five rows of tube feet.

INTERACTIVE QUESTION 33.10

List the key characteristics that distinguish the phylum Echinodermata.

a.

b.

c.

Chordates   Echinoderms and chordates have evolved independently for over 500 million years. Phylum Chordata contains two basal groups of invertebrates in addition to the vertebrates.

Word Roots

arthr- = jointed; **-pod** = foot (*arthropod:* a segmented ecdysozoan with a hard exoskeleton and jointed appendages)

brachi- = the arm (*brachiopod:* a marine lophophorate with a shell divided into dorsal and ventral halves; also called lamp shells)

choan- = a funnel; **-cyte** = cell (*choanocyte:* a flagellated feeding cell found in sponges; also called a collar cell)

cnid- = a nettle (*cnidocyte:* a specialized cell unique to cnidarians that functions in defense and prey capture)

cut- = the skin (*cuticle:* the exoskeleton of an arthropod; also a tough coat that covers the body of a nematode)

echin- = spiny; **-derm** = skin (*echinoderm:* a sessile or slow-moving marine deuterostome with a water vascular system; the group includes sea stars, brittle stars, sea urchins, feather stars, and sea cucumbers)

eury- = broad, wide; **-pter** = a wing, a feather, a fin (*eurypterid:* an extinct carnivorous chelicerate; also called a water scorpion)

exo- = outside (*exoskeleton:* a hard encasement on the surface of an animal that provides protection and points of attachment for muscles)

gastr- = stomach; **vascul-** = a little vessel (*gastrovascular cavity:* a central cavity with a single opening in the body of certain animals that functions in both the digestion and distribution of nutrients)

in- = without (*invertebrate:* an animal without a backbone)

meso- = the middle; **-hyl** = matter (*mesohyl:* a gelatinous region between the two layers of cells in a sponge)

meta- = change; **morph-** = shape (*complete metamorphosis:* the transformation of a larva into an adult that looks very different, and often functions very differently in its environment, from the larva)

nemat- = a thread; **-cyst** = a bag (*nematocyst:* in a cnidocyte of a cnidarian, a capsule-like organelle containing a coiled thread that when discharged can penetrate the body wall of prey)

os- = a mouth (*osculum:* a large opening in a sponge that connects the spongocoel to the environment)

parthen- = a virgin; **-genesis** = producing (*parthenogenesis:* a form of asexual reproduction in which females produce offspring from unfertilized eggs)

proto- = first; **nephr-** = the kidney (*protonephridia:* an excretory system, such as the flame bulb system of flatworms, consisting of a network of tubules lacking internal openings)

Structure Your Knowledge

1. Use the following evolutionary tree (which is based on molecular evidence) to help you review the diversity of animal phyla. Label clades **a.** through **e.** Then identify each phylum and list some key characteristics and common examples of each. The small sketches should help you identify the phyla. (Note that a sketch of a brachiopod represents the two lophophorate phyla.)

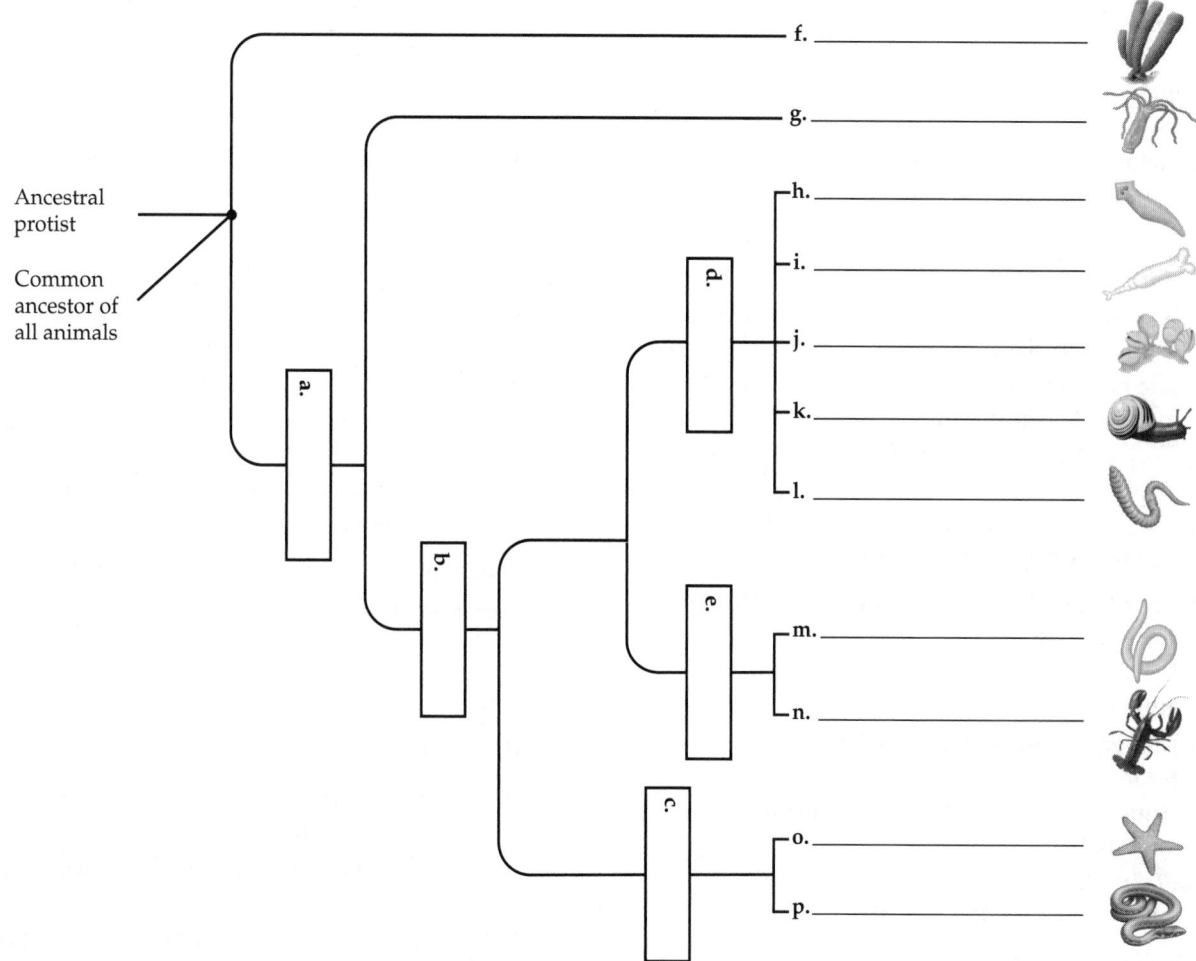

Test Your Knowledge

MULTIPLE CHOICE: *Choose the one best answer.*

1. Invertebrates include
   a. all animals except for those in the phylum Vertebrata.
   b. all animals without backbones.
   c. animals that have exoskeletons or hydrostatic skeletons, but not endoskeletons.
   d. all of the above.

2. Where would you look for a true coelom?
   a. in a mollusc   **c.** in a rotifer
   b. in a sponge   **d.** in a planarian

3. Which of the following lists provides the best description of the sponges?
   a. no real symmetry, diploblastic, cnidocytes for capturing prey
   b. radial symmetry, triploblastic, nematocysts
   c. no real symmetry, without true tissues, choanocytes for trapping food particles
   d. radial symmetry, osculum and spongocoel for filtering water

4. Which of the following organisms has an alimentary canal for digestion?
   a. flatworm   **c.** sea anemone
   b. hydra   **d.** bryozoan

5. The basal group in clade Eumetazoa is
   a. Syndermata.
   b. Platyhelminthes.
   c. Porifera.
   d. Cnidaria.

6. Hermaphrodites
   a. contain male and female sex organs but usually cross-fertilize.
   b. include sponges, earthworms, and many insects.
   c. include parthenogenic rotifers.
   d. Both **a** and **b** are correct.

7. Which of the following statements is *not* true of cnidarians?
   a. An alternation of medusa and polyp stage is common in clade Hydrozoa.
   b. They use a ring of tentacles armed with stinging cells to capture prey.
   c. They include corals, jellies, sponges, and sea anemones.
   d. They have a nerve net that coordinates contraction of microfilaments for movement.

8. Which of the following combinations of phylum and characteristics is *incorrect?*
   a. Brachiopoda—lophophore, stalked, marine animals with hinged shells
   b. Rotifera—crown of cilia, radial symmetry, parthenogenesis (in some)
   c. Annelida—segmentation, closed circulatory system, hydrostatic skeleton
   d. Echinodermata—radial anatomy in adults, endoskeleton, water vascular system

9. Which of the following pairings includes an excretory or osmoregulatory structure that is *correctly* matched with its groups?
   a. metanephridia—flatworms
   b. Malpighian tubules—cephalopods
   c. thin regions of cuticle and glands to regulate salt balance—crustaceans
   d. protonephridia—earthworms

10. Bivalves differ from other molluscs in that they
    a. are predaceous.
    b. have no heads and are suspension feeders.
    c. have shells.
    d. use a radula to feed as they burrow through sand.

11. In which of the following organisms would you expect to find a hard endoskeleton?
    a. *Strongylocentrocentrotus pupuratus*, a sea urchin
    b. *Caryophylia maculata*, a coral
    c. *Lactrodectus mactans*, a spider
    d. *Caulibugula dendrograpta*, an ectoproct

12. In which of the following organisms would you expect to find a book lung?
    a. *Achitina fulica*, a land snail
    b. *Ascaris lumbricoides*, a parasitic nematode
    c. *Hadronyche modesta*, a venomous spider
    d. *Adineta vaga*, a bdelloid rotifer

13. What do nematodes and arthropods have in common?
    a. They are both segmented.
    b. They are both pseudocoelomates.
    c. They include important members of plankton communities.
    d. They both have exoskeletons and undergo ecdysis (molting).

14. Which of the following structures is *not* associated with capturing prey?
    a. the mandibles of centipedes
    b. the chaetae of earthworms
    c. the tentacles of squid
    d. the cnidocytes of hydra

15. Which of the following statements is an *incorrect* description of parasitic animals?
    a. Ticks are bloodsucking parasites belonging to clade Arachnida.
    b. Some roundworms (Nematoda) are internal parasites of humans.
    c. Leeches are parasites that secrete hirudin to keep the host's blood from coagulating.
    d. Tapeworms are annelids that reproduce by shedding proglottids.

16. Which of the following phenomena is *not* considered a factor in the amazing radiation of insect groups?
    a. the evolution of wings and flight
    b. the acquisition of new *Hox* genes
    c. the evolution of mouthparts specialized for feeding on plants
    d. the radiation of flowering plants, providing new food sources

Chapter 34

The Origin and Evolution of Vertebrates

Key Concepts

34.1 Chordates have a notochord and a dorsal, hollow nerve cord

34.2 Vertebrates are chordates that have a backbone

34.3 Gnathostomes are vertebrates that have jaws

34.4 Tetrapods are gnathostomes that have limbs

34.5 Amniotes are tetrapods that have a terrestrially adapted egg

34.6 Mammals are amniotes that have hair and produce milk

34.7 Humans are mammals that have a large brain and bipedal locomotion

Framework

This chapter focuses on the phylogeny of chordates, introducing the derived characters that define the major clades. Fill in the phylogenetic tree of chordates on p. 245 as you work through this chapter. Fossil evidence and hypotheses about human ancestry are also described.

Chapter Review

There are more than 57,000 species of **vertebrates**, animals with a vertebral column, or backbone. This group is known for its *disparity*—its widely varying characteristics, such as body mass.

34.1 Chordates have a notochord and a dorsal, hollow nerve cord

The deuterostome clade of the bilaterians includes the echinoderms and **chordates**. The chordates include the cephalochordates, urochordates, and vertebrates.

Derived Characters of Chordates   Chordates have four structural trademarks, some of which may occur only during the embryonic stage: (1) a flexible rod called a **notochord** that provides skeletal support along the length of chordate embryos and in some adult chordates; (2) a dorsal, hollow nerve cord that develops into the brain and spinal cord; (3) **pharyngeal clefts** or grooves that often develop into **pharyngeal slits**, which open from the pharynx to the outside and function in suspension feeding, as gills, or as parts of the head; and (4) a muscular, post-anal tail containing skeletal elements and muscles.

Lancelets   Lancelets (Cephalochordata) are tiny marine animals that retain all four chordate characteristics into the adult stage. The suspension-feeding adults burrow backward into the sand and filter food particles through a mucous net secreted across the pharyngeal slits. A lancelet swims by coordinated contractions of serial muscles that flex the notochord from side to side. These muscle segments develop from blocks of mesoderm called *somites*, which are present in all chordate embryos.

Tunicates   Recent molecular studies indicate that **tunicates** (Urochordata) are more closely related to other chordates than are lancelets. In the saclike, sessile adult tunicate, water flows into the incurrent siphon, through pharyngeal slits where food is filtered by a mucous net, and out the excurrent siphon. Tunicates have only nine *Hox* genes, apparently having lost four of these developmental genes after branching off from other chordates.

Early Chordate Evolution   The ancestral chordate may have resembled a lancelet, the basal branch of chordates. Studies have shown that several *Hox* genes that organize the major regions of the vertebrate brain are expressed in the anterior end of the lancelet nerve cord in the same anterior-to-posterior pattern. The sequenced genome of tunicates indicates that whereas genes associated with the vertebrate heart and thyroid gland are present, genes involved with the long-range transmission of nerve impulses are found only in the vertebrate lineage. Tunicates also have embryonic cells that resemble *neural crest* cells, which are characteristic of vertebrates.

INTERACTIVE QUESTION   34.1

Name the animal shown in the following diagram and indicate the chordate group to which it belongs. Identify the labeled structures and indicate the four chordate characters.

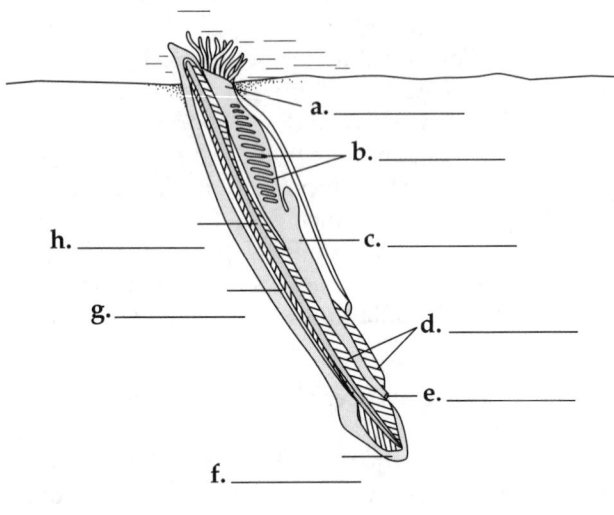

a. _____

b. _____

h. _____

c. _____

g. _____

d. _____

e. _____

f. _____

34.2 Vertebrates are chordates that have a backbone

Derived Characters of Vertebrates   Derived characters that distinguish vertebrates from other chordates include two or more sets of *Hox* genes instead of only one and duplicated families of genes for signaling molecules and transcription factors. These gene duplications may be associated with nervous system innovations and the more extensive skull and backbone of vertebrates. Also unique to vertebrates is the **neural crest**. These cells arise early in embryonic development and migrate widely throughout the body, giving rise to many different tissue types. In most vertebrates, the vertebrae enclose the spinal cord.

Hagfishes and Lampreys   The earliest vertebrates lacked jaws, although only two lineages are found today: the **hagfishes** and the **lampreys**. Both groups have rudimentary vertebrae. Molecular data indicate that these sister groups form a clade of living, jawless

vertebrates, the **cyclostomes**. The other major clade of vertebrates is the gnathostomes, vertebrates with jaws.

The hagfishes (clade Myxini) have a cartilaginous skull, a notochord that facilitates swimming, a small brain, and a mouth with tooth-like structures made of the protein keratin. Hagfishes are marine, mostly bottom-dwelling scavengers that produce slime to repulse predators and competing scavengers.

The lampreys (Petromyzontida) inhabit marine and freshwater environments. They are suspension feeders as larvae, and the adults of some species are blood-sucking parasites. The notochord is the main axial skeleton. A flexible sheath around the notochord has pairs of cartilaginous projections that partially enclose the nerve cord.

Early Vertebrate Evolution   Recently discovered 530 million-year-old fossils (from the Cambrian explosion) appear to be transitional stages. *Haikouella* resembled lancelets but had vertebrate characters such as eyes, a well-formed brain, and segmented muscles. It lacked a skull and ear organs. Another fossil from this period, *Myllokunmingia*, had a skull with ear and eye capsules but lacked vertebrae.

Early fossils of vertebrates date to 500 million years ago (mya). **Conodants** were soft-bodied vertebrates with barbed, *mineralized* hooks in the mouth, dental elements in the pharynx for processing food, and prominent eyes. Conodonts were abundant for more than 300 million years.

Vertebrates that had paired fins and an inner ear with semicircular canals (also present in lampreys) emerged during the Ordovician, Silurian, and Devonian periods. These animals lacked jaws but had a muscular pharynx and were armored with mineralized bone. They all disappeared by the end of the Devonian.

The origin of mineralization in vertebrates may be associated with the switch from suspension-feeding to scavenging and predation. The earliest known mineralized structures are conodont dental elements. The armor of jawless vertebrates was composed of small, tooth-like structures. The internal skeleton of vertebrates was originally cartilage, but mineralization of the skeleton developed between 470 and 430 mya.

INTERACTIVE QUESTION   34.2

List the derived characters of vertebrates.

INTERACTIVE QUESTION 34.3

The following phylogenetic hypothesis shows the major clades of chordates with some of the derived characters that define them. As you work through this chapter, fill in the seven large clades along the side and the name of each lineage.

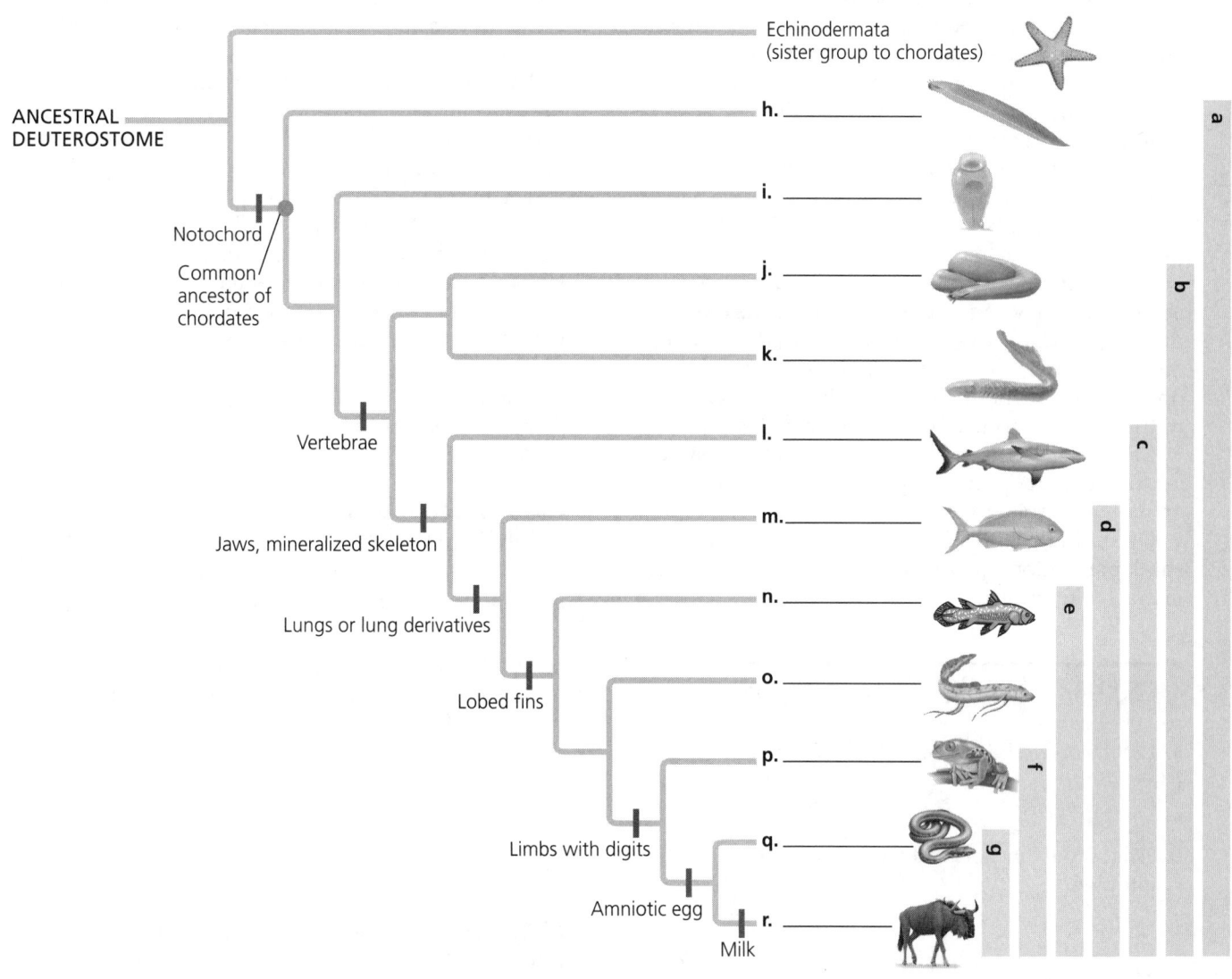

34.3 Gnathostomes are vertebrates that have jaws

Derived Characters of Gnathostomes   The hinged jaws of **gnathostomes** may have evolved from the skeletal rods of the anterior pharyngeal slits, leaving the remaining gill slits to function in gas exchange.

Other derived characters of gnathostomes include an additional duplication of *Hox* genes (now four sets), an apparent duplication of the whole genome, an enlarged forebrain associated with enhanced senses of smell and vision, and, in aquatic gnathostomes, a vibration-sensing **lateral line system**.

Fossil Gnathostomes   Beginning about 440 mya, gnathostomes became increasingly diverse. Their hinged jaws, paired fins supported by bony fin rays, and tail facilitated a predatory lifestyle. The earliest gnathostomes included the armored **placoderms**. The **acanthodians** also radiated during the Silurian and Devonian periods. Both groups became extinct, but by 420 mya three lineages of gnathostomes had diverged: chondrichthyans, ray-finned fishes, and lobe-fins.

Chondrichthyans (Sharks, Rays, and Their Relatives)   **Chondrichthyans** have a skeleton made predominantly of cartilage, which appears to be a derived condition.

The largest group of chondrichthyans consists of sharks, rays, and skates. Sharks swim using powerful muscles in the trunk and tail fin. The paired pectoral and pelvic fins enhance maneuverability. Swimming maintains buoyancy and moves water past the gills. Although most sharks are carnivores, the largest sharks and rays are suspension feeders.

Sharks have keen senses: sharp vision, nostrils, and skin regions in the head that can detect electric fields generated by nearby animals. The shark's body transmits sound waves to the inner ear.

Following internal fertilization, **oviparous** species of sharks lay eggs that hatch outside the mother; **ovoviviparous** species retain the fertilized eggs until they hatch; and in the few **viviparous** species, developing young are nourished by a yolk sac placenta, by nutrients in a uterine fluid, or by eating other eggs.

Rays, which propel themselves with their enlarged pectoral fins, are primarily bottom dwellers that feed on molluscs and crustaceans.

Overfishing has resulted in a dramatic decline in shark populations, particularly in regions where they interact with humans.

INTERACTIVE QUESTION   34.4

a. List the derived characters of gnathostomes.

b. What is the function of the *spiral valve* in a shark's intestine?

c. What is a **cloaca**?

Ray-Finned Fishes and Lobe-Fins   Nearly all vertebrates belong to a clade of gnathostomes called Osteichthyes, which historically included only the "bony fishes" but now includes the tetrapods. Almost all living **osteichthyans** have an ossified endoskeleton with a hard calcium-phosphate matrix.

Aquatic osteichthyans (informally called fishes) move water through the mouth and out between the gills by the contraction of muscles and movement of the protective bony flap called the **operculum**. Most fishes have a **swim bladder**, an air sac that controls

buoyancy. It appears that lungs arose in early osteichthyans (as a supplement to gas exchange in the gills) and later evolved into swim bladders in some lineages.

The flattened, bony scales that cover the skin of most fishes are coated with mucus to reduce drag. Fishes have a lateral line system. Most species are oviparous, with eggs that are fertilized externally.

Ray-finned fishes (Actinopterygii) originated during the Silurian period and are named for the bony rays that support their fins. They include most of the familiar modern fish.

The **lobe-fins** (Sarcopterygii), which also originated during the Silurian period, have muscular pectoral and pelvic fins supported by a series of rod-shaped bones that may have enabled these lobe-fins to swim and "walk" along the underwater substrate. Only three lobe-fin lineages survive: the coelacanths (Actinistia), which were once thought to be extinct; the lungfishes (Dipnoi), found in swamps in the Southern Hemisphere; and the tetrapods, found on land.

INTERACTIVE QUESTION   34.5

a. What three lineages of lobe-fins survive today?

b. Go back to the phylogenetic tree in Interactive Question 34.3 and make sure you have filled in the information up to the letter *o*.

34.4 **Tetrapods are gnathostomes that have limbs**

About 365 million years ago, the fins of some lobe-fins evolved into the limbs and feet of tetrapods.

Derived Characters of Tetrapods   **Tetrapods** have limbs that support their weight on land and feet with digits. Neck vertebrae enable greater head movement. The pelvic girdle is fused to the backbone, transferring forces generated by the hind legs to the rest of the body. The adults of almost all living tetrapods lack gills. Pharyngeal clefts formed during development give rise to parts of the ears, glands, and other structures.

The Origin of Tetrapods   A wide range of lobe-fins evolved during the Devonian, probably using stout fins to crawl through shallow waters and supplementing gas exchange with lungs. The fossil record documents how fins became progressively more limb-like. The first tetrapods appeared 365 mya. A diversity of tetrapods arose during the next 60 million years, most of which probably remained tied to the water.

Describe the 375-million-year-old "fishapod" fossil *Tik-taalik.* Is it considered to be a fish or a tetrapod?

Amphibians   **Amphibians** include salamanders (clade Urodela), frogs (clade Anura), and caecilians (clade Apoda).

Salamanders, which may be aquatic or terrestrial, move with a lateral bending of the body. Frogs are more specialized for moving on land, using their powerful hind legs for hopping. Tropical caecilians are nearly blind, wormlike (legless) burrowing animals.

Many frogs undergo a metamorphosis from a tadpole—the aquatic, herbivorous larval form with gills, a lateral line system, and a finned tail—to the carnivorous, terrestrial adult form with legs, lungs, and external eardrums. Some amphibians are exclusively aquatic or terrestrial. Amphibians are most abundant in damp habitats, and much of their gas exchange occurs across their moist skin.

Fertilization is generally external. Oviparous species lay their eggs in aquatic or moist environments. Some species show varying types of parental care. During breeding season, many frogs communicate via vocalizations. Some species migrate to specific breeding sites, using various forms of communication and navigation.

Amphibian populations worldwide have suffered dramatic declines in the last few decades. Causes for this decline include habitat loss, pollution, climate change, and infection by a chytrid fungus.

34.5 Amniotes are tetrapods that have a terrestrially adapted egg

Amniotes include reptiles (and birds) and mammals.

Derived Characters of Amniotes   The **amniotic egg** contains *extraembryonic membranes* that protect and nourish the embryo. Most reptiles and some mammals have shelled amniotic eggs. In most mammals, the embryo develops within the amnion inside the mother. Other terrestrial adaptations of the amniotes include a less permeable skin and the movement of the rib cage to ventilate the lungs.

Identify the four extraembryonic membranes in the following sketch of an amniotic egg. Identify the function of each membrane.

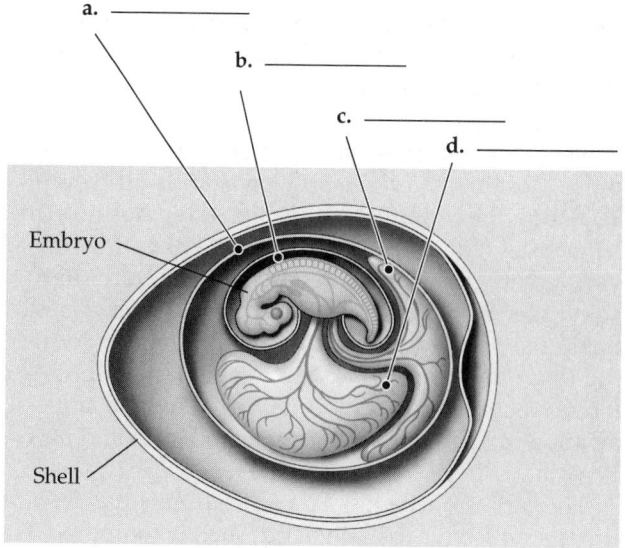

Early Amniotes   The common ancestor of living amphibians and amniotes likely lived about 350 mya. Over time, early amniotes expanded into drier and cooler environments.

Reptiles   The **reptile** clade includes a number of extinct groups and tuataras, lizards, snakes, turtles, crocodilians, and birds. Their derived characters include waterproof scales, and most lay their shelled amniotic eggs on land. Fertilization is internal.

Many reptiles are **ectothermic**—they absorb external heat rather than generating their own—but they may regulate their body temperature through behavioral adaptations. Birds are **endothermic**, warming their bodies with metabolic heat.

The oldest reptilian fossils resembled lizards and date from about 310 mya. Like all modern reptiles these were **diapsids**, characterized by two openings on each side of the skull through which jaw muscles pass. The two main lineages of diapsids are the **lepidosaurs**, which include tuataras, lizards, and snakes (and a number of now extinct huge marine reptiles), and the **archosaurs**, which produced the turtles and crocodilians and the extinct pterosaurs and dinosaurs.

Pterosaurs had wings formed from a collagen-strengthened membrane stretched between an elongated digit on the foreleg and the body. Two main branches of the highly diverse **dinosaurs** were the herbivorous ornithischians and the saurischians. The saurischians included long-necked giant quadrupeds and a group of bipedal carnivores—the **theropods**, which included the ancestors of birds.

Evidence indicates that many dinosaurs were active, agile, and in some species, social. Some dinosaurs brooded their eggs in nests. There is anatomical evidence that some were endotherms. At the end of the Cretaceous (66 mya), all dinosaurs except birds became extinct.

The lizardlike tuatara, representing one lineage of lepidosaurs, survives only on islands off New Zealand. The other living lineage of lepidosaurs, the squamates, consists of lizards and snakes. Lizards vary greatly in size. Snakes are limbless, although some species have vestigial pelvic and limb bones. Snakes are carnivorous, with adaptations for locating (heat-detecting and olfactory organs), killing (sharp teeth, which sometimes inject toxins), and swallowing prey (loosely articulated jaws and elastic skin).

The earliest turtle fossil dates from 240 mya. A turtle's protective (usually hard) shell is composed of upper and lower shields fused to the vertebrae, clavicles, and ribs. Recent evidence indicates that the lower portion of the turtle shell evolved first and the upper shell appeared later. Turtles are now classified as archosaurs, so they are more closely related to crocodilians and birds than they are to the lepidosaurs. Modern turtles are divided into two branches based on the way the neck is retracted into the shell.

Crocodiles and alligators, collectively called crocodilians, belong to an archosaur lineage. They now live in warm regions and are adapted to aquatic habitats.

INTERACTIVE QUESTION   34.8

The two main lineages of diapsids are archosaurs and lepidosaurs. List several representatives, both living and extinct, of these two lineages.

Birds are archosaurs whose features have become adapted to flight. Many of the derived characters of birds are adaptations that reduce weight. Feathers are extremely light and strong and shape the wing into an airfoil. Large pectoral muscles attached to a keel on the sternum flap the aerodynamic wings. The evolution of powered flight enhanced the abilities to hunt, forage, escape from predators, and migrate to favorable habitats.

Birds are endothermic; feathers help retain metabolic heat. The four-chambered heart and efficient respiratory system facilitate a high metabolic rate.

Birds have excellent vision. Their relatively large brains permit detailed visual processing and motor coordination. Birds display complex behaviors, particularly in courtship rituals and parenting. Eggs are brooded during development by the mother, the father, or both.

Recall that birds belong to the theropods, a group of bipedal saurischian dinosaurs. The fossils of feathered but flightless theropods have been found. Functions of early feathers may have been insulation, camouflage, and courtship displays. Feathered theropods had evolved into birds by 160 million years ago. *Archaeopteryx*, the earliest known bird, had feathered wings but retained clawed digits, teeth, and a long tail. Fossils from the Cretaceous period show a gradual loss of ancestral dinosaur features and the acquisition of bird innovations.

The clade Neornithes, which includes the 28 orders of living birds, appeared before 66 million years ago. The **ratites** include the ostrich, kiwi, and emu. These flightless birds lack a keeled breast-bone and enlarged pectoral muscles. The general body form of many flying birds is similar, with beaks adapted to different foods and variations in foot structure.

INTERACTIVE QUESTION   34.9

List several adaptations for flight found in birds.

34.6   Mammals are amniotes that have hair and produce milk

Derived Characters of Mammals   **Mammals** have mammary glands, which produce milk to nourish the young, and hair, which (along with a layer of fat under the skin) helps to insulate these endothermic animals. Active metabolism is provided for by an efficient respiratory system that uses a diaphragm to help ventilate the lungs and a circulatory system with a four-chambered heart.

Mammals generally have large brains. Extended parental care of the young provides them time to observe and learn survival skills. Mammalian teeth come in an assortment of shapes and sizes specialized for eating various foods.

Early Evolution of Mammals   Mammals are part of the amniote group known as **synapsids**, distinguished by the temporal fenestra, a single hole behind each eye socket. During the evolution of mammals, the jaw was remodeled and two former bones of the jaw joint became incorporated into the middle ear.

Synapsids were the dominant tetrapods during a time in the Permian period. Increasingly mammal-like synapsids emerged over a period of 100 million years. The first true mammals arose and diversified during the Jurassic period (200 to 145 mya). By 140 mya, the three major lineages of monotremes, marsupials, and eutherians had emerged. Mammals underwent an extensive radiation in the wake of the Cretaceous extinctions.

Monotremes   The platypus and echidnas (spiny anteaters) are **monotremes**, egg-laying mammals. Although egg-laying is an ancestral character for amniotes, monotremes have hair and produce milk for their young. Monotremes appear to have diverged from other mammals about 180 million years ago.

Marsupials   **Marsupials**, including opossums, kangaroos, and koalas, share derived characters with eutherians, including a high metabolic rate, nipples that provide milk, and giving birth to live young. A **placenta**, formed from the uterine lining and the extraembryonic membranes of the embryo, nourishes the developing embryo. A marsupial is born very early in development and completes its embryonic development attached to a nipple in a maternal pouch called a *marsupium*. Australia has been separated from other continents for about 66 million years, and its marsupials have diversified in isolation and filled the niches occupied by eutherian mammals in other parts of the world.

Eutherians (Placental Mammals)   **Eutherians** complete development attached to a placenta within the maternal uterus. Molecular and morphological data indicate that the major groups of living eutherians diverged between 60–100 mya.

Molecular systematics is helping to establish the phylogeny of eutherian orders, although there is still no consensus. There appear to be at least four main clades. One clade evolved in Africa and includes the elephants, manatees, hyraxes, and aardvarks. A second clade radiated in South America and includes the sloths, anteaters, and armadillos. The third and largest clade includes rabbits, rodents, and primates. The fourth clade includes many diverse groups: carnivores, even- and odd-toed hoofed herbivores, bats, some shrews and moles, and cetaceans (dolphins and whales).

Primates include the lemurs, tarsiers, monkeys, and apes (including humans). Derived characters include grasping hands and feet and flat nails on the digits. Compared to other mammals, primates have a large brain, short jaws, and forward-looking eyes with overlapping visual fields that enhance depth perception. They also exhibit complex social behavior and extensive parental care.

Many primate traits were shaped by natural selection in the tree-dwelling early primates. All primates—except humans—have a wide separation between the big toe and the other toes, allowing them to grasp branches with their feet. Monkeys and apes have an **opposable thumb**, which functions in a grasping "power grip" in monkeys and nonhuman apes but is adapted for precise manipulation in humans.

There are three main groups of living primates: the lemurs, lorises, and bush babies; the tarsiers; and the anthropoids. The **anthropoids** include monkeys and apes. The lemur group probably resembles early arboreal primates, and fossil evidence indicates that tarsiers are more closely related to anthropoids.

Both Old World and New World monkeys probably originated in Africa or Asia. New World monkeys, which colonized South America about 25 million years ago, are strictly arboreal. Most monkeys in both groups are diurnal and live in social bands.

The primates informally called apes include gibbons, orangutans, gorillas, chimpanzees and bonobos, and humans. Apes diverged from Old World monkeys about 30 mya. All nonhuman apes live in the Old World tropics. All living apes have relatively long arms, short legs, and no tails. Apes have a larger brain and more flexible behavior than do other primates.

INTERACTIVE QUESTION   34.10

a. List the derived characteristics of mammals.

b. Go back to the phylogenetic tree in Interactive Question 34.3 and make sure you have filled in all the information.

34.7   **Humans are mammals that have a large brain and bipedal locomotion**

Derived Characters of Humans   The species *Homo sapiens* is about 200,000 years old. Characters that distinguish humans from other apes are upright stance and bipedal locomotion; a larger brain with the capacity for language, symbolic thought, and artistic expression; and the ability to manufacture and use complex tools. Humans also have reduced jaws and a shorter digestive tract.

Although the genomes of humans and chimpanzees are 99% identical, recent studies show that they differ in the expression of regulatory genes, which may account for the many differences between the two groups.

The Earliest Hominins   **Paleoanthropology** is the study of human origins. Paleoanthropologists have identified approximately 20 extinct species of **hominins**—species that are more closely related to humans than to chimpanzees. The oldest of these, *Sahelanthropus*, lived about 6.5 mya. These early hominins shared some derived characters of humans, such as reduced canine teeth, relatively flat faces, and a more upright stance, as indicated by the location of the foramen magnum underneath the skull. The increasing bipedalism of early hominins is suggested by fossil pelvis, leg, and foot bones of 4.4-million-year-old *Ardipithecus ramidus*. Different human features have evolved at different rates.

Australopiths   Hominin diversity increased between 4 and 2 mya, and many of the hominins from this period are called australopiths, although their phylogeny is unresolved. Two lineages from that time include several "gracile" (slender) *Australopithecus* species and "robust" *Paranthropus* species that had sturdy skulls with powerful jaws and teeth. Australopiths retained small brains, comparable in size to those of chimpanzees.

Bipedalism   By about 10 million years ago, the Himalayan range had formed, the climate had become drier, and the forests of Africa and Asia had contracted, presenting a more extensive savannah habitat that may have influenced hominin evolution. But all recently discovered fossils of early hominins show signs of bipedalism, and these hominins lived in mixed forests and open woodlands. Some early hominins appear to have been able to switch between walking upright and climbing trees. About 1.9 mya, hominins that lived in more arid environments began to walk long distances on two legs.

Tool Use   The first evidence of complex tool use is 2.5-million-year-old cuts on animal bones, suggesting the use of stone tools. *Australopithecus garhi*, whose fossils were found nearby, had a relatively small brain.

Early Homo   Larger-brained fossils dating from 2.4 to 1.6 mya are the first to be placed in the genus *Homo*. Sometimes sharp stone tools are found with fossils of *Homo habilis*, or "handy man." Fossils from 1.9 to 1.5 mya are recognized by a number of paleoanthropologists as the species *Homo ergaster*, which had a larger brain, long legs and hip joints adapted for long-distance walking, relatively short and straight fingers, smaller teeth, and more sophisticated stone tools.

The sexual dimorphism of size difference was also less in *Homo ergaster* than in *Australopithecus*. In living primates, reduced sexual dimorphism is associated with more pair-bonding.

Homo erectus was the first hominin to migrate out of Africa. The oldest fossils of hominins outside of Africa date back 1.8 million years. Fossil evidence indicates that *H. erectus* became extinct between 200,000 and 70,000 years ago.

Neanderthals   *Homo neanderthalensis*, or Neanderthals, lived in Europe from about 350,000 years ago and later spread to the Near East, Central Asia, and Siberia. They became extinct between 28,000 and 40,000 years ago. They had large brains, buried their dead, and made tools from stone and wood. Genetic data indicate that the lineages leading to *H. sapiens* and Neanderthals diverged about 400,000 years ago. A 2010 analysis of the Neanderthal genome indicates that limited gene flow occurred between the two species and that their genomes were 99.7% identical. Neanderthals also interbred with another species, the Denisovans, which has been identified on the basis of DNA evidence.

Homo sapiens   The oldest known fossils of *Homo sapiens* have been found in Ethiopia and date from 195,000 and 160,000 years ago. DNA comparisons show that Europeans and Asians share a relatively recent common ancestor, with many African lineages branching off earlier. These analyses suggest a common African ancestor of all *H. sapiens*.

The oldest fossils outside Africa come from the Middle East and date back about 115,000 years. One or more waves of humans appear to have first spread into Asia, then to Europe and Australia. The date of arrival in the New World may be 15,000 years ago.

In 2015 a fully bipedal species with hands capable of fine motor control was identified and called *Homo naledi*. Unfortunately, it has not been possible to date this species, but it may offer new insights into human evolution.

Our understanding of the hominin story is constantly being updated as new discoveries are made. In 2004 a small hominin called *Homo floriensis* was found on the island Flores in Indonesia. Continuing studies of these fossils suggest that this species may have branched off before the origin of the clade that includes Neanderthals and humans, and its small stature and brain size could relate to its island habitat.

The rapid expansion of our species may be tied to changes in human cognition. The discovery of 77,000-year-old engravings and other works of art document the origin of sophisticated thought in humans.

INTERACTIVE QUESTION   34.11

List some of the derived characters of humans.

Word Roots

arch- = ancient (*archosaur:* member of the reptilian group that includes crocodiles, alligators, and dinosaurs, including birds)

cycl- = circle; **-stome** = mouth; (*cyclostome:* member of the vertebrate subgroup lacking jaws; includes hagfishes and lampreys)

di- = two (*diapsid:* member of an amniote clade distinguished by a pair of holes on each side of the skull; diapsids include the lepidosaurs and archosaurs)

din- = terrible; **-saur** = lizard (*dinosaur:* member of an extremely diverse clade of reptiles varying in body shape, size, and habitat)

endo- = inner; **therm-** = heat (*endothermic:* referring to organisms with bodies that are warmed by heat generated by their own metabolism)

eu- = good (*eutherian:* placental mammal; mammal whose young complete their embryonic development within the uterus, joined to the mother by the placenta)

gnath- = the jaw (*gnathostome:* member of the vertebrate subgroup possessing jaws)

homin- = man (*hominin:* a member of the human branch of the evolutionary tree; including *Homo sapiens* and our ancestors, a group of extinct species that are more closely related to us than to chimpanzees)

lepid- = a scale (*lepidosaur:* member of the reptilian group that includes lizards, snakes, and tuataras)

marsupi- = a bag, pouch (*marsupial:* a mammal, such as a koala, kangaroo, or opossum, whose young complete their embryonic development inside a maternal pouch called the marsupium)

mono- = one (*monotreme:* an egg-laying mammal, such as a platypus or echidna)

not- = the back; **-chord** = a string (*notochord:* a flexible rod, made of tightly packed mesodermal cells, that runs along the dorsal anterior-posterior axis of a chordate)

opercul- = a covering, lid (*operculum:* in aquatic osteichthyans, a protective bony flap that covers and protects the gills)

oste- = bone; **ichthy-** = fish (*osteichthyan:* member of a vertebrate clade with jaws and mostly bony skeletons)

ov- = an egg; **-parous** = bearing (*oviparous:* referring to a type of development in which young hatch from eggs laid outside the mother's body)

pale- = ancient; **anthrop-** = man; **-logy** = the science of (*paleoanthropology:* the study of human origins and evolution)

plac- = a plate (*placoderm:* a member of an extinct group of fishlike vertebrates that had jaws and were enclosed in a tough, outer armor)

pter- = a wing (*pterosaur:* winged reptile that lived during the Mesozoic era)

rat- = a raft (*ratite:* member of the group of flightless birds)

syn- = together (*synapsid:* member of an amniote clade distinguished by a single hole on each side of the skull; includes the mammals)

tetra- = four; **-pod** = foot (*tetrapod:* a vertebrate clade whose members have limbs with digits; includes mammals, amphibians, and birds and other reptiles)

vivi- = alive (*ovoviviparous:* referring to a type of development in which young hatch from eggs that are retained in the mother's uterus)

Structure Your Knowledge

1. Starting from primates, list the increasingly inclusive clades (from eutherians to chordates) and some of the derived characters that define each of these clades. Reviewing the phylogenetic tree on page 245 may help.
2. Describe several examples from vertebrate evolution that illustrate the common evolutionary theme of exaptation—that new adaptations usually evolve from preexisting structures.
3. How do various tetrapod groups meet the challenges of a terrestrial habitat?

Test Your Knowledge

FILL IN THE BLANKS

_____ 1. invertebrate chordate clade most closely related to vertebrates

_____ 2. the most basal chordate group

_____ 3. blocks of mesoderm along notochord that develop into muscles

_____ 4. clade of jawed vertebrates

_____ 5. flap over the gills of bony fishes

_____ 6. adaptation that allows reptiles to reproduce on land

_____ 7. egg-laying mammals

_____ 8. structure that helps mammals ventilate their lungs

_____ 9. group of primates that includes monkeys and apes

_____ 10. genus in which the fossil Lucy is pla

MULTIPLE CHOICE: *Choose the one best answer.*

1. Pharyngeal slits appear to have functioned first as
   a. suspension-feeding devices.
   b. gill slits for respiration.
   c. components of the jaw.
   d. mouth openings.

2. The flexible rod that provides support in chordate embryos is
   a. the dorsal, hollow nerve cord.
   b. the notochord.
   c. the spinal column.
   d. the somitic cord.

3. Which of the following living groups represents the basal lineage(s) of vertebrates?
   a. lancelets and tunicates
   b. sharks and rays
   c. hagfishes and lampreys
   d. coelacanths

4. Which of the following organisms are in the lobe-fin clade?
   a. lampreys
   b. sharks and rays
   c. ray-finned fishes
   d. tetrapods

5. Which of the following is a derived character of vertebrates?
   a. pharyngeal gill slits
   b. neural crest
   c. dorsal, hollow nerve cord
   d. amniotic egg

6. Which of the following groups is *not* in the same lineage as the others?
   a. lizards
   b. birds
   c. dinosaurs
   d. crocodilians

7. Among vertebrates, mineralized tissues first appeared in
   a. *Haikouella.*
   b. conodonts.
   c. lampreys.
   d. bony fishes.

8. Non-bird reptiles have much lower caloric needs than do birds of comparable size because they
   a. are ectothermic, whereas birds are endothermic.
   b. have waterproof scales instead of feathers.
   c. provide less parental care than do birds.
   d. have a less efficient respiratory system and thus a lower metabolic rate.

9. In which of the following organisms would you expect to find just two sets of *Hox* genes?
   a. *Branchiostoma lanceolatum*, a lancelet
   b. *Clavelina picta*, a tunicate
   c. *Eptatretus stoutii*, a hagfish
   d. *Carcharodon carcharias*, the great white shark

10. Oviparity is a reproductive strategy that
    a. is common in reptiles and some sharks.
    b. is necessary for vertebrates to reproduce on land.
    c. is characteristic of all flying vertebrates.
    d. protects the embryo inside the mother but uses the food resources of the egg.

Use the following choices to answer questions 11–13.
   a. archosaurs   c. dinosaurs
   b. diapsids   d. lepidosaurs

11. Which of these clades includes all the others?

12. Which of these clades includes the birds but not the crocodilians?

13. Which clade includes the tuataras and squamates (lizards and snakes)?

14. It is thought that feathers first arose as a means of maintaining body heat, but they have properties that allow them to form useful flight surfaces on the wings of birds. This change in function provides an example of:
    a. artificial selection.
    b. convergent evolution.
    c. exaptation.
    d. descent with modification.

15. Which of the following statements is supported by DNA studies?
    a. *Homo sapiens* originated in Africa.
    b. Erect posture preceded the enlargement of the brain in human evolution.
    c. The key differences between chimpanzees and humans appear to involve duplications of *Hox* genes.
    d. Neanderthals are more closely related to Asians than to Europeans.

Chapter 35

Vascular Plant Structure, Growth, and Development

Key Concepts

35.1 Plants have a hierarchical organization consisting of organs, tissues, and cells

35.2 Different meristems generate new cells for primary and secondary growth

35.3 Primary growth lengthens roots and shoots

35.4 Secondary growth increases the diameter of stems and roots in woody plants

35.5 Growth, morphogenesis, and cell differentiation produce the plant body

Framework

A rather large new vocabulary is needed to name the specialized cells and structures in a study of plant structure and growth. Focus your attention on how a plant's roots, stems, and leaves are specialized to function in absorption, support, transport, protection, and photosynthesis. The plant body is composed of dermal, vascular, and ground tissue systems. Apical meristems at the tips of roots and shoots create primary growth. The lateral meristems—vascular cambium and cork cambium—create secondary growth that adds girth to stems and roots. New techniques and model organisms such as *Arabidopsis* are enabling researchers to explore the molecular bases for plant growth, morphogenesis, and cellular differentiation.

Chapter Review

35.1 **Plants have a hierarchical organization consisting of organs, tissues, and cells**

A plant **organ** performs a specific function and is composed of several types of **tissues**, groups of **cells** with a common function.

Basic Vascular Plant Organs: Roots, Stems, and Leaves In vascular plants, the underground **root system** for obtaining water and minerals from the soil and an aerial **shoot system** of stems and leaves for absorbing light and CO_2 for photosynthesis are evolutionary adaptations to the dispersed resources in a terrestrial environment. *Photosynthates* from the shoot system nourish the roots. The shoot system includes vegetative shoots and reproductive shoots, which, in angiosperms, bear flowers.

The functions of **roots** include anchorage, absorption, and storage of food. The *primary root* emerges from a seed and branches to form **lateral roots**. A *taproot system* is found commonly in tall, large plants. The one main deep **taproot** anchors the plant and may store nutrients. Small plants usually have a shallower *fibrous root system* in which many small roots grow from the stem. Such roots are called *adventitious,* as they grow from an unusual location. These roots then form their own lateral roots, forming mats that are good for anchoring the plant and preventing soil erosion. Most monocots have fibrous root systems.

Most absorption of water and minerals occurs through tiny **root hairs**, extensions of epidermal cells that are clustered near root tips. Most plants also form *mycorrhizal associations* with soil fungi, which increase mineral absorption. Modified roots may serve various functions, including support, storage, and oxygen absorption.

A **stem** consists of alternating **nodes**, the points at which leaves are attached, and segments between nodes, called **internodes**. Located at the tip or apex of a shoot is an **apical bud**. **Axillary buds** are found in the upper angle (axil) between a leaf and the stem. Axillary buds may develop into lateral branches or flowers.

Modifications of shoots include stolons (horizontal shoots that enable asexual reproduction), horizontally growing underground rhizomes, and food storage structures (such as tubers).

Leaves, the main photosynthetic organs of most plants, usually consist of a flattened **blade** and a **petiole**, or stalk. Many monocots lack petioles. Monocot leaves usually have parallel major **veins** (the vascular tissue of leaves), whereas eudicot leaves have networks of branched veins arising from the *midrib*. Leaves may be simple or compound. Modified leaves may function in support, protection, storage, or reproduction.

INTERACTIVE QUESTION   35.1

Label the parts in the following diagram of a flowering plant.

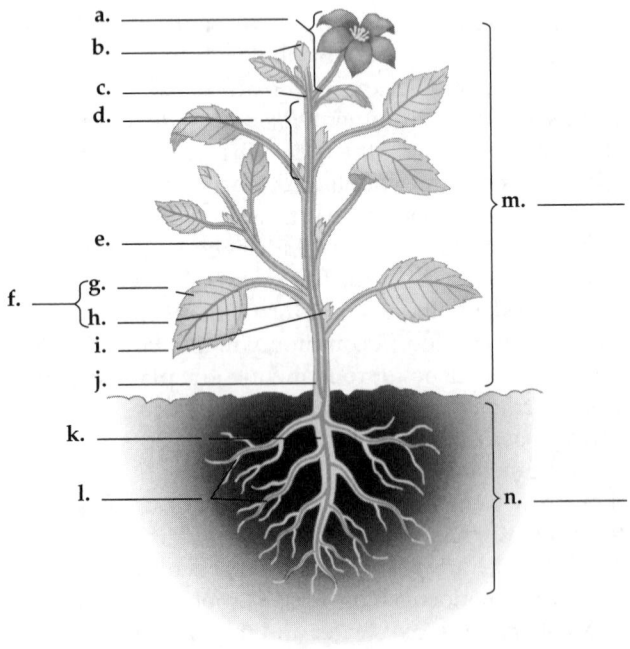

a. _____
b. _____
c. _____
d. _____
m. _____
e. _____
f. { g. _____
    h. _____
    i. _____
j. _____
k. _____
l. _____
n. _____

Dermal, Vascular, and Ground Tissues   These three **tissue systems** are continuous throughout the plant but have specific characteristics in each plant organ (root, stem, and leaf).

The **dermal tissue system** forms a protective outer layer. The **epidermis** is a single layer of tightly packed cells that covers nonwoody plants. The epidermis of leaves and most stems is covered with a **cuticle**, a waxy coating that prevents excessive water loss. **Guard cells** are specialized epidermal cells involved in regulating gas exchange and water loss. Older stems and roots of woody plants are covered by a protective **periderm**. **Trichomes** are outgrowths of the shoot epidermis that may protect against insects and disease or reduce water loss.

The **vascular tissue system** provides mechanical support and consists of the tissues **xylem** and **phloem**, which function in long-distance transport of water and minerals (xylem) and of sugars (phloem). In angiosperms, the **stele**, or vascular tissue of an organ, takes the form of a solid central **vascular cylinder** in roots but is arranged in *vascular bundles* in stems and leaves.

The **ground tissue system** contains cells that function in photosynthesis, support, and storage. Ground tissue internal to vascular tissue is called **pith**; ground tissue outside the vascular tissue is called **cortex**.

Common Types of Plant Cells   There are five main types of *differentiated* plant cells, which are structurally suited for their specific functions.

Parenchyma cells carry on most of a plant's metabolic functions, such as photosynthesis and food storage. They usually lack secondary walls and have large central vacuoles. Most parenchyma cells retain the ability to divide and differentiate into other types of plant cells.

Collenchyma cells lack secondary walls but have thickened primary walls. Strands or cylinders of collenchyma cells provide flexible support for young parts of the plant and elongate along with the plant.

Sclerenchyma cells have thick secondary walls strengthened with **lignin**. The protoplasts of these specialized supporting cells often die at maturity. **Fibers** are long, tapered cells that are usually grouped in strands. **Sclereids** are shorter and irregular in shape, with very thick, lignified cell walls.

The water-conducting cells of xylem also die at functional maturity, leaving behind their secondary walls, which have interspersed thinner pits. **Tracheids**, found in all vascular plants, are long, thin, tapered cells. Water passes through pits from cell to cell. **Vessel elements**, found mainly in angiosperms, are wider, shorter, and thinner walled, with perforations in their end walls. They align to form long tubes known as **vessels**. Both tracheids and vessel elements have lignin-strengthened walls.

In the phloem of angiosperms, sugars flow through chains of cells called **sieve-tube** elements, or sieve-tube members. These cells remain alive at functional maturity but lack nuclei, ribosomes, and vacuoles. Fluid flows through pores in the **sieve plates** in the end walls between cells. The nucleus and ribosomes of an adjacent **companion cell**, which is connected to a sieve-tube element by numerous plasmodesmata, serve both cells.

INTERACTIVE QUESTION 35.2

a. Which types of plant cells are dead at functional maturity?

b. Which types of plant cells lack nuclei at functional maturity?

35.2 Different meristems generate new cells for primary and secondary growth

Plants have tissues called **meristems**, which remain undifferentiated and can perpetually divide to form new cells. This division results in continuous or **indeterminate growth**. (Some tissues, such as leaves and thorns, stop growing when fully developed; this pattern of growth is said to be **determinate**.) **Apical meristems**, located at the tips of roots and shoots, produce growth in length, called **primary growth**. Herbaceous (nonwoody) plants usually undergo primary growth only. **Secondary growth** is an increase in diameter as new cells are produced by **lateral meristems**. The lateral meristem called **vascular cambium** produces secondary xylem and secondary phloem; **cork cambium** produces the protective periderm. Cells that remain to divide in a meristem are called *stem cells* (or *initials*), whereas cells that are displaced from the meristem become specialized in mature tissues including the **primary meristems**: *protoderm*, *ground meristem*, and *procambium*.

Flowering plants may be *annuals*, which complete their life cycle in a year or less; *biennials*, which have a life cycle spanning two years; or *perennials*, which live many years.

INTERACTIVE QUESTION 35.3

a. What is the difference between indeterminate growth and determinate growth?

b. What is the difference between primary and secondary growth?

35.3 Primary growth lengthens roots and shoots

Primary Growth of Roots   The apical meristem of the root tip is protected by a **root cap**, which secretes a lubricating polysaccharide slime. The *zone of cell division* includes the apical meristem and its derivatives. In the *zone of elongation*, cells lengthen to many times their original size, which pushes the root tip through the soil. As the zone of elongation grades into the *zone of differentiation* (or maturation), cells become specialized in structure and function.

In the vascular cylinder of most eudicot roots, xylem cells radiate from the center in spokes, with phloem located between them. The vascular tissue of a monocot may have pith, a central core of undifferentiated parenchyma cells, inside a ring of xylem and phloem.

The ground tissue consists of parenchyma cells in the cortex. The innermost layer of cortex is the one-cell-thick **endodermis**, which regulates the passage of materials into the vascular cylinder.

Lateral roots develop from the **pericycle**, the outer layer of cells in the vascular cylinder, and push through the cortex and epidermis.

INTERACTIVE QUESTION 35.4

Label the tissues in the following cross sections of a eudicot root and its vascular cylinder. Identify the functions of each of these tissues.

a. _____

b. _____

c. _____

500 μm

d. _____

e. _____

f. _____

g. _____

70 μm

Primary Growth of Shoots   The dome-shaped mass at the tip of the apical bud is the shoot apical meristem. The apical bud may exhibit **apical dominance**, inhibiting the growth of axillary buds. When axillary buds break dormancy, they develop into lateral shoots, or branches, each with its own apical bud, leaves, and axillary buds. Elongation of the shoot occurs by cell elongation within young internodes.

The epidermis of the dermal tissue system covers stems. In most eudicots, vascular bundles are arranged in a ring, with the ground tissues pith inside and cortex outside the ring. Xylem is located internal to the phloem in the vascular bundles. In most monocot stems, the vascular bundles are scattered throughout the ground tissue. A layer of collenchyma just beneath the epidermis and sclerenchyma fiber cells may strengthen stems during primary growth.

Leaf primordia along the sides of the shoot apical meristem give rise to leaves. The cuticle-covered epidermis of a leaf has tiny pores called **stomata** flanked by guard cells, which permit both gas exchange and evaporation of water from the leaf. The ground tissue, or **mesophyll**, consists of parenchyma cells containing chloroplasts. In many eudicot leaves, columnar *palisade mesophyll* is located above *spongy mesophyll*, which has loosely packed cells surrounding many air spaces.

Vascular tissue in the stem branches into the leaf and divides repeatedly, providing support and vascular tissue to the photosynthetic mesophyll. Each vein is enclosed in a ring of protective cells called a *bundle sheath*.

INTERACTIVE QUESTION   35.5

Name the indicated structures in the following diagram of a leaf and review their functions.

a. _____

b. _____

c. _____

d. _____

e. _____

f. _____

g. _____   i. _____

h. _____

35.4   Secondary growth increases the diameter of stems and roots in woody plants

The vascular cambium and cork cambium produce secondary growth, which occurs in all gymnosperms and in many eudicots but is rare in monocots.

The Vascular Cambium and Secondary Vascular Tissue   In a woody stem, vascular cambium forms a continuous cylinder of meristematic cells that produce secondary xylem to the inside of the cylinder and secondary phloem to the outside. Some cambial initials produce radial lines of parenchyma cells called *vascular rays*, which function in lateral transport of water and nutrients and in storage.

Wood is the accumulation of secondary xylem cells with thick, lignified walls. Annual growth rings in temperate regions result from the seasonal cycle of early or spring wood production (with larger, thinner cells) and late or summer wood production (thicker-walled cells). The study of tree growth rings (*dendrochronology*) can provide information about seasonal variations over the life of the tree. Scientists analyzing tree rings of Mongolian conifers from over the past five centuries have found evidence of warming temperatures during the 20th century.

In older trees, a central column of *heartwood* consists of older xylem with resin-filled cell cavities; the *sapwood* consists of secondary xylem that still functions in transport.

Older secondary phloem is sloughed off as bark; only that closest to the vascular cambium functions in sugar transport.

The Cork Cambium and the Production of Periderm   The epidermis splits off during secondary growth and is replaced by new protective tissues produced by the cork cambium, a meristematic cylinder that first forms in the outer cortex in stems and in the outer pericycle in roots. The cork cambium produces *cork cells,* which develop *suberin*-impregnated walls. The protective coat formed by a cork cambium and its tissues is a layer of periderm. As secondary growth continually splits the outer layers of periderm, new cork cambia develop. **Lenticels** are spaces between cork cells through which gas exchange occurs. **Bark** refers to phloem and periderm.

INTERACTIVE QUESTION   35.6

Place the letters of the following tissues in the blanks so that they are in the order in which they are located in a tree trunk going from the outside in.

A. primary phloem   **C.** primary xylem

B. secondary phloem   **D.** secondary xylem

E. pith   **G.** vascular cambium

F. cork cambium   **H.** cork

_____ _____ _____ _____ _____ _____ _____ _____

Evolution of Secondary Growth   Studies with *Arabidopsis* have suggested that the accumulation of weight by the stem may cue wood formation. The genetic regulation of both apical meristem and vascular cambium involves common genes, suggesting that these genes have been conserved by evolution.

35.5 Growth, morphogenesis, and cell differentiation produce the plant body

Development involves the processes of growth and differentiation. *Growth* is an increase in size. The organization of cells to create body form is called *morphogenesis. Differentiation* is the process by which cells become specialized for different functions. Plant development and structure are controlled by both genetic and environmental factors, resulting in *developmental plasticity* in response to local environments.

Model Organisms: Revolutionizing the Study of Plants   *Arabidopsis thaliana* has many traits of a model organism: It is a small, easily cultivated plant with a short generation time and prolific seed production. Its genome is small for a plant and was the first plant genome to be sequenced. *Arabidopsis* strains from various environments show significant differences that can be explored to understand evolutionary adaptations. In addition, *Arabidopsis* is particularly amenable to genetic manipulation, including the introduction of foreign genes (*transgenes*) and the production of targeted mutations using CRISPR-Cas technology.

INTERACTIVE QUESTION 35.7

Plant biologists use CRISPR-Cas technology to disrupt specific genes in *Arabidopsis*. How are the resulting "knockout mutants" being used?

Growth: Cell Division and Cell Expansion   The plane of cell division, which is usually perpendicular to the long axis of a dividing cell, is determined by the *preprophase band,* a ring of cytoskeletal microtubules that forms during late interphase.

Studies with the *tangled-1* mutant of maize, in which transverse cell divisions are normal but longitudinal divisions are aberrant, indicate that the generation of leaf shape does not depend solely on the orientation of cell division.

In *asymmetrical cell division,* one daughter cell receives more cytoplasm than the other, often leading to a key developmental event, such as the formation of guard cells. The first asymmetric division of a plant zygote establishes the **polarity** of plants, which have a root end and a shoot end. In the *gnom* mutant of *Arabidopsis,* the first division is not asymmetrical, and the resulting ball-shaped plant lacks leaves and roots.

About 90% of plant cell expansion is due to the uptake of water into the central vacuole. This economical means of cell elongation produces the rapid growth of shoots and roots. Cells expand in a direction perpendicular to the orientation of the cellulose microfibrils in the inner layers of the cell wall.

INTERACTIVE QUESTION 35.8

Review the role of microtubules in the orientation of plant cell division and expansion.

Morphogenesis and Pattern Formation   **Pattern formation** is the characteristic development of structures in specific locations. The fate of plant cells does not appear to be determined by *lineage-based mechanisms,* but rather by *position-based mechanisms.* Experiments demonstrate that a cell's fate is determined late in its development and depends on signaling from neighboring cells.

Animal cell fate, which depends on lineage-dependent mechanisms, involves transcription factors encoded by homeotic (*Hox*) genes. In many plants, a *Hox*-like gene called *KNOTTED-1* influences leaf morphology. Its overexpression in tomato plants results in "super-compound" leaves.

Gene Expression and Control of Cell Differentiation Cell-to-cell communication provides the positional information that results in the differential gene expression necessary for cell differentiation. For example, in the root epidermis of *Arabidopsis,* the homeotic gene *GLABRA-2* is expressed in epidermal cells that are in contact with only one underlying cortical cell. The gene is not expressed in those cells in contact with two cortical cells, and they differentiate into root hair cells.

Shifts in Development: Phase Changes   In a process known as **phase change**, the apical meristem can switch from one developmental stage, or *phase,* to another, such as from producing juvenile vegetative nodes and internodes to laying down adult nodes. A node's developmental phase is fixed when it is produced, and thus both juvenile and mature leaves can coexist along the shoots of a plant.

Genetic Control of Flowering   In the transition from vegetative growth to reproductive growth (production of a flower), **meristem identity genes** code for transcription factors that activate the genes involved in the conversion of indeterminate vegetative meristems to determinate floral meristems.

The four floral organs develop from floral primordia in four concentric whorls: sepals, petals, stamens, and carpels. Several **organ identity genes**, belonging to the *MADS-box* family, code for transcription factors that regulate floral pattern. Positional information influences their expression. Mutations of organ-identity genes result in the placement of one type of floral organ where another type would normally develop. The **ABC hypothesis** of flower formation identifies three classes of organ identity genes, each of which is expressed in two adjacent whorls.

INTERACTIVE QUESTION   35.9

The following diagram depicting the ABC hypothesis shows the active genes in each whorl and the resulting anatomy of a wild-type flower.

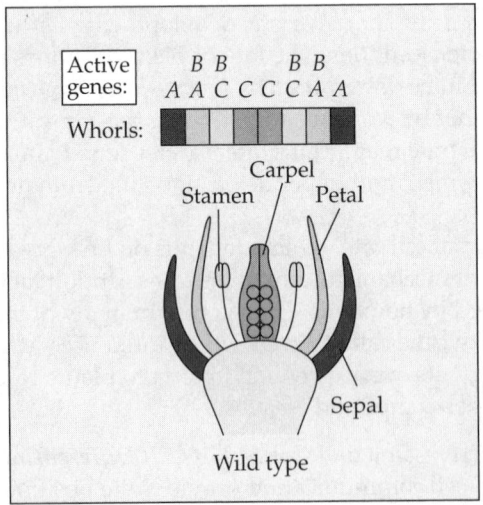

Wild type

a. Fill in the following table to indicate which organs are produced in the whorls in a normal flower. In a mutant that lacks a functional gene *A* ("mutant A"), what gene expression pattern and resulting flower

organ arrangement would be produced? (Remember that the lack of *A* activity removes the inhibition of gene *C*, and vice versa.)

Whorl	Genes Active	Organs in Normal Flower	Genes Active in Mutant A	Organs in Mutant A Flower
1.	A			
2.	AB			
3.	BC			
4.	C			

b. If you had a double-mutant plant that had no gene activity for *B* or *C*, what would the resulting flower look like?

Word Roots

apic- = the tip; **merist-** = divided (*apical meristem:* embryonic plant tissue, located in the tips of roots and the buds of shoots, whose dividing cells enable the plant to grow in length)

coll- = glue; **-enchyma** = an infusion (*collenchyma cell:* a flexible plant cell type that occurs in strands or cylinders that support young parts of the plant without restraining growth)

endo- = inner; **derm-** = skin (*endodermis:* in plant roots, the innermost layer of the cortex that surrounds the vascular cylinder)

epi- = over (*epidermis:* the dermal tissue system of nonwoody plants, usually consisting of a single layer of tightly packed cells)

inter- = between (*internode:* a segment of a plant stem between the points where leaves are attached)

meso- = middle; **-phyll** = a leaf (*mesophyll:* leaf cells specialized for photosynthesis)

peri- = around; **-cycle** = a circle (*pericycle:* the outermost layer in the vascular cylinder, from which lateral roots arise)

phloe- = the bark of a tree (*phloem:* vascular plant tissue consisting of living cells arranged into elongated tubes that transport sugar and other organic nutrients throughout the plant)

scler- = hard (*sclerenchyma cell:* a rigid, supportive plant cell type usually lacking a protoplast and possessing thick secondary walls strengthened by lignin)

trache- = the windpipe (*tracheid:* a long, tapered water-conducting cell found in the xylem of nearly all vascular plants; functioning tracheids are no longer living.)

vascul- = a little vessel (*vascular tissue system:* a transport system formed by xylem and phloem throughout a vascular plant)

xyl- = wood (*xylem:* vascular plant tissue consisting mainly of tubular dead cells that conduct water and minerals upward from the roots to the rest of the plant)

Structure Your Knowledge

1. How does the indeterminate growth pattern of plants lead to developmental plasticity? What is the adaptive advantage of such plasticity?
2. In the following cross section of a young stem, label the indicated structures. Is this a stem of a monocot or a eudicot? How can you tell? Does this stem illustrate primary growth or secondary growth? How can you tell?

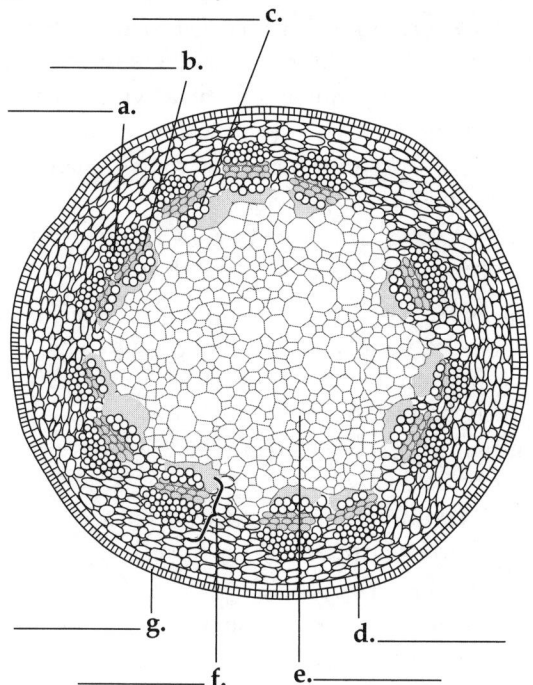

Test Your Knowledge

MULTIPLE CHOICE: *Choose the one best answer.*

1. Which of the following is *not* a reason that *Arabidopsis* is used as a model organism?
   a. It has a short generation time and is easy to grow in the laboratory.
   b. Its relatively small genome has been sequenced.
   c. Its ability to reproduce only by self-fertilization has facilitated genetic studies.
   d. The production of "knockout mutants" allows researchers to study the function of individual genes.

2. Which of the following is an *incorrect* completion of this sentence? Monocots typically have
   a. a taproot rather than a fibrous root system.
   b. leaves with parallel veins rather than branching venation.
   c. no secondary growth.
   d. scattered vascular bundles in the stem rather than bundles in a ring.

3. Mycorrhizal associations involve interactions between fungi and
   a. roots.
   b. stems.
   c. leaves.
   d. both stems and leaves.

4. What would be the effect if a mutant plant had sieve-tube elements with sieve plates that were thickened and much less porous?
   a. The lateral transport of nutrients through a woody stem would stop.
   b. The stems of these plants would be much stronger and the plants could grow taller.
   c. Transport of photosynthate from leaves to roots would be reduced.
   d. Transport through xylem would increase as companion cells transferred sugars through the pits in tracheids.

5. Axillary buds
   a. always develop into vegetative shoots.
   b. form at nodes in the angle where leaves join the stem.
   c. grow out from the pericycle layer to form lateral roots.
   d. are formed from intercalary meristems.

6. Which of the following structures provide structural support to a leaf?
   a. petioles
   b. vascular bundles
   c. suberin-impregnated cork cells
   d. epidermal cells with lignin-strengthened cell walls

7. Which of the following plant structures show determinate growth?
   a. taproots
   b. vegetative shoots
   c. adventitious roots
   d. eudicot leaves

8. Which of the following is *not* a primary meristem?
   a. protoderm
   b. periderm
   c. procambium
   d. ground meristem

9. Which of the following is *incorrectly* paired with the length of its life cycle?
   a. pine tree—perennial
   b. corn—biennial
   c. carrot—biennial
   d. marigold—annual

10. Secondary xylem and phloem are produced in a root by the
    a. pericycle.
    b. endodermis.
    c. vascular cambium.
    d. vascular cylinder.

11. Bark consists of
    a. secondary phloem.
    b. layers of periderm.
    c. cork cambium.
    d. all of the above.

12. You are a companion cell in an angiosperm. What are you a companion to?
    a. a tracheid
    b. a sclereid
    c. a sieve-tube element
    d. a vessel element

13. In what direction does a plant cell enlarge?
    a. toward its basal end as a result of positional information
    b. parallel to the orientation of the preprophase band of microtubules
    c. perpendicular to the orientation of cellulose microfibrils in the cell wall
    d. upward if it is in the stem, downward if it is in the root

14. Which of the following is *incorrectly* paired with its function?
    a. collenchyma—provide flexible support for growing stems
    b. lenticels—facilitate gas exchange in woody stems
    c. root hairs—protect the root tips as they push through soil
    d. periderm—protective layers of cork cells produced by cork cambium

15. Which of the following is essential to establishing the axial polarity of a plant?
    a. expression of different homeotic genes in the shoot and root meristems
    b. orientation of cortical microtubules perpendicular to the ground
    c. the asymmetric first division of the zygote
    d. the proper orientation of microtubules, which then orient cellulose microfibrils

16. The results from genetic studies of which of the following support the idea that plant form is not determined merely by the orientation of cell divisions?
    a. the *gnom* mutant of *Arabidopsis,* in which the first division of the zygote is symmetrical
    b. the *tangled-1* mutant of maize, with its aberrant longitudinal division and normal leaves
    c. the *KNOTTED-1* mutant in tomato plants, with its "super-compound" leaves
    d. the expression of the *GLABRA-2* gene in root epidermal cells that do not develop root hairs

17. Recall the ABC hypothesis for flower development. You have developed a mutant plant in which gene B is produced in all whorls of the flower. What do you expect the flowers to look like?
    a. two whorls of petals surrounding two whorls of stamens
    b. one whorl of sepals; one whorl of stamens; one whorl of carpels
    c. four whorls of sepals
    d. one whorl of petals; one whorl of stamens; two whorls of carpels

18. Morphogenesis and pattern formation in plants results from
    a. differences in the direction of cell expansion as well as positional information from neighboring cells.
    b. the order in which cells move from the zone of elongation to the zone of differentiation.
    c. lineage-based mechanisms in which cell fate is passed from cells to their progeny.
    d. the expression of meristem identity genes and then organ-identity genes.

Resource Acquisition and Transport in Vascular Plants

Key Concepts

36.1 Adaptations for acquiring resources were key steps in the evolution of vascular plants

36.2 Different mechanisms transport substances over short or long distances

36.3 Transpiration drives the transport of water and minerals from roots to shoots via the xylem

36.4 The rate of transpiration is regulated by stomata

36.5 Sugars are transported from sources to sinks via the phloem

36.6 The symplast is highly dynamic

Framework

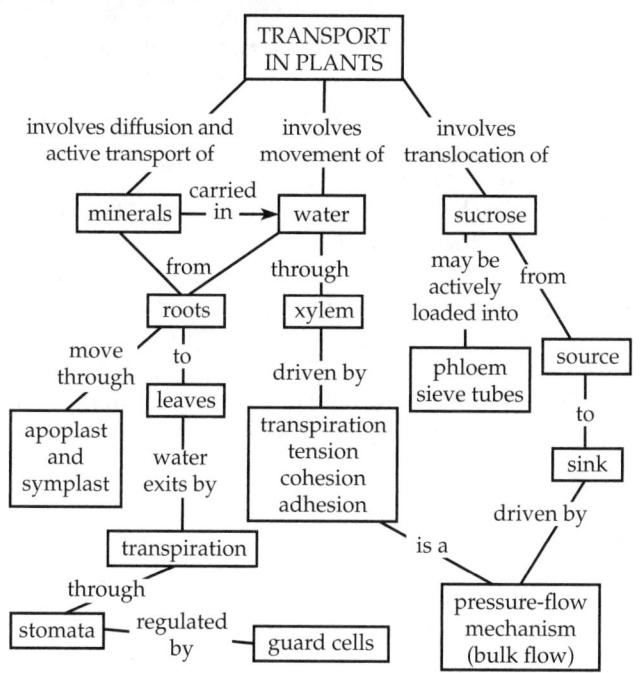

Chapter Review

36.1 Adaptations for acquiring resources were key steps in the evolution of vascular plants

Evolutionary adaptations to life on land involved the specialization of roots to absorb water and minerals, taller shoots with broad leaves to absorb CO_2 and light, and vascular tissue consisting of **xylem** and **phloem** to transport materials between roots and shoots.

Shoot Architecture and Light Capture The orientation of leaves and stems affects light capture. The extent of branching and stem elongation represents a compromise between energy spent growing tall to avoid shading versus energy spent branching and increasing photosynthetic surface area. Increased stem thickness usually accompanies growth in height.

Leaf size is often correlated with water availability. **Phyllotaxy**, the arrangement of leaves on the stem of a plant, may be one leaf per node (alternate, or spiral), two leaves per node (opposite), or multiple (whorled). Alternate phyllotaxy, with leaves emerging at a light-maximizing 137.5° angle, is most common in angiosperms.

The depth of the layers of leaves in a plant community affects productivity. When the layers are too deep, the shaded, nonproductive lower leaves of a plant may be shed through *self-pruning*. The *leaf area index* is the ratio of the total area of the top surfaces of the leaves to the area of ground covered by the plant or crop. In mature crops, the leaf area index may approach 7.

Open stomatal pores allow CO_2 into photosynthetic tissues, but they are also responsible for most water loss from plants: As a result, shoot adaptations are a compromise between maximizing photosynthetic efficiency and minimizing water loss.

Root Architecture and Acquisition of Water and Minerals Evidence indicates that plants have mechanisms that reduce the competition between their own roots for the same limited resources.

Mycorrhizae, symbiotic associations between roots and fungi, greatly increase the surface area for water and mineral absorption.

INTERACTIVE QUESTION   36.1

a. What leaf orientation works best in low light levels? In bright sunny conditions?

b. Which results in greater productivity: a high leaf area index or a low leaf area index?

c. How might root branching and physiology change in response to patches of nutrients such as nitrate?

36.2   Different mechanisms transport substances over short or long distances

The Apoplast and Symplast: Transport Continuums   The **apoplast** consists of plant cell walls, extracellular spaces, and interiors of dead cells, such as xylem vessels. The protoplasts of plant cells are connected by plasmodesmata, forming a cytoplasmic continuum called the **symplast**.

Transport of water and solutes within plant tissues can occur by three routes: the *symplastic route*, moving through plasmodesmata; the *apoplastic route*, moving along cell walls and extracellular spaces; and the *transmembrane route*, by crossing plasma membranes and cell walls as materials move from cell to cell. Substances may change routes during transit.

Short-Distance Transport of Solutes Across Plasma Membranes   Substances may move across the selectively permeable plasma membrane by passive or active transport, using protein carriers, ion channels, and cotransporters.

A plant proton pump moves H^+ out of the cell, generating an energy-storing proton gradient and a membrane potential. Cotransport with H^+ can move solutes such as sucrose and ions such as NO_3^- across the plasma membrane. Gated ion channels open or close in response to stimuli such as voltage, pressure, or chemicals.

Short-Distance Transport of Water Across Plasma Membranes   The diffusion of free water (not bound to solutes or surfaces) across a membrane is called **osmosis**. **Water potential** is a useful measurement for predicting the direction that water will move, and it takes into account both solute concentration and physical pressure. Free water will flow from a region of higher water potential to one of lower water potential, and this movement has the capacity to perform work.

Water potential, designated ψ *(psi)*, is measured in **megapascals** (MPa); 1 MPa is equal to about 10 atmospheres of pressure. The water potential of pure water in an open container under standard conditions is assigned the value of 0 MPa.

The combined effect of solute concentration and pressure is shown by the *water potential equation*: $\Psi = \Psi_S + \Psi_P$. This equation states that water potential is the sum of the solute potential and the pressure potential.

Solute potential (Ψ_S), also called *osmotic potential*, is proportional to the molarity of a solution. Solutes reduce the number of free water molecules and the capacity of water to do work. Solutes lower the water potential, and a solution's Ψ_S is always negative.

Pressure potential (Ψ_P) measures the physical pressure on a solution and can have a positive or negative value. A plant **protoplast** that is expanding due to the osmotic uptake of water pushes against the cell wall. This positive pressure is called **turgor pressure**. The water in hollow xylem cells is often under a negative pressure (tension).

A **flaccid** plant cell (Ψ_P = 0 MPA) bathed in a solution more concentrated than the cell will lose water by osmosis because the solution has a lower (more negative) Ψ than the Ψ_S of the cell. The cell's protoplast shrinks and pulls away from the cell wall (**plasmolysis**). A flaccid cell bathed in pure water has a Ψ that is lower than the Ψ of the water. Water will then enter the cell until enough turgor pressure builds up such that Ψ_P and Ψ_S are equal and opposite in magnitude, and $\Psi = 0$ both inside and outside the cell. Net movement of water will then stop.

Plant cells are usually **turgid**; they have a greater solute concentration than their extracellular environment and turgor pressure keeps them firm. Loss of turgor causes **wilting** of a plant's leaves and stems.

Water-specific transport proteins called **aquaporins** increase the rate of water diffusion across a membrane.

INTERACTIVE QUESTION 36.2

a. A flaccid plant cell has a water potential of −0.6 MPa. Fill in the water potential equation for this cell.

$$\psi_p =$$
$$+ \; \psi_s =$$
$$\overline{\psi \;\; = -0.6 \text{ MPa}}$$

b. The cell is then placed in a beaker of water ($\Psi = 0$ MPa). Fill in the equation for the cell after it reaches equilibrium in pure water. Explain what happens to the Ψ of this cell.

$$\psi_p =$$
$$+ \; \psi_s =$$
$$\overline{\psi \;\; =}$$

c. Explain what would happen to the same cell if it is placed in a solution that has a water potential of −0.8 MPa. Fill in the equation for the cell after it reaches equilibrium in the solution.

$$\psi_p =$$
$$+ \; \psi_s =$$
$$\overline{\psi \;\; =}$$

Long-Distance Transport: The Role of Bulk Flow   Long-distance transport in plants occurs by **bulk flow**, the movement of fluid driven by a pressure gradient. Sieve-tube elements lack most cellular organelles, and vessel elements and tracheids are dead (and empty) at maturity; both factors facilitate more efficient bulk flow.

36.3   **Transpiration drives the transport of water and minerals from roots to shoots via the xylem**

Absorption of Water and Minerals by Root Cells
Much of the absorption of water and mineral ions occurs along root tips, where root hairs are located. The soil solution flows along the hydrophilic walls of epidermal cells and moves along the apoplast into the root cortex, exposing a large surface area of plasma membrane for the uptake of water and minerals. Active transport allows cells to accumulate essential minerals.

Transport of Water and Minerals into the Xylem
Water and minerals that entered the symplast through a cortex or epidermal cell pass through plasmodesmata of cells of the **endodermis** and into the vascular cylinder, or stele. A ring of suberin around each endodermal cell, called the **Casparian strip**, prevents water and minerals from the apoplast from entering the stele without passing through a *selectively permeable* plasma membrane.

By a combination of diffusion and active transport, minerals move from endodermal cells to the apoplast and enter the nonliving xylem tracheids and vessel elements along with water.

INTERACTIVE QUESTION 36.3

Label the following diagram of a section of a root. Letters a and b refer to transport routes of water and minerals; letters c–i refer to cell layers or structures.

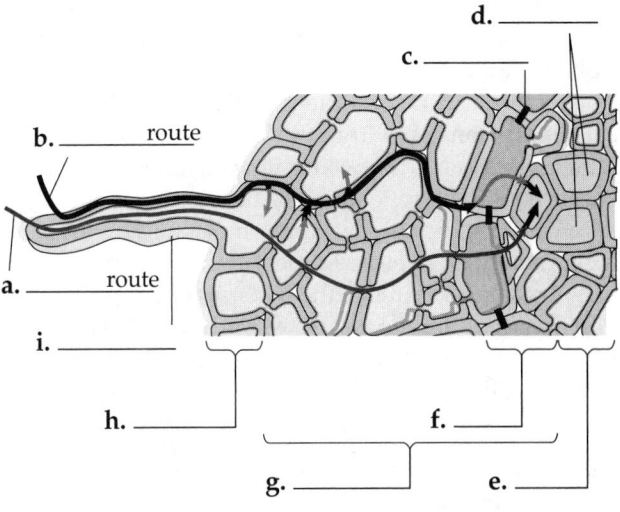

Bulk Flow Transport via the Xylem   The water and dissolved minerals of **xylem sap** move by bulk flow from the vascular cylinder of a root to the branching veins of leaves. Through the evaporative loss of water vapor from leaves, called **transpiration**, the plant loses a tremendous amount of water that must be replaced by water transported up from the roots.

At night, the transpiration rate is low, but minerals are still actively pumped into the vascular cylinder and prevented from leaking out by the endodermis. Water flowing from the cortex into this area of lower water potential produces **root pressure**, which pushes xylem sap upward. In some plants, root pressure may cause **guttation**, the exudation of water droplets from leaves when more water is forced up the xylem than is transpired by the plant.

According to the **cohesion-tension hypothesis**, transpiration creates a negative-pressure pull on xylem sap, which is transmitted from shoots to roots by the cohesion of water molecules within the xylem.

Water vapor from saturated air spaces within a leaf exits through stomata. The thin layer of water that coats the mesophyll cells lining the air spaces begins to evaporate. The adhesion of the remaining water to the hydrophilic walls causes a curvature in the air-water interface. Because of the high surface tension of water, this curvature produces tension in the water layer. This negative pressure draws water from the cells, which is replaced by water from the xylem. Water moves along a gradient toward more negative water potentials, from xylem to neighboring cells to air spaces to the drier air outside the leaf.

The transpirational pull on xylem sap is transmitted from the leaves to the root tips by the cohesiveness of water, which results from hydrogen bonding between water molecules. Adhesion of water molecules to the hydrophilic walls of the narrow xylem cells also contributes to overcoming the downward pull of gravity.

The upward transpirational pull on the cohesive sap creates tension within the xylem, lowering the water potential so that water flows passively from the soil, across the cortex, and into the vascular cylinder.

A break in the chain of water molecules by the formation of a water vapor pocket in a xylem vessel, called cavitation, breaks the transpirational pull. The vessel cannot function in transport unless the break is repaired or the chain of water molecules detours through adjacent vessels. In some trees, transfer of water from phloem to xylem can prevent cavitation.

Xylem Sap Ascent by Bulk Flow: A Review   The cohesion-tension mechanism results in the bulk flow of water and solutes from roots to leaves. Solar-powered tension produced by transpiration creates the pressure gradient that is responsible for long-distance transport of water and minerals.

INTERACTIVE QUESTION   36.4

Explain the contribution of each of the following phenomena to the long-distance transport of water.

a. Transpiration:

b. Tension:

c. Cohesion:

d. Adhesion:

36.4 The rate of transpiration is regulated by stomata

For photosynthesis to occur, leaves must exchange gases through the stomata and provide a large internal surface area for CO_2 uptake, both of which increase evaporative water loss.

Stomata: Major Pathways for Water Loss   A plant loses most of its water through stomata. Guard cells regulate the size of stomatal openings, which controls the rate of transpiration. Stomatal densities in many plant species are both genetically and environmentally influenced.

Mechanisms of Stomatal Opening and Closing
When the kidney-shaped guard cells of most angiosperms become turgid and swell, their radially oriented microfibrils cause them to bow outward, increasing the size of the gap between them.

Guard cells can actively accumulate potassium ions (K^+), which lowers water potential and leads to the osmotic inflow of water and an increase in turgor pressure. The flow of K^+ through specific membrane channels into the cell is coupled with the generation of a membrane potential by proton pumps that transport H^+ out of the cell. The exodus of K^+ (with water following) leads to a loss of turgor and stomatal closing.

Stimuli for Stomatal Opening and Closing   The opening of stomata at dawn is related to at least three factors. First, the illumination of blue-light receptors activates proton pumps; the proton gradient is then used to transport K^+ into the guard cells, and water follows passively. Second, the depletion of CO_2 within air spaces of the leaf as photosynthesis begins stimulates stomata to open. The third factor is a daily rhythm of opening and closing that is endogenous to guard cells. Cycles that have intervals of approximately 24 hours are called **circadian rhythms**.

Environmental stress can cause stomata to close during the day. Guard cells lose turgor when water is in short supply. Also, **abscisic acid (ABA)**, a hormone produced in roots and leaves in response to a lack of water, signals guard cells to close stomata.

Effects of Transpiration on Wilting and Leaf Temperature   When transpiration exceeds the water available, leaves wilt as cells lose turgor pressure. Transpiration also produces evaporative cooling, maintaining a cooler temperature for critical enzymes in leaf cells.

Adaptations That Reduce Evaporative Water Loss   Many **xerophytes**, plants adapted to arid climates, have leaves that are highly reduced; photosynthesis is carried out in thick, water-storing stems. Leaf cuticles may be thick, and stomata may be sheltered in depressions. A form of photosynthesis called crassulacean acid metabolism (CAM) is used by succulents of several plant families. These plants take in CO_2 at night, so their stomata can be closed during the day, reducing water loss.

Chapter 36: Resource Acquisition and Transport in Vascular Plants

INTERACTIVE QUESTION   36.5

What is meant by saying that plants face a photosynthesis-transpiration compromise? Explain how a hot, sunny day with a dry wind can affect this compromise in a plant.

36.5 Sugars are transported from sources to sinks via the phloem

Movement from Sugar Sources to Sugar Sinks
Translocation, the transport of photosynthetic products throughout the plant, occurs in the sieve-tube elements of phloem. **Phloem sap** may have a sucrose concentration as high as 30% and may also contain minerals, amino acids, and hormones.

Phloem sap flows from a **sugar source,** where it is produced by photosynthesis or the breakdown of starch, to a **sugar sink,** an organ that consumes or stores sugar. The direction of transport in any one sieve tube depends on the location of the source and sink connected by that tube, and the direction may change with the season or the needs of the plant.

In some species, sugar in the leaf moves through the symplast of the mesophyll cells to sieve-tube elements. In other species, sugar first moves through the symplast and then into the apoplast in the vicinity of sieve-tube elements and companion cells, which actively accumulate sugar. In some plants, companion (transfer) cells have specialized ingrowths of the cell wall that increase surface area for movement of solutes.

Phloem loading requires active transport in plants that accumulate high sugar concentrations in the sieve tubes. Proton pumps and the cotransport of sucrose through cotransport proteins along with the returning protons is the mechanism used for active transport. In sink tissues, the concentration gradient favors the diffusion of sugar out of sieve tubes because sugar is either used or converted into starch within sink cells.

INTERACTIVE QUESTION   36.6

Label the components of the following diagram of the mechanism involved in the active transport of sucrose into companion cells or sieve-tube elements. Which side of this diagram represents the inside of the cell?

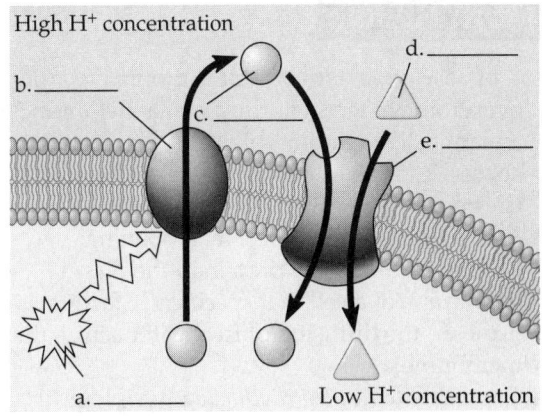

Bulk Flow by Positive Pressure: The Mechanism of Translocation in Angiosperms   The rapid movement of phloem sap from source to sink is driven by *pressure flow.* High solute concentration at the source lowers the water potential, and the resulting movement of water into the sieve tube produces positive pressure. At the sink end, the osmotic loss of water following the exodus of sucrose into the surrounding tissue results in a lower pressure. The difference in these pressures causes sap to move by bulk flow from source to sink. Innovative tests of this hypothesis using aphid stylets support it as the explanation for the flow of sap in the phloem of angiosperms.

In a phenomenon known as *self-thinning,* plants abort flowers, seeds, or fruits when they have more sinks than can be supported by sources.

36.6 The symplast is highly dynamic

Changes in Plasmodesmatal Number and Pore Size   Recent studies have shown that plasmodesmata are dynamic structures that can open and close rapidly, increase in number or size, and be eliminated as developmental needs change. Plant viruses produce *viral movement proteins* that mimic the cell's regulatory machinery that dilates plasmodesmata. Communication during plant development may occur within interconnected groups of cells called *symplastic domains.*

Phloem: An Information Superhighway   The systemic transport via phloem of certain macromolecules, such as flower-inducing signals or defense signals, helps integrate the functions of the whole plant.

Electrical Signaling in the Phloem   Some studies indicate that phloem can transmit electrical signals that affect gene transcription or other processes in other locations of the plant.

Word Roots

apo- = off, away; **-plast** = formed, molded (*apoplast:* everything external to the plasma membrane of a plant cell, including cell walls, extracellular spaces, and the lumen of xylem vessels and tracheids)

aqua- = water; **por-** = a pore, small opening (*aquaporin:* a channel protein in the plasma membrane of a cell that specifically facilitates osmosis, the diffusion of free water across the membrane)

circa- = a circle (*circadian rhythm:* a physiological cycle of about 24 hours that persists even in the absence of external cues)

endo- = within, inner; **-derm** = skin (*endodermis:* in plant roots, the innermost layer of the cortex that surrounds the vascular cylinder)

gutt- = a drop (*guttation:* the exudation of water droplets from leaves, caused by root pressure in certain plants)

mega- = large, great (*megapascal:* a unit of pressure equivalent to about 10 atmospheres of pressure)

myc- = a fungus; **rhiz-** = a root (*mycorrhiza:* a mutualistic association of plant roots and a fungus)

sym- = with, together (*symplast:* in plants, the continuum of cytoplasm connected by plasmodesmata between cells)

xer- = dry; **-phyte** = a plant (*xerophyte:* a plant adapted to an arid climate)

Structure Your Knowledge

1. Describe the ways in which solutes may move across the plasma membrane in plants.
2. Both xylem sap and phloem sap move by bulk flow in an angiosperm. Compare and contrast the mechanisms for the movement of each.

Test Your Knowledge

MULTIPLE CHOICE: *Choose the one best answer.*

1. If a plant has a phyllotaxy of alternate leaves that emerge at a 30° angle from each previous leaf, one can surmise that the
   a. leaves are most likely very broad, indicating that the plant grows where there is sufficient water.
   b. leaves are most likely very broad, indicating that the plant grows in low light levels.
   c. leaves are most likely very thin because due to emerging at such a close angle, they would have to be thin to avoid shading each other.
   d. the layers of leaves of the plant must be very thick with leaves spaced so close together, and lower leaves are most likely lost through self-pruning.

2. Shoot architecture represents a compromise between
   a. photosynthesis and cellular respiration.
   b. light capture and temperature sensitivity.
   c. water availability and photosynthetic efficiency.
   d. growth and reproduction.

3. Which of the following structures are *not* components of the symplast?
   a. sieve-tube elements
   b. xylem tracheids
   c. endodermal cells
   d. root hairs

4. Proton pumps in the plasma membranes of plant cells may
   a. contribute to the movement of ions through a cotransport mechanism.
   b. be coupled to the movement of K^+ into guard cells.
   c. drive the accumulation of sucrose in sieve-tube elements.
   d. be involved in all of the above.

5. Consider an enzymatic reaction that forms product at a rate of 50 µg/sec at 20°C, and a rate of 150 µg/sec at 30°. What is Q_{10} for this reaction?
   a. 50
   b. 3.0
   c. 0.67
   d. Can't tell; more information is required.

6. If a turgid plant cell placed in a solution in an open beaker becomes flaccid, which of the following statements is true?
   a. The water potential of the cell was initially higher than that of the solution.
   b. The water potential of the cell was initially equal to that of the solution.
   c. The pressure (Ψ_P) of the cell was initially lower than that of the solution.
   d. The Ψ_S of the cell was initially more negative than that of the solution.

7. Which of the following mechanisms facilitates the movement of NO_3^- into epidermal cells of the root?
   a. bulk flow of water into the root
   b. transport along with H^+ through a membrane cotransporter
   c. passage through selective channels, aided by the membrane potential created by proton pumps
   d. active transport through a nitrate pump

8. Aquaporins are
   a. cytoplasmic connections between cortical cells.
   b. pores through the ends of sieve-tube elements through which phloem sap flows.
   c. openings in the lower epidermis of leaves through which water vapor escapes.
   d. water-specific channels in membranes that speed the rate of osmosis.

9. Which of the following statements is true?
   a. Ψ is equal to 0 MPa when the cell is in pure water and is turgid.
   b. Ψ is positive in a plasmolyzed cell.
   c. Ψ_S is always positive.
   d. Ψ becomes greater as K^+ ions are actively moved into a plant cell.

10. The Casparian strip prevents water and minerals from entering the vascular cylinder through the
    a. plasmodesmata.
    b. endodermal cells.
    c. apoplast.
    d. xylem vessels.

11. The surface area of a root available for absorbing water and minerals is *not* increased by
    a. mycorrhizae.
    b. root hairs.
    c. cytoplasmic extensions of tracheids.
    d. the large surface area of cortical cells.

12. Guttation results from
    a. the pressure flow of sap through phloem.
    b. a water-vapor break in a column of xylem sap.
    c. root pressure causing water to flow up through xylem faster than it can be lost through transpiration.
    d. specialized transport cell structures that accumulate sucrose.

13. The formation of a curvature in the air-water interface along the cell walls of cells lining the air space of a leaf contributes to water transport by
    a. creating a more positive water potential than in the surrounding mesophyll cells.
    b. creating tension, thereby lowering pressure and the water potential of the leaf.
    c. increasing the adhesion of water molecules to the cell walls.
    d. increasing the rate of transpiration from the leaf.

14. Cohesion is a result of
    a. hydrogen bonding between water molecules.
    b. the pull on the water column as water evaporates from the surfaces of mesophyll cells.
    c. tension within the xylem caused by a negative pressure.
    d. attraction of water molecules to the hydrophilic walls of narrow xylem tubes.

15. You observe a CAM plant that has its stomata closed on a sunny day. How does the plant obtain CO_2?
    a. CO_2 is taken up through open stomata at night.
    b. CO_2 is transported up from the roots.
    c. CO_2 transport proteins move CO_2 into photosynthetic cells.
    d. This is a trick question! CAM plants do not require CO_2 for photosynthesis.

16. Your favorite houseplant is wilting. Which of the following statements describes the most likely cause and remedy for its declining condition?
    a. Water potential is too low; apply sugar water.
    b. The plant's cells are undergoing plasmolysis; water the plant.
    c. Cavitation has occurred; perform a xylem vessel bypass.
    d. Circadian rhythm has caused the stomata to open; place the plant in the dark.

17. Given the current increase in atmospheric CO_2 levels, which of the following changes would you predict?

    a. an increase in distribution of CAM plants, which more effectively take in CO_2 at night

    b. an increase in transpiration due to increased levels of photosynthesis

    c. a decrease in stomatal density due to increased CO_2 levels

    d. an increase in stomatal density to provide for increased levels of photosynthesis

18. Which of the following mechanisms explains the movement of sucrose from source to sink?

    a. osmotic movement of water into the sucrose-loaded sieve-tube elements, creating a higher pressure in the source than in the sink

    b. tension created by pressure differences in the source and the sink

    c. active transport of sucrose through the sieve-tube cells driven by proton pumps

    d. the hydrolysis of starch to sucrose in mesophyll cells, which raises their water potential and drives the bulk flow of sap to the sink

19. Which of the following would most likely result in self-pruning or self-thinning?

    a. the transport of viral particles facilitated by viral movement proteins through plasmodesmata

    b. the shading of leaves to the extent that respiration exceeds photosynthesis

    c. a lack of sufficient products of photosynthesis in sources to supply all sinks

    d. both **b** and **c**

20. Which of the following statements is *not* correct concerning the dynamic nature of the symplast?

    a. Proteins and RNAs may travel through symplastic domains to coordinate development within tissues.

    b. As a leaf matures from a sink to a source, some of its plasmodesmata may close, thereby stopping phloem unloading.

    c. Electrical signals traveling through the phloem can affect changes in gene transcription, respiration, photosynthesis, or hormone levels in other parts of the plant.

    d. Transpiration rates increase on warm, sunny days and decrease on cloudy, damp days.

Chapter 37

Soil and Plant Nutrition

Key Concepts

37.1 Soil contains a living, complex ecosystem

37.2 Plant roots absorb essential elements from the soil

37.3 Plant nutrition often involves relationships with other organisms

Framework

The fertility of soil is influenced by its texture and composition. Good agricultural practices incorporate soil and water conservation. The nutritional requirements of plants include essential macronutrients and micronutrients. The rhizosphere includes bacteria and fungi that are supported by plant roots and may enhance plant growth. Plants assimilate nitrogen made available by the decomposition of humus and from nitrogen fixation by bacteria. Mycorrhizae are important associations between fungi and plant roots that increase mineral and water absorption.

Chapter Review

37.1 Soil contains a living, complex ecosystem

Soil Texture The texture of soil depends on particle size, which varies from coarse sand to silt to fine clay. The formation of soil begins with the weathering of rock as frozen water in crevices mechanically fractures rock and soil acids break down rocks chemically. Plant roots accelerate these processes. **Topsoil** is a mixture of mineral particles from rock, **humus** (decomposing organic matter), and living organisms. Two other distinct soil layers, or **soil horizons**, are found under the topsoil layer.

Loams, containing almost equal parts of sand, silt, and clay, are often the most fertile soils, having enough fine particles to provide a large surface area for retaining water and minerals but enough coarse particles to provide air spaces for respiring roots.

Topsoil Composition Positively charged ions, such as K^+, Ca^{2+}, and Mg^{2+}, adhere to the negatively charged surfaces of soil particles and are not easily lost by *leaching*, the filtration of water through soil. In **cation exchange**, hydrogen ions displace positively charged mineral ions from soil particles, making the ions available for absorption. Soils with more clay and organic matter have a higher capacity for cation exchange. Negatively charged ions, such as nitrate (NO_3^-), phosphate ($H_2PO_4^-$), and sulfate (SO_4^{2-}), tend to leach away more quickly.

Humus builds a crumbly soil that retains water, provides good aeration, and supplies mineral nutrients.

The activities of the numerous soil inhabitants—bacteria, fungi, algae and other protists, insects, worms, nematodes, and plant roots—affect the physical and chemical properties of soil.

INTERACTIVE QUESTION 37.1

Describe the characteristics of a fertile soil.

Soil Conservation and Sustainable Agriculture As ancient farmers learned to fertilize the soil, humans were able to establish permanent residences. Soil mismanagement, however, leads to low productivity.

Irrigation can make farming possible in arid regions, but it places a huge drain on surface and underground water resources and leads to soil *salinization*. Excess water removal from *aquifers* results in *land subsidence*, the sinking of Earth's surface. *Drip irrigation,* which slowly releases water to the root zone, reduces some of these problems.

Agriculture removes mineral nutrients from the soil; **fertilization** replaces them. In industrialized nations, commercially produced fertilizers (usually containing nitrogen, phosphorus, and potassium) are used. Such fertilizers may rapidly leach from the soil, polluting streams and lakes. Manure, fishmeal, and compost (so-called "organic" fertilizers) contain organic material that slowly decomposes into inorganic nutrients.

The acidity of the soil affects cation exchange. It can also alter the chemical form of minerals and thus their ability to be absorbed by the plant. Managing soil pH is an important aspect of maintaining fertility.

A large amount of topsoil is lost to water and wind erosion each year. Agricultural use of cover crops, windbreaks, and terracing can minimize erosion, as can **no-till agriculture**, which minimally disturbs the soil.

The use of plants to extract heavy metals and other pollutants from contaminated soils is an emerging technology known as **phytoremediation**.

INTERACTIVE QUESTION   37.2

a. What is **sustainable agriculture**?

b. What is the N-P-K ratio?

c. How is soil pH related to toxic aluminum ions (Al^{3+})?

37.2   Plant roots absorb essential elements from the soil

Water makes up 80–90% of the weight of a fresh plant. Carbon, oxygen, and hydrogen—the components of carbohydrates such as cellulose—are the most abundant elements making up the dry weight of a plant. Inorganic substances absorbed from the soil account for only 4% of the dry mass of a plant.

Essential Elements   **Essential elements** are those required for a plant to complete its life cycle from a seed to the production of its own seeds. **Hydroponic culture** has been used to help identify the 17 elements that are essential in all plants.

Nine **macronutrients** are required by plants in relatively large amounts and include the six major elements of organic compounds as well as calcium, potassium, and magnesium.

Eight **micronutrients** are needed by plants in very small amounts, mainly as cofactors of enzymes.

INTERACTIVE QUESTION   37.3

List some functions of each of the following macronutrients.

a. magnesium

b. phosphorus

c. calcium

Symptoms of Mineral Deficiency   Mobile nutrients move to young, growing tissues, so deficiencies of those nutrients first appear in older parts of the plant. Symptoms of a given mineral deficiency may be distinctive enough to allow diagnosis by a plant physiologist or farmer. Soil and plant analysis can confirm a specific deficiency. Deficiencies of potassium, phosphorus, and especially nitrogen are most common.

INTERACTIVE QUESTION   37.4

a. What is *chlorosis?*

b. Where in a plant would you first expect to see a deficiency of a relatively immobile element?

Improving Plant Nutrition by Genetic Modification
Resistance to aluminum toxicity has been engineered into tobacco and papaya plants by introducing a citrate synthase gene from a bacterium. Citric acid secreted by roots binds to free aluminum ions, lowering the levels of toxic aluminum in the soil.

When a promoter that responds to a decrease in a nutrient is linked to a "reporter" gene in plants, a grower can tell when a deficiency of that nutrient is imminent and can then fertilize only in response to need.

37.3 Plant nutrition often involves relationships with other organisms

Plants form *mutualistic* relationships with many bacteria and fungi.

Bacteria and Plant Nutrition The **rhizosphere** is the layer of soil surrounding plant roots. It contains high populations of soil bacteria called **rhizobacteria**. Bacteria (or fungi) that live within the plant are called **endophytes**. Both rhizobacteria and endophytic bacteria are supported by plant secretions and, in turn, enhance plant growth by producing growth-stimulating hormones or antibiotics, by absorbing toxins, or by making nutrients more available. A metagenomic study found that the "species" composition of bacterial communities varied based on the location within or outside the root and whether the soil was clayey or porous.

The **nitrogen cycle** describes the movement of nitrogen in an ecosystem, which involves soil microbes, plants, decomposers, and the atmosphere. Through a process called ammonification, decomposers convert organic nitrogen in humus to ammonium (NH_4^+). *Nitrogen-fixing bacteria* convert atmospheric N_2 to NH_3. In the soil solution, ammonia forms ammonium (NH_4^+). Different types of *nitrifying bacteria* perform the two-step process of *nitrification*, converting ammonia to nitrate (NO_3^-), the form of nitrogen most readily absorbed by roots. Most plants incorporate nitrogen into amino acids or other organic compounds in the roots and then transport these compounds or NO_3^- through the xylem to shoots. Some nitrate is lost to the atmosphere by the action of denitrifying bacteria.

Some free-living soil bacteria and bacteria of the genus *Rhizobium*, which are closely associated with the roots of legumes, are able to convert atmospheric nitrogen into ammonia through the process of **nitrogen fixation**. *Nitrogenase* is an enzyme complex that reduces N_2 by adding H^+ and electrons to form NH_3. This process is energetically expensive, and nitrogen-fixing bacteria rely on decaying material or plant roots for carbohydrates.

Plants of the legume family have root swellings, called **nodules**, composed of plant cells with vesicles containing bacteria in a form called **bacteroids**. Nitrogen fixation requires an anaerobic environment. Lignified external layers of nodule cells limit gas exchange. In some nodules, the presence of oxygen-binding leghemoglobin keeps the concentration of free O_2 low and regulates the oxygen supply for the bacterial respiration required to provide ATP for nitrogen fixation.

Mutual signaling between roots and bacteria leads to root hair elongation and formation of an infection thread within a root hair. Vesicles containing bacteroids bud from the infection thread into dividing cortical and pericycle cells. These dividing cells fuse to form the nodule, which grows and develops vascular tissue connecting with the vascular cylinder. Bacteria in the nodules are supplied with carbohydrates, and nodule cells use the fixed nitrogen to make amino acids, which are then transported throughout the plant.

Legume seeds are often coated with their specific *Rhizobium* before planting. Rice farmers culture a water fern that has symbiotic cyanobacteria that fix nitrogen, improving the fertility of rice paddies.

INTERACTIVE QUESTION 37.5

Fill in the types of bacteria (a–d) that participate in the nitrogen nutrition of plants. Indicate the form (e) in which nitrogen is transported in xylem to the shoot system.

a. What is **crop rotation**?

b. What is "green manure"?

Fungi and Plant Nutrition   Most plants have symbiotic associations between their roots and fungi called **mycorrhizae**. The fungus receives sugar from the plant while providing a large surface area for the absorption of water and minerals (especially phosphate) for the plant. The fungus secretes growth factors that stimulate root growth and branching and antibiotics that help protect the plant from pathogens.

Especially common in woody plants, **ectomycorrhizae** form a dense sheath or mantle of mycelium over the root surface. Hyphae extending from the mantle provide a huge absorptive surface area. Hyphae also grow into extracellular spaces in the root cortex, facilitating exchange between plant and fungus.

Arbuscular mycorrhizae do not form a sheath but extend fine fungal hyphae into the soil. Hyphae also penetrate through the walls of cortex cells, forming tubes that invaginate the cell membrane. Some of these tubes form dense branched structures called arbuscles that facilitate nutrient exchange. Over 85% of plant species have arbuscular mycorrhizae.

Most plants form mycorrhizae when they grow in their natural habitat. When seeds are planted in foreign soil lacking their symbiotic fungal species, the resulting plants show signs of malnutrition. The invasive plant garlic mustard has been shown to interfere with the growth of arbuscular mycorrhizae on native trees.

Why is it thought that mycorrhizae were important in the colonization of land by the first plants?

Epiphytes, Parasitic Plants, and Carnivorous Plants   **Epiphytes** are plants that grow on the surface of another plant but do not take nourishment from it. Many parasitic plants produce roots that function as haustoria that siphon nutrients from their host plant.

Carnivorous plants living in acid bogs or other nutrient-poor soils obtain nitrogen and minerals by killing and digesting insects and other small animals that are caught in various types of traps formed from modified leaves.

Word Roots

arb- = tree; **myc-** = a fungus; **rhiz-** = a root (*arbuscular mycorrhiza:* association of a fungus with a plant root system in which the fungus causes the invagination of the plant cells' plasma membranes)

ecto- = outside (*ectomycorrhizae:* associations of a fungus with a plant root system in which the fungus surrounds the roots but does not cause invagination of the plant cells' plasma membranes)

endo- = inside; **-phyte** = a plant (*endophyte:* a fungus or bacteria that lives between the cells of a plant part without causing harm to the plant)

macro- = large (*macronutrient:* an essential element that an organism must obtain in relatively large amounts)

micro- = small (*micronutrient:* an essential element that an organism needs in very small amounts)

Structure Your Knowledge

1. Develop a concept map to organize your understanding of the nutritional requirements of plants.
2. What are the similarities and differences between root nodules and mycorrhizae? How is each beneficial to plants?

Test Your Knowledge

MULTIPLE CHOICE: *Choose the one best answer.*

1. The most fertile type of soil is usually
   a. sand, because its large particles allow room for air spaces.
   b. loam, which has a mixture of fine and coarse particles.
   c. clay, because the fine particles provide much surface area to which minerals and water adhere.
   d. humus, which is decomposing organic material.

2. Which of the following is not a principle of sustainable agriculture?
   a. prudent use of irrigation
   b. salinization of topsoil
   c. protection of soil from erosion
   d. restoration of degraded land

3. Which of the following is an example of phytoremediation?
   a. dusting legume seeds with specific *Rhizobium* strains to increase nodule formation
   b. inoculating seeds with fungal spores to ensure mycorrhizal formation
   c. using plants to remove toxic heavy metals from contaminated soils
   d. genetic engineering of plants to increase protein content

4. Which soil component is associated with higher cation exchange capability?
   a. clay
   b. humus
   c. sand
   d. silt

5. The effects of mineral deficiencies involving a nutrient that is relatively immobile will first be observed in
   a. older portions of the plant.
   b. new leaves and shoots.
   c. the root system.
   d. the color of the leaves.

6. Micronutrients
   a. may be cofactors in enzymes or components of cytochromes.
   b. are required in very minute quantities.
   c. can be identified using hydroponic culture techniques.
   d. All of the above are true.

7. Chlorosis is
   a. yellowing leaves due to decreased chlorophyll synthesis in response to a mineral deficiency.
   b. the uptake of the micronutrient chlorine by a plant, facilitated by endophytic bacteria.
   c. the production of chlorophyll within the thylakoid membranes of a plant.
   d. a contamination of glassware in hydroponic culture.

8. Plants require nitrogen primarily for synthesis of
   a. carbohydrates.
   b. proteins.
   c. proteins and nucleic acids.
   d. carbohydrates, lipids and proteins.

9. The insertion of a bacterial citrate synthase gene into tobacco and papaya plants
   a. is a form of phytoremediation that enables these genetically modified plants to remove toxic zinc from contaminated soils.
   b. increases the availability of nitrogen by promoting the growth of nitrogen-fixing cyanobacteria.
   c. enables these genetically modified plants to overproduce citric acid, which binds to and thus lowers the level of toxic aluminum ions in the soil.
   d. allows more precise fertilizer use, as this gene serves as a reporter gene that signals a nutrient deficiency before damage has occurred.

10. Nitrogenase
    a. is an enzyme complex that reduces atmospheric nitrogen to ammonia.
    b. is found in *Rhizobium* and other nitrogen-fixing bacteria.
    c. requires an anaerobic environment to catalyze the energy-expensive fixation of nitrogen.
    d. is or does all of the above.

11. "Smart plants"
    a. are capable of signaling when fruit is ripe.
    b. are capable of signaling when they require water.
    c. are capable of signaling when they are nutrient-deficient.
    d. are capable of reducing metabolism when water or nutrients are deficient.

12. Epiphytes
    a. have haustoria for anchoring to their host plants and obtaining water and nutrients.
    b. are symbiotic relationships between specialized leaves and specific fungi.
    c. live in poor soil and capture and digest insects to obtain nitrogen.
    d. grow on other plants but do not obtain nutrients from their hosts.

13. The nitrogen content of some agricultural soils may be improved by
    a. crop rotation with legumes that house nitrogen-fixing bacteria.
    b. arbuscular mycorrhizae that form mutualistic interactions with specific legumes.
    c. cation exchange that is facilitated when roots pump H^+ into the soil.
    d. the action of ammonifying, nitrifying, and denitrifying bacteria.

14. Using a technique called metagenomics, researchers found that
    a. all plant roots house endophytes.
    b. many plants exhibit some amount of heterotrophy when their roots take up and digest bacteria and yeasts.
    c. the "species" composition of bacterial communities vary the most between bacteria inside roots (endophytes) and soil bacteria in the rhizosphere.
    d. the composition of the soil (clayey or porous) does not affect the "species" composition of bacteria either inside or outside the rhizosphere.

15. What result would you predict if a gardener sterilizes all of her soil in an effort to reduce plant diseases?
    a. increased growth due to the destruction of disease-causing microbes
    b. increased plant growth due to lack of soil microbes draining away the photosynthetic output of the plants
    c. decreased growth due to lack of humus that provides texture to the soil and various nutrients
    d. decreased plant growth due to destruction of mutualistic associations with microbes in the rhizosphere

Angiosperm Reproduction and Biotechnology

Framework

This chapter describes the sexual and asexual reproduction of flowering plants. The flower produces spores by meiosis that grow into the haploid gametophyte stages of the life cycle: Microspores in the anther develop into male gametophytes in pollen grains, and a megaspore in the ovule produces an embryo sac. Pollination and the double fertilization of egg and polar nuclei are followed by the development of a seed with an embryo and endosperm, protected in a seed coat and housed within a fruit. Seed dormancy is broken following proper environmental cues and the imbibition of water.

Asexual reproduction allows successful plants to clone themselves. Agriculture makes extensive use of vegetative propagation with cuttings, grafts, and test-tube cloning. Plant biotechnologists are creating genetically modified (GM or transgenic) plants that have such traits as insect or disease resistance, herbicide tolerance, and improved nutritional value.

Chapter Review

Humans and their crop plants are an example of the mutually beneficial relationships common between angiosperms and animals.

38.1 Flowers, double fertilization, and fruits are key features of the angiosperm life cycle

Plants exhibit an alternation of generations between the dominant sporophytes, which produce spores, and gametophytes, which produce gametes.

Flower Structure and Function Flowers are reproductive shoots that usually contain four concentric whorls of modified leaves—**sepals, petals, stamens**, and **carpels**—which attach to a part of the stem called the **receptacle**. A carpel consists of a sticky **stigma** at the top of a slender **style**, which leads to an **ovary**. The ovary encloses one or more **ovules**. A flower may have a single carpel or multiple fused carpels; either may be referred to as a **pistil**. Stamens consist of a filament and an **anther**, which contains microsporangia (pollen sacs). Petals are generally more brightly colored than sepals and may attract pollinators. Sepals enclose and protect the unopened floral bud.

A **complete flower** has sepals, petals, stamens, and carpels. **Incomplete flowers** lack one or more of these floral organs. Some incomplete flowers may be sterile; others may be *unisexual*. Flowers may be single or in clusters called **inflorescences**. The various shapes, colors, and odors of flowers are generally adaptations for attracting specific pollinators.

INTERACTIVE QUESTION   38.1

Identify the flower parts in the following diagram.

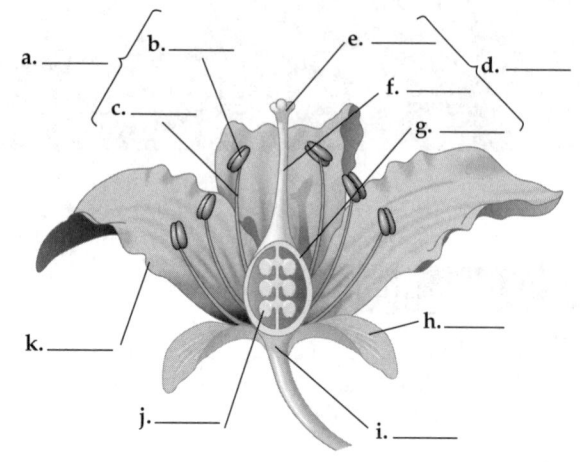

a. _____
b. _____
c. _____
e. _____
d. _____
f. _____
g. _____
h. _____
k. _____
j. _____
i. _____

Methods of Pollination   **Pollination**, the transfer of pollen from a stamen to a pistil, may be accomplished by wind, water, or animals, including bees, moths and butterflies, flies, birds, or bats. Many species of flowering plants and their specific pollinators illustrate **coevolution**, the evolution of reciprocal adaptations in interacting species. For instance, natural selection favors traits that enhance flower-pollinator mutualism, in which flowers provide nectar to specific insects while those insects efficiently pick up pollen and transfer it to other flowers. The close coevolutionary link between a plant and its pollinator may be a problem if climate change affects either species.

The Angiosperm Life Cycle: An Overview   The female gametophyte, also called an **embryo sac**, develops inside an ovule. Two *integuments* surround each megasporangium except at the *micropyle*. The *megasporocyte* in the megasporangium undergoes meiosis to form four haploid **megaspores**, only one of which survives. This megaspore grows and divides by mitosis three times, forming the female gametophyte, which typically consists of eight nuclei contained in seven cells. At the micropylar end of the embryo sac, an egg cell is lodged between two cells called synergids, which help attract the pollen tube. Three antipodal cells are at the other end; and two nuclei, called polar nuclei, are in a large central cell.

Within each microsporangium (pollen sac) in an anther, diploid cells called *microsporocytes* undergo meiosis to form four haploid **microspores**. A microspore divides once by mitosis to produce a *tube cell* and a *generative cell,* which moves into the tube cell.

The spore wall surrounding the cells thickens into the sculptured coat of the **pollen grain**.

The fusion of gametes is called **fertilization**. After the pollen grain lands on a receptive stigma, the tube cell begins to form the **pollen tube**. The generative cell divides to form two sperm cells. The pollen tube releases the sperm cells near the female gametophyte. In a process called **double fertilization**, one sperm fertilizes the egg to form the zygote, and the other combines with the polar nuclei to form a triploid nucleus, which will develop into a food-storing tissue called the **endosperm**.

INTERACTIVE QUESTION   38.2

a. Describe the male gametophyte.

b. Describe the female gametophyte.

c. What function does double fertilization serve?

Each ovule develops into a seed enclosed in the ovary, which develops into a fruit.

Seed Development and Structure: A Closer Look   The triploid nucleus formed by double fertilization develops into the endosperm, a multicellular mass rich in nutrients that are stored for later use by the seedling. In many eudicots, these food reserves may be transferred to the cotyledons before the seed matures.

The first mitotic division of the zygote creates a basal cell and a terminal cell. The basal cell divides to produce a thread of cells, called the suspensor, which anchors the embryo and transfers nutrients to it. The terminal cell divides to form a spherical proembryo, on which the cotyledons (two in eudicots, one in monocots) begin to form. The embryo elongates, and apical meristems develop at the apices of the embryonic shoot and root.

As it matures, the seed dehydrates and the embryo enters **dormancy**. The embryo and its food supply, the endosperm and/or enlarged cotyledons, are enclosed in a **seed coat** formed from the ovule integuments.

In a eudicot seed such as a bean, the embryo is an elongated embryonic axis attached to fleshy cotyledons. The axis below the cotyledonary attachment is called the **hypocotyl**; it terminates in the **radicle**, or embryonic root. The upper axis is the **epicotyl**; it leads to a shoot tip with a pair of leaves. Together, the epicotyl, young leaves, and shoot apical meristem are called the *plumule*.

The monocot seed found in grasses has a single thin cotyledon, called a *scutellum,* which absorbs nutrients from the endosperm during germination. A sheath called a **coleorhiza** covers the root, and a **coleoptile** encloses the young shoot.

INTERACTIVE QUESTION 38.3

Label the parts in the following diagrams of a bean seed and a corn seed.

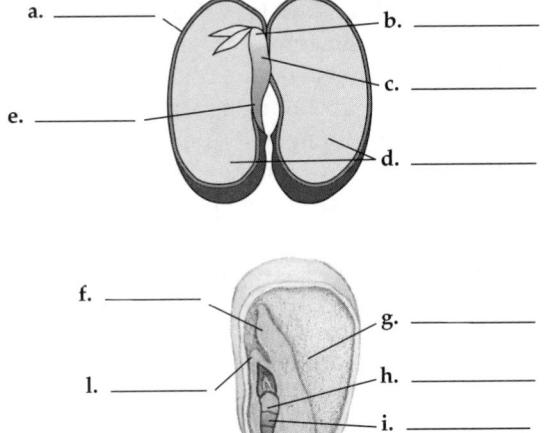

Dormancy increases the chances that the seed will germinate when and where the seedling has a good chance of surviving. The specific cues for breaking dormancy vary with the environment and may include heavy rain, intense heat from fires, cold, light, or chemical breakdown of the seed coat within an animal's digestive tract. The viability of a dormant seed may vary from a few days to decades or longer.

Sporophyte Development From Seed to Mature Plant   **Imbibition**, the absorption of water by the dry seed, causes the seed to expand, rupture its coat, and begin a series of metabolic changes. Enzymes digest stored compounds, and nutrients are sent to growing regions.

The radicle emerges from the seed first, followed by the shoot tip. In many eudicots, a hook that forms in the hypocotyl is pushed up through the ground, pulling the delicate shoot and cotyledons behind it. Light stimulates the straightening of the hook, and the first foliage leaves begin to perform photosynthesis. In some monocots, the shoot tip emerges through the coleoptile into the air.

The plant usually undergoes vegetative growth until the reproductive phase begins. A combination of environmental and internal cues triggers the transition of the shoot apical meristem to a *floral meristem,* which then produces the four floral organs.

INTERACTIVE QUESTION 38.4

Why is flowering usually delayed for some time after germination?

Fruit Structure and Function   The ovary of the flower develops into a **fruit**, which both protects and helps to disperse the seeds. Hormonal changes following fertilization cause the ovary to enlarge; its wall becomes the *pericarp,* or thickened wall of the fruit. Fruit usually does not develop if a flower has not been pollinated.

A **simple fruit** is derived from a single carpel or several fused carpels; an **aggregate fruit** results from a flower with more than one separate carpel; a **multiple fruit** develops from an inflorescence whose many ovaries fuse together to become one fruit. Other floral parts (such as the receptacle) may develop into a fruit. These fruits—apples, for example—are called **accessory fruits**.

Fruits usually ripen as the seeds are completing their development. Ripening of fleshy fruits involves changes that promote seed dispersal by animals.

INTERACTIVE QUESTION 38.5

a. List some ways in which fruits aid in seed dispersal.

b. What changes usually occur when a fleshy fruit ripens?

38.2 Flowering plants reproduce sexually, asexually, or both

Mechanisms of Asexual Reproduction   Many plant species produce genetically identical copies (clones) of themselves through **asexual reproduction**. Asexual reproduction is an extension of the indeterminate growth of plants in which meristematic tissues can grow indefinitely and parenchyma cells can divide and differentiate into specialized cells. A common type of vegetative reproduction is **fragmentation**, the formation of whole plants from parts of a parent plant. In some species, the root system gives rise to many adventitious shoots that develop into a clone with separate shoot systems. Some plants, such as dandelions, can produce seeds asexually, a process called **apomixis**.

Advantages and Disadvantages of Asexual and Sexual Reproduction   Advantages of asexual or **vegetative reproduction** include the production of clones of plants that are well suited to a certain environment and of progeny that are usually stronger than seedlings. Sexual reproduction generates variation in a population, an advantage when the environment changes. Seeds, which are almost always produced sexually, provide a means of dispersal to new locations and dormancy during harsh conditions.

INTERACTIVE QUESTION   38.6

a. What is an advantage of apomixis?

b. What is "selfing," and why might it be a desirable trait?

Mechanisms That Prevent Self-Fertilization   Self-fertilization may be prevented by temporal or structural adaptations in flowers. In **dioecious** species, flowers are either staminate or carpellate and cannot self-fertilize because they are on separate plants.

The ability of flowers to reject their own pollen or that of closely related individuals, called **self-incompatibility**, depends on genes called *S*-genes. A plant population may have dozens of alleles of the *S*-gene. Gametophytic self-incompatibility depends on the *S*-allele of the pollen grain. In some cases, RNA-hydrolyzing enzymes produced by the style destroy the RNA of developing pollen tubes that have a matching allele. In sporophytic self-incompatibility, a signal-transduction pathway in stigma cells blocks germination of pollen with matching sporophyte *S*-alleles.

Further research on mechanisms of self-incompatibility may allow plant breeders to manipulate crop species to assure hybridization.

Totipotency, Vegetative Reproduction, and Tissue Culture   Many plant cells are **totipotent**, able to divide and produce a clone of the original plant. The cloning of plants by humans is called **vegetative propagation**. New plants may develop from stem cuttings when adventitious roots develop from a mass of dividing, totipotent cells, called a **callus**, that forms

at the cut end of the shoot. Adventitious roots can also form from a node in the shoot fragment.

Twigs or buds of one plant can be grafted onto a plant of a different variety or closely related species. The plant that provides the root system is called the **stock**, and the twig is called the **scion**. Grafting can combine the best qualities of different plants.

In test-tube cloning, whole plants can develop from pieces of tissue from the parent plant. A single plant can be cloned into thousands of plants by subdividing the undifferentiated calluses as they grow in tissue culture. Stimulated by proper hormone balances, calluses sprout shoots and roots and develop into plantlets, which can be transferred to soil to develop.

Because apical meristems are often virus-free, tissue culture can be used to eliminate plant viruses from cloned plants. Also, foreign DNA may be inserted into individual plant cells, which then grow into genetically modified (GM) plants using *in vitro* culture.

INTERACTIVE QUESTION   38.7

How are vegetative reproduction and self-fertilization the same? How are they different?

38.3 **People modify crops by breeding and genetic engineering**

Almost all our crop species were first domesticated about 10,000 years ago. Natural hybridization between different species of plants is common, and humans have exploited the resulting genetic variation using artificial selection to develop and improve crops.

Plant Breeding   Breeders search for valuable traits in domesticated and wild species and hasten mutations by treating seeds or seedlings with mutagens. Wild plants with desirable traits are repeatedly crossed with the domesticated species until they are agriculturally suitable.

Plant Biotechnology and Genetic Engineering   Plant biotechnology refers both to the age-old use of plants to make products for human use and to the use of GM organisms in agriculture and industry. Genetic engineering allows plant biotechnologists to transfer genes between species, producing **transgenic** organisms.

Plant biotechnology can help increase crop yields to help alleviate the serious malnutrition of millions of people.

Cotton, maize, and potatoes have been engineered to contain genes from *Bacillus thuringiensis* that code for *Bt* toxin, reducing the need for spraying chemical insecticides. Some transgenic plants are more resistant to disease or have improved nutritional quality. Researchers have developed other transgenic crops that are resistant to the herbicide glyphosate, allowing farmers to "weed" crops without heavy tillage. Overuse of this technique has resulted in glyphosate-resistant weeds, and glyphosate may affect the gut microbiome.

Biomass from fast-growing plants, such as switchgrass and poplars, may help meet the world's energy demands. Enzymatic reactions break down plant cell walls into sugars, which are fermented and distilled to yield **biofuels**.

INTERACTIVE QUESTION 38.8

Biofuels are considered carbon neutral, even though the burning of biofuels releases CO_2, just as the burning of fossil fuels does. Explain.

The Debate over Plant Biotechnology   GM organisms (GMOs) may present an unknown risk to human health or the environment. One concern is the transfer of allergens to a food source, although there is no evidence that this occurs. Some GM crops may be safer than regular crops; *Bt* maize contains lower levels of a harmful toxin called fumonisin, which is produced by a fungus that infects insect-damaged crops, and work has been done in removing allergens from some crops.

Many ecologists are concerned about the effects of GM crops on nontarget organisms. A serious concern is the escape of herbicide- or disease-resistance genes through crop-to-weed hybridization that may create "superweeds." Various techniques are being developed to reduce the ability of transgenic crops to hybridize, such as developing male sterility so that transgenic pollen is not produced; engineering apomixis into the crop; introducing genes into chloroplast DNA so that those genes are not present in pollen; and developing flowers that self-pollinate and do not open so that pollen cannot escape.

Word Roots

apo- = away from; **-mixis** = mingling (*apomixis:* the ability to reproduce asexually through seeds without fertilization)

anth- = a flower (*anther:* the terminal pollen sac of a stamen, where pollen grains containing sperm-producing male gametophytes form)

carp- = a fruit (*carpel:* the ovule-producing reproductive organ of a flower, consisting of the stigma, style, and ovary)

cole- = a sheath; **rhiz-** = a root (*coleorhiza:* the covering of the young root of the embryo of a grass seed)

di- = two (*dioecious:* having male and female reproductive parts on different plants of the same species)

dorm- = sleep (*dormancy:* a condition typified by extremely low metabolic rate and a suspension of growth and development)

endo- = within (*endosperm:* a nutrient-rich tissue, formed by the union of a sperm with two polar nuclei during double fertilization that provides nourishment to the developing embryo in angiosperm seeds)

epi- = on, over (*epicotyl:* the embryonic axis above the point of attachment of the cotyledon[s] and below the first pair of miniature leaves)

hypo- = under (*hypocotyl:* the embryonic axis below the point of attachment of the cotyledon[s] and above the radicle)

mega- = large (*megaspore:* a spore that develops into a female gametophyte)

micro- = small (*microspore:* a spore that develops into a male gametophyte)

stam- = standing upright (*stamen:* the pollen-producing reproductive organ of a flower, consisting of an anther and filament)

Structure Your Knowledge

1. Draw a diagram of the major events in the life cycle of an angiosperm.
2. List the advantages and disadvantages of sexual and asexual reproduction in plants.
3. List some of the potential benefits and dangers of plant biotechnology.

Test Your Knowledge

FILL IN THE BLANKS

_____ 1. structure from which fruit typically develops

_____ 2. generation that produces spores by meiosis

_____ 3. fuel produced from organic matter

_____ 4. female gametophyte of angiosperms

_____ 5. embryonic root

_____ 6. embryonic axis above attachment point of cotyledon

_____ 7. protects a grass shoot as it breaks through the soil

_____ 8. twig or stem portion of a graft

_____ 9. spore that develops into a male gametophyte

_____ 10. mass of dividing cells at the cut end of a shoot

MULTIPLE CHOICE: *Choose the one best answer.*

1. A flower on a dioecious plant is
   a. complete.
   b. incomplete.
   c. staminate or carpellate.
   d. both **b** and **c**.

2. Which of the following is a correct pairing of cell type and chromosome complement?
   a. microspore, diploid
   b. zygote, triploid
   c. endosperm, triploid
   d. endosperm, haploid

3. Double fertilization involves fusion of sperm nuclei with
   a. the egg and the synergids.
   b. the egg and the polar nuclei.
   c. the egg and the antipodal cells.
   d. the synergids and the polar nuclei.

4. The terminal cell of an early plant embryo
   a. develops into the shoot apex of the embryo.
   b. forms the suspensor that anchors the embryo and transfers nutrients.
   c. develops into the endosperm when fertilized by a sperm nucleus.
   d. divides to form the proembryo.

5. Where does meiosis occur in an angiosperm?
   a. in the sporophytes
   b. in the gametophytes
   c. in the cotyledons
   d. in the coleoptiles

6. Which of the following is *not* true of endosperm?
   a. Its nutrients may be absorbed by the cotyledons in the seeds of eudicots.
   b. This triploid tissue originates from the joining of a diploid sporophyte cell and a haploid sperm.
   c. It is digested by enzymes in monocot seeds following imbibition.
   d. It develops in concert with the embryo as a result of double fertilization.

7. A seed consists of a(n)
   a. embryo, a seed coat, and a nutrient supply.
   b. embryo sac.
   c. gametophyte enclosed in a mature ovule.
   d. enlarged ovary.

8. Which of the following structures protects a bean shoot as it breaks through the soil?
   a. hypocotyl hook
   b. coleoptile
   c. coleorhiza
   d. seed coat

9. Which of the following is an example of the asexual process of fragmentation?
   a. You save some seeds from this year's jack-o-lantern to plant next year.
   b. Your strawberry plants send out long "runners" that produce a new plant where they touch the ground.
   c. Your friend likes your spider plant, so you give him a cutting from it to produce a new plant.
   d. Dandelion seeds have blown from your neighbor's scruffy yard into your carefully maintained lawn.

10. A white, highly scented flower is most likely pollinated by
    a. a moth.
    b. a bird.
    c. a bat.
    d. either **a** or **d** is likely.

11. In the plant embryo, the suspensor
    a. is produced by vertical mitotic divisions within the proembryo.
    b. connects the early root and shoot apexes.
    c. is analogous to the umbilical cord in mammals.
    d. is the point of attachment of the cotyledons.

12. Which of the following statements is *false*?
    a. Fly-pollinated flowers have reduced petals but large numbers of stamens to provide their pollinators with pollen.
    b. Insects pollinate approximately 65% of all flowering plants, and bees are the most important of these pollinators.
    c. The flowers of many temperate, wind-pollinated trees appear before the production of the leaves, which could interfere with pollen movement.
    d. Moths and bats, which are often active at night, are attracted to light-colored, fragrant flowers.

13. Which of the following is an example of coevolution?
    a. the dispersal of the sharp fruits of the puncture vine by animals
    b. the germination of the seeds of edible fruits following the breakdown of the seed coat in an animal's digestive tract
    c. the pollination of a yucca plant by a single species of moth, the appendages of which pack pollen onto the plant's stigma and that deposits its eggs in the flower's ovaries, where the larvae eat some seeds
    d. "nectar guides" that help bees (which can see ultraviolet light) locate the nectaries of flowers

14. Self-incompatibility provides a plant with a
    a. means of transferring pollen to another plant.
    b. means of destroying foreign pollen before it fertilizes the egg cell.
    c. biochemical block to self-fertilization so that cross-fertilization is assured.
    d. means of producing seeds without the need for fertilization.

15. Which of the following statements about fruit is *not* true?
    a. The primary functions of fruits include protection of dormant seeds and their dispersal away from the parent plant.
    b. Wind-dispersed fruits may have wings or "parachutes."
    c. Animals may help disperse seeds by "planting" fruits in underground caches.
    d. Most fruits develop from the ovule or sometimes the receptacle of a flower.

16. Clones formed when multiple shoots emerge from a single root system benefits a plant by
    a. providing a strong start for new plants.
    b. dispersing offspring to new habitats.
    c. providing a defense against viruses and other pathogens.
    d. providing a strong root system for genetically diverse shoot systems.

17. Which of the following techniques is being developed to reduce the likelihood that introduced genes for herbicide or insect resistance will escape to closely related weed species?
    a. engineering flowers to self-fertilize and remain closed so that pollen is not released
    b. breeding male sterility into transgenic plants so that they have no pollen to be transferred to nearby weeds
    c. engineering the gene of interest into chloroplast DNA, which is inherited from the maternal plant and is not transferred by pollen
    d. All of the above would reduce the risk of crop-to-weed transgene escape.

Chapter 39

Plant Responses to Internal and External Signals

Framework

Environmental stimuli and internal signals are linked by signal transduction pathways to cellular responses such as changes in gene expression and activation of enzymes. Plant hormones—auxin, cytokinins, gibberellins, abscisic acid, ethylene, brassinosteroids, jasmonates, and strigolactones—control growth and development as plants respond and adapt to their environments. Plant movements in response to environmental stimuli include phototropism, gravitropism, and thigmotropism. The biological clock of plants controls circadian rhythms, such as stomatal opening and sleep movements. Blue-light photoreceptors are involved in phototropism and stomatal opening. Phytochromes absorb red and far-red light and are involved in the de-etiolation process, seed germination, and the photoperiodic control of flowering. Plants have various physiological responses to environmental stresses, pathogens, and herbivores.

Chapter Review

Plants are able to sense and adaptively respond to their environments, generally by altering their patterns of growth and development.

39.1 Signal transduction pathways link signal reception to response

The growth pattern of a sprouting potato shoot growing in darkness, called **etiolation**, facilitates the shoot breaking ground. When the shoot reaches sunlight, stem elongation slows, leaves expand, the root system elongates, and chlorophyll production begins—all part of a process known as **de-etiolation** or greening.

Reception   Receptors are proteins, either built into the plasma membrane or located in the cell, that change shape in response to a specific stimulus. A *phytochrome* is the photoreceptor involved in de-etiolation.

Transduction   **Second messengers** are small molecules that amplify the signal from the receptor and transfer it to proteins that produce the specific response. Phytochrome signal transduction involves two pathways: the production of cyclic GMP and the opening of calcium channels. Cyclic GMP and Ca^{2+} activate specific protein kinases that in turn affect gene expression. Both pathways are required for complete de-etiolation.

Response   The response to a signal transduction pathway usually involves the activation of specific enzymes, either by modification of existing enzymes or by transcriptional regulation.

Post-translational modification of existing proteins usually involves phosphorylations. Cascades of protein kinase activations may lead to the phosphorylation of transcription factors and thus, ultimately, to changes in gene expression. Protein phosphatases are enzymes that dephosphorylate specific proteins, allowing signal pathways to be turned off when a signal is no longer present.

Several *specific transcription factors*, which bind to DNA and control the transcription of specific genes, are activated by protein kinases during phytochrome-induced de-etiolation. Transcription factors may function as activators or repressors of gene expression.

Enzymes activated or newly synthesized during de-etiolation include those involved in chlorophyll production and photosynthesis and others that affect growth through the regulation of plant hormones.

INTERACTIVE QUESTION 39.1

Answer the following questions concerning the process known as de-etiolation.

a. What is the signal, and what is the receptor?

b. What are the important steps in the transduction of this signal?

c. What is the plant's growth response? What types of proteins are involved?

39.2 Plant hormones help coordinate growth, development, and responses to stimuli

Hormones are generally considered to be molecules produced in small quantities that act via specific receptors in target tissues distant from their sources. Some signaling molecules thought of as *plant hormones* act only locally, and others occur at much higher concentrations than a typical hormone. Thus, many plant biologists use the term *plant growth regulators* to describe compounds that affect specific physiological processes in plants.

Depending on its concentration relative to other hormones, the site of action, and the developmental stage of the plant, each hormone can have multiple effects on growth and development.

A Survey of Plant Hormones A **tropism** is a growth response of plant organs toward or away from a stimulus. The growth of a shoot toward light is called positive **phototropism**. A shoot bends toward the light when illuminated from one side because of the greater elongation of cells on the darker side.

Darwin and his son observed that a grass seedling enclosed in its coleoptile would not bend toward light if its tip were removed or covered by an opaque cap. They postulated that a signal must be transmitted from the tip to the elongating region of the coleoptile. P. Boysen-Jensen demonstrated that the signal was a mobile substance, capable of being transmitted through a block of gelatin separating the tip from the rest of the coleoptile. The side of a coleoptile away from the light elongates due to a higher concentration of this growth-promoting chemical, which was named auxin.

Auxin refers to any substance that stimulates elongation of coleoptiles. The major auxin extracted from plants is indoleacetic acid (IAA). Auxin is transported from the shoot tip apical meristem (where it is synthesized) down the shoot. This *polar transport* involves auxin transporters located only at the basal ends of cells.

INTERACTIVE QUESTION 39.2

Fill in the blanks in this description of the *acid growth hypothesis*.

Auxin stimulates a plasma membrane's **a.** _____, resulting in an increased membrane potential and a **b.** _____ pH in the cell wall. This activates enzymes called **c.** _____, which break the cross-links between cellulose microfibrils. The increased membrane potential enhances ion uptake, which is followed by the **d.** _____. Thus, the two factors that cause the cell to elongate are **e.** _____ and **f.** _____. For continued growth and elongation, the cell must produce more **g.** _____ and **h.** _____; these processes are mediated by auxin effects on gene expression.

Auxin is important in the *pattern formation* of a developing plant. Auxin transported from shoot tips controls branching patterns. A reduced flow from a branch releases lateral buds below the branch from dormancy. Polar auxin transport also helps establish *phyllotaxy*, the arrangement of leaves. Peaks in auxin establish the sites of leaf primordium formation. Auxin flowing from the leaf margin also affects leaf venation patterns.

The auxin IBA affects lateral root formation and is used commercially to enhance formation of adventitious roots at the cut base of stems or leaves. Synthetic auxins, such as 2,4-D, are used as herbicides that kill eudicot (broadleaf) weeds with a hormonal overdose. Auxin produced by developing seeds promotes fruit growth; synthetic auxins can induce fruit development in greenhouses.

Cytokinins—so named because they stimulate cytokinesis—are modified forms of adenine. Zeatin is the most common naturally occurring cytokinin. Cytokinins act with auxin to stimulate cell division, and the ratio of cytokinins to auxin directs cell differentiation.

The control of apical dominance involves the interaction between auxin, which is transported down from the terminal bud and restrains axillary bud development, and cytokinins, which are transported up from the roots and stimulate bud growth. It now appears, however, that the polar flow of auxin stimulates the production of strigolactone, a hormone that represses bud growth.

Cytokinins can retard aging of some plant organs, because they stimulate RNA and protein synthesis, mobilize nutrients, and inhibit protein breakdown.

In the 1930s Japanese scientists determined that the fungus *Gibberella* secretes a chemical that causes the hyperelongation of rice stems, or "foolish seedling disease." More than 100 different naturally occurring **gibberellins** (GAs) have now been identified.

Gibberellins, which are produced by roots and young leaves, stimulate growth by enhancing both cell elongation and division. *Bolting,* the growth of an elongated floral stalk, is caused by a surge of gibberellins.

In many plants, both auxin and gibberellins contribute to fruit development. Gibberellins are applied in the production of Thompson seedless grapes.

The release of gibberellins from the embryo signals the seeds of many plants to break dormancy. GAs stimulate the production of enzymes that mobilize stored nutrients for a growing seedling.

The hormone **abscisic acid (ABA)** generally slows growth. The high concentration of ABA in maturing seeds inhibits germination and stimulates production of proteins that protect the seeds during dehydration. For dormancy to be broken in some seeds, ABA must be removed or inactivated, or the ratio of gibberellins to ABA must increase.

ABA also reduces drought stress. In a wilting plant, ABA causes stomata in the leaves to close. ABA may be produced in the roots in response to water shortage and then be transported to the leaves.

Plants produce the gas **ethylene** in response to stress and during fruit ripening and programmed cell death. Ethylene production may be induced by a high concentration of auxin.

The mechanical stress in a seedling pushing against an obstacle induces the production of ethylene. Ethylene then initiates a growth pattern called the **triple response**: slowing of stem elongation, thickening of the stem, and initiation of horizontal growth. As the effects of ethylene lessen, normal upward growth resumes.

Researchers have identified *Arabidopsis* mutants that are ethylene insensitive *(ein)*, ethylene overproducing *(eto)*, and that undergo the triple response in the absence of ethylene. In these latter constitutive triple-response *(ctr)* mutants, the ethylene signal transduction pathway is permanently turned on, even though ethylene is not present. The implicated gene in *ctr* mutants codes for a protein kinase, suggesting that the non-mutant kinase is a negative regulator of ethylene signal transduction. Binding of ethylene to the ethylene receptor may lead to the inactivation of the negative kinase, which allows the synthesis of the proteins involved in the triple response.

Senescence is the programmed death of plant cells, organs, or the whole plant. Apoptosis requires the synthesis of new enzymes that break down many cellular components, which the plant may salvage. Ethylene is almost always associated with this programmed cell death.

Deciduous leaf loss protects against winter desiccation. Before leaves abscise in the autumn, many of the leaves' compounds are stored in the stem. A change in the balance between auxin and ethylene initiates changes in the abscission layer located near the base of the petiole, including the production of enzymes that hydrolyze polysaccharides in cell walls.

Ethylene initiates the breakdown of cell walls and the conversion of starches to sugars associated with fruit ripening. In an example of positive feedback, ethylene triggers ripening, and ripening triggers more ethylene production. Many commercial fruits are ripened in huge containers perfused with ethylene gas.

Brassinosteroids are steroids that have effects very similar to those of auxin: They promote cell elongation and division, retard leaf abscission, and promote xylem differentiation. Identification of a brassinosteroid-deficient mutant of *Arabidopsis* helped to establish these compounds as nonauxin plant hormones.

Jasmonates, which include *jasmonate* (JA) and *methyl jasmonate* (MeJA), are fatty acid–derived molecules that play many roles in plant development and plant defense. Various physiological processes, ranging from nectar secretion to mycorrhizal symbiosis, are regulated by jasmonates.

Strigolactones stimulate seed germination, enhance mycorrhizal associations, and repress the growth of lateral buds in apical dominance. They were first identified as signals released by plant roots that stimulate germination of seeds of *Striga*, a rootless parasitic plant that is a serious agricultural pest in Africa.

Fill in the name of the hormone that is responsible for each of the following functions:

a. _____ promotes fruit ripening; initiates the triple response; is involved in apoptosis

b. _____ stimulate cell division, growth, and germination; are anti-aging compounds

c. _____ inhibits growth; maintains dormancy; closes stomata during water stress

d. _____ regulate a wide variety of functions, including plant defenses against herbivores and pathogens

e. _____ stimulates stem elongation, root branching, and fruit development; is involved in apical dominance

f. _____ promote cell elongation and division, and xylem differentiation; effects similar to auxin

g. _____ stimulate seed germination; help establish mycorrhizal associations; repress lateral bud growth

h. _____ promote stem elongation and seed germination; contribute to fruit set

39.3 Responses to light are critical for plant success

The effect of light on plant growth and development is called **photomorphogenesis**. An **action spectrum** is a graph of a physiological response across different wavelengths of light. An absorption spectrum shows the wavelengths of light a pigment absorbs. Close correlation between an action spectrum for a plant response and the absorption spectrum of a pigment may indicate that the pigment is the photoreceptor involved in the response. **Blue-light photoreceptors** and **phytochromes**, which absorb mostly red light, are the two major classes of light receptors.

Blue-Light Photoreceptors Molecular biologists have determined that plants use different types of pigments to detect blue light: *cryptochromes* are involved in inhibition of stem elongation when a seedling breaks ground, and *phototropin* is a protein kinase involved in phototropism and stomatal opening.

Phytochrome Photoreceptors Studies of lettuce seed germination in the 1930s determined that red light (wavelength: 660 nm) increased germination the most, and that far-red light (730 nm) inhibited germination. The effects of red and far-red light are reversible, and a seed's response is determined by the last flash of light it receives. Five phytochromes have been identified thus far in *Aribidopsis*.

The light-absorbing portion of phytochrome is photoreversible; the P_r form absorbs red light and is converted to the P_{fr} form, which is converted back to P_r when it absorbs far-red light. Their interconversion acts as a switch that controls various events in the life of the plant.

The signal that tells the plant that sunlight is present is the conversion of P_r (the phytochrome form the plant synthesizes) to P_{fr}. Subsequently, P_{fr} triggers many plant responses to light, such as breaking seed dormancy.

Because chlorophyll in leaves absorbs red light, the ratio of far-red to red light is increased in areas shaded by vegetation. Such changes in the *quality* of light allow the plant to change its behavior in response to shading.

Explain how phytochrome can also indicate the *quality* of light available to a plant. What effect might this have on a shaded tree?

Biological Clocks and Circadian Rhythms A **circadian rhythm** is a physiological cycle with about a 24-hour frequency. These rhythms persist, even when the organism is sheltered from environmental cues. Research indicates that the circadian clock is internal, although it is set (entrained) to a 24-hour period by daily environmental signals.

What are free-running periods?

The molecular mechanism of the biological clock may be negative-feedback loops involving cyclical changes in the transcription of genes for transcription factors that inhibit, after a time delay, the expression of their own genes.

Researchers have identified clock mutants of *Arabidopsis* by splicing the gene for luciferase to the promoter for genes for cyclically produced photosynthesis-related proteins. When the biological clock turned on the promoter for these genes, the plants glowed. Plants that glowed for an abnormal amount of time were clock mutants. Some of these mutants had defects in proteins that normally bind photoreceptors, suggesting that a light-dependent mechanism sets the biological clock.

The Effect of Light on the Biological Clock   Both blue-light photoreceptors and phytochrome can entrain the biological clock in plants. In darkness, the phytochrome ratio shifts toward P_r, in part because P_{fr} converts slowly to P_r in some plants, and in part because P_{fr} is degraded and new pigment is synthesized as P_r. When the sun rises, P_r is rapidly converted to P_{fr}, resetting the biological clock each day at dawn.

Photoperiodism and Responses to Seasons   Seasonal events in the life cycle of plants usually are cued by *photoperiod*, the interval of light exposure each day. A physiological response to photoperiod is called **photoperiodism**.

Researchers discovered that a variety of tobacco plant flowered only when the photoperiod was 14 hours or shorter. They termed it a **short-day plant**. **Long-day plants** flower when days are longer than a certain number of hours. The flowering of **day-neutral plants** is unaffected by photoperiod.

Researchers have determined that it is actually night length that controls flowering and other photoperiod responses. If the dark period is interrupted by even a few minutes of light, a short-day (long-night) plant will not flower. Photoperiodic responses thus depend on a critical night length.

Red light was found to be the most effective in interrupting a plant's perception of night length. A flash of red light breaks a dark period of sufficient length and prevents short-day plants from flowering, whereas a flash of red light during a dark period that is longer than the critical length will induce flowering in a long-day plant.

Some plants respond to photoperiod only after exposure to another environmental stimulus. **Vernalization** is the need for pretreatment with cold to induce flowering.

Leaves detect the photoperiod. The signal for flowering is called **florigen**. Recent research indicates that the gene *FLOWERING LOCUS T* (*FT*) produces the FT protein in the leaf; the protein then travels through the symplast to the shoot apical meristem and initiates flowering.

INTERACTIVE QUESTION   39.6

Indicate whether a short-day plant and a long-day plant would flower or would not flower under the indicated light conditions.

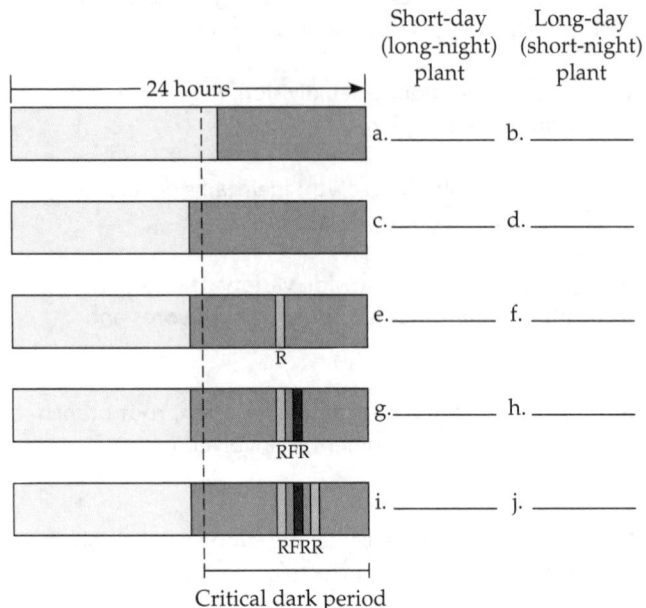

Critical dark period

How do these results demonstrate the red/far-red photoreversibility of phytochrome?

39.4   Plants respond to a wide variety of stimuli other than light

Gravity   Roots exhibit positive **gravitropism**, whereas shoots show negative gravitropism. According to one hypothesis, the settling of **statoliths**—plastids containing dense starch grains—in cells of the root cap triggers movement of calcium, which causes the lateral transport of auxin. Accumulating on the lower side of the growing root, auxin inhibits cell elongation, causing the root to curve downward. The settling of the protoplast and large organelles may distort the cytoskeleton and also signal gravitational direction.

Mechanical Stimuli   The stunting of growth in height and increase in girth of plants that are exposed to wind is an example of **thigmomorphogenesis**, changes in plant form in response to mechanical stimulation.

Most climbing plants have tendrils that coil rapidly around supports, exhibiting **thigmotropism** or directional growth in response to touch.

The sensitive plant *Mimosa* folds its leaves after being touched due to the rapid loss of turgor by cells in specialized motor organs called pulvini, located at the joints of the leaf. The message travels through the plant from the point of stimulation as the result of electrical impulses called **action potentials**. These electrical messages may be used in plants as a form of internal communication.

INTERACTIVE QUESTION 39.7

a. Name the types of tropisms that a stem exhibits.

b. Name and describe the growth mechanism that produces the coiling of a tendril.

Environmental Stresses   Plants respond to both **abiotic** environmental stresses and to **biotic** stresses such as pathogens and herbivores.

Mechanisms that reduce transpiration help a plant respond to water deficit. Guard cells lose turgor and stomata close. Abscisic acid, produced by leaves in response to water deficit, acts on guard cell membranes to keep stomata closed. Wilted leaves of grasses may roll up, further reducing transpiration. These responses, however, reduce photosynthesis.

Plants adapted to wet habitats may have aerial roots that provide oxygen to their submerged roots. When roots of other plants are in waterlogged soils, oxygen deprivation may stimulate ethylene production, causing some root cortical cells to undergo apoptosis, opening up air tubes within the roots.

In addition to the general toxicity of some ions, excess salts in the soil may lower the water potential of the soil solution, reducing water uptake by roots. Plants may respond to moderate soil salinity by producing organic solutes that lower the water potential of root cells, thus increasing water uptake. Special adaptations for dealing with high soil salinity have evolved in halophytes.

Transpiration creates evaporative cooling for a plant, but this effect may be lost on hot, dry days when stomata close. Above critical temperatures, plant cells produce **heat-shock proteins** that may provide temporary support to reduce protein denaturation.

Plants respond to cold stress by increasing the proportion of unsaturated fatty acids in membrane lipids, which maintain the fluidity of cell membranes.

Subfreezing temperatures cause ice to form in cell walls, lowering the extracellular water potential and causing cells to dehydrate. Plants adapted to cold winters have adaptations to deal with freezing stress, such as changing the solute composition of the cytosol. Plants may also produce *antifreeze proteins* that reduce the formation of ice crystals.

INTERACTIVE QUESTION 39.8

a. How might a plant exposed to a spell of very hot and dry weather survive this stress?

b. How might a plant respond to an unusually cold and wet fall?

39.5 Plants respond to attacks by pathogens and herbivores

Defenses Against Pathogens   The physical barrier of the plant's epidermis and/or periderm is the first line of defense against pathogenic viruses, bacteria, and fungi. Should pathogens enter the plant due to injuries or through openings such as stomata, the plant mounts two lines of immune defense.

PAMP-triggered immunity is triggered when **pathogen-associated molecular patterns (PAMPs)**, which are specific to certain pathogens, are recognized by a plant's Toll-like receptors. The resulting signaling pathways lead to the local production of antimicrobial *phytoalexins*, which have fungicidal and bactericidal properties. This innate immune system is similar to that present in fungi and animals. Toughening of the cell wall also occurs during PAMP-triggered immunity.

Some pathogens produce **effectors**, which shut down a plant's innate immune system. A second level of a plant's immune system is called *effector-triggered immunity*. A plant has hundreds of disease resistance (*R*) genes, which produce proteins that are activated by a specific effector. The binding of an effector to an R protein triggers signal transduction pathways that produce local and systemic responses.

In the **hypersensitive response**, plant cells modify their cell walls to seal off infected areas and produce enzymes that attack the cell walls of the pathogen. Finally, cells in the infected areas destroy themselves, producing lesions that restrict the spread of the pathogen.

A hypersensitive response also involves the release methylsalicylic acid, which travels throughout the plant, resulting in a **systemic acquired resistance**.

This signaling molecule is converted to **salicylic acid** in cells distant from the infection and activates nonspecific resistance to a diversity of pathogens for several days.

Defenses Against Herbivores   Plant defenses against **herbivory** have evolved at multiple levels. Molecular-level defenses include compounds that are distasteful or cause insects to molt prematurely and die. Cellular-level defenses include trichomes that hinder chewing insects and vacuoles that store irritating chemicals that deter herbivores. At the tissue level, leaves may be very hard to chew due to thickened sclerenchyma tissue. Organ-level defenses include spines and thorns, unattractive leaf shapes, and egg mimicry that discourages insects from laying eggs on leaves.

At the organismal level, plants may respond to mechanical damage by physiological changes that deter further attack, such as changing when their flowers open. Some plant species show coordinated behavior at the population level. Chemical signals released from a plant attacked by an herbivore may signal neighboring plants to become more resistant to attack. Or a population may synchronously release large numbers of seeds, a phenomenon called masting, which assures that at least some seeds escape herbivory. In a type of community-level defense, some plants recruit predators of their herbivores. A combination of eating damage by caterpillars and a compound in caterpillar saliva stimulates leaves to release compounds that attract parasitoid wasps. The wasps lay their eggs inside the caterpillar and then the developing wasp larvae eat their host.

INTERACTIVE QUESTION   39.9

a. In what ways may insects harm plants?

b. In what ways may insects benefit plants?

Word Roots

aux- = grow, enlarge (*auxin:* a term that primarily refers to indoleacetic acid, a natural plant hormone that has a variety of effects, including cell elongation, root formation, secondary growth, and fruit growth)

circ- = a circle (*circadian rhythm:* a physiological cycle of about 24 hours that persists even in the absence of external cues)

cyt- = cell; **-kine** = moving (*cytokinin:* any of a class of related plant hormones that retard aging and act in concert with auxin to stimulate cell division, influence cell differentiation, and control apical dominance)

gibb- = humped (*gibberellin:* any of a class of related plant hormones that stimulate growth in the stem and leaves, trigger the germination of seeds and breaking of bud dormancy, and [with auxin] stimulate fruit development)

hyper- = excessive (*hypersensitive response:* a plant's localized defense response to a pathogen, involving the death of cells around the site of infection)

phot- = light; **trop-** = turn, change (*phototropism:* growth of a plant shoot toward or away from light)

phyt- = a plant; **-chrome** = a colored substance (*phytochrome:* a type of light receptor that absorbs red light and regulates many plant responses, such as seed germination and shade avoidance)

stat- = standing, placed; **-lith** = a stone (*statolith:* a specialized plastid that contains dense starch grains and may play a role in detecting gravity)

thigm- = a touch; **morph-** = form; **-genesis** = origin (*thigmomorphogenesis:* a response in plants to chronic mechanical stimulation, resulting from increased ethylene production; an example is thickening stems in response to strong winds)

Structure Your Knowledge

1. List some agricultural uses of plant hormones.
2. Develop a concept map to illustrate your understanding of photoperiodism and the control of flowering. Remember to include the role of the biological clock.
3. Briefly describe the steps in the hypersensitive response and in systemic acquired resistance that result from a plant's encounter with a pathogen that produces effectors that one of a plant's R proteins can recognize.

Test Your Knowledge

TRUE OR FALSE: *Indicate T or F and then correct the false statements.*

_____   **1.** Abscisic acid is necessary for a seed to break dormancy.

_____   **2.** Cytokinins, synthesized in the root, seem to counteract apical dominance.

_____ 3. Thigmomorphogenesis is a growth response of a seedling that encounters an obstacle while pushing upward through the soil.

_____ 4. Action potentials are electrical impulses that may be used as internal communication in plants.

_____ 5. A physiological response to day length is called a circadian rhythm.

_____ 6. Vernalization is the pretreatment with cold before a plant flowers.

_____ 7. A plant's PAMP-triggered immunity involves the recognition by a Toll-like receptor of a pathogen's specific effector proteins and the triggering of a hypersensitive response and systemic acquired resistance.

MULTIPLE CHOICE: *Choose the one best answer.*

1. The body form within a species of plant may vary more than that within a species of animal because
   a. growth in animals is controlled by genes.
   b. plants respond adaptively to their environments by altering their patterns of growth and development.
   c. plant growth and development are governed by many hormones.
   d. plants can respond to environmental stress.

2. Which of the following is *not* a feature of the de-etiolation response?
   a. leaf expansion
   b. root elongation
   c. stem elongation
   d. chlorophyll production

3. Small molecules in the cell that amplify signals and transfer them to other effectors are known as
   a. phytochromes.
   b. transcription factors.
   c. hormones.
   d. second messengers.

4. According to the acid growth hypothesis,
   a. auxin stimulates membrane proton pumps.
   b. a lowered pH outside the cell activates expansins, which break cross-links between cellulose microfibrils.
   c. the membrane potential created by the proton pumps enhances ion uptake, increasing the osmotic movement of water into cells, which then elongate.
   d. all of the above are involved in cell elongation.

5. Which of the following situations would most likely stimulate the development of axillary buds?
   a. a large quantity of auxin traveling down from the shoot and a small amount of cytokinin produced by the roots
   b. a reduction of auxin traveling from the shoot and thus a reduced production of strigolactone
   c. a 1:1 ratio of gibberellins to auxin
   d. the absence of cytokinins caused by the removal of the terminal bud

6. Seeds of desert plants typically break dormancy when
   a. heavy rains wash abscisic acid out of the seed.
   b. the seeds undergo vernalization.
   c. the seeds undergo de-etiolation.
   d. the seeds undergo thigmomorphogenesis.

7. A circadian rhythm
   a. is seen in the oscillation of stomatal opening and production of photosynthetic enzymes in a 24-hour period.
   b. involves an internal biological clock that is not set by daily environmental signals.
   c. provides the signal for flowering.
   d. involves all of the above.

8. Which of the following is *not* a component of the signal transduction pathway involved in the de-etiolation process when a shoot breaks ground?
   a. reception of light by a phytochrome located in the cytosol
   b. inhibition of the triple response and initiation of phototropism as the shoot bends toward light
   c. production of second messengers such as cGMP and Ca^{2+}
   d. cascades of protein kinase activations that phosphorylate transcription factors

9. Injecting phytochrome into cells of the tomato mutant *aurea*
   a. causes the plant to undergo the triple response above ground when exposed to ethylene.
   b. causes the plant to undergo the triple response above ground without exposure to ethylene.
   c. helps researchers identify clock mutants.
   d. causes the plant to develop a normal de-etiolation response to light.

10. A flash of far-red light during a critical-length dark period will
    a. induce flowering in a long-day plant.
    b. induce flowering in a short-day plant.
    c. not influence flowering.
    d. increase the P_{fr} level suddenly.

11. The conversion of P_r to P_{fr}
    a. occurs slowly at night.
    b. may be the way in which plants sense daybreak and serves to entrain the biological clock.
    c. is the molecular mechanism responsible for the biological clock.
    d. occurs in flower buds and induces flowering.

12. Ethylene-induced destruction of root cortex cells in some plants is a response to
    a. heat stress.
    b. flooding.
    c. salt stress.
    d. cold stress.

13. Which of the following hormones would be sprayed on barley seeds to speed germination in the production of malt for making beer?
    a. abscisic acid
    b. cytokinin
    c. brassinosteroid
    d. gibberellin

14. You have grown seedlings of an *ein* mutant of *Arabidopsis*. What is their growth pattern in the absence of ethylene gas?
    a. They will behave like ordinary *Arabidopsis* seedlings.
    b. They will exhibit the triple response even without ethylene.
    c. They will produce inhibitors of ethylene synthesis.
    d. They will demonstrate stem elongation, but not other aspects of the triple response.

15. Many plants will flower in response to a specific
    a. photoperiod, which seems to be measured by the length of darkness to which the leaves of the plant are exposed.
    b. effector, which is produced in leaves and travels to a floral meristem.
    c. minimal temperature that appears to signal a seasonal change.
    d. combination of high levels of auxin and FT protein and low levels of cytokinin and strigolactone.

16. Which of the following is *not* a plant defense against herbivory?
    a. production of distasteful compounds
    b. physical defenses such as thorns
    c. initiation of a hypersensitive response with production of antimicrobial proteins
    d. release of volatile compounds that attract parasitoid wasps

17. Which of the following is *not* true of the hypersensitive response?
    a. It relies on recognition between a specific plant R protein and a pathogen effector.
    b. It enhances the production of ethylene, which serves as a signal molecule that is transported throughout the plant and activates systemic acquired resistance.
    c. It contains an infection by stimulating the cross-linking of cell wall molecules and the production of lignin.
    d. The plant cells involved in the defense undergo apoptosis, leaving lesions that indicate the site of the contained infection.

Chapter 40

Basic Principles of Animal Form and Function

Key Concepts

40.1 Animal form and function are correlated at all levels of organization

40.2 Feedback control maintains the internal environment in many animals

40.3 Homeostatic processes for thermoregulation involve form, function, and behavior

40.4 Energy requirements are related to animal size, activity, and environment

Framework

An animal's body is built from four basic tissues: epithelial, connective, muscle, and nervous. Organs function together in organ systems. Structure correlates with function in each hierarchical level of organization.

The internal environment is regulated by homeostatic negative feedback mechanisms. Thermoregulatory mechanisms include insulation, circulatory adaptations, evaporative cooling, behavior, and regulation of heat production. Endotherms use metabolic heat to maintain body temperature; ectotherms gain their heat mostly from the environment.

Metabolic rate, the amount of energy used in a unit of time, is higher for endothermic animals. The rate of energy use per gram of body mass is inversely related to body size.

Chapter Review

Anatomy is the study of an organism's structure; **physiology** is the study of biological function.

40.1 Animal form and function are correlated at all levels of organization

Evolution of Animal Size and Shape   Physical laws drove natural selection for a fusiform body shape in fast-swimming fishes, birds, and mammals, an example of convergent evolution. Physical laws also influence body size, as larger bodies require stronger skeletons and relatively larger muscle masses for locomotion.

Exchange with the Environment   Every cell must be in an aqueous medium to allow exchange across its plasma membrane. Animals with two-layered, saclike bodies or thin, flat bodies can maintain sufficient cellular contact with the environment for such exchange.

Animals with larger, thicker bodies, must provide extensively branched or folded surfaces for exchanging materials with the environment. A circulatory system often connects these exchange surfaces with the **interstitial fluid** bathing the body's cells.

INTERACTIVE QUESTION   40.1

List some advantages of a compact, complex body form.

Hierarchical Organization of Body Plans   **Tissues** are groups of similar cells with a common function. **Organs**, which perform specific functions, often consist of a layered arrangement of tissues. Groups of organs are integrated into **organ systems**, which carry out the main functions required for life.

Tissue Structure and Function   **Epithelial tissues** line the outer and inner surfaces of the body in sheets of tightly packed cells. A simple **epithelium** has one layer of cells; a stratified epithelium has multiple layers of cells. The shape of cells at the *apical* surface facing the lumen or the outside of an organ may be squamous (flat), cuboidal, or columnar. The opposite side of an epithelium is the *basal* surface.

INTERACTIVE QUESTION   40.2

Name the two types of epithelia illustrated below. One of these types lines the mouth; the other lines the intestines. Explain how each type is adapted to its location.

a. _____   b. _____

Connective tissue binds and supports other tissues. It is composed of relatively few cells suspended in an extracellular matrix of protein fibers embedded in a liquid, jellylike, or solid material. **Fibroblasts** secrete the protein fibers, and **macrophages** engulf foreign material and cellular debris.

Connective tissue fibers are of three types. *Collagenous fibers* are strong and flexible; *elastic fibers* stretch and provide resilience; and *reticular fibers* provide attachment to adjacent tissues.

The widespread *loose connective tissue* attaches epithelia to underlying tissues and holds organs in place.

Fibrous connective tissue, with its parallel collagenous fibers, is found in **tendons**, which attach muscles to bones, and in **ligaments**, which join bones to other bones at joints.

Bone is formed by *osteoblasts,* which deposit a collagen matrix. Calcium, magnesium, and phosphate ions form a hard mineral within the matrix. *Osteons* consist of concentric layers of matrix around a central canal containing blood vessels and nerves.

Cartilage is composed of collagenous fibers embedded in a rubbery substance; both are secreted by *chondrocytes*. Cartilage is a strong but flexible support material.

Adipose tissue pads and insulates the body and stores fuel in the form of fat. Each adipose cell contains a large fat droplet.

Blood has a liquid extracellular matrix called plasma. Erythrocytes (red blood cells) carry oxygen; leukocytes (white blood cells) function in defense; and cell fragments called platelets are involved in the clotting of blood.

INTERACTIVE QUESTION   40.3

Identify the type of connective tissue and the indicated structures or cells in the following two micrographs. Compare the matrix of these two connective tissues.

a. _____

b. _____

c. _____

d. _____

e. _____

f. _____

g. _____

The cells of **muscle tissue** contain filaments of actin and myosin. **Skeletal muscle**—also called *striated muscle*—is responsible for voluntary body movements. The contractile units called sarcomeres form the striations along these long, multinucleated cells. **Cardiac muscle**, forming the wall of the heart, is also striated, but its cells are branched and joined by intercalated discs. **Smooth muscle** is composed of spindle-shaped cells lacking striations. It is found in the walls of the digestive tract, blood vessels, and other internal organs, and produces involuntary movements.

Nervous tissue receives, processes, and transmits information. The **neuron**, or nerve cell, consists of a cell body and two types of extensions: dendrites, which receive impulses, and axons, which conduct impulses to other cells. **Glial cells (glia)** provide support for neurons.

Coordination and Control   The **endocrine system** and the **nervous system** coordinate the body's activities. Endocrine cells secrete **hormones** into the bloodstream. Body cells with matching receptors respond to these signaling molecules. Neurons deliver signals directly to target cells, which include other neurons, muscle cells, and various secretory cells.

INTERACTIVE QUESTION   40.4

Compare the speed of signaling and the duration of the response in the endocrine and nervous systems. How do the functions of these two communication systems differ?

40.2   Feedback control maintains the internal environment in many animals

Regulating and Conforming   For a given environmental variable, an animal may be a **regulator**, using mechanisms to control internal fluctuations, or a **conformer**, allowing internal conditions to vary with external changes.

Homeostasis   The maintenance of a "steady state" or relative internal balance is called **homeostasis**. Homeostatic control mechanisms involve a receptor or **sensor**, which detects a **stimulus** (a fluctuation above or below a **set point** for a variable), and a *control center*, which triggers a **response** that returns the variable to the set point.

Most homeostatic mechanisms operate by **negative feedback**, in which the response reduces the stimulus

and thus turns off the response. Variables may have a *normal range* within which they fluctuate instead of a single set point.

INTERACTIVE QUESTION   40.5

What is **positive feedback**? Give an example. Are positive feedback mechanisms usually involved in maintaining homeostasis? Explain.

Homeostatic mechanisms allow for *regulated change* in the body's internal environment when necessary. A **circadian rhythm** involves physiological changes that occur over a 24-hour cycle. In **acclimatization**, an animal's internal processes may gradually adjust to environmental changes.

40.3   Homeostatic processes for thermoregulation involve form, function, and behavior

Animals maintain body temperatures within an optimal range through the process of **thermoregulation**.

Endothermy and Ectothermy   Animals that use metabolic heat in generating body warmth are **endothermic** (birds and a few other reptiles, mammals, some fishes, and some insects). Animals that gain their heat mostly from the environment are **ectothermic** (amphibians, and the majority of invertebrates, fishes, and nonavian reptiles). Ectotherms require less food energy and tolerate wider internal temperature variations than do endotherms.

Variation in Body Temperature   A *poikilotherm* is an animal whose body temperature varies with changes in its environment; a *homeotherm* is an animal with a relatively stable body temperature regardless of environmental conditions.

INTERACTIVE QUESTION   40.6

Explain the following statement: Some ectotherms are homeotherms, and some even have higher body temperatures than those of endotherms.

Balancing Heat Loss and Gain   The flow of heat within an organism and between an organism and its environment is affected by the processes of *conduction*, the direct transfer of thermal motion (heat) between objects in contact; *convection*, the transfer of heat by the flow of air or water past a surface; *radiation*, the emission of electromagnetic waves by all objects; and *evaporation*, the loss of heat due to the conversion of the surface molecules of a liquid to a gas.

The mammalian **integumentary system** serves a thermoregulatory function. Insulation, in the form of hair, feathers, and fat layers in mammals and birds, reduces the rate of heat exchange with the environment. A very thick layer of blubber insulates marine mammals.

Vasodilation of superficial (surface) blood vessels increases both blood flow in the skin and the transfer of body heat to the environment. *Vasoconstriction* reduces blood flow and heat transfer.

In **countercurrent exchange**, common in aquatic mammals and birds, the close proximity of blood vessels servicing the extremities allows heat in arterial blood leaving the body core to be transferred to venous blood returning to the body core. Some sharks and fish that are powerful swimmers, as well as endothermic insects, use countercurrent heat exchangers to maintain high temperatures for muscles involved in swimming or flight.

Mammals and birds may use the cooling mechanism of evaporative heat loss across the skin and respiratory surfaces. Panting, sweating, and bathing may enhance evaporative cooling.

Behavioral responses contribute to thermoregulation in both endotherms and ectotherms. Many ectotherms may thermoregulate by orienting the body to the sun or moving to cool areas. The social organization of honeybees enables them to retain heat by huddling together and to cool the hive by bringing in water and fanning their wings.

Endotherms can vary *thermogenesis* to match heat loss. For example, they may increase the rate of metabolic heat produced by muscles through moving and shivering. Endothermic flying insects may warm up before taking off by contracting their powerful flight muscles.

Some mammals generate heat through *nonshivering thermogenesis*, a rise in metabolic rate and the production of heat instead of ATP. Some mammals have *brown fat* stores that are specialized for rapid heat production.

A few large reptiles use shivering to generate heat and become endothermic while incubating eggs.

Acclimatization in Thermoregulation   Birds and mammals often adjust the thickness of their insulating coats to meet seasonal changes. Ectotherms may adjust their cellular physiology (changes in enzymes, changes in membranes, or production of "antifreeze" proteins) to acclimatize to changes in environmental (and thus body) temperature.

Physiological Thermostats and Fever   In mammals, a group of neurons in the **hypothalamus** functions as a thermostat that triggers heat loss or gain mechanisms in response to a body temperature outside the normal range. Mammals and birds develop a *fever* in response to certain infections, apparently due to the resetting of the hypothalamic set point.

INTERACTIVE QUESTION   40.7

a. What cooling mechanisms are activated when the hypothalamus senses a body temperature above the set point?

b. What mechanisms are activated when body temperature falls below the set point?

40.4   **Energy requirements are related to animal size, activity, and environment**

An animal's **bioenergetics** involves the overall flow and transformation of energy—the input of energy in the form of food, its use in the body, and its return to the environment as heat or in waste products.

Energy Allocation and Use   Autotrophs produce organic fuel molecules using an external energy source; *heterotrophs* obtain chemical energy from food sources. Fuel molecules are used to generate ATP for cellular work and biosynthesis.

Quantifying Energy Use   The total energy an animal uses in a unit of time is its **metabolic rate**. Metabolic rate can be determined by measuring heat loss, the rate of oxygen consumption, the rate of CO_2 production, or the energy content of food and the energy lost in waste products.

Minimum Metabolic Rate and Thermoregulation The minimum metabolic rate for a nongrowing endotherm that is at rest, fasting, and nonstressed is called the **basal metabolic rate (BMR)**. The **standard metabolic rate (SMR)** is the metabolic rate of a resting, fasting, nonstressed ectotherm determined at a specific temperature.

Influences on Metabolic Rate   BMR increases with increasing body size. However, the energy required to maintain each gram of body weight is inversely related to body size. Smaller mammals have higher metabolic, respiration, and heart rates; have larger blood volumes; and require more food per unit of body mass.

Maximal metabolic rates occur during intense activity. Most terrestrial animals have an average daily rate of energy consumption that is 2 to 4 times BMR or SMR.

Species vary in the food energy they allocate for BMR or SMR, activity, growth, reproduction, and thermoregulation.

Torpor and Energy Conservation **Torpor** is a physiological state characterized by decreases in metabolism and activity that allow an animal to save energy while avoiding unfavorable environmental conditions.

INTERACTIVE QUESTION 40.8

Describe the following types of torpor. What challenges does each type help an animal meet?

a. Hibernation

b. *Estivation*

c. Daily torpor

Word Roots

con- = with; **vect-** = carried (*convection:* the mass movement of warmed air or liquid to or from the surface of a body or object)

counter- = opposite (*countercurrent exchange:* the exchange of a substance or heat between two fluids flowing in opposite directions)

ecto- = outside; **therm-** = heat (*ectothermic:* referring to organisms for which external sources provide most of the heat for temperature regulation)

endo- = inner (*endothermic:* referring to organisms that are warmed by heat generated by their own metabolism, This heat usually maintains a relatively stable body temperature higher than that of the external environment)

fibr- = a fiber (*fibroblast:* a type of cell in loose connective tissue that secretes extracellular protein fibers)

home- = same; **-stasis** = standing, posture (*homeostasis:* the steady-state physiological condition of the body)

inter- = between (*interstitial fluid:* the fluid filling the spaces between cells in most animals)

Structure Your Knowledge

1. Organs are composed of layers of several different tissues. Which of the four animal tissues would you expect to find in all organs? In what types of organs would you predict the remaining tissues to occur?

2. Explain the following statement: Endotherms can tolerate wider external temperature fluctuations; ectotherms may be able to tolerate wider internal temperature fluctuations.

3. An animal's energy budget may include basal (or standard) metabolism, reproduction, temperature regulation, growth, and activity. Describe some factors that lead to different energy allocations for different animals.

Test Your Knowledge

MULTIPLE CHOICE: *Choose the one best answer.*

1. Tunas, dolphins, penguins, and sharks all have a fusiform body shape. What is the best explanation for this similarity?
   a. They are all vertebrates and share homologous skeletal structures.
   b. The physical laws of hydrodynamics drove the natural selection for this shape.
   c. All aquatic animals have this shape, an example of convergent evolution.
   d. This shape is determined by the thermodynamics of maintaining body temperature in cold water.

2. Which of the following best describes smooth muscle?
   a. spindle-shaped cells; involuntary control
   b. striated, branching cells; involuntary control
   c. spindle-shaped cells connected by intercalated disks; voluntary control
   d. spindle-shaped striated cells; voluntary control

3. If a house were used to represent a human body, which bits would be considered epithelial tissues?
   a. aluminum siding, shingles, carpets, wallpaper
   b. water heater, sewer pipes, bathtub, toilet
   c. supporting members, 2×4s in the frame
   d. wiring, electrical outlets, telephone cables

4. Which of the following statements is *not* true of connective tissue?
   a. Some of its scattered cells secrete protein fibers.
   b. It includes such diverse tissues as bone, cartilage, adipose, and loose connective tissue.
   c. It forms the internal and external lining of many organs.
   d. It can have a matrix that is a liquid, a gel, or a solid.

5. Which of the following is an *incorrect* pairing?
   a. bone—osteoblasts embedded in a mineral matrix
   b. loose connective tissue—collagenous, elastic, and reticular fibers
   c. adipose tissue—loose connective tissue with fat-storing cells
   d. fibrous connective tissue—chondrocytes embedded in chondroitin sulfate

6. Consider an epithelium consisting of several cell layers. The outermost layer of this epithelium consists of flattened cells. You would properly classify this epithelium as
   a. cuboidal
   b. simple columnar
   c. simple squamous
   d. stratified squamous

7. The interstitial fluid of animals
   a. is the internal environment within cells.
   b. surrounds cells and provides for the exchange of nutrients and wastes.
   c. makes up blood plasma and is involved in transporting oxygen.
   d. is not necessary in aquatic vertebrates.

8. Each of the following is a form of connective tissue *except*
   a. cartilage
   b. blood
   c. bone
   d. glial cells

9. Which of the following would be *least* likely to be true of an animal that is a regulator?
   a. It can live in a variable climate because of its homeostatic mechanisms.
   b. It may have a larger geographic range than a conformer.
   c. Much of its energy budget can be allocated to reproduction.
   d. It may use behavioral as well as physiological mechanisms for responding to changing conditions.

10. Negative feedback loops are
    a. mechanisms that most commonly maintain homeostasis.
    b. activated only when a physiological variable falls below a set point.
    c. analogous to a radiator that heats a room.
    d. involved in maintaining contractions during childbirth.

11. Which of the following mechanisms dissipates heat?
    a. countercurrent exchange between vessels that service the extremities
    b. vasoconstriction of surface vessels
    c. vasodilation of surface vessels
    d. both **a** and **c**

12. In some hot climates it is customary to have a period of sleep, or at least reduced activity, during the hottest part of the day. This practice might be considered a form of
    a. thermogenesis.
    b. estivation.
    c. torpor.
    d. hibernation.

13. Which of the following animals would be likely to produce antifreeze compounds in its cells?
    a. a polar bear or an arctic wolf
    b. a hibernating mammal
    c. an estivating amphibian
    d. an ectotherm in a subzero environment

14. The rate of metabolic heat production can be increased by
    a. nonshivering thermogenesis.
    b. increased metabolic activity of brown fat.
    c. muscle contraction through shivering.
    d. all of the above.

15. Which of the following statements is true?
    a. BMR is always higher than SMR.
    b. BMR per kg of body mass is inversely related to body size.
    c. SMR is inversely related to body size.
    d. Overall metabolic rate is not correlated with body mass.

16. Which of the following animals would most likely have the highest metabolic rate?
    a. a bat
    b. a lizard
    c. a giant squid
    d. a whale

17. Which of the following animals would have the largest percentage of its energy budget available for growth?
    a. a deer mouse from a temperate forest
    b. a python from tropical India
    c. a human adult from a temperate climate
    d. a desert bird such as a roadrunner

18. For most terrestrial animals, the average daily rate of energy consumption is 2–4 times BMR or SMR. Which of the following animals has an unusually low average daily metabolic rate?
    a. a human in a developed country
    b. a wild turkey
    c. a domesticated turkey
    d. a desert snake

19. You have identified a gene that you think is involved in determining circadian rhythms. You isolate mRNA for this gene from the suprachiasmatic nuclei (SCN) of active and hibernating ground squirrels. Which of the following results would *not* support your hypothesis that this gene is a clock gene?
    a. In active animals, the RNA level was much higher during the day than during the night.
    b. In active animals, the RNA level was much higher during the night than during the day.
    c. In hibernating animals, the RNA levels in day and night were exactly reversed from what they were in active animals.
    d. In hibernating animals, the RNA levels were about equal during day and night.

Chapter 41

Animal Nutrition

Key Concepts

41.1 **An animal's diet must supply chemical energy, organic building blocks, and essential nutrients**

41.2 **Food processing involves ingestion, digestion, absorption, and elimination**

41.3 **Organs specialized for sequential stages of food processing form the mammalian digestive system**

41.4 **Evolutionary adaptations of vertebrate digestive systems correlate with diet**

41.5 **Feedback circuits regulate digestion, energy storage, and appetite**

Chapter Review

Nutrition is the process by which organisms take in and use food. **Herbivores** eat plants or algae; **carnivores** eat animals; and **omnivores** consume both plants and animals. Most animals opportunistically eat foods from different categories when they are available.

41.1 An animal's diet must supply chemical energy, organic building blocks, and essential nutrients

An animal's diet must provide fuel for ATP production via cellular respiration, organic carbon and nitrogen precursors for biosynthesis, and essential nutrients.

Essential Nutrients An animal's diet must also provide **essential nutrients**, which are molecules that an animal requires but cannot synthesize itself.

Framework

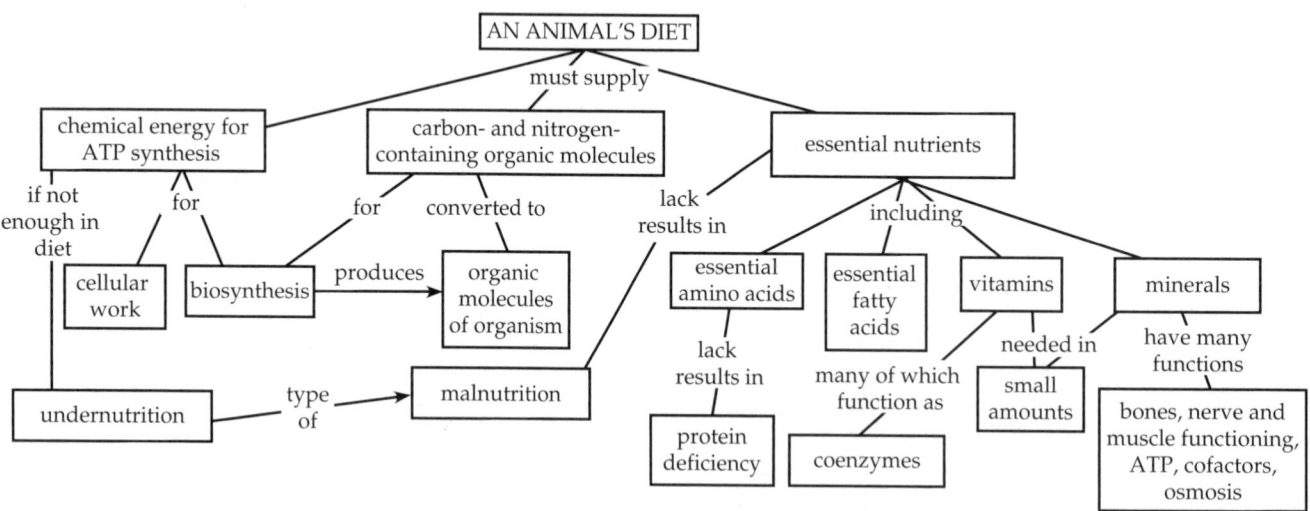

Eight of the 20 amino acids that are needed to make proteins are **essential amino acids** required in the diet of most animals (including adult humans). Meat, eggs, and cheese contain "complete" proteins with all essential amino acids in their proper proportions. Most plant proteins are "incomplete."

INTERACTIVE QUESTION 41.1

How can vegetarians obtain all the essential amino acids from their diet?

Animals can make many fatty acids. The **essential fatty acids**, such as linoleic acid, contain one or more double bonds. Such fatty acids are readily supplied in the diet from plant sources.

Vitamins are organic molecules required in very small amounts in the diet. Thirteen vitamins essential to humans have been identified. Water-soluble vitamins include the B complex, most of which function as coenzymes, and vitamin C, which is required for the production of connective tissue. The fat-soluble vitamins are vitamins A, incorporated into visual pigments; D, which aids in calcium absorption and bone formation; E, an antioxidant; and K, required for blood clotting.

Minerals are inorganic nutrients that are usually needed in small amounts. Some minerals are built into the structure of enzymes and other proteins. Iodine is needed by vertebrates to make the metabolism-regulating thyroid hormone. Sodium, potassium, and chloride are important in nerve and muscle function and osmotic balance.

Dietary Deficiencies *Malnutrition* results from a failure to obtain one or more essential nutrients or sufficient chemical energy. Undernourishment refers to a diet deficient in calories. With severe deficiency, the body breaks down its own proteins for energy, eventually causing irreversible damage or death.

Assessing Nutritional Needs Research on human dietary requirements involves many challenges. Much information comes from *epidemiology*, the study of health and disease among various populations (in this case, those with different dietary practices).

INTERACTIVE QUESTION 41.2

a. What is the cause of the most common type of malnutrition among humans?

b. Which two minerals do vertebrates require in relatively large amounts? Why?

41.2 Food processing involves ingestion, digestion, absorption, and elimination

Ingestion is the act of consuming food. Many aquatic animals are **filter feeders** (a type of suspension feeding), which strain small organisms or food particles from the water. **Substrate feeders** live in or on their food, eating their way through it. **Fluid feeders** suck fluids from a living plant or animal. Most animals are **bulk feeders**, which eat relatively large pieces of food.

Digestion splits food into small molecules. Mechanical digestion of food into smaller fragments often precedes chemical digestion, which is the *enzymatic hydrolysis* of bonds between components of large molecules. Bacteria often contribute to digestion.

In the last two stages of food processing, **absorption** and **elimination**, small molecules are taken into the cells of an animal, and the undigested remainder of the food leaves the digestive system.

INTERACTIVE QUESTION 41.3

Why must large molecules in food be digested into monomers or smaller components?

Digestive Compartments Breaking down food in specialized compartments protects animals from digesting themselves. In intracellular digestion, food vacuoles formed by phagocytosis or pinocytosis fuse with enzyme-containing lysosomes. Most animals break down their food, at least initially, by extracellular digestion within a separate compartment that connects to the external environment.

A **gastrovascular cavity** functions in both digestion and the transport of nutrients throughout the body. In the small cnidarian hydra, digestive enzymes secreted into the gastrovascular cavity begin food breakdown. Cells lining the cavity then engulf food particles for hydrolysis within food vacuoles. Undigested materials are expelled through the single opening (mouth) of the gastrovascular cavity.

Most animals have *complete digestive tracts* or **alimentary canals** with two openings, a mouth and an anus. Food, ingested through the mouth and pharynx, passes through an esophagus that may lead to a crop, stomach, or gizzard—organs specialized for storing or grinding food. In the intestine, food is chemically digested and nutrients are absorbed. Undigested material exits through the anus.

INTERACTIVE QUESTION   41.4

What are the advantages of an alimentary canal compared to a gastrovascular cavity?

41.3 Organs specialized for sequential stages of food processing form the mammalian digestive system

The mammalian digestive system consists of the alimentary canal and accessory glands that secrete digestive juices through ducts into the canal.

The Oral Cavity, Pharynx, and Esophagus   Physical and chemical digestion begins in the mouth, or **oral cavity**, where teeth grind food to expose a greater surface area to enzyme action. Saliva, released from **salivary glands**, has several components: **mucus**, which contains slippery glycoproteins that protect the mouth lining from abrasion and lubricate food for swallowing; buffers to neutralize acidity; antimicrobial agents; and **amylase**, an enzyme that begins the hydrolysis of starch and glycogen. The tongue tastes and manipulates food and pushes the food ball, or **bolus**, into the pharynx for swallowing.

The **pharynx**, or throat region, is the intersection leading to both the esophagus and the trachea. During swallowing, the larynx, the upper part of the respiratory tract, moves up so that the epiglottis covers and blocks the trachea.

Food moves down through the **esophagus** to the stomach, squeezed along by rhythmic waves of muscle contraction called **peristalsis**. Passage of food into the stomach is regulated by a muscular ringlike valve called a **sphincter**.

Digestion in the Stomach   In the expandable **stomach**, food is combined with **gastric juice**, producing a mixture called **chyme**.

Parietal cells within gastric glands in the stomach secrete hydrogen ions and chloride ions. The resulting hydrochloric acid (HCl) breaks down food tissues, kills bacteria, and denatures proteins. A **protease** called **pepsin**, which hydrolyzes specific peptide bonds in proteins, is synthesized and secreted by *chief cells* in an inactive form called **pepsinogen**. It is activated by HCl and by pepsin itself, which produces more active pepsin through positive feedback.

Mucus and rapid cell replacement help protect the stomach epithelium from digestion. Gastric ulcers, damaged areas of the stomach lining, have been linked to infection by the bacterium *Helicobacter pylori*.

Smooth muscles churn the contents of the stomach. Except when a bolus arrives, the opening from the esophagus is closed to prevent backflow of the acidic chyme. A sphincter regulates passage of chyme into the intestine.

INTERACTIVE QUESTION   41.5

Two types of macromolecules have been partially digested by the time chyme moves into the intestine. What are these molecules, where does this digestion take place, and what enzymes are involved?

a.

b.

Digestion in the Small Intestine   Digestive juices are mixed with chyme in the **duodenum**, the first section of the **small intestine**. The **pancreas** produces digestive enzymes and a bicarbonate-rich alkaline solution that offsets the acidity of the chyme. The **liver** produces **bile**, which is stored in the **gallbladder** until needed. Bile salts emulsify fats for easier digestion. Bile also contains pigments that are by-products of the breakdown of red blood cells in the liver. The lining of the duodenum secretes some enzymes; others are bound to the surfaces of epithelial cells.

Most digestion occurs in the duodenum. The *jejunum* and *ileum* function in nutrient absorption.

INTERACTIVE QUESTION   41.6

For each of the following molecules, describe their enzymatic digestion in the small intestine.

a. polysaccharides

b. polypeptides

c. nucleic acids

d. fats

Absorption in the Small Intestine   Large folds of the small intestine lining are covered with fingerlike projections called **villi**. Each epithelial cell of a villus has microscopic extensions called **microvilli**. This huge surface area for absorption is called the *brush border*. Nutrients may be actively or passively transported across the membranes.

Absorbed amino acids and sugars leave the epithelial cells and enter capillaries in the core of each villus. The nutrient-laden blood from the small intestine is carried directly to the liver by the **hepatic portal vein**. The liver regulates the nutrient content of the blood and detoxifies a variety of molecules, including foreign substances such as certain drugs.

Monoglycerides and fatty acids absorbed by epithelial cells are recombined to form triglycerides. These fats are coated with proteins, phospholipids, and cholesterol to make tiny globules called **chylomicrons**, which move out of epithelial cells by exocytosis and into a lymph vessel called a **lacteal** in the center of each villus. They are then transported by the lymphatic system to large veins.

Most of the water from drinking and secreted in digestive juices is reabsorbed by osmosis in the small intestine.

Processing in the Large Intestine   The small intestine leads into the **large intestine** at a T-shaped junction. One arm of this junction is a pouch called the **cecum**, which in humans has a fingerlike extension called the **appendix**. The other arm of the junction is the **colon**, which continues the reabsorption of water and processes the **feces**.

Huge numbers of mostly harmless bacteria live on unabsorbed organic matter in the colon. The feces contain a large proportion of intestinal bacteria, cellulose, and other undigested wastes. Feces are stored in the **rectum**, the final section of the large intestine, and expelled through the anus.

41.4 **Evolutionary adaptations of vertebrate digestive systems correlate with diet**

Dental Adaptations   Dentition, the type and arrangement of teeth, correlates with diet.

Stomach and Intestinal Adaptations   Vertebrates that eat large, sporadic meals may show highly expandable stomachs. Plant material is more difficult to digest than meat. The longer alimentary canal of herbivores facilitates the digestion of plant material.

Mutualistic Adaptations   The human digestive system may house from 10 to 100 trillion bacteria. Many intestinal bacteria are part of a mutualistic symbiosis with their host. Some produce nutrients such as vitamin K, biotin, and folic acid. Recent research has revealed more than 400 bacterial species in the **microbiome** found in the human digestive system. Comparisons of sequenced DNA collected from *H. pylori*-infected and uninfected individuals revealed that infection resulted in a drastic reduction in the diversity of the microbiome. Recent research has linked changes in the gut microbiome to a variety of conditions, including obesity.

Many herbivorous mammals have fermentation chambers filled with symbiotic bacteria and protists. These microorganisms, often housed in the cecum, digest cellulose to simple sugars and produce a variety of essential nutrients. Rabbits and some rodents ingest their feces (*coprophagy*) to capture the nutrients that their mutualistic bacteria have released in the large intestine.

Ruminants have an elaborate system involving several stomach chambers, regurgitation and rechewing of the cud, and digestion of symbiotic bacteria, which maximizes the nutrient yield of their grass diet.

INTERACTIVE QUESTION   41.7

Compare and contrast the dentition and alimentary canals of carnivores and herbivores.

41.5 Feedback circuits regulate digestion, energy storage, and appetite

Regulation of Digestion   The *enteric division* of the nervous system regulates peristalsis and release of digestive juices. Hormones, such as *gastrin*, released by the stomach, and *secretin* and *cholecystokinin (CCK)*, released by the duodenum, stimulate the release of digestive secretions from the stomach, pancreas, and gallbladder.

Regulation of Energy Storage   When an animal consumes more calories than are needed to meet its energy requirements, the excess can be stored in the liver and muscles as glycogen, or in adipose tissue as fat (when glycogen stores are full).

Glucose homeostasis is maintained by two pancreatic hormones: **insulin**, which promotes glycogen synthesis in the liver and glucose uptake into body cells, thus lowering blood glucose; and **glucagon**, which stimulates the liver to break down glycogen and release glucose, thus raising blood glucose. Scattered within the exocrine tissue of the pancreas are clusters of endocrine cells called pancreatic islets. Within each islet are *alpha cells*, which make glucagon, and *beta cells*, which make insulin.

In **diabetes mellitus**, the absence of insulin in the bloodstream or the loss of response to insulin in target tissues reduces glucose uptake by cells. The body must use fats for fuel, and acids from the breakdown of fats may lower blood pH. Glucose accumulates in the blood and is excreted in the urine, with an accompanying loss of water.

Type 1 diabetes, also known as insulin-dependent diabetes, is an autoimmune disorder in which pancreatic beta cells are destroyed. This type of diabetes is treated by regular injections of insulin. More common is *type 2 diabetes*, or non-insulin-dependent diabetes, characterized by reduced responsiveness of target cells to insulin. Lack of exercise and excess weight increase the risk of type 2 diabetes.

INTERACTIVE QUESTION   41.8

Complete the following concept map on the regulation of blood glucose levels by hormones of the pancreas.

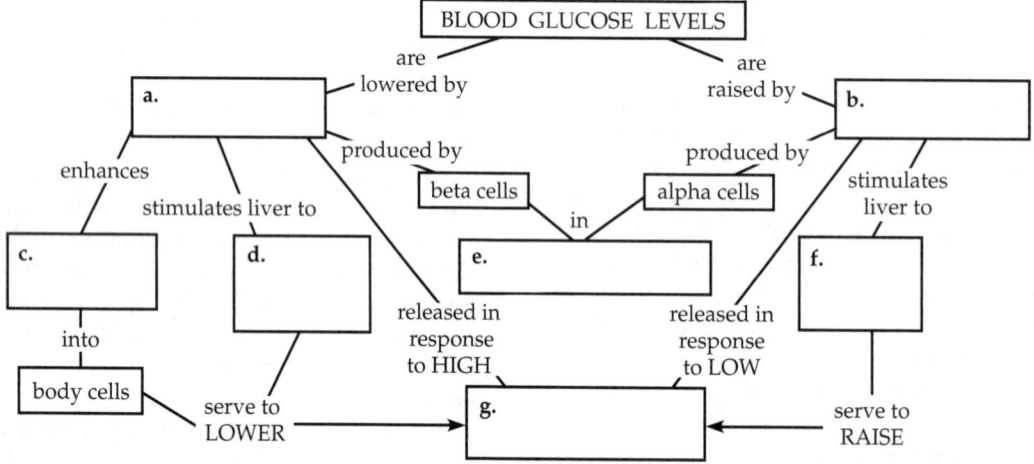

Regulation of Appetite and Consumption   *Overnourishment*, consuming more calories than the body needs, can lead to obesity. This excessive accumulation of fat is associated with a number of health problems.

Researchers have identified feedback circuits that control fat storage and metabolism. Appetite-regulating hormones act on the brain's "satiety center," which generates impulses that make one feel hungry or satiated. These hormones include *ghrelin*, which is secreted by the stomach and triggers hunger, as well as insulin and *PYY*, secreted by the small intestine after meals, which both suppress appetite. *Leptin*, produced by adipose tissue, also suppresses appetite and appears to play a role in regulating body fat.

Word Roots

chyl- = juice; **micro-** = small (*chylomicron:* a lipid transport globule composed of fats mixed with cholesterol and coated with proteins)

chym- = juice (*chyme:* the mixture of partially digested food and digestive juices formed in the stomach)

gastr- = stomach; **vascul-** = a little vessel (*gastrovascular cavity:* a central cavity with a single opening that functions in both digestion and nutrient distribution)

herb- = grass; **-vore** = one who eats (*herbivore:* an animal that mainly eats plants or algae)

micro- = small; **-villi** = shaggy hair (*microvilli:* the many fine, fingerlike projections of the epithelial cells in the lumen of the small intestine that increase its surface area)

omni- = all (*omnivore:* an animal that regularly eats animals as well as plants or algae)

peri- = around; **-stalsis** = a constriction (*peristalsis:* alternating waves of contraction and relaxation in smooth muscles that push food along the alimentary canal)

Structure Your Knowledge

1. Label the indicated structures in the following diagram of the human digestive system. Review the functions of these structures.

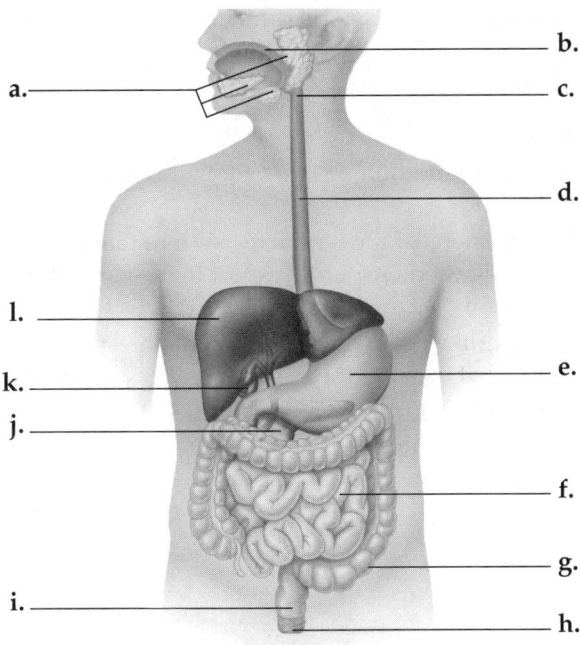

Test Your Knowledge

MATCHING: *Match the description with the correct enzyme, hormone, or other substance.*

_____ 1. enzyme that hydrolyzes peptide bonds in the stomach

_____ 2. hormone that stimulates secretion of gastric juice

_____ 3. hormone that stimulates release of bile and pancreatic enzymes; also inhibits peristalsis

_____ 4. lipid transport globule

_____ 5. hormone secreted by adipose cells that suppresses appetite

_____ 6. substances that emulsify fats

_____ 7. enzyme that hydrolyzes fats

_____ 8. hormone that is secreted by the stomach and triggers hunger

A. carboxypeptidase
B. bile salts
C. cholecystokinin (CCK)
D. chylomicron
E. gastrin
F. ghrelin
G. insulin
H. leptin
I. lipase
J. pepsin

MULTIPLE CHOICE: *Choose the one best answer.*

1. Which of the following is a *true* statement about vitamins?
   a. They may be produced by intestinal bacteria.
   b. Requirements for them are the same from one species to the next.
   c. They are inorganic nutrients, needed in small amounts, that may be required for enzyme function.
   d. Because they are water soluble, they must be replaced every day.

2. Corn (maize) and beans are staple items in the traditional cuisine of Mexico. What is the advantage of consuming corn and beans in the same meal?
   a. They are relatively inexpensive and nutritious.
   b. They can be grown successfully in many parts of Mexico.
   c. In combination they contain all the amino acids that are essential for adult humans.
   d. They contain certain bacteria that contribute to a healthy gut microbiome.

3. Which of the following substances is mismatched with its function?
   a. most B vitamins—coenzymes
   b. vitamin A—incorporated into visual pigment
   c. iron—component of thyroid hormones
   d. phosphorus—bone formation, nucleotide synthesis

4. You have a friend who maintains a strict vegan diet; that is, she eats no animal products whatsoever. She complains of lack of energy and numbness in her extremities. You suspect the problem may be a deficiency of
   a. vitamin A
   b. vitamin B_9
   c. vitamin B_{12}
   d. vitamin K

5. Substrate feeders such as earthworms
   a. eat mostly mineral substances.
   b. filter small organisms from damp earth.
   c. are herbivores and thus eat autotrophs.
   d. move and eat through their food.

6. In animals, a distinct advantage of extracellular digestion over intracellular digestion is that
   a. polymers are hydrolyzed to monomers by digestive enzymes.
   b. a greater surface area is available for the absorption of digested nutrients.
   c. larger pieces of food can be ingested and then digested.
   d. the products of extracellular digestion can be absorbed into all body cells, without the need for a transport system.

7. Why does salivary amylase not hydrolyze starch in the duodenum?
   a. Starch is completely hydrolyzed into maltose in the oral cavity.
   b. The acid pH of the stomach denatures salivary amylase, and pepsin begins hydrolyzing it.
   c. Salivary amylase is produced by salivary glands and never leaves the oral cavity.
   d. Pancreatic amylase is a more effective enzyme in the pH of the duodenum.

8. What is the advantage of mechanical digestion of food?
   a. It increases the surface area available for chemical digestion.
   b. It creates small particles that can be taken in to food vacuoles.
   c. It removes bits of food that cannot be broken down by chemical digestion.
   d. It prevents food from entering the glottis during swallowing.

9. Where is starch broken down?
   a. in the mouth
   b. in the stomach
   c. in the small intestine
   d. both in the mouth and in the small intestine

10. Heartburn is caused by
    a. leakage of acidic stomach contents into the esophagus.
    b. overexpansion of the stomach, which puts pressure on the heart.
    c. spasms of the pyloric sphincter of the stomach.
    d. damage to the stomach lining resulting from oversecretion of stomach acid.

11. Villi are
    a. folds in the stomach that allow the stomach to expand.
    b. extensions of the lymphatic system that pick up digested fats for transport to the circulatory system.
    c. fingerlike projections of the small intestine lining that increase the surface area for absorption.
    d. microscopic extensions of epithelial cells lining the small intestine that provide more surface area for digestion.

12. The hepatic portal vein
    a. supplies the capillaries of the intestines with oxygenated blood.
    b. carries absorbed nutrients to the liver for processing.
    c. carries blood from the liver to the heart.
    d. drains the lacteals of the villi.

13. You have just eaten a cheeseburger and french fries. Which of the following components of your lunch will be subject to chemical digestion in the stomach?
    a. starch in the fries and in the bun
    b. protein in the burger and the cheese
    c. fat in the burger, the fries, and the cheese
    d. starch, protein, and fat will all be chemically digested in the stomach

14. The most likely action of antidiarrheal medicine is to
    a. slow down peristalsis in the large intestine.
    b. kill *E. coli* in the intestine.
    c. decrease water reabsorption in the intestine.
    d. increase salt secretion into the feces.

15. Which of the following is *not* a common component of feces?
    a. intestinal bacteria
    b. cellulose
    c. saturated fats
    d. undigested materials

16. Ruminants
    a. get all of their nutrition from digested plant material.
    b. house microorganisms that digest cellulose in stomach chambers.
    c. eat their feces to obtain nutrients digested from cellulose by microorganisms.
    d. house symbiotic bacteria in a long cecum.

17. Which of the following statements is *false*?
    a. The average human has enough stored fat to supply calories for several weeks.
    b. An increase in leptin levels leads to an increase in appetite and weight gain.
    c. After glycogen stores are filled, excessive calories are stored as fat, regardless of their original food source.
    d. Carbohydrates and fats are preferentially used as fuel before proteins are used.

18. You and two friends are trying a cereal diet to lose weight. You each eat 30 g of cereal, six times a day. You choose an organic cereal with nuts, grains, and no added sugar that has 3 g of fat, 20 g of carbohydrates (4 g of which are nondigestible insoluble fiber), and 7 g of protein. Friend 1 chooses a high-fiber cereal with 1 g of fat, 27 g of carbohydrates (9 g of insoluble fiber), and 2 g of protein. Friend 2 chooses a sugary cereal with 0 g of fat, 28 g of carbohydrates (3 g of insoluble fiber), and 2 g of protein. Predict the order, from fastest to slowest, in which the three of you will lose weight.
    a. you (because you ate the fewest carbs), then Friend 1 (because 9 g of the carbohydrates eaten are not digestible), then Friend 2
    b. Friend 1 (who ate 9 g of insoluble fiber), then you, then your sugary Friend 2
    c. Friend 1, then Friend 2, then you (because each of your fat grams contains approximately two times the energy content of carbohydrate or protein)
    d. you (because proteins are not used as fuel, those 7 g of protein do not count), then Friend 2, then Friend 1

Chapter 42

Circulation and Gas Exchange

Key Concepts

42.1 Circulatory systems link exchange surfaces with cells throughout the body

42.2 Coordinated cycles of heart contraction drive double circulation in mammals

42.3 Patterns of blood pressure and flow reflect the structure and arrangement of blood vessels

42.4 Blood components function in exchange, transport, and defense

42.5 Gas exchange occurs across specialized respiratory surfaces

42.6 Breathing ventilates the lungs

42.7 Adaptations for gas exchange include pigments that bind and transport gases

Framework

This chapter surveys the circulatory and gas exchange systems found in animals, with special attention given to the human systems and the problems of cardiovascular disease. The following concept map organizes some of the chapter's key ideas.

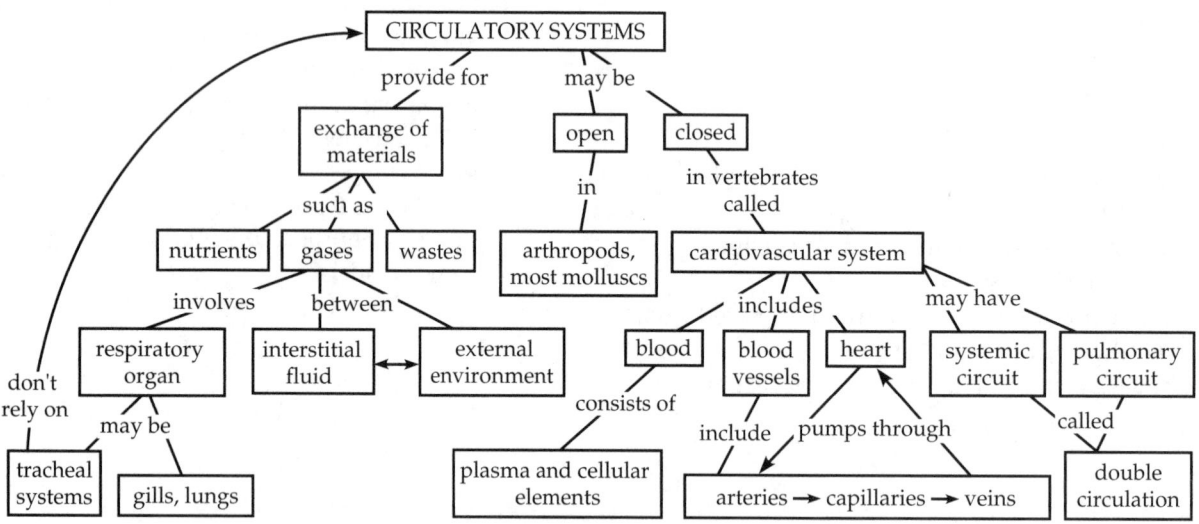

Chapter Review

42.1 Circulatory systems link exchange surfaces with cells throughout the body

Every cell must obtain O_2 and nutrients and dispose of CO_2 and metabolic wastes. Many of these substances can enter or leave a cell by **diffusion**, moving down their concentration gradient across the cell membrane. Some animals have body shapes and sizes that allow most body cells to be in contact with the environment for exchange of materials. Because diffusion is too slow a process across distances more than a few millimeters, most animals have body tissues that exchange materials with the environment and a circulatory system to service body cells.

Gastrovascular Cavities   The central **gastrovascular cavity** inside the two-cell-thick body wall of cnidarians serves for both the digestion and transport of materials. The internal fluid exchanges directly with the aqueous environment through the single opening. Flatworms also have gastrovascular cavities that branch throughout their thin, flat bodies.

Open and Closed Circulatory Systems   A circulatory system consists of a pumping **heart**, a circulatory fluid, and transport vessels. The circulatory system connects the interstitial fluid surrounding cells with organs that interface directly with the external environment.

Arthropods and some molluscs have an **open circulatory system**. A heart pumps **hemolymph** (the *interstitial fluid*) into sinuses (spaces) between organs, where chemical exchange occurs with body cells.

Cephalopods, annelids, and vertebrates have **closed circulatory systems**, in which **blood** remains in vessels and materials are exchanged between the blood and interstitial fluid bathing the cells.

INTERACTIVE QUESTION   42.1

a. How is the hemolymph different from blood?

b. Compare the advantages of open and closed circulatory systems.

Organization of Vertebrate Circulatory Systems   In the **cardiovascular system** of vertebrates, **arteries** carrying blood away from the heart branch into tiny **arterioles** within organs. Arterioles then divide into microscopic **capillaries**, which infiltrate tissues in networks called **capillary beds**. Capillaries converge to form **venules**, which meet to form the **veins** that return blood to the heart. In a few instances, portal veins connect two capillary beds without the blood first returning to the heart.

The vertebrate heart has one or two **atria**, which receive blood, and one or more **ventricles**, which pump blood out of the heart.

The ventricle of the two-chambered heart of a fish pumps blood to the gills, from which the oxygen-rich blood flows through a vessel to capillary beds in the other organs. Veins return the oxygen-poor blood to the atrium. Passage through two capillary beds reduces blood pressure and slows the flow of blood, but body movements help to maintain circulation.

In the circulatory systems of amphibians, reptiles, and mammals, the right side of the heart pumps blood to a *pulmonary circuit* (lungs) or *pulmocutaneous circuit* (lungs and skin), while the left side pumps the returning oxygen-rich blood to capillary beds in the body tissues (the **systemic circuit**). This double pumping assures a strong flow of oxygen-rich blood to the brain, muscles, and body organs.

In the three-chambered heart of amphibians, the single ventricle pumps blood into both the pulmocutaneous and systemic circuit. A ridge in the ventricle diverts most of the blood from the left atrium (returning from the lungs and skin), into the systemic circuit. When underwater, most blood flow to the lungs can be shut off.

The three-chambered heart of turtles, snakes, and lizards has a partially divided ventricle that helps to separate blood flowing to the lungs and to the body tissues. In crocodilians, the ventricles are completely divided, although connections between arteries of the pulmonary and systemic circuits allow the lungs to be bypassed when the animal is underwater.

Delivery of oxygen for cellular respiration is most efficient in birds and mammals, which, as endotherms, have high oxygen and energy demands. The left side of the large and powerful four-chambered heart receives and pumps oxygen-rich blood only, whereas the right side receives and pumps oxygen-poor blood.

INTERACTIVE QUESTION   42.2

a. Explain the difference between **single circulation** and **double circulation**.

b. Describe the difference between the circulatory system of amphibians and that of mammals.

42.2   Coordinated cycles of heart contraction drive double circulation in mammals

Mammalian Circulation   Review the circulation of blood through a mammalian cardiovascular system by completing Interactive Question 42.3.

INTERACTIVE QUESTION   42.3

In the following diagram of a mammalian circulatory system, label the indicated parts, color the vessels that carry oxygen-rich blood red, and then trace the flow of blood by numbering the circles from 1 to 11. Start numbering by assigning 1 to the right ventricle.

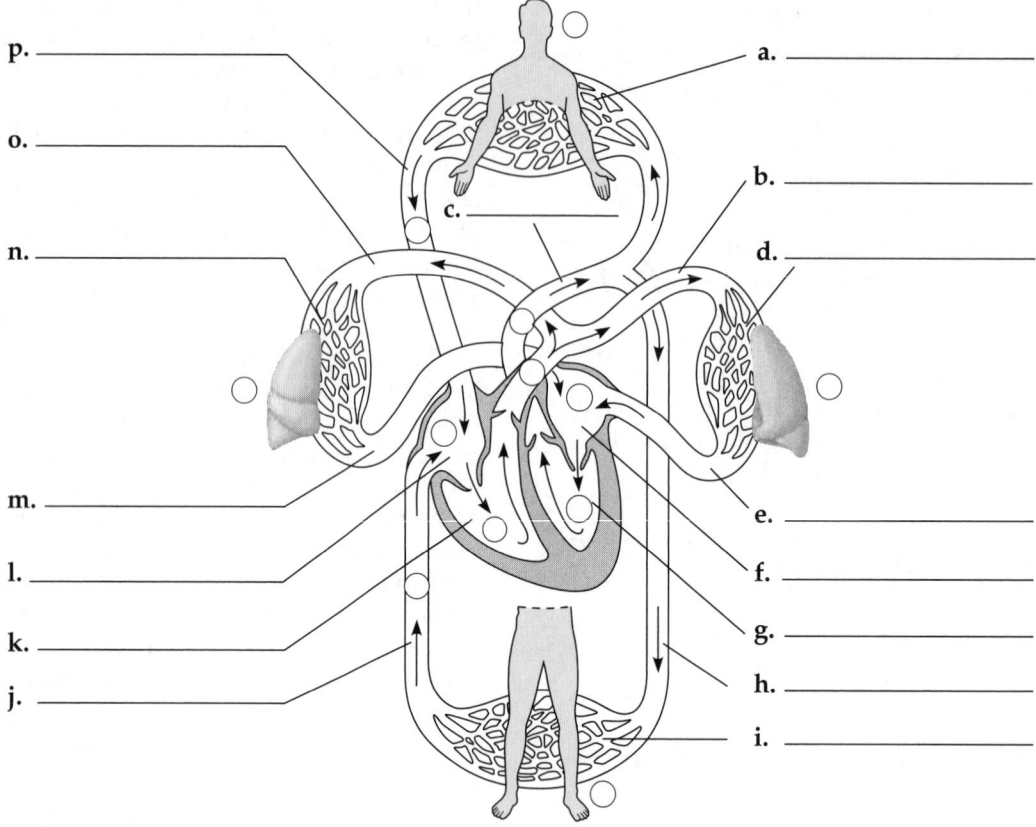

p. _____   a. _____

o. _____   b. _____

n. _____   c. _____   d. _____

m. _____   e. _____

l. _____   f. _____

k. _____   g. _____

j. _____   h. _____

                                                                              i. _____

The Mammalian Heart: A Closer Look   The human heart, located just beneath the sternum, is composed mostly of cardiac muscle. The atria have relatively thin muscular walls, whereas the ventricles have thicker walls. The **cardiac cycle** consists of **systole**, during which the cardiac muscle contracts and the chambers pump blood, and **diastole**, when the heart chambers are relaxed and filling with blood.

The **cardiac output**, or volume of blood pumped by each ventricle per minute, depends on both the **heart rate** and **stroke volume** (the quantity of blood pumped by each contraction).

Atrioventricular (AV) valves between each atrium and ventricle are snapped shut when blood is forced against them as the ventricles contract. The **semilunar valves** where the aorta and pulmonary artery exit the heart are forced open by ventricular contraction and close when the ventricles relax.

The heart sounds are caused in part by the recoil of blood against the closed heart valves. A **heart murmur** is the detectable hissing sound of blood leaking back through a defective valve.

Maintaining the Heart's Rhythmic Beat   The rhythm of heart contractions is coordinated by the **sinoatrial (SA) node**, or *pacemaker*, a region of specialized autorhythmic muscle cells located in the wall of the right atrium. The SA node initiates an electrical impulse that spreads via the intercalated disks of the cardiac muscle cells, and the two atria contract. After a 0.1-second delay, the **atrioventricular (AV) node**, located between the left and right atria, relays the impulse to the ventricles through bundle branches and Purkinje fibers. The electrical currents produced during the heart cycle can be detected by electrodes placed on the skin and recorded in an **electrocardiogram (ECG or EKG)**.

The SA node is controlled by sympathetic and parasympathetic nerves (which speed or slow the heart rate) and is influenced by hormones and body temperature.

INTERACTIVE QUESTION 42.4

a. Name the valve between an atrium and ventricle.

_____

b. Name the valve between a ventricle and the aorta or pulmonary artery. _____

c. Which node initiates atrial systole? _____

d. Which node initiates ventricular systole? _____

42.3 **Patterns of blood pressure and flow reflect the structure and arrangement of blood vessels**

Blood Vessel Structure and Function   The wall of an artery or vein consists of three layers: an outer connective tissue layer with elastic fibers that allow the vessel to stretch and recoil; a middle layer of smooth muscle and more elastic fibers; and an inner lining of **endothelium**, a single layer of flattened cells. The outer two layers are thicker in arteries than they are in veins. One-way valves in the larger veins assure that blood cannot flow backward away from the heart. Capillaries have only an endothelium.

Blood Flow Velocity   The enormous number of capillaries creates a large total cross-sectional area. Blood flow speeds up within venules and veins due to the decrease in total cross-sectional area.

Blood Pressure   Blood pressure, the force that drives blood from the heart to the capillary beds, is highest during ventricular systole **(systolic pressure)**. Resistance caused by the narrow openings of the arterioles impeding the exit of blood from arteries contributes to the swelling of the arteries during systole, which can be felt as the **pulse**. The recoiling of the stretched elastic arteries creates **diastolic pressure**, which maintains a continuous flow of blood into arterioles and capillaries.

   Vasoconstriction, a reduction in vessel diameter due to contraction of smooth muscles in arteriole walls, increases resistance and thus increases blood pressure. **Vasodilation** of arterioles as smooth muscles relax increases blood flow into the arterioles and thus lowers blood pressure. Nervous and hormonal signals control the production of the gas nitric oxide (NO) and the peptide endothelin, which induce vasodilation and vasoconstriction, respectively. Changes in cardiac output also affect blood pressure.

   Arterial blood pressure may be measured with a sphygmomanometer. The first number is the pressure during systole (when blood first spurts through the artery that was closed off by the cuff), and the second number is the pressure during diastole.

   Blood pressure is very low by the time blood exits the capillary beds. Contraction of the smooth muscle walls of venules and veins and the contraction of skeletal muscles between which veins are embedded force blood to flow back to the heart. Valves in veins prevent gravity from moving blood away from the heart.

INTERACTIVE QUESTION 42.5

Complete the following concept map to help you organize your understanding of blood pressure.

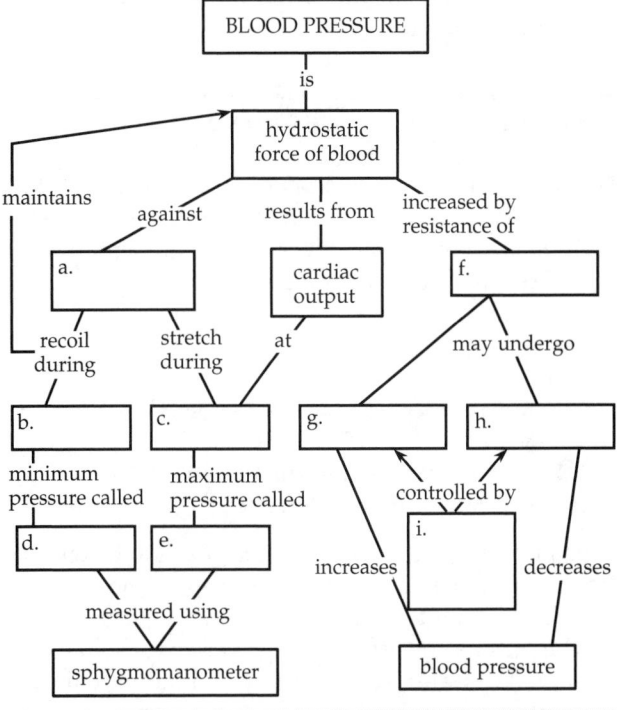

Capillary Function   Capillaries branch off thoroughfare channels that are direct connections between arterioles and venules. The distribution of blood to capillary beds varies with need and is regulated by nerve impulses, hormones, and local chemicals that control contraction of smooth muscles in arteriole walls and in *precapillary sphincters* at the entrance to capillary beds.

The exchange of substances between blood and interstitial fluid may occur by endocytosis and exocytosis by endothelial cells, by passive diffusion of small molecules across endothelial cells, and by movement through microscopic pores in the capillary wall. Water, sugars, salts, and urea may move through these openings, whereas blood cells and proteins are too large to fit through. Blood pressure forces fluid out of capillaries by bulk flow. The blood's *osmotic pressure* due to the proteins that remain within the capillary tends to counter this fluid movement out of the capillary.

Fluid Return by the Lymphatic System   The fluid lost from capillaries and any proteins that may have leaked out are returned to the blood through the **lymphatic system**. Fluid moves into tiny lymph vessels intermingled with blood capillaries. The fluid, called **lymph**, moves through lymph vessels containing one-way valves, mainly as a result of the movement of skeletal muscles, and is returned to the circulatory system via large veins in the neck. In **lymph nodes**, the lymph is filtered, and white blood cells attack viruses and bacteria.

INTERACTIVE QUESTION   42.6

Edema, swelling resulting from the accumulation of fluid in body tissues, can result from blockages that interfere with lymph flow. Explain how edema might also result from a severe dietary protein deficiency.

42.4   Blood components function in exchange, transport, and defense

Blood Composition and Function   Cells and cell fragments make up about 45% of the volume of blood; the rest is a liquid matrix called **plasma**.

The plasma consists of a large variety of solutes dissolved in water. The collective and individual concentrations of the dissolved ions of inorganic salts (electrolytes) are important to osmotic balance, pH levels, and the maintaining of ion levels in the interstitial fluid for proper functioning of muscles and nerves.

Plasma proteins function in the osmotic balance of blood and as buffers, antibodies, escorts for lipids, and clotting factors. Blood plasma from which clotting factors have been removed is called *serum*. Nutrients, metabolic wastes, gases, and hormones are transported in the plasma.

Erythrocytes (red blood cells) transport O_2. The red blood cells of mammals lack nuclei and generally have a biconcave shape. This shape and their small size create a large surface area of plasma membrane across which O_2 can diffuse. Erythrocytes are packed with **hemoglobin**, an iron-containing protein that binds O_2.

In **sickle-cell disease**, abnormal hemoglobin molecules cause erythrocytes to distort into a sickle shape, resulting in blocked vessels and other serious effects.

There are five major types of **leukocytes**, or white blood cells, all of which fight infections. Some white cells are phagocytes that engulf bacteria and cellular debris; lymphocytes give rise to cells that produce the immune response. Leukocytes are also found in interstitial fluid and the lymphatic system.

Platelets are cell fragments involved in blood clotting.

Red and white blood cells and platelets are produced from multipotent **stem cells** in red bone marrow. Erythrocytes circulate for about four months before they are replaced. The production of red blood cells is controlled by a negative feedback mechanism involving the hormone **erythropoietin (EPO)**, which is produced by the kidney in response to low O_2 supply in tissues. Recombinant EPO is used to treat *anemia*, a condition of low red blood cells or hemoglobin levels.

The clotting process usually begins when platelets, clumped together along a damaged endothelium, release clotting factors. By a series of steps involving a positive feedback loop, the enzyme thrombin converts the plasma protein fibrinogen to its active form, fibrin. The threads of fibrin form a patch. Hemophilia is caused by an inherited defect in any step of the complex clotting process.

Anticlotting factors normally prevent the clotting of blood in the absence of injury. A **thrombus** is a clot that occurs within a blood vessel and blocks the flow of blood.

INTERACTIVE QUESTION   42.7

Complete the following concept map of the components of vertebrate blood and their functions.

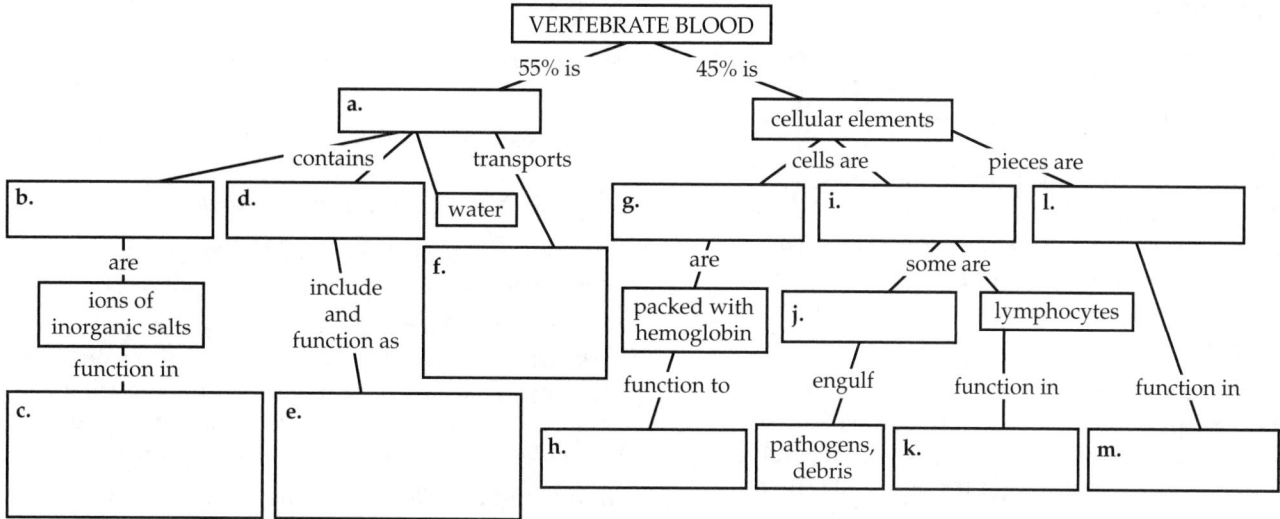

Cardiovascular Disease   In the chronic cardiovascular disease known as **atherosclerosis**, fatty deposits called plaques stiffen and narrow arteries. Cholesterol is carried to body cells for inclusion in cell membranes in particles called **low-density lipoproteins (LDLs)**. **High-density lipoproteins (HDLs)** return excess cholesterol to the liver. A high LDL/HDL ratio is a risk factor for atherosclerosis.

Injury to or infection of the lining of an artery can lead to *inflammation*. Leukocytes are attracted and take up lipids. The plaque grows and may obstruct the artery or rupture, forming a thrombus.

A **heart attack**, or *myocardial infarction,* is the death of cardiac muscle due to blockage of coronary arteries; a **stroke**, the death of nervous tissue in the brain, results from blockage or rupture of arteries in the head.

Statins are drugs used to lower LDL levels in individuals at risk of cardiovascular disease. High levels of C-reactive protein (CRP), indicative of acute inflammation, also indicate disease risk.

Hypertension, or high blood pressure, is thought to damage the endothelium and initiate plaque formation, promoting atherosclerosis and increasing the risk of heart attack and stroke. This condition can be easily diagnosed and controlled by drugs, diet, and exercise.

INTERACTIVE QUESTION   42.8

What lifestyle choices have been correlated with increased risk of cardiovascular disease?

42.5   **Gas exchange occurs across specialized respiratory surfaces**

Gas exchange, or respiration, is the uptake of O_2 from and the discharge of CO_2 to the environment.

Partial Pressure Gradients in Gas Exchange   In a mixture of gases, the contribution of a particular gas to the total gas pressure is the **partial pressure** for that gas. At sea level, the partial pressure of O_2, which makes up 21% of the atmosphere, is 160 mm Hg (0.21 × 760 mm, which is atmospheric pressure at sea level). A gas undergoes net diffusion from a region of higher partial pressure to a region of lower partial pressure. When a gas is at equilibrium between air and water, the partial pressure is the same in both media; the *concentration* of the gas differs, however, because of different solubilities of the gas in air and water.

Respiratory Media   Air is a respiratory medium with plentiful O_2. Water has much less O_2 and greater density and viscosity. Thus gas exchange in water is often more energetically demanding and often involves adaptations for increased efficiency.

Respiratory Surfaces   The respiratory surface, where gas exchange with the respiratory medium occurs, must be moist, thin, and have a large surface area for exchange.

In animals with simple body forms, gases can easily diffuse in and out of body cells. In earthworms and some amphibians, the skin, supplied with a network of capillaries, serves as a respiratory organ. Most animals with denser bodies have a branched or folded respiratory surface.

Gills in Aquatic Animals   Gills are outfoldings of the body surface that increase surface area for gas exchange. Their structure ranges from simple bumps on echinoderms to more complex localized structures. To maintain partial pressure gradients of O_2 and CO_2, gills usually require **ventilation**, or movement of the respiratory medium across the respiratory surface.

In the gills of a fish, blood flows through capillaries in a direction opposite that of the flow of water. This arrangement sets up a **countercurrent exchange**, in which the O_2 diffusion gradient favors the movement of O_2 into the blood along the entire length of the capillary.

Tracheal Systems in Insects   **Tracheal systems** are tiny air tubes that branch throughout the body and come into contact with nearly every cell. Thus, an insect's open circulatory system does not function in O_2 and CO_2 transport. In large insects, rhythmic muscle contractions ventilate tracheal systems.

Lungs   **Lungs** are invaginated respiratory surfaces restricted to one location. A circulatory system transports gases between the lungs and the body.

The lungs of mammals are located in the thoracic cavity. Air entering through the nostrils is filtered, warmed, humidified, and smelled in the nasal cavity. Air passes through the pharynx and enters the respiratory tract through the glottis. The **larynx** moves up and tips the epiglottis over the glottis when food is swallowed. The larynx functions to produce sound in most mammals as exhaled air vibrates a pair of vocal folds (called vocal cords in humans). The **trachea**, or windpipe, usually branches into two **bronchi**, which then branch repeatedly into **bronchioles** within each lung. Ciliated, mucus-coated epithelium lines the major branches of the respiratory tree.

Alveoli are clustered air sacs encased in webs of capillaries at the tips of the tiniest bronchioles. Gas exchange takes place across the thin, moist epithelium of each alveolus. These tiny sacs are coated with **surfactants**, which decrease surface tension and help keep the sacs from sticking shut.

INTERACTIVE QUESTION   42.9

What causes *respiratory distress syndrome (RDS)* in infants?

42.6   Breathing ventilates the lungs

Vertebrate lungs are ventilated by **breathing**, the alternate inhalation and exhalation of air.

How an Amphibian Breathes   A frog uses **positive pressure breathing**. It lowers the floor of the oral cavity, drawing air into the expanding cavity; raising the floor of the oral cavity then pushes air into the lungs. Exhalation occurs primarily through elastic recoil.

How a Bird Breathes   Birds have air sacs that act as bellows to maintain air flow through the lungs. The air sacs and lungs are ventilated through a circuit that includes a one-way passage through tiny, gas-exchange channels called *parabronchi*. Two cycles of inhalation and exhalation are required to pass air through the entire system, in which fresh air and stale air do not mix.

How a Mammal Breathes   Mammals ventilate their lungs by **negative pressure breathing**. Contraction of the rib muscles and the **diaphragm** expands the chest cavity and increases the volume of the lungs. The reduction in air pressure in this increased volume allows atmospheric pressure to force air into the lungs. Relaxation of the rib muscles and diaphragm compresses the lungs, increasing the pressure and forcing air out.

Tidal volume is the volume of air inhaled and exhaled by an animal during normal breathing. The maximum volume during forced breathing is called **vital capacity**. The **residual volume** is the air that remains in the alveoli and lungs after forceful exhaling.

Control of Breathing in Humans   Breathing is under automatic regulation by neural circuits that form a pair of *breathing control centers* in the medulla oblongata. (The pons, near the medulla, also plays a role in regulating breathing.) A negative-feedback mechanism involving sensors in the lungs and control circuits in the medulla prevent overexpansion of the lungs.

When sensors in the medulla detect a drop in the pH of the cerebrospinal fluid, which is caused by an increase in blood CO_2 concentration, its control circuits increase the depth and rate of breathing. Although breathing centers respond primarily to CO_2 levels, they are alerted by oxygen sensors in the aorta and carotid arteries that react to severe deficiencies of O_2.

INTERACTIVE QUESTION   42.10

For the following animals, indicate the respiratory medium, respiratory surface, and means of ventilation.

Animal	Respiratory Medium	Respiratory Surface	Means of Ventilation
Fish	a.		
Grasshopper	b.		
Frog	c.		
Human	d.		

42.7 Adaptations for gas exchange include pigments that bind and transport gases

Coordination of Circulation and Gas Exchange
Blood entering the capillaries of the lungs has a lower P_{O_2} and a higher P_{CO_2} than does the air in the alveoli, so O_2 diffuses into the capillaries and CO_2 diffuses out. In the tissue capillaries, pressure differences favor the diffusion of O_2 out of the blood into the interstitial fluid and of CO_2 into the blood.

Respiratory Pigments   In most animals, O_2 is carried in the blood or hemolymph by **respiratory pigments**, which consist of a metal bound to a protein. Copper is the oxygen-binding component in the blue respiratory pigment *hemocyanin* common in arthropods and many molluscs.

Hemoglobin, contained in erythrocytes, is the respiratory pigment of almost all vertebrates. Hemoglobin is composed of four subunits, each of which has a co-factor called a heme group with iron at its center. The binding of O_2 to the iron atom of one subunit induces a change in shape of the other subunits, and their affinity for O_2 increases. Likewise, the unloading of the first O_2 lowers the other subunits' affinity for O_2.

The dissociation curve for hemoglobin shows the relative amounts of O_2 bound to hemoglobin under varying O_2 partial pressures. In the steep part of this S-shaped curve, a slight change in partial pressure will cause hemoglobin to load or unload a substantial amount of O_2. Because of the **Bohr shift** in response to the lower pH of active tissues, hemoglobin releases more O_2 in such tissues.

Most CO_2 is transported in blood as bicarbonate ions. CO_2 enters red blood cells, where it first combines with H_2O to form carbonic acid (catalyzed by carbonic anhydrase) and then dissociates into H^+ and HCO_3^-. Bicarbonate moves into the plasma for transport. The H^+ binds to hemoglobin and other proteins, which minimizes changes in pH during the transport of CO_2. In the lungs, the diffusion of CO_2 out of the blood shifts the equilibrium in favor of the conversion of HCO_3^- back to CO_2, which is unloaded from the blood.

INTERACTIVE QUESTION   42.11

The following graph shows dissociation curves for hemoglobin at two different pH values. Explain the significance of these two curves.

Respiratory Adaptations of Diving Mammals
Special adaptations have enabled some air-breathing mammals to make sustained underwater dives. Weddell seals store twice the amount of O_2 per kilogram of body weight as do humans, mostly by having a larger volume of blood and a higher concentration of **myoglobin**, an oxygen-storing muscle protein.

Oxygen-conserving adaptations during a dive include decreasing the heart rate, rerouting blood to the brain and essential organs, and restricting blood supply to the muscles, which use fermentation to produce ATP during long dives. These diving abilities reflect both short-term physiological responses to the environment and long-term adaptations that have evolved through natural selection.

Word Roots

alveol- = a cavity (*alveolus:* one of the dead-end air sacs where gas exchange occurs in a mammalian lung)

atri- = a vestibule; **ventricul-** = ventricle (*atrioventricular node:* a region of specialized heart muscle tissue between the left and right atria where electrical impulses are delayed before spreading to both ventricles and causing them to contract)

cardi- = heart; **vascul-** = a little vessel (*cardiovascular system:* the closed circulatory system characteristic of vertebrates)

counter- = opposite (*countercurrent exchange:* exchange of a substance or heat between two fluids flowing in opposite directions)

endo- = inner (*endothelium:* the simple squamous layer of cells lining the lumen of blood vessels)

erythr- = red; **poiet-** = produce (*erythropoietin:* a hormone that stimulates the production of erythrocytes. It is secreted by the kidney when body tissues do not receive enough oxygen)

hem- = blood; **-lymph** = clear fluid (*hemolymph:* in invertebrates with an open circulatory system, the body fluid that bathes tissues)

leuk- = white; **-cyte** = cell (*leukocyte:* a blood cell that functions in fighting infections; also called a white blood cell)

my- = muscle (*myoglobin:* an oxygen-storing, pigmented protein in muscle cells)

semi- = half; **lun-** = moon (*semilunar valve:* a valve located at each exit of the heart, where the aorta leaves the left ventricle and the pulmonary artery leaves the right ventricle)

thromb- = a clot (*thrombus:* a fibrin-containing clot that forms in a blood vessel and blocks the flow of blood)

Structure Your Knowledge

1. Identify the labeled structures in the following diagram of a human heart. Draw arrows to trace the flow of blood.

2. Trace the path of a molecule of O_2 from its inhalation into the nasal cavity to its delivery to the kidney.

Test Your Knowledge

MULTIPLE CHOICE: *Choose the one best answer.*

1. A gastrovascular cavity
   a. is a digestive and circulatory chamber in cnidarians and annelids.
   b. pumps fluid through short vessels from which the fluid spreads throughout the body.
   c. functions in both digestion and distribution of nutrients.
   d. has a single opening for ingestion and elimination, but a separate opening for gas exchange.

2. Certain inherited disorders interfere with the functioning of cilia. How would the respiratory system be affected by the loss of ciliary action?
   a. Diffusion would slow down because cilia in the alveoli would no longer produce air currents.
   b. Blood flow would be reduced because cilia would no longer drive the blood through the alveolar capillaries.
   c. The delicate lung tissue would dehydrate because ciliary action would no longer move water through the lungs.
   d. The lungs would become clogged with mucus and debris.

3. In a system with double circulation,
   a. blood is usually pumped at two separate locations as it circulates through the body.
   b. hemolymph circulates both through a pumping blood vessel and through body sinuses.
   c. blood is pumped to the gas exchange organ, and returning blood is then pumped to the systemic circuit.
   d. there are always two ventricles and often two atria.

4. Which of the following statements is the best explanation for the presence of four-chambered hearts in both birds and mammals?
   a. They shared a common ancestor that had a four-chambered heart.
   b. They are the only vertebrates with double circulation, which requires four heart chambers.
   c. They are both endotherms, and the evolution of efficient circulatory systems supported the high metabolic rate of endotherms.
   d. This is an example of convergent evolution because animals that obtain their O_2 from air require both a pulmonary circuit and a systemic circuit.

5. During diastole,
   a. blood fills the atria and passively flows into the ventricles.
   b. the elastic recoil of the arteries maintains hydrostatic pressure on the blood.
   c. semilunar valves are closed, but atrioventricular valves are open.
   d. all of the above are occurring.

6. Which of the following blood vessels carries deoxygenated blood in human adults?
   a. the aorta
   b. the pulmonary veins
   c. the pulmonary arteries

7. As a general rule, blood leaving the right ventricle of a mammal's heart will pass through how many capillary beds before it returns to the right ventricle?
   a. one
   b. two
   c. three
   d. one or two, depending on whether it passes through a portal vein

8. The nurse tells you that your blood pressure is 112/70. The "70" refers to
   a. your heart rate.
   b. the velocity of blood during diastole.
   c. the systolic pressure from ventricular contraction.
   d. the diastolic pressure from the recoil of the arteries.

9. Blood flows more slowly in arterioles than in arteries because arterioles
   a. have thoroughfare channels to venules that are often closed off.
   b. collectively have a larger cross-sectional area than do arteries.
   c. must provide opportunity for exchange with the interstitial fluid.
   d. are narrower than arteries.

Use the following to answer questions 10–13.
   1. vena cava
   2. left ventricle
   3. pulmonary vein
   4. right atrium
   5. aorta
   6. pulmonary capillaries

10. Which of the following sequences represents the flow of blood through the human body? (Obviously, many structures are not included.)
   a. 1-5-3-6-2-4   c. 1-4-3-6-2-5
   b. 1-2-6-3-4-5   d. 1-4-6-3-2-5

11. In which vessel is blood pressure the highest?
    a. 1   c. 4
    b. 3   d. 5

12. In which vessel is the velocity of blood flow the lowest?
    a. 1   c. 4
    b. 3   d. 6

13. Of these structures, which has the thickest muscle layer?
    a. 1   c. 4
    b. 2   d. 5

14. Exchange between the blood and the tissues of the body occurs in
    a. arterioles.
    b. venules.
    c. capillaries.
    d. portal veins.

15. Which of the following processes would tend to reduce the blood supply to a capillary bed?
    a. release of EPO (erythropoietin)
    b. relaxation of precapillary sphincters
    c. vasodilation
    d. release of endothelin

16. Which of the following changes would you expect following an increase in the concentration of blood proteins?
    a. the swelling of body tissues (edema)
    b. the swelling of lymph glands
    c. an increase in the outward movement of fluid at the arterial end of capillaries
    d. a decrease in lymph flow through the lymphatic system

17. In which of the following vessels would you expect to find valves?
    a. arteries
    b. veins
    c. lymph vessels
    d. both b and c

18. In preparation for a big date, you have cut yourself while shaving. (Your chromosome complement will largely determine the location of the injury.) Which of the following will be most directly involved with stopping the flow of blood?
    a. phagocytic cells
    b. leukocytes
    c. platelets
    d. fibroblasts

19. In countercurrent exchange of respiratory gases,
    a. the flow of fluids in opposite directions maintains a favorable diffusion gradient along the length of an exchange surface.
    b. O_2 is exchanged for CO_2.
    c. double circulation keeps oxygenated and deoxygenated blood separate.
    d. O_2 moves from a region of high partial pressure to one of low partial pressure, but CO_2 moves in the opposite direction.

20. The tracheae of insects
    a. are stiffened with chitinous rings and lead into the lungs.
    b. ramify along capillaries of the circulatory system for gas exchange.
    c. are highly branched and come into contact with almost every cell to effect gas exchange.
    d. provide an extensive, fluid-filled system of tubes for gas exchange.

21. What do the alveoli of mammalian lungs, the gill filaments of fish, and the tracheoles of insects all have in common?
    a. They use a circulatory system to transport absorbed O_2.
    b. Their respiratory surfaces are invaginated.
    c. They function using countercurrent exchange.
    d. They have a large, thin surface area for gas exchange.

22. *Most* of the CO_2 in the blood is carried
    a. as CO_2 dissolved in plasma.
    b. as CO_2 bound to the heme group of hemoglobin.
    c. as CO_2 bound to the globin portion of hemoglobin.
    d. as HCO_3^- dissolved in the plasma.

23. When you hold your breath, which of the following leads to the urge to breathe?
    a. falling O_2
    b. rising CO_2
    c. falling blood pH
    d. both b and c

24. Which of the following mechanisms is *not* involved in increasing the respiration rate?
    a. stretch receptors in the lungs
    b. impulses from the breathing centers in the medulla
    c. severe deficiencies of oxygen
    d. a drop in the pH of cerebrospinal fluid

25. Which of the following statements is the best explanation for why birds can fly over the Himalayas but most humans require bottled oxygen to climb these mountains?

    a. Birds are much smaller and require less O_2.

    b. With a one-way flow of air and efficient ventilation, the lungs of birds extract more O_2 from the air they breathe.

    c. The circulatory system of birds is much more efficient in delivering O_2 to body tissues.

    d. The vital capacity of bird lungs, relative to their size, is much greater than that of humans.

26. Why is more O_2 unloaded from hemoglobin in an actively metabolizing tissue than in a resting tissue, even at the same P_{O_2}.

    a. At the lower pH of an actively metabolizing tissue, hemoglobin has a lower affinity for O_2.

    b. Hemoglobin binds four O_2 due to a shape change caused by the lower pH of active tissues.

    c. More O_2 is bound to myoglobin at lower O_2 concentrations.

    d. More CO_2 is drawn out of tissue cells as it is converted to bicarbonate in red blood cells.

Chapter 43

The Immune System

Key Concepts

43.1 In innate immunity, recognition and response rely on traits common to groups of pathogens

43.2 In adaptive immunity, receptors provide pathogen-specific recognition

43.3 Adaptive immunity defends against infection of body fluids and body cells

43.4 Disruptions in immune system function can elicit or exacerbate disease

Framework

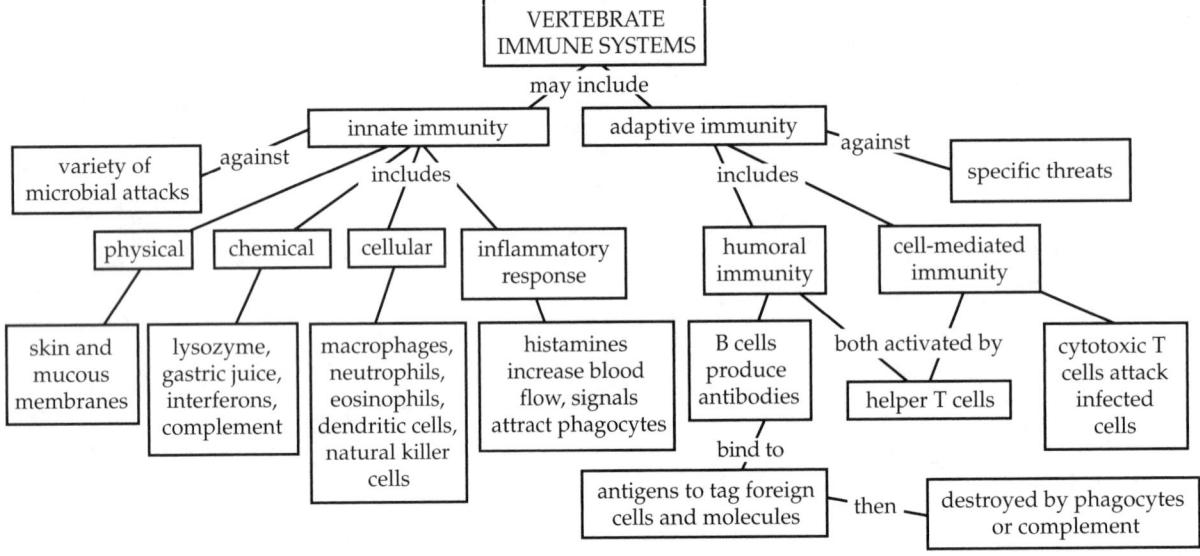

Chapter Review

The **immune system** defends against **pathogens**—disease-causing agents such as bacteria, viruses, and fungi. *Molecular recognition* is the key attribute of the immune system that enables it to identify internal invaders that are nonself. **Innate immunity** includes both external physical barriers and internal defenses. The immune cells involved in innate immunity have a small group of receptor proteins that recognize broad groups of pathogens and can trigger a rapid defensive response. **Adaptive immunity**, also called acquired immunity, is a line of defense in vertebrates in which immune cells react specifically to pathogens. A vast array of adaptive immune receptors enables recognition and response to specific pathogens, and the response is initially slower but enhanced upon reexposure to a pathogen.

43.1 In innate immunity, recognition and response rely on traits common to groups of pathogens

Innate Immunity of Invertebrates The insect exoskeleton composed of chitin provides a protective barrier against pathogens. The intestine is also lined with chitin, and it contains **lysozyme**, which attacks microbial walls. *Hemocytes* are circulating cells that can destroy bacteria by **phagocytosis** or release chemicals such as *antimicrobial peptides* that kill fungi and bacteria. Different classes of pathogens bind with recognition proteins secreted by immune cells. The complex then activates distinct Toll receptors that initiate pathways for the production of antimicrobial peptides effective against that group of pathogens.

INTERACTIVE QUESTION 43.1

a. Briefly describe the insect defense system against infecting RNA viruses.

b. Why does this system not attack the RNA of normal insect cells?

Innate Immunity of Vertebrates In mammals, skin and the mucous membranes lining the digestive, respiratory, urinary, and reproductive tracts provide barrier defenses. The ciliated epithelial lining of the respiratory tract sweeps *mucus* containing trapped microbes upward and out. Lysozyme present in tears, saliva, and mucus kills bacteria. Skin secretions maintain a low pH, which discourages the colonization of many bacteria. The acidity of gastric juice kills most microbes that reach the stomach.

Cellular innate defenses involve specific **Toll-like receptors (TLRs)** that recognize molecules that are common to a set of pathogens, such as double-stranded RNA from certain viruses or lipopolysaccharides on the surface of many bacteria. As in invertebrates, binding to such a receptor triggers phagocytosis.

Neutrophils are phagocytic cells that circulate in the blood. **Macrophages** are larger phagocytic cells that migrate through the body or remain permanently in various organs, such as the spleen and other organs of the lymphatic system.

The primary role of **dendritic cells**, located in tissues in contact with the environment, is to stimulate adaptive immunity. *Eosinophils* attack multicellular parasitic invaders with destructive enzymes. **Natural killer cells** recognize abnormal surface proteins on some virus-infected or cancerous cells and release chemicals that trigger cell death.

Various antimicrobial peptides and proteins are released in response to pathogen recognition. Virus-infected cells produce **interferons**, which stimulate neighboring cells to produce substances that inhibit viral reproduction in those cells. Some white blood cells produce an interferon that activates macrophages.

The **complement system** is a group of about 30 proteins in the blood plasma that, when activated by contact with microbes, may lyse cells, trigger inflammation, or assist in adaptive defenses.

In response to physical injury or pathogen entry, signaling molecules released from the affected area trigger an **inflammatory response**, characterized by redness, swelling, and heat. *Mast cells* in connective tissue release **histamine**, a chemical that triggers dilation and leakiness of blood vessels; activated macrophages release *cytokines*, signaling molecules that recruit neutrophils to the damaged area. Vasodilation and signaling proteins result in the congregation of phagocytes and antimicrobial compounds. *Pus* is an accumulation of fluid containing white blood cells, dead microbes, and cell debris.

A systemic (whole-body) response to an infection may include an increase in the number of circulating white blood cells and a fever. Fevers, triggered by substances released by activated macrophages, may stimulate phagocytosis and speed tissue repair. *Septic shock* is a life-threatening condition resulting from an overwhelming systemic inflammatory response.

Evasion of Innate Immunity by Pathogens Certain bacteria may evade the innate immune system with an outer capsule that interferes with recognition and phagocytosis or by resisting breakdown and growing within the host's cells.

INTERACTIVE QUESTION   43.2

Complete the following table that summarizes the functions of the cells and chemicals of the innate defense mechanisms of mammals.

Cells or Compounds	Functions
Neutrophils	a.
Macrophages	b.
Eosinophils	c.
Dendritic cells	d.
Natural killer cells	e.
Mast cells	f.
Histamine	g.
Lysozyme	h.
Interferons	i.
Complement system	j.
Cytokines	k.

43.2 In adaptive immunity, receptors provide pathogen-specific recognition

Lymphocytes are the key cells of adaptive immunity. **T cells** migrate from the bone marrow to the **thymus** to mature, whereas **B cells** remain in the bone marrow and mature there.

B cells and T cells become activated upon binding with an **antigen**, often a protein or polysaccharide protruding from the surface of a pathogen or a toxin secreted by bacteria. Each B cell or T cell has thousands of identical membrane-bound **antigen receptors** that enable it to recognize a specific antigen.

The small region of an antigen to which an antigen receptor binds is called an **epitope**.

Antigen Recognition by B Cells and Antibodies   Each Y-shaped antigen receptor on a B cell consists of four polypeptide chains: two identical **light chains** and two identical **heavy chains**, linked together by disulfide bridges. Both heavy and light chains have *variable (V) regions* at the ends of the two arms of the Y, which together form two identical antigen-binding sites. The *constant (C) regions* of the molecule vary little among receptors on different B cells. The constant region of the heavy chains includes an anchoring transmembrane region and a cytoplasmic tail.

Activated B cells give rise to cells that secrete **antibodies**, or **immunoglobulins (Ig)**, which are soluble antigen receptors that defend against pathogens. The particular amino acid sequences of the variable regions of an antibody or B cell receptor accounts for the highly specific binding with a particular epitope.

Antigen Recognition by T Cells   A T cell antigen receptor, also anchored by a transmembrane region, consists of one α *chain* and one β *chain*, linked by a disulfide bridge. The variable regions at the tip of the molecule form a single antigen-binding site. T cells recognize small fragments of antigens displayed on the surface of cells. The displaying protein is called a **major histocompatibility complex (MHC) molecule**.

Host cells that are infected by or take in a pathogen cleave antigens into small peptides. In a process called **antigen presentation**, an MHC molecule binds with an *antigen fragment* within the cell and then displays that fragment on the cell's surface, where a T cell receptor can recognize the specific antigen–MHC molecule complex.

INTERACTIVE QUESTION   43.3

Describe the differences between the antigens that B cell antigen receptors recognize and the antigens that T cell antigen receptors recognize.

B Cell and T Cell Development   The four major characteristics of adaptive immunity include the immense diversity of antigen receptors, the self-tolerance of the system, the huge increase in B cells and T cells specific for an antigen following activation, and *immunological memory*, an enhanced response to a previously encountered foreign molecule.

Each person may have more than 1 million different B cell antigen receptors and 10 million different T cell antigen receptors. The genes coding for this remarkable diversity have numerous coding segments that are randomly and permanently rearranged. For example, in the Ig (immunoglobulin) light-chain gene, there are 40 variable *(V)* gene segments and five joining *(J)* gene segments. The *J* segments are followed by a single *C* segment, which codes for the constant region. Early in B cell development, one *V* gene segment is randomly linked to one *J* segment by an enzyme complex called *recombinase*, producing one of 200 possible light chains. These light chains combine with heavy chains, which have an even greater number of possible recombinations. Transcription, splicing, and then translation of such permanently rearranged genes produces receptors in different B cells (and T cells) that provide an enormous diversity of antigen-binding sites.

In the bone marrow and thymus, respectively, the antigen receptors of maturing B and T cells are tested for self-reactivity. Lymphocytes with receptors specific for the body's own molecules are either inactivated or destroyed by *apoptosis*. This critical *self-tolerance* means that normally the body contains no mature lymphocytes that react against self components.

When an antigen interacts with antigen receptors on B cells or T cells that are specific for epitopes of that antigen, those particular lymphocytes are activated to divide repeatedly into a clone of identical cells. Some daughter cells develop into short-lived **effector cells**, either *plasma cells*, which secrete antibodies, or helper T cells and cytotoxic T cells. Other cells in the clone develop into long-lived **memory cells**. By this process of **clonal selection**, a small number of cells is selected by their interaction with a specific antigen to produce thousands of identical cells keyed to that particular antigen.

The body mounts a **primary immune response** upon a first exposure to an antigen. About 10 to 17 days are required for selected lymphocytes to proliferate and yield the maximum response. Should the body reencounter the same antigen, the resulting **secondary immune response** is more rapid, effective, and prolonged. The long-lived T memory cells and B memory cells are responsible for this immunological memory. The secondary immune response provides long-term protection against a previously encountered pathogen.

43.3 Adaptive immunity defends against infection of body fluids and body cells

The **humoral immune response** involves the production of antibodies that help destroy toxins and pathogens in the blood and lymph. The **cell-mediated immune response** involves specialized T cells that identify and destroy infected body cells.

Helper T Cells: Activating Adaptive Immunity
Both humoral and cell-mediated immune responses are triggered by **helper T cells**. Helper T cells recognize specific antigens displayed by **antigen-presenting cells**.

Most body cells have only class I MHC molecules, but antigen-presenting cells have both class I and class II MHC molecules. These antigen-presenting cells display the antigen fragments in class II MHC molecules. A helper T cell antigen receptor binds to the antigen fragment, and a surface protein called CD4 enhances the binding by attaching to the class II MHC molecule. Signaling between the two cells via cytokines results in the proliferation of activated helper T cells. These helper T cells secrete cytokines that activate both B cells involved in humoral immunity and cytotoxic T cells, which are part of cell-mediated immunity.

The interactions of dendritic cells and macrophages with helper T cells result in a clone of activated helper T cells. A B cell presents antigens to an already activated helper T cell, which in turn activates the B cell.

INTERACTIVE QUESTION   43.5

Label the components in the following diagram that shows a helper T cell being activated by interaction with an antigen-presenting cell and the helper T cell's central role in activating both humoral and cell-mediated immunity.

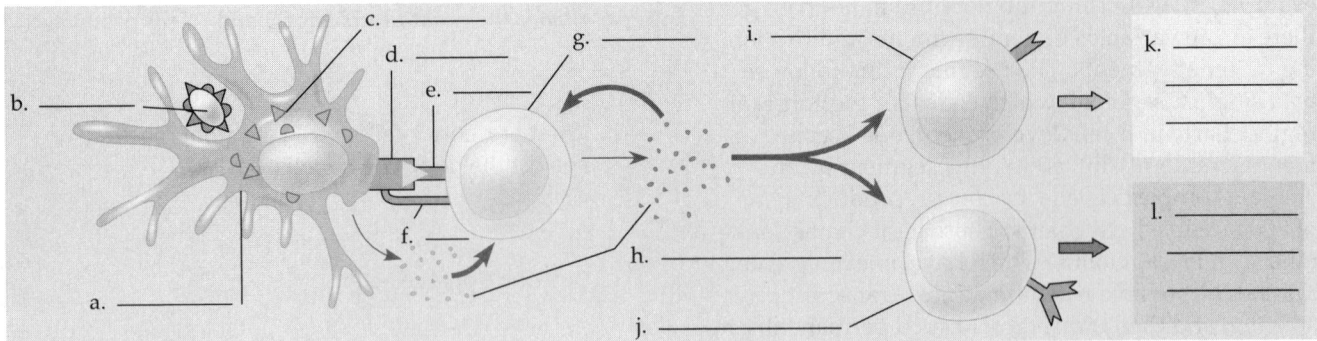

B Cells and Antibodies: A Response to Extracellular Pathogens   Upon first binding antigen, a B cell takes in a few foreign molecules by receptor-mediated endocytosis and presents antigen fragments in its class II MHC molecules to a helper T cell that had been activated by binding with a macrophage or dendritic cell displaying the same antigen. Aided by both direct contact with a helper T cell and the cytokines it secretes, an activated B cell proliferates into memory B cells and antibody-secreting plasma cells.

INTERACTIVE QUESTION   43.6

Describe each of the following mechanisms by which antibodies mediate the disposal of antigens.

a. *neutralization*

b. *opsonization*

c. *complement* activation and formation of *membrane attack complex*

When antibodies bind to viral proteins that appear on the surface of an infected cell, natural killer cells may be recruited.

There are five types of constant regions, creating five *classes* of antibodies. One is the membrane-bound B cell antigen receptor. The other four classes are soluble antibodies, which can be found in various body fluids.

Cytotoxic T Cells: A Response to Infected Cells
Activation of **cytotoxic T cells** requires signals from nearby helper T cells and interaction with an antigen-presenting cell. An infected cell produces fragments of foreign proteins and displays them in its class I MHC molecules. A cytotoxic T cell receptor (with the aid of accessory protein CD8) recognizes and binds to a class I MHC-fragment-complex. The T cell then secretes proteins that destroy the infected cell.

Summary of the Humoral and Cell-Mediated Immune Responses   Adaptive immunity defends against pathogens in the blood and lymph with antibodies secreted by B cells (humoral immune response) and against intracellular pathogens and cancer cells with cytotoxic T cells (cell-mediated immune response). Both responses depend on antigen-presenting cells activating helper T cells, which secrete cytokines that activate cytotoxic T cells and B cells.

INTERACTIVE QUESTION   43.7

a. What does an activated helper T cell release, and what do these substances do?

b. What does a cytotoxic T cell attached to an infected body cell release, and what do these substances do?

Immunization An immune response can be stimulated through **immunization** (or *vaccination*), the artificial introduction of antigens. Preparations of antigens called *vaccines* may contain inactivated toxins, weakened or killed pathogens, or genes for microbial proteins. The immune response triggered by the vaccine results in adaptive immunity and memory cell formation. Through the use of vaccines, many diseases have been controlled, and one (smallpox) has been eliminated.

Active and Passive Immunity **Active immunity** results when the body responds to an infection and develops immunological memory specific to that pathogen. In **passive immunity**, temporary immunity is provided by antibodies supplied through the placenta to a fetus (IgG antibodies), through breast milk (IgA antibodies), or through the introduction of antibodies from an immune individual.

Antibodies as Tools A technique for making **monoclonal antibodies** derived from a single clone of B cells can supply quantities of identical antibodies, which can be used in medical diagnosis and treatment.

Immune Rejection The immune response to the carbohydrates on red blood cells that determine ABO blood groups must be considered in blood transfusions. Antibodies to blood group antigens other than the individual's own antigens arise in response to normal bacterial flora. These antibodies circulate in the blood plasma and will produce a devastating reaction to transfused blood cells that possess matching antigens.

Because human MHC molecules are produced from a large number of genes and allelic forms, MHC proteins are highly variable and likely to differ among individuals. When tissues are transplanted, donor MHC molecules will be recognized as foreign and result in an immune response and the phenomenon of rejection. The use of closely related donors, as well as drugs that suppress immune responses, helps to minimize rejection.

INTERACTIVE QUESTION 43.8

Individuals with blood type O produce neither A nor B antigens on their red blood cells. Briefly explain why such individuals are sometimes called "universal donors" with regard to blood transfusions.

43.4 Disruptions in immune system function can elicit or exacerbate disease

Exaggerated, Self-Directed, and Diminished Immune Responses Allergies are hypersensitivities to certain antigens called *allergens*. IgE antibodies created in response to initial exposure to an allergen may bind to mast cells. When later allergens bind and cross-link to these cell surface antibodies, the mast cells release histamines. The resulting inflammatory response may include sneezing, a runny nose, and difficulty in breathing due to smooth muscle contractions. Antihistamines are drugs that reduce symptoms by blocking receptors for histamine. *Anaphylactic shock* is a severe allergic response in which the rapid release of inflammatory chemicals from mast cells causes abrupt dilation of peripheral blood vessels and a drop in blood pressure, as well as constriction of bronchioles.

Sometimes the immune system turns against self, leading to **autoimmune diseases**, such as *lupus*, *rheumatoid arthritis*, type 1 diabetes mellitus, and multiple sclerosis. These diseases result from a loss of self-tolerance and are influenced by gender, genetics, and environment.

Exercising to exhaustion and psychological stress have been shown to impair immune system function.

An immunodeficiency is the inability of the immune system to protect against pathogens. A genetic or developmental defect in the immune system is called an *inborn immunodeficiency*; a defect that arises later in response to chemical or biological agents is called an *acquired immunodeficiency*. In the rare congenital disease known as severe combined immunodeficiency (SCID), lymphocytes (and therefore adaptive immune responses) are lacking. Acquired immunodeficiency may be the result of drugs used to treat autoimmune diseases or to suppress transplant rejection.

Evolutionary Adaptations of Pathogens That Underlie Immune System Avoidance Some viruses and parasites evade the body's defenses through *antigenic variation*, a change in their surface epitopes so that the immune system cannot develop an immunological memory of the pathogen. Examples include the parasite that causes sleeping sickness and the influenza virus.

Some viruses enter an inactive state called *latency*, in which viral DNA remains in a host cell's nucleus. Should conditions change, the virus can replicate and spread to neighboring cells or to new hosts.

The infectious agent responsible for **acquired immunodeficiency syndrome (AIDS)** is the **human immunodeficiency virus (HIV)**. HIV infects helper T cells (by binding with surface CD4 accessory protein), macrophages, and brain cells. HIV RNA is reverse-transcribed into DNA, which is integrated into the host cell genome, from which it directs production of new viral particles. HIV escapes the immune system through both antigenic variation (when it mutates rapidly during replication) and latency (when it integrates into host cell DNA). As HIV infection kills helper T cells, both humoral and cell-mediated immune responses are impaired. Individuals with AIDS are highly susceptible to opportunistic infections and cancers that take advantage of a suppressed immune system.

HIV is transmitted by the transfer of body fluids, such as blood, semen, and breast milk, containing viral particles or infected cells. Although new drug combinations are unable to cure HIV infection, they are slowing the progression to AIDS.

INTERACTIVE QUESTION   43.9

a. Why is AIDS such a deadly disease?

b. Why has AIDS proved so difficult to prevent and cure?

Cancer and Immunity   The incidence of certain cancers increases when the immune system is impaired. A healthy immune system may be able to recognize and defend against the 15–20% of human cancers that involve viruses. Vaccines are now available against hepatitis B virus, which is linked to liver cancer, and human papillomavirus (HPV), which is associated with cervical cancer.

Word Roots

anti- = against; **-gen** = producing substance (*antigen:* a substance that elicits an immune response by binding to receptors of B cells or T cells)

epi- = over; **top-** = place (*epitope:* a small, accessible region of an antigen to which an antigen receptor or antibody binds)

immun- = safe, free; **glob-** = globe, sphere (*immunoglobulin:* a protein secreted by plasma cells that binds to a particular antigen; also called an antibody)

macro- = large; **-phage** = one that eats (*macrophage:* a phagocytic cell present in many tissues that functions in innate immunity by destroying microbes, and in adaptive immunity as an antigen-presenting cell)

neutr- = neutral; **-phil** = loving (*neutrophil:* the most abundant type of white blood cell; neutrophils are phagocytic and tend to self-destruct as they destroy foreign invaders)

Structure Your Knowledge

This chapter contains a wealth of information that is probably fairly new to you. If you take a little time and pull out the key players of the immune system and organize them first into very basic concept clusters and then develop more interrelated concept maps, you will find that this information is both understandable and fascinating.

1. Create a concept map outlining adaptive immunity; include the cells involved in the humoral and cell-mediated responses and their functions.
2. Describe the structure of an antibody molecule, and relate this structure to its function.
3. Although innate and adaptive immunity were presented separately, these defense mechanisms interact in several ways. Describe a few of the chemical and cellular components they share.

Test Your Knowledge

MULTIPLE CHOICE: *Choose the one best answer.*

1. Which of the following statements best describes an insect's immune system?
   a. Insects rely on the barrier defense of an exoskeleton and lining of the intestine, both composed of chitin.
   b. Hemocytes can carry out phagocytosis of bacteria and foreign substances.
   c. Insects can produce different antimicrobial peptides in response to the binding of molecules from a broad class of pathogens to their Toll receptors.
   d. All of the above are part of an insect's innate immunity.

2. Which of the following defense mechanisms is *incorrectly* paired with its function?
   a. fever—may stimulate phagocytosis
   b. histamine—causes blood vessels to dilate
   c. cytokines—attract phagocytes in the inflammatory response
   d. lysozyme—attacks the cell wall of viruses

3. Which of the following cells would release interferon?
   a. a macrophage that has become an antigen-presenting cell
   b. an injured endothelial cell of a blood vessel
   c. a cell infected by a virus
   d. a mast cell that has bound an antigen

4. The process of clonal selection accounts for
   a. the ability of natural killer cells to detect and eliminate cancerous cells.
   b. the production of large numbers of lymphocytes with identical antigenic specificity.
   c. the ability of the cellular innate defenses to attack foreign cells.
   d. the antigenic specificity of lymphocytes.

5. Which of the following statements correctly describes the main difference between innate immunity and adaptive immunity?
   a. Innate immunity responds only to free pathogens in a localized area; adaptive immunity responds only to pathogens that have entered body cells.
   b. Innate immunity involves only leukocytes, whereas adaptive immunity involves only lymphocytes.
   c. Innate immunity relies on phagocytes to destroy pathogens, whereas adaptive immunity does not involve phagocytes.
   d. Innate immunity recognizes molecules common to a set of pathogens, whereas adaptive immunity reacts to specific microbes on the basis of their unique antigens.

6. Which of the following destroys a target cell by phagocytosis?
   a. a neutrophil
   b. a cytotoxic T cell
   c. a natural killer cell
   d. complement proteins

7. Antibodies are proteins
   a. or polysaccharides usually found on the surface of invading bacteria.
   b. embedded in T cell membranes.
   c. circulating in the blood that may tag foreign cells for phagocytosis or destruction by complement.
   d. that are predominantly found in lymph nodes and other organs of the lymphatic system.

8. Neutrophils are recruited to a site of inflammation by
   a. histamine released from mast cells.
   b. cytokines released from macrophages.
   c. antibodies produced by plama cells.
   d. interferon produced by infected body cells.

9. A secondary immune response is more rapid and effective than a primary immune response because
   a. the second response is an active immunity, whereas the primary one was a passive immunity.
   b. helper T cells are available to activate B cells and T cells.
   c. memory cells respond to the pathogen and rapidly proliferate into effector cells.
   d. chemical signals cause the more rapid accumulation of phagocytic cells.

10. How does the vertebrate immune system come to produce so many different antibodies?
    a. Vertebrates inherit millions of different antibody genes.
    b. The genes for the polypeptide chains in an antibody are generated by recombination among many gene sequences.
    c. Meiotic recombination and independent assortment lead to new antibody variations.
    d. The immune system creates new B cells each time it is exposed to a new antigen.

11. Major histocompatibility complex molecules
    a. are involved in the ability to distinguish self from nonself.
    b. may trigger T cell responses after transplant operations.
    c. are a collection of cell surface proteins that present antigen fragments on infected cells.
    d. are or do all of the above.

12. In opsonization,
    a. antibodies coat microorganisms and help phagocytes bind to and engulf the foreign cell.
    b. a set of complement proteins lyses a hole in a foreign cell's membrane.
    c. antibodies coat binding sites of viruses.
    d. a flood of histamines is released and may result in anaphylactic shock.

13. A transfusion of type A blood into a person who has type AB blood would result in

    a. the recipient's anti-A antibodies reacting with the donated red blood cells.

    b. enzymes in the recipient's blood cleaving the A carbohydrates from the donated red blood cells.

    c. probably no reaction in that the recipient has neither anti-A nor anti-B antibodies.

    d. the introduced blood cells being destroyed by innate defense mechanisms.

14. Plasma cells are committed to large-scale production of antibodies for export. You would expect these cells to contain

    a. many mitochondria.

    b. an extensive network of rough endoplasmic reticulum.

    c. an extensive network of smooth endoplasmic reticulum.

    d. multiple nucleoli.

15. Which of the following best describes the relationship between antibodies and complement?

    a. They are both coded for by genes that have hundreds of alleles.

    b. They are both involved in innate defenses.

    c. Antibodies bound to antigens on a pathogen's membrane may activate complement proteins to form a membrane attack complex.

    d. Complement proteins tag foreign cells; antibodies disable these marked cells by neutralization.

16. In an adaptive immune response, a dendritic cell

    a. binds to an accessory protein on cytotoxic T cells to activate their production of perforin.

    b. destroys antibody-labeled cells by phagocytosis.

    c. activates humoral and cell-mediated immunity by releasing interferons after engulfing a virus.

    d. presents peptide antigens of an engulfed pathogen in its class II MHC molecules to helper T cells, and releases cytokines.

17. Which of the lists in choices a–d places the following steps in the helper T cell activation of cell-mediated and humoral immunity in the correct order?

    1. Macrophage and helper T cell secrete cytokines.

    2. Macrophage engulfs pathogen and presents antigen in class II MHC.

    3. Plasma cells secrete antibodies, and cytotoxic T cells attack cells with class I MHC molecule-antigen complex.

    4. Helper T cell receptor recognizes class II MHC molecule-antigen complex.

    5. Activated B cells produce plasma cells and memory cells, and activated T cells produce cytotoxic T cells and memory cells.

       a. 1, 3, 5, 2, 4

       b. 2, 4, 1, 5, 3

       c. 5, 2, 4, 1, 3

       d. 2, 1, 4, 5, 3

18. What is the key advantage to the use of monoclonal antibodies as medicines?

    a. Because we already have antibodies in our blood these molecules are not recognized as foreign.

    b. They are specific for a single antigen, or even for a specific epitope on an antigen.

    c. Because they are produced by living cells, they are more natural and healthful than artificially synthesized drugs.

    d. They are inexpensive to produce, so they can be widely distributed.

19. Which of the following statements about cell-mediated immunity is *incorrect?*

    a. It may be induced by vaccination as a form of active immunity.

    b. It protects against pathogens that have invaded body cells as well as against cancer cells.

    c. It is mounted by lymphocytes that have matured in the thymus.

    d. It requires recognition of class II MHC molecule–antigen complexes on infected body cells.

20. In which of the following circumstances would a B cell display antigens to a T cell?

   a. A B cell takes in a few antigen molecules and displays them in class II MHC molecules to activated helper T cells.

   b. A B cell engulfs bacteria and displays bacterial peptide antigens in class II MHC molecules to a helper T cell.

   c. After being infected by a virus, a B cell displays viral peptides it has synthesized in its class I MHC molecules to a cytotoxic T cell.

   d. Both **a** and **c** are possible ways that a B cell could display antigens to a T cell.

21. Which of the following conditions is *not* considered a disease or malfunction of the immune system?

   a. MHC-induced transplant rejection

   b. SCID (severe combined immunodeficiency)

   c. lupus, multiple sclerosis, and type I diabetes

   d. allergic anaphylactic shock

22. From which of the following conditions would an AIDS patient be *least* likely to suffer?

   a. Kaposi's sarcoma or other cancers

   b. rheumatoid arthritis

   c. pneumonia

   d. yeast infections of mucous membranes

Chapter 44

Osmoregulation and Excretion

Key Concepts

44.1 Osmoregulation balances the uptake and loss of water and solutes

44.2 An animal's nitrogenous wastes reflect its phylogeny and habitat

44.3 Diverse excretory systems are variations on a tubular theme

44.4 The nephron is organized for stepwise processing of blood filtrate

44.5 Hormonal circuits link kidney function, water balance, and blood pressure

Framework

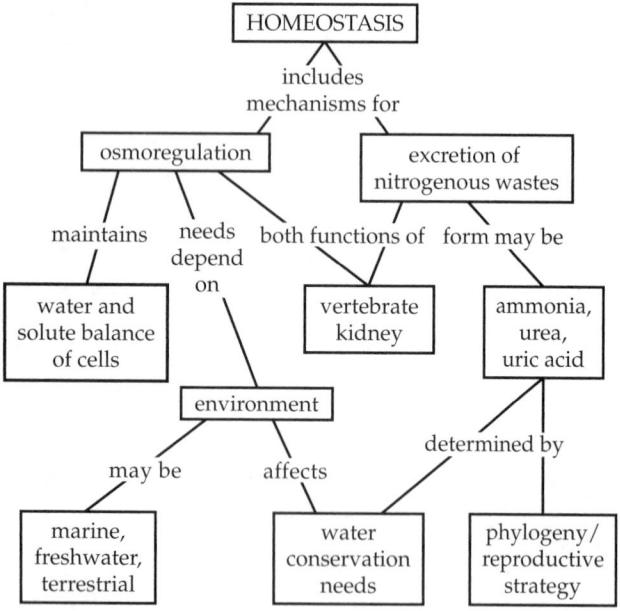

Chapter Review

Animals are able to tolerate the osmoregulatory challenges in their external environments by regulating solute concentrations and water balance

(**osmoregulation**) in their internal environments. Animals must also dispose of *nitrogenous* (nitrogen-containing) wastes through the process of **excretion**.

44.1 Osmoregulation balances the uptake and loss of water and solutes

Osmosis and Osmolarity Osmosis is the diffusion of free water across a selectively permeable membrane separating two solutions that differ in **osmolarity**—total solute concentration expressed as moles of solute per liter (mOsm/L). *Isoosmotic* solutions are equal in osmolarity, and no net osmosis occurs between them. There is a net flow of water from a *hypoosmotic* (more dilute) to a *hyperosmotic* (more concentrated) solution.

Osmotic Challenges and Mechanisms **Osmoconformers** are isoosmotic with their marine environment. **Osmoregulators** regulate their internal osmolarity. Osmoregulators must get rid of excess water if they live in a hypoosmotic habitat, or they must take in water to offset osmotic loss if they inhabit a hyperosmotic environment.

Most animals are *stenohaline*, able to tolerate only small changes in external osmolarity. Animals that are *euryhaline* can survive large differences in their osmotic environments.

Most marine invertebrates are osmoconformers, although they may regulate specific solutes. Many marine vertebrates are osmoregulators. Marine bony fishes gain salt from food and from the large quantities of seawater they must drink to replace the water they lose by osmosis. Ingested salts are excreted through their gills and in the scanty urine produced by their kidneys.

A shark's osmolarity is slightly higher than that of seawater because it maintains in its tissues a high concentration of urea as well as trimethylamine oxide (TMAO), which protects proteins from the damaging effects of urea. Water that enters a shark's body by osmosis or in food, and salt that diffuses in through its gills, are disposed of in urine.

Unlike a marine bony fish, a freshwater bony fish certainly doesn't "drink like a fish." In fact, it hardly drinks at all. Like all freshwater animals, it constantly takes in water by osmosis and loses salts by diffusion.

To compensate, freshwater fishes excrete large quantities of dilute urine and gain salts from food or by active uptake across the gills.

Some animals are capable of **anhydrobiosis**, surviving dehydration, or *desiccation*, in a dormant state. Anhydrobiotic roundworms produce large quantities of trehalose, a sugar that replaces water around membranes and proteins during dehydration. Many insects survive freezing by producing trehalose.

Adaptations to prevent dehydration in terrestrial animals include water-impervious coverings and behavioral adaptations such as nocturnal lifestyles. Water is lost in urine and feces and through evaporation and is gained by drinking and eating moist food and by metabolic production during cellular respiration.

Energetics of Osmoregulation   Osmoregulation is energetically costly because animals must actively transport solutes to maintain osmotic gradients needed to gain or lose water. The energy cost is related to the difference in osmolarity between an animal and its environment, and the body fluid composition of many animals is adapted to the salinity of their habitats.

INTERACTIVE QUESTION   44.1

Indicate whether the following animals are isoosmotic, hyperosmotic, or hypoosmotic to their environment and identify their osmoregulatory mechanisms.

Animal	Osmotic Relation to Environment	Osmoregulatory Mechanisms
a. Marine invertebrate		
b. Shark		
c. Marine bony fish		
d. Freshwater bony fish		

Transport Epithelia in Osmoregulation   Animals regulate the solute composition of the fluid that bathes body cells, whether it is hemolymph or interstitial fluid. The movement of specific solutes across a **transport epithelium** is involved in both osmotic regulation and the disposal of metabolic wastes. Transport epithelia are usually arranged in tubular networks that provide large surface areas for exchange. Nasal glands in marine birds utilize countercurrent exchange in removing salt from the blood and producing a concentrated salt secretion.

44.2   An animal's nitrogenous wastes reflect its phylogeny and habitat

Forms of Nitrogenous Waste   **Ammonia**, a small and toxic molecule, is produced when proteins and nucleic acids are metabolized and their amino groups removed. Aquatic animals can excrete nitrogenous wastes as ammonia because it easily diffuses through membranes into the surrounding water.

Most terrestrial animals and many marine species excrete **urea**, a much less toxic compound than ammonia. Urea can be tolerated in a more concentrated form and excreted with less loss of water. Ammonia and carbon dioxide are combined in the vertebrate liver to produce urea. Both conversion of ammonia to urea and conversion of urea to uric acid are energetically expensive processes.

Insects, land snails, and many reptiles, including birds, produce **uric acid**, a compound of low solubility in water that can be excreted as a semisolid with very little water loss. Its synthesis, however, is energetically expensive.

The Influence of Evolution and Environment on Nitrogenous Wastes   The form of an animal's nitrogenous wastes relates to its habitat. Terrestrial turtles excrete mainly uric acid, whereas aquatic turtles excrete both urea and ammonia. The mode of reproduction influenced which form of nitrogenous waste product evolved in various animal groups; this is particularly true for amniotes that lay shelled eggs.

a. Of two animals of the same size, which would produce more nitrogenous waste:

an endotherm or an ectotherm?

an herbivore or a carnivore?

Explain your answers.

b. Vertebrates that produce shelled eggs excrete _____.

Mammals produce _____.

What adaptive advantage do these types of nitrogenous wastes provide?

44.3   **Diverse excretory systems are variations on a tubular theme**

Excretory Processes   During **filtration**, water and small solutes are forced out of the blood or body fluids across a transport epithelium into an excretory tubule, generally by hydrostatic pressure. Two mechanisms transform this **filtrate** to urine: Valuable solutes are returned from the filtrate by selective **reabsorption**, and excess ions and toxins are added to the filtrate by selective **secretion**. The osmotic movement of water into or out of the filtrate follows the direction in which solutes are pumped. The processed filtrate is released from the body as urine by excretion.

Survey of Excretory Systems   The **protonephridia** of flatworms are branched systems of closed tubules. Water and solutes from the interstitial fluid pass into flame bulbs at the ends of the tubules, propelled by cilia. Urine is excreted into the external environment. The protonephridia of freshwater flatworms are primarily osmoregulatory. In parasitic flatworms, however, protonephridia function mainly in the excretion of nitrogenous wastes.

Metanephridia are found in most annelids. They occur in pairs in each segment of an earthworm. An open ciliated funnel collects coelomic fluid, which then moves through a tubule encased in capillaries. The transport epithelium of the tubule reabsorbs most solutes, which then reenter the blood. Hypoosmotic urine, consisting of nitrogenous wastes and excess water absorbed by osmosis, exits to the outside.

Malpighian tubules are excretory organs in insects and other terrestrial arthropods. The initial filtration step common to excretory processes does not occur in this system. These blind sacs, which open into the digestive tract, are lined by transport epithelia that secrete solutes from the hemolymph into the tubule. The fluid passes into the rectum, where most of the solutes are pumped back into the hemolymph, and water follows by osmosis. In this water-conserving system, uric acid is eliminated as dry matter along with the feces.

The numerous excretory tubules of most vertebrates are associated with a network of capillaries and arranged into compact kidneys. The kidneys and the structures that carry urine out of the body constitute the vertebrate excretory system.

Which of the three excretory processes is *least* selective?

Exploring the Mammalian Excretory System (Figure 44.12)   Blood enters each of the pair of bean-shaped **kidneys** through a renal artery and leaves by way of a renal vein. Urine exits each kidney through a **ureter** and is temporarily stored in the **urinary bladder**. Urine exits the body through the **urethra**.

The outer **renal cortex** and inner **renal medulla** of the kidney are packed with excretory tubules and blood vessels. After the filtrate is processed and most water is reabsorbed into the blood, urine collects in the inner **renal pelvis**. In the human kidney, 85% of the **nephrons** (the functional units of the kidney) are **cortical nephrons** located almost entirely in the renal cortex. **Juxtamedullary nephrons** extend into the renal medulla and enable the formation of hyperosmotic urine.

A nephron consists of a long tubule, at the blind end of which is a cuplike **Bowman's capsule** enclosing a ball of capillaries called the **glomerulus**. Blood pressure forces fluid from the blood into Bowman's capsule. The resulting filtrate passes through the **proximal tubule**; the **loop of Henle**, consisting of a descending limb and an ascending limb; and the **distal tubule**. **Collecting ducts** receive processed filtrate from many nephrons and pass it to the renal pelvis.

An *afferent arteriole* supplying each nephron subdivides to form the capillaries of the glomerulus, which then converge to form an *efferent arteriole*. This arteriole divides into a second network of **peritubular capillaries** that surrounds the proximal and distal tubules. The **vasa recta** includes descending and ascending capillaries that parallel the loop of Henle.

INTERACTIVE QUESTION   44.4

a. What substances are removed from the blood during filtration?

b. What substances remain in the capillaries?

44.4 The nephron is organized for stepwise processing of blood filtrate

From Blood Filtrate to Urine: A Closer Look   The numbered steps in the following discussion correspond to the figure in Interactive Question 44.5.

1. Reabsorption and secretion both take place in the proximal tubule. Salt (NaCl) moves from the filtrate into the cells of the proximal tubule, which then actively transport Na^+ across the membrane into the interstitial fluid; Cl^- is transported passively out to balance the positive charge; and water follows by osmosis. Salt and water diffuse from the interstitial fluid into the peritubular capillaries. Glucose, amino acids, K^+, and the buffer bicarbonate (HCO_3^-) are reabsorbed either actively or passively. The transport epithelium secretes H^+, ammonia (as a buffer to the H^+), and drugs and poisons (processed by the liver and delivered via peritubular capillaries) into the filtrate.

2. The cells of the descending limb of the loop of Henle have numerous **aquaporin** proteins and are thus freely permeable to water. They are not permeable to NaCl or other small solutes. The interstitial fluid is increasingly hyperosmotic toward the inner medulla, so water exits by osmosis, leaving behind a small volume of filtrate with a high solute concentration.

3. The cells of the ascending limb of the loop of Henle are not permeable to water but are permeable to salt, which diffuses out of the thin lower segment of the loop, increasing the osmolarity of the medulla. NaCl is actively transported out of the thick upper portion of the ascending limb. As salt leaves, the filtrate becomes less concentrated, but filtrate volume is not increased.

4. The distal tubule regulates K^+ secretion and NaCl reabsorption, and it helps to regulate pH by secreting H^+ and reabsorbing HCO_3^-.

5. The pair of human kidneys processes about 180 L of filtrate each day to produce about 1.5 L of urine. As the filtrate moves along the collecting duct through the osmotic gradient of the medulla, more and more water exits by osmosis through aquaporin channels. Urea diffuses out of the lower portion of the duct and helps to maintain the osmotic gradient in the medulla. When producing dilute urine, the transport epithelium actively reabsorbs NaCl and withdrawal of aquaporin channels from the plasma membrane prevents water from following by osmosis.

Review these steps by completing Interactive Question 44.5.

INTERACTIVE QUESTION　44.5

In the following diagram of a nephron and collecting duct, label the parts on the indicated lines. Review what is happening in steps 1 to 5 and label the arrows to indicate the movement of NaCl, water, nutrients, K^+, HCO_3^-, H^+, NH_3, and urea into or out of the tubule.

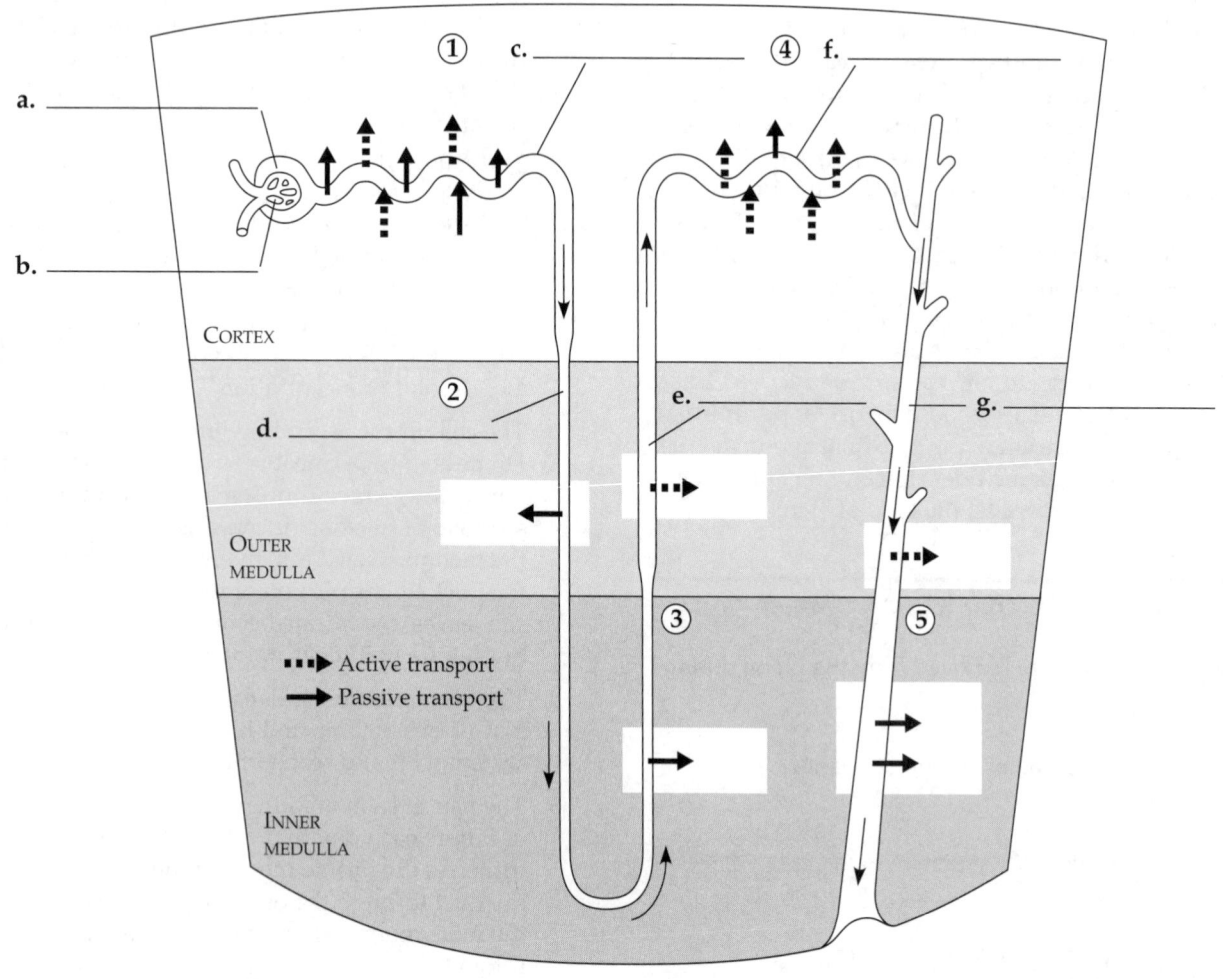

Solute Gradients and Water Conservation The osmolarity gradient in the renal medulla involves energy-requiring active transport by the loops of Henle and collecting ducts and enables the human kidney to produce urine up to four times as concentrated as blood and normal interstitial fluid.

Filtrate leaving Bowman's capsule has an osmolarity of about 300 mOsm/L, the same as blood. Both water and salt are reabsorbed in the proximal tubule; thus, the volume decreases but the osmolarity remains the same. During the filtrate's movement down the descending limb of the loop of Henle, water exits by osmosis, and the filtrate becomes more concentrated. Salt, which is now in high concentration in the filtrate, diffuses out as the filtrate moves up the salt-permeable but water-impermeable ascending limb—helping to create the osmolarity gradient.

Through a **countercurrent multiplier system**, the thick segment of the ascending limb expends energy to actively transport NaCl from the filtrate and generate the osmotic gradient in the medulla.

The vasa recta also consists of descending and ascending vessels. As blood flows down through the osmotic gradient in the inner medulla, water leaves the descending vessel by osmosis and salt enters; as the ascending vessel carries blood back up toward the cortex, water reenters and salt diffuses out. Thus the capillaries can carry nutrients and other supplies to the medulla without disrupting the osmolarity gradient.

The filtrate is hypoosmotic to body fluids after salt is pumped out of the ascending limb. The filtrate makes one final pass through the medulla, this time in the collecting duct, which is permeable to water and urea but not to salt. Water flows out by osmosis, and as the filtrate becomes more concentrated, urea diffuses into the interstitial fluid, adding to the high osmolarity of the inner medulla. The resulting concentrated urine may achieve an osmolarity as high as 1,200 mOsm/L, isoosmotic to the interstitial fluid of the inner medulla.

Adaptations of the Vertebrate Kidney to Diverse Environments Modifications of nephron structure and function correspond to osmoregulatory requirements in various habitats. Mammals in dry habitats, needing to conserve water and excrete hyperosmotic urine, have exceptionally long loops of Henle. Mammals that spend much of their time in fresh water have nephrons with relatively short loops. The case study of the vampire bat underscores the flexibility possible in mammalian kidney function.

Birds also have juxtamedullary nephrons, but their shorter loops of Henle prevent them from producing urine as hyperosmotic as that of mammals. The production of uric acid is their main means of water conservation. The kidneys of other reptiles have only cortical nephrons, and these animals conserve water through reabsorption in the cloaca and the production of uric acid as their nitrogenous waste.

Freshwater fishes are hyperosmotic to their environment and must excrete large quantities of very dilute urine. Salts are reabsorbed from the filtrate. When in fresh water, frogs and other amphibians actively absorb certain salts across their skin and excrete a dilute urine. When on land, water is reabsorbed from the urinary bladder.

The kidneys of marine bony fishes have fewer and smaller nephrons and excrete very little urine. They function mainly to eliminate Ca^{2+}, Mg^{2+}, and sulfate ions taken in by drinking seawater. Excess Na^+ and Cl^- are secreted by *chloride cells* in the gills.

INTERACTIVE QUESTION 44.6

a. What type of nephron enables the production of hyperosmotic urine?

b. Which animal groups have this type of nephron?

44.5 Hormonal circuits link kidney function, water balance, and blood pressure

Homeostatic Regulation of the Kidney Through nervous and hormonal controls, the mammalian kidney regulates urine volume and osmolarity to maintain the animal's water and salt balance and its blood pressure and blood volume.

When blood osmolarity rises above a set point of 285-295 mOsm/L, osmoreceptor cells in the hypothalamus trigger the release of **antidiuretic hormone (ADH)**, also called *vasopressin*. ADH binds to receptors on cells in the collecting ducts, leading to the insertion of additional aquaporin proteins in their plasma membranes. These water channels increase water permeability of the collecting ducts, and thus water reabsorption from the urine. As blood osmolarity declines, negative feedback decreases the release of ADH from the pituitary. When blood osmolarity is low, little ADH is released, resulting in the production of a large volume of dilute urine, called diuresis.

Alcohol inhibits ADH release and can cause dehydration. Mutations that affect ADH production, ADH receptor action, or aquaporin formation can cause severe dehydration in the condition *diabetes insipidus*.

The *renin-angiotensin-aldosterone system (RAAS)* is a homeostatic feedback circuit that maintains adequate blood pressure and blood volume. The **juxtaglomerular apparatus (JGA)**, located around the afferent arteriole leading to the glomerulus, responds to a drop in blood pressure or volume by releasing renin. The activity of this enzyme leads to the conversion of the plasma protein angiotensinogen to *angiotensin II*. Angiotensin II raises blood pressure (by constricting arterioles) and stimulates the adrenal glands to release *aldosterone*. This hormone stimulates Na^+ and water reabsorption in the distal tubules and collecting duct. Some drugs that treat hypertension target the enzyme angiotensin converting enzyme (ACE), which functions in the activation of angiotensin II.

Both ADH and the RAAS increase water reabsorption, but in response to different problems. ADH counters dehydration by responding to an increase in blood osmolarity. The RAAS responds to a loss of both fluid and salts by sensing a drop in blood volume or pressure.

Atrial natriuretic peptide (ANP) has a mechanism opposite to that of the RAAS. ANP, released by the atria of the heart in response to increased blood volume and pressure, inhibits the release of renin from the JGA, inhibits NaCl reabsorption by the collecting ducts, and reduces the release of aldosterone from the adrenal glands.

INTERACTIVE QUESTION   44.7

The following concept map may look a little overwhelming, but working through it should help organize your understanding of the homeostatic regulation of the kidney.

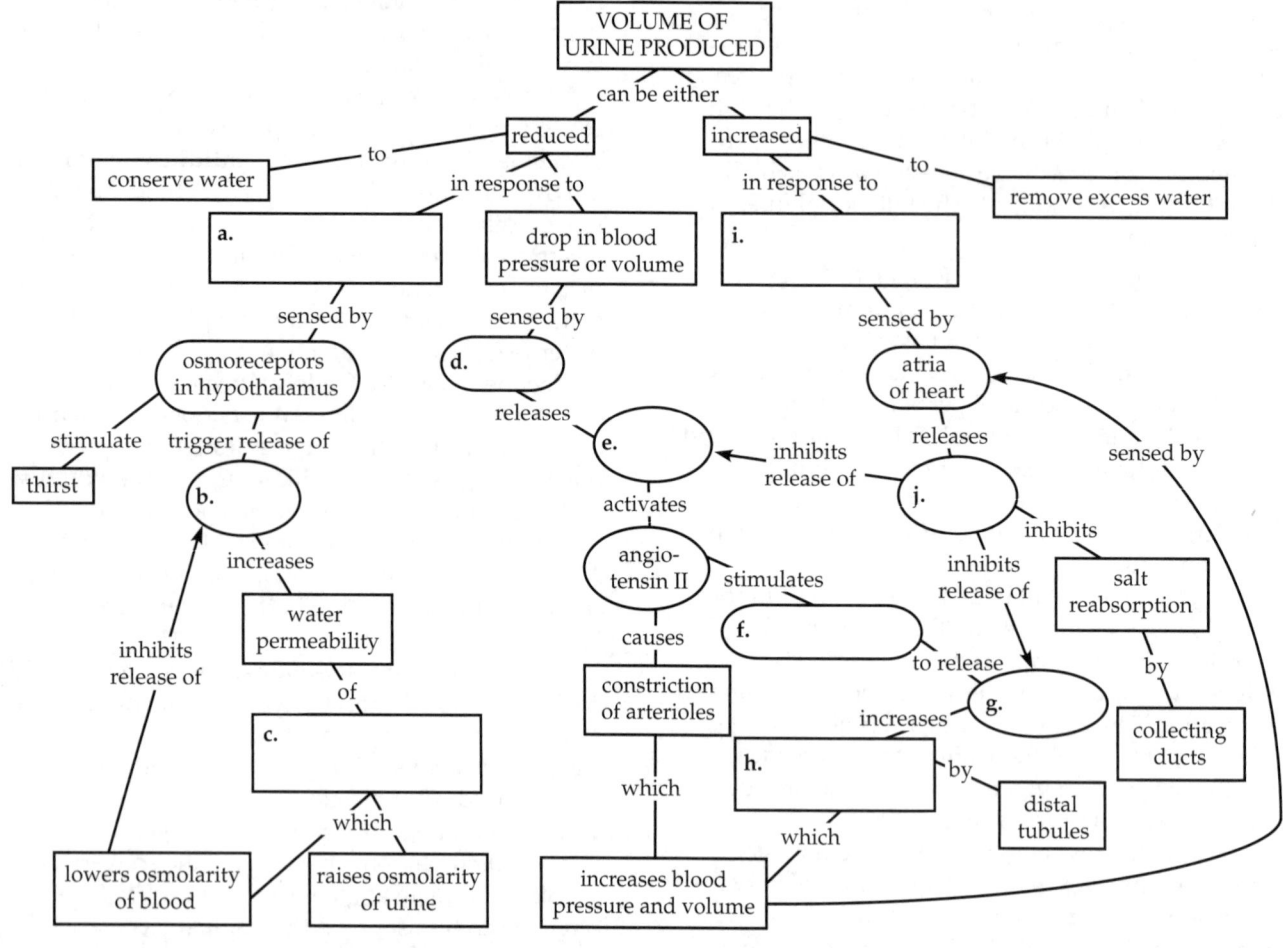

Word Roots

an- = without, **hydr-** = water; **bi-** = life (*anhydrobiosis:* a dormant state involving loss of almost all body water)

glomer- = a ball (*glomerulus:* a ball of capillaries surrounded by Bowman's capsule in the nephron; is the site of filtration in the kidney)

juxta- = near to (*juxtaglomerular apparatus:* a specialized tissue in nephrons that releases the enzyme renin in response to a drop in blood pressure or volume)

meta- = after; **nephr-** = kidney (*metanephridium:* an excretory organ found in many invertebrates that consists of tubules connecting ciliated internal openings to external openings)

osm- = pushing (*osmoregulation:* regulation of solute concentrations and water balance by a cell or an organism)

peri- = around (*peritubular capillaries:* the network of tiny blood vessels surrounding the proximal and distal tubules in the kidney)

proto- = first (*protonephridium:* an excretory system, such as the flame bulb system of flatworms, consisting of a network of tubules lacking internal openings)

ren- = a kidney; **angi-** = a vessel; **tens-** = stretched (*renin-angiotensin-aldosterone system:* a hormone cascade pathway that helps regulate blood pressure and volume)

vasa = a vessel; **recta** = straight (*vasa recta:* the capillary system in the kidney that serves the loop of Henle)

Structure Your Knowledge

1. Provide the following information concerning the production of urine in mammals.
   a. Substances filtered:
   b. Substances secreted:
   c. Substances reabsorbed:

2. After referring to the diagram in Interactive Question 44.5, describe the two-solute model that enables the production of hyperosmotic urine.

Test Your Knowledge

MULTIPLE CHOICE: *Choose the one best answer.*

1. Contractile vacuoles would most likely be used to expel excess water from unicellular eukaryotes that
   a. live in a freshwater environment.
   b. live in a marine environment.
   c. are internal parasites.
   d. are hypoosmotic to their environment.

2. Transport epithelia are responsible for
   a. pumping water across a membrane.
   b. forming an impermeable boundary at an interface with the environment.
   c. the movement of solutes in osmoregulation or excretion.
   d. the passive transport of H^+ and HCO_3^- for the regulation of pH.

3. Which of the following organisms would be most likely to produce the sugar trehalose?
   a. a desert mammal
   b. a euryhaline fish
   c. a shark
   d. an anhydrobiotic roundworm

4. A freshwater fish would be expected to
   a. pump salt out through salt glands in the gills.
   b. produce copious quantities of dilute urine.
   c. retain urea and TMAO in its tissues.
   d. do all of the above.

5. The production of uric acid is advantageous to birds because uric acid
   a. has low toxicity and can be safely stored as an embryo develops in an egg.
   b. contributes to the production of the egg shell and can thus serve two purposes.
   c. takes less energy to produce than urea and is nontoxic.
   d. Both **a** and **b** are correct.

6. Which of the following statements is *incorrect*?
   a. The Malpighian tubules of insects secrete nitrogenous wastes and solutes from the hemolymph into the tubules, which empty into the digestive tract. Most solutes are reabsorbed in the rectum and water follows by osmosis.
   b. The metanephridia of earthworms function primarily in osmoregulation; ammonia diffuses out from blood vessels lining their thin, moist skin.
   c. The main function of the protonephridia of freshwater flatworms is osmoregulation.
   d. Because their habitat is isoosmotic, the main function of the protonephridia of parasitic flatworms is excretion of nitrogenous wastes.

7. The correct sequence of structures through which urine flows in mammals is
   a. collecting duct, renal pelvis, urethra, bladder, ureter
   b. renal pelvis, collecting duct, urethra, bladder, ureter
   c. renal pelvis, ureter, collecting duct, bladder, urethra
   d. collecting duct, renal pelvis, ureter, bladder, urethra

8. The osmotic gradient in the renal medulla is largely produced by the diffusion of _____ out of the collecting ducts.
   a. sodium and chloride ions
   b. potassium ions
   c. urea
   d. both **a** and **c** contribute to the medullary gradient

9. The mechanism for the filtration of blood within the nephron involves
   a. the active transport of Na$^+$ and glucose, followed by osmosis.
   b. both active and passive secretion of ions, toxins, and NH$_3^+$ into the tubule.
   c. high hydrostatic pressure of blood forcing water and small molecules out of the capillary.
   d. a lower osmotic pressure in Bowman's capsule compared to that in the glomerulus.

10. The process of secretion in the formation of urine ensures that
    a. drugs and toxins are removed from the blood.
    b. urine is hyperosmotic to the blood.
    c. nutrients and water are returned to the blood.
    d. excess ions are removed from the filtrate.

11. The peritubular capillaries
    a. form the ball of capillaries inside Bowman's capsule.
    b. intertwine with the proximal and distal tubules and exchange solutes with and reabsorb water from the interstitial fluid.
    c. produce a countercurrent flow of blood through the medulla to supply nutrients without interfering with the osmolarity gradient.
    d. rejoin to form the efferent arteriole.

12. Which of the following sections of the mammalian nephron is *incorrectly* paired with its function?
    a. Bowman's capsule and glomerulus—filtration of blood
    b. proximal tubule—secretion of ammonia and H$^+$ into the filtrate and transport of glucose and amino acids out of tubule
    c. descending limb of loop of Henle—diffusion of urea out of the filtrate
    d. ascending limb of loop of Henle—diffusion and pumping of NaCl out of the filtrate

13. Which of the following animals is hyperosmotic to its environment?
    a. a marine shark
    b. a marine bony fish
    c. a freshwater bony fish
    d. both **a** and **c** are hyperosmotic to their environments

14. Ethanol inhibits antidiuretic hormone (ADH) secretion. After drinking a mug of beer, you would therefore expect:
    a. to produce less urine.
    b. to produce a more concentrated urine.
    c. to secrete more Na$^+$ in your urine.
    d. to have to empty your bladder more frequently than usual.

15. Which of the following stimuli causes the atria of the heart to release ANP (atrial natriuretic peptide)?
    a. a rise in blood pressure
    b. a drop in blood pressure
    c. a drop in blood pH
    d. a rise in blood osmolarity

16. Which of the following is an effective mechanism for a drug used to treat hypertension?
    a. increase the production of ADH
    b. block the production or release of ANP
    c. inhibit an enzyme involved in producing angiotensin II
    d. stimulate the insertion of aquaporins in cells lining the collecting ducts

17. Which of the following would occur while a vampire bat is consuming a large blood meal?
    a. an increase in the release of ANP from the atria of the heart
    b. diuresis resulting from a decrease in water reabsorption
    c. a decrease in the production of aldosterone by the adrenal glands
    d. an increase in the release of ADH

18. You are working on a research project to modify the human organism by increasing the length of the loop of Henle in the kidney. If you succeed, what will be the effect on human physiology?
    a. Humans will be able to reabsorb more sodium in the kidney and will therefore be able to eat less salt.
    b. Humans will be able to secrete more potassium ions into the urine and will need to eat fewer bananas.
    c. Humans will be able to produce a copious dilute urine and could therefore survive better in an aquatic environment.
    d. Humans will be able to produce a more concentrated urine and will be able to survive very dry conditions without becoming dehydrated.

Chapter 45

Hormones and the Endocrine System

Key Concepts

45.1 Hormones and other signaling molecules bind to target receptors, triggering specific response pathways

45.2 Feedback regulation and coordination with the nervous system are common in hormone pathways

45.3 Endocrine glands respond to diverse stimuli in regulating homeostasis, development, and behavior

Framework

This chapter introduces the intricate system of chemical control and communication within animals. Endocrine cells (or neurosecretory cells) produce hormones that regulate the activity of other cells and organs. Steroid hormones bind with receptors within their target cells and influence gene expression. The signal transduction pathways of most polypeptide and amine hormones involve binding with cell-surface receptors and triggering metabolic reactions in target cells.

The hypothalamus and pituitary gland play coordinating roles by integrating the nervous and endocrine systems and producing many tropic hormones that control the synthesis and secretion of hormones in other endocrine glands and organs.

Chapter Review

An animal **hormone** is a signaling molecule that is released into the extracellular fluid, circulates through the hemolymph or blood, and elicits a specific response from *target cells* (that is, cells with matching receptors). Coordination and communication among the specialized parts of complex animals are achieved by the **endocrine system**, which produces hormones that regulate many biological processes, and the **nervous**

system, which conveys messages along neurons that regulate other neurons, muscle cells, and endocrine cells.

45.1 Hormones and other signaling molecules bind to target receptors, triggering specific response pathways

Intercellular Communication   There are five forms of signaling that use secreted molecules. In endocrine signaling, hormones travel via the circulatory system throughout the body. Hormones maintain homeostasis and regulate growth, development, and reproduction.

Local regulators are molecules that reach their target cells by diffusion. In **paracrine** signaling, secreted molecules act on neighboring cells; in **autocrine** signaling, the molecules act on the secreting cell itself. Local regulators include cytokines (communication between immune cells), growth factors (promote cell division), and prostaglandins. **Prostaglandins** help induce labor and promote inflammation, the sensation of pain, and platelet aggregation in clot formation. Aspirin and ibuprofen inhibit the synthesis of prostaglandins and thus reduce these effects.

In *synaptic signaling*, **neurotransmitters** are secreted by neurons at synapses with other neurons or muscle cells. In *neuroendocrine signaling*, specialized neurons called neurosecretory cells secrete **neurohormones**, which travel through the bloodstream to target cells.

Pheromones are signaling molecules that are released into the environment and communicate between individuals of the same species.

Chemical Classes of Local Regulators and Hormones   Prostaglandins are modified fatty acids; cytokines and growth factors are polypeptides. **Nitric oxide (NO)** is a gas that can function as a neurotransmitter or as a local regulator that produces vasodilation, increasing blood flow.

Hormone classes include polypeptides, steroids, and amines (synthesized from amino acids). Polypeptides and most amines are water soluble. Steroids and other nonpolar hormones are lipid soluble.

Cellular Response Pathways   Water-soluble hormones are released by exocytosis, travel through the bloodstream, and bind to cell-surface receptors, triggering a cytoplasmic response or a change in gene expression. Lipid-soluble hormones travel in the bloodstream bound to transport proteins, enter target cells, and bind to a receptor in the cytoplasm or nucleus, triggering changes in gene transcription.

Binding of water-soluble hormones to plasma membrane receptors initiates **signal transduction**, which converts extracellular signals to specific intracellular responses. For example, after **epinephrine** (also called *adrenalin*) is released by the adrenal glands in response to stress, it binds to G protein-coupled receptors on liver cells, initiating a signal transduction pathway that produces cAMP as a *second messenger*, which activates a protein kinase; the result is glycogen breakdown and the release of glucose into the bloodstream.

Steroid hormones bind to cytoplasmic receptors, and the hormone-receptor complex then moves into the nucleus, where it stimulates transcription of specific genes. Thyroxine, vitamin D, and other nonsteroid, lipid-soluble hormones bind with receptors inside the nucleus that then stimulate transcription.

INTERACTIVE QUESTION   45.1

Label the parts of the following diagrams, indicating which parts represent a water-soluble hormone and which a lipid-soluble hormone.

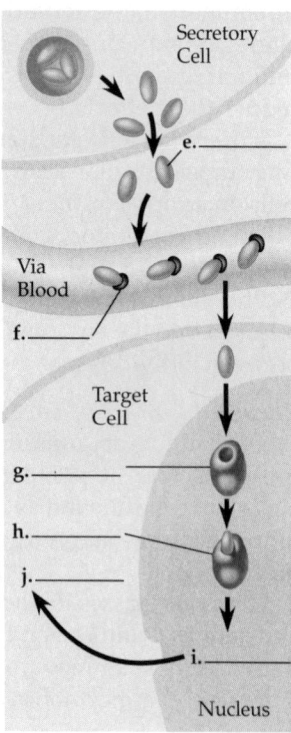

Multiple Effects of Hormones   A given signal can have different effects on different target cells as a result of the types of receptors (for example, the α and β epinephrine receptors) or the specific signal transduction pathways and effector proteins present in the cell.

Endocrine Tissues and Organs   Some endocrine cells are located in organs that are part of other organ systems. **Endocrine glands** are ductless secretory organs composed of groups of endocrine cells. The secretions of *exocrine glands* are carried by ducts to their point of action.

45.2   Feedback regulation and coordination with the nervous system are common in homone pathways

Simple Endocrine Pathways   In a *simple endocrine pathway*, endocrine cells secrete a hormone in response to a stimulus; the hormone travels in the bloodstream to target cells, where signal transduction produces a physiological response. For example, endocrine cells of the duodenum secrete the hormone *secretin* in response to a low pH caused by the entry of acidic stomach contents. Target cells in the pancreas release bicarbonate, which raises the pH in the duodenum.

Simple Neuroendocrine Pathways   In a *simple neurohormone pathway*, a neurosecretory cell secretes a neurohormone in response to a stimulus received by a sensory neuron. The circulating neurohormone triggers a specific response in target cells. For example, in response to nerve signals from the nipples, the hypothalamus triggers the release of the neurohormone **oxytocin** from the posterior pituitary gland, which stimulates the mammary glands to secrete milk.

Feedback Regulation   In a feedback loop called **negative feedback**, the response to a stimulus reduces the stimulus, and the pathway shuts off. In **positive feedback**, the response to a stimulus reinforces the stimulus, leading to a further increase in response. The stimulation of milk secretion and oxytocin's role in inducing contraction of uterine muscles during birth are two examples of positive feedback regulation. Negative feedback is typical of pathways involved in maintaining homeostasis. In many instances, a pair of pathways, each counteracting the other, constitutes the control system. As an example, insulin lowers elevated blood glucose levels, while glucagon raises low blood glucose levels.

Compare negative and positive feedback in the regulation of hormonal pathways and give an example of each.

Coordination of Endocrine and Nervous Systems   In insects, development and metamorphosis are usually controlled through an integration of the endocrine and nervous systems. Neurosecretory cells in the brain produce prothoracicotropic hormone (PTTH), which stimulates the prothoracic glands to secrete *ecdysteroid*, which triggers molting and metamorphosis. Juvenile hormone promotes retention of larval characteristics during molting. When its level is high, ecdysteroid-induced molting produces larger larval stages; only after the level of juvenile hormone has decreased does molting result in a pupa.

In vertebrates, the **hypothalamus**, situated in the lower brain, plays a key role in integrating the endocrine and nervous systems. In response to nerve signals it receives from throughout the body, the hypothalamus sends endocrine signals to the pituitary gland located at its base.

The **pituitary gland** has two discrete parts that develop separately and have different functions. The **posterior pituitary** stores and secretes two neurohormones (ADH and oxytocin) delivered through axons of neurosecretory cells of the hypothalamus. The **anterior pituitary** produces and secretes hormones in response to hormonal signals from the hypothalamus.

Antidiuretic hormone (ADH), or *vasopressin*, increases water retention in the kidneys and plays a role in social behavior. In addition to regulating milk secretion and uterine contractions, oxytocin influences behaviors related to maternal care, pair bonding, and sexual activity.

Many of hormones secreted by the anterior pituitary regulate other endocrine cells or glands. The anterior pituitary itself is regulated by *releasing hormones* and *inhibiting hormones* secreted by neurosecretory cells of the hypothalamus. These hypothalmic hormones are released into capillaries at the base of the hypothalamus and travel via a short portal vessel to capillary beds in the anterior pituitary. For example, *prolactin-releasing hormone* stimulates the release of **prolactin**, which stimulates milk production.

In a *hormone cascade pathway*, the hypothalamus secretes a hormone that stimulates release of an anterior pituitary hormone, which then acts on a target endocrine tissue, stimulating the release of a hormone that produces a response.

What is a **tropic hormone**? Give some examples.

Thyroid Regulation: A Hormone Cascade Pathway
Thyroid hormone regulates metabolic rate in mammals and helps to maintain normal blood pressure, heart rate, muscle tone, digestion, and reproductive functions. In response to a lowered level of thyroid hormone in the blood, the hypothalamus secretes thyrotropin-releasing hormone (TRH), which stimulates the anterior pituitary to secrete thyroid-stimulating hormone (TSH), also called thyrotropin. TSH stimulates the **thyroid gland** to release thyroid hormone. Negative feedback regulates this pathway: High levels of thyroid hormone inhibit the secretion of TRH and TSH.

The thyroid gland produces two hormones, *thyroxine* (T_4) and *triiodothyronine* (T_3). Explain how a lack of iodine in the diet may result in goiter, an enlarged thyroid gland.

Hormonal Regulation of Growth   **Growth hormone (GH)** has both tropic and nontropic effects on growth. It signals the liver to release *insulin-like growth factors (IGFs)*, which stimulate bone and cartilage growth. It also acts to raise blood glucose levels. Gigantism and acromegaly are human growth disorders caused by hypersecretion of GH. Pituitary dwarfism can now be treated with genetically engineered human GH (HGH).

45.3 **Endocrine glands respond to diverse stimuli in regulating homeostasis, development, and behavior**

Parathyroid Hormone and Vitamin D: Control of Blood Calcium   In response to a drop in blood Ca^{2+} level below a set point, the four **parathyroid glands** secrete **parathyroid hormone (PTH)**. PTH stimulates the reabsorption of Ca^{2+} in the kidney and its release from bone. PTH also activates vitamin D, which then increases Ca^{2+} uptake from food in the intestines. If blood calcium rises above the set point, the thyroid gland secretes **calcitonin**, a hormone that lowers blood calcium levels. In humans, calcitonin appears to be required only during the bone growth of childhood.

Adrenal Hormones: Response to Stress   In mammals, the **adrenal glands** consist of two different glands: the outer adrenal *cortex* composed of endocrine cells and the central adrenal *medulla* derived from neural tissue.

The adrenal medulla produces **norepinephrine** (noradrenaline) and epinephrine (adrenaline)—both of which are *catecholamines*, a class of hormones synthesized from the amino acid tyrosine. Epinephrine and norepinephrine, released in response to positive or negative short-term stress, increase the availability of energy sources by stimulating glycogen hydrolysis in skeletal muscle and the liver, glucose release from liver cells, and fatty acid release from fat cells. These hormones increase the rate and volume of the heartbeat, dilate bronchioles in the lungs, and influence the contraction or relaxation of smooth muscles to increase blood supply to the heart, brain, and skeletal muscles while reducing the supply to other organs. Nerve signals from the hypothalamus stimulate the release of epinephrine and norepinephrine from the adrenal medulla. The multiple effects of epinephrine in varying tissues are achieved through differences in receptor types and cytoplasmic responses.

The adrenal cortex responds to endocrine signals released in response to long-term stress. A releasing hormone from the hypothalamus causes the anterior pituitary to release adrenocorticotropic hormone (ACTH). This tropic hormone stimulates the adrenal cortex to secrete *corticosteroids*.

Glucocorticoids promote the synthesis of glucose from noncarbohydrates (such as muscle proteins) and thus increase energy supplies during stress. Glucocorticoids may be used to treat inflammatory diseases such as arthritis, but their side effects can be dangerous.

Mineralocorticoids affect salt and water balance. *Aldosterone* functions in ion and water homeostasis and in the body's response to severe stress.

INTERACTIVE QUESTION   45.5

Describe how the adrenal gland is signaled by the hypothalamus and responds to short-term and long-term stress.

a. Short-term stress:

b. Long-term stress:

Sex Hormones   Sex hormones affect growth, development, and reproductive cycles and behaviors. The three major categories of gonadal steroid sex hormones—androgens, estrogens, and progestins—are present in different proportions in males and females.

The testes primarily synthesize **androgens**, such as **testosterone**, which determine the sex of the developing embryo and stimulate development of the male reproductive system. At puberty, androgens produce male secondary sex characteristics.

In females, **estrogens**, in particular **estradiol**, regulate the maintenance of the reproductive system and the development of female secondary sex characteristics. In mammals, **progestins**, including **progesterone**, help prepare and maintain the uterus for the growth of an embryo.

In a hormone cascade pathway, a hypothalamic releasing hormone (GnRH) controls the anterior pituitary secretion of the gonadotropins follicle-stimulating hormone (FSH) and luteinizing hormone (LH), which then control the synthesis of estrogens and androgens.

An *endocrine disruptor* is a foreign molecule that disrupts a hormone pathway. The synthetic estrogen diethylstilbestrol (DES), once prescribed for problem pregnancies, was shown to affect the reproductive system of female offspring. Other potential endocrine disruptors, such as bisphenol A, are under study.

Hormones and Biological Rhythms   The **pineal gland**, located near the center of the mammalian brain, secretes **melatonin**, which regulates functions related to light and seasonal changes in day length and to reproduction. The secretion of melatonin at night is controlled by the suprachiasmatic nuclei (SCN) in the hypothalamus, a biological clock that receives input from the retina. Nightly increases in melatonin may play an important role in promoting sleep.

Evolution of Hormone Function   Hormone functions may change among evolutionary lineages. In many vertebrates, thyroid hormone is involved in the regulation of general metabolism, but in amphibians it has become associated with changes during metamorphosis. The hormone prolactin has effects ranging from milk production in mammals, to delay of metamorphosis in amphibians, to osmoregulation in fishes. The diversity of actions in different vertebrate species suggests that prolactin is an ancient hormone whose functions diversified during the evolution of vertebrate groups.

Melanocyte-stimulating hormone (MSH) regulates the activity of pigment-containing cells in the skin of most vertebrates, and it also functions in hunger and metabolism in mammals. Drugs that inhibit MSH receptors in the brain may be a potential treatment for the wasting condition called cachexia, which is associated with AIDS, tuberculosis (TB), and late-stage cancer.

Word Roots

andr- = male; **-gen** = producing substance (*androgen:* any steroid hormone, such as testosterone, that stimulates the development and maintenance of the male reproductive system and secondary sex characteristics)

anti- = against; **diur-** = urinate (*antidiuretic hormone:* a hormone produced in the hypothalamus and released from the posterior pituitary that promotes water retention by the kidneys)

epi- = above, over (*epinephrine:* a catecholamine that, when secreted as a hormone by the adrenal medulla, mediates "fight-or-flight" responses to short-term stresses; also called adrenaline)

melan- = black (*melatonin:* a hormone secreted by the pineal gland involved in regulation of biological rhythms and sleep)

neur- = nerve (*neurohormone:* a molecule that is secreted by a neuron, travels in body fluids, and acts on specific target cells)

para- = beside, near (*parathyroid glands:* four endocrine glands, embedded in the surface of the thyroid gland, that secrete parathyroid hormone, which raises blood calcium levels)

pro- = before; **lact-** = milk (*prolactin:* a hormone produced by the anterior pituitary gland with diverse effects in different vertebrates; it stimulates milk production in mammals)

Structure Your Knowledge

1. Briefly describe the two general mechanisms (depending on the location of the receptor) by which hormones trigger responses in target cells.

2. What is a hormone cascade pathway? Describe how thyroid regulation illustrates such a pathway.

Test Your Knowledge

MATCHING: *Write the letter of the hormone (A–L) and the letter of the gland or organ that produces it (a–k) in the blank next to the description of each hormone's action. Choices may be used more than once, and not all choices must be used.*

Hormones
A. ACTH
B. androgen
C. ADH
D. calcitonin
E. epinephrine
F. glucocorticoid
G. gonadotropin
H. insulin
I. melatonin
J. oxytocin
K. PTH
L. thyroxine

Gland or Organ
a. adrenal cortex
b. adrenal medulla
c. anterior pituitary
d. hypothalamus
e. pancreas
f. parathyroid gland
g. pineal
h. posterior pituitary
i. testis
j. thymus
k. thyroid

Hormone	Gland or Organ	Hormone Action
_____	_____	1. functions in biological clock and seasonal activities
_____	_____	2. breaks down muscle protein for conversion to glucose
_____	_____	3. decreases blood sugar through glycogen storage in liver
_____	_____	4. stimulates development of male reproductive system
_____	_____	5. stimulates adrenal cortex to synthesize corticosteroids
_____	_____	6. increases available energy, heart rate and metabolism in response to stress
_____	_____	7. regulates general metabolism
_____	_____	8. raises blood calcium levels
_____	_____	9. increases reabsorption of water by kidney
_____	_____	10. stimulates contraction of uterus, milk secretion

MULTIPLE CHOICE: *Choose the one best answer.*

1. Which of the following statements is *not* accurate?
   a. Not all hormones are secreted by endocrine glands.
   b. Hormones are transported through the circulatory system to their destinations.
   c. Target cells have specific hormone receptors.
   d. Steroid hormones often function as neurotransmitters.

2. The best description of the difference between pheromones and hormones is that
   a. pheromones are small, volatile molecules; hormones are large, hydrophilic molecules.
   b. pheromones are a form of neural communication; hormones are a form of chemical communication.
   c. pheromones are signaling molecules that function between organisms; hormones communicate among parts within an organism.
   d. pheromones are local regulators; hormones circulate in blood or hemolymph.

3. Ecdysteroid is
   a. a steroid hormone produced in insects that promotes retention of larval characteristics.
   b. secreted by prothoracic glands in insects and triggers molting and the development of adult form.
   c. responsible for color changes in amphibians.
   d. involved in metamorphosis in amphibians.

4. The activation of helper T cells during an immune response by cytokines released by antigen-presenting cells and helper T cells themselves is an example of
   a. endocrine signaling.
   b. autocrine signaling.
   c. paracrine signaling.
   d. both b and c.

5. The anti-inflammatory and pain-relieving effects of ibuprofen are due to
   a. inhibition of the synthesis of the local regulator prostaglandin.
   b. inhibition of the release of NO by pain-sensing neurons and blood vessels.
   c. stimulation of the release of glucocorticoids by the adrenal cortex.
   d. activation of pain-reducing centers of the hypothalamus.

6. The finding that MSH (melanocyte-stimulating hormone) causes frog skin cells to darken when added to the interstitial fluid but has no effect on color when injected into the cells supports which of the following ideas?
   a. MSH is a tropic hormone.
   b. MSH is a neurohormone.
   c. MSH is a steroid hormone.
   d. MSH binds to cell surface receptors.

7. Which of the following hormones is secreted by the thyroid gland?
   a. calcitonin
   b. TSH
   c. FSH
   d. glucagon

8. Steroid hormones
   a. influence gene expression.
   b. act through second messengers.
   c. bind to cell-surface receptors.
   d. do not require specific receptor molecules.

9. Which of the following situations is an example of positive feedback?
   a. In response to growth hormone, the liver produces insulin-like growth factors, which promote skeletal growth.
   b. The neurotransmitter acetylcholine causes skeletal muscle to contract, heart muscle to relax, and cells of the adrenal medulla to secrete epinephrine.
   c. Prostaglandins released from placental cells promote muscle contraction during childbirth, and those muscle contractions stimulate further prostaglandin release.
   d. Elevated levels of stress result in neural stimulation of the adrenal medulla and hormonal stimulation of the adrenal cortex.

10. Hypothalamic releasing and inhibiting hormones act on
    a. the posterior pituitary.
    b. the anterior pituitary.
    c. the adrenal cortex.
    d. both a and b.

11. Injury to the posterior pituitary gland may result in the production of a copious, dilute urine. Which hormone is likely involved?

    a. oxytocin   c. ADH

    b. calcitonin   d. MSH

12. The anterior pituitary

    a. stores and releases oxytocin and ADH produced by the hypothalamus.

    b. receives releasing and inhibiting hormones from the hypothalamus through portal vessels connecting capillary beds.

    c. produces releasing and inhibiting hormones.

    d. produces neurohormones in response to nervous stimulation.

13. The release of milk during nursing is promoted by

    a. oxytocin.   c. prolactin.

    b. progesterone.   d. estrogen.

14. Which of the following hormones is *not* involved in increasing the blood glucose concentration?

    a. epinephrine

    b. glucocorticoids

    c. ACTH (adrenocorticotropic hormone)

    d. insulin

15. Which of the following statements about epinephrine is *not* true?

    a. It is secreted by the adrenal medulla.

    b. It dilates bronchioles in the lungs.

    c. Its release is stimulated by ACTH.

    d. It is part of the fight-or-flight response to stress.

Choose from the following hormones to answer questions 16–20.

    a. ACTH

    b. prolactin

    c. oxytocin

    d. estrogen

    e. insulin

16. Which hormone is a tropic hormone?

17. Which hormone is produced as the result of a hormone cascade pathway?

18. Which hormone is part of a simple neuroendocrine pathway?

19. Which hormone has diverse effects in different vertebrate species?

20. Which hormone is part of an antagonistic pair of hormones?

Chapter 46

Animal Reproduction

Framework

This chapter covers the patterns and mechanisms of animal reproduction. In sexual reproduction, fertilization may occur externally or internally. Development of the zygote may take place internally within the female; externally in a moist environment; or in a protective, resistant egg. Placental mammals provide nourishment as well as shelter for the embryo, and they nurse their young.

The human reproductive system is described, including its organs, glands, and hormones; gamete formation; and sexual response. The chapter also covers pregnancy and birth, contraception, and reproductive technologies.

Chapter Review

46.1 Both asexual and sexual reproduction occur in the animal kingdom

In **sexual reproduction**, two meiotically formed haploid gametes fuse to form a diploid **zygote**, which develops into an offspring. A female gamete is a relatively large, nonmotile **egg**; male gametes are small, motile **sperm**. In **asexual reproduction**, a single individual produces offspring, with no fusion of egg and sperm.

Mechanisms of Asexual Reproduction Many invertebrates can reproduce asexually by *budding*, in which a new individual grows out from the parent's body; by **fission**, in which a parent is separated into two individuals of equal size; or by *fragmentation*, in which the body is broken into several pieces, each of which develops into a complete animal. *Regeneration*, the regrowth of body parts, is necessary for a fragment to develop into a new organism. In **parthenogenesis**, eggs develop without being fertilized.

Variation in Patterns of Sexual Reproduction In **hermaphroditism**, found in some sessile, burrowing, and parasitic animals, each individual has functioning male and female reproductive systems. Mating between two individuals results in fertilization of both. In oysters and some fishes, individuals reverse their sex during their lifetime.

Reproductive Cycles Periodic reproductive cycles may be linked to favorable conditions or energy supplies. A combination of environmental and hormonal cues controls the timing of **ovulation**, the release of eggs.

Climate change is causing a mismatch between some animal migrations, which are cued by seasonal changes in daylight, and the new plant growth, cued by temperature, that provides nutrition for successful rearing of offspring.

Animals may reproduce sexually, asexually, or both. The freshwater crustacean *Daphnia* produces eggs that develop by parthenogenesis, as well as eggs that are fertilized. Environmental conditions determine whether reproduction is asexual or sexual. In a few species of parthenogenetic lizards, mating behaviors and ovulation are mediated by changing levels of female hormones.

Sexual Reproduction: An Evolutionary Enigma Asexual reproduction potentially produces many more offspring, yet most eukaryotic species reproduce sexually. The shuffling of genes in sexual reproduction may allow more rapid adaptation to changing conditions or allow harmful genes to be eliminated more easily.

a. What adaptive advantages may asexual reproduction provide?

b. What adaptive advantages may sexual reproduction provide?

c. If a hermaphrodite self-fertilizes, is that asexual or sexual reproduction? Explain.

46.2 Fertilization depends on mechanisms that bring together sperm and eggs of the same species

Fertilization is the union of sperm and egg. In *external fertilization*, eggs and sperm are shed, and fertilization occurs in the environment. External fertilization occurs almost exclusively in moist habitats, where the gametes and developing zygote are not in danger of desiccation. When individuals release their gametes at the same time, it is called *spawning*.

Internal fertilization involves the placement of sperm in or near the female reproductive tract so that the egg and sperm unite internally. Behavioral cooperation, as well as copulatory organs and sperm receptacles, are required for internal fertilization.

A *pheromone* is a small, volatile, or water-soluble molecule that influences the behavior of an organism other than the one that releases it. In some cases, pheromones function as mate attractants.

List three mechanisms that may help to ensure that gamete release is synchronized when fertilization is external.

a.

b.

c.

Ensuring the Survival of Offspring Species with internal fertilization usually produce fewer gametes, but a larger fraction of zygotes survive.

The shells of the eggs of birds and other reptiles protect the embryo from water loss or injury. Embryos of marsupial mammals complete development in the mother's pouch, attached to a mammary gland. Embryos of eutherian (placental) mammals develop in the uterus, nourished by the mother's blood through the placenta. Parental care of young is common in mammals and birds and even found in some invertebrates.

Gamete Production and Delivery Precursor cells for eggs or sperm are often established early in embryogenesis and later *amplified* in number. The simplest reproductive systems lack **gonads** to produce gametes. In polychaete worms, eggs or sperm develop from cells lining the coelom.

Most insects have separate sexes and complex reproductive systems. Sperm develop in the testes, are stored in the seminal vesicles, and are ejaculated during copulation. Eggs are produced in the ovaries, fertilized in the uterus, and then expelled from the body. Females may have **spermathecae**, or sperm-storing sacs, from which sperm may be released when conditions favor successful reproduction.

Most animals mate with more than one partner. In some species, males have evolved mechanisms to increase their reproductive success by interfering with other males or by decreasing the receptivity of female partners.

With the exception of mammals, many vertebrates have a common opening—a **cloaca**—for the digestive, excretory, and reproductive systems. Mammals generally have a separate opening for the digestive tract. The reproductive and excretory openings are separate in most female mammals.

Mechanisms that influence which sperm are successful in sequential matings have evolved in some male and female insects.

46.3 Reproductive organs produce and transport gametes

Human Male Reproductive Anatomy The external male reproductive organs are the scrotum and penis. Sperm are produced in the highly coiled **seminiferous tubules** of the **testes**. Sperm production requires a cooler temperature than the core body temperature of most mammals, so the **scrotum** suspends the testes below the abdominal cavity.

Sperm pass from a testis into the coiled **epididymis**, in which they mature and gain motility. During **ejaculation**, sperm are propelled through the **vas deferens**, into a short *ejaculatory duct*, and out through the **urethra**, which runs through the penis.

Three sets of accessory glands add secretions to the sperm to form **semen**. The **seminal vesicles** contribute an alkaline fluid containing mucus, fructose (an energy source for the sperm), a coagulating enzyme, ascorbic acid, and prostaglandins. The **prostate gland** produces a secretion that contains anticoagulant enzymes and citrate. The *bulbourethral glands* produce a mucus that neutralizes urine remaining in the urethra.

The **penis** is composed of spongy tissue that becomes engorged with blood during sexual arousal, producing an erection that facilitates insertion of the penis into the vagina. Some mammals have a baculum, a bone that helps stiffen the penis. A fold of skin called the **prepuce**, or foreskin, covers the head, or **glans**, of the penis.

Human Female Reproductive Anatomy   The external reproductive structures in human females include the clitoris and two sets of labia. The **ovaries** contain many **follicles**, each of which contains an **oocyte** surrounded by cells that nourish and protect the developing egg during its development.

The egg cell is swept by cilia into the **oviduct**, or fallopian tube, through which it travels to the **uterus**. The uterine lining, the **endometrium**, is richly supplied with blood vessels. The neck of the uterus, called the **cervix**, opens into the **vagina**, the elastic-walled birth canal and repository for sperm during copulation.

The external genitalia are collectively called the **vulva**. Two pairs of skin folds, the thick outer **labia majora** and the thin inner **labia minora**, enclose the separate openings of the vagina and urethra. The vaginal opening is partly covered by a membrane called the *hymen*. The **clitoris** is composed of erectile tissue supporting a glans, or head, covered by the prepuce.

A **mammary gland**, located within a breast, is composed of fatty tissue and small milk-secreting sacs that drain into ducts that open at the nipple. Breast tissue is found in both sexes, but milk is generally produced only by females.

INTERACTIVE QUESTION   46.3

Label the indicated structures in the following diagrams of the human male and female reproductive systems.

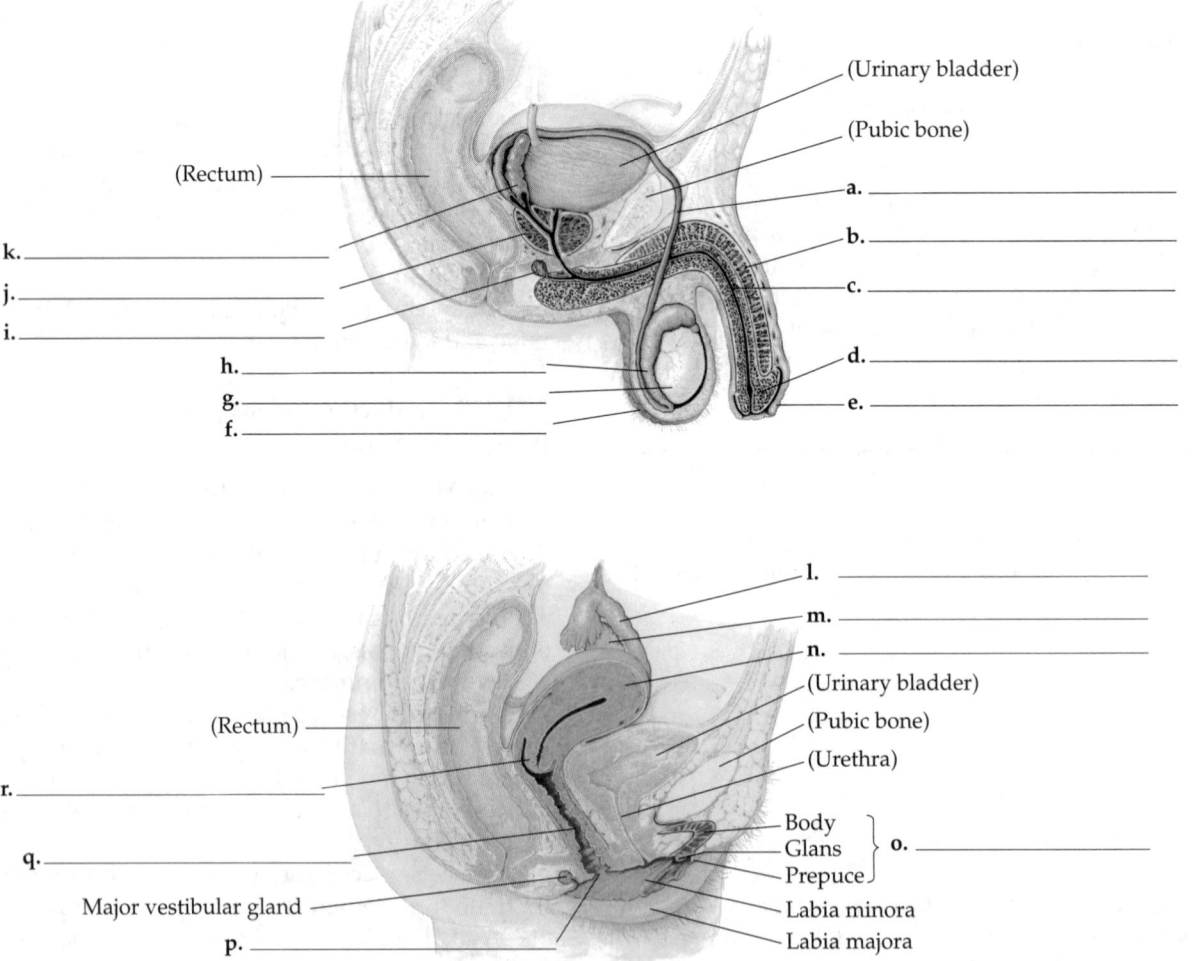

Gametogenesis (Figure 46.11) The production of gametes, or **gametogenesis**, differs in females and males. **Spermatogenesis** occurs continuously in the seminiferous tubules of the testes after puberty. Primordial germ cells in male embryos differentiate into spermatogonial stem cells. These cells divide mitotically to form **spermatogonia**, which also divide mitotically to form spermatocytes. Spermatocytes undergo meiosis to produce four spermatids. The haploid nucleus of a sperm is contained in a head, tipped with an **acrosome**, which contains enzymes that help the sperm penetrate the egg. Mitochondria provide ATP for movement of the flagellum, or tail.

Oogenesis begins in the female embryo. **Oogonia**, produced from primordial germ cells, divide and differentiate into **primary oocytes**, which are arrested in prophase I of meiosis before birth. Beginning at puberty, follicle-stimulating hormone (FSH) periodically stimulates a follicle to resume growth. Its enclosed primary oocyte finishes meiosis I and develops into a **secondary oocyte**, which is arrested in metaphase II. In humans, meiosis is completed if a sperm penetrates the oocyte. Meiotic cytokinesis is unequal, producing one large egg and up to three small haploid polar bodies that disintegrate.

Following ovulation, the follicle forms a corpus luteum, which secretes progesterone and estradiol. If fertilization does not occur, the corpus luteum degenerates.

INTERACTIVE QUESTION 46.4

List three important ways in which spermatogenesis differs from oogenesis.

a.

b.

c.

46.4 **The interplay of tropic and sex hormones regulates mammalian reproduction**

Gonadotropin-releasing hormone (GnRH) from the hypothalamus directs the anterior pituitary to produce **follicle-stimulating hormone (FSH)** and **luteinizing hormone (LH)**. These tropic hormones (*gonadotropins*) regulate gametogenesis and sex hormone production in both males and females. The steroid sex hormones produced by the gonads include *androgens* (primarily **testosterone**), *estrogens* (chiefly **estradiol**), and **progesterone**.

Androgens are responsible for the embryonic development of male primary sex characteristics. At puberty, androgens direct formation of secondary sex characteristics, such as voice deepening, pubic and facial hair growth, and muscle growth. In females, estrogens stimulate breast and pubic hair development at puberty. Sex hormones influence sexual behaviors in both sexes.

Hormonal Control of the Male Reproductive System FSH acts on *Sertoli cells*, which nourish developing sperm. LH acts on *Leydig cells*, which secrete testosterone and other androgens, promoting spermatogenesis. Testosterone inhibits production of GnRH, FSH, and LH; *inhibin*, produced by Sertoli cells, reduces FSH secretion. These feedback mechanisms maintain proper levels of androgen production.

INTERACTIVE QUESTION 46.5

Fill in the blanks in the following description of the control of male reproductive hormones.

a. _____, produced by the hypothalamus, regulates the release of gonadotropic hormones from the anterior pituitary.

b. _____ stimulates Sertoli cells of the seminiferous tubules, which nourish sperm.

c. _____ stimulates the production of **d.** _____ by Leydig cells.

e. The production of GnRH, LH, and FSH is regulated by _____.

Hormonal Control of Female Reproductive Cycles

In females, the endometrium thickens to prepare for the implantation of the embryo and then is shed in a process called **menstruation** if fertilization does not occur. The human female reproductive cycle involves the integration of the **ovarian cycle**, which involves follicle growth and ovulation, and the **uterine cycle** (or **menstrual cycle**), which involves these cyclic changes in the uterus.

The ovarian cycle starts with the release from the hypothalamus of GnRH, which stimulates the anterior pituitary to secrete small amounts of FSH and LH. In this *follicular phase*, FSH stimulates follicular growth, and the cells of the follicle secrete estradiol. The low level of estradiol inhibits the release of FSH and LH.

When the secretion of estradiol by the growing follicle rises sharply, the hypothalamus is stimulated to increase GnRH output, which results in a rise in LH and FSH release. The increase in LH induces maturation of the follicle, which enlarges and forms a bulge near the ovary surface. Ovulation occurs about a day after the LH surge, with the rupture of the follicle and adjacent ovary wall.

In the *luteal phase*, LH stimulates the transformation of the residual follicular tissue into the corpus luteum, which secretes estradiol and progesterone. The rising level of these hormones exerts negative feedback on the hypothalamus and pituitary, inhibiting secretion of LH and FSH and thereby preventing another egg from maturing. Without LH to maintain it, the corpus luteum disintegrates, which reduces the levels of estradiol and progesterone. This drop releases the inhibition of the hypothalamus and pituitary, and FSH secretion stimulates the growth of new follicles and the start of the next ovarian cycle.

The uterine (menstrual) cycle is controlled by hormones secreted by the ovaries. Estradiol secreted by the growing follicle causes the endometrium to begin to thicken in the *proliferative phase*, which correlates with the follicular phase of the ovarian cycle. After ovulation, estradiol and progesterone from the corpus luteum stimulate continued development of the endometrium and of glands that secrete a nutrient fluid. This *secretory phase* of the uterine cycle correlates with the luteal phase of the ovarian cycle.

If fertilization and implantation do not occur, the rapid drop in ovarian hormones caused by the disintegration of the corpus luteum reduces blood supply to the endometrium and begins its disintegration, leading to the *menstrual flow phase* of the uterine cycle. By convention, the uterine and ovarian cycles begin with the first day of menstruation.

Endometriosis is a condition in which some endometrial tissue becomes **ectopic**—that is, located outside the uterus. The cyclic swelling and breakdown of this tissue causes pain and bleeding into the abdomen.

Menopause, the cessation of ovulation and menstruation, occurs in older women as the ovaries become less responsive to FSH and LH.

Humans and some other primates have menstrual cycles. Other mammals have **estrous cycles**, during which the endometrium thickens but is reabsorbed (not shed) if fertilization does not occur. Females are receptive to sexual activity only during estrus, or "heat," the period surrounding ovulation.

Human Sexual Response   The human sexual response cycle includes two types of physiological reactions: *vasocongestion*, increased blood flow to a tissue, and *myotonia*, increased muscle tension. The excitement phase involves vasocongestion and vaginal lubrication, preparing the vagina and penis for *coitus*, or sexual intercourse. The plateau phase continues vasocongestion and myotonia, and breathing rate and heart rate increase. In both sexes, *orgasm* is characterized by rhythmic, involuntary contractions of reproductive structures. In males, emission moves semen into the urethra, and ejaculation occurs when the urethra contracts and semen is expelled. In the resolution phase, vasocongested organs return to normal size, and muscles relax.

INTERACTIVE QUESTION   46.6

In the following diagram of the human female reproductive cycle, label the lines indicating the levels of the gonado-tropic and the ovarian hormones in the blood and the phases of the ovarian and uterine cycles.

46.5 In placental mammals, an embryo develops fully within the mother's uterus

Conception, Embryonic Development, and Birth   The 2–5 mL of ejaculated semen contains hundreds of millions of sperm. Semen initially coagulates and then liquifies as the sperm start swimming through the female reproductive tract.

Fertilization, or **conception**, occurs in the oviduct. Cleavage produces a ball of cells by the time the embryo reaches the uterus. The **blastocyst**, a hollow sphere of cells, implants in the endometrium about 7 days after conception.

The carrying of one or more embryos in the uterus is called **pregnancy** or **gestation**. In humans, pregnancy averages 266 days (38 weeks).

Human gestation can be divided into three *trimesters*. During the first trimester, the embryo secretes *human chorionic gonadotropin (hCG)*, which maintains the corpus luteum's secretion of progesterone and estrogens.

After implantation, the outer layer of the blastocyst, the **trophoblast**, grows and mingles with the endometrium, helping to form the **placenta**, a disk-shaped organ that is the site of waste removal and gas and nutrient exchange between maternal blood and embryonic capillaries.

Splitting of the embryo in the first month can produce *monozygotic* or identical twins. In contrast, independent fertilization of two eggs can lead to *dizygotic* or fraternal twins. Many embryos may stop developing due to chromosomal or developmental abnormalities.

Organogenesis proceeds so that by the eighth week the embryo has all the rudimentary structures of the adult and is called a **fetus**. In the mother, high progesterone levels stimulate growth of the uterus and the maternal part of the placenta, the cessation of menstrual cycling, and breast enlargement.

During the second trimester, hCG declines, the corpus luteum degenerates, and the placenta secretes its own progesterone, which maintains the pregnancy. The fetus grows rapidly and is quite active. The third trimester is a period of rapid fetal growth.

Labor, a series of strong contractions of the uterus, leads to birth. Hormones (oxytocin and estradiol) and local regulators (prostaglandins) regulate labor. Through a positive feedback system, uterine contractions stimulate the release of oxytocin, which stimulates further contractions.

Lactation is unique to mammals. Prolactin secretion by the anterior pituitary stimulates milk production, and oxytocin triggers the release of milk during nursing.

INTERACTIVE QUESTION   46.7

List the functions of each of the following hormones and the structures that secrete them:

a. human chorionic gonadotropin

b. progesterone

c. oxytocin

Maternal Immune Tolerance of the Embryo and Fetus   Current research is attempting to explain why a mother does not reject an embryo, which has paternal as well as maternal chemical markers. Changes in immune system regulation appear to occur during pregnancy.

Contraception and Abortion   **Contraception**, the deliberate prevention of pregnancy, can be accomplished by several methods: preventing release of egg or sperm, preventing fertilization, or preventing implantation. Complete abstinence from sexual intercourse is the most effective means of birth control. *Natural family planning* is based on refraining from intercourse during the period in which conception is most likely: a number of days before and after ovulation. *Coitus interruptus*, or withdrawal, is an unreliable method of preventing pregnancy.

Barrier methods of contraception include *condoms* and *diaphragms*, which, when used in conjunction with spermicidal foam or jelly, present a physical and chemical barrier to fertilization. Condoms are the only form of birth control that offers some protection against *sexually transmitted diseases (STDs)*, also called *sexually transmitted infections (STIs)*.

IUDs (intrauterine devices) interfere with fertilization and implantation. The release of gametes may be prevented by hormonal contraception, as in **birth control pills**, and by sterilization. Most birth control pills are combinations of synthetic estrogen and progestin, which act by negative feedback to prevent the release of GnRH by the hypothalamus and thus of FSH and LH by the pituitary. This results in both a cessation of follicle development and ovulation. Progestin-only injections or minipills alter the cervical mucus such that it blocks sperm entry to the uterus. Cardiovascular problems are a potential effect of combination birth control pills.

Tubal ligation in women and **vasectomy** in men are sterilization procedures that permanently prevent gamete release.

Abortion is the termination of a pregnancy. Spontaneous abortion, or *miscarriage*, occurs in as many as one-third of all pregnancies. The drug mifepristone (RU486) blocks progesterone receptors in the uterus and can be used to terminate a pregnancy within the first 7 weeks.

INTERACTIVE QUESTION 46.8

List the three general types of birth control methods, and cite examples of each. Indicate which examples are most likely to prevent pregnancy, and which are least likely to do so.

a.

b.

c.

Modern Reproductive Technologies Some genetic diseases and developmental defects can be detected while the fetus is in the uterus, using such techniques as amniocentesis, chorionic villus sampling, ultrasound imaging, and a new technique that identifies fetal cells in the mother's blood, which can then be tested.

Some cases of infertility—the inability to conceive offspring—are the result of inflammatory disorders caused by STDs. Hormone therapy may increase gamete production, and surgeries may correct blocked ducts. Reproductive technology can help couples unable to conceive. *In vitro* **fertilization (IVF)** involves the removal of eggs from a woman, fertilization within a culture dish, and implantation of the developing embryo in the uterus. If sperm are defective or low in number, a sperm nucleus may be directly injected into an oocyte.

Word Roots

a- = not, without (*asexual reproduction:* the generation of offspring from a single parent, without the fusion of gametes; offspring are usually genetically identical to the parent)

acr- = tip; **-some** = body (*acrosome:* a vesicle in the tip of a sperm containing enzymes that help the sperm reach the egg)

blast- = produce; **-cyst** = sac, bladder (*blastocyst:* the blastula stage of mammalian embryonic development, consisting of an inner cell mass, a cavity, and an outer layer, the trophoblast)

contra- = against (*contraception:* the deliberate prevention of pregnancy)

endo- = inside (*endometrium:* the inner lining of the uterus, which is richly supplied with blood vessels)

epi- = above, over (*epididymis:* a coiled tubule adjacent to the testes, where sperm are stored)

o- = egg; **-genesis** = producing (*oogenesis:* the process in the ovary that results in the production of female gametes)

parthen- = a virgin (*parthenogenesis:* asexual reproduction in which females produce offspring from unfertilized eggs)

-theca = a cup, case (*spermatheca:* in many insects, a sac in the female reproductive system in which sperm are stored)

vas- = a vessel; **-ectomy** = cut out (*vasectomy:* the cutting and sealing of each vas deferens to prevent sperm from entering the urethra)

Structure Your Knowledge

1. Use a simple flow diagram to trace the path of a human sperm from the point of production to the point of fertilization; then briefly describe the functions of both the structures the sperm passes through and the associated glands.

2. Answer the following questions concerning the reproductive cycle of the human female.
   a. What does GnRH do?
   b. What does FSH stimulate?
   c. What causes the spike in LH level?
   d. What does this LH surge induce?
   e. What does LH maintain during the luteal phase?
   f. What inhibits the secretion of LH and FSH?
   g. What allows LH and FSH secretion to begin again?

3. Describe how birth control pills work. How does the drug RU486 (mifepristone) function?

Test Your Knowledge

FILL IN THE BLANKS

_____ 1. type of asexual reproduction in which a new individual grows while attached to the parent's body

_____ 2. small, volatile chemicals that may act as mate attractants

_____ 3. development of egg without fertilization

_____ 4. individual with functioning male and female reproductive systems

_____ 5. common opening of digestive, excretory, and reproductive systems in nonmammalian vertebrates

_____ 6. reproductive cycle in which thickened endometrium is shed if fertilization and implantation do not occur

_____ 7. procedure in which oocytes are fertilized in a culture dish and then implanted in the uterus

_____ 8. common duct for urine and semen in mammalian males

_____ 9. filling of a tissue with blood due to increased blood flow

_____ 10. period when ovulation and menstruation cease in human females

MULTIPLE CHOICE: *Choose the one best answer.*

1. Asexual reproduction leads to
   a. haploid offspring from diploid parents.
   b. diploid offspring from haploid parents.
   c. genetically identical offspring.
   d. genetically diverse offspring.

2. Which of the following animals is *least* likely to be hermaphroditic?
   a. an earthworm
   b. a barnacle (sessile crustacean)
   c. a tapeworm
   d. a grasshopper

3. External fertilization is *least* likely to be associated with
   a. spermatheca.
   b. large numbers of gametes and zygotes.
   c. behavioral interaction.
   d. moist environments.

4. What are the products of meiosis I during spermatogenesis?
   a. 2 spermatogonia
   b. 2 primary spermatocytes
   c. 2 secondary spermatocytes
   d. 4 secondary spermatocytes

5. In the developing mammalian oocyte, meiosis is initially halted at
   a. prophase I.
   b. metaphase I.
   c. prophase II.
   d. metaphase II.

6. In which location does fertilization usually take place in a human female?
   a. ovary
   b. oviduct
   c. uterus
   d. vagina

7. Which of the following statements *correctly* describes how the production of human sperm and eggs differs?
   a. The meiotic production of gametes occurs before a female is born but does not begin until puberty in males.
   b. Meiosis occurs in the testes of males but in the oviducts of females.
   c. Primary oocytes stop dividing by mitosis before birth, whereas male primary spermatocytes continue to divide throughout life.
   d. Meiosis is an uninterrupted process in males, whereas in females it resumes when a follicle matures and is only completed when a sperm penetrates the egg.

8. What is the role of the corpus luteum in pregnancy?
   a. It makes testosterone.
   b. It stimulates ovulation.
   c. It makes estradiol and progesterone.
   d. It stimulates uterine contractions during labor.

9. The secretory phase of the uterine cycle
   a. begins with falling levels of estradiol and progesterone.
   b. corresponds with the luteal phase of the ovarian cycle.
   c. corresponds with the follicular phase of the ovarian cycle.
   d. occurs when the endometrium begins to disintegrate and menstrual flow occurs.

Choose from the following hormones to answer questions 10–14.

    a. estradiol

    b. progesterone

    c. LH (luteinizing hormone)

    d. FSH (follicle-stimulating hormone)

    e. hCG (human chorionic gonadotropin)

10. Which hormone stimulates ovulation and the development of the corpus luteum?

11. Which hormone is produced by the developing follicle and initiates thickening of the endometrium?

12. Which hormone is produced by the embryo during the first trimester and is necessary for maintaining a pregnancy?

13. Which hormone stimulates Leydig cells to make testosterone, which in turn stimulates sperm production?

14. Which hormone is produced by the corpus luteum and later by the placenta and is responsible for maintaining a pregnancy?

15. FSH was named for its function in the female reproductive system. Which cells in the male reproductive system respond to FSH?

    a. follicle cells   **c.** spermatogonia

    b. Sertoli cells   **d.** Leydig cells

16. Examples of birth control methods that prevent the production or release of gametes are

    a. hormonal contraception and sterilization.

    b. birth control pills and IUDs.

    c. condoms and diaphragms.

    d. the progestin minipill and RU486.

17. Certain maternal diseases, drugs, alcohol, and radiation are most dangerous to embryonic development

    a. during the first 2 weeks, when the embryo has not yet implanted and spontaneous abortion may occur.

    b. during the first 3 months, when organogenesis is occurring.

    c. during the first and second trimesters, when the embryonic liver is not yet filtering toxins.

    d. during the second and third trimesters, when the most rapid growth is occurring.

Chapter 47

Animal Development

Key Concepts

47.1 Fertilization and cleavage initiate embryonic development

47.2 Morphogenesis in animals involves specific changes in cell shape, position, and survival

47.3 Cytoplasmic determinants and inductive signals regulate cell fate

Framework

Fertilization initiates physical and molecular changes in the egg cell. Early embryonic development includes cleavage to form a blastula, gastrulation, and organogenesis. The amount of yolk in the egg and the phylogeny of the animal determine how these processes occur in any particular animal group.

Development and the differentiation of cells result from the control of gene expression by both cytoplasmic determinants and induction. Pattern formation depends on the positional information a cell receives within a developing structure. Developmental biologists are gradually unraveling some of the mechanisms underlying the complex processes of animal development.

Chapter Review

A common sequence of developmental stages, a common set of regulatory genes, and basic mechanisms of development are shared across many animal species. Biologists use molecular genetics, classical embryology, and **model organisms**, which are easy to study in the lab, to study development.

47.1 Fertilization and cleavage initiate embryonic development

Fertilization **Fertilization** is the union of a haploid egg and sperm to form a diploid zygote. Successful fertilization requires that (1) sperm penetrate any protective layers around the egg, (2) molecules on the sperm surface bind to matching egg receptors, and (3) changes at the egg surface prevent **polyspermy**, or entry of multiple sperm.

In the external fertilization of sea urchins, the jelly coat surrounding the egg releases molecules that attract the sperm. When a sperm comes in contact with the jelly coat, the **acrosome** at the tip of the sperm discharges hydrolytic enzymes. This **acrosomal reaction** allows the *acrosomal process* to elongate through the jelly coat. Fertilization within the same species is assured when proteins on the tip of the acrosomal process bind to specific receptor proteins extending from the egg's plasma membrane.

In response to the fusion of sperm and egg plasma membranes, ion channels open and sodium ions flow into the egg. The resulting depolarization of the membrane prevents other sperm from fusing with the egg, providing a *fast block to polyspermy*. The sperm nucleus then enters the egg cytoplasm.

Membrane fusion also initiates the *cortical reaction*. In response to a high concentration of Ca^{2+} released from the endoplasmic reticulum, cortical granules located in the outer cortex release their contents by exocytosis. The secreted enzymes and other macromolecules clip off sperm-binding receptors and cause the vitelline layer (formed from the egg's extracellular matrix) to elevate and harden into the fertilization envelope, which functions as a *slow block to polyspermy*.

The rise in Ca^{2+} concentration also increases cellular respiration and protein synthesis, known as egg activation. Injecting calcium into an egg that lacks a nucleus artificially activates the egg, indicating that cytoplasmic proteins and mRNAs are sufficient for egg activation.

After the sperm nucleus fuses with the egg nucleus, DNA replication begins in preparation for the cell division that begins the development of the embryo.

In mammalian fertilization, a sperm migrates through the follicle cells released with the egg to the **zona pellucida**, the extracellular matrix of the egg. Complementary binding of a sperm to a receptor in the zona pellucida induces an acrosomal reaction, facilitating passage to the egg.

Binding of the sperm triggers the release of enzymes from cortical granules. This cortical reaction causes alterations of the zona pellucida and functions as a slow block to polyspermy.

Cleavage   The rapid cell divisions of **cleavage** parcel the egg cytoplasm into smaller cells called **blastomeres**. These first divisions usually produce a **blastula**—a cluster of cells surrounding a fluid-filled cavity called the **blastocoel**.

In frogs and many other animals, cleavage is affected by the distribution of **yolk**, which is much more concentrated toward the **vegetal pole** of the egg than toward the **animal pole**. A *cleavage furrow* marks each cell division. The first division of a frog egg bisects the gray crescent, a lighter region opposite the site of sperm entry. The first two cleavage divisions are vertical, extending from the animal pole to the vegetal pole; the third division is equatorial or perpendicular to the polar axis. Yolk impedes cell division, and the plane of division is shifted, resulting in smaller cells being produced in the animal hemisphere. In the frog blastula, the blastocoel forms in the animal hemisphere due to unequal cell division.

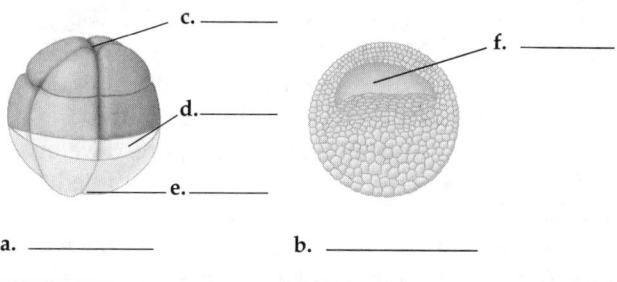
In species in which eggs have little yolk, blastomeres are of equal size and the blastocoel forms centrally. The complete cleavage of eggs with small or moderate amounts of yolk is said to be *holoblastic*.

The large amount of yolk in the eggs of birds, other reptiles, many fishes, and insects results in a type of incomplete cleavage called *meroblastic*.

Divisions of a fertilized bird egg are restricted to the small disk of cytoplasm on top of the yolk. The blastomeres of the resulting cap of cells (called the blastoderm) separate into upper and lower layers, forming a cavity between them comparable to a blastocoel.

In the yolk-rich eggs of most insects, the zygote nucleus undergoes repeated divisions, and the resulting nuclei migrate to the egg surface. Plasma membranes eventually form around the nuclei, creating a blastula made of cells surrounding a mass of yolk.

The RNA and proteins of the egg regulate initial development. After cleavage, the nuclei in the resulting smaller blastomeres are able to produce enough RNA to support each cell's metabolism and program further development.

47.2 **Morphogenesis in animals involves specific changes in cell shape, position, and survival**

Morphogenesis, the development of body shape, occurs through **gastrulation** and **organogenesis**.

Gastrulation   The movement of cells into the blastula produces a two- or three-layered embryo called a **gastrula**. The embryonic **germ layers** produced by gastrulation are the outer **ectoderm** and the inner **endoderm** that lines the embryonic digestive tract. Bilateral animals also have a middle germ layer called **mesoderm**. Adult tissues and organs derive from one or more of these cell layers.

In gastrulation in a sea urchin, cells at the vegetal plate flatten and then buckle inward. This invagination deepens to form a narrow pouch called the archenteron, which will become the digestive tube. The opening to the archenteron, the blastopore, develops into the anus. The tip of the archenteron is pulled across the blastocoel, where it fuses with the ectoderm to form the mouth. Unlike protostomes, in which the mouth develops from the blastopore, sea urchins and chordates are deuterostomes—the mouth develops from this second opening.

Gastrulation in the frog begins with a group of cells forming a small crease on the dorsal side. The part above the crease is called the **dorsal lip** of the blastopore. In a process called involution, surface cells roll over the dorsal lip and move inside, becoming organized into the endoderm and mesoderm. The blastopore eventually forms a circle surrounding a plug of yolk-filled cells. Gastrulation produces an endoderm-lined archenteron surrounded by mesoderm; the outer layer of the gastrula is composed of ectoderm.

INTERACTIVE QUESTION   47.3

Label the parts in the following diagrams of gastrulation in a sea urchin and a frog.

In the chick embryo, a linear thickening called the *primitive streak* is the site of gastrulation. Cells of the upper layer, the *epiblast*, move inward through the primitive streak, forming the mesoderm and the endoderm. Cells remaining in the epiblast become ectoderm. The *hypoblast* cells of the original lower layer form part of the yolk sac and connecting stalk.

Because mammalian eggs contain little yolk, cleavage is holoblastic. The **blastocyst** consists of an outer epithelium, called the **trophoblast**, surrounding a cavity into which protrudes a cluster of cells, called

the **inner cell mass**. These inner cells develop into the embryo. The trophoblast secretes enzymes that enable the blastocyst to embed in the uterine lining (endometrium) and then extends projections that will contribute to the fetal portion of the placenta.

The inner cell mass forms a flat disk with an upper *epiblast* and lower *hypoblast*. In gastrulation, cells from the epiblast move through a primitive streak to form the mesoderm and endoderm. Four **extraembryonic membranes** develop from and support the embryo.

INTERACTIVE QUESTION   47.4

Label the indicated cell layers in the following diagram of a human blastocyst implanting in the endometrium.

Developmental Adaptations of Amniotes   Reptiles (including birds) and mammals are **amniotes**, meaning that they provide the aqueous environment necessary for embryonic development within a fluid-filled sac surrounded by a membrane called the amnion. Extraembryonic membranes are essential to development within the egg of reptiles (and monotremes) and in the uterus of mammals. The chorion is the site of gas exchange in shelled eggs and forms part of the placenta in mammals. The allantois stores wastes in an egg and forms blood vessels in a mammalian umbilical cord. The yolk sac encloses the yolk in reptilian eggs and is the site of blood cell formation in mammals.

Organogenesis   During organogenesis, organs begin to develop from the embryonic germ layers. *Neurulation* begins the formation of the vertebrate brain and spinal cord. In a frog embryo, the **notochord**—the dorsal rod characteristic of chordate embryos—forms from dorsal mesoderm. Through **induction**, signals from nearby tissues cause the ectoderm above the developing notochord to form a thickened *neural plate*. The neural plate curves inward and rolls into a **neural tube**, from which the central nervous system will develop. In human development, a failure of the posterior neural tube to close properly results in the disabling birth defect called *spina bifida*.

In vertebrate embryos, a band of cells called the **neural crest** separates from the side of the neural tube. These neural crest cells migrate to form peripheral nerves as well as teeth and parts of the bones of the skull.

Mesoderm along the sides of the notochord separates into a series of blocks called **somites**. Mesenchyme (migratory) cells from the somites give rise to vertebrae and to the muscles associated with the vertebral column and ribs, arranged in segmented fashion.

Organogenesis in the chick proceeds in a fashion similar to that in the frog embryo. The borders of the embryo fold down and join, forming a three-layered tube attached by a stalk to the yolk.

In invertebrates, many of the same mechanisms drive organogenesis: cell migrations, inductive interactions, and cell shape changes. Although the neural tube forms on the ventral side of an insect embryo, the formation of the tube from ectoderm and the signaling pathways involved are similar to those of vertebrates.

INTERACTIVE QUESTION 47.5

Label the indicated structures in the following diagram of organogenesis in a frog embryo.

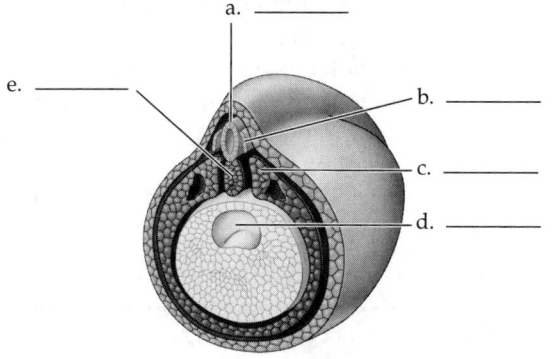

The Cytoskeleton in Morphogenesis Changes in cell shape usually involve reorganization of the cytoskeleton. Extension of microtubules elongates a cell, and then contraction of actin filaments at one end of the cell creates a wedge shape, initiating invaginations and evaginations.

During neurulation in vertebrates, elongation of cells to form the neural plate is mediated by microtubules. Contraction of microfilaments at the apex of these cells then creates the wedge-shaped cells associated with neural tube formation.

By **convergent extension**, cells of a tissue layer elongate and crawl between each other, causing the sheet of cells to extend and narrow.

Embryonic cells use their cytoskeleton to extend and retract cellular protrusions in amoeboid movement. Transmembrane glycoproteins called *cell adhesion molecules* help regulate movement by enabling cell interactions. Secreted molecules in the *extracellular matrix (ECM)* help direct migrating cells.

Apoptosis is a common developmental feature in which cells or whole tissues die and are engulfed by surrounding cells. This type of *programmed cell death* may occur because tissues are no longer needed (a tadpole's tail), because more cells are produced than needed (neurons survive only if they make functional connections with each other), or because structures that are part of an ancestral developmental program are later eliminated (webbing between mammalian digits).

47.3 Cytoplasmic determinants and inductive signals regulate cell fate

As a result of differences in gene expression during development, cells and groups of cells become committed to particular fates (**determination**) and then become specialized in structure and function (**differentiation**).

Fate Mapping In the 1920s, W. Vogt developed a **fate map** for amphibian embryos by labeling regions of the blastula with dye to determine where specific cells showed up in later developmental stages. One type of fate mapping is cell lineage analysis, in which individual blastomeres are marked and their mitotic descendants followed.

The developmental history, or *lineage*, of every cell has been determined for the transparent nematode *Caenorhabditis elegans*. *Germ cells* give rise to eggs or sperm, and their fate is determined early in development. The first cleavage division in *C. elegans* partitions cytoplasmic determinants known as *P granules* in the posterior cell. Further divisions result in P granules ending up in germ cells in the adult worm's gonad.

Studies of programmed cell death during *C. elegans* development helped identify cell death pathways present in many animals.

Axis Formation In bilateral animals, the three body axes (anterior-posterior, dorsal-ventral, and left-right) are established early in development. In the frog egg, the animal hemisphere has melanin granules in the cortex, whereas the vegetal hemisphere contains the yellow yolk. The animal-vegetal axis determines the anterior-posterior (head-tail) axis of the embryo.

Fertilization results in a rotation of the animal pole cortex toward the point of sperm entry, exposing the gray crescent that marks the future dorsal side. This *cortical rotation* leads to inductive interactions that influence the development of dorsal and ventral structures.

In chicks, gravity may be involved in establishing the anterior-posterior axis while the egg moves down the oviduct, and pH differences may determine the dorsal-ventral axis of the blastoderm. In mammals, polarity becomes obvious only after cleavage, but the location of sperm and egg nuclei before fusion may play a role. In insects, morphogen gradients establish the body axes.

The left-right axis is fixed once the other two axes are determined. Vertebrates have a left-right asymmetry of several internal organs.

Restricting Developmental Potential   By definition, fertilized eggs are **totipotent**; that is, they are capable of giving rise to all cell types. Working with frog embryos, H. Spemann found that the *developmental potential* of the first two blastomeres is not restricted. Because the first cleavage division bisects the gray crescent, cytoplasmic determinants in this region are divided equally between the two blastomeres.

Mammalian blastomeres remain totipotent through the eight-cell stage, although this may be partly due to a separated cell's ability to respond to its environment. The progressive restriction of a cell's developmental potency is characteristic of animal development. Experimental manipulation shows that the tissue-specific fates of cells of a late gastrula are usually fixed.

INTERACTIVE QUESTION   47.6

If a researcher divides an eight-cell stage of a sea urchin embryo vertically into two four-cell groups, both develop into normal larvae. If the division is made horizontally, abnormal larvae develop. Explain this experimental result.

Cell Fate Determination and Pattern Formation by Inductive Signals   Cells influence the development of other cells through signals that usually switch on sets of genes.

Using transplantation experiments with amphibian embryos, H. Mangold and H. Spemann established that the dorsal lip acts as an "organizer" of the embryo because of its influence in early organogenesis. The molecular basis of induction by *Spemann's organizer* appears to be tied to the inactivation of bone morphogenetic protein 4 (BMP-4), an event that allows cells to make dorsal structures on the dorsal side of the embryo. Proteins similar to BMP-4 and its inhibitors appear to play roles in development in the fruit fly.

Pattern formation is the ordering of cells and tissues into characteristic structures in the proper locations in the animal. Cells and their offspring differentiate based on **positional information**, molecular cues that indicate the cells' positions with respect to body axes.

A limb bud of a chick consists of a core of mesoderm beneath a layer of ectoderm. Researchers have identified two regions that provide positional information that direct limb development.

Cells of the **apical ectodermal ridge (AER)** at the tip of the limb bud secrete fibroblast growth factor (FGF). These growth signals appear to promote limb bud outgrowth and patterning along the proximal-distal axis.

The **zone of polarizing activity (ZPA)**, mesodermal tissue located at the posterior attachment of the bud, appears to communicate position along the anterior-posterior axis in a developing limb. Cells of the ZPA secrete a protein called Sonic hedgehog. Hedgehog proteins have been found to influence development in many different organisms.

A cell's developmental history shapes that cell's response to positional information relating to its location in a developing organ. Thus, patterns of *Hox* gene expression first determine whether cells will become part of a wing or part of a leg; then positional cues indicate what part of the wing or leg the cells will help form.

INTERACTIVE QUESTION   47.7

What would be the effect of implanting cells genetically engineered to produce and secrete Sonic hedgehog to the anterior attachment of a limb bud?

Cilia and Cell Fate   Recent research indicates that the stationary primary cilia, or *monocilia*, found on eukaryotic cells play important developmental roles through their reception of signaling proteins, including Sonic hedgehog.

Kartagener's syndrome, which stems from a defect in the function of motile cilia, is characterized by recurrent respiratory tract infections, immotile sperm, and often, a reversal of the placement of internal organs, called *situs inversus*. It appears that during embryonic development, cilia generate a leftward flow of fluid that initiates the normal left-right asymmetry of organs.

Word Roots

acro- = the tip (*acrosomal reaction:* the discharge of hydrolytic enzymes from the acrosome when a sperm approaches or contacts an egg)

arch- = ancient, beginning (*archenteron:* the endoderm-lined cavity, formed during gastrulation, that develops into the digestive tract)

blast- = bud, sprout; **-pore** = a passage (*blastopore:* in a gastrula, the opening of the archenteron that typically develops into the anus in deuterostomes and the mouth in protostomes)

ecto- = outside; **-derm** = skin (*ectoderm:* the outermost of the three primary germ layers in animal embryos)

endo- = within (*endoderm:* the innermost of the three primary germ layers in animal embryos)

extra- = beyond (*extraembryonic membrane:* one of four membranes located outside the embryo that support the developing embryo in reptiles and mammals)

fertil- = fruitful (*fertilization:* the union of haploid gametes to produce a diploid zygote)

gastr- = stomach, belly (*gastrulation:* a series of cell and tissue movements in which the blastula-stage embryo folds inward, producing a three-layered embryo, the gastrula)

hol- = whole (*holoblastic:* referring to a type of cleavage in which there is complete division of the egg)

mer- = a part (*meroblastic:* referring to a type of cleavage in which there is incomplete division of a yolk-rich egg, characteristic of avian development)

mes- = middle (*mesoderm:* the middle primary germ layer in an animal embryo)

not- = the back; **-chord** = a string (*notochord:* a dorsal, flexible rod made of tightly packed mesodermal cells that runs along the anterior-posterior axis of a chordate)

som- = body (*somite:* one of a series of paired blocks of mesoderm just lateral to the notochord in a vertebrate embryo)

troph- = nourish (*trophoblast:* the outer epithelium of a mammalian blastocyst; forms the fetal part of the placenta)

zona = a belt; **pellucid-** = transparent (*zona pellucida:* the extracellular matrix surrounding a mammalian egg)

Structure Your Knowledge

1. Create a flowchart that shows the sequence of key events in the fertilization of a sea urchin egg and indicate the functions of each of these events.

2. Fill in the following table, briefly describing these stages of development in sea urchin, frog, bird, and mammalian embryos.

Animal	Cleavage	Blastula	Gastrulation
Sea urchin			
Frog			
Bird			
Mammal			

Test Your Knowledge

MULTIPLE CHOICE: *Choose the one best answer.*

1. The fact that an eunucleated sea urchin egg can be activated by the injection of Ca^{2+} is evidence that
   a. development can proceed parthenogenetically.
   b. inactive mRNA and other proteins had been stockpiled during development of the egg.
   c. the cortical reaction is not necessary for development to proceed.
   d. development is a cytoplasmic process that does not require genetic regulation.

2. The formation of the fertilization envelope in a sea urchin egg is part of the process known as
   a. cleavage.
   b. neurulation.
   c. the slow block to polyspermy
   d. the fast block to polyspermy

3. The blastocoel
   a. develops into the archenteron or embryonic gut.
   b. is a fluid-filled cavity in the blastula.
   c. opens to the exterior through a blastopore.
   d. Both **b** and **c** are correct.

4. In a frog embryo, gastrulation
   a. is slowed by the large amount of yolk.
   b. proceeds by involution as cells roll over the dorsal lip of the blastopore.
   c. produces a blastocoel that is displaced into the animal hemisphere.
   d. occurs along the primitive streak.

5. Which of the following body structures is *incorrectly* paired with the embryonic germ layer from which it arose?
   a. muscles—mesoderm
   b. central nervous system—ectoderm
   c. lining of gut—endoderm
   d. notochord—ectoderm

6. The primitive streak of mammalian embryos is analogous to the
   a. dorsal lip of frog embryos.
   b. blastoderm of bird embryos.
   c. multinucleated stage of fruit fly embryos.
   d. archenteron of sea urchin embryos.

7. Which of the following phenomena is a result of apoptosis during development?
   a. the birth defect spina bifida
   b. the left-right asymmetry of some vertebrate organs
   c. the concentration of P granules in the formation of germ cells in *C. elegans*
   d. the removal of webbing between mammalian digits

8. Somites are
   a. blocks of mesoderm circling the coelom.
   b. clumps of cells from which the notochord arises.
   c. serially arranged mesoderm blocks lateral to the notochord in a vertebrate embryo.
   d. structures arising from neural crest cells.

9. To which portion of the developing embryo do cells of the trophoblast contribute?
   a. ectoderm
   b. mesoderm
   c. endoderm
   d. None of the above. Trophoblast cells contribute only to extraembryonic structures.

Use the following diagram of a chick embryo for questions 10–13.

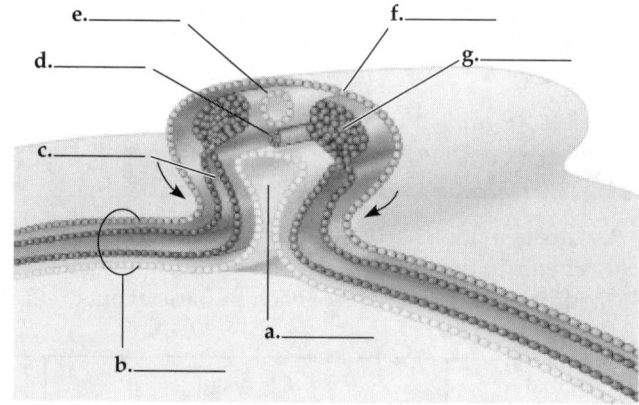

10. Which line points to the notochord?

11. Which line points to the archenteron?

12. To what structure does line b point?
   a. the primitive streak through which gastrulation occurred
   b. the hypoblast
   c. the mesodermal lining of the coelom
   d. the developing extraembryonic membranes

13. Which structures are derived from ectoderm?
    a. e and f
    c. f and g
    b. d and e
    d. f

14. During gastrulation in the frog embryo, a new cavity is formed. This cavity is
    a. the archenteron.
    b. the blastocoel.
    c. the gastrocoel.
    d. the brain cavity.

15. Gastrulation and organogenesis in mammals most closely resemble those processes in
    a. sea urchins, because the eggs of both groups have little yolk and cleavage is holoblastic.
    b. chicks, because both birds and mammals share a common reptilian ancestor.
    c. frogs, because gastrulation takes place through the dorsal lip of the blastopore in both groups.
    d. neither chicks nor frogs, because mammals have some embryonic cells that develop into extraembryonic membranes.

16. The first cleavage in frog eggs bisects the gray crescent. What is the developmental significance of the gray crescent?
    a. The gray crescent marks the site of the future mouth.
    b. The neural tube will be derived from cells containing gray crescent cytoplasm.
    c. Cells containing gray crescent cytoplasm will form germ cells.
    d. The gray crescent marks the future dorsal side of the embryo.

17. In their transplantation experiments with frog embryos, Mangold and Spemann found that moving a part of the dorsal lip to another location would result in a second gastrulation, and even in the development of a doubled, face-to-face tadpole. The best explanation for these results is
    a. morphogenetic movements.
    b. totipotency of cells.
    c. induction by a dorsal lip "organizer."
    d. fate map determination.

18. Pattern formation is most directly influenced by
    a. positional information a cell receives from gradients of molecular signals.
    b. differentiation of cells that then migrate into developing organs.
    c. the migration of cells guided by molecules secreted into the ECM.
    d. gastrulation and the formation of the three tissue layers.

19. Cells of the zone of polarizing activity (ZPA) secrete
    a. Sonic hedgehog, which communicates location along the anterior-posterior axis.
    b. cell adhesion molecules that hold migrating cells together as they form tissues of the leg.
    c. fibroblast growth factors that promote limb bud outgrowth.
    d. positional information that causes the limb bud to elongate by convergent extension.

20. How might *situs inversus* be related to infertility?
    a. Both involve defects in monocilia that interfere with cell signaling.
    b. Both involve asymmetrical arrangements of organs that interfere with their functioning.
    c. Both involve defects in the structure of cilia or flagella.
    d. Both involve the inability of cells to propel themselves through a liquid medium.

Neurons, Synapses, and Signaling

Framework

This chapter describes the transmission of nervous impulses along and between neurons. Ion channels establish the membrane potential of a neuron, and the rapid flow of ions through voltage-gated channels creates the rapid depolarization and repolarization of an action potential. The release of neurotransmitters at synapses converts electrical signals to chemical signals that pass information to receiving cells.

Chapter Review

Neurons transmit long-distance electrical signals throughout the body and usually communicate between cells using short-distance chemical signals. The higher-order processing of nervous signals may involve clusters of neurons called **ganglia** or more structured groups of neurons organized into a **brain**.

48.1 Neuron structure and organization reflect function in information transfer

Neuron Structure and Function   A neuron consists of a **cell body**, which contains the nucleus and

organelles, as well as numerous extensions. The highly branched, short **dendrites** (along with the cell body) *receive* signals from other neurons; the single, longer **axon** *transmits* signals to other cells. Signals that travel down an axon originate from a region of the cell body called the axon hillock.

Information is transmitted to another cell at a junction called a **synapse**. At the branching ends of axons are *synaptic terminals*, which usually release **neurotransmitters** that relay signals to another neuron, a muscle, or a gland cell. The transmitting cell is called the *presynaptic cell*; the receiving cell is the *postsynaptic cell*.

The very numerous supporting cells called **glial cells**, or **glia**, give structural integrity and physiological support to neurons.

INTERACTIVE QUESTION   48.1

Label the indicated structures on the following diagram of a neuron. Indicate the direction of impulse transmission. What happens at part e?

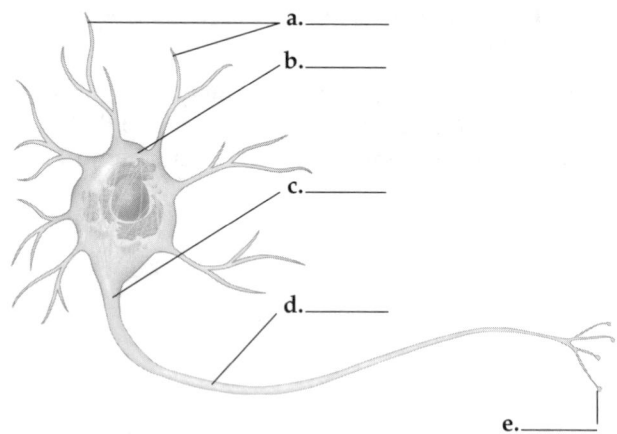

a. _____
b. _____
c. _____
d. _____
e. _____

Introduction to Information Processing The three stages of information processing are sensory input, integration, and motor output. Sensors detect external stimuli or internal conditions. **Sensory neurons** transmit this information to the brain or ganglia where **interneurons** integrate it. **Motor neurons** carry impulses that stimulate muscle contraction or glandular activity.

A **central nervous system (CNS)**, found in many animals, includes the neurons involved in integration. The **peripheral nervous system (PNS)** consists of neurons that carry information to and from the CNS. Bundles of axons outside the CNS are called **nerves**.

48.2 Ion pumps and ion channels establish the resting potential of a neuron

Due to the separation of charges, a voltage, or **membrane potential**, exists across a cell's plasma membrane. In a typical neuron at rest, the membrane potential—called the **resting potential**—is between −60 and −80 mV.

Formation of the Resting Potential In most neurons, the concentration of K^+ is higher inside the cell than outside and the Na^+ concentration is greater outside than inside. The **sodium-potassium pump** maintains these gradients, using the energy of ATP hydrolysis to move three Na^+ out of the cell for every two K^+ it pumps in.

Ion channels that are *selectively permeable* allow specific ions to diffuse across the membrane. The diffusion of K^+ out of a cell through the many open potassium channels (sometimes called leak channels) results in a buildup of negative charge within the neuron, creating the membrane potential. As negative charges inside the cell increase, they attract positively charged potassium ions, slowing their movement out of the cell.

Modeling the Resting Potential Ions diffuse through channels down their concentration gradients until their chemical gradient is balanced by the electrical gradient across the membrane. The membrane voltage at this equilibrium, or **equilibrium potential (E_{ion})**, is determined for a single ion with a 1+ charge at 37°C by the Nernst equation: $E_{ion} = 62$ mV (log ([ion]$_{outside}$/[ion]$_{inside}$)). The equilibrium potential for K^+ (E_K) across a membrane is −90 mV. Using the concentration gradient for Na^+, E_{Na} is +62 mV (the inside of the membrane is more positive than outside).

Neurons at rest have more open K^+ channels than open Na^+ channels, and the resting potential is closer to E_K than to E_{Na}.

INTERACTIVE QUESTION 48.2

a. What is the principal cation inside the cell? Outside the cell?

b. Which side of the membrane has a negative charge?

c. What change in the permeability of the cell's membrane to K^+ and/or Na^+ would cause the cell's membrane potential to shift from −70 mV to −80 mV?

48.3 Action potentials are the signals conducted by axons

You have learned that ungated potassium and sodium ion leak channels create a neuron's resting potential. But neurons also have **gated ion channels**, which open or close in response to stimuli and alter the membrane potential. Electrophysiologists measure membrane potential by placing microelectrodes connected to a voltage recorder inside and outside a cell.

Hyperpolarization and Depolarization A stimulus that opens gated potassium channels will result in **hyperpolarization**, as K^+ flows out and the membrane potential moves toward E_K (−90 mV). When gated sodium channels open and Na^+ flows in, a **depolarization** occurs as the inside of the cell becomes less negative and shifts toward E_{Na} (+62 mV).

Graded Potentials and Action Potentials A shift in membrane potential can result in a transient **graded potential**, whose magnitude is proportional to the strength of the stimulus. An **action potential** is a massive change in membrane voltage that has a constant magnitude and can spread to adjacent regions of the membrane. **Voltage-gated ion channels** open or close in response to a change in membrane potential. When a depolarization opens voltage-gated sodium channels, Na^+ inflow increases depolarization and causes more sodium channels to open. Once this positive feedback results in depolarization of a neuron to a certain membrane voltage called the **threshold**, an action potential is triggered. The action potential is an *all-or-none* event, always creating the same positive voltage spike once the threshold is reached.

Generation of Action Potentials: A Closer

Look Both Na$^+$ and K$^+$ voltage-gated channels are involved in an action potential. Sodium channels open rapidly in response to depolarization. After a very short time, these channels become inactivated when a loop of the channel protein blocks the opening. Potassium channels open slowly in response to depolarization, but remain open throughout the action potential.

As a stimulus depolarizes the membrane, Na$^+$ channels open; the influx of Na$^+$ causes further depolarization, which opens more channels. If the depolarization reaches the threshold, an action potential is triggered. The continuing positive-feedback cycle of Na$^+$ influx during the *rising phase* brings the membrane potential close to E_{Na}. The inactivation of voltage-gated Na$^+$ channels and the opening of voltage-gated K$^+$ channels rapidly bring the membrane potential back toward E_K during the *falling phase*.

During the *undershoot*, the membrane's permeability to K$^+$ is higher than at rest, and the continued outflow of K$^+$ temporarily hyperpolarizes the membrane. During the **refractory period**, which occurs while the Na$^+$ channels remain inactivated, the neuron cannot respond to another stimulus.

INTERACTIVE QUESTION 48.3

The following diagram shows the changes in voltage-gated channels during an action potential. Label the channels and inactivation loop, the ions, and the components of the graph. Name and describe the five phases of the action potential. Place numbers on the graph to show where each phase is occurring.

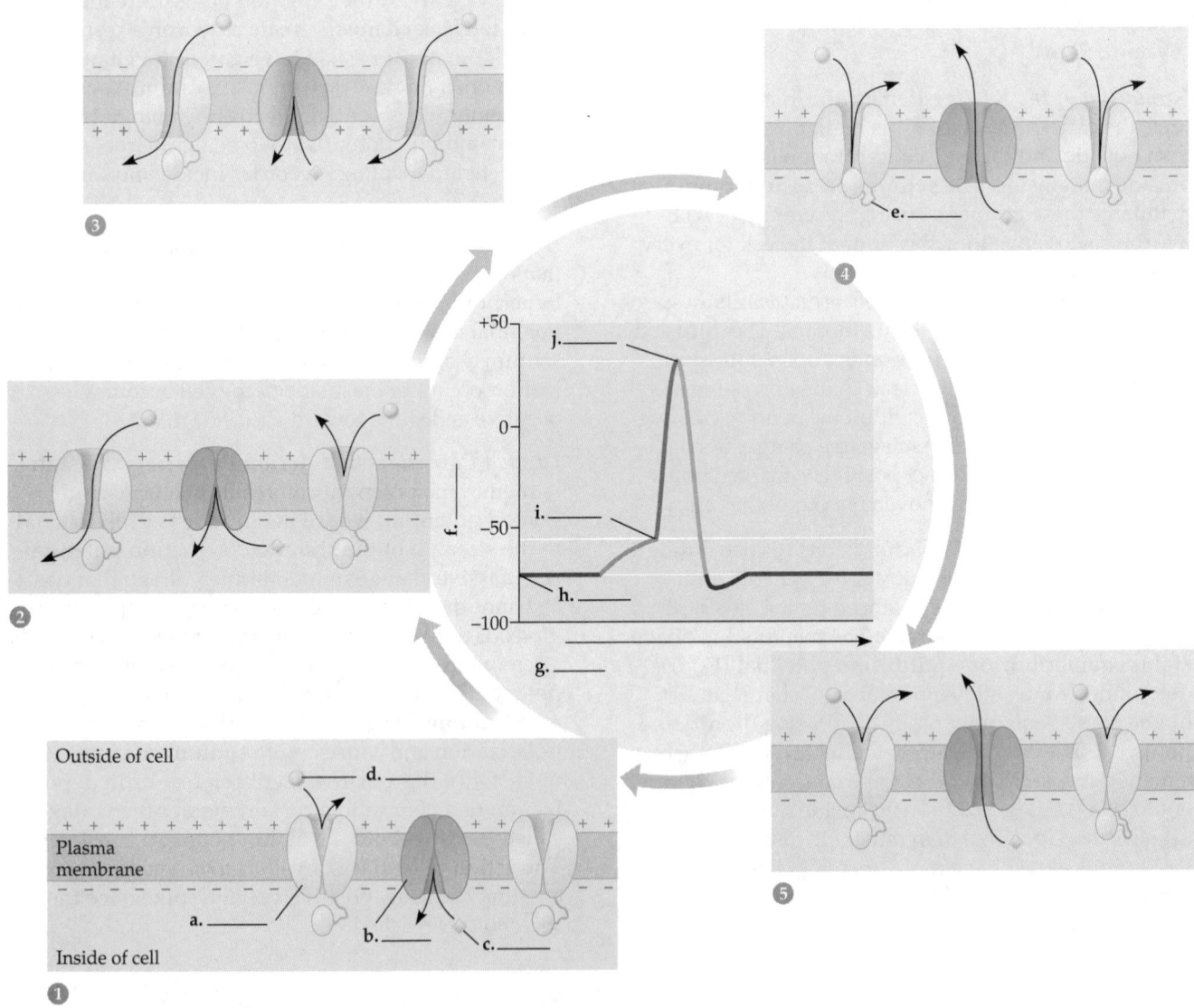

Conduction of Action Potentials Na$^+$ influx in the rising phase depolarizes adjacent sections of the membrane, bringing them to threshold. Local depolarizations and action potentials across the membrane result in the propagation of serial action potentials along the length of the neuron. Because of the brief refractory period, the action potential is propagated in one direction only.

Action potentials last only about 1–2 milliseconds. The frequency of action potentials increases with the intensity of the stimulus.

Resistance to current flow is inversely proportional to the cross-sectional area of the conducting "wire." The greater the axon diameter, the faster action potentials are propagated. Some invertebrates, such as squid, have giant axons that conduct impulses very rapidly.

In vertebrates, **oligodendrocytes** (in the CNS) and **Schwann cells** (in the PNS) insulate axons in multiple layers of membranes called a **myelin sheath**. Voltage-gated ion channels are concentrated in the **nodes of Ranvier**, small gaps between successive Schwann cells. Action potentials can be generated only at these nodes, and a nerve impulse "jumps" from node to node, resulting in a faster mode of transmission known as **saltatory conduction**.

INTERACTIVE QUESTION 48.4

Return to the diagram of a neuron in Interactive Question 48.1 and draw in a myelin sheath. Label a Schwann cell and node of Ranvier.

48.4 Neurons communicate with other cells at synapses

At electrical synapses, an electrical current flows directly from cell to cell via gap junctions. Electrical synapses are found in the giant axons of squid and lobsters and in many neurons of the vertebrate brain.

Most synapses are chemical synapses, which involve the release of neurotransmitters, which are stored in *synaptic vesicles* in synaptic terminals. The depolarization of the presynaptic membrane opens voltage-gated calcium channels in the membrane. The influx of Ca^{2+} causes the synaptic vesicles to fuse with the presynaptic membrane and release neurotransmitter into the narrow *synaptic cleft*.

Generation of Postsynaptic Potentials Binding of neurotransmitter to a receptor on a **ligand-gated ion channel** (often called an *ionotropic receptor*) in the postsynaptic cell membrane allows ions to cross the membrane, creating a graded *postsynaptic potential*. If both Na$^+$ and K$^+$ are able to diffuse through the channel, the net inflow of positive charge depolarizes the membrane, creating an **excitatory postsynaptic potential (EPSP)** that brings the membrane potential closer to threshold. If binding of neurotransmitter opens K$^+$ or Cl$^-$ channels, the membrane hyperpolarizes, producing an **inhibitory postsynaptic potential (IPSP)**.

INTERACTIVE QUESTION 48.5

Identify the components of this chemical synapse following the depolarization of the synaptic terminal.

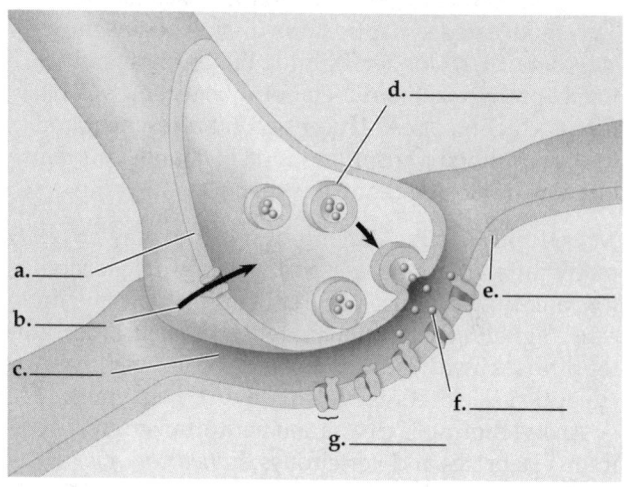

Summation of Postsynaptic Potentials Postsynaptic potentials are graded potentials. Their magnitude depends on the number of neurotransmitter molecules that bind to receptors and the distance a synapse is from the axon hillock. The membrane potential of the axon hillock at any given time is determined by the sum of all EPSPs and IPSPs.

INTERACTIVE QUESTION   48.6

a. _____ *summation* occurs with repeated release of neurotransmitter from a synaptic terminal before the postsynaptic potential returns to its resting potential.

b. _____ *summation* occurs when several different presynaptic terminals, usually from different neurons, release neurotransmitter simultaneously.

Termination of Neurotransmitter Signaling   To return a synapse to its resting state, neurotransmitters must be removed from the synaptic cleft. This may occur through diffusion of the neurotransmitter out of the cleft, enzymatic degradation of the neurotransmitter, or reuptake of the neurotransmitter by the presynaptic neuron.

Modulated Signaling at Synapses   Some neurotransmitters bind to *metabotropic receptors*, which trigger a signal transduction pathway in the postsynaptic cell. This modulated synaptic transmission begins more slowly but lasts longer. Binding of neurotransmitter to a G protein-coupled metabotropic receptor leads to the production of cAMP as a second messenger and to the amplified phosphorylation of channel proteins, opening or closing many ion channels.

Neurotransmitters   A neurotransmitter may have many different ionotropic and metabotropic receptors and may produce different effects in postsynaptic cells. Signaling is terminated when the neurotransmitter diffuses away, is taken up and repackaged into synaptic vesicles, or is broken down by enzymes.

Acetylcholine is a common neurotransmitter in invertebrates and vertebrates. In *neuromuscular junctions*, acetylcholine released from a motor axon produces an EPSP in a muscle cell. The enzyme acetylcholinesterase breaks down the neurotransmitter in the synaptic cleft. The same ionotropic receptor in the PNS and CNS binds nicotine, producing its physiological and psychological stimulant effect.

Aside from acetylcholine, the rest of the more than 100 identified neurotransmitters fall into four groups: amino acids, biogenic amines, neuropeptides, and gases.

The amino acids **glutamate** and **gamma aminobutyric acid (GABA)** function as neurotransmitters in the CNS. Glutamate, the most common CNS neurotransmitter, is excitatory and plays a key role in long-term memory formation. GABA is the most common inhibitory transmitter in the brain. Glycine is an inhibitory amino acid in parts of the CNS outside the brain.

Biogenic amines are neurotransmitters derived from amino acids. **Norepinephrine**, derived from tyrosine, is an excitatory neurotransmitter in the autonomic nervous system. It also acts as a hormone, as does the related biogenic amine *epinephrine*. **Dopamine**, also derived from tyrosine, and **serotonin**, made from tryptophan, affect sleep, mood, attention, and learning. Imbalances of these neurotransmitters have been associated with several nervous system disorders.

Neuropeptides are short chains of amino acids that function via G protein-coupled metabotropic receptors. *Substance P* is an excitatory neurotransmitter that functions in pain perception. **Endorphins** are neuropeptides that decrease pain perception. Endorphins are produced in the brain during physical or emotional stress, and their effects include pain reduction, euphoria, and other physiological effects. Opiates bind to endorphin receptors in the brain.

Some neurons use nitric oxide (NO) and carbon monoxide as neurotransmitters. The release of NO triggers relaxation of smooth muscle cells and vessel dilation. CO produced by neurons in the brain affects the release of hypothalamic hormones.

INTERACTIVE QUESTION   48.7

Acetylcholine stimulates skeletal muscle contraction but inhibits or slows cardiac muscle contraction. Explain how this neurotransmitter has such opposite effects.

Word Roots

bi- = life; **-genic** = producing, produced by (*biogenic amine*: a neurotransmitter derived from an amino acid)

dendr- = tree (*dendrite*: one of usually numerous, short, highly branched extensions of a neuron that receive signals from other neurons)

de- = down, out (*depolarization*: a change in a cell's membrane potential such that the inside of the membrane is made less negative relative to the outside)

endo- = within (*endorphin*: any of several hormones produced in the brain that inhibit pain perception)

glia = glue (*glia*: cells of the nervous system that support, regulate, and augment the functions of neurons)

hyper- = over, above, excessive (*hyperpolarization*: a change in a cell's membrane potential such that the inside of the membrane becomes more negative relative to the outside)

inter- = between (*interneurons*: an association neuron; a nerve cell within the central nervous system that forms synapses with sensory and/or motor neurons and integrates sensory input and motor output)

neur- = nerve; **trans-** = across (*neurotransmitter: a molecule that is released from the synaptic terminal of a neuron at a chemical synapse, diffuses across the synaptic cleft, and binds to the postsynaptic cell, triggering a response*)

olig- = few, small (*oligodendrocyte: a type of glial cell that forms insulating myelin sheaths around the axons of neurons in the central nervous system*)

salt- = leap (*saltatory conduction: rapid transmission of a nerve impulse along an axon resulting from the action potential jumping from one node of Ranvier to another, skipping the myelin-sheathed regions of membrane*)

syn- = together (*synapse: the junction where a neuron communicates with another cell across a narrow gap via a neurotransmitter or an electrical coupling*)

Structure Your Knowledge

1. Develop a flowchart or diagram or write a description of the sequence of events in the (1) creation and propagation of an action potential and (2) in the transmission of this action potential across a chemical synapse.

Test Your Knowledge

MULTIPLE CHOICE: *Choose the one best answer.*

1. Which of the following nervous system components is *incorrectly* paired with its function?
   a. axon hillock—region of neuron where an action potential originates
   b. Schwann cells—produce a myelin sheath around axons in the PNS
   c. synapse—space between presynaptic and postsynaptic cells into which neurotransmitter is released
   d. synaptic terminal—receptor that is part of an ion channel that is keyed to a specific neurotransmitter

2. Which of the following best describes interneurons?
   a. may connect sensory and motor neurons
   b. more common in the PNS than the CNS
   c. typically have more axons than dendrites
   d. All of the above are accurate descriptions.

3. During a neuron's resting state
   a. there are more potassium ions outside the axon than inside it.
   b. there are more sodium ions outside the axon than inside it.
   c. ions are distributed equally across the axonal membrane.
   d. ions move freely across the membrane.

4. Which of the following contribute(s) to the resting potential of a neuron?
   a. the Na^+/K^+ pump
   b. K^+ leak channels
   c. voltage-gated Na^+ channels
   d. both **a** and **b**

5. An action potential results from a dramatic depolarization of the plasma membrane of a neuron. This depolarization involves an
   a. influx of sodium ions.
   b. influx of potassium ions.
   c. efflux of sodium ions.
   d. influx of electrons.

6. The threshold of a membrane
   a. is midway between E_K and E_{Na}.
   b. will, once it is reached, open voltage-gated channels and permit the rapid outflow of sodium ions.
   c. is the depolarization that is needed to generate an action potential.
   d. is a graded potential that is proportional to the strength of a stimulus.

7. How is an increase in the strength of a stimulus communicated by a neuron?
   a. The spike of the action potential reaches a higher voltage.
   b. The frequency of action potentials generated along the neuron increases.
   c. The length of an action potential (the duration of the rising phase) increases.
   d. All action potentials are the same, so the nervous system cannot discriminate between stimuli of different strengths.

8. After the rapid depolarization of an action potential, the fall in the membrane potential occurs due to the
   a. delay in the opening of voltage-gated potassium channels.
   b. closing of potassium and sodium channels.
   c. refractory period, during which the membrane is hyperpolarized.
   d. opening of voltage-gated potassium channels and the closing of sodium inactivation loops.

9. Nodes of Ranvier are
   a. gaps where Schwann cells abut and action potentials are generated.
   b. neurotransmitter-containing vesicles located in the synaptic terminals.
   c. the parts of neurons where action potentials are initiated.
   d. clusters of receptor proteins located on the postsynaptic membrane.

10. Movement of an action potential in only one direction along a neuron is a function of
    a. saltatory conduction.
    b. the pathway from dendrite to axon.
    c. the refractory period, when sodium channels are still inactivated.
    d. the localized depolarization of the surrounding membrane.

11. Signal transmission is faster in myelinated axons because
    a. these axons are thinner and thus present less resistance to voltage flow.
    b. these axons use electrical synapses rather than chemical synapses.
    c. the action potential can jump from node to node along the insulating myelin sheath.
    d. these axons are thicker and thus present less resistance to voltage flow.

12. Which of the following statements concerning chemical synapses is *not* true?
    a. Synaptic terminals at the ends of branching axons contain synaptic vesicles, which enclose the neurotransmitter.
    b. The influx of sodium when an action potential reaches the presynaptic membrane causes synaptic vesicles to release their neurotransmitter into the cleft.
    c. The binding of neurotransmitter to receptors on the postsynaptic membrane usually opens (or sometimes closes) ligand-gated ion channels.
    d. Neurotransmitter is often rapidly degraded in the synaptic cleft.

13. What would be the direct result if a toxin blocked Ca^{2+} voltage-gated ion channels in neurons?
    a. No effect because Na^+ and K^+ voltage-gated channels were not blocked.
    b. No IPSP could be created, although EPSP would still be generated.

c. Synaptic vesicles could not initiate the process of exocytosis, and transmission across chemical synapses would cease.
d. No action potentials could be generated; death would quickly follow.

14. A neurotransmitter that caused potassium ions to leave the postsynaptic cell would cause
    a. an EPSP.
    b. an IPSP.
    c. an action potential in the postsynaptic cell.
    d. the death of the postsynaptic cell.

15. Consider a motor neuron that stimulates a muscle fiber to contract. If you wanted to enhance the contraction of the muscle fiber you could
    a. increase the amount of acetylcholine released from the neuron.
    b. decrease the amount of acetylcholinesterase in the space between the neuron and the fiber.
    c. decrease the number of acetylcholine receptors on the muscle fiber plasma membrane.
    d. either **a** or **b**

16. Which of the following pairs is mismatched?
    a. amino acid—glutamate
    b. neuropeptide—endorphin
    c. biogenic amine—dopamine
    d. steroid—acetylcholine

17. The main effect of the neurotransmitter GABA in the CNS is to
    a. increase pain.
    b. create inhibitory postsynaptic potentials.
    c. affect sleep, mood, and learning.
    d. decrease pain and induce euphoria.

18. If the binding of a neurotransmitter to its receptor opens Cl^- channels, what would be the effect on the postsynaptic cell? (Cl^- is in higher concentration outside the cell.)
    a. It would hyperpolarize, producing an IPSP.
    b. It would depolarize and form an EPSP but probably not generate an action potential.
    c. It would initiate a signal transduction pathway as part of modulated signaling.
    d. Its membrane potential would not change because neither sodium nor potassium channels opened.

Chapter 49

Nervous Systems

Key Concepts

49.1 Nervous systems consist of circuits of neurons and supporting cells

49.2 The vertebrate brain is regionally specialized

49.3 The cerebral cortex controls voluntary movement and cognitive functions

49.4 Changes in synaptic connections underlie memory and learning

49.5 Many nervous system disorders can be explained in molecular terms

Framework

This chapter describes the organization of the vertebrate nervous system and the functional specialization of its components. The branches of the peripheral nervous system include the sensory afferent neurons and the efferent motor system and autonomic nervous system. The central nervous system includes the spinal cord, which mediates reflexes and transmits information, and the regionally specialized brain, which regulates homeostatic functions, integrates sensory information, and controls voluntary movement and cognitive functions.

Chapter Review

The estimated 10^{11} neurons of the human brain communicate in complex circuits. New techniques of "brainbow" staining of neurons and powerful imaging techniques of brain activity are enhancing the study of the structure and function of the brain.

49.1 Nervous systems consist of circuits of neurons and supporting cells

Mechanisms to sense and respond to changes in the environment evolved in prokaryotes. Modification of this ability allowed for the development of communication between the cells of multicellular organisms.

Cnidarians have a simple diffuse *nerve net* of interconnected, individual neurons. In the nervous system of more complex animals, axons of multiple neurons are bundled into **nerves**.

Cephalization, the concentration of sensory neurons and interneurons at the anterior end of the body, evolved in bilaterally symmetrical animals. In many animals, a **central nervous system (CNS)** consists of neurons that integrate information. A **peripheral nervous system (PNS)** consists of neurons that carry information to and from the CNS. Flatworms have the simplest CNS, consisting of a small brain and two longitudinal nerve cords. Annelids and arthropods have a more complicated brain and a ventral nerve cord with segmentally arranged clusters of neurons called **ganglia**. Vertebrates have a brain and a dorsal spinal cord; nerves and ganglia make up the PNS.

INTERACTIVE QUESTION 49.1

The organization of an organism's nervous system correlates with how the animal interacts with its environment. How do the different nervous systems of clams and squid illustrate this generalization?

Glia The nervous systems of vertebrates and most invertebrates include **glial cells**, or **glia**. Glia in the vertebrate CNS include ciliated ependymal cells, which line the ventricles; microglia, which protect against pathogens; and oligodendrocytes, which function in axon myelination. **Astrocytes** are supporting glia in the CNS with various functions. Some facilitate information transfer at synapses and increase blood flow to active neurons. Astrocytes help form the *blood–brain barrier*, which restricts the passage of most substances from the blood into the brain. Schwann cells myelinate axons in the PNS. *Radial glia* form tracks that guide the embryonic growth of neurons. Radial glia and astrocytes can act as stem cells.

Organization of the Vertebrate Nervous System

The CNS develops from the embryonic dorsal hollow nerve cord. Spaces in the brain called *ventricles* are continuous with the narrow **central canal** of the spinal cord, and all are filled with *cerebrospinal fluid*. This fluid, formed by filtration of arterial blood, carries out circulatory functions.

Neuron cell bodies primarily make up the **gray matter** of the brain and spinal cord. **White matter** is made up of bundles of axons. The spinal cord carries information to and from the brain and controls **reflexes**, which are automatic responses to stimuli.

The Peripheral Nervous System

Sensory information is carried to the CNS along *afferent* PNS neurons. *Efferent* neurons of the PNS carry instructions to muscles, glands, and endocrine cells. Most nerves contain axons of both afferent and efferent neurons.

Functionally, the PNS has two efferent components. The **motor system** carries signals to skeletal muscles, whose movement is largely under conscious control. The **autonomic nervous system** maintains involuntary control over smooth and cardiac muscles.

The autonomic nervous system is separated into three divisions. The **enteric division** includes networks of neurons involved in the control of the digestive tract, pancreas, and gallbladder. The **sympathetic division** accelerates heart rate and metabolic rate, generating energy and arousing an organism for action. The **parasympathetic division** carries signals that enhance self-maintenance activities that conserve energy, such as digestion and slowing the heart rate.

Parasympathetic nerves exit from the base of the brain or the end of the spinal cord and have ganglia near an internal organ. Sympathetic nerves exit midway along the spinal cord and form synapses in ganglia just outside the spinal cord. The *preganglionic neurons* of both divisions have cell bodies in the CNS and release acetylcholine as a neurotransmitter. Parasympathetic *postganglionic neurons* release acetylcholine; sympathetic postganglionic neurons release norepinephrine.

INTERACTIVE QUESTION   49.2

Complete the following concept map to help you review the organization of the peripheral nervous system.

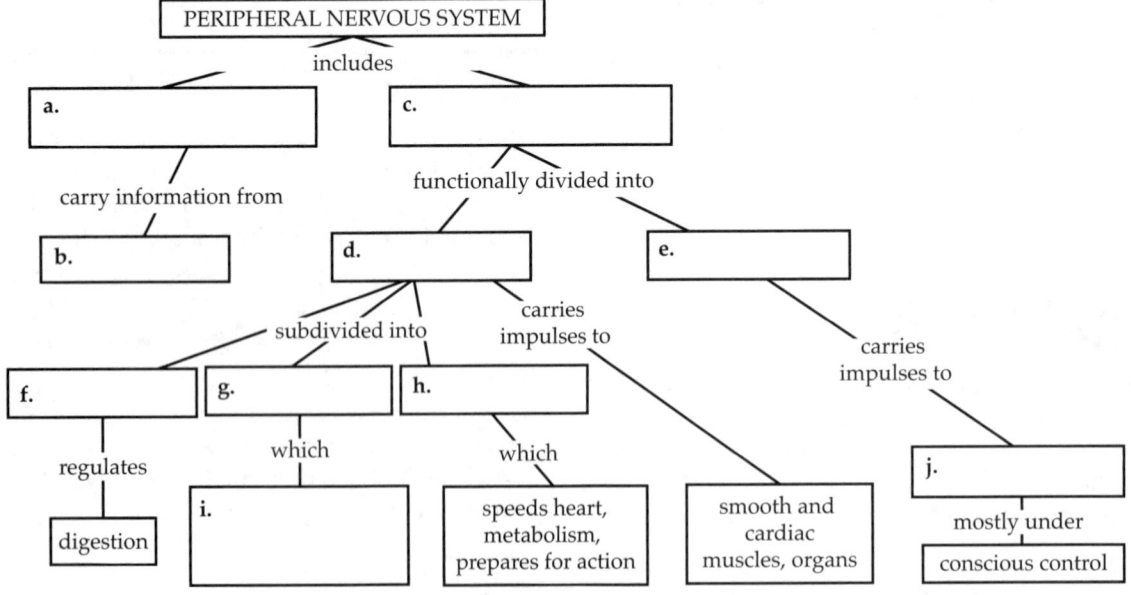

49.2 The vertebrate brain is regionally specialized

The vertebrate brain has three main regions. The activities of the **forebrain** (containing the *olfactory bulbs* and *cerebrum*) include processing olfactory input, regulating sleep, learning, and complex processing. The **midbrain** routes sensory input. The **hindbrain**, part of which forms the *cerebellum*, controls blood circulation and other involuntary activities and coordinates motor activities. The relative sizes of particular brain regions, which vary among different animal groups, relate to the importance of a particular brain function.

INTERACTIVE QUESTION 49.3

a. How does the size of the cerebellum differ between a lamprey and a ray-finned fish (such as a tuna differ), and how does that difference correlate with the activity level of each?

b. What two evolutionary trends are apparent in the brains of birds and mammals?

Exploring the Organization of the Human Brain (Figure 49.11) As a human embryo develops, the forebrain, midbrain, and hindbrain become evident as three anterior bulges of the neural tube. These brain regions subsequently subdivide: the forebrain becomes the telencephalon (anterior) and diencephalon (posterior), and the hindbrain becomes the metencephalon (anterior) and the myelencephalon (posterior). The midbrain does not divide but is now known as the mesencephalon.

The **brainstem** is made up of the mesencephalon, the **pons** (from the metencephalon), and the **medulla oblongata** or *medulla* (from the myelencephalon). The metencephalon also gives rise to the cerebellum.

The forebrain develops into the diencephalon and the telencephalon, which becomes the **cerebrum**. During the second and third months of development, the outer portion of the cerebrum, called the cortex, rapidly expands over much of the brain.

The cerebrum is divided into right and left **cerebral hemispheres**. The **cerebral cortex** is involved in perception, voluntary movement, and learning. The right side of the cortex receives information from and controls the movement of the left side of the body, and vice versa. Communication between the two hemispheres travels through the **corpus callosum**, a thick band of axons. The *basal nuclei*, embedded in the inner white matter, are important in planning and learning movements.

The diencephalon gives rise to the thalamus, hypothalamus, and epithalamus. The **thalamus** is the major sorting center for sensory information going to the cerebrum. The **hypothalamus** is the major brain region for homeostatic regulation. It produces the posterior pituitary hormones and the releasing hormones that control the anterior pituitary. The *epithalamus* includes the pineal gland, which produces melatonin, and one of several clusters of capillaries that produce cerebrospinal fluid.

The cerebellum is involved in learning and remembering motor skills, as well as in coordination and error checking during motor and perceptual functions. The cerebellum integrates information from the auditory and visual systems with sensory input from the joints and muscles as well as motor commands from the cerebrum to provide coordination of movements and balance.

The midbrain receives and integrates sensory information, such as that involved in hearing and visual reflexes, and relays it to specific regions of the forebrain. The pons and medulla transfer information between the PNS and the rest of the brain. They help coordinate large-scale movements, and most axons of motor neurons cross in the medulla, so that the right side of the brain controls much of the movement of the left side of the body, and vice versa. The medulla contains control centers for such homeostatic functions as breathing, swallowing, heart and blood vessel actions, and digestion. The pons functions with the medulla in some of these activities.

Arousal and Sleep Arousal is a state in which an individual is aware of the external world. While asleep, an individual does not consciously perceive external stimuli, although electroencephalograms (EEGs) record active brain waves. Sleep and dreaming may be involved in consolidating learning and memory. Sleep and arousal are partly controlled by the *reticular formation*, a diffuse network of neurons that controls the timing of rapid eye movement (REM) sleep and filters sensory information going to the cerebral cortex.

Biological Clock Regulation Most organisms appear to have daily cycles of biological activity. Studies show that an internal timekeeper, called the **biological clock**, is important for maintaining these circadian rhythms. In mammals, the **suprachiasmatic nucleus (SCN)**, located in the hypothalamus, functions as the biological clock. Sensory neurons provide visual information to the SCN to keep the clock synchronized with natural cycles of light and dark.

Emotions   The amygdala, hippocampus, and sections of the thalamus form a ring around the brainstem called the *limbic system*. This system interacts with sensory and other areas of the cerebrum in generating emotions, both basic emotions, such as laughing and crying, and emotional feelings associated with the survival-related functions of the brainstem, such as aggression, feeding, and sexuality. The **amygdala** functions in creating emotional memories.

Functional Imaging of the Brain   Functional magnetic resonance imaging (fMRI) is used to identify active parts of the brain in the study of emotion and cognition.

INTERACTIVE QUESTION   49.4

Identify the structures (a–g) in the following illustration of the human brain. Then match the functions (1–7) to these structures.

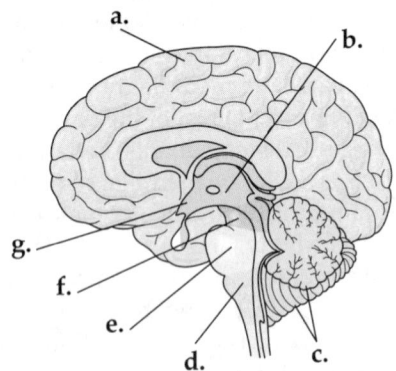

Structure	Function
a. _____	_____
b. _____	_____
c. _____	_____
d. _____	_____
e. _____	_____
f. _____	_____
g. _____	_____

Functions

1. Coordinates balance and movement

2. Aids the medulla in some functions; conducts information between the brain and spinal cord

3. Sorts and relays information to the cerebrum

4. Regulates breathing, heart rate, and digestion

5. Integrates sensory and motor information; is center for learning, emotion, memory, and perception

6. Produces hormones; functions in homeostatic regulation

7. Sends sensory information to the forebrain; is involved in hearing and visual reflexes

---

49.3   The cerebral cortex controls voluntary movement and cognitive functions

Sensory areas, *association areas*, and *motor areas* are located in the cortex. Each side of the cerebral cortex is divided into four lobes—frontal, temporal, occipital, and parietal—each with specialized functions.

Information Processing   Sensory information from sensory organs and individual *somatosensory* receptors in the body, which provide touch, pressure, temperature, and positional information, is directed to primary sensory areas located in different lobes of the cortex. Association areas near the primary sensory areas integrate ("associate") the different aspects of sensory inputs. The prefrontal cortex may generate motor response plans, which the primary motor cortex directs to the brainstem and spinal cord, from which action potentials travel in motor neurons to skeletal muscles.

Language and Speech   Studies of brain injuries and functional imaging of brain activity show that several areas in different lobes are involved in understanding and generating language. *Broca's area* in the left frontal lobe is involved in speech generation; *Wernicke's area* in the left temporal lobe is involved in speech comprehension. Other areas are involved in reading, speaking, and attaching meaning to words.

Lateralization of Cortical Function   Lateralization results in the two hemispheres differing in function. The left hemisphere usually becomes specialized for language, math, and logic, whereas the right hemisphere is involved in pattern recognition, spatial perception, and nonverbal thinking.

Frontal Lobe Function   Injury to the frontal lobe often results in flawed decision making and diminished emotional responses.

Evolution of Cognition in Vertebrates   The outer region of the cerebrum, consisting of six layers of neurons, has a large surface area and is highly convoluted in primates and cetaceans. These convolutions were once thought to be required for cognition, the perception and reasoning associated with knowledge. The now-documented cognitive ability of birds is linked to the *pallium*, the outer portion of the avian brain, which contains clusters of nuclei.

What are the functions of the primary somatosensory cortex, located along the front of the parietal lobe, and the primary motor cortex, located along the rear of the frontal lobe? Describe the spatial distribution of neurons in these two areas.

49.4 Changes in synaptic connections underlie memory and learning

During embryonic development, neurons that reach their proper targets receive growth-supporting factors; those that do not undergo programmed cell death. A developing neuron forms numerous synapses; although some of these synapses are stabilized and retained into adulthood, more than half of them are lost.

Neuronal Plasticity The nervous system can be remodeled in response to its activity, a capacity called **neuronal plasticity**. A high level of activity at a synapse may lead to the recruitment of additional synaptic terminals from the presynaptic neuron. A lack of activity may lead to the loss of connections with a neuron. Signals at a synapse can also be strengthened when several synapses are active at the same time on a postsynaptic cell.

The developmental disorder *autism* may involve a disruption of activity-dependent synapse remodeling.

Memory and Learning Human memory consists of **short-term memory**, information held briefly, and **long-term memory**, the retention of this information for later recall. Transferring information from short-term memory, which involves temporary links in the hippocampus, to long-term memory requires the making of more permanent connections within the cerebral cortex. Long-term memories become integrated with existing knowledge.

Skill memories are formed by repetition of motor activities and do not require conscious recall of specific steps. Learning and remembering skills and procedures seem to involve nerve cells making new connections. Remembering facts and numbers may involve changes in the strength of existing neuronal connections.

Long-Term Potentiation In **long-term potentiation (LTP)**, the strength of synaptic transmission is increased. A burst of action potentials causes glutamate release from a presynaptic cell, which binds with NMDA receptors that are blocked by Mg^{2+}. When the postsynaptic membrane has been depolarized by activity at nearby synapses, NMDA receptors are no longer blocked, Ca^{2+} enters the cell, and AMPA receptors are inserted in the postsynaptic membrane. After LTP is established, glutamate released by the presynaptic neuron binds to and activates both NMDA and AMPA receptors, triggering strong postsynaptic potentials and initiating action potentials. LTP may be involved in learning and memory storage.

Action potentials are initiated more readily in synapses exhibiting long-term potentiation. Explain why it initially takes at least two signaling neurons to establish LTP in a synapse.

49.5 Many nervous system disorders can be explained in molecular terms

Research efforts seek to identify genetic and environmental components of nervous system disorders for both prevention and development of better drug treatments.

Schizophrenia **Schizophrenia** is characterized by psychotic episodes involving hallucinations and delusions. The disease has both environmental and genetic components. Relatives of people with schizophrenia have higher risks of developing the disease. Evidence indicates that schizophrenia involves neuronal pathways using dopamine, although glutamate signaling may also be involved.

Depression Depressive illnesses include **major depressive disorder**, which involves a persistent low mood with changes in sleep, appetite, and energy; and **bipolar disorder**, which involves extreme swings of mood. Drug treatments for depression often increase the activity of biogenic amines.

The Brain's Reward System and Drug Addiction The neuronal circuit called the *reward system* involves neurons of the *ventral tegmental area (VTA)*, located near the base of the brain, that release dopamine at synapses in regions of the cerebrum, including the *nucleus accumbens*. Drug addiction is the compulsive consumption of drugs. Addictive drugs such as cocaine, heroin, nicotine, and amphetamines affect the reward system in different ways—by stimulating dopamine-releasing VTA neurons, blocking the removal of dopamine from synapses, or decreasing the activity of inhibitory neurons in the pathway. Long-lasting changes in the reward circuitry result in increased craving for the drug.

Alzheimer's Disease   **Alzheimer's disease** is dementia characterized by confusion and memory loss. This age-related, progressive disease involves the death of neurons in large areas of the brain. Postmortem examination reveals neurofibrillary tangles (bundles of the protein tau, which normally supports nutrient movement along axons) and amyloid plaques (aggregates of β-amyloid, an insoluble peptide that accumulates outside neurons and appears to trigger death of neighboring neurons).

Parkinson's Disease   **Parkinson's disease** is a progressive motor disorder characterized by difficulty in movement, rigidity, and muscle tremors. The death of neurons in the midbrain that normally release dopamine in the basal nuclei leads to the symptoms of this disease. Brain surgery, deep-brain stimulation, and drugs such as L-dopa (a dopamine precursor) are used to manage the symptoms. A potential cure may be to implant dopamine-secreting neurons in the midbrain or basal ganglia.

Future Directions   The BRAIN (Brain Research Through Advancing Innovative Neurotechnologies) Initiative began in 2014. Its aim is to promote development of new technologies that will increase our understanding of the brain and how neuronal activity results in thought and behavior.

Word Roots

astr- = a star; **-cyte** = cell (*astrocyte:* a glial cell with diverse functions, including providing structural support for neurons, regulating the interstitial environment, facilitating synaptic transmission, and assisting in regulating the blood supply to the brain)

auto- = self (*autonomic nervous system:* an efferent branch of the vertebrate peripheral nervous system that regulates the internal environment; consists of the sympathetic, parasympathetic, and enteric divisions)

hypo- = below (*hypothalamus:* the ventral part of the vertebrate forebrain; functions in maintaining homeostasis, especially in coordinating the endocrine and nervous systems)

para- = near (*parasympathetic division:* one of three divisions of the autonomic nervous system; generally enhances body activities that gain and conserve energy)

supra- = above, over (*suprachiasmatic nuclei:* a group of neurons in the hypothalamus of mammals that functions as a biological clock)

Structure Your Knowledge

1. List the location and functions of the following important brain nuclei or functional systems.
   a. reticular formation
   b. suprachiasmatic nucleus
   c. basal nuclei
   d. limbic system
   e. amygdala and hippocampus

Test Your Knowledge

MULTIPLE CHOICE: *Choose the one best answer.*

1. Which of the following animals is *incorrectly* paired with the description of its nervous system?
   a. sea star (echinoderm)—modified nerve net, central nerve ring with radial nerves
   b. hydra (cnidarian)—ring of ganglia leading to multiple nerve nets
   c. annelid worm—brain, ventral nerve cord with segmental ganglia
   d. vertebrate—central nervous system of brain and spinal cord and peripheral nervous system

2. Which type of glial cell has the same role in the CNS that Schwann cells have in the PNS?
   a. oligodendrocytes
   b. ependymal cells
   c. astrocytes
   d. microglia

3. Which of the following pairs is correct for the parasympathetic nervous system?
   a. preganglionic, acetylcholine; postganglionic, norepinephrine
   b. preganglionic, norepinephrine; postganglionic, norepinephrine
   c. preganglionic, norephinephrine; postganglionic, acetylcholine
   d. preganglionic, acetylcholine; postganglionic, acetylcholine

4. The amygdala and hippocampus
   a. control biorhythms and are found in the thalamus.
   b. are the parts of the reticular formation found in the midbrain.
   c. are nuclei in the midbrain involved in hearing and vision.
   d. are found in the inner cortex and are involved in memory storage.

5. A nerve that innervates the triceps muscle in your arm is part of the
   a. central nervous system.
   b. afferent division of the peripheral nervous system.
   c. efferent division of the peripheral nervous system.
   d. the sympathetic nervous system.

6. Which of the following statements provides evidence indicating that schizophrenia, bipolar disorder, and major depression have a genetic as well as an environmental component?
   a. All three conditions seem to involve an excess of dopamine and can be treated with the same drugs.
   b. If one identical twin has one of these disorders, the other twin has a much higher risk, indicating that genes and the environment both contribute to these disorders.
   c. The genes for these disorders have been identified, and they share a common sequence.
   d. Identical twins raised separately do not share these disorders; those raised in the same home are both likely to be affected by the same disorder.

7. Long-term potentiation seems to be involved in
   a. memory storage and learning.
   b. enhanced response of a presynaptic neuron resulting from strong depolarizations due to bursts of action potentials.
   c. the attachment of emotion to events.
   d. the release of the excitatory neurotransmitter glutamate, which opens voltage-gated channels.

8. The white matter of the spinal cord is composed of
   a. the ventricles.
   b. myelinated sheaths of axons.
   c. cerebrospinal fluid.
   d. motor and interneuron cell bodies.

9. What does lateralization of the cerebrum refer to?
   a. Visceral functions are controlled by the hindbrain, whereas thinking and language are centered in association areas of the cortex.
   b. The right side of the brain controls the left side of the body, and vice versa.
   c. The primary somatosensory cortex receives sensory information that is mapped onto one region of the cerebral cortex, whereas the primary motor cortex localized in another region controls motor output to the body.
   d. The left hemisphere usually controls verbal and analytic ability, whereas the right hemisphere is specialized for nonverbal thinking and pattern recognition.

10. A function of the reticular formation is to
    a. keep you awake during a lecture.
    b. coordinate movement and balance.
    c. produce emotions.
    d. control visceral functions such as breathing and heart rate.

11. After an unfortunate experience as a child, you have developed a fear of clowns. Which brain region is involved in associating your early memory with the emotion of fear?
    a. cerebrum
    b. Broca's area
    c. amygdala
    d. reticular formation

12. Which of the following is *not* associated with the brain's reward system?
    a. the neurotransmitter dopamine
    b. amygdala
    c. ventral tegmental area
    d. nucleus accumbens

13. Signals passing between the medulla and the cerebral hemispheres are transmitted by the
    a. cerebellum.
    b. hypothalamus.
    c. pons.
    d. reticular formation.

14. Which of the following statements correctly describes neuronal plasticity?
    a. Excesses or deficiencies of neurotransmitters such as dopamine are implicated in most nervous system disorders.
    b. Several different regions of the left hemisphere are involved in understanding and generating speech.
    c. Certain glia form tracks along which newly formed neurons migrate during development.
    d. The number and activity of synapses can change in the process of learning and developing long-term memories.

Sensory and Motor Mechanisms

Framework

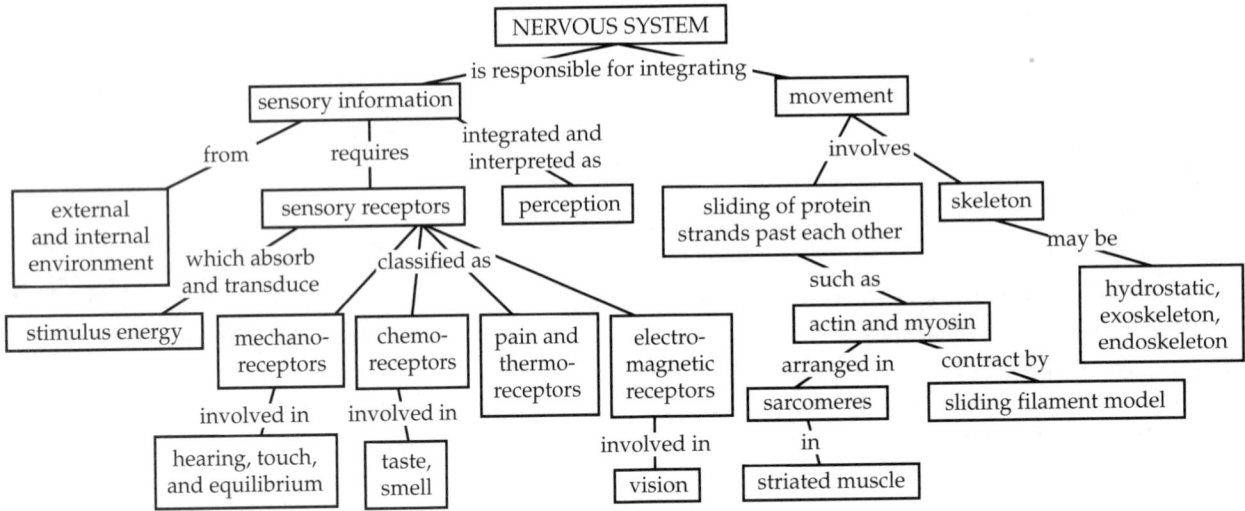

Chapter Review

50.1 Sensory receptors transduce stimulus energy and transmit signals to the central nervous system

Stimuli are forms of energy that a sensory receptor converts to a change in membrane potential, which can result in action potentials sent to the central nervous system (CNS). Integration of sensory input by the brain leads to motor responses.

Sensory Reception and Transduction **Sensory reception** is the detection of energy from a particular stimulus by sensory cells, which are usually modified neurons or non-neuronal cells that regulate neurons. Sensory cells may occur singly or in groups within sensory organs. The term **sensory receptor** can refer to an organ, cell, or structure within a cell that detects specific stimuli from the external or internal environment.

Sensory transduction is the conversion of a physical or chemical stimulus into a **receptor potential**, a graded change in the membrane potential of a receptor cell that is proportional to the strength of the stimulus. Receptor potentials result from the opening or closing of ion channels in response to a stimulus.

Transmission Sensory information is transmitted to the CNS as action potentials. The magnitude of a receptor potential correlates with either the frequency of action potentials generated (when the sensory receptor is a neuron) or the quantity of neurotransmitter released (when a non-neuronal sensory receptor synapses with a sensory neuron). Many sensory neurons generate action potentials spontaneously at a low rate, and reception of a stimulus changes the frequency of action potentials.

Integration of sensory information may begin with the summation of graded potentials from receptors. Sensory information is further integrated within complex receptors and the CNS.

Perception Impulses arriving from sensory neurons are routed to different parts of the brain that interpret them, producing **perceptions** of various stimuli.

Amplification and Adaptation The strengthening of the energy of a stimulus, either by accessory structures of sense organs or as part of signal transduction pathways, is called **amplification**. Ongoing stimulation may lead to **sensory adaptation** in which continuous stimulation results in a decline in sensitivity of the receptor cell.

Mechanoreceptors respond to the mechanical energy of pressure, touch, stretch, motion, and sound. Bending or stretching of external (or internal) cell structures alters the permeability of ion channels, leading to depolarization or hyperpolarization.

Vertebrate stretch receptors monitor the length of skeletal muscles. Dendrites of sensory neurons wound around certain muscle fibers are stimulated by stretching of the muscles. In humans, dendrites of sensory neurons function as touch receptors; the type of stimulus to which they respond best depends on their location and the structure of their connective tissue covering.

Chemoreceptors include both general receptors that monitor total solute concentration and specific receptors that respond only to a single kind of important molecule.

Electromagnetic receptors respond to various forms of electromagnetic energy, such as visible light, electricity, and magnetic fields. Specific proteins are associated with sensors that detect Earth's magnetic field in a variety of organisms.

In humans, **thermoreceptors** in the skin and hypothalamus respond to heat or cold and send information to the body's thermostat. Researchers have found that the receptor proteins for capsaicin, a component of hot peppers, also respond to hot temperatures. Mammals have a number of thermoreceptors that respond to specific temperature ranges.

Pain receptors (nociceptors) in humans include naked dendrites in the epidermis. These receptors respond to excess heat, pressure, or chemicals. Prostaglandins released from injured tissues increase pain by increasing nociceptor sensitivity.

INTERACTIVE QUESTION 50.1

List and describe the four functions common to all sensory pathways.

a.

b.

c.

d.

Types of Sensory Receptors Sensory receptors can be grouped into five categories.

INTERACTIVE QUESTION 50.2

How do aspirin and ibuprofen reduce pain?

50.2 **In hearing and equilibrium, mechanoreceptors detect moving fluid or settling particles**

Sensing of Gravity and Sound in Invertebrates Most invertebrates have **statocysts**, which often consist of a layer of ciliated receptor cells lining a chamber containing **statoliths**, dense granules or grains of sand. The stimulation of mechanoreceptors beneath the statoliths provides positional information to the animal.

Most insects sense sounds with body hairs that vibrate in response to sound waves of specific frequencies. A tympanic membrane stretched over an internal air-filled chamber provides a vibration-sensing organ in some insect species.

Hearing and Equilibrium in Mammals   In mammals and most terrestrial vertebrates, the sensory organs for hearing and balance are in the ear. In *hearing*, pressure waves in the air are transduced into nerve impulses that the brain perceives as sound.

The ear consists of three regions: The external pinna and the auditory canal make up the **outer ear**. The **tympanic membrane** (eardrum), which separates the outer ear from the **middle ear**, transmits sound pressure waves to three small bones—the malleus (hammer), incus (anvil), and stapes (stirrup)—which conduct the waves to the **inner ear** by way of a membrane called the **oval window**. The **Eustachian tube**, connecting the pharynx and the middle ear, equalizes pressure within the middle ear. The inner ear contains fluid-filled channels. The **semicircular canals** function in equilibrium, and the **cochlea** functions in hearing.

The coiled cochlea has two large, connected, fluid-filled canals—an upper vestibular canal and a lower tympanic canal—separated by the smaller cochlear duct. The **organ of Corti**, located on the basilar membrane that forms the floor of the cochlear duct, contains sensory receptors called **hair cells**. Their hairs extend into the cochlear duct, and some attach to the tectorial membrane, which overhangs the organ of Corti.

Pressure waves in air—transmitted by the tympanic membrane, the three bones of the middle ear, and the oval window—produce pressure waves in the perilymph, which travel from the vestibular canal through the tympanic canal and dissipate when they strike the **round window**. These pressure waves vibrate the basilar membrane, bending the hairs of its receptor cells against the tectorial membrane. When bent in one direction, the hair cells depolarize, increasing neurotransmitter release and the frequency of action potentials in sensory neurons that travel in the auditory nerve to the brain. When bent in the other direction, hair cells hyperpolarize, which reduces neurotransmitter release and the generation of action potentials.

Volume is a result of the amplitude, or height, of the sound wave; a stronger wave bends the hairs more and results in more action potentials. *Pitch* is related to the frequency of sound waves. Different regions of the basilar membrane vibrate in response to different frequencies, and neurons associated with the vibrating region transmit action potentials to those auditory regions of the cortex where a particular pitch is perceived.

INTERACTIVE QUESTION   50.3

Label the parts in the following diagram of the human ear.

1. Which of these structures are filled with air?

2. Which of these structures are filled with fluid?

Within the inner ear are two chambers, the *utricle* and *saccule*, and three semicircular canals responsible for detection of body position and movement. Hair cells in the utricle and saccule project into a gelatinous material containing calcium carbonate particles called *otoliths*. These heavy particles exert a pull on the hairs in the direction of gravity; different head angles stimulate different hair cells, thereby altering the information about head angle or linear acceleration traveling through the vestibular nerve to the brain. Clusters of hair cells in the swellings at the base of the semicircular canals project into a gelatinous mass called the cupula and sense angular motion in any direction by detecting the pressure of the moving semicircular canal fluid against the cupula.

Hearing and Equilibrium in Other Vertebrates   In fish, sound-wave vibrations pass through the skeleton of the head to the inner ears, within which hair cells are stimulated by the movement of otoliths. Some fishes have a series of bones that transmit vibrations from the swim bladder to the inner ear.

Clusters of hair cells with sensory hairs embedded in a gelatinous cap or cupula are contained in the

lateral line system of fishes and aquatic amphibians. Water moving through a tube along both sides of the body bends the cupula and stimulates the hair cells, enabling the fish to perceive its own movement, water currents, and the pressure waves generated by other moving objects.

The inner ear is the organ of hearing and equilibrium in terrestrial vertebrates. In amphibians, sound vibrations are conducted by a tympanic membrane on the body surface to a single middle ear bone to the inner ear. A cochlea has evolved in birds and other reptiles, as well as in mammals.

50.3 The diverse visual receptors of animals depend on light-absorbing pigments

Evolution of Visual Perception All light detectors involve **photoreceptors** containing light-absorbing pigment molecules. Photoreceptors of flatworms, annelids, arthropods, and vertebrates share common developmental genes, indicating that photoreceptors were present in the earliest bilaterian animals.

Most invertebrates have light-detecting organs. The ocelli or eyespots of planarians detect light direction and enable the animal to navigate a path away from a light source.

The **compound eye** of insects, crustaceans, and some polychaete worms contains up to thousands of light detectors called **ommatidia**, each with its own lens. The differing intensities of light entering the many ommatidia from different angles produce a visual image. The compound eyes of insects are adept at detecting movement and color, and some detect ultraviolet radiation.

Some jellies, polychaetes, spiders, and many molluscs have a **single-lens eye**, in which light is focused through a movable single lens onto a layer of photoreceptors. The **iris** changes the diameter of the **pupil** to adjust the amount of light entering the eye.

The Vertebrate Visual System The human eye is surrounded by a mucous membrane called the conjunctiva, a tough, outer connective tissue layer called the sclera, and a thin, pigmented inner layer called the choroid. At the front of the eye, the sclera becomes the transparent *cornea*, and the choroid forms the colored *iris*, which regulates the amount of light entering through the pupil. The **retina**, the innermost layer of the eyeball, contains neurons and photoreceptors. The optic nerve attaches to the eye at the optic disk, forming a "blind spot" on the retina.

The transparent **lens** focuses an image onto the retina. The *aqueous humor* fills the anterior eye cavity; jellylike *vitreous humor* fills the posterior cavity.

Rods and **cones** are the photoreceptors in the retina. Several rods or cones relay information to each *bipolar*

cell, several of which synapse with *ganglion cells*, whose axons form the optic nerve. *Horizontal* and *amacrine cells* in the retina help to integrate visual information.

Rods are more light sensitive and enable night vision, whereas cones distinguish colors.

INTERACTIVE QUESTION 50.4

Label the parts of the human eye in the following illustration.

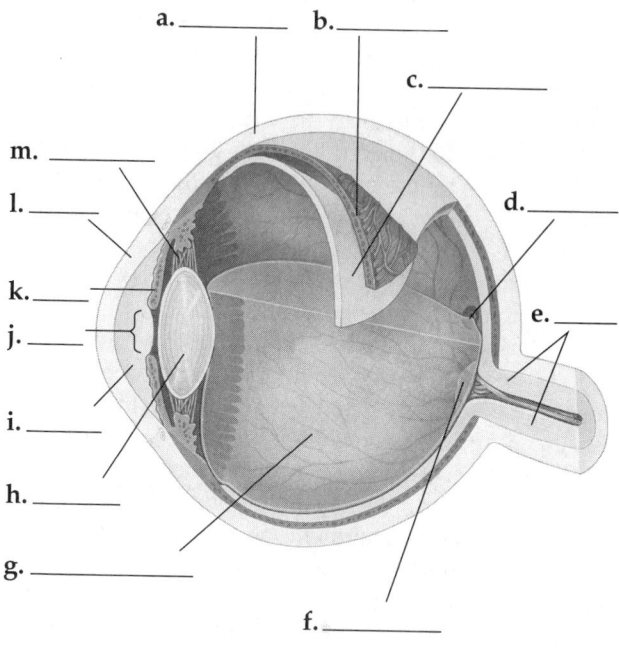

a. _____ b. _____

c. _____

m. _____

l. _____ d. _____

k. _____

j. _____ e. _____

i. _____

h. _____

g. _____

f. _____

Both rods and cones have *visual pigments* embedded in a stack of disks in the outer segment of each cell. The light-absorbing molecule **retinal** is bound to a membrane protein called an **opsin**. The visual pigment in rod cells is called **rhodopsin**. When retinal absorbs light, it changes from the *cis* isomer to the *trans* isomer, activating its opsin protein.

Light-activated rhodopsin activates a membrane-bound G protein called transducin, which then activates an enzyme that detaches cGMP bound to Na^+ channels in the plasma membrane. Sodium channels close, hyperpolarizing the cell. Except in bright light when rhodopsin remains active, enzymes convert retinal to its original shape, and the signal transduction pathway shuts off.

Visual processing begins in the retina, where rods and cones synapse with bipolar cells. In the dark, rods and cones are depolarized and continually release glutamate, which, depending on the postsynaptic receptor, either depolarizes or hyperpolarizes bipolar cells. When rods and cones hyperpolarize after absorbing light, glutamate release ceases. In response, the bipolar cells reverse their polarization, either becoming hyperpolarized or depolarized, depending on the receptor.

Signals from rods and cones may follow a pathway directly from photoreceptors to bipolar cells to ganglion cells. Other pathways may involve *lateral inhibition* by horizontal cells, which is a form of integration that enhances contrast. Lateral inhibition also occurs as amacrine cells relay information from one bipolar cell to several ganglion cells, as well as at other levels of visual processing in the brain. Each ganglion cell receives information from a number of rods and cones, forming a *receptive field*.

The left and right optic nerves meet at the *optic chiasma* at the base of the cerebral cortex. Information from the left visual field of both eyes travels to the right side of the brain, whereas what is sensed in the right field of view goes to the left side.

Most axons of the ganglion cells go to the *lateral geniculate nuclei*, where neurons lead to the *primary visual cortex* in the cerebrum. Neurons also carry information to other visual processing and integrating centers in the cortex. At least 30% of the cerebral cortex may be devoted to creating the perception of what we see.

Most fishes, amphibians, and reptiles have good color vision. The relative proportion of rods and cones correlates with the activity pattern of a mammal. In humans, red, green, and blue cones, each with its own type of opsin that binds with retinal to form a *photopsin*, are named for the color of light they optimally absorb. As the genes for the red and green pigments are on the X chromosome, color blindness is more common in males. Squirrel monkeys have only one X-linked opsin gene, which can have either an allele for red or an allele for green (as in humans, their blue opsin gene is on an autosome). Researchers used gene therapy to inject the missing allele into the retina of male monkeys, enabling them to distinguish red from green and demonstrating that visual neural circuits can form or be activated even in adults.

You focus on close objects by accommodation, in which ciliary muscles contract, causing suspensory ligaments to slacken and the elastic lens to become rounder. In the human eye, rods are most concentrated toward the edge of the retina, whereas the center of the visual field, the **fovea**, is filled with cones.

INTERACTIVE QUESTION   50.5

The *receptive field* of a ganglion cell includes the rods and/or cones that supply information to it.

a. Would a large or a small receptive field produce the sharpest image?

b. Where on the retina are the smallest receptive fields found?

50.4 **The senses of taste and smell rely on similar sets of sensory receptors**

For terrestrial animals, the sense of **gustation** (taste) detects chemicals called **tastants** present in a solution, whereas **olfaction** (smell) detects airborne **odorants**. Both rely on chemoreceptors.

Sensory hairs containing taste receptors are found on the feet and mouthparts of insects. Insects also have olfactory hairs, usually located on their antennae.

Taste in Mammals   Taste cells recognize five types of tastants, producing the taste perceptions—sweet, sour, salty, bitter, and umami. **Taste buds**, scattered on the tongue and mouth, contain receptor cells for all five tastes. Individual taste cells express a single receptor type. Experiments indicate that taste perception depends on which sensory neurons are activated: A sweet taste cell reprogrammed to express a receptor for a bitter molecule caused mice to be attracted to that bitter chemical.

A G protein-coupled receptor (GPCR) is involved with sweet, umami, and bitter tastes. Humans have more than 30 different bitter receptors, but only one type of sweet and umami receptor. The sour tastant receptor is a transient receptor protein (TRP), similar to capsaicin and the other thermoreceptor proteins. The taste receptor for salt is a sodium channel.

Smell in Humans   Olfactory receptor cells are neurons that line the upper part of the nasal cavity. Binding of an odorant to a GPCR protein, called an odorant receptor (OR), on an olfactory cilium triggers a signal transduction pathway, which opens ion channels. The membrane depolarizes and sends action potentials to the olfactory bulb of the brain. Approximately 2% of the genes in the human genome are OR genes.

INTERACTIVE QUESTION   50.6

For what purposes do animals use their senses of gustation and olfaction?

50.5 **The physical interaction of protein filaments is required for muscle function**

Muscle cell contraction involves the interaction between **thin filaments**, which consist of two coiled strands of actin proteins, and **thick filaments**, which are arrays of myosin molecules.

Vertebrate Skeletal Muscle   Vertebrate **skeletal muscle** consists of a bundle of multinucleated muscle cells, called fibers, running parallel to the length of the muscle. Each fiber contains bundles of **myofibrils**, each containing thin and thick filaments.

The regular arrangement of filaments produces repeating light and dark bands; thus, skeletal muscle is also called *striated muscle*. The repeating sections are called **sarcomeres**, which are the contractile units of skeletal muscle. Thin filaments attach to the Z lines at the border of a sarcomere and project toward the center of the sarcomere. Thick filaments lie in the center of a sarcomere, anchored in the M lines.

According to the **sliding-filament model**, the thick and thin filaments do not change length during muscle contraction but simply slide past each other, increasing their area of overlap.

The mechanism for the sliding of filaments is the hydrolysis of ATP by the globular head of a myosin molecule, which then changes to a high-energy form that binds to actin and forms a cross-bridge. The myosin head then bends back to its low-energy configuration and pulls the attached thin filament toward the center of the sarcomere. When a new molecule of ATP binds to the myosin head, it breaks its bond to actin, and the cycle repeats with the myosin head attaching to an actin molecule farther along on the thin filament.

Energy for muscle contraction is stored in creatine phosphate, which regenerates ATP, and in glycogen, which is broken down to glucose. Glucose can be used to generate ATP by glycolysis or aerobic respiration.

INTERACTIVE QUESTION   50.7

Identify the components in the following diagram of a section of a skeletal muscle fiber. Explain how this diagram will look different when the muscle fiber is fully contracted, and briefly describe the mechanism for this change.

When a muscle fiber is at rest, the myosin-binding sites of the actin molecules are blocked by the regulatory protein **tropomyosin**, which is associated with another set of regulatory proteins, the **troponin complex**. When troponin binds to Ca^{2+}, the regulatory proteins change shape, and myosin-binding sites on the thin filament are exposed and muscle contraction can occur.

Calcium ions are actively transported into the **sarcoplasmic reticulum (SR)** of a muscle cell. When an action potential of a motor neuron causes the release of acetylcholine into the synapse with a muscle fiber, the fiber depolarizes and an action potential spreads into the **transverse (T) tubules**, infoldings of the muscle cell's plasma membrane. The action potential opens Ca^{2+} channels in the sarcoplasmic reticulum, releasing Ca^{2+} into the cytosol. Calcium binding with troponin exposes myosin-binding sites, and contraction begins. Contraction ends when motor neuron input stops, the sarcoplasmic reticulum pumps Ca^{2+} back out of the cytosol, and the tropomyosin–troponin complex again blocks the binding sites.

The response of a single muscle fiber to an action potential of a motor neuron is a brief twitch. Graded contraction of whole muscles may result from involving additional fibers. A single motor neuron and all the muscle fibers it innervates make up a **motor unit**. The strength of a muscle contraction depends on the size and number of motor units involved. Muscle force (tension) can be increased by the activation of additional motor units, called *recruitment*. Prolonged contraction can cause muscle fatigue due to depletion of ATP and the loss of ion gradients required for depolarization.

Repeated, rapid-fire action potentials produce summation and a state of smooth and sustained contraction called **tetanus**.

There are different types of muscle fibers with distinct properties. Oxidative fibers, which rely mostly on aerobic respiration, have many mitochondria, a good blood supply, and **myoglobin**, a pigment that extracts oxygen from the blood and stores it. Glycolytic fibers primarily use glycolysis to produce ATP. **Fast-twitch fibers** produce rapid and powerful contractions. **Slow-twitch fibers**, common in muscles that must sustain long contractions, have less sarcoplasmic reticulum and pump calcium more slowly; thus, calcium remains in the cytosol longer and twitches last longer.

INTERACTIVE QUESTION   50.8

a. Identify the three distinct types of skeletal muscle fibers.

b. Which type of fiber has few mitochondria, has little myoglobin, and produces brief, powerful contractions?

c. Which type of fiber would you expect to be present in the highest proportion in postural muscles?

Other Types of Muscle   Vertebrate **cardiac muscle** fibers are found in the heart. Ion channels in the plasma membrane of a cardiac muscle cell produce rhythmic depolarizations that generate action potentials without nervous input. Cardiac muscle fibers are connected by *intercalated disks* through which action potentials spread to all the cells of the heart. These action potentials last longer than those of skeletal muscle fibers.

The scattered arrangement of actin and myosin filaments in **smooth muscle** accounts for its nonstriated appearance. Smooth muscle fibers lack T tubules, a troponin complex, and a well-developed sarcoplasmic reticulum. Calcium ions enter the cell in response to an action potential and bind to calmodulin, which activates an enzyme that phosphorylates myosin. Contractions are slower and may be triggered by the autonomic nervous system or by the muscle cells themselves.

Invertebrates have muscle cells similar to the skeletal and smooth muscle cells of vertebrates. The flight muscles of insects are capable of independent and rapid contraction.

50.6   Skeletal systems transform muscle contraction into locomotion

Skeletons function in support, protection, and movement. Because muscles can only contract, antagonistic muscle pairs are required to move body parts in opposite directions.

Types of Skeleton Systems   Fluid under pressure in a closed body compartment creates a **hydrostatic skeleton**. As muscles change the shape of the fluid-filled compartment, the animal elongates or moves. Earthworms and other annelids move by **peristalsis**, using rhythmic waves of contractions of circular and longitudinal muscles.

Typical of molluscs and arthropods, **exoskeletons** are hard coverings deposited on the surface of animals. The *cuticle* of an arthropod contains fibrils of the polysaccharide **chitin** embedded in a protein matrix. Where protection is most needed, this flexible support is hardened by organic compounds and calcium salts.

The supporting elements of an **endoskeleton** are embedded in the soft tissues of the animal. Sponges have endoskeletons composed of inorganic needle-like structures or protein fibers, and echinoderms have hard plates composed of magnesium carbonate and calcium carbonate beneath the skin.

The endoskeletons of vertebrates are composed of cartilage and/or bone. The bony portion of the skeleton is built and remodeled by the combined actions of *osteoblasts*, which deposit the bone matrix, and *osteoclasts*, which break down and resorb the matrix. The mammalian skeleton consists of more than 200 bones, including a skull, vertebral column, rib cage, the pectoral and pelvic girdles, and the limb bones. Joints, where bones are connected by ligaments, allow for flexibility in body movement. Tendons attach muscles to bones.

INTERACTIVE QUESTION   50.9

List an advantage and a disadvantage of each of the following types of skeletons, and indicate in which animal groups each type is found.

a. hydrostatic skeleton

b. exoskeleton

c. endoskeleton

Body proportions of small and large animals vary, partly as a result of the physical relationship between the diameter of a support (strength increases with the square of its diameter) and the strain produced by increasing weight (weight increases with the cube of its dimensions). Posture is an important structural factor in supporting body weight. Most of the weight of large land mammals is supported by muscles and tendons, which hold the legs relatively straight and directly under the body.

Types of Locomotion   Locomotion, the active movement from place to place, requires energy to overcome the forces of friction and gravity. Animals move to obtain food, escape from danger, and find mates.

Walking, running, hopping, or crawling on land requires the animal to support itself and move against gravity; however, air offers little resistance to movement. Powerful leg muscles and strong skeletal support are important for supporting an animal and propelling it forward. A running or hopping animal may store some energy in its tendons to facilitate the next step or hop. Maintaining balance is another requirement for upright movement. A crawling animal must overcome the friction produced by its contact with the ground.

Overcoming gravity is less of a problem for swimming animals due to the buoyancy of water. Most fast swimmers have a fusiform shape that reduces friction or drag caused by the density of water. Animals may swim by using their legs as oars, using water to jet-propel themselves, or by undulating their bodies and tails from side to side or up and down.

A flying animal uses its wings to provide enough lift to overcome the force of gravity. Flying animals are relatively light to decrease the effects of gravity and streamlined to decrease air friction. Wings of various designs function as airfoils and attach to light-weight, fusiform bodies.

The energy cost of locomotion depends on the mode of locomotion and the body mass of an animal.

INTERACTIVE QUESTION 50.10

a. Which mode of locomotion is the most energy efficient per distance traveled? What problems does this mode of locomotion present, and what adaptations facilitate this type of movement?

b. Which mode is the most energetically expensive per distance traveled? What problems does this mode of locomotion present, and what adaptations facilitate this type of movement?

Word Roots

chem- = chemical (*chemoreceptor:* a sensory receptor that responds to a chemical stimulus)

electr- = electricity (*electromagnetic receptor:* a receptor of electromagnetic energy, such as visible light, electricity, or magnetism)

endo- = within (*endoskeleton:* a hard skeleton buried within the soft tissues of an animal)

exo- = outside (*exoskeleton:* a hard encasement on the surface of an animal that provides protection and points of attachment for muscles)

hydr- = water (*hydrostatic skeleton:* a skeletal system composed of fluid held under pressure in a closed body compartment)

inter- = between (*intercalated disk:* a specialized junction between cardiac muscle cells that provides direct electrical coupling between the cells)

mechan- = an instrument (*mechanoreceptor:* a sensory receptor that detects physical deformation in the body's environment associated with pressure, touch, stretch, motion, and sound)

my- = muscle; **fibr-** = fiber (*myofibril:* a longitudinal bundle in a muscle cell [fiber] that contains thin filaments of actin and regulatory proteins and thick filaments of myosin)

noc- = harm (*nociceptor:* a sensory receptor that responds to noxious or painful stimuli)

olfact- = smell (*olfaction:* the sense of smell)

ommat- = the eye (*ommatidium:* one of the facets of the compound eye of arthropods and some polychaete worms)

peri- = around; **-stalsis** = a constriction (*peristalsis:* a type of movement on land produced by rhythmic waves of muscle contraction passing from front to back, as in many annelids)

phot- = light (*photoreceptor:* an electromagnetic receptor that detects the radiation known as visible light)

rhod- = red (*rhodopsin:* a visual pigment consisting of retinal and opsin)

sarc- = flesh; **-mere** = a part (*sarcomere:* the fundamental, repeating unit of striated muscle, delimited by the Z lines)

semi- = half (*semicircular canals:* a three-part chamber of the inner ear that functions in maintaining equilibrium)

stat- = standing; **-cyst** = a pouch (*statocyst:* a type of mechanoreceptor that functions in equilibrium in invertebrates by use of statoliths, which stimulate hair cells in relation to gravity)

therm- = heat (*thermoreceptor:* a receptor stimulated by either heat or cold)

trans- = across (*transmission:* the passage of a nerve impulse along an axon)

trop- = turn, change (*tropomyosin*: the regulatory protein that blocks the myosin-binding sites on actin molecules)

tympan- = a drum (*tympanic membrane*: another name for the eardrum, the membrane between the outer ear and the middle ear)

Structure Your Knowledge

1. Describe the path of a light stimulus from where it enters the vertebrate eye to its transmission as an action potential through the optic nerve.

2. Arrange the following structures in an order that enables you to describe the passage of sound waves from the external environment to the perception of sound in the brain.

round window	basilar membrane
oval window	cerebral cortex
pinna	vestibular and tympanic canals
auditory nerve	tectorial membrane
cochlea	tympanic membrane
auditory canal	malleus, incus, stapes
cochlear duct	organ of Corti

3. Trace the sequence of events in muscle contraction from an action potential in a motor neuron to the relaxation of the muscle.

Test Your Knowledge

MULTIPLE CHOICE: *Choose the one best answer.*

1. Sensory perception is the
   a. reception of a stimulus by a sensory cell.
   b. conversion of stimulus energy into a receptor potential by a sensory receptor or sensory neuron.
   c. interpretation of a stimulus according to the region of the brain that receives the nerve impulse.
   d. amplification and transmission of sensory data to the CNS.

2. Some flowers look different to bees than they do to us because bees can see a color called "bee purple" that our eyes can't detect. Bees can see 'bee purple' because their eyes are sensitive to
   a. ultraviolet light.
   b. infrared light.
   c. polarized light.
   d. X-rays.

3. The function of the three bones of the middle ear is to
   a. respond to different frequencies of sound waves.
   b. transmit pressure waves to the oval window.
   c. equalize pressure within the inner ear by communicating through the Eustachian tube.
   d. sense rotation in three different planes (*x, y,* and *z* axes).

4. The pitch of a sound is determined by the
   a. area in the brain that receives the signal.
   b. section of the basilar membrane that vibrates.
   c. degree to which hair cells are bent and the resulting increase in action potentials in the sensory neuron.
   d. Both **a** and **b** are true.

5. Rotation of the head of a mammal is sensed by
   a. the organ of Corti located in the utricle and saccule.
   b. changes in the action potentials of hair cells in response to bending against the tectorial membrane.
   c. the bending of hair cells embedded in a cupula in a semicircular canal in response to the movement of internal fluid.
   d. hair cells in response to movement of fluid in the coiled cochlea.

6. Fish can perceive pressure waves from moving objects with
   a. a lateral line system containing clusters of hair cells embedded in a gelatinous cap.
   b. semicircular canals in which sensory hairs are embedded in a gelatinous cap.
   c. inner ears, within which sensory hairs are stimulated by the movement of otoliths.
   d. a series of bones that transmit vibrations from the swim bladder to the inner ear.

7. What do the lateral line of fish, statocysts in invertebrates, and the cochlea of your ear all have in common?
   a. They are used to sense sound or pressure waves.
   b. They are organs of equilibrium.
   c. They use hair cells as mechanoreceptors.
   d. They use granules to stimulate their receptor cells.

8. The compound eye of insects and crustaceans
   a. is composed of thousands of prisms that focus light onto a few photoreceptor cells.
   b. cannot sense colors but may detect infrared and ultraviolet radiation.
   c. contains many ommatidia, each of which contains a lens that focuses light from a tiny portion of the field of view.
   d. All of the above are true.

9. The eyes of individuals with *achromatopsia* contain only rods. Such people would
   a. lack color vision.
   b. be red-green color-blind.
   c. possess normal night vision.
   d. Both a and c are true.

10. Which of the following structures is *incorrectly* paired with its function?
    a. cones—respond to different light wavelengths, producing color vision
    b. horizontal cells—produce lateral inhibition of receptor cells and bipolar cells
    c. bipolar cells—relay impulses between rods or cones and ganglion cells
    d. ganglion cells—synapse with the optic nerve at the optic disk or blind spot

11. The molecule rhodopsin
    a. dissociates when struck by light and is functional only in the dark.
    b. is more sensitive to light than are photopsins and is involved in black-and-white vision in dim light.
    c. activates a signal transduction pathway that increases a rod cell's permeability to sodium.
    d. is or does all of the above.

12. The absorption of light by rhodopsin in a rod cell leads to
    a. the opening of sodium channels caused by the binding of cGMP.
    b. a hyperpolarized membrane and a decrease in the release of neurotransmitter.
    c. the bleaching of rhodopsin, which increases the sensitivity of rod cells to light.
    d. the closing of sodium channels and an increase in the release of neurotransmitter.

13. A receptive field in the retina involves a *single*
    a. bipolar cell.
    b. amacrine cell.
    c. horizontal cell.
    d. ganglion cell.

14. Humans have the most different receptors for
    a. bitter.
    b. umami.
    c. sour.
    d. salty.

15. When striated muscle fibers contract,
    a. the Z lines are pulled closer together.
    b. the thin filaments become shorter.
    c. the thick filaments become thicker.
    d. both a and b occur.

16. Venom from the black widow spider (*Lactrodectus mactans*) causes massive release of acetylcholine from neurons. How would muscles be affected by this venom?
    a. The muscles would exhibit spastic paralysis; that is, they would be massively contracted and unable to relax.
    b. The muscles would exhibit flaccid paralysis; that is, they would be floppy and unable to contract.
    c. The muscles would be unable to utilize ATP.
    d. The muscles would be unaffected because acetylcholine is a neurotransmitter used only for communication between neurons.

17. The role of ATP in muscle contraction is to
    a. form cross-bridges between thick filaments and thin filaments.
    b. release the myosin head from actin when it binds and to provide energy when hydrolyzed to form myosin's high-energy form.
    c. remove the tropomyosin–troponin complex from blocking the binding sites on actin.
    d. bend the cross-bridge and pull the thick filaments toward the center of the sarcomere.

18. How does calcium affect muscle contraction?
    a. It is released from the T tubules in response to an action potential and initiates contraction.
    b. The binding of acetylcholine opens calcium channels in the plasma membrane, creating an action potential that travels down the T tubules.
    c. It binds to tropomyosin and helps to stabilize cross-bridge formation.
    d. Its binding to troponin causes tropomyosin to move away from the myosin-binding sites on the actin filament.

19. A motor unit is
    a. a sarcomere that extends from one Z line to the next Z line.
    b. a motor neuron and the muscle fibers it innervates.

c. the muscle fibers that make up a muscle.

d. the antagonistic set of muscles that flex and extend a body part.

20. Which of the following characteristics does *not* pertain to cardiac muscle?

a. scattered arrangement of actin and myosin filaments

b. intercalated disks that spread action potentials between cells

c. striations

d. ability to generate action potentials without nervous input

21. Smooth muscle contracts and relaxes relatively slowly because

a. the only ATP available is supplied by fermentation.

b. its contraction is stimulated by hormones, not motor neurons.

c. it does not have a well-developed sarcoplasmic reticulum, and Ca^{2+} enters the cell through the plasma membrane during an action potential.

d. it is composed exclusively of slow-twitch muscle fibers.

22. Hydrostatic skeletons are used for movement by all of the following animals *except*

a. cnidarians.

b. arthropods.

c. nematodes.

d. annelids.

23. When you pull up your lower arm and "make a muscle" in your biceps, you are

a. contracting a flexor.

b. relaxing a flexor.

c. contracting an extensor.

d. contracting a tendon.

24. Which of the following combinations of animal size and mode of locomotion has the lowest energy cost (calories/kilogram·meter)?

a. small and swimming

b. small and running

c. large and flying

d. large and swimming

Chapter 51

Animal Behavior

Key Concepts

51.1 Discrete sensory inputs can stimulate both simple and complex behaviors

51.2 Learning establishes specific links between experience and behavior

51.3 Selection for individual survival and reproductive success can explain diverse behaviors

51.4 Genetic analyses and the concept of inclusive fitness provide a basis for studying the evolution of behavior

Framework

This chapter introduces the complex and fascinating subject of animal behavior.

Behaviors can range from simple fixed-action patterns in response to specific stimuli to problem solving in novel situations. The parameters of behavior are controlled by genetics and thus are acted upon by natural selection. Behavioral ecology focuses on the ultimate cause of reproductive fitness, which can be used to interpret foraging behavior, mating patterns, and altruistic behavior. Sociobiology extends evolutionary interpretations to human social behavior.

Chapter Review

A **behavior** is an action performed under the direction of the nervous system. Behavior is essential for survival and reproduction, and thus subject to natural selection.

51.1 Discrete sensory inputs can stimulate both simple and complex behaviors

Niko Tinbergen's set of questions to guide behavioral studies emphasizes the importance of both *proximate causation*, the immediate cause of a behavior in terms of the stimuli that trigger it, the mechanisms that produce it, and how experience modifies a response, and *ultimate causation*, which concerns the benefit to survival and reproduction and the evolutionary basis of the behavior. **Behavioral ecology** is the study of the ecological and evolutionary basis of animal behavior.

Fixed Action Patterns   A **fixed action pattern** is a sequence of unlearned behaviors that, once begun, is usually carried through to completion. It is triggered by a simple external stimulus called a **sign stimulus**.

Migration   Many animals use environmental cues to guide regular, long-distance changes in location called **migrations**. Some migratory animals alter their orientation to the sun during the day with the aid of the *circadian clock*. Some animals have specific receptors that allow them to navigate using Earth's magnetic field.

Behavioral Rhythms   Many behaviors follow a circadian rhythm mediated by the biological clock. *Circannual rhythms* are linked to the yearly cycle of seasons and are influenced by changing lengths of daylight and darkness. The mating behavior of fiddler crabs is linked to the lunar cycle, which corresponds to times of greatest tidal movement and the subsequent dispersal of larvae.

Animal Signals and Communication   **Communication** between animals involves the transmission and reception of special stimuli called **signals**. Courtship often involves a *stimulus-response chain*, in which each response is the stimulus for the next behavior. Modes of communication may be visual, auditory, chemical, or tactile, depending on the lifestyle and sensory specializations of a species.

Pheromones are chemical signals commonly used by mammals and insects in reproductive behavior to attract mates and to trigger specific courtship behaviors. Pheromones can also function as alarm signals.

INTERACTIVE QUESTION   51.1

Why is most communication among mammals olfactory and auditory, whereas communication among birds is primarily visual and auditory?

51.2   Learning establishes specific links between experience and behavior

Behavior that is performed virtually the same by all individuals in a population, regardless of developmental or environmental differences, is developmentally fixed and called **innate behavior**.

Experience and Behavior   In a **cross-fostering study**, young of one species are raised by adults of another species. Such studies examine how an animal's social and physical surroundings influence behavior. Male California mice provide extensive parental care and are always highly aggressive toward other mice. When cross-fostered in nests of white-footed mice, California mice showed reduced aggression and reduced parental behavior. Early experiences that alter parental behavior extend this environmental influence to the next generation.

Human **twin studies** involving identical twins raised apart or in the same household have been used to study various behavioral disorders in humans.

Learning   **Learning** is the modification of behavior as a result of specific experiences.

Imprinting is characterized by a limited **sensitive period** during which learning may occur and involves a behavioral response to an individual or object that is usually long lasting. The ability to respond is innate; the environment provides the imprinting stimulus. Imprinting on parents may be required for future *pair-bonding* with a mate of the proper species.

The establishment of a memory of the spatial structure of the environment is called **spatial learning**. Animals may learn and use a particular set of landmarks to find their way within their area.

More complicated than a set of learned landmarks, **cognitive maps** are neural representations of the spatial relationships of objects in an animal's surroundings. Evidence for cognitive maps comes from research with nutcrackers, birds that are able to retrieve their hidden food stores.

In **associative learning**, animals learn to associate one stimulus with another. In *classical conditioning*, an arbitrary stimulus is associated with a particular outcome. *Operant conditioning* refers to trial-and-error learning through which an animal associates a behavior with a reward or punishment. An animal's nervous system enables the association of behaviors with those stimuli that are most likely to occur in the animal's natural habitat.

Cognition refers to an animal's ability to learn using awareness, reasoning, recollection, and judgment. Research with bees indicates that they are capable of categorizing environmental objects as "same" or "different."

Problem solving, the ability to achieve an end in the face of obstacles, is most often observed in mammals, especially in primates and dolphins. Such behavior has also been documented in some bird species.

Development of learned behaviors may occur over a short period of time or in distinct stages. For example, white-crowned sparrows appear to have a 50-day sensitive period in which they memorize the song of their species. In a second learning phase, a juvenile bird sings a subsong, which gradually improves, apparently as the bird compares its own singing with the memorized song. The sparrow then sings this "crystallized" song for the rest of its life.

Many animals use the behavior of others as information in problem solving. Vervet monkeys have an innate ability to give alarm calls in response to threatening objects. They learn to discriminate in their calls by observing other members of the group and by receiving social confirmation of their calls. **Social learning**, learning through the observation of others, forms the basis of **culture**—a system of information transfer that involves teaching and/or social learning and influences behavior in a population.

INTERACTIVE QUESTION   51.2

Indicate the type of learning illustrated by the following examples:

a. Ewes will adopt and nurse a lamb shortly after they give birth to their own lamb but will butt and reject a foreign lamb introduced a day or two later.

b. A dog whose early "accidents" were cleaned up with paper towels accompanied with harsh discipline hides any time a paper towel is used in the household.

c. In experiments, honeybees have the ability to distinguish between patterns that are the same or different.

d. Tinberger showed that female digger wasps use landmarks to locate their hives.

e. Young chimpanzees learn to crack oil palm nuts with two stones by observing others.

51.3 Selection for individual survival and reproductive success can explain diverse behaviors

Evolution of Foraging Behavior **Foraging** is behavior involved with searching for, recognizing, obtaining, and consuming food. Natural selection is expected to refine behaviors that improve foraging. Laboratory studies of *Drosophila meanogaster* have documented an evolutionary change in foraging path length in populations of high or low density. The frequency of the *for^R* (rover) allele increased in high-density populations, whereas the frequency of the *for^S* (sitter) allele increased in low-density populations.

According to the **optimal foraging model**, feeding behaviors maximize energy intake over energy expenditure and the risk of being eaten while foraging. Mule deer appear to forage more in open areas where food may be slightly less available but the deer are less likely to fall prey to mountain lions.

INTERACTIVE QUESTION 51.3

Explain how the study of whelk-eating crows supports the optimal foraging model.

Mating Behavior and Mate Choice *Mating systems* vary among species. Many species have *promiscuous* mating, in which no strong pair-bonds form between mates. Longer-lasting relationships may be **monogamous** or **polygamous**. Polygamous relationships may be *polygynous* (one male and many females), or *polyandrous* (one female and many males). Monogamous species are less likely to exhibit *sexual dimorphism*.

The needs of offspring are an important factor in the evolution of reproductive patterns. If young require more food than one parent can supply, a male may increase his reproductive fitness by helping to care for offspring rather than going off in search of more mates. With mammals, the female often provides all the food, and males are often polygynous.

Certainty of paternity also influences mating behavior and parental care. With internal fertilization, the acts of mating and egg laying or birth are separated, and paternity is less certain than when eggs are fertilized externally.

INTERACTIVE QUESTION 51.4

Exclusive male parental care is observed much more frequently in species with external fertilization. Explain how such behavior could evolve.

Sexual selection may be *intrasexual*, involving competition among members of one sex for mates, or *intersexual*, in which mates are chosen by one sex on the basis of particular characteristics.

Female choice in stalk-eyed fruit flies has been a selection factor in the evolution of long eyestalks, which correlate with male health. Experiments involving feather ornaments added to zebra finch parents suggest that females imprint on their fathers, and that mate choice may play a role in the evolution of ornamentation in male zebra finches.

In **mate-choice copying**, individuals may copy the mate choice of others in a population. Experiments with guppies have shown that females will preferentially mate with males whom they have observed engaged in courtship with other females.

Agonistic behavior involves an often-ritualized contest to determine which competitor gains access to a resource, such as food or a mate.

In studying mating behavior, behavioral ecologists apply **game theory**, which evaluates strategies in situations in which the outcome depends on both an individual's strategy and the strategies of others. Researchers studying the coexistence of three male phenotypes of the side-blotched lizard found that the reproductive success of each phenotype depended on the frequency of the other phenotypes. Aggressive orange throat males defend large territories with many females; blue throat males defend smaller territories and fewer females; and yellow throat males mimic females and use "sneaky" tactics to obtain matings.

Starting with a high abundance of orange throats in a side-blotched lizard population, which type of male will tend to increase its mating success and increase in frequency next, and which type of male will then replace this second type after it increases in numbers? Explain your answer.

Britain. Offspring of these new migrants had similar migratory orientations, as documented in funnel cage studies.

Why do experiments examining behavioral variations among natural populations often raise and test animals in the laboratory?

51.4 Genetic analyses and the concept of inclusive fitness provide a basis for studying the evolution of behavior

Genetic Basis of Behavior   The gene called *fru* (for *fruitless*) in fruit flies is a master regulatory gene that controls the sex-specific development of the nervous system and thus the male courtship ritual. Males lacking a functional *fru* gene fail to court, and females expressing a male *fru* gene court other females.

Male prairie voles form pair-bonds, help care for the young, and are aggressive to intruders. Male meadow voles do not show these behaviors. The neurotransmitter **antidiuretic hormone (ADH)**, or **vasopressin**, is released during mating. The gene that codes for vasopressin receptors is highly expressed in the brains of prairie voles but not in those of meadow voles. Male meadow voles with inserted prairie vole receptor genes developed more vasopressin receptors in their brains and showed mating behaviors similar to those of male prairie voles.

Genetic Variation and the Evolution of Behavior
Closely related species often exhibit behavioral differences. Variations in behavior between populations within a species may correlate with variations in the environment.

Most laboratory-born garter snakes from coastal areas (where banana slugs are an abundant source of prey) ate slugs when they were offered, whereas few laboratory-born snakes from inland populations ate the slugs. The researcher proposed that snakes in coastal areas with the ability to recognize slugs by chemoreception had higher fitness, leading to the evolution of this difference in prey selection behavior.

Field observations and laboratory studies have documented recent changes in migratory behavior in European blackcaps. Instead of migrating southwest and then south to Africa for the winter, some German blackcaps are migrating west to Great

Altruism   Many social behaviors are selfish, benefiting one individual's reproductive success at the expense of others. Selflessness, or **altruism**, is behavior that reduces an animal's fitness while increasing the fitness of other individuals.

Inclusive Fitness   Natural selection favors traits that increase reproductive success, thereby propagating the genes for those traits. W. Hamilton explained altruistic behavior in terms of **inclusive fitness**, the ability of an individual to pass on its genes by producing its own offspring and by helping close relatives produce their offspring.

The three variables involved in altruism are: benefit (B) to the recipient of an altruistic act is the average number of additional offspring as a result of the act; cost (C) to the altruist is the average number of fewer offspring produced as a result of the act; and the **coefficient of relatedness** (r), the average fraction of genes shared between the recipient and the altruist. **Hamilton's rule** states that natural selection will favor altruistic behavior if $rB > C$. **Kin selection** is the term for the natural selection of altruistic behavior that enhances the reproductive success of related individuals.

Studies show that most cases of altruistic behavior involve close relatives, such as females in Belding's squirrel populations and worker bees in a hive, and thus improve the individual's inclusive fitness.

When altruistic behavior involves nonrelated animals, the explanation offered is **reciprocal altruism**; there is no immediate benefit for the altruistic individual, but some future benefit may occur if the helped animal "returns the favor." Although "cheaters" could obtain a large benefit, behavioral ecologists, using game theory, propose a behavioral strategy called *tit for tat*. If cheating is immediately retaliated against, reciprocal altruism may evolve and persist in a population.

INTERACTIVE QUESTION 51.7

According to kin selection theory, would an individual be more likely to exhibit altruistic behavior toward a parent, a sibling, or a first (full) cousin? Explain your answer in terms of the coefficient of relatedness and Hamilton's rule.

Evolution and Human Culture **Sociobiology** relates evolutionary theory to social behavior and to human culture. Due to our capacity for learning, human behavior appears to be quite plastic. Our structured societies—with their codification of acceptable behaviors and the exclusion of some behaviors that might otherwise enhance an individual's fitness—may be a unique characteristic separating humans and other animals.

Word Roots

mono- = one; **-gamy** = reproduction (*monogamous:* a type of relationship in which one male mates with just one female)

poly- = many (*polygamous:* a type of relationship in which an individual of one sex mates with several of the other)

soci- = a companion (*sociobiology:* the study of social behavior based on evolutionary theory)

Structure Your Knowledge

1. How does the nature-versus-nurture controversy apply to the study of behavior?

2. How does the concept of evolutionary fitness apply to all aspects of behavior?

Test Your Knowledge

MULTIPLE CHOICE: *Choose the one best answer.*

1. Behavioral ecology is the
   a. study of the behavior of animals, focusing on stimulus and response.
   b. application of human emotions and thoughts to other animals.
   c. study of animal behavior from an ecological and evolutionary perspective.
   d. behavioral study of ecology.

2. Proximate causes
   a. explain the evolutionary significance of a behavior.
   b. are immediate causes of behavior such as environmental stimuli.
   c. are environmental, whereas ultimate causes are genetic.
   d. show that nature is more important than nurture.

3. The behavior that maximizes an animal's energy intake-to-expenditure ratio is called
   a. optimal foraging.
   b. Hamilton's rule.
   c. a fixed-action pattern.
   d. cognition and learning.

4. Which of the following descriptions is an example of a fixed action pattern?
   a. a crane in a captive-breeding program imprinting on its human caregiver
   b. a male stickleback chasing a red-bellied object from its territory
   c. the claw-waving behavior of a male fiddler crab
   d. a digger wasp returning to its nest with the aid of landmarks

5. The honeybee waggle dance communicates the location of food resources relative to
   a. the location of the sun.
   b. the Earth's magnetic field.
   c. the north-south compass axis.
   d. the North Star.

6. A sensitive period
   a. is the time right after birth.
   b. usually follows the reception of a sign stimulus.
   c. is a limited time in which imprinting can occur.
   d. is the period during which birds can learn to fly.

7. In operant conditioning, a(n)
   a. animal improves its performance of a fixed-action pattern.
   b. animal learns as a result of associating a benefit or harm with an action.
   c. bird can learn the song of a related species if it hears only that song.
   d. irrelevant stimulus can elicit a response because of its association with a normal stimulus.

8. Which of the following types of intraspecies communication signal is best suited to a fruit fly's stimulus-response chain during mating behavior?
   a. an auditory signal
   b. a visual signal
   c. a chemical or a tactile signal
   d. All of the above are suited.

9. You have discovered a new primate species in which there is very little difference between males and females. You would correctly predict that this species is
   a. monogamous.
   b. promiscuous.
   c. polyandrous.
   d. polygynous.

10. Which of the following examples of behavior provides evidence of animal cognition?
   a. ravens pulling up string to obtain a food item attached to it
   b. trained honeybees that can discriminate between the concepts of "same" and "different" by matching colors or patterns
   c. a biology student using various resources to study for an exam
   d. All of the above show evidence of information processing and animal cognition.

11. In frogs, the male releases sperm onto the eggs as they are deposited, thus ensuring paternity. What behavior would you expect to find in frog species?
   a. Males are often responsible for parental care of eggs.
   b. Males defend a harem of several females.

   c. Females are highly ornamented, and are selected by males for mating.
   d. Males select females because of pheromone signals.

12. A crow that aids its parents in raising its siblings is increasing its
   a. survival success.
   b. altruistic behavior.
   c. inclusive fitness.
   d. coefficient of relatedness.

13. According to the concept of kin selection,
   a. an animal would be more likely to aid a stranger if the "kindness" could be reciprocated.
   b. an animal would aid its parent before it would help its sibling.
   c. animals are more likely to choose close relatives as mates.
   d. examples of altruism usually involve close relatives and increase an animal's inclusive fitness.

14. Which of the following behaviors would arise from a tit-for-tat social strategy?
   a. You perform random acts of kindness regardless of how you are treated.
   b. You only help out individuals who have previously helped you.
   c. You help anybody, but stop helping those who don't treat you well in return.
   d. You don't help anyone who isn't related to you.

15. The cross-fostering of California mice in white-footed mice nests provides evidence for
   a. the genetic control of aggression and parenting behavior.
   b. the relationship between the distribution of vasopressin receptors and parenting behavior.
   c. an imprinting period during which behaviors related to aggression and parenting are set.
   d. the influence of the early social environment on the expression of aggressive and parental behaviors.

16. A popular song from 1911 was called "I Want a Girl (Just Like the Girl that Married Dear Old Dad)." The title of this song suggests
   a. that human males are strictly monogamous.
   b. that human males choose mates based on good genes.
   c. that human males choose mates based on mate-choice copying.
   d. that human males choose mates based on imprinting.

UNIT 8 Ecology

Chapter 52

An Introduction to Ecology and the Biosphere

Framework

This chapter describes the scope of ecology, global climate patterns, the major terrestrial and aquatic biomes, and the factors that help determine the distribution of species.

Chapter Review

Ecology is the study of the interactions of organisms with each other and with their environment. Rigorous experimental designs are used to investigate complex ecological questions. These questions may focus on different levels of the biological hierarchy, outlined as follows.

Organismal ecology, which may include the disciplines of behavioral, physiological, and evolutionary ecology, considers the adaptations of an organism to its environment.

Population ecology is concerned with the factors that affect the size of **populations**, which are groups of individuals of the same species occupying a particular area.

A **community** includes the populations of different species in an area; **community ecology** looks at how species' interactions such as predation and competition affect community structure.

An **ecosystem** includes all the organisms in an area and the abiotic factors with which they interact, and **ecosystem ecology** addresses such topics as the flow of energy and chemical cycling.

A **landscape** or seascape consists of several connected ecosystems in a region; **landscape ecology** is the study of the flow of energy, materials, and organisms across ecosystems.

The **biosphere** includes all of Earth's ecosystems and landscapes. **Global ecology** is the study of the effect of regional energy and material exchanges on organisms across the biosphere.

52.1 Earth's climate varies by latitude and season and is changing rapidly

The **climate**, or prevailing weather conditions of an area, is influenced by temperature, precipitation, sunlight, and wind.

Global Climate Patterns   The absorption of solar radiation heats the atmosphere, land, and water, setting patterns for temperature variations, air circulation, and water evaporation that cause latitudinal variations in climate. The shape of Earth creates seasonal variations in day length and temperature, which increase with latitude. The **tropics**, located between latitudes 23.5° north and 23.5° south, receive the greatest amount of and least variation in solar radiation.

The global circulation of air begins as intense solar radiation near the equator causes warm, moist air to rise and release its water, producing the characteristic wet tropical climate, the arid conditions around 30° north and south as dry air descends, the fairly wet although cool climate at about 60° latitude as air rises again, and the cold and rainless climates of the polar regions. Air flowing along Earth's surface produces global wind patterns, such as the cooling trade winds blowing from east to west in the tropics and the prevailing westerlies in temperate zones.

Regional and Local Effects on Climate   The changing angle of the sun over the year produces wet and dry seasons around 20° north and 20° south latitude. Seasonal wind pattern changes affect ocean currents, sometimes causing upwellings of cold, nutrient-rich water.

Coastal areas are generally wetter than inland areas, and large bodies of water moderate the climate. Water warmed at the equator flows as the Gulf Stream toward the North Atlantic and warms the coast of western Europe. A *Mediterranean climate* is created when cool, dry ocean breezes warm as they cross land, absorbing moisture and creating a hot, rainless climate.

INTERACTIVE QUESTION   52.1

Describe how mountains affect local climate with respect to the following three factors:

a. sunlight

b. temperature

c. rainfall

Microclimate   Climate also varies on a very small scale. Differences in **abiotic** factors created by open spaces, or even fallen logs or large stones affect the local distributions of organisms. **Biotic** factors, or living organisms, also influence the distribution and abundance of organisms.

Global Climate Change   Scientists predict that increasing concentrations of CO_2 and other greenhouse gases will warm the Earth 1–6°C by the end of the 21st century. Such **climate change** will have great

effects on the distributions of plants and animals. Fossil pollen deposits have documented the rates of northward expansion of the ranges of various tree species following the last continental glaciers. As their geographic limits change with climatic warming, seed dispersal may not be rapid enough for some species of plants to move into new suitable habitat as their former ranges become inhospitable. Changes in distribution have already been documented for many plant and animal groups.

52.2   The distribution of terrestrial biomes is controlled by climate and disturbance

Biomes are major types of ecosystems characterized by the predominant vegetation or (in aquatic biomes) by the physical environment.

Climate and Terrestrial Biomes   A **climograph** plots annual mean temperature and rainfall for a region; generally these values correlate with the distribution of various biomes. Overlaps of biomes on a climograph indicate the importance of variation in the seasonal patterns of rainfall and temperatures.

General Features of Terrestrial Biomes   Biomes are usually named for their predominant vegetation and major climatic features. Each biome also has characteristic microorganisms, fungi, and animals. The area of transition between biomes is called an **ecotone**.

Terrestrial biomes have vertical stratification, such as the layers in a forest from upper **canopy**, low-tree layer, shrub understory, herbaceous plant ground layer, forest floor (litter), to root layer. Vertical stratification of vegetation provides diverse habitats for animals.

Species composition of any one biome varies locally. Often, convergent evolution has produced a superficial resemblance of unrelated species.

Disturbance and Terrestrial Biomes   A **disturbance** is an event such as a fire, a storm, or human activity that can modify a community and alter resource availability. The patchiness of most biomes results from disturbances. Grasslands, savannas, chaparral, and many coniferous forests are maintained by the periodic disturbance of fire. Urban and agricultural communities now cover a large portion of Earth's landmass.

Exploring Terrestrial Biomes   Let's review the main characteristics of the eight terrestrial biomes.

Tropical forests occur in equatorial and subequatorial regions. Variations in rainfall result in **tropical dry forests**, where rainfall is seasonal, and **tropical rain forests**, where rainfall is more constant and abundant. Temperatures are uniformly high. The tropical rain forest has pronounced vertical stratification, including emergent trees above a closed canopy and abundant epiphytes. Animal diversity is higher than in any other terrestrial biome. Agriculture and development are destroying many tropical forests.

Characterized by low and variable precipitation, **deserts** occur around 30° north and south latitude and in the interior of continents. Deserts may be hot or cold, and temperature varies seasonally and daily. Plants may use C_4 and CAM photosynthesis and have reduced leaf surface area, water storage adaptations, and protective spines and toxins. Desert animals have physiological and behavioral adaptations to dry conditions. Irrigated agriculture and urbanization are now common in deserts, reducing natural biodiversity.

Savannas are equatorial and subequatorial grasslands with scattered trees and long dry seasons. Fires are common, and the dominant grasses and forbs (small nonwoody plants) are fire-adapted, drought tolerant, and able to withstand grazing. Large grazing mammals and their predators are common, although insects are the dominant herbivores. Cattle ranching and overhunting have reduced the large-mammal populations of savannas.

Chaparral, common along coastlines in midlatitude regions, has cool, rainy winters and hot, dry summers. The dominant vegetation—evergreen shrubs and small trees—is maintained by and adapted to periodic fires. The high plant diversity present also includes many grasses and herbs. Browsing animals and small mammals and birds are common. Urbanization and agriculture have reduced areas of chaparral.

Temperate grasslands are maintained by fire, seasonal drought, and grazing by large mammals. Winters are generally cold and dry; summers are hot and wet. Grasses and forbs are dominant plant forms in these areas. In addition to grazers, burrowing mammals are common in grasslands. Soils are deep and fertile, and most North American grasslands have been converted to farmland.

The largest terrestrial biome, the **northern coniferous forest**, or *taiga*, is characterized by harsh winters with heavy snowfall and periodic droughts. Summers may be hot. Birds and large mammals are common animals. Old-growth stands of coniferous trees are rapidly being logged.

Characterized by broad-leaved deciduous trees, **temperate broadleaf forests** grow in midlatitude regions in the Northern Hemisphere, with smaller areas on other continents. Temperate broadleaf forests have distinct vertical layers. Winters are cold and summers hot and humid. During the cold winters, many mammals hibernate and birds migrate. Humans have heavily logged these forests, clearing land for agriculture and development.

Tundra, covering large areas of the Arctic, is characterized by long, cold winters; short, cool summers; and dwarfed or herbaceous vegetation. A layer of frozen soil called permafrost restricts root growth. Migratory large mammals and birds are common. The *alpine tundra*, found at all latitudes on very high mountains, has similar plant communities. Mineral and oil extraction are becoming common in areas of arctic tundra.

INTERACTIVE QUESTION 52.2

Temperature and precipitation are two of the key factors that influence the vegetation found in a biome. On the following climograph, label the North American biomes (arctic and alpine tundra, northern coniferous forest, desert, temperate grassland, temperate broadleaf forest, and tropical forest) represented by each plotted area of temperature and precipitation.

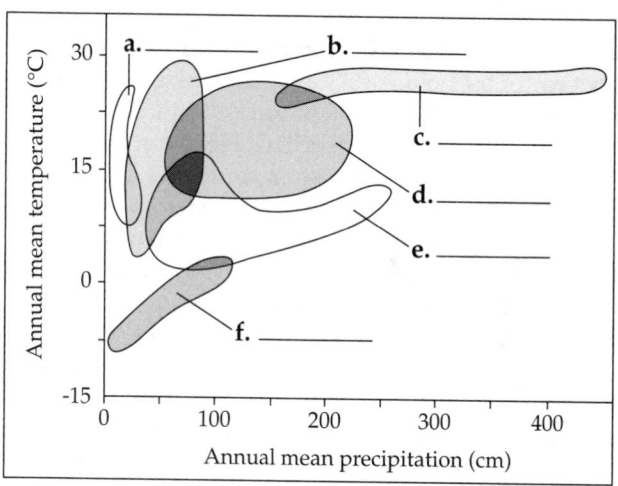

52.3 Aquatic biomes are diverse and dynamic systems that cover most of Earth

One of the chemical differences in aquatic biomes is salt concentration—less than 0.1% for freshwater biomes versus an average of 3% for marine biomes. Three-fourths of Earth is covered by oceans, which influence global rainfall, climate, and wind patterns. Marine algae and photosynthetic bacteria produce a large portion of the world's O_2 and consume enormous amounts of CO_2. Freshwater biomes are influenced by the surrounding terrestrial biome.

Zonation in Aquatic Biomes Many aquatic biomes are layered both vertically and horizontally. The

photic zone receives sufficient light for photosynthesis, whereas little light penetrates into the lower **aphotic zone**. The **pelagic zone** includes the waters of the photic and aphotic zone. The very deep ocean region is called the **abyssal zone**. The bottom substrate, called the **benthic zone**, is home to organisms collectively called **benthos**. Settling **detritus** (dead organic material) provides food for many benthic species.

In the ocean and in most lakes, a narrow layer of abrupt temperature change called a **thermocline** separates warmer surface waters from the cold bottom layer. In temperate lakes, seasonal temperature changes produce the **turnover** of water that sends oxygenated water to the bottom and brings nutrient-rich water to the surface.

Exploring Aquatic Biomes   Let's review the main characteristics of the eight aquatic biomes.

Oligotrophic lakes are often deep, nutrient-poor, and generally oxygen-rich. The nutrient-rich waters of **eutrophic lakes** may be seasonally depleted of oxygen in deeper regions due to decomposition. Rooted and floating plants are found in the **littoral zone** close to shore; phytoplankton inhabit the **limnetic zone**. Heterotrophs include floating zooplankton, benthic invertebrates, and fishes. Dumping of municipal wastes and runoff from fertilized lands can cause algal blooms, oxygen depletion, and fish kills.

Wetlands, defined as areas covered with water often enough that they support plants adapted to water-saturated soil, are among the most productive of biomes. Water and soils may be periodically oxygen-poor due to high rates of decomposition. Topography creates *basin, riverine,* and *fringe wetlands*. Heterotrophs include diverse invertebrates, birds, and others. Wetlands filter nutrients and chemical pollutants and reduce flooding. Up to 90% of European wetland areas have been lost to draining and filling.

Streams and rivers are flowing habitats with physical and chemical characteristics that vary from the headwaters to the mouth. Oxygen levels are high in turbulently flowing water and low in murky, warm waters. Adjacent vegetation contributes to nutrient content. Unpolluted rivers and streams are home to diverse fishes and invertebrates. Human impact on streams and rivers includes pollution and damming.

Where a freshwater river meets the ocean, an **estuary** is formed. Salinity varies both spatially and daily with the rise and fall of tides. Salt marsh grasses and algae are the major producers. Estuaries serve as feeding and breeding areas for marine invertebrates, fish, and waterfowl. Pollution, filling, and dredging have extensively disrupted these highly productive biomes.

INTERACTIVE QUESTION   52.3

In the following table, indicate with a + or a − whether the listed biomes are relatively high or low in oxygen level, nutrient content, and productivity.

Biome	Oxygen Level	Nutrient Content	Productivity
Oligotrophic lake	a.		
Eutrophic lake	b.		
Headwater stream	c.		
Turbid river	d.		
Wetlands	e.		
Estuary	f.		

In **intertidal zones**, the daily cycle of tides exposes the shoreline to variations in water, nutrients, and temperature, and to the mechanical force of wave action. Rocky intertidal communities are vertically stratified, with diverse marine algae and animals adapted to attaching firmly to the hard substrate. Sandy intertidal zones are home to burrowing worms, clams, and crustaceans. Oil pollution and construction of protective barriers have disrupted many intertidal areas.

The water of the **oceanic pelagic zone** is typically nutrient-poor but oxygen-rich. In temperate and high-latitude areas, seasonal mixing of waters brings nutrients to surface waters. Phytoplankton in the photic region provide half of global photosynthetic productivity. They are grazed on by numerous types of zooplankton. Free-swimming animals include squid, fishes, and marine mammals. Overfishing and waste dumping have damaged the Earth's oceans.

Found in the photic zone of clear tropical waters, **coral reefs** are highly diverse and productive biomes. The structure of the reef is produced by the calcium carbonate skeletons of the coral (diverse cnidarians) and serves as a substrate for red and green algae. The coral animals are nourished by symbiotic unicellular algae. Overfishing, coral collecting, pollution, and global warming are destroying coral reefs.

The **marine benthic zone** consists of the ocean floor. The shallow benthic areas of the **neritic zone** have diverse seaweeds and algae, as well as numerous invertebrates and fishes. The rest of the marine benthic zone receives no sunlight, and nutrients fall as detritus from the waters above. Inhabitants of the abyssal zone are adapted to cold and to high water pressure. Chemoautotrophic prokaryotes form the basis of a collection of organisms adapted to the hot, low-oxygen environment surrounding deep-sea **hydrothermal vents**. Overfishing has reduced many benthic fish populations.

INTERACTIVE QUESTION   52.4

Different marine environments can be classified on the basis of light penetration, distance from shore, and open water versus bottom. Match the following zones to their corresponding numbers on the diagram:

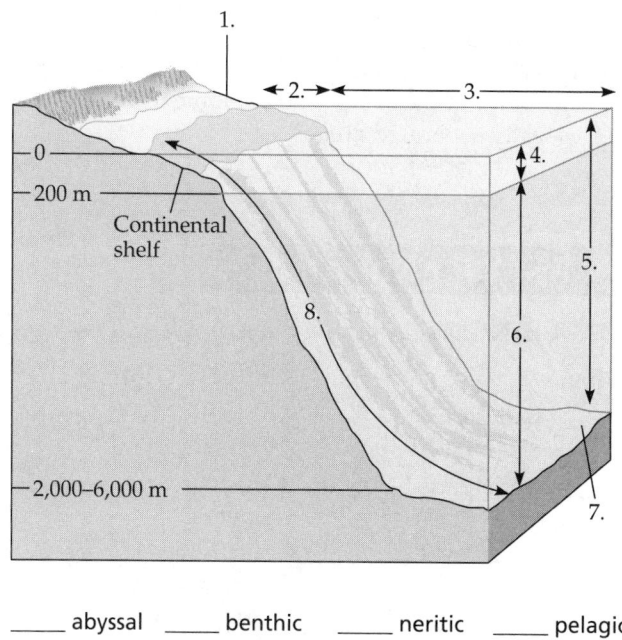

| _____ abyssal | _____ benthic | _____ neritic | _____ pelagic |
| _____ aphotic | _____ intertidal | _____ oceanic | _____ photic |

52.4 Interactions between organisms and the environment limit the distribution of species

The distribution of species reflects ecological and evolutionary interactions with both biotic and abiotic factors. Ecologists often work through a series of four logical questions to determine what factors limit the geographic distribution of a species.

Dispersal and Distribution   Does dispersal limit a species distribution? **Dispersal** is the movement of

individuals away from their original area. The cattle egret is an example of a species that has naturally extended its range through dispersal. Occasionally, such long-distance dispersal can lead to the adaptive radiation of many new species.

Transplants of a species can indicate whether dispersal limits its distribution. A successful transplant shows that the *potential* range of a species is larger than its *actual* range. Species that are introduced to new areas, either purposely or accidentally, often disrupt their new ecosystem.

Biotic Factors   Is the distribution of a species limited by biotic factors such as predation, pathogens, parasitism, or competition? Removal experiments can test whether other species limit the distribution of a species. A study showed that sea urchins limited the abundance and distribution of seaweeds.

Abiotic Factors   Is a species' geographic distribution influenced by abiotic factors? Temperature is an important environmental factor because of its effects on metabolism. Most organisms function best within a narrow range of environmental temperatures.

The availability of water in different habitats can vary greatly. The distribution of organisms reflects their ability to obtain sufficient water and avoid desiccation in terrestrial habitats. Oxygen concentrations can be low in deep waters, in sediments rich in organic matter, and in flooded soils.

The ability of organisms to osmoregulate determines their distribution in freshwater, saltwater, or other high-salinity habitats.

Light energy drives almost all ecosystems. The quantity of sunlight is a limiting factor in aquatic environments and on forest floors. Excessive light intensity at high elevations and in deserts also limits the survival of plants and animals.

Soils, which vary in their physical structure, pH, and mineral composition, affect the distribution of plants and, in turn, the distribution of animals. Substrate composition in aquatic environments influences water chemistry and the types of organisms that can inhabit those areas.

52.5 Ecological change and evolution affect one another over long and short periods of time

Ecological changes alter the selective pressures on populations, leading to evolutionary change. Evolutionary changes alter the role of a population in an ecosystem. Ecological and evolutionary change influence each other over time scales from millions of years, to centuries, to decades.

INTERACTIVE QUESTION 52.5

Describe the four factors that ecologists examine to understand the geographic distribution of a species.

a.

b.

c.

d.

Word Roots

a- =without; **bio-** = life (*abiotic:* nonliving; referring to physical and chemical properties of an environment)

abyss- = deep, bottomless (*abyssal zone:* the part of the ocean's benthic zone between 2,000 m and 6,000 m deep)

benth- = the depths of the sea (*benthic zone:* the bottom surface of an aquatic environment)

estuar- = the sea (*estuary:* the area where a freshwater stream or river merges with the ocean)

eu- = good, well; **troph-** = food, nourishment (*eutrophic lake:* a lake with a high rate of biological productivity and nutrient cycling)

hydr- = water; **therm-** = heat (*deep-sea hydrothermal vent:* a dark, hot, oxygen-deficient environment associated with volcanic activity on or near the seafloor; the producers are chemoautotrophic prokaryotes)

inter- = between (*intertidal zone:* the shallow zone of ocean adjacent to land and between the high- and low-tide lines)

limn- = a lake (*limnetic zone:* in a lake, the well-lit, open surface waters far from shore)

littor- = the seashore (*littoral zone:* in a lake, the shallow, well-lit waters close to shore)

micro- = small (*microclimate:* climate patterns on a very fine scale, such as the specific climatic conditions underneath a log)

oligo- = small, scant (*oligotrophic lake:* a nutrient-poor, clear lake with few phytoplankton)

pelag- = the sea (*pelagic zone:* the open-water component of aquatic biomes)

phot- = light (*aphotic zone:* the part of an ocean or lake beneath the photic zone, where light does not penetrate sufficiently for photosynthesis to occur)

clin- = slope (*thermocline:* a narrow stratum of abrupt temperature change in the ocean and in many temperate-zone lakes)

Structure Your Knowledge

1. **a.** Define ecology.
   b. How does ecology relate to evolutionary biology?
2. **a.** What are biomes?
   b. What accounts for the similarities in life forms found in the same type of biome in geographically separated areas?

Test Your Knowledge

MATCHING: *Match the biotic description with its biome.*

Biome	Biotic Description
_____ 1. desert	A. broad-leaved deciduous trees
_____ 2. chaparral	B. lush growth, vertical layers
_____ 3. savanna	C. evergreen shrubs, fire-adapted vegetation
_____ 4. northern coniferous forest	D. scattered trees, grasses, and forbs
_____ 5. temperate broadleaf forest	E. tall stands of cone-bearing trees
_____ 6. temperate grassland	F. low shrubby vegetation, lichens
_____ 7. tropical rain forest	G. grasses adapted to fire and drought
_____ 8. tundra	H. widely scattered shrubs, cacti, and euphorbs

MULTIPLE CHOICE: *Choose the one best answer.*

1. Which level of ecology considers the effects of predation, parasitism, and competition on species distribution?
   a. landscape
   b. community
   c. ecosystem
   d. population

2. Oceans and lakes have a moderating effect on the climate of adjacent land areas. Ultimately, this effect results from
   a. the large surface area of oceans and lakes.
   b. reflection of sunlight from water onto land.
   c. the polarity of covalent bonds in water molecules.
   d. the saturation of adjacent land with water.

3. The ample rainfall of the tropics and the arid areas around 30° north and south latitudes are caused by
   a. ocean currents that flow clockwise in the northern hemisphere and counterclockwise in the southern hemisphere.
   b. the global circulation of air initiated by intense solar radiation near the equator that produces wet and warm air.
   c. the heavier rain on the windward side of mountain ranges and the drier climate on the leeward side.
   d. the location of tropical rain forests and deserts.

4. Which of the following would be most affected if Earth were not tilted on its axis relative to its orbit around the sun?
   a. the number of hours in a day
   b. the number of days in a year
   c. the distinct seasons in temperate and polar regions
   d. the amount of solar radiation that the tropics receives

5. Which of the following statements reflects a concern about the effects of global climate change on tree species?
   a. Trees may not be able to disperse fast enough to reach new habitats that meet their climatic requirements.
   b. Warmer temperatures may speed tree growth, producing trees that are too tall and spindly.
   c. The additional CO_2 in the atmosphere may actually increase photosynthetic rates and prove beneficial to tree growth.

   d. Most trees will be able to evolve adaptations to deal with climate changes because temperatures are rising very slowly.

6. A rain shadow would be expected to appear
   a. on the windward side of a mountain.
   b. on the leeward side of a mountain.
   c. on both the windward and leeward sides of a mountain.
   d. on top of a mountain.

7. Which of the following biomes has the highest species diversity?
   a. savannah
   b. temperate broadleaf forest
   c. tropical forest
   d. temperate grasslands

8. Many plant species have adaptations for dealing with the periodic fires typical of
   a. a savanna.
   b. a chaparral.
   c. a temperate grassland.
   d. all of the above.

9. The best way to explain how two communities can have the same annual mean temperature and rainfall but very different biota and characteristics is that the communities
   a. are found at different altitudes.
   b. are composed of species that have very low dispersal rates.
   c. are found on different continents.
   d. have different seasonal temperatures and patterns of rainfall throughout the year.

10. Which of the following aquatic zones is *incorrectly* paired with its description?
    a. pelagic zone—area of open water
    b. abyssal zone—very deep benthic region of the ocean
    c. intertidal zone—shallow coastal area of the ocean
    d. limnetic zone—shallow waters close to shore in a lake

11. In which of the following biomes would you expect to find organisms with adaptations for tolerating changes in salinity?
    a. desert
    b. coral reef
    c. deep-sea hydrothermal vent
    d. estuary

12. Where would you find the coldest water in a lake that has frozen over in winter?
    a. at the bottom of the lake
    b. just beneath the ice at the surface
    c. in a thermocline separating layers of warmer water at the surface and the bottom of the lake
    d. everywhere—the water would be uniformly cold throughout the lake

13. Upwellings in oceans
    a. support reef communities.
    b. occur over deep-sea hydrothermal vents.
    c. bring nutrient-rich water to the surface.
    d. are most common in tropical waters, where they bring oxygen-rich water to the surface.

14. Which of the following environments would be most likely to be associated with low oxygen levels?
    a. marine benthic zone
    b. wetland
    c. coral reef
    d. oligotrophic lake

15. An experiment that removed a species from an area would enable an ecologist to test whether
    a. competition or predation from that species was limiting the local distribution of other species.
    b. dispersal ability was limiting the range of the species.
    c. abiotic factors were limiting the dispersal of the species.
    d. habitat selection behavior had limited the distribution of the species.

Chapter 53

Population Ecology

Key Concepts

53.1 Biotic and abiotic factors affect population density, dispersion, and demographics

53.2 The exponential model describes population growth in an idealized, unlimited environment

53.3 The logistic model describes how a population grows more slowly as it nears its carrying capacity

53.4 Life history traits are products of natural selection

53.5 Density-dependent factors regulate population growth

53.6 The human population is no longer growing exponentially but is still increasing rapidly

Framework

Chapter Review

Population ecology is the study of the influence of the abiotic and biotic environment on population size, density, and distribution.

53.1 Biotic and abiotic factors affect population density, dispersion, and demographics

A **population** is a localized group of individuals of the same species. Members of a population use the same resources and have a high probability of interacting and breeding with each other. Ecologists define a population's boundaries based on the type of organism and the research question being asked.

Density and Dispersion   The number of individuals per unit area or volume is a population's **density**; the pattern of spacing of those individuals is referred to as **dispersion**.

   Scientists often use one of a variety of sampling techniques to estimate population density. Indirect indicators, such as burrows or nests, also may be used. In the **mark-recapture method**, a population

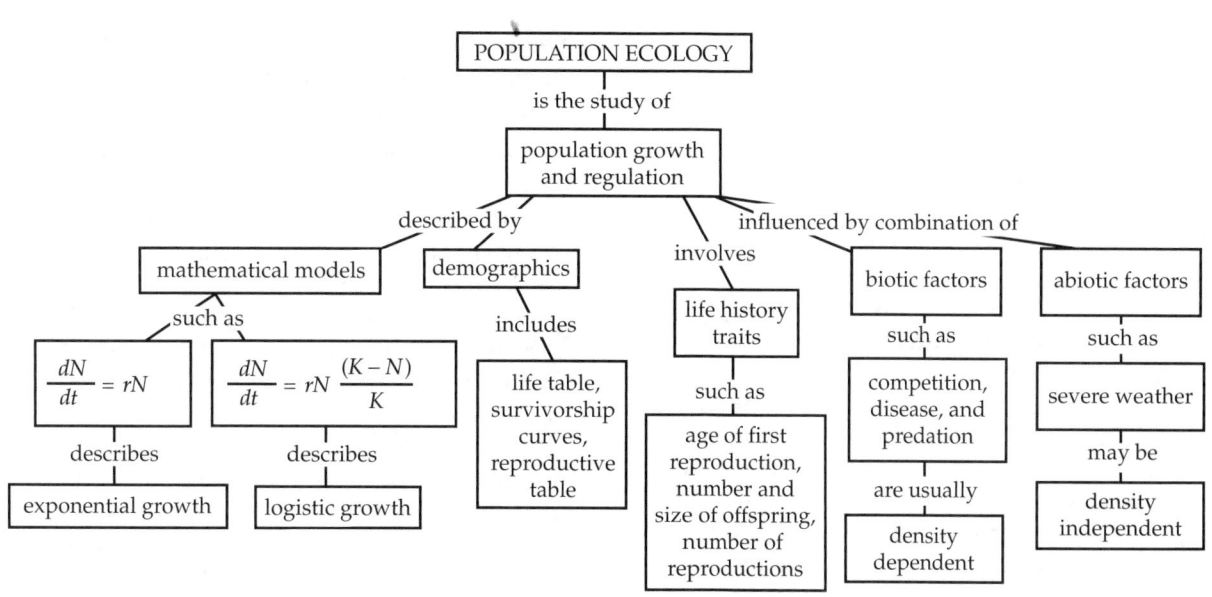

POPULATION ECOLOGY

is the study of

population growth and regulation

described by

influenced by combination of

mathematical models

such as

$$\frac{dN}{dt} = rN$$

describes

exponential growth

$$\frac{dN}{dt} = rN \frac{(K - N)}{K}$$

describes

logistic growth

demographics

includes

life table, survivorship curves, reproductive table

involves

life history traits

such as

age of first reproduction, number and size of offspring, number of reproductions

biotic factors

such as

competition, disease, and predation

are usually

density dependent

abiotic factors

such as

severe weather

may be

density independent

is sampled twice. First, animals are trapped and marked (or identified), and released. Animals are then captured or sampled again. The proportion of the total second trapping that is marked indicates the proportion of the total population that was initially marked. The calculation uses the equation $x/n = s/N$ (where x is recaptured marked animals; n is number in second sampling; s is the number of animals initially marked and released; and N is estimated population size). Solving for $N = sn/x$.

INTERACTIVE QUESTION   53.1

In a mark-recapture study, an ecologist traps, marks, and releases 25 voles in a small wooded area. A week later she traps 30 voles, 10 of which are marked. What is her estimate of the vole population in that area?

Changes in population density reflect both additions of members through birth (including all forms of reproduction) and **immigration**, and removal of members through death (mortality) and **emigration**.

What types of dispersal patterns may be found in a population's geographic range? *Clumping* may indicate a heterogeneous environment or social interactions between individuals. *Uniform* distribution may be related to competition for resources and result from interactions between individuals. **Territoriality**, the defense of a bounded physical space, can lead to uniform dispersion. *Random* spacing indicates the absence of strong interactions among individuals or a fairly uniform habitat.

Demographics   The study of the vital statistics of a population, such as birth and death rates, is called **demography**.

A **life table** presents age-specific survival data for a population. It can be constructed by following a **cohort** of organisms from birth to death and calculating the proportion of the cohort surviving in each age group.

A **survivorship curve** shows the number or proportion of members of a cohort still alive at each age. Survivorship curves are often based on a convenient beginning cohort of 1,000 individuals and typically use a logarithmic scale on the y-axis and a relative scale on the x-axis, so that species with different life spans can be compared on the same graph. There are three general types of survivorship curves. Type I has low mortality during early and middle age and a

rapid increase with old age. In a Type II curve, death rate is relatively constant throughout the life span. A Type III curve has high initial mortality, with the few offspring that survive likely to reach adulthood. Many species show intermediate or more complex survivorship patterns.

INTERACTIVE QUESTION   53.2

Identify the types of survivorship curves in the following graph. For each curve, describe the characteristics of some representative species.

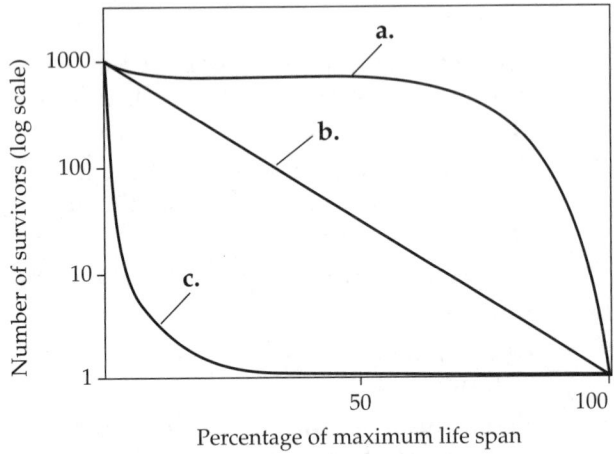

In sexually reproducing species, demographers usually follow only the reproduction of females through the generations. The age-specific reproductive rate for a sexually reproducing species is the average number of female offspring for females in a specific age class.

53.2 **The exponential model describes population growth in an idealized, unlimited environment**

Changes in Population Size   A small population in a very favorable environment may exhibit unrestricted growth. Ignoring immigration and emigration, the change in population size during a specific time period is equal to the number of births (B) minus the number of deaths (D). If we define R as the difference between the number of births and the number of deaths ($B - D$) then the change in population size is $\Delta N/\Delta t = R$.

It is more useful to consider changes in population size over time on a *per capita* basis. If we define $r_{\Delta t}$ as the per capita rate of increase, then $R = r_{\Delta t}N$. Substituting in the equation in the paragraph above gives $\Delta N/\Delta t = r_{\Delta t}N$. We can express this relationship using differential calculus as $dN/dt = rN$.

Exponential Growth Ecologists define *r* as the **intrinsic rate of increase**; under ideal conditions a population may grow according to the equation $dN/dt = rN$, and is said to exhibit **exponential population growth**. When expressed graphically, exponential population growth produces a J-shaped growth curve. The larger the population (*N*) becomes, the faster the population grows. Periods of exponential growth may occur in some populations that exploit an unfilled environment or rebound from a catastrophic event.

INTERACTIVE QUESTION 53.3

Define *r*. How does *r* relate to exponential growth?

53.3 The logistic model describes how a population grows more slowly as it nears its carrying capacity

A population may grow exponentially for only a short time before its increased density limits the resources available to its members. The **carrying capacity** (*K*) is the maximum population size that a particular environment can maintain. Crowding and resource limitations may lead to decreased per capita birth rates and increased per capita death rates.

The Logistic Growth Model The per capita rate of increase decreases from its maximum at low population size to zero as carrying capacity is reached. The mathematical model of **logistic population growth** is $dN/dt = rN(K - N)/K$. The equation includes the expression $(K - N)/K$ to reflect the impact of increasing *N* on the per capita rate of increase as the population approaches the carrying capacity.

When *N* is small, $(K - N)/K$ is close to 1, and growth is approximately exponential (*rN*). As population size approaches the carrying capacity, the $(K - N)/K$ term becomes a smaller fraction, and per capita rate of increase is small. When *N* reaches *K*, the term $(K - N)/K$ is 0, and the population stops growing. The logistic model produces a sigmoid (S-shaped) growth curve, and maximum increase in population numbers occurs when *N* is intermediate in size.

INTERACTIVE QUESTION 53.4

Label the following exponential and logistic growth curves and write the equation associated with each curve. What is *K* for the population shown in curve *b*?

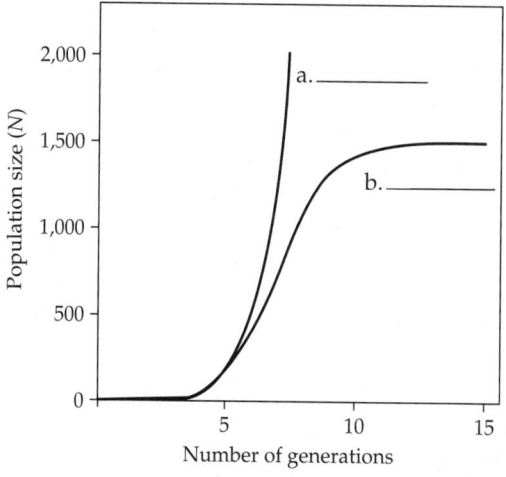

The Logistic Model and Real Populations Some laboratory populations of small animals and microorganisms show logistic growth. Natural populations often overshoot and then fluctuate around carrying capacity.

The logistic model makes the assumption that any increase in population size will have a negative effect on population growth. Population growth may slow because birth rates decrease, death rates increase, or both.

53.4 Life history traits are products of natural selection

The **life history** of an organism involves traits that reflect tradeoffs between survival and reproduction.

Diversity of Life Histories Life history variables include the age at first reproduction, how often an organism reproduces, and the number of offspring produced in each reproductive episode.

Some species put all their reproductive resources into a single effort, called **semelparity**, or big-bang reproduction. Other species follow the strategy of **iteroparity**, making repeated reproductive efforts over a span of time. Two factors thought to be important in the selection for and evolution of these two reproductive strategies are the survival rate of young offspring and the probability that an adult will survive to reproduce more than once.

"Trade-offs" and Life Histories   Because organisms have a finite energy budget, they cannot maximize all life history traits simultaneously. The costs of reproduction often include a reduction in survival. The production of large numbers of offspring is related to the selective pressures of high mortality rates of offspring in uncertain environments or from intense predation. Parental investments in the physical size of offspring (or seeds) and extended care increase the survival chances of offspring.

Natural selection will favor different life history traits at different population densities. Populations at high density, close to their carrying capacity, may experience **K-selection** or density-dependent selection for traits such as competitive ability and efficient resource use. In environments in which population density is low, **r-selection** (or density-independent selection) would favor traits that maximize population growth, such as the production of numerous, small offspring.

INTERACTIVE QUESTION   53.5

a. Explain why the life history of an organism cannot be to reproduce early and often, to have large numbers of offspring at a time, and to live a long life.

b. Indicate whether the following life history traits would be considered to be r-selected or K-selected:

early age at first reproduction; many small offspring produced

few, relatively large offspring produced every year

53.5   Density-dependent factors regulate population growth

Population Change and Population Density   A birth rate or a death rate that does not change as population density changes is said to be **density independent**. The death rate is **density dependent** if it rises with increasing population density; the birth rate is density dependent if it falls with increasing population density. An equilibrium density may be reached in a population as long as birth rate or death rate or both are density dependent.

Mechanisms of Density-Dependent Population Regulation   As population density increases, density-dependent decreases in birth rate and increases in death rate operate through negative feedback to slow or stop population growth.

Competition for resources such as nutrients or food may limit reproductive output. The availability of territorial space may be the limiting resource for some animals. Increased population densities affect the transmission rate of diseases. Predation may be a density-dependent factor when a predator feeds preferentially on a prey population that has reached a high density. The accumulation of toxic metabolic wastes may also be a limiting factor for some organisms.

Density-dependent regulation may also involve intrinsic factors. Studies of mice have shown that even when food or shelter is not limiting, aggressive interactions and hormonal changes inhibit reproduction and increase mortality at high population density.

Population Dynamics   All populations show fluctuations in numbers. The study of **population dynamics** looks at the biotic and abiotic factors that cause them. Fluctuations in populations of some large mammals, such as moose on Isle Royale, may be linked to the severity of winter, the availability of food, or predation.

The 10-year cycles in the densities of snowshoe hares and lynx in northern forests have been studied to determine whether winter food shortages or predator overexploitation causes the cyclic collapses in hare populations. Experiments that excluded predators from an area support the hypothesis that it is mainly excessive predation that drives hare cycles.

Emigration may increase with population density. A group of populations occupying separated suitable habitats may be linked through immigration and emigration to form a **metapopulation**. Emigrants may serve to repopulate habitats in which local populations become extinct. Studies of the Glanville fritillary have shown that a butterfly's genetic makeup may influence its likelihood to colonize new locations.

INTERACTIVE QUESTION   53.6

a. List some density-dependent factors that may limit population growth.

b. List some abiotic factors that may cause population fluctuations.

53.6 The human population is no longer growing exponentially but is still increasing rapidly

The Global Human Population Historically, human population growth has often exceeded exponential growth. After 1650, it took 200 years for the global population to double to 1 billion. In the next 80 years, it doubled to 2 billion; it doubled again to 4 billion in the next 45 years; and now, some 38 years later, it is more than 7 billion. The rate of growth has begun to slow, partly due to diseases such as AIDS and voluntary population control, although population ecologists project a population of 8.1–10.6 billion by 2050.

Population stability can be reached in one of two ways: Zero population growth can be achieved either by a balance of high birth rates and high death rates, or low birth rates and low death rates. The movement from the first configuration to the second is called the **demographic transition**. Death rates declined rapidly in most developing countries after 1950, but birth rate decline has been more uneven—from rapid in China to slower in sub-Saharan Africa and India. The annual population growth rate is regionally variable; it is near equilibrium in industrialized nations. Less industrialized countries, where 80% of the world's population lives, contribute the most to the global growth rate.

Human population growth is unique in that it can be consciously controlled by voluntary contraception and family planning. The key to the demographic transition is reduced family size. In many cultures, women are receiving more education and delaying marriage and reproduction, thus slowing population growth.

The **age structure** of a population affects present and future growth and can be predictive of future social conditions and needs.

Infant mortality and *life expectancy at birth* vary among human populations, with mortality being much higher and life expectancy lower in less industrialized countries.

INTERACTIVE QUESTION 53.7

What do the age structure pyramids in Figure 53.24 in the textbook predict for the future population growth of Afghanistan and Italy?

Global Carrying Capacity Estimates of Earth's carrying capacity have varied greatly. These estimates involve different assumptions, such as the logistic equation, the amount of habitable land, or food as the limiting factor.

The concept of an **ecological footprint** takes into account multiple human needs and the land and water area required to meet those needs and absorb wastes. A *global hectare* (gha) is an area of land or water of average productivity. This value has been used to estimate that a sustainable ecological footprint is 1.7 gha. In the United States, a typical person, with an ecological footprint of about 8 gha, is using an unsustainable share of Earth's resources.

Impact may also be based on the amount of energy used. A typical person in the United States consumes roughly 30 times the energy consumed by an individual in central Africa.

The ultimate carrying capacity of Earth may be determined by food and water supplies, the availability of space or nonrenewable resources, degradation of the environment, or several interacting factors. When and how we reach zero population growth is an issue of great social and ecological consequence.

Word Roots

co- = together (*cohort*: a group of individuals of the same age in a population)
dem- = people; **-graphy** = writing (*demography*: the study of the vital statistics of populations, especially birth and death rates)
iter- = to repeat; **-parity** = to beget (*iteroparity*: reproduction in which adults produce offspring over many years; also known as repeated reproduction)
semel- = once (*semelparity*: reproduction in which an organism produces all of its offspring in a single event; also known as big-bang reproduction)

Structure Your Knowledge

1. Create a concept map to organize your understanding of the exponential and logistic equations—the mathematical models of population growth.
2. What is the best collection of life history traits that would maximize reproductive success?

Test Your Knowledge

MULTIPLE CHOICE: *Choose the one best answer.*

1. In an area with a heterogeneous distribution of suitable habitats, the dispersion pattern of a population is probably
   a. clumped.
   b. uniform.
   c. random.
   d. dense.

2. Which of the following statements about life tables is *not* true?
   a. They show the age-specific death rate for a population.
   b. They are used to predict logistic growth.
   c. They can be used to construct survivorship curves.
   d. They are often constructed by following a cohort from birth to death.

3. In a species showing a Type III survivorship curve, you would expect to see
   a. many, small offspring.
   b. few, large offspring.
   c. significant parental care of offspring.
   d. long life expectancy at birth.

4. The middle of the S-shaped growth curve in the logistic growth model
   a. shows that at middle densities, individuals of a population do not affect each other.
   b. is best described by the term rN.
   c. is the point where birth rate starts to decline.
   d. is the period when the population growth rate is the highest.

5. In the logistic growth equation, which term represents the intrinsic rate of increase?
   a. N
   b. r
   c. K
   d. dt

6. The term $(K - N)/K$
   a. is the carrying capacity for a population.
   b. is greatest when K is very large.
   c. is zero when population size equals carrying capacity.
   d. increases in value as N approaches K.

7. The carrying capacity for a population is estimated at 500, the population size is currently 400, and r is 0.1. What is dN/dt?
   a. 0.01   c. 8
   b. 0.8   d. 40

8. In order to maintain the largest sustainable fish harvest, fishing efforts should
   a. take only postreproductive fish.
   b. maintain the population close to its carrying capacity.
   c. reduce the population to a very low number to take advantage of exponential growth.
   d. maintain the population density close to ½ K.

9. Immigration and emigration are likely to play a role in population dynamics in
   a. metapopulations.
   b. exponential growth.
   c. the demographic transition.
   d. small, isolated populations.

10. Chimpanzee females typically first give birth at about 15 years of age. They nearly always have a single offspring, and they do not give birth again for 5–6 years. You would correctly categorize chimpanzees as
    a. K-selected
    b. r-selected.
    c. semelparous.
    d. both **a** and **b**

11. Which of the following factors is *not* a density-dependent factor limiting a population's growth?
    a. increased specialization by a predator
    b. intraspecific competition
    c. a stress syndrome that alters hormone levels
    d. a very early fall frost

12. Consider a population experiencing negative growth; that is, the population size is decreasing. Which of the following statements *must* be true?
    a. birth rate (B) is decreasing
    b. death rate (D) is increasing
    c. the value of r is negative
    d. K is greater than N

13. You have introduced a small number of individuals into a new environment with unlimited resources. What pattern of growth would you expect to see in the early days of your experiment?
    a. K-selection
    b. exponential growth
    c. logistic growth
    d. Type III growth

In questions 14–18, use the following choices to indicate how these factors would be affected by the described changes.

    a. an increase

    b. a decrease

    c. no change

    d. the effect cannot be predicted

14. For a population regulated by density-dependent factors, what change would occur in clutch size or seed crop size with increased population density?

15. In the study of European kestrels described in the textbook (Figure 53.14), what effect did reducing the brood size by transferring chicks have on the survivorship of the parents in the following winter?

16. In a population showing exponential growth, what change would occur in dN/dt with an increase in N?

17. As a population approaches zero population growth, what change would occur in the per capita rate of increase (r)?

18. For a population showing exponential growth, how would r change if the carrying capacity of its environment were increased?

19. The human population is growing at such a fast rate because

    a. the age structure of many countries is highly skewed toward younger ages.

    b. the death rate has greatly decreased since the Industrial Revolution.

    c. fertility rates in many developing countries are above the replacement level of 2.1 children per female.

    d. All of the above are true.

20. The demographic transition is the gradual shift from

    a. a Type III survivorship curve to a Type I curve.

    b. iteroparity to semelparity.

    c. an age structure skewed toward the younger ages to an even age distribution.

    d. high birth rates and high death rates to low birth rates and low death rates.

21. An ecological footprint is an estimate of

    a. the carrying capacity of a nation.

    b. the number of offspring an adult produces and the resulting demand on resources.

    c. the land and water area needed per person to meet the current demand on resources.

    d. how much energy is used to produce food for a vegetarian versus a meat eater.

Chapter 54

Community Ecology

Key Concepts

54.1 Community interactions are classified by whether they help, harm, or have no effect on the species involved

54.2 Diversity and trophic structure characterize biological communities

54.3 Disturbance influences species diversity and composition

54.4 Biogeographic factors affect community diversity

54.5 Pathogens alter community structure locally and globally

Framework

Communities are composed of populations of various species that may interact through competition, predation, herbivory, symbiosis, or facilitation. The structure of a community—its species composition and relative abundance—is determined by these interactions between species. Disturbances keep most communities in a state of nonequilibrium. Species diversity relates to community size and geographic location. Identifying the hosts and vectors of pathogens helps to combat zoonotic diseases.

Chapter Review

A collection of populations of different species living close enough to interact is called a biological **community**.

54.1 Community interactions are classified by whether they help, harm, or have no effect on the species involved

Interspecific interactions occur between the different species living in a community. The positive or negative effect of these interactions on the survival and reproduction of the interacting species can be signified by + or − signs.

Competition **Competition** is a −/− interaction that results when populations of two species use the same limited resource. Such interactions negatively affect both populations.

G. F. Gause's laboratory experiments showed that two species of *Paramecium* that rely on the same limited resource could not coexist in the same community. The **competitive exclusion** principle predicts that the less efficient competitor will be eliminated locally.

An organism's **ecological niche** is described as its role in an ecosystem—its use of biotic and abiotic resources. The competitive exclusion principle holds that two species with identical niches cannot coexist permanently in a community.

Resource partitioning involves slight variations in niche that allow ecologically similar species to coexist. Due to competition, a species' *realized niche* might be smaller than its *fundamental niche*. **Character displacement** of some morphological trait or in resource use enables closely related sympatric (geographically overlapping) species to avoid competition. When these species are allopatric (geographically separate), their differences may be much smaller.

INTERACTIVE QUESTION 54.1

When two spiny mouse species coexist, *Acomys cahirinus* is nocturnal, whereas *A. russatus* is active during the day. When all *A. cahirinus* were removed from a research site, diurnal *A. russatus* mice became nocturnal. How were resources partitioned between these species when they coexisted, and what do these results indicate about the niche of *A. russatus*?

Exploitation Interactions in which one organism benefits while the other is negatively affected (+/−) are described by the term **exploitation**.

Predation involves a predator killing and eating prey. Predator adaptations include acute senses, speed and agility, camouflage coloration, and physical structures such as claws, fangs, teeth, and stingers.

Animals can defend against predation by hiding, fleeing, or forming herds or schools. Mechanical and chemical defenses discourage predation. Some animals can synthesize toxins or passively accumulate them from the food they eat. Bright, conspicuous, **aposematic coloration** in prey species "warns" predators that the prey possesses chemical defenses. Prey may be camouflaged by **cryptic coloration**.

Prey may use mimicry to exploit the warning coloration of other species through **Batesian mimicry** (in which a harmless or palatable species mimics a venomous or unpalatable species) or **Müllerian mimicry** (in which venomous or unpalatable species resemble each other). Predators may use mimicry to "bait" their prey.

INTERACTIVE QUESTION 54.2

Name the type of mimicry described in each of the following descriptions:

a. harmless species resembling a poisonous or distasteful species

b. mutual imitation by two or more unpalatable species

In **herbivory**, an herbivore eats parts of a plant or alga. Most herbivores are small invertebrates such as insects, which may have chemical sensors that recognize nontoxic and nutritious plants. Herbivores may have teeth or digestive systems adapted for processing vegetation.

Plants may defend themselves with mechanical devices, such as thorns, or chemical compounds. Distasteful or toxic chemicals include such well-known compounds as strychnine, nicotine, tannins, and various spices.

In **parasitism**, a **parasite** obtains its nourishment from its **host**. Parasites that live within a host are called **endoparasites**; those that feed on the surface of a host are called **ectoparasites**. Parasitoid insects lay eggs on or in hosts on which their larvae then feed.

Parasites may have complex life cycles with a number of hosts. Parasites can have a substantial effect on the density of their host population.

Positive Interactions   Situations in which at least one species benefits and neither is harmed (+/+ or +/0) are **positive interactions**.

In **mutualism**, interactions between species benefit both participants (+/+). In *obligate mutualism*, at least one species is unable to survive on its own; in *facultative mutualism*, both species can survive alone. Mutualistic interactions may involve the coevolution of related adaptations in both species.

In **commensalism**, only one member appears to benefit from the interaction (+/0). Examples include "hitchhiking" species and species that feed on food incidentally exposed by another species.

INTERACTIVE QUESTION 54.3

Name and give examples of the interspecific interactions symbolized in the table.

	Interspecific Interaction	Examples
− / −	**a.**	
+ / +	**b.**	
+ / 0	**c.**	
+ / −	**d.**	
+ / −	**e.**	
+ / −	**f.**	

54.2 Diversity and trophic structure characterize biological communities

Species Diversity   The **species diversity** of a community is determined both by **species richness**, the number of different species present, and by **relative abundance**, the proportional abundance of the different species.

A widely used quantitative diversity index is **Shannon diversity** (H), which is based on species richness and relative abundance:

$$H = - (p_A \ln p_A + p_B \ln p_B + p_C \ln p_C + \ldots)$$

where p is the proportion of each species (A, B, C, and so on) in the community and ln is the natural log. Thus Shannon diversity (H) for a community is the negative sum of $p \ln p$ for all species present.

INTERACTIVE QUESTION   54.4

Tide pool 1 has three species of sea urchins with the following numbers: A = 8, B = 6, C = 6; tide pool 2 has four species of sea urchins with the following numbers: A = 14, B = 2, C = 2, D = 2.

a. Compute the Shannon diversity index of the sea urchin communities in the two pools.

b. Which pool has the greater species richness? Which community is the more diverse according to the Shannon diversity index?

Estimating the number and relative abundance of species requires various sampling techniques and may be difficult due to the rarity of most species in a community. The diversity and richness of bacterial communities may be determined using molecular tools such as DNA barcodes.

Diversity and Community Stability   Long-term experiments on plant diversity have shown that increased diversity correlates with the increased productivity (producing more **biomass**) and stability of a community. Experiments have indicated that more diverse communities use more of the available resources, making it more difficult for an **invasive species** to become established outside its native range.

Trophic Structure   The feeding relationships in a community determine its **trophic structure.** A **food chain** shows the transfer of food energy from one trophic level to the next: from producers to herbivores (primary consumers) to carnivores (secondary, tertiary, or quaternary consumers). The decomposers in a community feed on all trophic levels.

A **food web** diagrams the complex trophic relationships within a community. The complicated connections of a food web arise because many consumers feed at various trophic levels. Food webs can be simplified by grouping species into functional groups (such as primary consumers) or by isolating partial food webs.

Within a food web, each food chain usually consists of five or fewer links. According to the **energetic hypothesis**, food chains are limited by the inefficiency of energy transfer from one trophic level to the next. Only about 10% of the biomass of one trophic level is converted to the biomass of the next. The increasing

size of animals at successive trophic levels may also limit food chain length.

Species with a Large Impact   A **dominant species** in a community has the greatest abundance or largest biomass and is a major influence on the presence and distribution of other species. A species may become dominant due to its more competitive use of resources or its success at avoiding predation or disease. The removal of a dominant species from a community may adversely affect any species that relied exclusively on that species, but its role may quickly be filled by other species.

A **keystone species** has a large impact on community structure as a result of its ecological role, rather than its abundance. Paine's study of a predatory sea star demonstrated its role in maintaining species richness in an intertidal community by reducing the density of mussels, a highly competitive prey species.

Ecosystem engineers or "foundation species" influence community structure by changing the physical environment.

Bottom-Up and Top-Down Controls   Arrows can be used to indicate the effect an increase in the biomass in one trophic level has on another trophic level. $V \rightarrow H$ indicates that an increase in vegetation (V) increases the number of herbivores (H); $V \leftarrow H$ indicates that an increase in herbivores decreases vegetation biomass. $V \leftrightarrow H$ means that each trophic level is affected by changes in the other. According to the **bottom-up model** of community organization, $N \rightarrow V \rightarrow H \rightarrow P$, an increase in mineral nutrients (N) yields an increase in biomass at each succeeding trophic level: vegetation, herbivores, and predators (P). The **top-down model**, $N \leftarrow V \leftarrow H \leftarrow P$, assumes that predation controls community organization, with a series of +/− effects cascading down the trophic levels. According to this *trophic cascade model*, an increase in predator abundance decreases herbivore abundance, which results in increased vegetation and lowered nutrient levels.

INTERACTIVE QUESTION   54.5

Many freshwater communities appear to be organized along the top-down model. What actions might ecologists take if they wanted to use **biomanipulation** to control excessive algal blooms in a lake with four trophic levels (algae, zooplankton, primary predator fish, and top predator fish)?

54.3 Disturbance influences species diversity and composition

Traditionally, biological communities were viewed as existing in a state of equilibrium maintained by interspecific interactions. The ability of a community to reach and maintain this *climax community* with its relatively constant species composition is known as *stability*. The **nonequilibrium model**, in contrast, emphasizes that communities are constantly changing as a result of **disturbances**. Disturbances such as fire, drought, storms, overgrazing, or human activities may change resource availability and reduce or eliminate some populations.

Characterizing Disturbance   According to the **intermediate disturbance hypothesis**, moderate intensity or frequency of disturbance may result in greater species diversity than either low or high levels of disturbance. Small-scale disturbances may enhance environmental patchiness, helping to maintain species diversity. Human prevention of some natural disturbances, such as small fires, may lead to large-scale disturbances, such as large, destructive fires, to which species may not be adapted.

Ecological Succession   The sequential transition in species composition in a community, usually following a severe disturbance, is known as **ecological succession**. If no soil was originally present, as on a new volcanic island or on the moraine left by a retreating glacier, the process is called **primary succession** and may take hundreds or thousands of years. A series of colonizers usually begins with autotrophic and heterotrophic prokaryotes and protists and then moves through lichens, mosses, grasses, shrubs, and trees until the community reaches its prevalent form of vegetation.

Early colonizers may either *facilitate* the arrival of other species by improving the environment or *inhibit* the establishment of later species. Or species may be independent in their colonization and *tolerate* the conditions created by early species.

Secondary succession occurs when a community is disrupted by fire, logging, or farming, but the soil remains intact. Herbaceous species may colonize first, followed by woody shrubs and eventually trees.

INTERACTIVE QUESTION   54.6

Describe the effects of the alder stage on soil fertility during the succession following the retreat of a glacier.

Human Disturbance   Human activities such as converting land to agriculture, logging, clearing for urban development, and ocean trawling have altered the structure of communities all over the world. Because human disturbances are usually severe, a common result is a reduction in species diversity.

54.4 Biogeographic factors affect community diversity

Latitudinal Gradients   Surveys of plant and animal species have documented much greater numbers of species in tropical habitats than in temperate and polar regions. Tropical communities are older, partly because of their longer growing season and partly because they have not had to "start over" after glaciation, as has been the case several times for many polar and temperate communities.

Solar radiation and precipitation are important climatic explanations for the latitudinal gradient in diversity. **Evapotranspiration**, the amount of water evaporated from soil and transpired by plants, is determined by solar energy, temperature, and water availability. *Potential evapotranspiration* is determined by just solar radiation and temperature. Evapotranspiration rates have been shown to correlate with species richness for both plants and animals.

INTERACTIVE QUESTION   54.7

Why would the fact that tropical communities are "older" than temperate or polar communities contribute to greater species diversity?

Area Effects   A **species-area curve** represents the correlation between the geographic size of a community and the number of species in it. In general, the larger the area, the greater is the diversity of habitats and the greater the species richness. Use of such curves in conservation biology can enable scientists to predict how a loss of habitat may affect diversity.

A mathematical description of the species-area relationship is $S = cA^z$, where S is the number of species, c is a constant, and A is the area of the habitat. The exponent z is the slope of the line through the data points on a log-log plot and indicates how species number should increase as area increases. The value of z is usually between 0.2 and 0.4.

Island Equilibrium Model   Any habitat surrounded by a significantly different habitat is considered an island and allows ecologists to study factors that affect species diversity. R. MacArthur and E. O. Wilson developed an *island equilibrium model*, which states that the size of the island and its closeness to the mainland (or source of dispersing species) are important variables directly correlated with the number of species present. When the rates of immigration and extinction become equal, species diversity reaches an equilibrium, although species composition may continue to change.

INTERACTIVE QUESTION 54.8

a. How do the rates of immigration and extinction change as the number of species on an island increases?

b. In what physical ways would an island with a high number of species probably differ from one with a low number of species?

54.5 Pathogens alter community structure locally and globally

Pathogens are disease-causing microorganisms, viruses, viroids, or prions. They can be particularly virulent when introduced into a new habitat, where host species have not yet evolved resistance or where natural controls are lacking.

Pathogens and Community Structure Pathogens have altered terrestrial ecosystems when dominant tree species are killed and are changing the community structure in coral reef communities. Global travel and trade contribute to the spread of pathogens.

Community Ecology and Zoonotic Diseases Many human diseases are caused by **zoonotic pathogens**, which are disease-causing agents transferred to humans from other animals, often by means of a **vector** such as ticks or mosquitoes. Understanding species interactions and identifying both hosts and vectors help to combat and control the spread of such diseases.

INTERACTIVE QUESTION 54.9

Why are ecologists trapping and testing migrating birds in Alaska?

Word Roots

ecto- = outer (*ectoparasite*: a parasite that feeds on the external surface of a host)

endo- = inner (*endoparasite*: a parasite that lives within a host)

herb- = grass; **-vory** = eating (*herbivory*: an interaction in which an organism eats parts of a plant or alga)

inter- = between (*interspecific competition*: competition for resources between individuals of two or more species when resources are in short supply)

mutu- = reciprocal (*mutualism*: a symbiotic relationship in which both participants benefit)

Structure Your Knowledge

1. Complete the following concept map to organize your understanding of the important factors that structure a community.

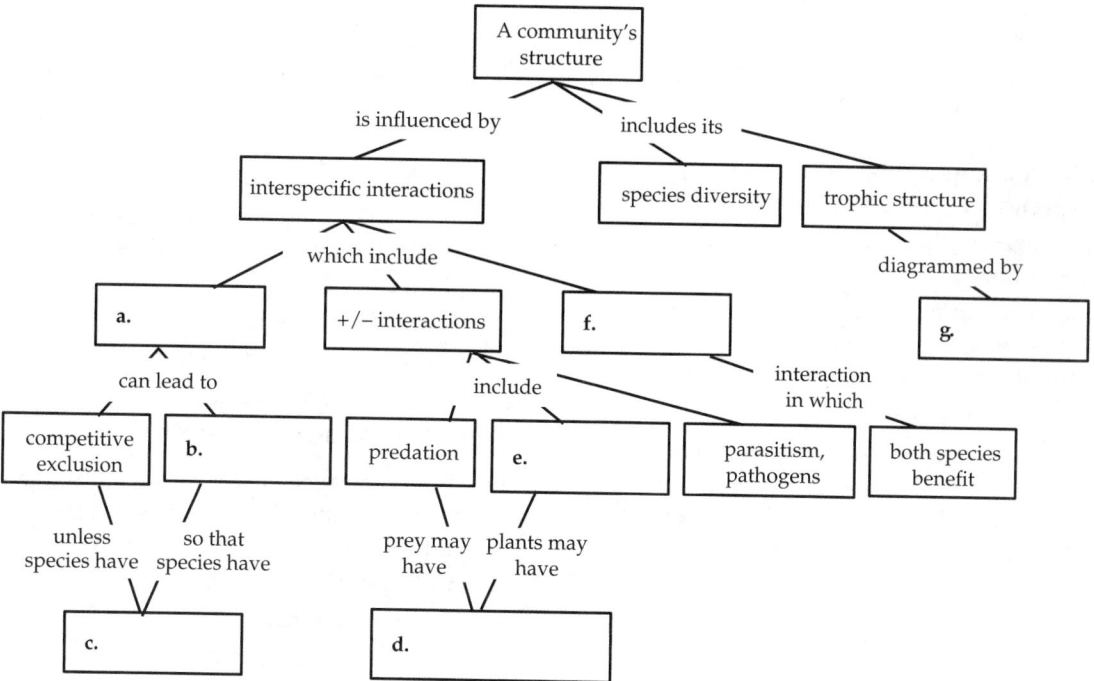

2. Community ecologists develop models or hypotheses to describe community structure and the factors that contribute to such structure. Briefly explain the following models described in this chapter.
   a. Competitive exclusion
   b. Energetic hypothesis
   c. Bottom-up model
   d. Top-down (trophic cascade) model
   e. Nonequilibrium model
   f. Intermediate disturbance hypothesis
   g. Island equilibrium model

Test Your Knowledge

MULTIPLE CHOICE: *Choose the one best answer.*

1. Two allopatric species of Galápagos finches have beaks of similar size. The observation that significant differences in beak size are evident when the two species occur on the same island is an example of
   a. competitive exclusion.
   b. species diversity.
   c. commensalism.
   d. character displacement.

2. The greater blue-ringed octopus (*Hapalochlaena lunulata*) is highly venomous and displays bright blue rings on its skin when threatened. These bright rings are a form of
   a. aposematic coloration.
   b. cryptic coloration.
   c. Batesian mimicry.
   d. Müllerian mimicry

3. Two other species of octopus (*H. maculosa* and *H. fasciata*) are also venomous and also display blue rings when threatened. The presence of a common display in these species is an example of
   a. aposematic coloration.
   b. cryptic coloration.
   c. Batesian mimicry
   d. Müllerian mimicry

4. Two species, A and B, occupy adjoining environmental patches that differ in several abiotic factors. When species A is experimentally removed from a portion of its patch, species B colonizes the vacated area and thrives. When species B is experimentally removed from a portion of its patch, species A does not successfully colonize the area. From these results you might conclude that
   a. both species A and species B are limited to their range by abiotic factors.
   b. species A is limited to its range by competition, and species B is limited by abiotic factors.
   c. both species are limited to their range by competition.
   d. species A is limited to its range by abiotic factors, and species B is limited to its range because it cannot successfully compete with species A.

5. Two shops in the same block are selling coffee and donuts, and business is poor for each party. One shop decides to specialize, selling gourmet coffee with donuts as a minor side line. The other shop sells specialty donuts and pastries and rather ordinary coffee. Both businesses thrive. The changes in the two shops are a model for
   a. competitive exclusion.
   b. resource partitioning.
   c. exploitation.
   d. mutualism.

6. A palatable (good-tasting) prey species may defend against predation by
   a. Müllerian mimicry.
   b. Batesian mimicry.
   c. cryptic coloration.
   d. either b or c.

7. Most humans are host to mites of the genus *Demodex* that inhabit hair follicles. The mites feed on skin cells and oils, but do not cause harm to their host; indeed, most people are unaware of the presence of these mites. The relationship between *Demodex* and humans is an example of
   a. parasitism.   c. commensalism.
   b. mutualism.   d. predation.

8. The species richness of a community refers to the
   a. relative numbers of individuals in each species.
   b. number of different species found in the community.
   c. species diversity of that community.
   d. community's stability or ability to persist through disturbances.

9. You have a friend who is a strict vegan; that is, he eats exclusively plant material, with no animal products whatsoever. In terms of trophic levels, you would correctly describe your friend as a
   a. primary producer.
   b. primary consumer.
   c. secondary consumer.
   d. tertiary consumer.

10. Ecologists survey the tree species in two forest plots. Plot 1 has six different species, and 95% of all trees belong to just one species. Plot 2 has five different species, each of which is represented by approximately 20% of the trees. Compared with plot 1, plot 2 has
    a. higher species richness.
    b. greater species diversity.
    c. lower species richness.
    d. Both b and c are correct.

11. When one species was removed from a tide pool, the pool's species diversity was significantly reduced. The removed species was probably
    a. the dominant species.
    b. a potent parasite.
    c. an ecosystem engineer.
    d. a keystone species.

12. Invasive species often reach a large biomass because
    a. they are better competitors than native species.
    b. they are usually producers and not top predators.
    c. they often lack natural predators or pathogens.
    d. their superior ability to disperse enables them to spread to new niches.

13. Why do most food chains consist of only three to five links?
    a. There are only five trophic levels: producers; primary, secondary, and tertiary consumers; and decomposers.
    b. Most communities are controlled bottom-up by mineral nutrient supply.
    c. According to the energetic hypothesis, the inefficiency of energy transfer from one trophic level to the next limits the number of links.
    d. According to the trophic cascade model, increasing the biomass of top trophic levels causes a decrease in the biomass of lower levels, so that the top levels are limited.

14. According to the top-down (trophic cascade) model of community control, which trophic level would you *decrease* if you wanted to *increase* the vegetation level in a community?
    a. nutrients
    b. vegetation
    c. secondary consumers (carnivores)
    d. tertiary consumers

15. During succession, inhibition by early species
    a. may prevent the achievement of a stable community.
    b. results from the frequent disturbances that often eliminate early colonizers.
    c. may slow down the successful colonization by other species.
    d. may involve changes in nitrogen levels or accelerated accumulation of humus.

16. According to the nonequilibrium model,
    a. chance events such as disturbances play major roles in the structure and composition of communities.
    b. species composition in a community is always in flux as a result of human interventions.
    c. food chains are limited to a few links because long chains are more unstable in the face of environmental disturbances.
    d. the communities with the most diversity have the least stability or resistance to change.

17. Which of the following shows a *mismatch* between organisms and their community roles?
    a. beavers—ecosystem engineers
    b. black rush *Juncus* in a salt marsh—facilitators
    c. sea stars in an intertidal community—keystone predators
    d. alder and *Dryas* (a mat-forming shrub)—inhibitors

18. An island that is small and far from the mainland, in contrast to a large island close to the mainland, would be expected to
    a. have lower species diversity.
    b. be in an earlier successional stage.
    c. have a higher rate of colonization but a higher rate of extinction.
    d. have a lower rate of colonization and a lower rate of extinction.

19. A major explanation for the decline in species richness along an equatorial-polar gradient is the correlation of high levels of solar radiation and precipitation with diversity. Which of the following factors is also thought to contribute to the high species richness of tropical communities?
    a. the inverse relationship between diversity and evapotranspiration
    b. the greater age of these communities (longer growing season and fewer climatic setbacks), providing more time for speciation events
    c. the larger area of the tropics and corresponding richness predicted by the species-area curve
    d. the lack of disturbances in tropical areas

20. Which of the following descriptions best describes a zoonotic pathogen?
    a. a pathogen that affects insects
    b. a pathogen that requires a vector to spread from animal to animal
    c. a disease-causing agent that is transmitted to humans from other animals
    d. a pathogen that is common in zoos due to the unnatural habitat provided for animals

Chapter 55

Ecosystems and Restoration Ecology

Key Concepts

55.1 Physical laws govern energy flow and chemical cycling in ecosystems

55.2 Energy and other limiting factors control primary production in ecosystems

55.3 Energy transfer between trophic levels is typically only 10% efficient

55.4 Biological and geochemical processes cycle nutrients and water in ecosystems

55.5 Restoration ecologists return degraded ecosystems to a more natural state

Framework

Chapter Review

An **ecosystem** consists of all the organisms in a given area and the abiotic factors with which they interact. Most ecosystems are powered by energy from sunlight, which is transformed to chemical energy by autotrophs, passed to heterotrophs in the organic compounds of food, and continually dissipated in the form of heat. Chemical elements are cycled among the abiotic and biotic components of the ecosystem as autotrophs incorporate them into organic compounds and the processes of metabolism and decomposition return them to the soil, air, and water.

55.1 Physical laws govern energy flow and chemical cycling in ecosystems

Conservation of Energy The first law of thermodynamics states that energy can neither be created nor destroyed, only transferred or transformed. Energy flows through most ecosystems from its input as solar

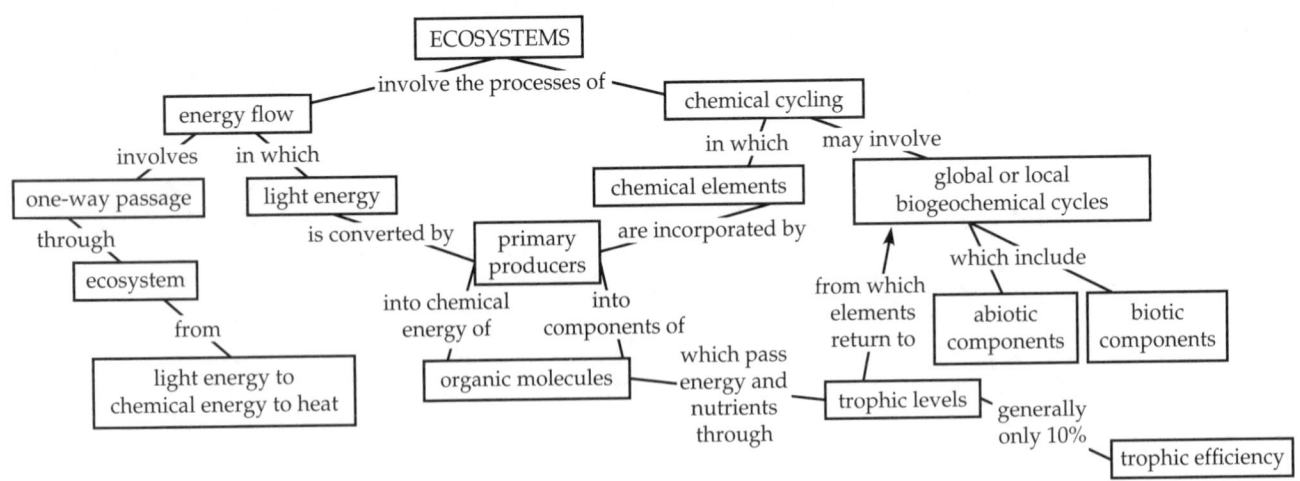

radiation to its conversion into chemical energy to its dissipation as heat. The second law of thermodynamics states that every energy exchange increases entropy. Thus, in each energy conversion, some energy is lost as heat.

Conservation of Mass The **law of conservation of mass** states that matter cannot be created or destroyed. Ecosystem ecologists measure the inputs and outputs of elements as well as chemical cycling within ecosystems.

INTERACTIVE QUESTION 55.1

What may happen if the input of a mineral nutrient into an ecosystem is less than the output of that element from the ecosystem?

Energy, Mass, and Trophic Levels Most **primary producers**, or autotrophs, use light energy to synthesize organic compounds, which they use as fuel and building materials. Heterotrophs depend on autotrophs for their organic compounds. **Primary consumers** are herbivores; **secondary consumers** are carnivores that eat herbivores. **Tertiary consumers** eat other carnivores.

Detritivores, or **decomposers**, consume **detritus**, which includes organic wastes, fallen leaves, and dead organisms. Detritivores often form a link between producers and consumers as they, themselves are eaten. Fungi and prokaryotes are important detritivores in most ecosystems, converting organic materials from all trophic levels to inorganic compounds that can be recycled by primary producers.

INTERACTIVE QUESTION 55.2

Compare the movement of energy and chemical elements in ecosystems.

55.2 **Energy and other limiting factors control primary production in ecosystems**

In most ecosystems, **primary production** is the amount of light energy converted by an ecosystem's photoautotrophs to chemical energy during a defined period of time. When the primary producers are chemoautotrophs, the energy input is chemical.

Ecosystem Energy Budgets Only a small portion of incoming solar radiation strikes photosynthetic organisms, and, of that, only about 1% is converted to chemical energy. Nevertheless, worldwide photosynthetic production is about 150 billion metric tons of organic material per year.

Net primary production (NPP), which represents the chemical energy available to consumers in an ecosystem, is equal to an ecosystem's **gross primary production (GPP)** minus the energy used by autotrophs in their cellular respiration (R_a):

$$NPP = GPP - R_a$$

NPP is usually about half of GPP.

Net primary production can be expressed as energy per unit area per unit time ($J/m^2 \cdot yr$) or as new biomass produced, expressed as dry weight of vegetation ($g/m^2 \cdot yr$). *Standing crop* is the total biomass of photosynthetic organisms in an ecosystem.

Because photosynthesizing vegetation absorbs more wavelengths of visible light, scientists can estimate primary production using data from satellites that compare the wavelengths of visible and near-infrared light reflected from Earth's surface. Primary production and the contribution to Earth's total production vary by ecosystem.

Net ecosystem production (NEP), which reflects *total biomass accumulation* in an ecosystem, deducts the total energy used in respiration by all autotrophs and heterotrophs (R_T) from gross primary production:

$$NEP = GPP - R_T$$

Scientists estimate NEP by measuring the net flux of CO_2 from terrestrial ecosystems or the net flux of CO_2 or O_2 in oceans. Researchers using oxygen sensors have found greater phytoplankton productivity in nutrient-poor regions of the oceans than previously thought.

Primary Production in Aquatic Ecosystems The depth to which light penetrates and the availability of nutrients affect primary production in oceans and lakes. Nitrogen or phosphorus is often the **limiting nutrient** in marine ecosystems. Nutrient enrichment experiments can identify limiting nutrients that may cause algal "blooms." Such blooms may result in oxygen depletion and marine "dead zones." The micronutrient iron limits phytoplankton growth in the Sargasso Sea. Conversely, marine environments in which *upwelling* brings nutrient-rich water to the surface are extremely productive.

In freshwater lakes, nutrient limitation also affects production. **Eutrophication**, the rapid growth of cyanobacteria and algae in response to increased nutrients, has been linked to phosphorus pollution from sewage and fertilizer runoff.

Primary Production in Terrestrial Ecosystems

There are usually positive correlations between primary production and precipitation. Another predictor of primary production—*evapotranspiration*, the amount of water evaporated from a landscape and transpired by plants—reflects both precipitation and temperature.

Nutrients affect primary production, and nitrogen and/or phosphorus are the limiting nutrients in many terrestrial ecosystems. Plant adaptations that increase uptake of nutrients include symbiotic relationships between roots and nitrogen-fixing bacteria, and mycorrhizal associations with fungi that provide phosphorus and other elements to plants. Plant root hairs increase absorptive surface area, and roots release substances that increase the availability of nutrients.

Because primary production is related to temperature and precipitation, it is sensitive to the effects of climate change. Droughts have increased wildfires and insect infestation of forests, resulting in decreased NPP. Climate change may also affect the release of carbon from an ecosystem.

INTERACTIVE QUESTION   55.3

a. List some ecosystems with high primary production.

b. List some ecosystems with low primary production.

c. The oceans have low primary production yet contribute as much global net primary production as terrestrial ecosystems. Explain.

d. Areas of the Antarctic Ocean are often more productive than tropical seas, even though they are colder and receive lower light intensity. Explain.

55.3   Energy transfer between trophic levels is typically only 10% efficient

The production of new biomass from the energy contained in consumers' food during a given time is called **secondary production**.

Production Efficiency   Herbivores consume only a fraction of the plant material produced; they cannot digest all they eat, and much of the energy they do absorb is used for cellular respiration. Only the chemical energy stored as growth or in offspring is available as food to secondary consumers. The proportion of assimilated food energy (not including losses in feces) that is used for net secondary production (growth and reproduction) is a measure of the efficiency of energy transformation: **production efficiency** = (net secondary production × 100%)/ assimilation of primary production. Production efficiencies vary from 1–3% for endothermic birds and mammals, to 10% for fishes, to around 40% for insects and microorganisms.

Trophic Efficiency and Ecological Pyramids   **Trophic efficiency** is the percentage of the production of one trophic level that makes it to the next level; it typically is about 10%. An *energy pyramid* shows this multiplicative loss of energy.

A *biomass pyramid* illustrates the standing crop biomass of organisms at each trophic level. This pyramid usually narrows rapidly from producers to the top trophic level. Some aquatic ecosystems have inverted biomass pyramids in which zooplankton (consumers) outlive and outweigh the highly productive, but heavily consumed, phytoplankton. The production pyramid for such an aquatic ecosystem, however, is normal in shape.

INTERACTIVE QUESTION   55.4

a. Why is production efficiency higher for fishes than for birds and mammals?

b. Assuming a 10% trophic efficiency (transfer of energy to the next trophic level), approximately what proportion of the chemical energy produced by photosynthesis in an ecosystem makes it to a tertiary consumer?

55.4   Biological and geochemical processes cycle nutrients and water in ecosystems

Decomposition and Nutrient Cycling Rates   Temperature and the availability of water influence decomposition rates; thus, nutrient cycling times vary in different ecosystems. Nutrients cycle rapidly in a tropical rain forest; the soil contains only about 10% of the ecosystem's nutrients. Decomposition is slower in temperate forests, and 50% of the ecosystem's organic material may be found in detritus and soil. Decomposition rates are slow where oxygen is limited, such as in peat lands and aquatic sediments.

Biogeochemical Cycles   Chemical elements are passed between abiotic and biotic components of ecosystems through **biogeochemical cycles**. Gaseous forms of carbon, oxygen, sulfur, and nitrogen have global cycles. Heavier elements, such as phosphorus, potassium, and calcium, have a more local cycle, especially in terrestrial ecosystems.

Ecologists study the movement of elements by tracking added radioactive isotopes or by following naturally occurring nonradioactive isotopes through ecosystems.

Exploring Water and Nutrient Cycling (Figure 55.15)
The water cycle involves evaporation from oceans and evapotranspiration from land, condensation of water vapor into clouds, and precipitation. Runoff and groundwater flow return water to the oceans.

In the carbon cycle, producers use CO_2 from the atmosphere in photosynthesis, producing organic compounds that are used by producers and consumers. Organisms release CO_2 in cellular respiration. Reservoirs of carbon include fossil fuels, dissolved carbon compounds in oceans, plant and animal biomass, CO_2 in the atmosphere, and sedimentary rocks such as limestone. Fossil fuel combustion is increasing atmospheric CO_2.

Plants require nitrogen in the form of ammonium (NH_4^+) or nitrate (NO_3^-). Animals obtain nitrogen in organic form from plants or other animals. The major reservoir is the atmosphere, and nitrogen enters ecosystems through *nitrogen fixation* when lightning, volcanic activity, or soil bacteria convert N_2 into compounds that can be assimilated by plants. Some bacteria reduce nitrate and return N_2 to the atmosphere through denitrification. Fertilizers and the planting of legume crops now add more nitrogen to ecosystems than do natural processes. Large quantities of reactive nitrogen gases are released to the atmosphere by human activities.

Weathering of rock adds phosphorus to the soil in the form of PO_4^{3-}, which is absorbed by plants. Organic phosphate is transferred from plants to consumers and returned to the soil through the action of decomposers or in animal excretion. Soil particles bind phosphate, keeping it available locally for recycling. Sedimentary rocks of marine origin are the largest reservoirs.

INTERACTIVE QUESTION   55.5

What is the biological importance of water, carbon, nitrogen, and phosphorus?

Case Study: Nutrient Cycling in the Hubbard Brook Experimental Forest   A team of scientists has looked at nutrient cycling in the Hubbard Brook forest ecosystem since 1963. The mineral budget for each of six valleys was determined by measuring the input of key nutrients in rainfall and their outflow through the creek that drained each watershed. Most minerals were recycled within the forest ecosystem.

The effect of deforestation on nutrient cycling was measured for three years in a valley that was completely logged. Compared with a control area, water runoff from the deforested valley increased 30–40%. Nitrate increased in concentration in the creek 60-fold, removing this critical soil nutrient and contaminating the water.

INTERACTIVE QUESTION   55.6

a. In which natural ecosystem do nutrients cycle the fastest? Why?

b. In which natural ecosystems do nutrients cycle slowly? Why?

c. What is the effect of loss of vegetation on nutrient cycling?

55.5 **Restoration ecologists return degraded ecosystems to a more natural state**

Areas degraded by farming, mining, or environmental pollution are often abandoned. In order to speed the successional processes involved in a community's recovery, restoration ecologists attempt to identify and manipulate the factors that most limit recovery time. In some cases, the physical structure of a site must first be reconstructed.

Bioremediation   **Bioremediation** uses prokaryotes, fungi, or plants to detoxify polluted ecosystems. Some plants are able to extract metals from contaminated soils; harvesting those plants removes the metals from the ecosystem. Some prokaryotes can metabolize dangerous elements to less soluble forms.

Biological Augmentation   **Biological augmentation** uses organisms to add essential substances to degraded ecosystems. Restoration ecologists may also reintroduce animals to restored sites.

Many restoration ecology projects are taking place around the world.

***Ecosystems*: A Review**   The concepts of Chapters 53, 54, and 55 are reviewed in Figure 55.20.

INTERACTIVE QUESTION   55.7

Contrast bioremediation and biological augmentation and give an example of each.

Word Roots

bi- = life; **ge-** = Earth (*biogeochemical cycles:* the various chemical cycles involving both biotic and abiotic components of ecosystems)

detrit- = wear off; **-vore** = one who eats (*detritivore:* a consumer that derives its energy and nutrients from nonliving organic material)

Structure Your Knowledge

1. What is primary production? What factors limit primary production in aquatic ecosystems? In terrestrial ecosystems?
2. What is secondary production? How does it relate to trophic efficiency?
3. What processes mediate the interconversion of the nutrients contained in organic materials and inorganic materials?

Test Your Knowledge

MULTIPLE CHOICE: *Choose the one best answer.*

1. The inefficiency of energy transformations is related to the
   a. first law of thermodynamics.
   b. second law of thermodynamics.
   c. third law of thermodynamics.
   d. law of conservation of mass.

2. Which of the following statements about ecosystems is true?
   a. Energy is recycled through the trophic structure.
   b. Energy is usually converted from sunlight by primary producers, passed to secondary producers in the form of organic compounds, and lost to detritivores in the form of heat.
   c. Chemicals are recycled between the biotic and abiotic sectors, whereas energy makes a one-way trip through the food web and is eventually dissipated as heat.
   d. There is a continuous process by which energy is lost as heat, and chemical elements leave the ecosystem through runoff.

3. Primary production is
   a. equal to the standing crop of an ecosystem.
   b. limited by light, nutrients, and moisture in all ecosystems.
   c. the amount of light energy converted to chemical energy per unit time in an ecosystem.
   d. inverted in some aquatic ecosystems.

4. The open ocean and tropical rain forest are the two largest contributors to Earth's net primary production because
   a. both have high rates of net primary production.
   b. both cover huge surface areas of Earth.
   c. nutrients cycle rapidly in these two ecosystems.
   d. the ocean covers a huge surface area and the tropical rain forest has a high rate of production.

5. Production in terrestrial ecosystems is affected by
   a. temperature and light intensity.
   b. availability of nutrients.
   c. availability of water.
   d. all of the above.

6. Which of the following statements concerning net ecosystem production and net primary production is *false?*
   a. NEP will always be less than NPP because it takes the respiration of heterotrophs and decomposers into account.
   b. If measurements indicate that more CO_2 enters an ecosystem than leaves, NEP would be positive.
   c. If measurements indicate that more O_2 leaves an ecosystem than enters, NEP would be negative.
   d. An increase in the rate of decomposition in an ecosystem would lower NEP, but NPP would probably remain the same or perhaps increase.

7. Secondary production is
   a. measured by the standing crop.
   b. the rate of biomass production in consumers represented by growth and reproduction.
   c. 10% less than primary production.
   d. the gross primary production minus the energy used for respiration.

8. Most food chains have only three to five links. The number of links in a food chain is limited by
   a. trophic efficiency.
   b. production efficiency.
   c. gross primary production.
   d. net ecosystem production.

9. Some authors have suggested that humans worldwide should consume insects rather than herbivores such as cattle or sheep. Insects have a high
   a. gross primary production.
   b. net ecosystem production.
   c. production efficiency.
   d. metabolic rate.

10. Which of the following organisms would you predict to have the highest production efficiency?
    a. lion
    b. caterpillar
    c. snail
    d. deer

11. If a bird eats seeds with an energy content of 400 J, produces feces that contain 200 J, and uses 190 J for respiration, what is this bird's production efficiency?
    a. 0.5%
    b. 5%
    c. 25%
    d. 47.5%

12. Biogeochemical cycles are global for elements
    a. that are found in the atmosphere.
    b. that are found mainly in the soil.
    c. such as carbon, nitrogen, and phosphorus.
    d. Both **a** and **c** are correct.

13. The atmosphere is the main reservoir for
    a. water (as water vapor).
    b. carbon (as carbon dioxide).
    c. nitrogen (as nitrogen gas).
    d. phosphorus (as volatile phosphates).

14. Which of these processes is *incorrectly* paired with its description?
    a. nitrification—deposition of nitrate in the soil by lightning and volcanism
    b. nitrogen fixation—reduction of atmospheric nitrogen to ammonia
    c. denitrification—return of N_2 to air; occurs when denitrifying bacteria metabolize nitrate
    d. ammonification—decomposition of organic compounds to ammonium

15. Clear-cutting tropical forests yields agricultural land with limited productivity because
    a. it is too hot in the tropics for most food crops.
    b. the tropical forest regrows rapidly and chokes out agricultural crops.
    c. few of the ecosystem's nutrients are stored in the soil; most are in the forest trees.
    d. decomposition rates are high but primary production is low in the tropics.

16. Plants, prokaryotes, and fungi are useful in bioremediation because some of them can
    a. destroy toxic elements.
    b. accumulate toxic elements.
    c. convert toxic elements to harmless forms.
    d. encourage native animals to recolonize the area.

17. Which of the following actions is an example of bioaugmentation?
    a. adding fertilizer to degraded soils
    b. restoring the natural flow of river channels
    c. planting nitrogen-fixing lupines in soils disturbed by mining
    d. adding ethanol to stimulate the growth of uranium-reducing bacteria in contaminated groundwater

Conservation Biology and Global Change

Framework

Biodiversity at the genetic, species, and ecosystem level is crucial to human welfare. This chapter explores the threats to biodiversity and several of the approaches to preserving it. Conservation biologists focus on the population size, genetic diversity, and habitat needs of endangered species and on establishing and managing nature reserves that are often in human-dominated landscapes. Human activities are causing global change through nutrient enrichment, the release of toxins, and increasing CO_2 levels in the atmosphere. Ecological research and our biophilia may help to achieve the goal of sustainable development—the long-term perpetuation of human societies *and* the ecosystems that support them.

Chapter Review

Conservation biology integrates all areas of biology in the effort to sustain ecosystem processes and biodiversity.

56.1 Human activities threaten Earth's biodiversity

Three Levels of Biodiversity Current high rates of extinction are many time higher than the typical extinction rates in the fossil record. This high extinction rate is tied to human activities and threatens Earth's biological diversity. At the first level of diversity, loss of the genetic diversity within and between populations lessens a species' adaptive potential.

A second level of biodiversity is species diversity. An **endangered species**, according to the U.S. Endangered Species Act, is one that is "in danger of extinction throughout all or a significant portion of its range." A **threatened species** is defined as one that is likely to become endangered. There are many well-documented examples of recent extinctions and endangered species in most taxonomic groups.

The third level of biodiversity is ecosystem diversity. A loss of a species can negatively affect other species in the ecosystem. Human activities have altered many ecosystems.

Biodiversity and Human Welfare There are both moral and practical reasons for preserving biodiversity. Humans experience *biophilia*, a sense of connection to all forms of life. A loss of biodiversity is a loss of the genetic potential held in the genomes of species, such as genes for traits that could improve crops. Biodiversity is a natural resource that can provide medicines, fibers, and food.

Humans depend on Earth's ecosystems. **Ecosystem services** include such things as purification of air and water, detoxification and decomposition of wastes, nutrient cycling, flood control, pollination of crops, and soil creation and preservation. The monetary value of ecosystem services is huge.

Threats to Biodiversity The greatest threat to biodiversity is habitat loss or fragmentation, caused by agriculture, urban development, forestry, mining, pollution, and global climate change. According to the International Union for Conservation of Nature and Natural Resources (IUCN), habitat destruction is implicated in 73% of species that have been designated as extinct, endangered, vulnerable, or rare. Fragmentation of natural habitats is a common occurrence and almost always leads to species loss. Both terrestrial ecosystems and aquatic habitats have been damaged.

Introduced species, sometimes called exotic species, compete with or prey upon native species. Humans have transplanted thousands of species, intentionally and unintentionally, with huge economic costs in terms of damage and control efforts.

Overharvesting involves harvesting wild plants or animals at rates higher than the populations' abilities to reproduce. Species of large animals with low reproductive rates and species on small islands are particularly vulnerable. Overfishing, particularly using new harvesting techniques, has drastically reduced populations of many commercially important fish species. The tools of molecular genetics enable conservation biologists to determine whether tissues have come from threatened or endangered species.

Global change, the fourth major threat to biodiversity and ranging from regional to global levels, includes alterations in climate, atmospheric chemistry, and ecological systems. The burning of fossil fuels and wood, for example, releases oxides of sulfur and nitrogen, which form sulfuric and nitric acid in the atmosphere. These acids contribute to *acid precipitation*, defined as rain, snow, sleet, or fog with a pH less than 5.2. Drifting emissions from factories create acid precipitation, which has damaged forests and fish populations in lakes across North America and Europe.

56.2 Population conservation focuses on population size, genetic diversity, and critical habitat

Small-Population Approach According to the small-population approach, the inbreeding and genetic drift characteristic of a small population may draw it down an **extinction vortex**, in which the loss of genetic variation leads to smaller and smaller numbers until the population becomes extinct. Low genetic diversity, however, does not always doom a population.

The greater prairie chicken provides an example of an averted extinction vortex. As agriculture fragmented their habitat, the number of prairie chickens in Illinois declined from millions in the 19th century to 50 in 1993. A comparison with DNA from museum specimens indicated decreased genetic variation in the threatened population. After importing birds from larger populations in other states, researchers noted an increase in egg viability, and the Illinois population rebounded.

Computer models that integrate many factors are used to estimate **minimum viable population (MVP)**, the minimum population size necessary to sustain a population.

The **effective population size** (N_e) is based on a population's breeding potential and can be determined by a formula that includes the number of individuals that breed and the sex ratio of the population: $N_e = (4N_f N_m)/(N_f + N_m)$. Alternative formulas take into account life history or genetic factors. Conservation efforts should be based on maintaining the minimum number of *reproductively active* individuals needed to prevent extinction.

INTERACTIVE QUESTION 56.1

Give an example of how each of the following threats to biodiversity has reduced population numbers or caused extinctions.

a. habitat destruction

b. introduced species

c. overharvesting

d. global change

INTERACTIVE QUESTION 56.2

Is the effective population size larger or smaller than the actual number of individuals in the population? Explain.

Population viability analysis uses MVP to predict the long-term viability of a population—the probability of survival over a particular period of time. M. Shaffer performed a population viability analysis as part of a long-term study of grizzly bears in Yellowstone National Park. He estimated that the minimum viable population size for threatened grizzly bear populations was 100 bears. Current estimates of the Yellowstone grizzly population is about

400. The effective population size, however, is only 25% of the total population size, or 100 bears. Genetic analyses indicate that the genetic variability within this population is low.

INTERACTIVE QUESTION 56.3

Explain the basic premise of the small-population approach. What conservation strategy is recommended for preserving small populations?

Declining-Population Approach　The emphasis of the declining-population approach is to identify populations that may be declining, determine the environmental factors that caused the decline, and then recommend corrective measures.

Several factors have driven the red-cockaded woodpecker into decline. Logging and agriculture have fragmented its mature pine forest habitats, and strict fire control has resulted in excessively thick undergrowth around the pine trees. Management strategies now include protection of some longleaf pine forests, controlled fires, and the excavation of breeding cavities in unoccupied habitats to encourage establishment of new breeding groups.

INTERACTIVE QUESTION 56.4

Describe the process that the declining-population approach follows in the conservation of endangered species.

Weighing Conflicting Demands　Preserving habitat for endangered species often conflicts with the economic and recreational desires of humans. Keystone species exert more influence on community structure and ecosystem processes, and prioritizing the species to be saved on the basis of their ecological role may be crucial to the survival of whole communities.

56.3　Landscape and regional conservation help sustain biodiversity

Conservation efforts increasingly are directed at sustaining the biodiversity of whole communities and ecosystems. A goal of landscape ecology is to make biodiversity conservation a part of ecosystem management and land-use planning.

Landscape Structure and Biodiversity　Landscape dynamics are important to conservation efforts because species often use more than one ecosystem or live on borders between ecosystems. The physical features of the landscape constitute its *structure*. Landscapes include ecosystems separated by boundaries or *edges*, which have their own sets of physical conditions. The increase in edge communities due to human fragmentation of habitats may serve to reduce biodiversity as edge-adapted species become predominant. The long-term Biological Dynamics of Forest Fragments Project has shown that species adapted to forest interiors decline the most when habitat fragments are small.

Movement corridors are narrow strips or clumps of habitat that connect isolated patches. Artificial corridors are sometimes constructed when habitat patches have been separated by major human disruptions.

INTERACTIVE QUESTION 56.5

What are some potential benefits of corridors? How might they be harmful?

Establishing Protected Areas　Currently, about 7% of Earth's land area has been set aside as reserves. **Biodiversity hot spots**—small areas with very high concentrations of endemic, threatened, and endangered species—are good choices for nature reserves. Global change will complicate the use of the hot-spot approach to conservation.

Even though nature reserves provide islands of protected habitat, the nonequilibrium model of natural disturbances applies to them. Patch dynamics, edges, and corridor effects must be considered in the design and management of reserves.

New information on the requirements for minimum viable population sizes indicates that most national parks and reserves are much too small—the area needed to sustain a population is usually much larger than the legal boundary set in a reserve.

What factors would favor the creation of larger, extensive reserves? What factors favor smaller, unconnected reserves?

Zoned reserves have protected core areas surrounded by buffer zones in which the human social and economic climate is stable and activities are regulated to promote the long-term viability of the protected zones. Costa Rica has established eight zoned reserves, but deforestation has continued in some buffer zones. The Florida Keys National Marine Sanctuary, a zoned marine reserve established in 1990, has increased marine life and generated revenue from recreational divers.

Urban Ecology   Cities now contain more than half of Earth's people and are predicted to continue to grow. **Urban ecology** looks at species preservation and ecological processes within cities.

56.4 Earth is changing rapidly as a result of human actions

Conservation efforts are complicated by ongoing habitat changes caused by human actions.

Nutrient Enrichment   The harvesting of crops removes nutrients that would otherwise be recycled in the soil. Once the organic and inorganic reserves of soils become depleted, crops require the addition of fertilizers. The addition of nitrogen fertilizers, increased legume cultivation, and burning of fossil fuels has more than doubled Earth's supply of fixed nitrogen.

Nitrogen in excess of the **critical load**, the amount of added nutrient that can be absorbed by plants without damaging the ecosystem, can contaminate groundwater, degrade lakes and rivers, and drain into the ocean. Elevated nitrogen levels create phytoplankton blooms. When these phytoplankton die, their decomposition lowers oxygen levels, creating "dead zones" in coastal waters. Eutrophication of lakes due to nutrient runoff can kill fish and harm other organisms.

Toxins in the Environment   Humans release a large variety of toxic chemicals into the environment. Organisms absorb these toxins from food and water and may retain them within their tissues. In a process known as **biological magnification**, the concentration of such compounds increases in each successive link of the food chain. Chlorinated hydrocarbons (such as DDT) and polychlorinated biphenyls (or PCBs) have been implicated in endocrine system problems in many animal species.

Over-the-counter and prescription drugs and drugs given to farm animals are toxins increasingly found in rivers and lakes. Sex steroids such as estrogens may alter sexual differentiation in fish and disrupt their reproduction.

Many toxic chemicals dumped into ecosystems are nonbiodegradable; others, such as mercury, may become more harmful as they react with other environmental factors.

Greenhouse Gases and Climate Change   The concentration of CO_2 in the atmosphere has been increasing since the Industrial Revolution as a result of the combustion of fossil fuel and deforestation. The addition of CO_2 and other gases can result in long-term directional alteration in global climate known as **climate change**. Through a phenomenon known as the **greenhouse effect**, CO_2, methane, water vapor, and other greenhouse gases in the atmosphere absorb infrared radiation reflected from Earth and re-reflect it back, thus warming Earth.

Scientific models predict a doubling of CO_2 levels and a temperature rise of 3°C by the end of the twenty-first century. In addition, the geographic distribution of precipitation is predicted to change. The correlation between CO_2 levels and temperatures in prehistoric times supports these models. Areas in the far north have already shown warming, and the coverage of Artic sea ice is declining.

To predict the impact of increasing temperatures, ecologists study the effects on vegetation of previous global warming trends. Because many plants disperse slowly, their ranges may not shift rapidly enough to survive global climate change. Some plant and animal species have already shown changes in distribution as a result of increasing temperatures.

Controlling the levels of CO_2 emissions in increasingly industrialized societies is a huge international challenge.

List some of the ways by which we may slow global warming.

Depletion of Atmospheric Ozone   A layer of ozone molecules (O_3) in the stratosphere absorbs damaging ultraviolet radiation. This layer has been gradually thinning since the mid-1970s, largely as a result of the accumulation of breakdown products of chlorofluorocarbons in the atmosphere. The dangers of ozone depletion may include increased incidence of skin cancer and cataracts and unpredictable effects on phytoplankton, crops, and natural ecosystems.

56.5 Sustainable development can improve human lives while conserving biodiversity

Sustainable Development   Economic development that meets the needs of people today without limiting the ability of future generations to meet their needs is known as **sustainable development**. The Ecological Society of America endorses a research agenda, the Sustainable Biosphere Initiative, that encourages studies of global change, biodiversity, and maintenance of the productivity of natural and artificial ecosystems. An important goal is developing the ecological knowledge necessary to make intelligent and responsible decisions concerning Earth's resources.

Partnerships between the government, nongovernment organizations (NGOs), and citizens have contributed to the success of conservation in Costa Rica. Living conditions in the country have improved, as evidenced by a decrease in infant mortality, an increase in life expectancy, and a high literacy rate.

The Future of the Biosphere   E. O. Wilson calls our attraction to Earth's diversity of life and our affinity for natural environments biophilia. By coming to know and understand nature through the study of biology, we may become better able to appreciate and preserve the processes and diversity of the biosphere.

Word Roots

bi- = life (*biodiversity hot spot:* a relatively small area with numerous endemic species and a large number of endangered and threatened species)

Structure Your Knowledge

1. Describe the four major threats to biodiversity.
2. How does the loss of biodiversity threaten human welfare?
3. What do edges and movement corridors have to do with habitat fragmentation?

Test Your Knowledge

MULTIPLE CHOICE: *Choose the one best answer.*

1. According to the Endangered Species Act, a threatened species is a(n)
   a. endemic species that is found nowhere else in the world.
   b. species that is found in disturbed habitats.
   c. species that is in danger of extinction in all or a large part of its range.
   d. species that is likely to become endangered.

2. Ecosystem services include all of the following, *except*
   a. pollination of crops.
   b. production of antibiotics and drugs.
   c. purification of air and water.
   d. decomposition of wastes.

3. The most serious threat to biodiversity is
   a. competition from introduced species.
   b. environmental toxins.
   c. habitat destruction.
   d. overharvesting.

4. Some grassland and conifer forest reserves have effective fire-prevention programs. The most likely result of such programs is
   a. an increase in species diversity because fires are prevented.
   b. a change in community composition because fires are natural disturbances that maintain the community structure.
   c. the preservation of endangered species in the area.
   d. succession to a deciduous forest.

5. According to the small-population approach, the most important remedy for preserving an endangered species is to
   a. establish a large nature reserve around its habitat.
   b. control the populations of its natural predators.
   c. determine the reason for its decline.
   d. encourage dispersal or introduction of new individuals, thus increasing genetic diversity.

6. Which of the following characteristics is typical of biodiversity hot spots?
   a. a large number of endemic species
   b. a high rate of habitat degradation
   c. a large land or aquatic area
   d. large populations of migratory birds

7. The smallest population that can sustain its numbers is
   a. the declining population.
   b. the effective population.
   c. the minimum viable population.
   d. the vortex population.

8. Movement corridors are
   a. strips or clumps of habitat that connect isolated habitats.
   b. the routes taken by migratory animals.
   c. the areas forming the boundary or edge between two ecosystems.
   d. buffer zones for human traffic that promote the long-term viability of protected areas.

9. You are studying a species in which a small number of males control harems of females for reproduction. You estimate that the population consists of 500 individuals, but only 50 males and 200 females reproduce. What is the effective population size?
   a. 500
   b. 250
   c. 160
   d. 150

10. The focus of the declining-population approach to conservation is to
    a. predict a species' minimum viable population size.
    b. transplant members from other populations to increase genetic diversity.
    c. perform a population viability analysis to predict the long-term viability of a population in a particular habitat.
    d. determine the cause of a species' decline and take remedial action.

11. Given their limited resources, conservation biologists working to preserve biodiversity should assign the highest priority to
    a. the northern spotted owl.
    b. declining keystone species in a community.
    c. a commercially important species.
    d. endangered and threatened vertebrate species.

12. The finding of harmful levels of DDT in grebes (fish-eating birds) following years of trying to eliminate bothersome gnat populations in a lakeshore town is an example of
    a. eutrophication.
    b. biological magnification.
    c. nutrient enrichment.
    d. increasing resistance to pesticides.

13. Thinning of the ozone layer is a result of CFCs introduced into the environment. Although global release of CFCs has been dramatically reduced, the ozone layer is recovering very slowly because
    a. chlorine from CFCs persists in the atmosphere, continuing to break down ozone.
    b. CFCs don't break down in the atmosphere.
    c. other compounds with the same effects as CFCs are being released.
    d. greenhouse gas emissions also damage the ozone layer.

14. Sustainable development
    a. uses nature reserves to save endangered species so that the rest of Earth's ecosystems can be used for human development.
    b. is economic development that meets the needs of people today without limiting the ability of people to meet their needs in the future.
    c. is the research agenda of the Sustainable Biosphere Initiative to study global change, biodiversity, and Earth's ecosystems.
    d. sustains the productivity of agricultural ecosystems through the use of chemical and organic fertilizers and irrigation.

15. Which of the following statements reflects a concern about the effects of global climate change on tree species?
    a. The increased ozone levels may damage leaf cells, reducing photosynthetic rates.
    b. Warmer temperatures may speed tree growth, producing trees that are too tall and spindly.
    c. The combination of acid precipitation and worsening droughts may threaten many tree species.
    d. Trees may not be able to disperse fast enough to reach new habitats that meet their climatic requirements.

Answer Section

Chapter 1: Evolution, the Themes of Biology, and Scientific Inquiry

INTERACTIVE QUESTIONS

1.1 **a.** The biosphere includes all life on Earth and the environments inhabited by life—most regions of land and water, the sediments and rocks below Earth's surface, and the atmosphere up to several kilometers.
b. An ecosystem includes all the living organisms in an area and the nonliving parts of the environment with which they interact.
c. A community consists of all the organisms inhabiting a particular ecosystem. Each diverse form of life is called a *species*.
d. A population includes all the individuals of a single species in an area.
e. An organism is an individual living entity.
f. Organs, found in more complex organisms, are body parts that perform a particular function. Organ systems are groups of organs that perform a larger function.
g. Tissues are groups of cells that collectively perform a specialized function. Several tissues make up an organ.
h. Cells are the fundamental units of life.
i. Organelles are the functional components of cells.
j. Molecules are composed of two or more atoms. Molecules are the chemical units that make up living organisms.

1.2 The order of nucleotides in a DNA molecule that makes up a gene "spells" the instructions, which are transcribed into RNA and then translated into a protein with a unique shape and specific function.

1.3 Both energy and chemical nutrients are transferred within an ecosystem from producers to consumers to decomposers. Energy, however, flows through ecosystems, entering as solar energy, being transformed to chemical energy, and exiting as heat. Chemical nutrients are recycled as photosynthesizing plants absorb water, minerals, CO_2, and sunlight, and produce sugar and release O_2; consumers eat plants and other organisms; and decomposers return nutrients to the ecosystem.

1.4 These organisms are characterized to a large extent by their mode of nutrition. Plants are photosynthetic, fungi absorb their nutrients from decomposing organic material, and animals ingest other organisms.

1.5 Most species tend to produce more offspring than can survive. Organisms with heritable traits best suited to the environment will tend to survive and leave more offspring. Over time, favorable adaptations will accumulate in a population. New species may arise as small populations are exposed to different environments and natural selection favors different traits.

1.6 **a.** The control groups were the camouflaged mice models placed in their native habitat. In each habitat, the experimental groups were non-camouflaged mice models.
b. The color of the models is the independent variable and the rate of predation is the dependent variable.
c. This experiment could not control for the number of predators in each area and thus for the total number of attacks on non-camouflaged mice. By presenting data as the predation rate or proportion of total attacks, the effect of this variable was eliminated.

1.7 **a.** A hypothesis is less broad in scope than a theory; it is a tentative explanation for a smaller set of observations. A theory can generate many testable hypotheses and is supported by a large body of evidence. Both hypotheses and theories are revised when new data fail to support their predictions.
b. Science seeks to understand natural phenomena; technology is the practical use of scientific knowledge.

SUGGESTED ANSWERS TO STRUCTURE YOUR KNOWLEDGE

1. **a.** With every increase in the biological hierarchy, from molecules to ecosystems, the organization and interactions of component parts lead to new, emergent properties of the dynamic system. The structure and function of biological components are correlated. Cells are the basic units of life. They come in two distinct forms: prokaryotic and eukaryotic.
    b. DNA is the molecule of inheritance; it codes information for proteins, which direct the functions of a cell, and is passed on from one generation to the next.
    c. All organisms require energy. Energy flows through ecosystems from sunlight to chemical energy in producers and consumers and then escapes as heat. Chemical nutrients are recycled in an ecosystem.
    d. An organism interacts with other organisms, and these interactions may be mutually beneficial, or one or both species may be harmed. Organisms also interact with the physical environment, and the environment is affected by the organisms that live there. Organisms regulate their internal environment, usually through negative feedback mechanisms that match supply to demand.
    e. Evolution explains the unity and diversity of life. All organisms are modified descendants of common ancestors. Darwin's theory of natural selection leading to unequal reproductive success accounts for the adaptation of populations to Earth's varying environments.

ANSWERS TO TEST YOUR KNOWLEDGE

Multiple Choice:

1. c	4. a	7. d	10. d
2. c	5. b	8. a	11. a
3. b	6. c	9. c	12. b

Chapter 2: The Chemical Context of Life

INTERACTIVE QUESTIONS

2.1 calcium, phosphorus, potassium, sulfur, sodium, chlorine, magnesium

2.2 neutrons; 15, 15, 16, 31 daltons

2.3 absorb; loses

2.4 **a.** Hydrogen, $_1$H   **c.** Oxygen, $_8$O

b. Carbon, $_6$C   **d.** Sodium, $_{11}$Na

   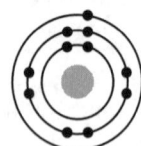

2.5 **a.** protons
   b. atomic number
   c. element
   d. neutrons
   e. mass number or atomic mass
   f. isotopes
   g. electrons
   h. electron shells
   i. valence shell

2.6 **a.** H = 1, O = 2, N = 3, C = 4

b. H:H   Ö::Ö   :N:::N: H:C:H (with H above and below)

2.7 **a.** Nonpolar; even though nitrogen has a high electronegativity, the three pairs of electrons are shared equally between the two N atoms because each atom has an equally strong attraction for the electrons.
   b. Polar; nitrogen is more electronegative than hydrogen and pulls the shared electrons in each covalent bond closer to itself.
   c. Nonpolar; carbon and hydrogen have similar electronegativities and share electrons fairly equally between them.
   d. The bond is polar because oxygen is more electronegative than carbon; the C—H bonds are relatively nonpolar.

2.8 **a.** $CaCl_2$
   b. Ca^{2+} is the cation.

2.9

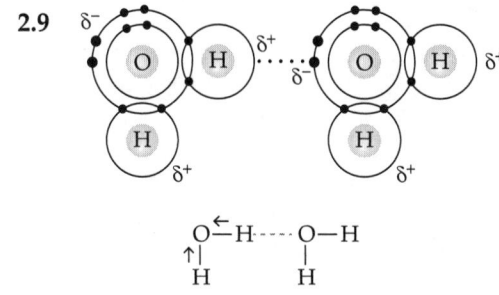

$$O \xleftarrow{} H \cdots O - H$$

2.10 Water's two covalent bonds with H are spread apart at a 104.5° angle due to the hybridization of the *s* and 3 *p* orbitals.

2.11 6, 6, 6

SUGGESTED ANSWERS TO STRUCTURE YOUR KNOWLEDGE

1.

Particle	Charge	Mass	Location
Proton	+1	1 dalton	nucleus
Neutron	0	1 dalton	nucleus
Electron	−1	negligible	orbitals in electron shells

2. **a.** The atoms of each element have a characteristic number of protons in their nuclei, referred to as the *atomic number*. In a neutral atom, the atomic number also indicates the number of electrons. The *mass number* is equal to the number of protons and neutrons in the nucleus.

b. The *valence*, an indication of bonding capacity, usually equals the number of unpaired electrons that an atom has in its valence shell.

c. The valence of an atom is most related to the chemical behavior of an atom because it is an indication of the number of bonds the atom will make to reach a filled valence shell.

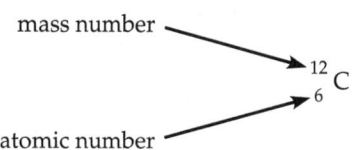

3. In nonpolar covalent bonds, electrons are equally shared between two atoms. Polar covalent bonds form when a more electronegative atom pulls the shared electrons closer to it, producing a partial negative charge associated with that portion of the molecule and a partial positive charge associated with the atom from which the electrons are pulled. In the formation of ions, an electron is completely pulled away from one atom and transferred to another, creating negatively and positively charged ions (anions and cations). These oppositely charged ions may be attracted to each other in an ionic bond.

ANSWERS TO TEST YOUR KNOWLEDGE

Multiple Choice:

1. b	**7.** b	**13.** d	**19.** b	**25.** c
2. a	**8.** b	**14.** a	**20.** d	**26.** c
3. d	**9.** c	**15.** c	**21.** b	**27.** a
4. b	**10.** c	**16.** d	**22.** d	**28.** d
5. d	**11.** a	**17.** d	**23.** b	
6. d	**12.** c	**18.** c	**24.** d	

Chapter 3: Water and Life

INTERACTIVE QUESTIONS

3.1

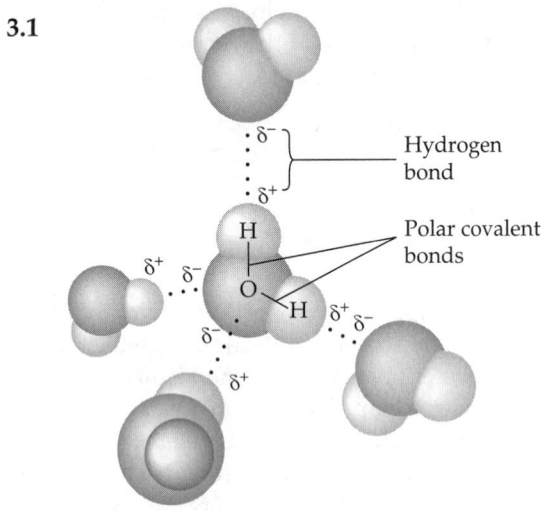

Hydrogen bond

Polar covalent bonds

3.2 **a.** polar water molecules
b. absorbed
c. released
d. specific heat
e. heat of vaporization
f. evaporative cooling
g. solar heat
h. rain

3.3 **a.** The molecular mass of $C_3H_6O_3$ is 90 daltons, the combined atomic masses of its atoms. A mole of lactic acid = 90 g. A 0.5 M solution would require ½ mol or 45 g.
b. ½ mol is ½ Avogadro's number of molecules: 3.01×10^{23}.

3.4

[H^+]	[OH^-]	pH	Acidic, Basic, or Neutral?
10^{-8}	10^{-6}	8	basic
10^{-7}	10^{-7}	7	neutral
10^{-1}	10^{-13}	1	acidic

3.5 carbonic acid   bicarbonate   hydrogen ion
$$H_2CO_3 \;\rightleftharpoons\; HCO_3^- \;+\; H^+$$
H^+ donor H^+ acceptor
(acid)   (base)

a. Bicarbonate acts as a base to accept excess H^+ ions when the pH starts to fall; the reaction moves to the left.
b. When the pH rises, H^+ ions are donated by carbonic acid, and the reaction shifts to the right.

3.6 **a.** $CO_2 + H_2O \rightleftharpoons H_2CO_3 \rightleftharpoons HCO_3^- + H^+$
Increasing [CO_2] will drive these reactions to the right, increasing [H^+].
b. A lower pH means an increasing [H^+], which will drive this reaction ($HCO_3^- \rightleftharpoons CO_3^{2-} + H^+$) to the left, thus decreasing [CO_3^{2-}].
c. $CO_3^{2-} + Ca^{2+} \rightleftharpoons CaCO_3$. With less CO_3^{2-} available to react with Ca^{2+}, calcification rates would be expected to decrease.

SUGGESTED ANSWERS TO STRUCTURE YOUR KNOWLEDGE

1. **a.** Cohesion, adhesion
   b. A water column is pulled up through plant vessels.
   c. Heat is absorbed or released when hydrogen bonds break or form. Water absorbs or releases a large quantity of heat for each degree of temperature change.
   d. High heat of vaporization
   e. Solar heat is dissipated from tropical seas.
   f. Hydrogen bonds in ice keep water molecules apart, making ice less dense.
   g. Floating ice insulates bodies of water so they don't freeze solid.
   h. Versatile solvent
   i. Polar water molecules surround and dissolve ionic and polar solutes.

2.

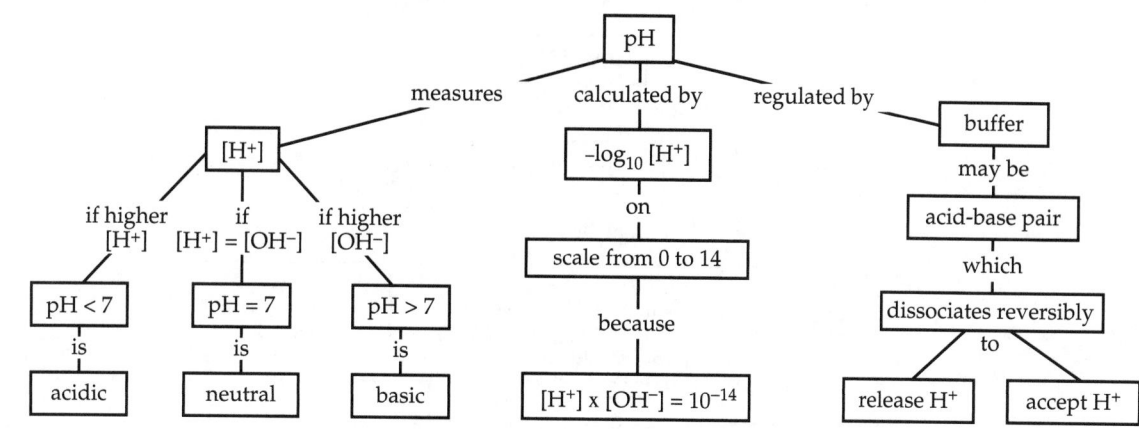

ANSWERS TO TEST YOUR KNOWLEDGE

Multiple Choice:

1. b	**4.** c	**7.** c	**10.** d	**13.** c	**16.** a
2. a	**5.** a	**8.** d	**11.** b	**14.** c	**17.** d
3. d	**6.** b	**9.** a	**12.** b	**15.** a	**18.** c

Chapter 4: Carbon and the Molecular Diversity of Life

INTERACTIVE QUESTIONS

4.1 The abiotic synthesis of organic compounds from inorganic molecules would be a first step in the origin of life. A variety of organic compounds, including amino acids, were produced in Miller's apparatus, which attempted to simulate the physical and chemical conditions on early Earth.

4.2 Ethanol and dimethyl ether are *structural isomers*, which have the same number and kinds of atoms but a different bonding sequence and very different properties. Maleic acid and fumaric acid are *cis-trans isomers* whose C=C bonds fix the spatial arrangement of the molecule. Maleic acid is the *cis* isomer; the carboxyl groups are on the same side of the double bond. Although the *enantiomers* L- and D-lactic acid look similar in a flat representation of their structures, they are not superimposable due to the arrangement of four different atoms or groups around an asymmetric carbon.

4.3

SUGGESTED ANSWERS TO STRUCTURE YOUR KNOWLEDGE

1.

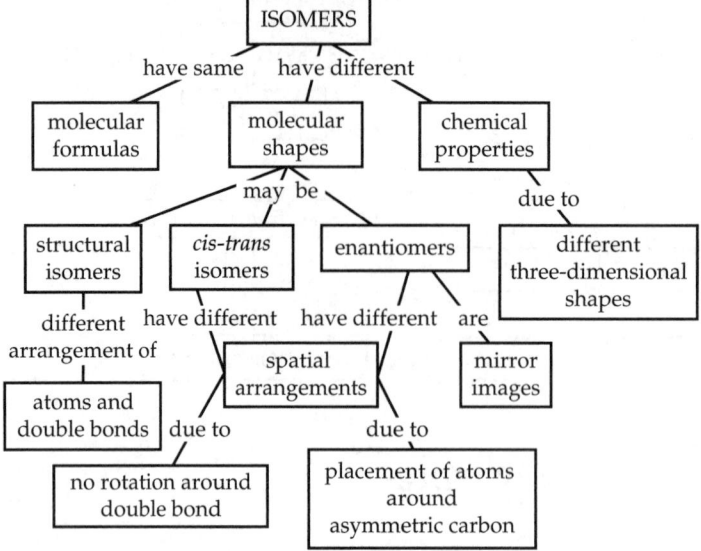

2.

Chemical Group	Molecular Formula	Name and Characteristics of Compounds Containing Group
Hydroxyl	—OH	Alcohols; polar group
Carbonyl	$>C=O$	Aldehyde or ketone; polar group
Carboxyl	—COOH	Carboxylic acid; donate H^+
Amino	—NH$_2$	Amines; basic, accept H^+
Sulfhydryl	—SH	Thiols; cross-links stabilize protein structure
Phosphate	—OPO$_3{}^{2-}$	Organic phosphates; involved in energy transfers, adds negative charge
Methyl	—CH$_3$	Methylated compounds; addition may alter expression of genes

ANSWERS TO TEST YOUR KNOWLEDGE

Multiple Choice:

1. b	**5.** d	**8.** d	**11.** b
2. c	**6.** a	**9.** b	**12.** d
3. b	**7.** c	**10.** c	**13.** a
4. a			

Matching:

1. a, c	**4.** e	**7.** b, f	**10.** a, d
2. b, f	**5.** e	**8.** e	**11.** e
3. a, d, e	**6.** a, c, d, e	**9.** a, c	**12.** c

Chapter 5: The Structure and Function of Large Biological Molecules

INTERACTIVE QUESTIONS

5.1 dehydration reactions/removal; hydrolysis/ addition

5.2 **a.** hydroxyl
       b. carbonyl

c. aldose or aldehyde sugar (carbonyl at end of carbon skeleton)
d. ketose or ketone sugar (carbonyl within carbon skeleton)
e. rings
f. -ose

5.3

Glucose   Fructose

Sucrose

5.4
a. monosaccharides
b. (CH_2O)
c. energy compounds
d. carbon skeletons, monomers
e. disaccharides
f. glycosidic linkages
g. polysaccharides
h. glycogen
i. animals
j. starch
k. cellulose
l. chitin

5.5
a. fats, triacylglycerides
b. phospholipids
c. glycerol
d. fatty acids
e. unsaturated (has some $C{=}C$ bonds)

f. saturated (no $C{=}C$ bonds)
g. phosphate group
h. cell membranes
i. steroids
j. animal cell membrane component (cholesterol), hormones

5.6 **a.**

alanine   serine

Peptide bond

dipeptide

b. Serine's R group is polar; alanine's R group is nonpolar.
c. a *polypeptide backbone*; an R group

5.7
a. hydrogen bond
b. hydrophobic interactions and van der Waals interactions
c. disulfide bridge
d. ionic bond

These interactions between R groups produce tertiary structure.

5.8

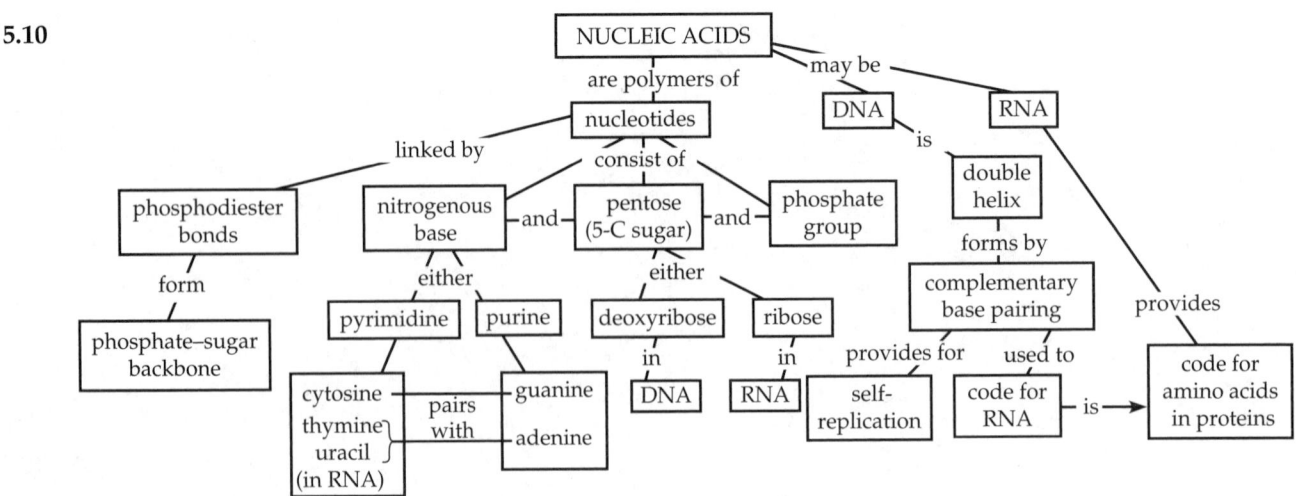

5.9 a.

b. pyrimidine; base has single ring
c. DNA; 2' carbon in sugar is lacking O

5.10

SUGGESTED ANSWERS TO STRUCTURE YOUR KNOWLEDGE

1. The primary structure of a protein is the specific, genetically coded sequence of amino acids in a polypeptide chain. The secondary structure involves the coiling (α helix) or folding (β pleated sheet) of the protein, stabilized by hydrogen bonds along the polypeptide backbone. The tertiary structure involves interactions between the side chains (R groups) of amino acids and produces a characteristic three-dimensional shape for a protein. Quaternary structure occurs in proteins composed of more than one polypeptide.

2. a. amino acid (glycine)
   b. fatty acid
   c. nitrogenous base, purine (adenine)
   d. glycerol
   e. phosphate group
   f. sugar (pentose, ribose)
   g. sugar (triose)

1. b, d	3. c, e, f	5. c	7. e, f
2. a	4. f, g	6. a	8. b

ANSWERS TO TEST YOUR KNOWLEDGE

Matching:

1. A	4. C	7. B	9. A
2. B	5. C	8. C	10. A
3. D	6. D		

Multiple Choice:

1. b	7. a	13. b	19. a
2. d	8. d	14. c	20. c
3. a	9. a	15. c	21. b
4. b	10. a	16. b	22. c
5. c	11. c	17. b	
6. d	12. d	18. a	

Chapter 6: A Tour of the Cell

INTERACTIVE QUESTIONS

6.1 a. the study of cell structure
   b. the internal fine structure of cells
   c. the three-dimensional surface topography of a specimen
   d. Light microscopy enables the study of living cells and may introduce fewer artifacts than do TEM and SEM.

6.2 a. a phospholipid bilayer with the hydrophobic tails clustered in the interior and the phosphate heads facing the hydrophilic outside and inside of the cell; proteins embedded in and attached to the membrane.
   b. area is proportional to a linear dimension squared; 10^2, or 100 times the surface area; volume is proportional to a linear dimension cubed; 10^3, or 1,000 times the volume.
   c. Smaller cells have a greater ratio of surface area to volume. Narrow, elongated shapes and *microvilli,* which are long, thin projections of the plasma membrane, also increase the surface area relative to the volume of a cell.

6.3 The genetic instructions for specific proteins are transcribed from DNA into messenger RNA (mRNA), which then passes into the cytoplasm to complex with ribosomes, where it is translated into the primary structure of proteins. Ribosomal subunits are synthesized in nucleoli from rRNA (transcribed from DNA) and proteins imported from the cytoplasm.

6.4 a. smooth ER—in different cells may house enzymes that synthesize lipids (including steroids), metabolize carbohydrates, detoxify drugs and alcohol, or store and release calcium ions in muscle cells
   b. nuclear envelope—double membrane that encloses nucleus; pores regulate passage of materials
   c. rough ER—attached ribosomes produce proteins that enter cisternae; produces secretory proteins and membranes
   d. transport vesicle—carries products of ER and Golgi apparatus to various locations
   e. Golgi apparatus—processes products of ER; makes polysaccharides; packages products in vesicles targeted to specific locations
   f. plasma membrane—selective barrier that regulates passage of materials into and out of the cell
   g. lysosome—houses hydrolytic enzymes that digest macromolecules

6.5

Mitochondrion

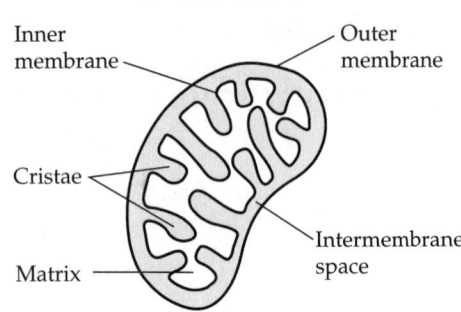

Inner membrane
Outer membrane
Cristae
Intermembrane space
Matrix

Chloroplast

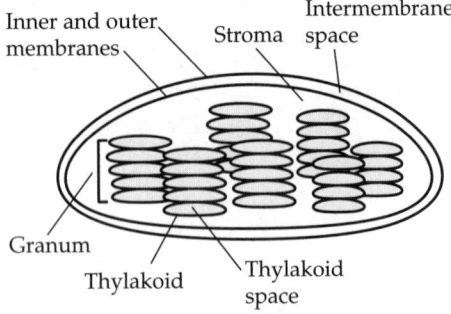

Inner and outer membranes
Stroma
Intermembrane space
Granum
Thylakoid
Thylakoid space

6.6   Peroxisomes do not bud from the endomembrane system, but grow by incorporating proteins and lipids made in the cytosol, the ER, or the peroxisome itself; they increase in number by dividing.

6.7   **a.**   hollow tube, formed from columns of tubulin dimers; 25-nm diameter
b.   cell shape and support (compression resistant), tracks for moving organelles, chromosome movement in cell division, beating of cilia and flagella
c.   two twisted chains of actin subunits; 7-nm diameter
d.   muscle contraction, maintain and change cell shape, amoeboid movement, cytoplasmic streaming
e.   supercoiled fibrous proteins of keratin family; 8–12 nm diameter
f.   reinforce cell shape (tension bearing); anchor nucleus; form nuclear lamina

6.8

Two adjacent plant cells

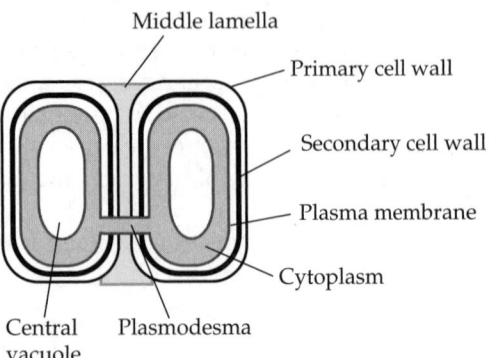

Middle lamella
Primary cell wall
Secondary cell wall
Plasma membrane
Cytoplasm
Central vacuole
Plasmodesma

6.9   **a.**   proteoglycan complex
b.   collagen
c.   fibronectin
d.   integrin
e.   microfilament

6.10   **a.**   See answer for 6.8.
b.   structural support; signaling between environment and cell via fibronectins, integrins, and the cytoskeleton (may affect gene expression), orientation of cell movements

SUGGESTED ANSWERS TO STRUCTURE YOUR KNOWLEDGE

1.   **a.**   nucleus, chromosomes, centrioles, microtubules (move chromosomes), microfilaments (pinch apart cells)
b.   nucleus, chromosomes, DNA → mRNA (to ribosomes) → enzymes and other proteins
c.   mitochondria
d.   ribosomes, rough and smooth ER, Golgi apparatus, transport vesicles
e.   smooth ER (peroxisomes also detoxify substances)
f.   lysosomes, food vacuoles, vesicles enclosing damaged organelles
g.   peroxisomes
h.   cytoskeleton: microtubules, microfilaments, intermediate filaments; extracellular matrix
i.   cilia and flagella (microtubules), microfilaments (actin) in muscles and amoeboid movement
j.   plasma membrane, transport vesicles
k.   desmosomes, tight and gap junctions, ECM

2.   **a.**   structural support, middle lamella glues cells together
b.   storage of inorganic ions, growth via cell expansion
c.   photosynthesis, production of sugars
d.   starch storage
e.   cell-cell communication

3.   **a.**   rough endoplasmic reticulum
b.   smooth endoplasmic reticulum
c.   chromatin
d.   nucleolus
e.   nuclear envelope
f.   nucleus
g.   ribosomes
h.   Golgi apparatus
i.   plasma membrane
j.   mitochondrion
k.   lysosome
l.   cytoskeleton
m.   intermediate filament

n. microtubule

o. microfilaments

p. microvilli

q. peroxisome

r. centrosome (contains a pair of centrioles)

s. flagellum

4.

ANSWERS TO TEST YOUR KNOWLEDGE

Multiple Choice:

1. b	**4.** b	**7.** d	**10.** c	**13.** b	**16.** c	**19.** d	**22.** d
2. a	**5.** d	**8.** b	**11.** d	**14.** a	**17.** a	**20.** c	**23.** a
3. a	**6.** b	**9.** d	**12.** c	**15.** b	**18.** b	**21.** b	

Chapter 7: Membrane Structure and Function

INTERACTIVE QUESTIONS

7.1 **a.** hydrophilic phosphate head
b. hydrocarbon tail (hydrophobic)
c. phospholipid bilayer (hydrophilic heads and hydrophobic tails)
d. hydrophobic region of integral protein
e. hydrophilic region of integral protein
f. peripheral protein (hydrophilic)

7.2 **a.** In hybrid human/mouse cells, membrane proteins rapidly intermix (Frye and Edidin, 1970).
b. The cell membranes of the cold lake population would have a higher proportion of unsaturated fatty acids in their phospholipids than that of the population living in warm temperatures.

7.3 transport, enzymatic activity, signal transduction, cell-cell recognition, intercellular joining, attachment to cytoskeleton and ECM (providing support and communication)

7.4 Ions and polar molecules such as glucose are impeded by the hydrophobic center of the plasma membrane's lipid bilayer.

7.5 Side A initially has fewer free water molecules; side B initially has more. More water molecules are clustered around the glucose in the 1 *M* solution than around the fructose and sucrose, whose combined concentration is 0.9 *M*. Water will move by osmosis from side B to side A.

7.6 **a.** The unicellular eukaryote will gain water from its hypotonic environment. It has a plasma membrane that is less permeable to water and contractile vacuoles that expel excess water.
b. Animal cells fare best in isotonic environments, in which they neither gain nor lose water. Plant cells are healthiest in hypotonic environments, in which the inward movement of water creates turgor pressure, providing mechanical support to the cell.

7.7 Although it speeds diffusion and requires transport proteins, facilitated diffusion is still passive transport because the solute is moving down its concentration gradient; the process is driven by the concentration gradient and not by energy expended by the cell.

7.8 Three Na^+ are pumped out of the cell for every two K^+ pumped in, resulting in a net movement of positive charge from the cytoplasm to the extracellular fluid.

7.9   **a.** Human cells use receptor-mediated endocytosis to take in cholesterol, which is used for the synthesis of other steroids and membranes.

b. In people with this disease, LDL receptor proteins in the plasma membrane are defective, and low-density lipoprotein particles cannot bind and be transported from the blood into the cell.

SUGGESTED ANSWERS TO STRUCTURE YOUR KNOWLEDGE

1.

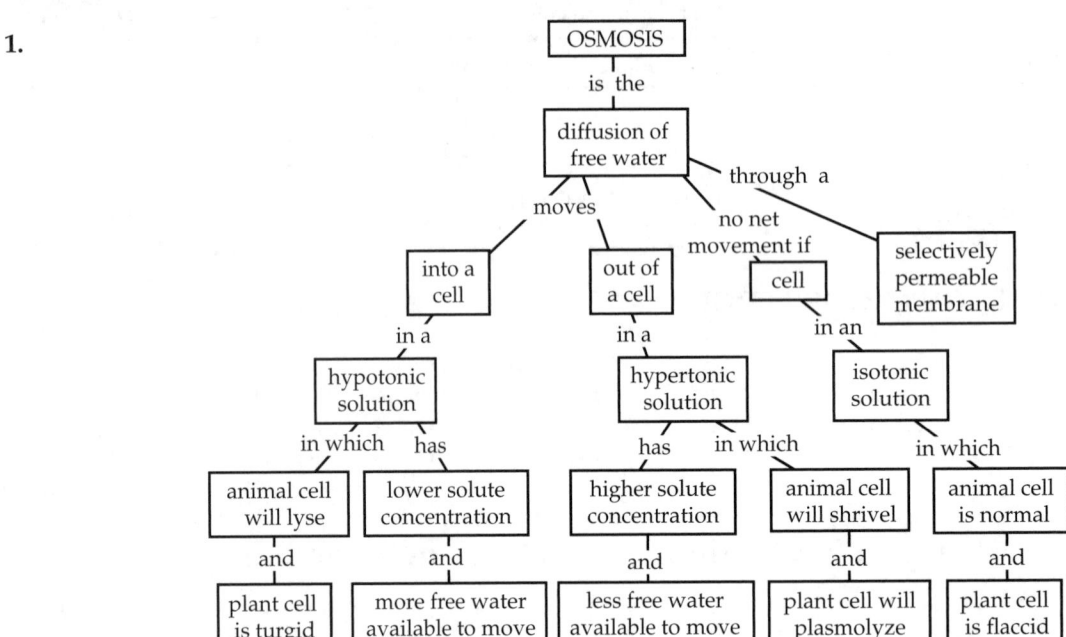

2.   **a.** II represents facilitated diffusion. The solute is moving through a transport protein and down its concentration gradient. The cell does not expend energy in this transport because it is driven by the concentration gradient. Polar molecules and ions may move by facilitated diffusion.

b. III represents active transport because the solute is clearly moving against its concentration gradient and the cell is expending energy derived from ATP hydrolysis to drive this transport against the gradient.

c. In order to diffuse through the lipid bilayer, solute molecules must be hydrophobic (nonpolar) or very small polar molecules.
d. I and II. Both diffusion through the lipid bilayer and facilitated diffusion are considered passive transport because the solute moves down its concentration gradient and the cell does not expend energy in the process.

ANSWERS TO TEST YOUR KNOWLEDGE

Multiple Choice:

1. b	**3.** d	**5.** d	**7.** d	**9.** d	**11.** d**	**13.** d	**15.** d	**17.** a	**19.** a
2. c	**4.** b	**6.** c	**8.** d*	**10.** b	**12.** c	**14.** a	**16.** b	**18.** c	**20.** b

*Explanation for answer to question 8: This problem involves both osmosis and diffusion. Although the solutions are initially equal in molarity, glucose will diffuse down its concentration gradient until it reaches dynamic equilibrium with a 1.5 *M* concentration on both sides. The increasing solute concentration on side A will cause water to move into this side, and the water level will rise.

**Explanation for answer to question 11: As the solute in the solution crosses the cell membrane, it increases the concentration of solutes within the cell, reducing the hypertonicity of the solution. As the solute reaches an equal concentration inside and outside the cell, it no longer causes osmotic changes in the cell.

Chapter 8: An Introduction to Metabolism

INTERACTIVE QUESTIONS

8.1
  a. capacity to cause change
  b. kinetic
  c. motion
  d. potential
  e. position
  f. conserved or constant
  g. created nor destroyed
  h. first
  i. transferred or transformed
  j. entropy
  k. second

8.2

	Stability	Work Capacity	Spontaneous Change?	Equilibrium
System with high free energy	low	high	yes	moves toward
System with low free energy	high	low	no	is at or near

8.3

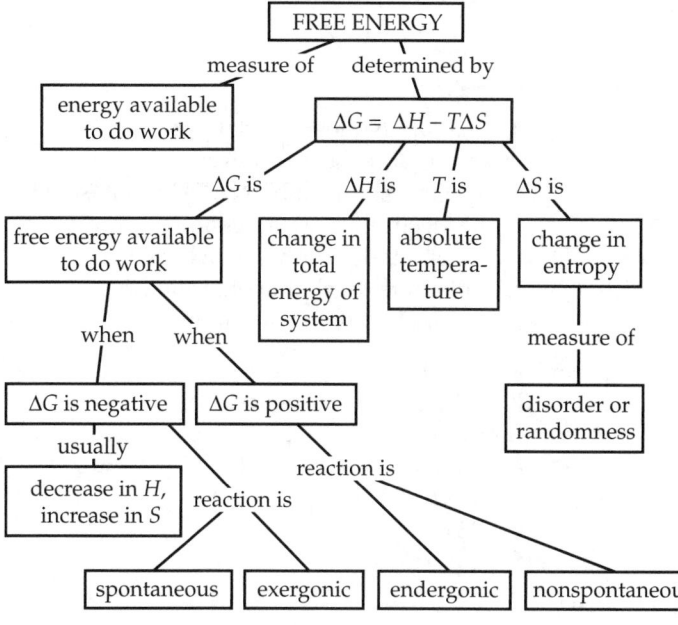

8.4
  a. adenine
  b. ribose
  c. triphosphate group (three phosphate groups)

d. A hydrolysis reaction breaks the terminal phosphate bond and releases a molecule of inorganic phosphate: ATP + H_2O → ADP + $\textcircled{P}_i$ + energy.

e. The negatively charged phosphate groups are crowded together, and their mutual repulsion makes this area unstable. The chemical change to a more stable state of lower free energy accounts for the relatively high release of energy.

8.5 Energy coupling is the use of an exergonic process to drive an endergonic one. The ATP cycle itself illustrates this coupling in that energy-releasing processes of catabolism are coupled with energy-consuming process of cellular work. But each side is also an example: The endergonic synthesis of ATP from ADP and $\textcircled{P}_i$ is coupled to the exergonic reactions of catabolism (left side), and the exergonic hydrolysis of ATP to ADP and $\textcircled{P}_i$ provides the immediate energy source for cellular work (right side).

8.6
  a. free energy
  b. transition state
  c. E_A (free energy of activation) without enzyme
  d. E_A with enzyme
  e. ΔG of reaction

8.7

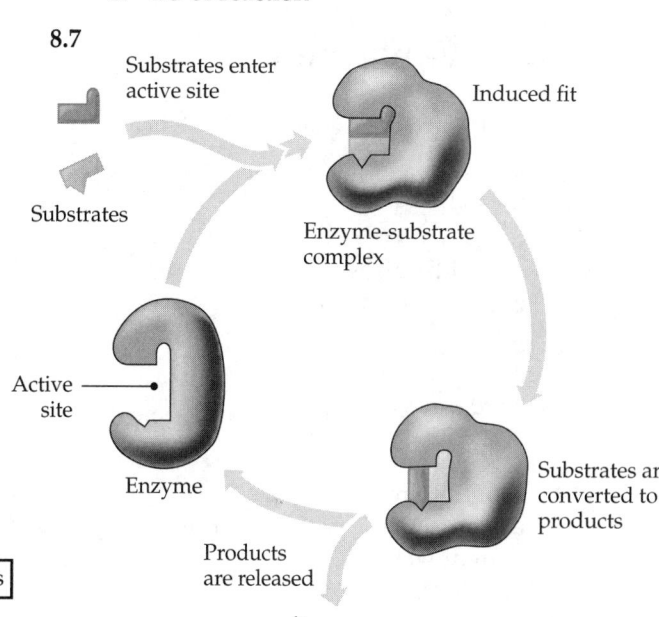

8.8   A competitive inhibitor would mimic the shape of the substrates and compete with them for the active site. A noncompetitive inhibitor would be a shape that could bind to another site on the enzyme molecule and would change the shape of the enzyme such that the active site functions less effectively.

8.9   ATP would act as an inhibitor to catabolic pathways, slowing the breakdown of fuel molecules if the supply of ATP exceeds demand. If ATP supplies drop, ADP would act as an activator of these catabolic enzymes and more ATP would be produced.

SUGGESTED ANSWERS TO STRUCTURE YOUR KNOWLEDGE

1.   Metabolism is the totality of chemical reactions that take place in living organisms. To create and maintain the structural order required for life necessitates an input of free energy—from sunlight for photosynthetic organisms and from energy-rich food molecules for other organisms. A cell couples catabolic, exergonic reactions ($-\Delta G$) with anabolic, endergonic reactions ($+\Delta G$), using ATP as the primary energy shuttle between the two.

2.   Enzymes are essential for metabolism because they lower the activation energy of the specific reactions they catalyze and allow those reactions to occur extremely rapidly at a temperature conducive to life. By regulating the enzymes it produces, a cell can regulate which of the myriad of possible chemical reactions take place at any given time. Metabolic control also occurs through allosteric regulation and feedback inhibition. The compartmental organization of a cell facilitates a cell's metabolism.

ANSWERS TO TEST YOUR KNOWLEDGE

Fill in the Blanks:

1.   metabolism
2.   anabolic
3.   potential
4.   heat
5.   entropy
6.   activation energy

7.   competitive inhibitors
8.   coenzymes
9.   feedback inhibition
10.   phosphorylated intermediate

Multiple Choice:

1. c	5. d	9. c	13. a	17. c
2. c	6. d	10. c	14. d	18. b
3. b	7. a	11. d	15. b	19. c
4. b	8. a	12. d	16. d	20. a

Chapter 9: Cellular Respiration and Fermentation

INTERACTIVE QUESTIONS

9.1 $C_6H_{12}O_6$; 6 CO_2; energy (ATP + heat)

9.2   **a.**   oxidized
   b.   reduced
   c.   donates (loses)
   d.   oxidizing agent
   e.   accepts (gains)

9.3   **a.** O_2
   b.   glucose
   c.   Some is stored in ATP and some is released as heat.

9.4   **a.**   electron carrier (or acceptor). It is a coenzyme that works with enzymes called dehydrogenases.
   b.   NADH

9.5   **a.**   glycolysis: glucose $\rightarrow$ pyruvate (considered part of cellular respiration in this textbook)
   b.   oxidation of pyruvate and citric acid cycle
   c.   oxidative phosphorylation: electron transport and chemiosmosis
   d.   electrons carried by NADH (and FADH$_2$) to the electron transport chain.
   e.   substrate-level phosphorylation
   f.   substrate-level phosphorylation
   g.   oxidative phosphorylation

9.6   **a.**   2 ATP
   b.   2 three-carbon sugars (glyceraldehyde-3-phosphate)
   c.   2 NAD$^+$ + 4 e$^-$ + 4H$^+$
   d.   2 NADH + 2H$^+$
   e.   4 ATP
   f.   2 pyruvate

9.7   **a.**   pyruvate
   b. CO_2
   c.   NADH + H$^+$
   d.   coenzyme A

    e. acetyl CoA
    f. oxaloacetate
    g. citrate
    h. $NADH + H^+$
    i. CO_2
    j. CO_2
    k. $NADH + H^+$
    l. GTP (may make ATP)
    m. $FADH_2$
    n. $NADH + H^+$

9.8 **a.** intermembrane space
      b. inner mitochondrial membrane
      c. mitochondrial matrix
      d. electron transport chain
      e. NADH
      f. NAD^+
      g. $FADH_2$
      h. $2 H^+ + \frac{1}{2} O_2$
      i. H_2O
      j. chemiosmosis

    k. ATP synthase
    l. $ADP + ⓟ_i$
    m. ATP

9.9 **a.** -2
      b. 4
      c. citric acid cycle
      d. 26 or 28
      e. 32
      f. 2
      g. 6
      h. 2
      i. 2

9.10 Respiration yields up to 16 times more ATP than does fermentation. By oxidizing pyruvate to CO_2 and passing electrons from NADH (and $FADH_2$) through the electron transport chain, respiration can produce a maximum of 32 ATP compared to the 2 net ATP that are produced by fermentation.

SUGGESTED ANSWERS TO STRUCTURE YOUR KNOWLEDGE

1. Use Interactive Questions 9.5, 9.6, 9.7, and 9.8 to help you review these pathways.

2.

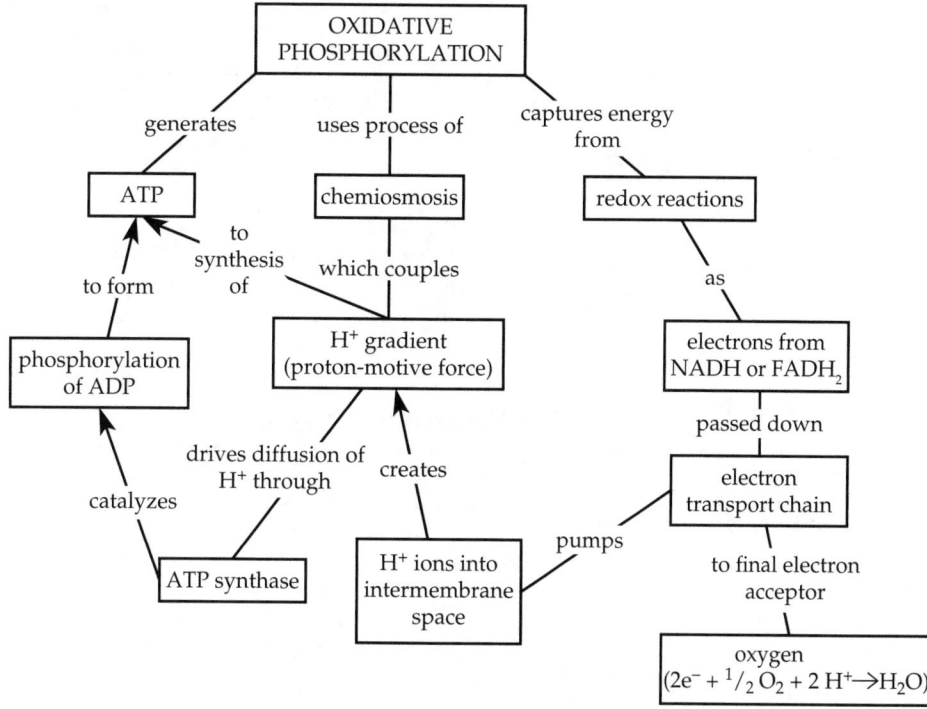

3.

Process	Brief Description	Main Inputs	Main Outputs
Glycolysis	Oxidation of glucose to 2 pyruvate with reduction of 2 NAD^+ to NADH, production of 2 ATP net.	glucose 2 ATP 2 NAD^+ 4 ADP + $\textcircled{P}_i$	2 pyruvate 4 ATP (2 net) 2 NADH + 2 H^+ 2 H_2O
Pyruvate to acetyl CoA and citric acid cycle	Oxidation of pyruvate to acetyl CoA, which combines with oxaloacetate → citrate. Citrate is cycled back to oxaloacetate as redox reactions produce NADH and $FADH_2$ and CO_2 is released. ATP is formed by substrate-level phosphorylation.	2 pyruvate 2 oxaloacetate	6 CO_2 8 NADH 2 $FADH_2$ 2 ATP
Oxidative phosphorylation	NADH and $FADH_2$ transfer electrons to electron transport chain. In a series of redox reactions, H^+ is pumped into intermembrane space, and electrons pass to O_2. Proton-motive force drives H^+ through ATP synthase to make ATP.	10 NADH 2 $FADH_2$ H^+ + O_2	H_2O 28 ATP (max)
Fermentation	Anaerobic catabolism: glycolysis followed by oxidation of NADH to NAD^+ so glycolysis can continue. Pyruvate is either reduced to ethyl alcohol and CO_2 or to lactate.	See glycolysis above 2 pyruvate 2 NADH	2 ATP 2 NAD^+ 2 ethanol and 2 CO_2 or 2 lactate

ANSWERS TO TEST YOUR KNOWLEDGE

Multiple Choice:

1. a	**4.** c	**7.** d	**10.** a	**13.** d	**16.** b	**19.** c	**22.** c
2. b	**5.** d	**8.** b	**11.** a	**14.** d	**17.** c	**20.** b	
3. d	**6.** b	**9.** c	**12.** a	**15.** d	**18.** d	**21.** c	

Chapter 10: Photosynthesis

INTERACTIVE QUESTIONS

10.1 **a.** outer membrane
  b. granum
  c. inner membrane
  d. thylakoid space
  e. thylakoid
  f. stroma

10.2 **a.** light
  b. H_2O
  c. light reactions in thylakoid membranes
  d. O_2
  e. ATP

  f. NADPH
  g. CO_2
  h. Calvin cycle in stroma
  i. $[CH_2O]$ (sugar)

10.3 The two y-axes should be labeled, "Absorption of light by chlorophyll *a*" and "Rate of photosynthesis." The solid line is the absorption spectrum; the dotted line is the action spectrum. Some wavelengths of light, particularly in the blue and the yellow-orange range, result in a higher rate of photosynthesis than would be indicated by the absorption of those wavelengths by chlorophyll *a*. These differences are partially accounted for by accessory pigments, such as chlorophyll *b* and carotenoids, which absorb light energy from different wavelengths and make that energy available to drive photosynthesis.

10.4 A photosystem contains light-harvesting complexes of pigment molecules (chlorophyll *a*, chlorophyll *b*, and carotenoids) bound to particular proteins and a reaction center complex, which includes a pair of chlorophyll *a* molecules (P680 or P700) and a primary electron acceptor.

10.5
  a. photosystem II
  b. photosystem I
  c. water (H_2O)
  d. oxygen ($\frac{1}{2} O_2$)
  e. P680, reaction-center chlorophyll *a*
  f. primary electron acceptor
  g. electron transport chain
  h. photophosphorylation by chemiosmosis
  i. ATP
  j. P700, reaction-center chlorophyll *a*
  k. primary electron acceptor
  l. $NADP^+$ reductase
  m. NADPH

  ATP and NADPH provide the chemical energy and reducing power for the Calvin cycle.

10.6
  a. Ferredoxin (Fd) passes the electrons to the cytochrome complex in the electron transport chain, from which they return to $P700^+$.
  b. Electrons from P680 are not passed to P700. Without the oxidizing agent $P680^+$, water is not split. Fd does not pass electrons to $NADP^+$ reductase to form NADPH.
  c. Electrons do pass down the electron transport chain, and the energy released by their "fall" drives chemiosmosis.

10.7
  a. in the thylakoid space (pH of about 5)
  b. (1) transport of protons into the thylakoid space as Pq transfers electrons to the cytochrome complex; (2) protons from the splitting of water remain in the thylakoid space; (3) removal of H^+ in the stroma during the reduction of $NADP^+$.

10.8
  a. carbon fixation
  b. reduction
  c. regeneration of CO_2 acceptor (RuBP)
  d. $3 CO_2$
  e. ribulose bisphosphate (RuBP)
  f. rubisco
  g. 3-phosphoglycerate
  h. $6 ATP \rightarrow 6 ADP$
  i. 1,3-bisphosphoglycerate
  j. $6 NADPH \rightarrow 6 NADP^+$
  k. $6 \, ℗_i$
  l. glyceraldehyde-3-phosphate (G3P)
  m. G3P
  n. glucose and other organic compounds
  o. $3 ATP \rightarrow 3 ADP$

  9 ATP and 6 NADPH are required to synthesize one G3P.

10.9 Photorespiration may be an evolutionary relic from the time when there was little O_2 in the atmosphere and the ability of rubisco to distinguish between O_2 and CO_2 was not critical. Photorespiration appears to protect plants from damaging products of the light reactions that build up when the Calvin cycle slows due to a lack of CO_2.

10.10
  a. in the bundle-sheath cells
  b. Carbon is initially fixed into a four-carbon compound in the mesophyll cells by PEP carboxylase. When this compound is broken down in the bundle-sheath cells, CO_2 is maintained at a high enough concentration that rubisco does not accept O_2 and cause photorespiration.
  c. ATP is used to convert pyruvate, returning from the bundle-sheath cells, to PEP in the mesophyll cells. Bundle-sheath cells produce this ATP using PS I and cyclic electron flow. (They lack PS II and thus do not generate O_2, which would compete with CO_2 for binding to rubisco.)

SUGGESTED ANSWERS TO STRUCTURE YOUR KNOWLEDGE

1.

2.

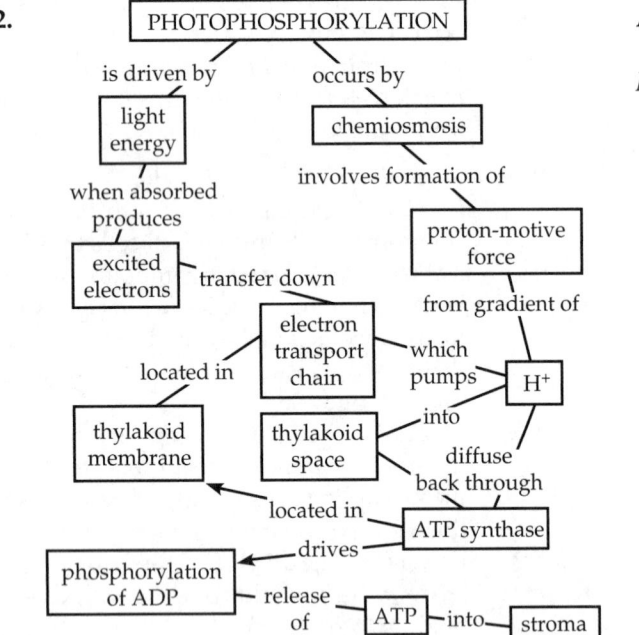

ANSWERS TO TEST YOUR KNOWLEDGE

Multiple Choice:

1. c	**7.** d	**13.** d	**19.** d	**25.** d
2. a	**8.** c	**14.** a	**20.** c	**26.** b
3. c	**9.** b	**15.** d	**21.** c	
4. a	**10.** b	**16.** b	**22.** a	
5. b	**11.** a	**17.** d	**23.** a	
6. b	**12.** c	**18.** a	**24.** b	

Chapter 11: Cell Communication

INTERACTIVE QUESTIONS

11.1 Plant hormones may reach their target cells by traveling through vessels, between cells via plasmodesmata, or even through the air as a gas.

11.2 The G protein is also a GTPase that hydrolyzes its bound GTP to GDP and inactivates itself. It then dissociates from the enzyme it had activated, and that enzyme returns to its original state.

11.3
  a. signaling molecules (ligands)
  b. α helix in the membrane
  c. phosphates
  d. 6 ATP → 6 ADP
  e. tyrosines
  f. activated relay proteins
  g. different cellular responses

11.4 Large, hydrophobic signaling molecules cannot cross the plasma membrane. They bind to cell-surface receptors such as GPCRs or ligand-gated ion channels. Small, hydrophobic signaling molecules can diffuse across the plasma membrane and bind with receptors within the cytoplasm or nucleus of target cells.

11.5
  a. A protein kinase transfers a phosphate group from ATP to a protein; adding a charged phosphate group causes a shape change that usually activates the protein.
  b. A phosphorylation cascade is a series of protein kinase relay molecules that sequentially phosphorylate the next kinase in the pathway.
  c. A protein phosphatase is an enzyme that removes a phosphate group from a protein, usually inactivating the protein. Protein phosphatases effectively shut down signaling pathways when the initial signal is no longer present.

11.6
  a. signaling molecule (first messenger)
  b. G protein-coupled receptor
  c. activated G protein (GTP bound)
  d. adenylyl cyclase
  e. ATP
  f. cAMP (second messenger)
  g. protein kinase A
  h. phosphorylation cascade to cellular response

11.7
  a. signaling molecule
  b. G protein
  c. membrane phospholipid
  d. IP_3
  e. Ca^{2+}
  f. endoplasmic reticulum

11.8
  a. Signaling molecules reversibly bind to receptors, and when they leave a receptor, the receptor reverts to its inactive form. The concentration of signaling molecules influences how many are bound at any time, with a threshold number of receptors with bound molecules required for the response to occur.
  b. Activated G proteins are inactivated when the GTPase portion of the protein converts GTP to GDP.
  c. Phosphodiesterase converts cAMP to AMP, thus damping out this second messenger.
  d. Protein phosphatases remove phosphate groups from activated proteins. The balance of active protein kinases and active phosphatases regulates the activity of many proteins.

11.9
  a. The basic mechanism for programmed cell death evolved early in the evolution of eukaryotes.
  b. Programmed cell death in humans is important in the normal development of the nervous system, fingers, and toes; for normal functioning of the immune system; and to prevent the development of cancerous cells (by the internal triggering of apoptosis in cells with DNA damage).

SUGGESTED ANSWERS TO STRUCTURE YOUR KNOWLEDGE

1. Cell signaling is essential for communication between cells. Unicellular organisms use signals to relay environmental or reproductive information. Communication in multicellular organisms allows for the development and coordination of specialized cells. Extracellular signals control the crucial activities of cells, such as cell division, differentiation, metabolism, and gene expression.

2. Cell signaling occurs through signal transduction pathways that include reception, transduction, and response. First a signaling molecule binds to a specific receptor. The message is transduced as the receptor

activates a protein that may relay the message through a sequence of activations, finally leading to the specific cellular response.

3. The sequence described in the previous answer can result in the activation of cellular proteins. When the signal is transduced to activate a transcription factor, however, the cellular response is a change in gene expression and the production of new proteins.

4. In a phosphorylation cascade, each step in the pathway activates multiple substrates of the next step, thus amplifying the original message to produce potentially millions of activated proteins and thus a large cellular response to a few signals.

ANSWERS TO TEST YOUR KNOWLEDGE

Multiple Choice:

1. b	4. b	7. a	10. a	13. b
2. d	5. d	8. c	11. c	14. c
3. a	6. c	9. d	12. d	15. b

Chapter 12: The Cell Cycle

INTERACTIVE QUESTIONS

12.1 a. 46
  b. 23
  c. 92

12.2 a. Growth—most organelles and cell components are produced continuously throughout these subphases.
  b. DNA synthesis and chromosome duplication

12.3 Refer to main text Figure 12.7 for chromosome diagrams (although you were asked to draw only four chromosomes, and the text shows six).
  a. G_2 of interphase
  b. prophase
  c. prometaphase
  d. metaphase
  e. anaphase
  f. telophase and cytokinesis
  g. duplicated centrosomes (with centrioles)
  h. chromatin (duplicated but uncondensed chromosomes)
  i. nuclear envelope
  j. nucleolus
  k. early mitotic spindle
  l. aster
  m. nonkinetochore microtubules
  n. metaphase plate
  o. spindle
  p. cleavage furrow
  q. nuclear envelope forming

12.4 a. MPF is a complex of cyclin and Cdk (cyclin-dependent kinase) that initiates mitosis by phosphorylating proteins and other kinases.

  b. MPF concentration is high as it triggers the onset of mitosis but is reduced at the end of mitosis because it depends on the concentration of cyclin in the cell. The Cdk level is constant throughout the cell cycle, but the level of cyclin varies because active MPF starts a process that degrades cyclin. Thus, MPF regulates its own level and can only become active when sufficient cyclin accumulates after being synthesized in the S and G_2 phases of the next interphase.

SUGGESTED ANSWERS TO STRUCTURE YOUR KNOWLEDGE

1. a. interphase
  b. telophase
  c. anaphase
  d. metaphase

2. Interphase: 90% of cell cycle; growth and DNA replication
  • G_1 phase: The chromosome consists of a long, thin chromatin fiber made of DNA and associated proteins. Growth and metabolic activities occur.
  • S phase—synthesis of DNA: The chromosome is duplicated; two exact copies, called sister chromatids, are produced and held together tightly at their centromeres and by cohesins along their length. Growth and metabolic activities continue.
  • G_2 phase: Growth and metabolism continue.
  • Mitotic phase: cell division consisting of mitosis and cytokinesis
  • Prophase: The sister chromatids, held together by sister chromatid cohesion and

at the centromere, become tightly coiled and condensed.
- Prometaphase: Kinetochore fibers from opposite ends of the mitotic spindle attach to the kinetochores of the sister chromatids; the chromosome moves toward midline.
- Metaphase: The centromere of the chromosome is aligned at the metaphase plate along with the centromeres of the other chromosomes.
- Anaphase: Cohesins are cleaved and the sister chromatids separate (now considered individual chromosomes) and move to opposite poles of the cell.
- Telophase: The chromatin fiber of the chromosome uncoils and is surrounded by reforming nuclear membrane.

3.

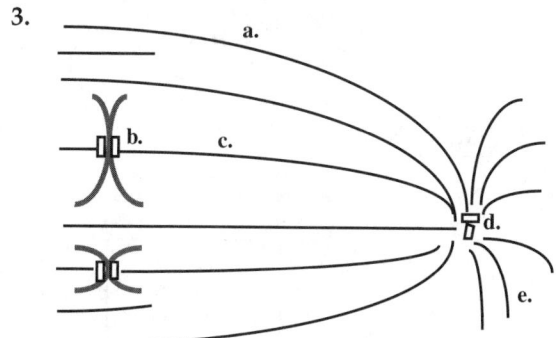

a. nonkinetochore microtubule: elongate cell by lengthening and "walking" past microtubules from the opposite pole

b. kinetochore: protein and DNA structure in region of centromere where microtubules attach

c. kinetochore microtubule: move chromosomes to metaphase plate and separate chromosomes as motor proteins of kinetochores "walk" toward opposite poles and the microtubules disassemble

d. centrosome and centrioles: region of mitotic spindle formation; organizes microtubules

e. aster: radiating spindle fibers (in animal cells)

ANSWERS TO TEST YOUR KNOWLEDGE

Fill in the Blanks:

1. G_0
2. anaphase
3. prophase
4. cytokinesis
5. S phase
6. metaphase
7. telophase
8. prophase
9. prometaphase
10. G_1 phase

Multiple Choice:

1. d	4. d	7. c	10. a	13. d
2. b	5. b	8. a	11. d	14. c
3. c	6. a	9. b	12. d	15. b

Chapter 13: Meiosis and Sexual Life Cycles

INTERACTIVE QUESTIONS

13.1 **a.** 14; 7
b. 28; 1
c. 56. Sister chromatids are produced when a chromosome duplicates before cell division. They are joined at the centromere and attached along their lengths. Nonsister chromatids are found on different chromosomes of a homologous pair (homologs).

13.2 **a.** meiosis
b. fertilization
c. zygote
d. mitosis
e. gametes
f. fertilization
g. zygote
h. meiosis
i. spores
j. gametes
k. fertilization
l. 2n
m. zygote
n. meiosis

13.3 The diploid number for this animal cell (2n) is 6.
a. metaphase II
b. prophase I
c. anaphase I
d. metaphase I
e. anaphase II
f. telophase I, beginning of cytokinesis
Proper sequence: b, d, c, f, a, e.

13.4   **a.** 2^{23}, approximately 8.4 million
b.   about 70 trillion ($2^{23} \times 2^{23}$)
c.   This number of combinations does not take into account the additional variation of recombinant chromosomes produced by crossing over.

SUGGESTED ANSWERS TO STRUCTURE YOUR KNOWLEDGE

1.   **a.**   sister chromatids
b.   centromere
c.   pair of homologous chromosomes
d.   nonsister chromatids

2.   **a.**   chromosome duplication; sister chromatids attached at centromere and by sister chromatid cohesion along their lengths

b.   Synapsis of homologous pairs (held by synaptonemal complex); crossing over (exchange of corresponding DNA segments) is evident at chiasmata; chromosomes condense.
c.   Homologous pairs line up independently at metaphase plate (orientation of maternal and paternal homologs is random).
d.   Homologous pairs of chromosomes separate, and homologs move toward opposite poles; sister chromatids remain attached at centromere.
e.   Haploid set of chromosomes, each consisting of two sister chromatids, aligns at metaphase plate; sister chromatids are not identical due to crossing over.
f.   Sister chromatids separate and move to opposite poles as individual chromosomes.

3.

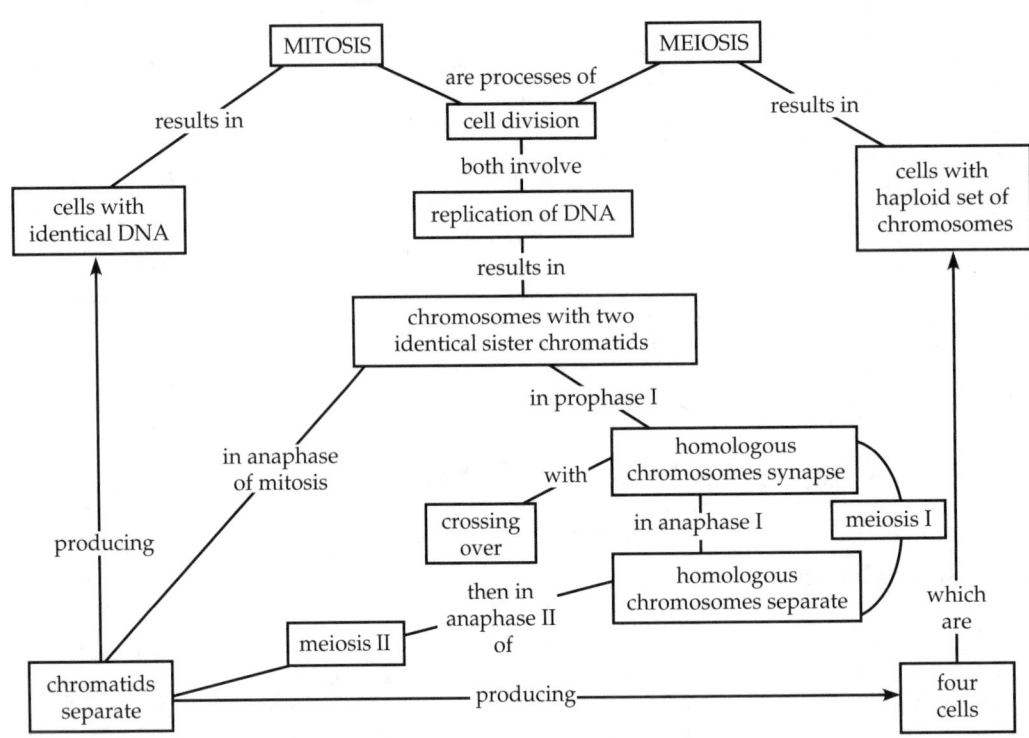

ANSWERS TO TEST YOUR KNOWLEDGE

Multiple Choice:

1. a	**4.** c	**7.** b	**10.** d	**13.** c	**16.** b	**19.** d	**22.** c
2. d	**5.** b	**8.** b	**11.** c	**14.** a	**17.** b	**20.** b	
3. b	**6.** a	**9.** d	**12.** a	**15.** b	**18.** c	**21.** d	

Chapter 14: Mendel and the Gene Idea

INTERACTIVE QUESTIONS

14.1
  a. R
  b. r
  c. F_1 generation
  d. Rr
  e. F_2 generation
  f. R
  g. r
  h. Rr ◯
  i. Rr ◯
  j. rr 〰
  k. 3 round:1 wrinkled
  l. 1 RR:2Rr:1rr

14.2
  a. all tall (Tt) plants, because the tall plant can only contribute a T allele and the dwarf plant can only contribute a t allele
  b. 1:1 tall (Tt) to dwarf (tt), because the tall plant can contribute either a T allele or a t allele
  You can use a Punnett square to determine the expected outcome of a testcross, but a shortcut is to use only one row or column for the recessive individual's gametes, as they produce only one type of gamete. For example, in the text Figure 14.7 you could use only one column for sperm because all sperm contain the recessive p allele.

14.3
  a. all tall purple plants
  b. $TtPp$
  c. TP, Tp, tP, tp
  d.

Sperm

	$\frac{1}{4}$ⓉP	$\frac{1}{4}$Ⓣp	$\frac{1}{4}$ⓣP	$\frac{1}{4}$ⓣp
$\frac{1}{4}$ⓉP	$TTPP$	$TTPp$	$TtPP$	$TtPp$
$\frac{1}{4}$Ⓣp	$TTPp$	$TTpp$	$TtPp$	$Ttpp$
$\frac{1}{4}$ⓣP	$TtPP$	$TtPp$	$ttPP$	$ttPp$
$\frac{1}{4}$ⓣp	$TtPp$	$Ttpp$	$ttPp$	$ttpp$

Eggs

  e. 9 tall purple:3 tall white:3 dwarf purple: 1 dwarf white
  f. 12:4 or 3:1 tall to dwarf; 12:4 or 3:1 purple to white

14.4
  a. Consider the outcome for each gene as a monohybrid cross. The probability that a cross of $Aa \times Aa$ will produce an $A\_$ offspring is ¾. The probability that a cross of $Bb \times bb$ will produce a $B\_$ offspring is ½. The probability that a cross of $cc \times CC$ will produce a $C\_$ offspring is 1. To have all of these events occur simultaneously, multiply their probabilities: ¾ × ½ × 1 = ⅜.
  b. First determine what genotypes would fill the requirement of at least two dominant traits. Offspring could be $A\_bbC\_$, $aaB\_C\_$, or A_B_C. The genotype A_B_cc is not possible. Can you see why?

Probability of $A\_bbC\_$ = ¾ × ½ × 1 = ⅜
Probability of $aaB\_C\_$ = ¼ × ½ × 1 = ⅛
Probability of $A\_B\_C\_$ = ¾ × ½ × 1 = ⅜

The probability of offspring showing at least two dominant traits is the sum of these independent probabilities, or ⅞.
  c. There is only one type of offspring that can show only one dominant trait ($aabbCc$). Can you see why? Its probability is ¼ × ½ × 1 = ⅛. As you may notice, this would have been an easier way to determine the answer to **b**. With problems such as these, always consider what the question is asking from as many different angles as you can.

14.5
  a. A: $I^A I^A$ and $I^A i$
  b. B: $I^B I^B$ and $I^B i$
  c. AB: $I^A I^B$
  d. O: ii

14.6 First determine the possible phenotypes of this cross and the genotypes that would produce each one. Remember that using a blank line facilitates genotype determination ($M\_B\_$ is easier than writing out $MMBB$, $MMBb$, $MmBB$, and $MmBb$). Then treat each gene as a separate monohybrid cross to determine the probabilities for each gene; multiply those probabilities to determine a particular genotype's probability. The ratio of offspring from this cross would be 9:3:4, a common ratio when one gene is epistatic to another.

Phenotype	Genotype	Ratio
Black	$M\_B\_$	¾ × ¾ = 9/16
Gray	M_bb	¾ × ¼ = 3/16
White	$mm\_\_$	¼ × 1 = ¼ or 4/16

14.7 **a.** The parental cross produced 25-cm-tall F_1 plants, all *AaBbCc* plants with 3 units of 5 cm added to the base height of 10 cm.

b. As a general rule, in the polygenic inheritance of a quantitative character the number of phenotypes resulting from a cross of heterozygotes equals the number of alleles involved plus one. In this case, 6 alleles (*AaBbCc*) + 1 = 7. So, there will be seven different phenotypes in the F_2 among the 64 possible combinations of the eight types of F_1 gametes. (See why you wouldn't want to go through a Punnett square to figure that out!) These phenotypes will range from six dominant alleles (40 cm), five dominant (35 cm), four dominant (30 cm), and so on, to all six recessive alleles (10 cm), yielding a total of seven phenotypes.

14.8 **a.** This trait is recessive. If it were dominant, it would be impossible to have children with albinism from two unaffected (and thus homozygous recessive) parents.

b. father *Aa*; mother *Aa*, because neither parent has albinism and they have affected offspring (*aa*)

c. mate #1 *AA* (probably); mate #2 *Aa*; grandson #4 *Aa*

d. The genotype of son #3 could be *AA* or *Aa*. If his wife is *AA*, then he could be *Aa* (because both his parents are carriers), and the recessive allele never would be expressed in his offspring. Even if he and his wife were both carriers (heterozygotes), there would be a $^{243}/_{1024}$ ($\frac{3}{4} \times \frac{3}{4} \times \frac{3}{4} \times \frac{3}{4} \times \frac{3}{4}$) or 24% chance that all five children would be normally pigmented.

14.9 **a.** $\frac{1}{4}$

b. $\frac{2}{3}$. Of offspring with a normal phenotype, $\frac{2}{3}$ would be predicted to be heterozygotes and thus carriers of the recessive allele.

14.10 Both sets of prospective grandparents must have been carriers. The prospective parents do not have the disorder, so they are not homozygous recessive. Thus, each has a $\frac{2}{3}$ chance of being a heterozygote carrier. The probability that both parents are carriers is $\frac{2}{3} \times \frac{2}{3} = \frac{4}{9}$; the chance that two heterozygotes will have a recessive homozygous child is $\frac{1}{4}$. The overall chance that a child will inherit the disease is $\frac{4}{9} \times \frac{1}{4} = \frac{1}{9}$. The fact that the first two children are unaffected does not establish the genotype of the parents. Thus the third child would also have a $\frac{1}{9}$ chance of inheriting the disorder. Should the third child have the disease, however, this would establish that both parents are carriers, and the chance that a subsequent child would have the disease is now estimated at $\frac{1}{4}$.

SUGGESTED ANSWERS TO STRUCTURE YOUR KNOWLEDGE

1. Mendel's law of segregation occurs in anaphase I, when alleles on homologous chromosomes move to opposite poles of the cell. The two cells formed from this division have one-half the number of chromosomes and one copy of each gene (although sister chromatids still have to separate in anaphase II). Mendel's law of independent assortment relates to the lining up of attached homologous chromosomes at the equatorial plate in a random fashion during metaphase I. Genes on different chromosomes will assort independently into gametes.

2. *Aa* = 2 different gametes   *AaBb* = 4 gametes   *AaBbCc* = 8 gametes   *AABbCc* = 4 gametes   A general formula is 2^n, where *n* is the number of gene loci that are heterozygous. Note that in the last example, *AABbCc*, all gametes will contain a dominant *A* allele.

3. Always look to the F_1 heterozygote resulting from a cross of true-breeding parents. If alleles show complete dominance/recessiveness, then the heterozygote will have a phenotype identical to that of one parent. In incomplete dominance, the F_1 phenotype will be intermediate between that of the parents. If codominant, the phenotype of both alleles will be exhibited in the heterozygote.

ANSWERS TO TEST YOUR KNOWLEDGE

Matching:

1. l	**4.** b	**7.** k	**10.** f
2. e	**5.** j	**8.** h	**11.** i
3. g	**6.** a	**9.** c	**12.** d

Multiple Choice:

1.	c	**5.**	a	**9.**	c	**13.**	b
2.	b	**6.**	c	**10.**	d	**14.**	a
3.	b	**7.**	d*	**11.**	d	**15.**	b
4.	b	**8.**	d**	**12.**	b	**16.**	c

*There are three different ways to get this outcome: HHT, HTH, THH. Each outcome has a probability of ⅛.

**Remember that only homozygotes are true-breeding. There are four possible homozygous genotypes: *AABB*, *aabb*, *AAbb*, and *aaBB*, with a probability of ¹⁄₁₆ for each one. Use the rule of addition to obtain the answer.

ANSWERS TO GENETICS PROBLEMS

1. White alleles are dominant to yellow alleles. If yellow were dominant, then you should be able to get white squash from a cross of two yellow heterozygotes.

2. **a.** ¼ [½ (to get *AA*) × ½ (*bb*)]
   b. ⅛ [¼ (*aa*) × ½ (*BB*)]
   c. ½ [1 (*Aa*) × ½ (*Bb*) × 1 (*Cc*)]
   d. ¹⁄₃₂ [¼ (*aa*) × ¼ (*bb*) × ½ (*cc*)]

3. Because flower color shows incomplete dominance, use symbols such as C^R and C^W for those alleles. There will be six phenotypic classes in the F₂ instead of the normal four classes found in a 9:3:3:1 ratio. You could find the answer with a Punnett square, but multiplying the probabilities of the monohybrid crosses is more efficient.

Tall red	$T\_C^RC^R$	¾ × ¼ = ³⁄₁₆
Tall pink	$T\_C^RC^W$	¾ × ½ = ⅜ or ⁶⁄₁₆
Tall white	$T\_C^WC^W$	¾ × ¼ = ³⁄₁₆
Dwarf red	$tt\ C^RC^R$	¼ × ¼ = ¹⁄₁₆
Dwarf pink	$tt\ C^RC^W$	¼ × ½ = ⅛ or ²⁄₁₆
Dwarf white	$tt\ C^WC^W$	¼ × ¼ = ¹⁄₁₆

4. Determine the possible genotypes of the mother and child. Then find the blood groups for the father that could not have resulted in a child with the indicated blood group.
   a. no groups exonerated
   b. B or O
   c. A or O
   d. AB only

5. The parents are *CcBb* and *Ccbb*. Right away you know that all *cc* offspring will die and no *BB* black offspring are possible because one parent is *bb*. Only four phenotypic classes are possible. Determine the proportion of each type by applying the rule of multiplication.

Lethal (*cc_ _*)	¼ of all offspring die
Normal brown (*CCBb*)	¼ × ½ = ⅛
Normal white (*CCbb*)	¼ × ½ = ⅛
Deformed brown (*CcBb*)	½ × ½ = ¼
Deformed white (*Ccbb*)	½ × ½ = ¼
Ratio of viable offspring:	1:1:2:2

6. The father's genotype must be *Pp* because polydactyly is dominant and he has had one normal child. The mother's genotype is *pp*. The chance of the next child having normal digits is ½ or 50% because the mother can only donate a *p* allele and there is a 50% chance that the father will donate a *p* allele.

7. The genotypes of the puppies were ⅜ *B_S_*, ⅜ *B_ss*, ⅛ *bbS_*, and ⅛ *bbss*. Because recessive traits show up in the offspring, both parents had to have had at least one recessive allele for both genes. Black:chestnut occurs in a 6:2 or 3:1 ratio, indicating a heterozygous cross. Solid:spotted occurs in a 4:4 or 1:1 ratio, indicating a cross between a heterozygote and a homozygous recessive. Parental genotypes were *BbSs* × *Bbss*.

8. The 1:1 ratio of the first cross indicates a cross between a heterozygote and a homozygote. The second cross indicates that the hairless hamsters cannot have a homozygous genotype, because then only hairless hamsters would be produced. Let us say that hairless hamsters are *Hh* and normal-haired are *HH*. The second cross of heterozygotes should yield a 1:2:1 genotype ratio. The 1:2 ratio indicates that the homozygous recessive genotype may be lethal, with embryos that are *hh* never developing.

9. **a.** Tail length appears to be a quantitative character. A general rule for the number of alleles in a cross involving a quantitative character is one less than the number of phenotypic classes. Here there are five phenotypic classes, and thus four alleles or two gene pairs. In this cross, the phenotypic ratio is approximately 1:4:6:4:1. The ratio base of 16 also indicates a dihybrid cross. Each dominant gene appears to add 5 cm in tail length. The 15-cm pigs are double recessives; the 35-cm pigs are double dominants.
   b. A cross of a 15-cm pig and a 30-cm pig would be equivalent to *aabb* × *AABb* (or *aabb* × *AaBB*). Offspring would have either one or two dominant alleles and would have tail lengths of 20 or 25 cm.

10. **a.** The black parent would be CC^{ch}. The Himalayan parent could be either C^hC^h or C^hc. First, list the alleles in order of dominance: C, C^{ch}, C^h, c. To approach this problem you should write down as much of the genotype as you can be sure about for each phenotype involved. The parents were black ($C\_$) and Himalayan ($C^h\_$). The offspring genotypes would be black ($C\_$) and Chinchilla ($C^{ch}\_$). Right away you can tell that the black parent must have been CC^{ch} because half of the offspring are Chinchilla; C^{ch} is dominant over C^h and c and could not have been present in the Himalayan parent. The genotype of this parent could be either C^hC^h or C^hc. A test-cross with a white rabbit would be necessary to determine this genotype.

b. You cannot definitely determine the genotype of the parents in the second cross. You can, however, eliminate some genotypes. The black parent could not be CC or CC^{ch}, because both the C and C^{ch} alleles are dominant to C^h, and Himalayan offspring were produced. The Chinchilla parent could not be $C^{ch}C^{ch}$ for the same reason. One or both parents had to have C^h as their second allele; one, but not both, might have c as their second allele.

11. **a.** First figure out possible genotypes: $\_\_E\_$ = golden (any B combination with at least one E), B_ee = black, $bbee$ = brown. All you know at the start is that both golden parents are $\_E\_$. Because you see non-golden offspring, you know that both parents had to be heterozygous for Ee, and at least one was heterozygous for Bb (to get black and brown offspring). Because black and brown are in a 1:1 ratio (like a testcross ratio), then one hamster was Bb and the other was bb. You need to consider the results from the second cross to know which hamster was Bb.

b. For the second cross, you now know that the black hamster 3 is $B?ee$ and the golden hamster 2 is $?bEe$. A ratio of approximately 3:1 black to brown looks like the results of a monohybrid cross, so both parents must be Bb. So hamster 3 is $Bbee$, and hamster 2 must be $BbEe$. That means that hamster 1 must be $bbEe$.

12.

(At least one of these grandparents must have been Tt)

13. Because the parents were true-breeding for two characters, the F_1s would be dihybrids. Because all F_1s had red terminal flowers, those two traits must be dominant, and their genotypes could be represented as $RrTt$. One would predict an F_2 phenotypic ratio of 9 red terminal:3 red axial:3 white terminal: 1 white, axial. If 100 offspring were counted, one would expect approximately 19 ($\frac{3}{16} \times 100$) plants with red axial flowers.

14. Because the F_2 generation is a ratio based on 16, one can conclude that two genes are involved in the inheritance of this flower color character. The parent plants were probably double homozygotes ($AABB$ for red and $aabb$ for white). The all-red F_1s must be $AaBb$, and the red color must require at least one dominant allele of both genes. The 9:6:1 ratio of the F_2s indicates that 9 plants had at least one dominant allele of both genes to produce the red color; 1 plant was doubly homozygous recessive and thus white; and the 6 pale purple were caused by having at least one dominant allele of only one of the two genes involved in flower color (A_bb or $aaB\_$).

15. Albinism (b) is a recessive trait; black (B) is dominant. First cross: parents $BB \times bb$; offspring all Bb (black coat). The black guinea pig in the second cross is a heterozygote. Second cross: parents $Bb \times bb$; offspring ½ Bb and ½ bb (7 Bb and 5 bb in this cross).

16. **a.** $PPLl \times PPLl$, $PPLl \times PpLl$, or $PPLl \times ppLl$
b. $ppLl \times ppLl$
c. $PPLL \times$ any of the 9 possible genotypes or $PPll \times ppLL$
d. $PpLl \times Ppll$
e. $PpLl \times PpLl$

17. Because they already have an affected child, you know that the parents' blood genotypes are I^Ai and I^Bi and that both are carriers for the disease allele. The probability of having a second child with O type blood and the disease is ¼ × ¼, or $\frac{1}{16}$.

18. 9 $B\_A\_$ (agouti):3 B_aa (black):3 $bbA\_$ (white):1 $bbaa$ (white). Overall, 9 agouti: 3 black:4 white.

Chapter 15: The Chromosomal Basis of Inheritance

INTERACTIVE QUESTIONS

15.1

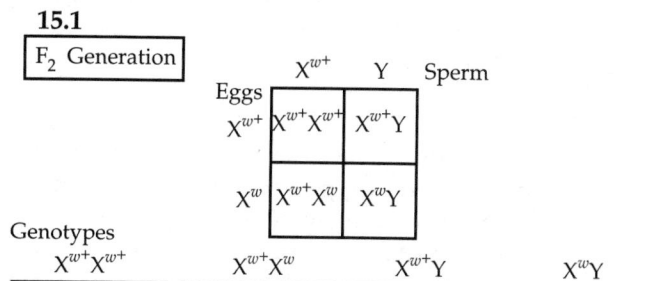

15.2 The gene is X-linked, so a good notation is X^N, X^n, and Y so that you will remember that the Y does not carry the gene. Capital N indicates normal sight. Genotypes are:
1. $X^N Y$
2. $X^N X^n$
3. $X^N X^N$ or $X^N X^n$ (probably $X^N X^N$ because four sons are $X^N Y$)
4. $X^N X^n$
5. $X^N Y$
6. $X^n Y$
7. $X^N X^n$

15.3 **a.** tall, purple-flowered and dwarf, white-flowered
b. tall, white-flowered and dwarf, purple-flowered

15.4 If linked genes have their loci close together on the same chromosome, they travel together during meiosis and *more* parental offspring are produced. Recombinants are the result of crossing over between nonsister chromatids of homologous chromosomes.

15.5 Solving a linkage problem is often a matter of trial and error. Sometimes it helps to lay out the loci with the greatest distance between them and fit the other genes between or on either side by adding or subtracting map unit distances.

```
_____ 15 _____
m ___ 9 ___ j ___ 6 ___ l ___ 6 ___ k
         _____ 12 _____
```

15.6 **a.** An organism with a trisomy ($2n + 1$) has an extra copy of one chromosome, usually caused by a nondisjunction during meiosis. A triploid organism ($3n$) has an extra set of chromosomes, possibly caused by a total nondisjunction in gamete formation.

b. A trisomy would probably disrupt the genetic balance more than having a complete extra set of chromosomes.

15.7 translocation (O-P-Q), deletion (D-E-F), and inversion (J-I-H)

15.8 Aneuploidies of sex chromosomes appear to upset genetic balance less, perhaps because relatively few genes are located on the Y chromosome and extra X chromosomes are inactivated as Barr bodies.

SUGGESTED ANSWERS TO STRUCTURE YOUR KNOWLEDGE

1. Genes that are not linked assort independently, and the ratio of offspring from a testcross with a dihybrid heterozygote should be 1:1:1:1 (*AaBb* × *aabb* gives *AaBb*, *Aabb*, *aaBb*, *aabb* offspring and a recombination frequency of 50%—50% of offspring are recombinants). Genes that are linked and do not cross over should produce a 1:1 ratio in such a testcross (*AaBb* and *aabb* because a heterozygote derived from a P generation cross of *AABB* × *aabb* produces only *AB* and *ab* gametes). If crossovers between distant genes almost always occur, the heterozygote will produce equal quantities of *AB*, *Ab*, *aB*, and *ab* gametes, and the genotype ratio of offspring will be 1:1:1:1, the same as it is for unlinked genes. Because each 1% recombination frequency is equal to 1 map unit, this measurement ceases to be meaningful at relative distances of 50 or more map units. However, crosses with intermediate genes on the chromosome could establish both that the genes are on the same chromosome and that they are a certain map unit distance apart.

2. A cross between a mutant female fly and a normal male should produce all normal females (who get a wild-type allele from their father) and all mutant male flies (who get the mutant allele on the X chromosome from their mother). $X^m X^m \times X^{m+} Y$ produces $X^{m+} X^m$ and $X^m Y$ offspring (normal females and mutant males).

3. The inactivation of two X chromosomes in XXX women would leave them with one genetically active X, as in women with the normal number of chromosomes (thus not upsetting normal genetic balance). To test whether this is the case, observing cells stained for DNA under a microscope should reveal two Barr bodies in XXX women.

4. The serious phenotypic effects that are associated with these chromosomal alterations indicate that normal development and functioning are dependent on genetic balance. Most genes appear to be vital to an organism's existence, and extra copies of genes upset genetic balance. Inversions and translocations, which do not disrupt the balance of genes, can alter phenotype because of the effect of neighboring genes on gene functioning.

ANSWERS TO TEST YOUR KNOWLEDGE

Multiple Choice:

1.	b	**5.**	d	**9.**	c	**13.**	d	**17.**	d
2.	a	**6.**	d	**10.**	c	**14.**	b	**18.**	d
3.	a	**7.**	c	**11.**	d	**15.**	a	**19.**	a
4.	a	**8.**	b	**12.**	b	**16.**	d	**20.**	c

ANSWERS TO GENETICS PROBLEMS

1. **a.** The trait is recessive and probably X-linked. Two sets of unaffected parents in the second generation have offspring with the trait, indicating that it must be recessive. More males than females display the trait. Females #3 and #5 must be carriers because their father has the trait, and they each pass it on to a son.
   b. Using the symbols X^T for the dominant allele and X^t for the recessive:
   1. X^tY
   2. X^TX^t
   3. X^TX^t
   4. X^TY
   5. X^TX^t
   6. X^TX^t or X^TX^T
   7. X^TX^t
   c. Female #6 has a brother who has the trait and her mother's father had the trait, so her mother must be a carrier of the trait, meaning there is a ½ probability that #6 is a carrier. If she

mates with a phenotypically normal male, none of her daughters will show the trait (0 probability). Her sons have a ½ chance of having the trait if she is a carrier. There is a ½ × ½ = ¼ probability that her sons will have the trait. There is a ½ chance that a child would be male, so the probability of an affected child is ½ × ¼, or ⅛.

2. c, a, b, d

3. The genes appear to be linked because the parental types appear most frequently in the offspring. Recombinant offspring represent 10 out of 40 total offspring for a recombination frequency of 25%, indicating that the genes are 25 map units apart.

4. One of the mother's X chromosomes carries the recessive lethal allele. One-half of male fetuses would be expected to inherit that chromosome and spontaneously abort. Assuming an equal sex ratio at conception, the ratio of girl to boy children would be 2:1, or six girls and three boys.

5. **a.** For female chicks to be black, they must have received a recessive allele from the male parent. If all female chicks are black, the male parent must have been Z^bZ^b. If the male parent was homozygous recessive, then all male offspring will receive a recessive allele, and the female parent would have to be Z^BW to produce all barred males.
   b. For female chicks to be both black and barred, the male parent must have been Z^BZ^b. If the female parent were Z^BW, only barred male chicks would be produced. To get an equal number of black and barred male chicks, the female parent must have been Z^bW.

6. ¼ for a color-blind daughter (½ chance that child will be female × ½ chance of a homozygous recessive genotype); ½ for first son

Chapter 16: The Molecular Basis of Inheritance

INTERACTIVE QUESTIONS

16.1 a. They grew T2 with radioactive sulfur to tag phage proteins.
b. They grew T2 with radioactive phosphorus to tag phage DNA.
c. Radioactivity was found in the liquid, indicating that the phage protein did not enter the bacterial cells.

d. In the samples with the labeled DNA, most of the radioactivity was found in the bacterial cell pellet.
e. They concluded that viral DNA is injected into the bacterial cells and serves as the hereditary material for viruses.

16.2 a. sugar-phosphate backbone
b. 3′ end of chain

c. hydrogen bonds
d. cytosine (pyrimidine)
e. guanine (purine)
f. adenine (purine)
g. thymine (pyrimidine)
h. 5' end of chain
i. nucleotide
j. deoxyribose
k. phosphate group
l. 3.4
m. 0.34
n. 2

16.3

| DNA molecules | Density bands |

First generation
(grown on light
^{14}N medium)

Second generation
(grown on light
^{14}N medium)

16.4 The phosphate end of each strand is the 5' end, and the hydroxyl group extending from the 3' carbon of the sugar marks the 3' end. See Figure 16.7b or 16.14 in the text.

16.5 a. single-strand binding protein
b. DNA pol III
c. leading strand
d. 5' end of parental strand
e. 3' end
f. helicase
g. RNA primer
h. primase
i. DNA pol III
j. Okazaki fragment
k. DNA pol I (replacing primer)
l. lagging strand
m. DNA ligase

16.6

Last fragment

RNA primer

5'
3'

5'
3'

No DNA nucleotides can be added after the RNA primer is removed because no 3' end is available for DNA polymerase to add to. Thus, the lagging strand is shorter. In the next round of replication, this now parental strand will provide a shorter template and thus yield a shorter DNA molecule. Further rounds of replication will continue to produce shorter DNA molecules. Also see text Figure 16.20.

16.7 Nucleosomes (10-nm fiber of nucleosomes and linker DNA); 30-nm fiber; looped domains (300-nm fiber); coiling and folding of looped domains into highly condensed metaphase chromosome.

SUGGESTED ANSWERS TO STRUCTURE YOUR KNOWLEDGE

1. Watson and Crick used Franklin's X-ray diffraction photo to deduce that DNA was a helix, 2 nm wide, with nitrogenous bases stacked 0.34 nm apart, and making a full turn every 3.4 nm. Franklin had concluded that the sugar-phosphate backbones were on the outside of the helix with the bases extending inside. Using molecular models of wire, Watson and Crick experimented with various arrangements and finally paired a purine base with a pyrimidine base, which produced the proper diameter. Specificity of base pairing (A with T and C with G) is ensured by hydrogen bonds.

2. Replication bubbles form where proteins recognize specific base sequences and open up the two strands. *Helicase*, an enzyme that works at the replication fork, unwinds the helix and separates the strands. *Single-strand binding proteins* support the separated strands while replication takes place. *Topoisomerase* eases the twisting ahead of the replication fork by breaking, untwisting, and rejoining DNA strands. *Primase* synthesizes a primer of 5–10 RNA bases to start the new strand. After a proper base pairs up on the exposed template, *DNA polymerase III* joins the nucleotide to the 3' end of the new strand. On the lagging strand, short Okazaki fragments are formed by primase and DNA pol III (again moving 5' → 3'). *DNA polymerase I* replaces the primer with DNA, adding to the 3' end of the leading strand or fragments. *Ligase* joins the 3' end of one fragment to the 5' end of its neighbor. Proofreading enzymes check for mispaired bases, and nucleases, DNA polymerase, ligase, and other enzymes repair damage or mismatches.

ANSWERS TO TEST YOUR KNOWLEDGE

Multiple Choice:

1. b	4. c	7. a	10. c	13. d	16. b	19. b	21. b
2. a	5. b	8. d	11. d	14. a	17. b	20. a	22. d
3. b	6. b	9. c	12. d	15. d	18. a		

Chapter 17: Gene Expression: From Gene to Protein

INTERACTIVE QUESTIONS

17.1 **a.** RNA differs from DNA in the following three ways: The sugar component of its nucleotides is ribose, rather than deoxyribose; uracil (U) replaces thymine as one of its nitrogenous bases; and RNA is usually single-stranded.
b. DNA $\xrightarrow{\text{transcription}}$ RNA $\xrightarrow{\text{translation}}$ protein

17.2 Met Pro Asp Phe Lys stop

17.3 **a.** Initiation: Transcription factors bind to the promoter and facilitate the binding of RNA polymerase II, forming a transcription initiation complex; RNA polymerase II separates DNA strands and transcription begins at the start point.
b. Elongation: RNA polymerase II moves along the DNA strand, connecting RNA nucleotides that have paired to the DNA template to the 3′ end of the growing RNA strand.
c. Termination: After polymerase transcribes past a polyadenylation signal sequence, the pre-mRNA is cut and released.

17.4 A 5′ cap consisting of a modified guanine nucleotide is added to the 5′ UTR. A poly-A tail consisting of up to 250 adenine nucleotides is attached to the 3′ UTR. Spliceosomes have cut out the introns and spliced the exons together.

17.5 **a.** A tRNA molecule translates the nucleotide sequence of a codon on mRNA into a specific amino acid.
b.

DNA Triplet 3′ → 5′	mRNA Codon 5′ → 3′	Anticodon 3′ → 5′	Amino Acid
TAC	AUG	UAC	methionine
GGA	CCU	GGA	proline
TTC	AAG	UUC	lysine
ATC	UAG	AUC	stop

17.6 1. Codon recognition: An elongation factor (not shown) helps an aminoacyl tRNA into the A site, where its anticodon base-pairs to the mRNA codon; hydrolysis of GTP increases accuracy and efficiency.
2. Peptide bond formation: Ribosome catalyzes peptide bond formation between the amino end of the new amino acid and the carboxyl end of the polypeptide held in the P site.
3. Translocation: The empty tRNA in the P site is moved to the E site and released; the tRNA now holding the polypeptide is moved from the A to the P site, taking the mRNA with it; GTP is required.
4. Termination: A release factor binds to the stop codon in the A site and promotes hydrolysis of the bond between the polypeptide and the tRNA in the P site. Free polypeptide is released and leaves through the exit tunnel. Ribosomal subunits and other assembly components separate. GTP is required.
a. amino end of growing polypeptide
b. aminoacyl tRNA
c. large subunit
d. 5′ end of mRNA
e. small subunit
f. peptide bond formation
g. A site
h. E site
i. release factor
j. stop codon
k. P site
l. free polypeptide

17.7 A ribosome that is translating an mRNA that codes for a secretory or membrane protein will become bound to the ER when an initial signal peptide on the polypeptide it is synthesizing is bound by an SRP, which then attaches to an ER receptor protein and translocation complex.

17.8 **a.** Silent: a nucleotide-pair substitution producing a codon that still codes for the same amino acid.

b. Missense: a nucleotide-pair substitution or frameshift mutation that results in a codon for a different amino acid.

c. Nonsense: a nucleotide-pair substitution or frameshift mutation that creates a stop codon and prematurely terminates translation.

d. Frameshift: an insertion or deletion of one, two, or more than three nucleotides that disrupts the reading frame and creates extensive missense and nonsense mutations.

SUGGESTED ANSWERS TO STRUCTURE YOUR KNOWLEDGE

1. Messenger RNA (mRNA) carries the code that specifies an amino acid sequence from DNA to ribosomes, where the RNA's sequence of nucleotides is translated into a polypeptide. Transfer RNA (tRNA) carries a specific amino acid to its position in a polypeptide based on matching its anticodon to an mRNA codon. Ribosomal RNA (rRNA) makes up about two-thirds of a ribosome and has specific binding and catalytic functions. Small RNAs that are part of spliceosomes play a catalytic role in splicing pre-mRNA.

The versatility of RNA molecules stems from their ability to base-pair with other RNA or DNA molecules, to form specific three-dimensional shapes by base-pairing within itself, and to act as a catalyst.

2.

	Transcription	Translation
Template	DNA	RNA
Location	nucleus (cytoplasm in prokaryotes)	cytoplasm; ribosomes can be free or attached to ER
Molecules involved	RNA nucleotides, DNA template strand, RNA polymerase, transcription factors	amino acids; tRNA; mRNA; ribosomes; ATP; GTP; enzymes; initiation, elongation, and release factors
Enzymes involved	RNA polymerases, spliceosomes (ribozymes)	aminoacyl-tRNA synthetase, ribosomal enzymes (ribozymes)
Control—start and stop	transcription factors locate promoter region with TATA box and start point; polyadenylation signal sequence	initiation factors, initiation sequence (AUG); stop codons, release factor
Product	primary transcript (pre-mRNA)	polypeptide
Product processing	RNA processing: 5' cap and poly-A tail, splicing of pre-mRNA—introns removed by small RNAs in spliceosomes	spontaneous folding, disulfide bridges, signal peptide removed, cleaving, quaternary structure, modification with sugars, etc.

3. The genetic code consists of the RNA triplets that code for amino acids. The order of nucleotides in these codons is specified by the sequence of nucleotides in DNA, which is transcribed into the codons found on mRNA and translated into their corresponding sequence of amino acids. There are 64 possible mRNA codons created from the four nucleotides used in the triplet code (4^3).

Redundancy of the code refers to the fact that several triplets may code for the same amino acid. Often these triplets differ only in the third nucleotide. The wobble phenomenon explains the fact that there are only about 45 different tRNA molecules that pair with the 61 possible codons (three codons are stop codons). The third nucleotide of many tRNAs can pair with more than one type of nucleotide. Because of the redundancy of the genetic code, these wobble tRNAs still place the correct amino acid in position.

The genetic code is nearly universal; each codon codes for exactly the same amino acid in almost all organisms. (Some exceptions have been found.) This universality points to an early evolution of the code in the history of life and the evolutionary relationship of all life on Earth.

4.

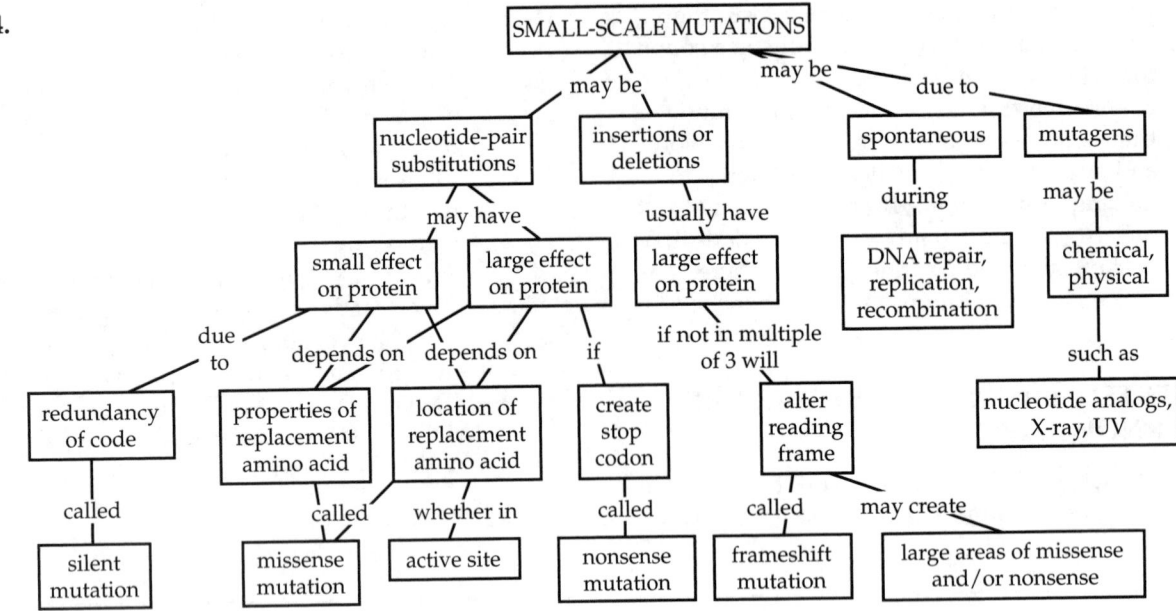

ANSWERS TO TEST YOUR KNOWLEDGE

Multiple Choice:

1. b	**4.** a	**7.** b	**10.** a	**13.** c	**16.** d	**19.** a	**22.** d	**25.** e	**27.** d
2. a	**5.** c	**8.** c	**11.** d	**14.** b	**17.** d	**20.** a	**23.** d	**26.** a	**28.** b
3. a	**6.** b	**9.** c	**12.** c	**15.** c	**18.** b	**21.** d	**24.** c		

Chapter 18: Regulation of Gene Expression

INTERACTIVE QUESTIONS

18.1 **a.** regulatory gene (*lacI*)
   b. promoter
   c. operator
   d. genes coding for enzymes
   e. operon
   f. RNA polymerase
   g. active repressor
   h. inducer (allolactose)
   i. mRNA for enzymes for lactose utilization

18.2 **a.** anabolic; corepressor; on; inactive
   b. catabolic; inducers; off; active

18.3 **a.** Barr body—one X chromosome is compacted in each cell
   b. Removing acetyl groups would decrease transcription because it would make genes in the nucleosome less accessible.

18.4 **a.** distal control elements in enhancer
   b. activators

   c. DNA-bending protein
   d. promoter
   e. mediator proteins
   f. general transcription factors
   g. TATA box
   h. RNA polymerase

18.5 **a.** Many genes code for several different proteins, depending on which exons are combined. Thus, the expression of a particular gene may depend on the regulatory proteins involved in alternate RNA splicing that are present within a cell.
   b. Regulatory proteins may bind to sequences in the 5′ or 3′ UTR and block attachment of ribosomes, thereby decreasing gene expression. Sequences in the 3′ UTR may affect the length of time an mRNA remains intact, thereby either increasing or decreasing gene expression.

18.6　**a.**　A single-stranded miRNA forms a complex with proteins. The complex binds to mRNA with complementary base sequences, and the mRNA is degraded (if the miRNA and mRNA bases are complementary all along their length) or translation is blocked (if the match between bases is less complete).
b.　Evidence indicates that they can do both. Using the same mechanism as miRNAs, siRNAs affect translation by degrading mRNA or blocking translation. Experiments have shown that siRNAs may silence transcription by changing chromatin structure.

18.7　**a.**　A cell is said to be determined when its developmental fate is set. Its series of gene activations and inactivations has set it on the path to express the genes for tissue-specific proteins. When it produces these proteins and develops its characteristic structure, the cell has become differentiated.
b.　The action of MyoD must depend on a combination of regulatory proteins, some of which may be lacking in the cells that it is not able to transform into muscle cells.

18.8　*Bicoid* mRNA was shown to be localized at one end of the unfertilized egg; later in development, Bicoid protein occurred in a gradient that was most concentrated in the anterior cells of the embryo. Also, injection of *bicoid* mRNA into various regions of early embryos caused anterior structures to form at those sites.

18.9　**a.**　Mutations may result in (1) more copies of the gene being present than normal (amplification); (2) translocation or transposition, which may bring the gene under the control of a more active promoter or control element; or (3) a point mutation in either a control element that increases gene expression or in the gene that creates a more active or resilient protein.
b.　Tumor-suppressor proteins may function in repair of damaged DNA, control of cell adhesion, or in signaling pathways that inhibit the cell cycle.

SUGGESTED ANSWERS TO STRUCTURE YOUR KNOWLEDGE

1.　**a.**　operons
　　b.　promoter
　　c.　operator
　　d.　negative control
　　e.　repressor
　　f.　inactive
　　g.　active
　　h.　corepressor
　　i.　inducer
　　j.　anabolic
　　k.　activate
　　l.　inactivate
　　m.　catabolic
　　n.　cAMP
　　o.　lack of glucose

2.　**a.**　DNA packing into nucleosomes; histone tail acetylation increases transcription, whereas removal of acetyl groups or methylation of tails decreases transcription; methylation of DNA may be involved in the long-term inactivation of genes; ncRNAs may promote heterochromatin formation.
b.　Specific transcription factors (activators) bind with control elements in enhancers, then interact with mediator proteins and the promoter region to form a transcription initiation complex; repressors can inhibit transcription.
c.　alternative splicing of primary RNA transcript, 5′ cap and poly-A tail added
d.　Nucleotide sequences in the 3′ UTR affect the lifespan of mRNA, and miRNAs and siRNAs target mRNA for degradation.
e.　Repressor proteins and miRNA or siRNA may prevent translation; activation of initiation factors begins translation of mRNA stockpiled in the egg.
f.　protein processing by cleavage or modification; transport to target location; selective degradation by proteosomes of proteins marked with ubiquitin

3.　Most cytoplasmic determinants are mRNA for transcription factors that are divided by the first few mitotic divisions. They are present in the cells, and their translated product can enter the nucleus and regulate transcription. Inducers must communicate between cells. They are often proteins that bind to cell-surface receptors and initiate a signal transduction pathway involving a cascade of enzyme activations, usually leading to the activation of transcription factors within the target cell.

ANSWERS TO TEST YOUR KNOWLEDGE

Multiple Choice:

1. b	**5.** d	**9.** b	**13.** d	**17.** c	**21.** a
2. c	**6.** b	**10.** a	**14.** c	**18.** d	**22.** d
3. d	**7.** c	**11.** d	**15.** a	**19.** d	**23.** a
4. a	**8.** a	**12.** b	**16.** a	**20.** c	**24.** b

Chapter 19: Viruses

INTERACTIVE QUESTIONS

19.1 1. Phage attaches to host cell and injects its DNA.
2. Phage DNA forms a circle. Certain factors determine which cycle is entered.
3. New phage DNA and proteins are synthesized and self-assemble into phages.
4. Bacterium lyses, releasing phages.
5. Phage DNA integrates into bacterial chromosome, becoming a prophage.
6. Bacterium reproduces, passing prophage to daughter cells.
7. Large population of infected bacteria forms.
8. Occasionally, prophage exits bacterial chromosome and begins lytic cycle.

a. phage DNA
b. bacterial chromosome
c. new phages
d. prophage
e. replicated bacterial chromosome with prophage

19.2 RNA → DNA → RNA; viral reverse transcriptase, host RNA polymerase

19.3 Viral particles spread easily through plasmodesmata, the cytoplasmic connections between plant cells. As there are no cures for plant viral diseases, reducing the spread of infection and breeding resistant varieties are the best approaches.

SUGGESTED ANSWERS TO STRUCTURE YOUR KNOWLEDGE

1.

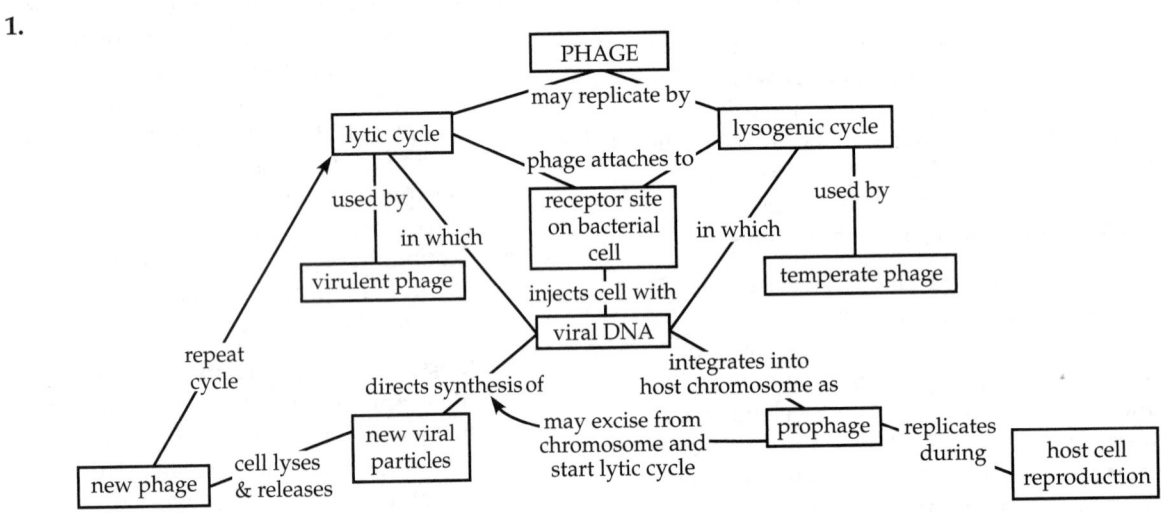

2. a. DNA
    b. RNA
    c. protein capsid
    d. host cell
    e. bacterium
    f. lytic or lysogenic cycle
    g. animal
    h. membranous envelope
    i. reverse transcriptase
    j. plant
    k. plasmids or transposons

ANSWERS TO TEST YOUR KNOWLEDGE

Multiple Choice:

1. b	**4.** d	**7.** c	**10.** d
2. c	**5.** a	**8.** a	**11.** b
3. c	**6.** d	**9.** b	**12.** a

Chapter 20: DNA Tools and Biotechnology

INTERACTIVE QUESTIONS

20.1 Third-generation sequencing will sequence a long, single-stranded DNA molecule on its own, without the need to synthesize a complementary strand or stitch together shorter DNA fragments to obtain a long sequence.

20.2
a. bacterial chromosome
b. plasmid
c. gene of interest
d. chromosome source of "foreign" DNA
e. recombinant plasmid
f. recombinant bacterium
g. clone of recombinant bacteria

1. Gene of interest is inserted into plasmid.
2. Recombinant plasmid is put into bacterial cell.
3. Recombinant bacteria are grown in culture to form a clone of cells containing cloned gene.

Copies of the cloned gene can be used in basic research or genetic engineering. The protein product of the gene may be harvested from bacteria and used in various applications.

20.3 The third sequence, because it has the same sequence running in opposite directions. The enzyme would probably cut between G and A, producing AATT and TTAA sticky ends.

20.4
a. The PCR primers may be synthesized to include a restriction site that will mark each end of the DNA fragment and match a site in the cloning plasmid, thus facilitating the production of recombinant plasmids.
b. A gene for antibiotic resistance will allow cells that have taken up a plasmid to be identified.
c. There is a limit to the number of accurate copies that can be made due to the accumulation of relatively rare copying errors. Large quantities of a gene are better prepared by DNA cloning in cells.

20.5
a. If similar sequences occur in known genes in other species, the function of a gene may be inferred.
b. RT-PCR: The mRNA from different tissue samples could be isolated. Reverse transcriptase would be used to make cDNA of all the mRNA, and PCR using specific primers could amplify only the gene of interest. Running the samples on a gel would show bands only in the tissues that were

expressing the gene. The use of fluorescent dyes avoids the need for electrophoresis and can provide quantitative data.
c. DNA microarray assay: As in RT-PCR, mRNA is isolated from different tissues, and cDNA is made and labeled with fluorescent dye. The cDNA is applied to a microarray (single-stranded DNA fragments of the genes of an organism arranged in a grid). The different cDNA will hybridize with the genes that were expressed in the tissue. The intensity with which the hybridized spots fluoresce indicates the relative amount of mRNA that was in the tissue. RNA-seq can also be used to sequence and then compare all the cDNA samples made from the isolated mRNA in different tissues.
d. By analyzing the phenotypes of the worms that developed from these embryos, they could study the effects of each gene on the process of development.

20.6 DNA methylation and histone acetylation help to regulate gene expression. An adult cell must have these epigenetic changes in its chromatin reprogrammed in order to support normal gene expression during development. The DNA of many cloned embryos has been found to have more methyl groups than that of normal embryos of an equivalent stage.

20.7 The major difficulty is to assure that proper control mechanisms are present so that the gene is expressed at the proper time, in the proper place, and to the proper degree. Insertion of the therapeutic gene must not harm other cell functions. Ethical considerations include whether genetic engineering should be done on any human cells or, in particular, on germ cells, which would influence the genetic makeup of future generations.

20.8 For each blood sample, PCR amplifies specific STRs, using fluorescently labeled primers that flank each STR region. The resulting fragments are separated by gel electrophoresis. Their location on the gel indicates their length and thus the number of repeats. Suspect 2 should be charged with the crime, because the length of his markers matches those found at the scene. Suspect 1's genetic profile does not match.

SUGGESTED ANSWERS TO STRUCTURE YOUR KNOWLEDGE

1. Agricultural applications: (1) improvement of the genomes of agricultural plants and animals to improve quality and yield, (2) "pharm" animals to produce pharmaceutical proteins, (3) development of plant varieties that have genes for resistance to diseases, herbicides, and insects

    Medical applications: (1) diagnosis of genetic and infectious diseases, (2) treatment of genetic disorders or other diseases, (3) production of insulin, human growth factor, TPA, and other useful products

2. **a.** Single-stranded DNA fragments are sequenced using either dideoxy chain termination, sequencing by synthesis, or "third-generation" sequencing. The faster sequencing by synthesis uses PCR to amplify fragments, multiwell plates that allow synthesis of millions of fragments in parallel, and high-throughput computer analysis. Use: determine nucleotide sequence of DNA fragments, stitch together fragments into whole chromosome or genome sequences

    b. Bacterial enzymes that cut DNA at restriction sites, creating "sticky ends" that can base-pair with other fragments. Uses: make recombinant DNA, produce DNA fragments used for many other techniques

    c. Mixture of molecules applied to gel in electric field; molecules separate, moving at different rates due to charge and size. Uses: separate DNA fragments into pattern of distinct bands

    d. DNA is mixed with heat-stable DNA polymerase, nucleotides, and primers having complementary sequences for targeted DNA section and repeatedly heated to separate, cooled to pair with primers and replicate. Use: rapidly produce multiple copies of a gene or section of DNA *in vitro*

    e. Fluorescently labeled single-stranded DNA or mRNA used to base-pair with complementary sequence of DNA or RNA. Uses: identify bands on gels; *in situ* hybridization;

    DNA microarray assay; diagnose infectious diseases

    f. mRNA is isolated; cDNA is synthesized by reverse transcriptase and DNA polymerase; cDNA is amplified using primers specific for the gene; gel electrophoresis shows band in samples that had mRNA from that gene. Use: determine whether a particular gene is expressed in a sample

    g. RNA samples are used to generate cDNAs, which are then sequenced. Uses: provides quantitative data regarding gene expression, including relative levels of alternatively spliced mRNAs

    h. Changes introduced in sequence of a cloned gene; mutated gene is inserted into a cell; phenotype of mutant is studied. Use: technique for determining function of a gene

    i. Double-stranded RNAs that match a gene sequence are inserted into cell; they trigger breakdown or block translation of that gene's mRNA. Use: silence the expression of genes to study their functions and the interactions between genes

3. The complementary base pairing of single strands of DNA enables plasmids and foreign DNA cut by the same endonuclease to match up and be joined by ligase; the DNA primers used in PCR to bind with flanking ends of a region of DNA that is to be copied; and nucleic acid probes to bind to sequences in DNA microarrays or *in situ* hybridization. Complementary base pairing also allows the sequencing of DNA as monitors determine the complementary nucleotides that sequentially bind in the replication of a new DNA strand.

ANSWERS TO TEST YOUR KNOWLEDGE

Multiple Choice:

1. c	**5.** d	**9.** a	**13.** c	**17.** a
2. d	**6.** b	**10.** c	**14.** c	**18.** c
3. d	**7.** d	**11.** b	**15.** b	**19.** c
4. b	**8.** a	**12.** d	**16.** a	**20.** c

Chapter 21: Genomes and Their Evolution

INTERACTIVE QUESTIONS

21.1 Microbial species—metagenomics enables the sequencing of a mixture of DNA (a *metagenome*) from an environmental sample without the need to grow each species separately in the lab.

21.2 Comparing nucleotide sequences may reveal matches with other genes of known function or reveal common amino acid domains for which a function is known and can identify similarities with closely or distantly related species that help trace a species' evolutionary history.

21.3 **a.** the archaean *Archaeoglobus fulgidus* (1,130 genes/Mb); humans (7/Mb)
b. corn, *Zea mays* (32,000 genes); the bacterium *Haemophilus influenzae* (1,700)
c. *Paris japonica* (a plant) (149,000 Mb); *H. influenzae* (1.8 Mb)
d. less than 21,000; alternative splicing of exons can allow each gene to code for more than one polypeptide; post-translational processing can also alter polypeptides

21.4 They are first transcribed into RNA; thus, the original retrotransposon remains in place. Reverse transcriptase (coded for by a retrotransposon gene) converts the RNA to DNA, which is inserted into another site in the genome.

21.5
1. D; 1.5
2. I; 20
3. H; 5
4. B; 44
5. E; 10
6. G; 17
7. C; 15
8. F; 5-6
9. A; 3

21.6 The lysozyme gene, which codes for a bacterial infection-fighting enzyme, was present in the last common ancestor of birds and mammals. After their lineages split, the gene underwent a duplication event in the mammalian lineage, and a copy of the lysozyme gene evolved into a gene coding for a protein involved in milk production.

21.7 **a.** Exon shuffling can happen through an error in meiotic recombination. This could occur by an uneven crossing over (as when crossing over occurs between two homologous but misaligned transposable elements), or by the inclusion or tagging along of an exon with a transposable element, which moves a copy of the exon to a new location.
b. Exons often code for domains of a protein. Providing an extra domain or a new domain to a protein may enhance or change its function.

21.8 The homeobox codes for a DNA-binding homeodomain, whereas other domains specific to each regulatory gene interact with transcription factors to recognize particular enhancers or promoters and thus control different batteries of developmental genes.

SUGGESTED ANSWERS TO STRUCTURE YOUR KNOWLEDGE

1. Many non-protein-coding sequences are highly conserved across species, indicating that they perform some important, but as yet unidentified, function. Much of this DNA is transcribed into RNA molecules, some of whose functions are being discovered.
   a. Transposable elements include transposons, which are DNA segments, and retrotransposons, which are produced from RNA transcribed into DNA. These transposable elements are very common and move DNA to new locations in the genome.
   b. *Alu* elements are about 300 nucleotides long and make up about 10% of the human genome. Many are transcribed into RNA; some of this RNA may be involved in regulation of gene expression.
   c. *L1* sequences are long retrotransposons that rarely move about. When active during development, they may help regulate gene expression.
   d. Simple sequence DNAs are highly repetitive tandem sequences found at centromeres and the telomeres of a chromosome. They appear to have structural functions in organizing chromatin. Sequences with fewer repetitions are called STRs (short tandem repeats) and are used in genetic profiling.
   e. Pseudogenes are remnants of genes that are no longer functional because mutations have altered their regulatory sequences.

2. Chromosome duplication provides additional copies of genes that may undergo mutation

and produce new proteins. Chromosomal rearrangements such as duplications, inversions, and translocations may create reproductive barriers between populations that lead to the formation of new species. Errors during meiosis can lead to duplications of genes or exons, or shuffling of exons, providing genetic material that may take on related or novel functions. Transposable elements can facilitate recombination between different chromosomes, can disrupt genes or control elements, and can carry genes or exons to new locations.

ANSWERS TO TEST YOUR KNOWLEDGE

Multiple Choice:

1. d	**4.** a	**7.** a	**10.** c	**13.** b
2. c	**5.** d	**8.** d	**11.** d	**14.** d
3. b	**6.** b	**9.** a	**12.** b	**15.** c

Chapter 22: Descent with Modification: A Darwinian View of Life

INTERACTIVE QUESTIONS

22.1 **a.** 1. E f
        2. A b
        3. B e
        4. C d, g
        5. F a
        6. D c
        7. G g

   b. a, f, d, e, b, g, c

22.2 Observation 1: Members of a population vary in their inherited traits.
Observation 2: All species can produce more offspring than the environment can support, and many offspring do not survive.

22.3 **a.** When a new antibiotic is developed, it is effective at killing most pathogens. A few, however, may have an inherent resistance, such as an enzyme like penicillinase that destroys penicillin. As the continued use of the new antibiotic selects against bacteria without an antibiotic-resistance gene, the resistant bacteria can multiply rapidly and the "beneficial" gene becomes more common.
   b. Resistance to multiple antibiotics can evolve if bacteria that are already resistant to one antibiotic are exposed to a different one. The new antibiotic, again, will probably kill most bacteria but "select" any that happen to have a variation that provides resistance to it. The exchange of genes between different strains and even different species of bacteria may also contribute to the evolution of multi-drug-resistant strains.

22.4 **a.** biogeography
   b. fossil record
   c. homologies
   d. island species and mainland species or neighboring island species
   e. ancestral and transitional forms
   f. homologous structures
   g. functions and form
   h. molecular comparisons
   i. DNA sequences and proteins
   j. descent from a common ancestor

SUGGESTED ANSWERS TO STRUCTURE YOUR KNOWLEDGE

1. The two main components of Darwin's evolutionary theory are that all of life has descended from a common ancestor and that this evolution has been the result of natural selection. The mechanism of natural selection is based on several key observations and inferences. Members of a population often vary in their inherited traits. The overproduction of offspring in conjunction with limited resources leads to the unequal reproductive success of those organisms best suited to the local environment. The increased survival and reproduction of the best-adapted individuals in a population leads to the gradual accumulation of favorable characteristics in a population.

2. Individuals interact with their environment and, as a result of their genetic endowment, may leave more or fewer offspring than other members of the population. The genotype of an individual does not, however, change over time. The differential reproduction of individuals in a population means that the alleles associated with higher reproductive success will become more numerous in the population with successive generations. This change in allele frequencies is evolution.

ANSWERS TO TEST YOUR KNOWLEDGE

Multiple Choice:

1. c	**4.** d	**7.** c	**10.** a	**13.** b	**16.** c
2. a	**5.** d	**8.** c	**11.** d	**14.** b	**17.** b
3. a	**6.** d	**9.** b	**12.** d	**15.** d	

Chapter 23: The Evolution of Populations

INTERACTIVE QUESTIONS

23.1 **a.** mutation
b. sexual reproduction
c. Prokaryotes and viruses have very short generation times, and a new beneficial mutation can increase in frequency rapidly in an asexually reproducing population. Although mutations are the source of new alleles, they are so infrequent that their contribution to genetic variation in a large, diploid population is minimal. However, the production and union of gametes produce zygotes with fresh combinations of alleles in each generation.

23.2 **a.** The frequencies of genotypes are 0.49 *BB*, 0.42 *Bb*, and 0.09 *bb* ($^{98}/_{200}$, $^{84}/_{200}$, $^{18}/_{200}$).
b. 0.7 *B*; 0.3 *b*. The 98 *BB* mice contribute 196 *B* alleles to the gene pool, and the 84 *Bb* mice contribute 84 *B* alleles. These 84 *Bb* mice also contribute 84 *b* alleles, and the 18 *bb* mice contribute 36 *b* alleles. Of a total of 400 alleles, 280 are *B* and 120 are *b*. Allele frequencies are 0.7 *B* and 0.3 *b*. Another way to determine allele frequencies is from genotype frequencies: Add the frequency of the homozygous dominant genotype and ½ the frequency of the heterozygote to determine the frequency of *p*. For *q*, add the homozygous recessive frequency and ½ the heterozygote frequency. Or determine *q* from $1 - p$.

23.3 0.7; 0.3; 0.49; 0.42; 0.09

23.4 **a.** 0.36; 0.48; 0.16. Plug *p* (0.6) and *q* (0.4) into the expanded binomial: $p^2 + 2pq + q^2 = 1$.
b. 0.6; 0.4. Add the frequencies of *AA* + ½ *Aa* to get *p*. Add the frequencies of *aa* + ½ *Aa* to get *q*. Alternatively, to determine *q*, take the square root of the homozygous recessive frequency if you are sure the population is in Hardy-Weinberg equilibrium. The frequency of *p* is then $1 - q$.
c. These results would suggest that the population is not in Hardy-Weinberg equilibrium and thus is evolving.

23.5 **a.** natural selection
b. genetic drift
c. gene flow
d. better reproductive success
e. small population
f. founder effect
g. bottleneck effect
h. genetic variation between populations

23.6 **a.** Diploidy—As the frequency of the sickle-cell allele decreases, homozygous recessives are less likely to occur and be selected against, while the allele can remain hidden from selection in heterozygotes.
b. Heterozygote advantage—Heterozygotes are protected from the most severe effects of malaria and have a selective advantage in areas where malaria is a major cause of death.

SUGGESTED ANSWERS TO STRUCTURE YOUR KNOWLEDGE

1. **a.** The Hardy-Weinberg equilibrium states that allele and genotype frequencies within a population will remain constant from one generation to the next as long as only Mendelian segregation and recombination of alleles are involved. This equilibrium requires five conditions: The population is large, mutation is negligible, migration is negligible, mating is random, and no natural selection occurs. The Hardy-Weinberg equilibrium provides a null hypothesis to enable researchers to test for evolution.
b. $p^2 + 2pq + q^2 = 1$. In the Hardy-Weinberg equation, *p* and *q* refer to the frequencies of two alleles in the gene pool. The frequency of homozygous offspring is ($p \times p$) or p^2 and ($q \times q$) or q^2. Heterozygous individuals can be formed in two ways, depending on whether the egg or sperm carries the *p* or *q* allele, and their frequency is equal to $2pq$.

2. Genetic variation is retained within a population by diploidy and balancing selection. Diploidy masks recessive alleles from selection when they occur in the heterozygote. Thus, less adaptive or even harmful alleles are maintained in the gene pool and are available should the environment change. Balancing selection maintains two or more phenotypic forms in a population. In heterozygote advantage, two alleles will be retained in stable frequencies within the gene pool. Frequency-dependent selection, in which the more common phenotype is selected against by predators or other factors, is another type of balancing selection.

ANSWERS TO TEST YOUR KNOWLEDGE

Multiple Choice:

1. d	**5.** b	**9.** d	**13.** a	**17.** c	**20.** c
2. a	**6.** a	**10.** d	**14.** c	**18.** d	**21.** c
3. d	**7.** c	**11.** b	**15.** a	**19.** d	
4. b	**8.** c	**12.** c	**16.** a		

Chapter 24: The Origin of Species

INTERACTIVE QUESTIONS

24.1 **a.** post-; reduced hybrid fertility
b. pre-; gametic isolation
c. pre-; mechanical isolation
d. pre-; temporal isolation
e. post-; hybrid breakdown
f. pre-; behavioral isolation
g. post-; reduced hybrid viability
h. pre-; habitat isolation

24.2 **a.** reproductive isolation
b. morphological
c. ecological

24.3 **a.** 20 chromosomes ($4n = 20$)
b. 24 chromosomes ($2n = 24$)

24.4 **a.** In allopatric speciation, a new species forms while geographically isolated from the parent population. In sympatric speciation, some reproductive barrier isolates the gene pool of a subgroup of a population within the same geographic range as that of the parent population.
b. Reproductive barriers may evolve as a by-product of the genetic change associated with the isolated population's adaptation to a new environment, genetic drift, or sexual selection. Sympatric speciation by polyploidy is common in plants. Sexual selection or a change in resource use may reproductively isolate a subset of an animal population.

24.5 **a.** Reinforcement: when hybrid offspring are less viable or reproduce less successfully, the reproductive barriers between the two species may strengthen. Traits that prevent or discourage crossbreeding will increase in frequency, whereas traits that allow for mating with the other species will decrease in frequency as the individuals with those traits produce fewer viable offspring.
b. Fusion: If hybrids are successful in mating with each other and with the parent species, reproductive barriers may weaken. Increased gene flow between the two species may eventually lead to their fusion into a single species.
c. Stability: Hybrids continue to be produced between the two species in the area of their overlap, but the gene pools of both parent species remain distinct. Several cases of such stability among hybridizing species have been documented.

24.6 The total time between speciation events has been estimated to range from 4,000 to 40 million years (with an average of 6.5 million years). Thus, even if the process of speciation is relatively rapid once it begins, it may be millions of years before that new species gives rise to another new species. The biodiversity of Earth may take a very long time to recover from the current increased extinction rate.

SUGGESTED ANSWERS TO STRUCTURE YOUR KNOWLEDGE

1. Speciation, which leads to an increase in biological diversity, is the process through which one species splits into two or more species. Microevolution involves changes in the gene pool of a population as a result of either chance events or natural selection. If the makeup of an isolated gene pool changes enough, microevolution may lead to speciation.

2. Reproductive barriers do not develop in order to create new species; they develop as a side consequence of the genetic changes that occur as a species adapts to a new environment or undergoes its own genetic drift. Gene flow between two species tends to break down reproductive barriers between them. Because gene flow is much less likely between allopatric species, one would predict that reproductive barriers are more likely to evolve. In the case of sympatric speciation by polyploidy, however, reproductive barriers arise in one generation.

3. The term "punctuated equilibria" describes the common pattern seen in the fossil record in which species appear relatively suddenly and then seem to change little for the rest of their existence. Thus, evolution seems often to occur in spurts of relatively rapid change interspersed with long periods of stasis.

ANSWERS TO TEST YOUR KNOWLEDGE

Multiple Choice:

1. b	**5.** b	**9.** d	**13.** d
2. d	**6.** a	**10.** b	
3. c	**7.** b	**11.** b	
4. a	**8.** a	**12.** c	

Chapter 25: The History of Life on Earth

INTERACTIVE QUESTIONS

25.1 Reproduction that faithfully passes genetic information on to offspring is a characteristic of life. But a cell needs metabolic machinery to provide the energy and building blocks necessary to replicate its genetic information and reproduce. Thus both self-replicating molecules that encode genetic information for metabolism and the resulting metabolic processes are required for life to have evolved.

25.2 17,190 years. Carbon-14 has a half-life of 5,730 years. In 5,730 years the ratio of C-14 to C-12 would be reduced by ½; in 11,460 years it would be reduced by ¼; and in 17,190 years it would be reduced by ⅛ (three half-lives; $3 \times 5,730$).

25.3 The inner membranes of both mitochondria and chloroplasts have enzymes and transport systems homologous to those in the plasma membranes of modern prokaryotes. They replicate by a splitting process similar to that of some prokaryotes. They contain a single, circular DNA molecule that is not associated with histones. The ribosomes of mitochondria and plastids are more similar to prokaryotic ribosomes than to the cytoplasmic ribosomes of eukaryotic cells.

25.4 The change in nutrient procurement from grazing, filter-feeding, and scavenging to predation: predators with claws and other prey-capturing adaptations as well as prey with heavy body armor appeared in the fossil record during the Cambrian explosion.

25.5 Marsupials migrated to Australia through South America and Antarctica while the continents were still joined. When Pangaea broke up, marsupials diversified on isolated Australia.

25.6 a. Early mammals were restricted in size and diversity because they were probably outcompeted or eaten by the larger and more diverse dinosaurs. After dinosaurs (except for birds) became extinct, mammals diversified and filled the vacated niches.
b. The rise in large predators in the Cambrian explosion, the diversification of terrestrial plants and tetrapods, the adaptive radiation of insects that ate or pollinated plants

c. The Hawaiian Archipelago is a series of relatively young, isolated, and physically diverse islands whose thousands of endemic species are examples of adaptive radiation following multiple colonization and speciation events.

25.7 a. This study provided experimental evidence that a specific change in the sequence of a developmental gene could produce a major evolutionary change—the suppression of legs associated with the origin of a six-legged insect from a crustacean-like ancestor with many legs.
b. The coding sequence of the developmental gene was identical in the two populations; the gene was not expressed in the ventral spine region of developing lake sticklebacks. Thus, the loss of ventral spines results from a change in gene regulation, not from a change in the gene itself.

25.8 a. The successful species that last the longest before extinction and generate the most new species will determine the direction of an evolutionary trend, just as the individuals that produce the most offspring determine the direction of adaptation in a population.
b. Both microevolution and macroevolution are primarily driven by natural selection resulting from interactions between organisms and their environments. Evolutionary trends are ultimately dictated by environmental conditions; if conditions change, an evolutionary trend may end or change direction.

SUGGESTED ANSWERS TO STRUCTURE YOUR KNOWLEDGE

1. The abiotic synthesis of small organic molecules; the joining of small molecules into macromolecules; the packaging of these molecules into membrane-bound protocells; the origin of self-replicating molecules, probably RNA, that made inheritance and natural selection possible.

2. **a.** Hadean eon
   b. Archaean eon
   c. Proterozoic eon
   d. prokaryotes
   e. atmospheric oxygen
   f. single-celled eukaryotes

g. multicellular eukaryotes
h. animals
i. colonization of land
j. Phanerozoic eon
k. Paleozoic
l. Mesozoic
m. Cenozoic
n. humans

3. *Plate tectonics:* When continents form one supercontinent, many habitats are destroyed or changed. As continents change latitude, their climates either warm or cool. The separation of continents creates major geographic separation events, making possible allopatric speciations. Indeed, continental drift helps explain much of biogeography.
Mass extinctions result in the loss of many evolutionary lineages. The resulting empty ecological roles, however, may be exploited by species that survived extinction.
Adaptive radiations are multiple speciation events that fill newly formed or newly emptied niches. Thus, adaptive radiations may follow mass extinctions, the colonization of new regions, or the evolution of novel adaptations. The adaptive radiation of one group may provide new resources (such as food or habitat) that support the radiation of other groups.

4. New morphological forms may evolve as a result of mutations in developmental genes or changes in the regulation of such genes. Such changes may affect the rate or timing of development or the spatial arrangement of body parts (e.g., *Hox* genes).

ANSWERS TO TEST YOUR KNOWLEDGE

Multiple Choice:

1.	b	4.	d	7.	a	10.	c	13.	a	16.	b
2.	a	5.	b	8.	c	11.	d	14.	b	17.	a
3.	b	6.	d	9.	c	12.	a	15.	d	18.	c

Chapter 26: Phylogeny and the Tree of Life

INTERACTIVE QUESTIONS

26.1. a. 1
       b. A
       c. B and C; D and E

26.2. Adaptations to different environments may lead to large morphologic differences within related groups, and convergent evolution can produce similar structures in unrelated organisms in response to similar selective pressures. Examples are the varied Hawaiian silversword plants and the similar Australian and North American burrowing moles.

26.3.

26.4. The four-chambered heart of birds and mammals is analogous, not homologous. It evolved independently in the two groups. An abundance of evidence supports the hypothesis that birds and mammals evolved from different reptilian ancestors.

26.5. a. rRNA genes
       b. mtDNA (mitochondrial DNA)

26.6. Assuming that genes have a fairly constant rate of mutation, in a protein that has a crucial function and whose amino acid sequence is central to that function, mutations that are harmful would be quickly removed from the population. Therefore, fewer mutations would be neutral and remain in the genome. If the exact sequence of amino acids in a protein is less critical to survival, more mutations would be neutral and remain. Thus, in the same amount of time, the sequence of the second gene will change more than the sequence of the more crucial gene.

26.7. a. Eukarya
       b. Archaea
       c. Bacteria
       d. endosymbiosis of proteobacterial ancestor of mitochondria
       e. endosymbiosis of cyanobacterial ancestor of chloroplasts and other plastids

SUGGESTED ANSWERS TO STRUCTURE YOUR KNOWLEDGE

1.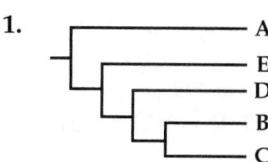

2. In order to reconstruct the evolutionary history of a group of organisms, systematists consider morphological and molecular homologies as well as evidence from the fossil record. Molecular homologies may be identified using computer programs that search for and align common sequences in homologous sections of DNA. Statistical analysis can help sort distant molecular homologies from coincidental homoplasies. Using the techniques of cladistics, the ingroup of organisms to be classified is compared with an outgroup, and the ingroup members are compared with each other. Shared ancestral characters and shared derived characters are used to determine the sequence of evolutionary divergence.

Computer programs can determine the most parsimonious tree (requiring the fewest evolutionary changes). When choosing among several phylogenetic hypotheses, the principle of maximum likelihood states that the most likely tree must take into account rules about how DNA changes over time.

3. A molecular clock is based on the observations that certain regions of DNA evolve (accumulate base changes) at a steady rate. By plotting the number of differences in DNA sequences between groups known to have diverged at specific times, scientists can apply this rate of molecular evolution to infer the timing of other evolutionary divergences.

ANSWERS TO TEST YOUR KNOWLEDGE

Multiple Choice:

1. c	5. d	9. d	13. a	17. d	21. c
2. b	6. a	10. d	14. b	18. a	
3. d	7. c	11. c	15. c	19. b	
4. c	8. a	12. c	16. b	20. d	

Chapter 27: Bacteria and Archaea

INTERACTIVE QUESTIONS

27.1　**a.** spheres (cocci), rods (bacilli), or spirals
　　　b. 0.5–5 μm in diameter
　　　c. cell wall made of peptidoglycan; gram-negative bacteria also have outer lipopolysaccharide membrane; archaea lack peptidoglycan; sticky capsule for adherence; fimbriae and pili for attachment or genetic exchange
　　　d. flagella; about half of prokaryotes show taxis to stimuli
　　　e. infoldings of plasma membrane may function in cellular respiration; thylakoid membranes in cyanobacteria
　　　f. circular DNA molecule with little associated protein found in nucleoid; may have plasmids with other genes
　　　g. binary fission; rapid population growth

27.2　**a.** circular chromosome (with less protein than eukaryotic linear chromosomes have)
　　　b. F plasmid, R plasmid
　　　c. mutation

　　　d. transformation
　　　e. naked DNA
　　　f. transduction
　　　g. bacteriophage (phage)
　　　h. conjugation
　　　i. F$^+$ or HRF cell and F$^-$ cell
　　　j. adaptation to environment/evolution

27.3　photoautotroph—uses light energy and CO_2 to synthesize organic compounds; the cell is a heterocyst, specialized for nitrogen fixation.

27.4　**a.** cyanobacteria
　　　b. Metagenomics is a technique of analyzing genetic material sampled directly from the environment. Before its development, researchers could study only those species that could be cultured in the lab. This technique has led to the discovery of new species and even new branches in prokaryotic phylogeny.

27.5　Chemoautotrophic bacteria form the basis of the food chain in vent communities, which can consist of hundreds of eukaryotic species.

SUGGESTED ANSWERS TO STRUCTURE YOUR KNOWLEDGE

1. Prokaryotes are adapted to live in the diverse habitats on (and in) Earth. They have a wide range of metabolic abilities, including the unique nutritional categories of chemoautotrophy and photoheterotrophy. Their genetic diversity, along with their short generation time, huge populations, and mechanisms of genetic exchange, enable rapid adaptation to changing environments. The formation of biofilms consisting of multiple species and their many symbiotic associations are further examples of the remarkable abilities of prokaryotes.

2. Decomposers—recycle nutrients; bioremediation: sewage treatment, clean up oil spills. Nitrogen fixers—provide nitrogen to the soil by bacteria in nodules on legume roots or by cyanobacteria.

Mutualism—intestinal bacteria, some digest food and produce vitamins.
Biotechnology—production of antibiotics, hormones, and many useful products.

ANSWERS TO TEST YOUR KNOWLEDGE

Fill in the Blanks:

1.	cocci	6.	endospore
2.	nucleoid	7.	biofilms
3.	Gram stain	8.	exotoxins
4.	fimbriae	9.	bioremediation
5.	taxis	10.	anaerobic respiration

Multiple Choice:

1. b	4. c	7. d	10. d	13. b
2. a	5. a	8. c	11. a	
3. c	6. a	9. b	12. d	

Chapter 28: Protists

INTERACTIVE QUESTIONS

28.1 The four membranes indicate secondary endosymbiosis of a green alga (whose chloroplasts originated from the primary endosymbiont, a cyanobacterium). The inner two membranes probably belonged to the green alga's chloroplast. The third membrane could be from the engulfed alga's plasma membrane, and the outer membrane from the heterotrophic eukaryote's food vacuole. The vestigial nucleus indicates that this process occurred relatively recently, and its genetic sequences point to a green algal origin. In more ancient secondary endosymbioses, many components of the engulfed alga have been lost.

28.2 Frequent global changes in the surface proteins of these parasites enable them to evade the host's immune system.

28.3 **a.** Control of malaria is complicated by the development of insecticide resistance in *Anopheles* mosquitoes, which spread the disease; drug resistance in *Plasmodium*; the sequestering of the parasite within human liver and blood cells; and the ability of the parasite to change its surface proteins.
b. The apicoplast appears to have originated by secondary endosymbiosis of a red algal cell, which had originally obtained its plastid from a cyanobacterium. Thus, drugs targeting metabolic pathways of the apicoplast, which is prokaryotic in origin, may not affect human pathways.

28.4 *Paramecium* is in the clade of ciliates in the subgroup Alveolata in the supergroup SAR.
a. cilia
b. oral groove, leading to cell mouth
c. food vacuoles, combine with lysosomes
d. contractile vacuole
e. release of wastes at region that functions as an anal pore
f. macronucleus
g. micronucleus

28.5 **a.** The calcareous tests of forams are long-lasting fossils in marine sediments and sedimentary rocks. Foram fossils are useful in correlating ages of sedimentary rocks in different areas.
b. Evidence indicates that chromatophores were derived from a cyanobacterium different from the one from which all other plastids were derived—a remarkable second case of primary endosymbiosis.

28.6 Size and complexity have increased via the formation of colonies; cell division and differentiation to produce true multicellular forms; and repeated nuclear divisions to produce a multinucleated "supercell."

28.7 a. amoebozoans
      b. opisthokonts
      c. slime molds
      d. tubulinids
      e. entamoebas
      f. nucleariids
      g. choanoflagellates

28.8 Higher sea surface temperatures increase the differences between warm and cold layers of water, reducing the upwelling of cold, nutrient-rich waters from below that contributes to the growth of marine producers. A reduction in this base of marine food webs would affect all consumers in the ecosystem, including important fish species. If diatom populations decrease, less CO_2 would be "pumped" from the atmosphere to the deep-ocean floor.

SUGGESTED ANSWERS TO STRUCTURE YOUR KNOWLEDGE

1. a. cyanobacterium
    b. primary endosymbiosis
    c. red alga
    d. green alga
    e. secondary endosymbiosis
    f. dinoflagellates
    g. stramenopiles
    h. euglenids

ANSWERS TO TEST YOUR KNOWLEDGE

Matching:

1. C	3. B	5. H	7. A
2. F	4. G	6. E	8. D

Multiple Choice:

1. c	4. c	7. a	10. a	13. b
2. c	5. d	8. d	11. b	14. c
3. d	6. b	9. c	12. d	

Chapter 29: Plant Diversity I: How Plants Colonized Land

INTERACTIVE QUESTIONS

29.1 Benefits: bright, unfiltered sunlight; plentiful CO_2 and soil minerals. Challenges: lack of structural support and relative scarcity of water.

29.2 a. gametes
      b. mitosis
      c. zygote
      d. mitosis
      e. sporophyte
      f. spores
      g. meiosis
      h. gametophyte

29.3 Although lycophytes and monilophytes are both seedless plants, they do not form a monophyletic group. Monilophytes (ferns) share a more recent common ancestor with seed plants than with lycophytes. Thus, these paraphyletic groups do not include all of the descendants of their most recent common ancestor (would have to include seed plants).

29.4 a. gametophyte
      b. archegonia
      c. antheridia
      d. swim
      e. sporophyte

      f. meiosis
      g. capsule (sporangium)
      h. gametophyte

29.5 Lignified xylem provides the mechanical support for aerial growth. Vascular tissues convey water and minerals to photosynthesizing leaves and transport sugars throughout the plant. Roots both absorb water and minerals and anchor the plant. Tall plants could outcompete shorter plants for access to sunlight, and their spores could disperse farther.

29.6 a. meiosis
      b. spore
      c. gametophyte
      d. antheridium
      e. sperm
      f. archegonium
      g. egg
      h. fertilization
      i. zygote
      j. gametophyte
      k. young sporophyte
      l. mature sporophyte
      m. sporangia (in sori on undersides of leaves)

      The gametophyte (upper portion of the diagram between spore and fertilization) is haploid; the sporophyte (lower portion) is diploid.

SUGGESTED ANSWERS TO STRUCTURE YOUR KNOWLEDGE

1. The adaptations of plants to a terrestrial habitat include an embryo retained in the female gametophyte that is protected and nourished, sporopollenin-walled resistant spores produced in multicellular sporangia, apical meristems that allow plants to continue to grow and exploit terrestrial resources, a waterproof cuticle, and stomata for gas exchange. The bryophytes have rhizoids that help anchor them. Vascular plants have vascular tissue strengthened with lignin, which allows them to grow tall and enables water and nutrient transport between roots and leaves.

2. a. land plants
    b. nonvascular plants (bryophytes)
    c. vascular plants
    d. seedless vascular plants
    e. seed plants
    f. liverworts
    g. mosses
    h. hornworts
    i. monilophytes (ferns, horsetails, whisk ferns)
    j. gymnosperms
    k. angiosperms

    1. origin of land plants (about 470 mya)
    2. origin of vascular plants (about 425 mya)
    3. origin of extant seed plants (about 305 mya)

ANSWERS TO TEST YOUR KNOWLEDGE

Multiple Choice:

1. d	4. a	7. a	10. a	13. a
2. d	5. b	8. a	11. b	14. c
3. c	6. c	9. d	12. d	15. c

Chapter 30: Plant Diversity II: The Evolution of Seed Plants

INTERACTIVE QUESTIONS

30.1 Resistant, airborne, or animal-carried pollen grains can transport the sperm-producing male gametophyte, even over long distances, to the female gametophyte. A germinating pollen tube delivers sperm directly to an ovule, eliminating the need for a moist environment for sperm to swim to reach eggs.

30.2 a. integument (2n)
      b. megaspore (n)
      c. megasporangium (2n)
      d. seed coat (from integument) (2n)
      e. food supply (from female gametophyte tissue) (n)
      f. embryo (new sporophyte) (2n)

30.3 a. The smaller pollen cones have many scalelike structures holding microsporangia. Microsporocytes undergo meiosis to produce microspores that develop into pollen grains, which enclose the male gametophyte. Sperm cells are eventually discharged from a pollen tube.
      b. Ovulate cones have scales that hold two ovules, each containing a megasporangium. A megasporocyte undergoes meiosis to form four megaspores. The surviving megaspore develops into a multicellular female gametophyte containing a few archegonia, each with an egg.

30.4 a. sepal
      b. petal
      c. anther (top of stamen)
      d. stigma (top of carpel)
      e. ovary
      f. ovule with megasporangium
      g. microsporangium
      h. microsporocytes
      i. meiosis
      j. microspore (n)
      k. pollen grain (containing *male gametophyte* with generative cell and tube cell)
      l. two sperm
      m. pollen tube
      n. style of the carpel
      o. surviving megaspore (n)
      p. embryo sac (*female gametophyte*)
      q. central cell with two nuclei
      r. egg
      s. fertilization
      t. endosperm nucleus (3n)
      u. zygote

v. seed coat
w. endosperm
x. embryo
y. germinating seed

30.5

Characteristic	Monocot	Eudicot
Cotyledons	one	two
Leaf venation	parallel	netlike
Stems	scattered vascular tissue	vascular tissue in ring
Roots	fibrous	taproot
Pollen	one opening	three openings
Floral organs	multiples of three	multiples of four or five

SUGGESTED ANSWERS TO STRUCTURE YOUR KNOWLEDGE

1. The key evolutionary adaptations to a terrestrial habitat are seeds and pollen grains. Seed plants have reduced male and female gametophytes, which develop from spores retained within the micro- and megasporangia of the sporophyte. The female gametophyte is protected and nourished in an ovule retained on the sporophyte plant. Nonswimming sperm are protected and dispersed in pollen. Seeds can remain dormant and provide for the protection, dispersal, and nourishment of the embryo.

2. Adaptations of angiosperms include flowers as enclosed reproductive structures, animal pollinators, and fruit for protection and dispersal of seeds.

3. **a.** *Amborella*
   b. water lilies
   c. star anise and relatives
   d. magnoliids
   e. monocots
   f. eudicots

ANSWERS TO TEST YOUR KNOWLEDGE

True or False:

1. False—add *or algae.*
2. True
3. False—change *gymnosperms* to *angiosperms.*
4. True
5. True
6. False—change *fruit* to *seed.*
7. True
8. False—change *angiosperms* to *gymnosperms.*

Multiple Choice:

1. c		**4.** c		**7.** a		**10.** b
2. b		**5.** c		**8.** b		**11.** a
3. c		**6.** d		**9.** c		**12.** d

Chapter 31: Fungi

INTERACTIVE QUESTIONS

31.1 The fungus provides phosphate ions and other minerals to the plant; the plant supplies the fungus with organic nutrients.

31.2 **a.** network of hyphae, typical body form
b. cross-walls between cells, with pores
c. multinucleated hyphae, lack septa
d. cytoplasmic fusion of two parent hyphae
e. nuclei from two different parents exist together in hyphae
f. hyphae with two different nuclei paired up in cells
g. fusion of nuclei from two parents to form diploid state

31.3 **a.** nucleariids
b. fungi
c. chytrids
d. other fungi (zygomycetes, glomeromycetes, ascomycetes, basidiomycetes)

31.4 1. ascomycete
2. zygomycete
3. basidiomycete

a. conidia (asexual spores)
b. ascocarp
c. ascospores (sexual spores)
d. asci
e. sporangia (forms asexual spores)
f. zygosporangium (forms sexual spores)
g. basidiocarp
h. basidium
i. basidiospores (sexual spores)

SUGGESTED ANSWERS TO STRUCTURE YOUR KNOWLEDGE

1. **a.** Chytrids include aquatic and soil decomposers or parasites; their key feature is flagellated zoospores.
    b. Zygomycetes include molds (such as the black bread mold) and parasites. Reproduction involves resistant zygosporangia in the sexual stage and asexual spores in sporangium on aerial hyphae.
    c. Almost all glomeromycetes form mycorrhizae. An estimated 90% of all plants have these arbuscular mycorrhizae.
    d. Ascomycetes (sac fungi) include yeasts, molds, morels, and mycorrhizal species. Reproduction involves dikaryotic hyphae in an ascocarp, karyogamy in asci, and ascospores, as well as asexual spores called conidia at the tips of hyphae.
    e. Basidiomycetes (club fungi) include mushrooms, shelf fungi, rusts, and smuts. They have long-lasting dikaryotic hyphae. Reproduction involves a fruiting body called a basidiocarp and basidiospores produced in basidia.

2. The role of fungi as decomposers of organic material is central to the recycling of chemicals between living organisms and their physical environment. Parasitic fungi have economic and health effects on humans. Plant pathogens cause extensive loss of food crops and trees. Mycorrhizae are found on the roots of almost all vascular plants. These mutualistic associations increase the efficiency with which plants absorb nutrients such as phosphorus. Endophytes living inside leaves benefit plants by deterring herbivores and protecting against heat and pathogens. Fungi are also partners in lichens, which are able to colonize bare rock and contribute to soil formation.

ANSWERS TO TEST YOUR KNOWLEDGE

Fill in the Blanks:

1. nucleariids
2. opisthokonts
3. chitin
4. enzymes
5. endophyte
6. mycorrhizae
7. chytrids
8. form arbuscular mycorrhizae

Multiple Choice:

1. a	4. d	7. d	10. c	13. a
2. c	5. c	8. a	11. d	14. c
3. b	6. b	9. b	12. e	15. b

Chapter 32: An Overview of Animal Diversity

INTERACTIVE QUESTIONS

32.1 a. A larva is a free-living, sexually immature stage in some animal life cycles that may differ from the adult in habitat, nutrition, and body form.
b. A juvenile resembles the adult but is not sexually mature; a larva differs from the adult form.
c. A larva transforms to a juvenile by **metamorphosis**.

32.2 (1) The emergence of new predator–prey relationships (increased locomotion and protective shells) may have triggered various evolutionary adaptations. (2) The accumulation of atmospheric oxygen may have supported the active metabolisms of mobile and larger animals. (3) The origin of the *Hox* complex of developmental genes and other regulatory sequences may have facilitated the evolution of the diverse body plans that appear during the Cambrian explosion. All three of these may have played a role in the Cambrian explosion.

32.3 Sponges have no symmetry to their bodies, no true tissues, and no digestive tract—and thus no body cavity between it and the body wall. Cnidarians have radial symmetry and two germ layers: ectoderm and endoderm. They also have no coelom. Bilateral animals have bilateral symmetry, three germ layers (mesoderm is the third) that form true tissues, and most have a body cavity between their complete digestive tract and body wall.

32.4

	Protostome	Deuterostome
Cleavage	spiral and determinate	radial and indeterminate
Coelom formation	split in solid mass of mesoderm	mesodermal buds from archenteron
Blastopore fate	mouth; second opening for anus	anus; second opening for mouth

SUGGESTED ANSWERS TO STRUCTURE YOUR KNOWLEDGE

1. **a.** Porifera (sponges): asymmetrical; lack true tissues

   b. Eumetazoa: "true" animals; have true tissues

   c. Cnidaria: radial symmetry; diploblastic

   d. Bilateria: bilateral symmetry; triploblastic

   e. Deuterostomia: echinoderms, chordates

   f. Lophotrochozoa: ectoprocts, molluscs, and annelids (and others)

   g. Ecdysozoa: nematodes and arthropods

ANSWERS TO TEST YOUR KNOWLEDGE

Multiple Choice:

1. d	**3.** a	**5.** b	**7.** a	**9.** a
2. a	**4.** c	**6.** d	**8.** b	

Chapter 33: An Introduction to Invertebrates

INTERACTIVE QUESTIONS

33.1 **a.** Flagellated choanocytes, which line the spongocoel, create water flow through a sponge. They trap and engulf food particles.

b. Amoebocytes wander through the mesohyl. They digest food and distribute nutrients, produce skeletal fibers, and may become other types of cells as needed.

33.2 **a.** polyp
b. medusa
c. mouth/anus
d. tentacle
e. mesoglea
f. gastrovascular cavity

33.3 **a.** Planarians have an extensively branched gastrovascular cavity; food is drawn in and undigested wastes are expelled through the mouth at the tip of a muscular pharynx.

b. Tapeworms are bathed in predigested food in their host's intestine.

33.4 **a.** 1. muscular foot for movement
2. visceral mass containing internal organs
3. mantle that may secrete shell and enclose visceral mass, forming a mantle cavity that houses gills

b. 1. Snails use their foot to creep slowly and their radula to rasp up algae.

2. Clams are sedentary suspension feeders that use siphons to draw water over the gills, where food is trapped in mucus and swept by cilia to the mouth.

3. Squid are active predators that capture prey with tentacles and crush it with jaws, aided by a poison in their saliva.

33.5 **a.** septum
b. intestine
c. metanephridium
d. chaetae
e. fused ventral nerve cords
f. dorsal and ventral blood vessels
g. longitudinal and circular muscles

33.6 Earthworms have longitudinal and circular muscle layers and a segmented body cavity that enable them to move by alternately elongating and contracting body sections and anchoring to the substrate with chaetae. Nematodes are not segmented and have only longitudinal muscles. These pull against the hydrostatic skeleton of the roundworm's tapered body, producing thrashing movements.

33.7 **a.** The exoskeleton provided protection from desiccation and physical support on land.

b. Pairs of jointed appendages have become modified for walking, feeding, sensing, mating, or defense.

c. Sensory organs include eyes, olfactory receptors, and antennae specialized for touch and smell.

33.8 **a.** Chelicerae are anterior feeding appendages modified as pincers or fangs.

b. Many spiders trap their prey in webs they have spun from abdominal gland secretions. Spiders kill prey with poison-equipped, fang-like chelicerae. After secreting digestive juices, they suck up their liquefied food.

33.9 **a.** The habitat of crustaceans is mostly aquatic, whereas insects are mostly terrestrial.

b. Crustaceans mostly swim or are planktonic; insects fly and walk.

c. Respiration in crustaceans is accomplished by gills or across the body surface; insects have an internal tracheal system.

d. Excretion in crustaceans takes place across thin areas of the cuticle; insects have Malpighian tubules.

e. Crustacean appendages include mandibles and multiple mouthparts, walking legs on the thorax, and abdominal appendages. Insects have mandibles and modified mouthparts for chewing, piercing, or sucking; three pairs of walking legs on the thorax; and many insects have one or two pairs of wings (extensions of cuticle).

33.10 a. radial anatomy as adults, usually five spokes; sessile or slow moving; bilaterally symmetrical larvae

b. endoskeleton of calcareous plates

c. water vascular system with tube feet for locomotion and feeding

SUGGESTED ANSWERS TO STRUCTURE YOUR KNOWLEDGE

1. **a.** Eumetazoa
    b. Bilateria
    c. Deuterostomia
    d. Lophotrochozoa
    e. Ecdysozoa
    f. Porifera (sponges); lack true tissues; choanocytes used in filter feeding
    g. Cnidaria (hydras, jellies, sea anemones, corals); diploblastic; radial symmetry; gastrovascular cavity; polyp and medusa forms; cnidocytes
    h. Platyhelminthes (flatworms: planaria, flukes, tapeworms); acoelomate; gastrovascular cavity

i. Syndermata (rotifers and acanthocephalans); pseudocoelomate; in rotifers: alimentary canal; crown of cilia around mouth; parthenogenesis (in some)

j. Lophophorates (brachiopods and ectoprocts); coelomate; lophophore feeding structure; exoskeleton or bivalve shell

k. Mollusca (clams, snails, squids); coelomates; muscular foot; visceral mass; mantle; most with hard shell

l. Annelida (segmented worms: Errantia and Sedentaria, which includes earthworms and leeches); coelomates; segmented body wall and organs

m. Nematoda (roundworms); pseudocoelomate; unsegmented, cylindrical worms; ecdysis

n. Arthropoda (chelicerates such as horseshoe crabs and arachnids [spiders, ticks, and mites]; myriapods such as millipedes and centipedes; and pancrustaceans such as crustaceans and insects); coelomate; segmented body; jointed appendages; exoskeleton

o. Echinodermata (starfishes, sea urchins); coelomate; endoskeleton; water vascular system

p. Chordata (vertebrates, two invertebrate basal groups); coelomates; segmentation

ANSWERS TO TEST YOUR KNOWLEDGE

Multiple Choice:

1.	b	**5.**	d	**9.**	c	**13.**	d
2.	a	**6.**	a	**10.**	b	**14.**	b
3.	c	**7.**	c	**11.**	a	**15.**	d
4.	d	**8.**	b	**12.**	c	**16.**	b

Chapter 34: The Origin and Evolution of Vertebrates

INTERACTIVE QUESTIONS

34.1 lancelet, Cephalochordata
  a. mouth
  b. *pharyngeal slits
  c. digestive tract
  d. segmental muscles
  e. anus
  f. *post-anal tail
  g. *dorsal, hollow nerve cord
  h. *notochord

  * Chordate characters

34.2 The derived characters of vertebrates include two or more sets of *Hox* genes and duplications of genes for signaling molecules and transcription factors; more extensive skull and vertebrae (backbone) that encloses the spinal cord; neural crest.

34.3 a. Chordata
  b. Vertebrata
  c. Gnathostomata
  d. Osteichthyes
  e. lobe-fins
  f. Tetrapoda
  g. Amniota
  h. lancelets (Cephalochordata)
  i. tunicates (Urochordata)

j. hagfishes (Myxini)
k. lampreys (Petromyzontida)
l. sharks, rays (Chondrichthyes)
m. ray-finned fishes (Actinopterygii)
n. coelacanths (Actinistia)
o. lungfishes (Dipnoi)
p. frogs, salamanders (Amphibia)
q. turtles, snakes, crocodiles, birds (Reptilia)
r. mammals (Mammalia)

34.4 **a.** The derived characters of gnathostomes include hinged jaws, lateral line system, enlarged forebrain with better olfaction and vision, four sets of *Hox* genes, and apparent duplication of the entire genome.
b. The spiral valve increases the area for absorption and slows the passage of food through the short shark intestine.
c. A cloaca is a common opening for the reproductive tract, excretory system, and digestive tract.

34.5 **a.** Coelacanths, lungfishes, and tetrapods
b. See answer to Interactive Question 34.3.

34.6 *Tiktaalik* has the fish characters of scales, fins, and gills and lungs, but the tetrapod characters of neck, ribs, flat skull, eyes on top of the skull, and bones in its front fins with the basic tetrapod pattern of humerus, radius and ulna, and wrist bones. It did not have limbs with digits and thus is considered to be a lobe-fin, not a tetrapod.

34.7 **a.** chorion—functions in gas exchange
b. amnion—encases embryo in fluid, prevents dehydration, and cushions shocks
c. allantois—stores metabolic wastes, functions with chorion in gas exchange
d. yolk sac—contains yolk; blood vessels transport stored nutrients into embryo

34.8 Archosaurs include the extinct dinosaurs and pterosaurs as well as crocodilians and birds. Lepidosaurs include extinct huge marine reptiles, tuatara, and the squamates (lizards and snakes).

34.9 Adaptations for flight include lightweight strong skeleton; reduced weight due to absence of teeth and some organs; feathers that shape wings into airfoil; strong keel on sternum to which flight muscles attach; high metabolic rate supported by efficient circulatory system (four-chambered heart), respiratory system, and endothermy;

well-developed visual and motor areas of brain.

34.10 **a.** Derived traits of mammals include hair; milk produced by mammary glands; endothermic with active metabolism; diaphragm to ventilate lungs; four-chambered heart; larger brains, increased learning capacity; differentiated teeth; and remodeled jaw
b. See answer to Interactive Question 34.3.

34.11 increased brain size; change in jaw shape and dentition; bipedal posture; reduced size difference between the sexes; capacity for language and symbolic thought; manufacture and use of complex tools

SUGGESTED ANSWERS TO STRUCTURE YOUR KNOWLEDGE

1. Primates: grasping hands and feet, flat nails, moveable thumb, forward-facing eyes, large brain
Eutherians: placental mammals; young complete development within uterus
Mammals: mammary glands that produce milk, hair, diaphragm, four-chambered heart
Amniotes: amniotic egg with four extraembryonic membranes, less permeable skin, use of rib cage to breathe
Tetrapods: limbs with digits, neck vertebrae, pelvic girdle fused to backbone
Lobe-fins: muscular fins supported by bones
Osterichthyans: bony skeleton, lungs or lung derivatives
Gnathostomes: hinged jaws, four sets of *Hox* genes, mineralized skeleton, enlarged forebrain with enhanced senses
Vertebrates: vertebral column surrounding nerve cord; *Hox* gene duplication
Chordates: notochord; dorsal, hollow nerve cord; post-anal tail; pharyngeal clefts or slits

2. (1) Pharyngeal slits used for suspension feeding became adapted for gas exchange; (2) skeletal supports for gills became adapted for use as hinged jaws; (3) lungs of bony fish were transformed to air bladders, aiding in buoyancy; (4) fleshy fins supported by skeletal elements developed into limbs for terrestrial animals; (5) feathers, probably originally used for insulation, came to be used for flight; (6) dexterous hands important for an arboreal life, evolved to manipulate tools.

3.   Amphibians remain in damp habitats, burrowing in mud during droughts; some secrete foamy protection for eggs laid on land. The amniote egg provides an aquatic environment for the developing embryo. Non-bird reptiles have scaly, waterproof skin and behavioral adaptations to modulate changing temperatures. Endothermy of birds and mammals maintains stable body temperature in the more variable terrestrial climate. Gestation protects a developing embryo. Tetrapods use limbs for locomotion on land.

ANSWERS TO TEST YOUR KNOWLEDGE

Fill in the Blanks:

1. Urochordata (tunicates)
2. Cephalochordata (lancelets)
3. somites
4. gnathostomes
5. operculum
6. amniotic egg
7. monotremes
8. diaphragm
9. anthropoids
10. *Australopithecus*

Multiple Choice:

1. a	**4.** d	**7.** b	**10.** a	**13.** d
2. b	**5.** b	**8.** a	**11.** b	**14.** c
3. c	**6.** a	**9.** c	**12.** c	**15.** a

Chapter 35: Vascular Plant Structure, Growth, and Development

INTERACTIVE QUESTIONS

35.1
a. reproductive shoot (flower)
b. apical bud
c. node
d. internode
e. vegetative shoot
f. leaf
g. blade
h. petiole
i. axillary bud
j. stem
k. taproot
l. lateral roots
m. shoot system
n. root system

35.2
a. sclerenchyma (fibers and sclereids), tracheids, and vessel elements
b. sieve-tube elements (and the cells listed in a.)

35.3
a. Plants exhibit indeterminate growth, continuing to grow as long as they live. Animals, as well as some plant organs such as leaves and flowers, have determinate growth and stop growing after reaching a certain size.
b. Primary growth is the elongation of shoots, branches, and roots and results from cell division within apical meristems. Secondary growth is produced by lateral meristems (vascular cambium and cork cambium) and results in the thickening of roots and shoots of woody plants

35.4
a. epidermis (dermal)—root hairs for absorption
b. cortex (ground)—transport of minerals and water from root hairs to stele; intercellular spaces allow *extracellular* diffusion of water, minerals, and oxygen; food storage
c. vascular cylinder (vascular)—transport
d. endodermis—regulates water and mineral movement into stele
e. pericycle—origin of lateral roots
f. xylem—water and mineral transport
g. phloem—nutrient transport

35.5
a. cuticle—waxy covering that reduces evaporative water loss
b. upper epidermis—tightly packed protective outer covering of cells
c. palisade mesophyll—photosynthesis
d. spongy mesophyll—photosynthesis; loosely packed cells surrounding air spaces
e. guard cells—regulate opening and closing of stomata
f. xylem—transport of water and minerals from roots
g. phloem—transport of carbohydrates from leaves to roots and growing parts of plant
h. vein—vascular tissue surrounded by ring of bundle-sheath cells that regulate movement of materials into and out of vein
i. stoma—pore that allows exchange of CO_2 and O_2 between air spaces of leaf and outside

35.6 H, F, A, B, G, D, C, E

35.7 They are helping to determine each gene's function and eventually to track the complex genetic and biochemical pathways involved in plant development.

35.8 Microtubules of the preprophase band determine the orientation of cell division. Microtubules in the outer cytoplasm orient the cellulose microfibrils in cell walls, which then determine the direction in which cells can expand.

35.9 **a.**

Whorl	Genes Active	Organs in Normal Flower	Genes Active in Mutant A	Organs in Mutant A Flower
1	A	sepals	C	carpels
2	AB	petals	BC	stamens
3	BC	stamens	BC	stamens
4	C	carpels	C	carpels

b. Gene *A* would be expressed alone in each whorl, producing a flower with four whorls of sepals.

SUGGESTED ANSWERS TO STRUCTURE YOUR KNOWLEDGE

1. Except for leaves and flowers, which stop growing when they reach maturity (determinate growth), plant roots and shoots continue to grow throughout the life of a plant. Although certain characteristic forms have been favored by natural selection in species adapted to particular environments, the specific environment in which a plant grows greatly influences its individual body form. For example, a plant may grow taller or produce different shapes of leaves depending on its light exposure. This developmental plasticity helps compensate for a plant's immobility.

2. This drawing is a eudicot stem, indicated by the ring of vascular bundles and the central pith. It represents primary growth because there is no layer of vascular cambium between the layers of xylem and phloem and no evidence of layers of secondary xylem or phloem.
   a. sclerenchyma
   b. phloem
   c. xylem
   d. cortex
   e. pith
   f. vascular bundle
   g. epidermis

ANSWERS TO TEST YOUR KNOWLEDGE

Multiple Choice:

1. c	4. c	7. d	10. c	13. c	16. b
2. a	5. b	8. b	11. d	14. c	17. a
3. a	6. b	9. b	12. c	15. c	18. a

Chapter 36: Resource Acquisition and Transport in Vascular Plants

INTERACTIVE QUESTIONS

36.1 **a.** At low light intensities, horizontal leaf orientation would maximize exposure to sunlight. In bright sun, a vertical orientation would shield leaves from too high a light intensity and allow light to penetrate to lower leaves.
b. Plant productivity would increase with an increasing leaf area index as more leaf area is available to capture sunlight. If this index gets too high, however, then lower branches and leaves are too shaded to photosynthesize, and they may die through a process of self-pruning.
c. The branching pattern of roots can increase to take advantage of the local availability of nitrate, as can the production of nutrient transporters within the root cells.

36.2 **a.** $\Psi_P = 0$, $\Psi_S = -0.6$, $\Psi = -0.6$ MPa
b. $\Psi_P = 0.6$, $\Psi_S = -0.6$, $\Psi = 0$ MPa. The turgor pressure of the cell that develops with movement of water into the cell finally offsets the solute potential of the cell. No net osmosis occurs, and the water potentials of both cell and solution are equal and = 0 MPa.
c. $\Psi_P = 0$, $\Psi_S = -0.8$, $\Psi = -0.8$ MPa. The cell would plasmolyze, losing water until the solute potential of the cell would equal that of the bathing solution (both -0.8 MPa). There would be no turgor pressure.

36.3 **a.** symplastic
b. apoplastic
c. Casparian strip
d. xylem vessels
e. vascular cylinder (stele)
f. endodermis
g. cortex

h. epidermis
i. root hair

36.4 **a.** The loss of water vapor from the air spaces through the stomata causes evaporation from the water film coating the mesophyll cells, leading to the negative pressure that draws water up from the roots.
b. Evaporation from the water film increases the surface tension of the curving air-water interface. This tension, or negative pressure, produces a gradient of water potentials that extends from the leaf to the root.
c. Water molecules hold together due to hydrogen bonding, and a pull on one water molecule is transmitted throughout the column of water.
d. Water molecules adhere to the hydrophilic walls of narrow xylem cells, helping to support the column of water against gravity.

36.5 In order to provide sufficient CO_2 for photosynthesis, stomata must be open, and large amounts of water will be lost to transpiration. A sunny, windy, dry day will increase the rate of transpiration, perhaps to the point that insufficient water is available in the soil. Due to a lack of water, guard cells may lose turgor (and also be signaled by abscisic acid) and stomata will close, slowing transpiration. In reducing excessive water loss, however, leaf cells now receive insufficient CO_2, and the photosynthetic rate will decline.

36.6 **a.** ATP
b. proton pump
c. H^+
d. sucrose
e. cotransporter

The lower portion of the diagram represents the inside of the cell.

SUGGESTED ANSWERS TO STRUCTURE YOUR KNOWLEDGE

1. Solutes may move across membranes by passive transport when they move down their concentration or electrochemical gradient. Transport proteins, which speed this passive transport, may be specific carrier proteins or selective channels. Active transport usually involves a proton pump that creates a membrane potential as H^+ ions are moved out of the cell. Cations such as K^+ may now move through specific channels down their electrochemical gradient. In cotransport, the inward diffusion of H^+ through a cotransporter moves a solute against its concentration gradient.

2. The cohesion-tension hypothesis explains the ascent of xylem sap. Tension created by the evaporation of water and the resulting curvature of the air-water interface lowers the water potential in the leaf and sets up the bulk flow of water. The cohesion of water molecules transmits the pull resulting from transpiration throughout the water column, and the adhesion of water to the hydrophilic walls of the xylem vessels aids the flow.

A pressure-flow mechanism also explains the movement of phloem sap, but the plant must expend energy to create the differences in water potential that drive translocation. The active accumulation of sucrose in sieve-tube elements lowers water potential at the source end, resulting in an inflow of water. The removal of sugar from the sink end of a phloem tube is followed by the osmotic loss of water. The resulting difference in pressure between the source and sink end of the phloem tube causes the bulk flow of phloem sap.

ANSWERS TO TEST YOUR KNOWLEDGE

Multiple Choice:

1. c	**5.** b	**9.** a	**13.** b	**17.** c
2. c	**6.** a	**10.** c	**14.** a	**18.** a
3. b	**7.** b	**11.** c	**15.** a	**19.** d
4. d	**8.** d	**12.** c	**16.** b	**20.** d

Chapter 37: Soil and Plant Nutrition

INTERACTIVE QUESTIONS

37.1 A loam, which is a mixture of sand, silt, and clay, has enough large particles to provide air spaces and enough small particles to retain water and minerals. Cations bind to soil particles and are made available to roots by cation exchange. Humus in a soil helps to retain water and provide a steady supply of mineral nutrients.

37.2 **a.** Sustainable agriculture includes the variety of conservation-minded, environmentally safe, and profitable farming methods that permit the use of soil as a renewable resource.
b. The N-P-K ratio is the percentage of nitrogen, phosphorus, and potassium that is in a fertilizer.
c. A soil pH of 5 or lower makes Al^{3+} more soluble, increasing absorption of these toxic ions that stunt root growth.

37.3 **a.** Mg: component of chlorophyll, cofactor and activator of enzymes
b. P: component of nucleic acids, phospholipids, ATP, some coenzymes
c. Ca: component of cell walls, membrane function, signal transduction

37.4 **a.** Chlorosis is the yellowing of leaves due to a lack of chlorophyll, often indicative of a deficiency of magnesium when found in older leaves. Deficiencies of iron or manganese can produce chlorosis in young leaves.
b. A deficiency in an immobile element would be evident in the younger, growing parts of a plant.

37.5 **a.** nitrogen-fixing bacteria
b. ammonifying bacteria
c. nitrifying bacteria
d. denitrifying bacteria
e. nitrate and nitrogenous organic compounds

37.6 **a.** Legume crops, which increase nitrogen in the soil, are often followed the next growing season with a nonlegume crop.
b. "Green manure" is a legume crop that is plowed under to improve soil fertility by adding both humus and nitrogen.

37.7 The fossil record documents mycorrhizal associations in the earliest land plants. These symbiotic associations may have been an important adaptation in helping plants to colonize nutrient-poor soils and providing carbohydrates for fungi.

SUGGESTED ANSWERS TO STRUCTURE YOUR KNOWLEDGE

1.

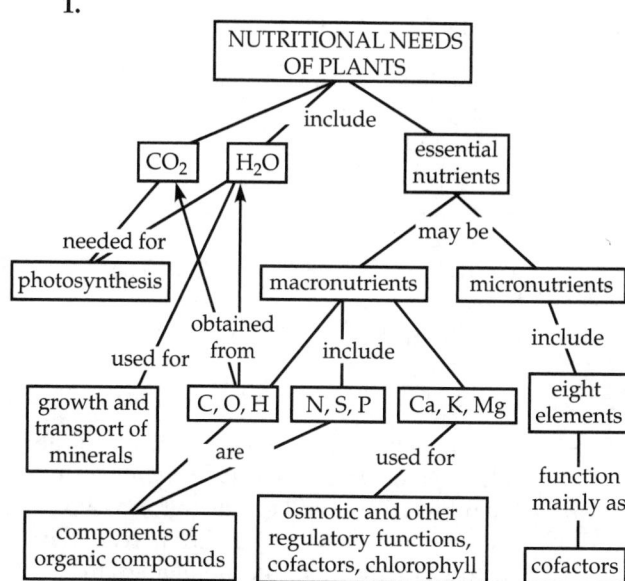

2. Both these types of symbiotic relationships involve chemical recognition between specific plant species and bacterial or fungal species. Root nodules are mutualistic associations between legumes and nitrogen-fixing bacteria of the genus *Rhizobium*. The bacteria receive nourishment from the plant and provide the plant with fixed nitrogen. Mycorrhizae are mutualistic associations between the roots of most plants and various species of fungi. The fungi stimulate root growth, increase surface area for absorption, and secrete antibiotics that may help protect the plant. In return, the fungi receive nourishment from the plant.

ANSWERS TO TEST YOUR KNOWLEDGE

Multiple Choice:

1. b	**4.** a	**7.** a	**10.** d	**13.** a
2. b	**5.** b	**8.** c	**11.** c	**14.** c
3. c	**6.** d	**9.** c	**12.** d	**15.** d

Chapter 38: Angiosperm Reproduction and Biotechnology

INTERACTIVE QUESTIONS

38.1 **a.** stamen
  b. anther
  c. filament
  d. carpel
  e. stigma
  f. style
  g. ovary
  h. sepal
  i. receptacle
  j. ovule
  k. petal

38.2 **a.** The male gametophyte consists of a tube cell and a generative cell (which moves into the tube cell and will divide to form two sperm). These cells and the spore wall that encloses them constitute a pollen grain.
  b. The female gametophyte is the embryo sac, often containing seven cells (one of which is an egg) and eight nuclei.
  c. Double fertilization conserves resources by allowing for nutrient development only when fertilization has occurred.

38.3 **a.** seed coat
  b. epicotyl
  c. hypocotyl
  d. cotyledons with endosperm
  e. radicle
  f. cotyledon (scutellum)
  g. endosperm
  h. epicotyl
  i. hypocotyl
  j. radicle
  k. coleorhiza
  l. coleoptile

38.4 Vegetative growth before flowering allows the plant to photosynthesize and grow, creating adequate resources to support flowering.

38.5 **a.** Buoyant fruits may be dispersed by water; winged fruits may "fly" seeds to new destinations; seeds in edible fruits may be dispersed in feces; prickly seeds may stick to animals, and some animals may "plant" seeds in their underground caches.
  b. Hormonal interactions induce softening of the pulp, a change in color, and an increase in sugar content.

38.6 **a.** Apomixis allows for the dispersal value of seeds, which in this case are clones of the parent plant.
  b. Selfing is self-fertilization, in which a plant's sperm fertilizes its own eggs. Although offspring are not as genetically diverse, selfing is an advantage in certain crop plants because it ensures that each ovule will develop into a seed. Also, if a plant is well adapted to a relatively constant environment, it may be advantageous to produce offspring with a similar genetic endowment.

38.7 In both vegetative reproduction and self-fertilization, offspring have only one parent. But in vegetative reproduction, all offspring are genetically identical clones. Self-fertilization involves meiosis and random fertilization of gametes. Thus, whatever genetic variation is present in a diploid parent plant is recombined in offspring, producing more genetic variation in offspring.

38.8 Burning of fossil fuels releases CO_2 that was removed from the atmosphere millions of years ago. CO_2 that is released by the burning of biofuels is absorbed for photosynthesis as new biofuel crops grow.

SUGGESTED ANSWERS TO STRUCTURE YOUR KNOWLEDGE

1. This is one version of the life cycle of an angiosperm. See also Figure 38.6 in the textbook.

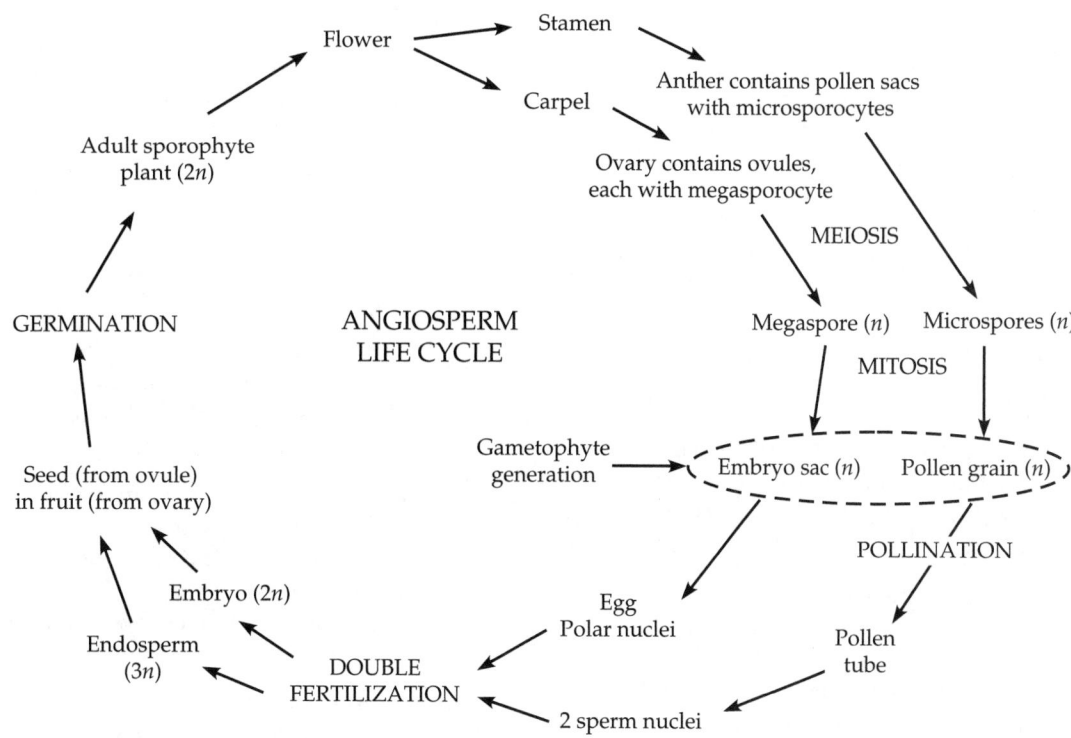

2. *Sexual:* Advantages include increased genetic variability, which provides the potential for adaptation to changing conditions, and the dispersal and dormancy capabilities provided by seeds. Disadvantages are that the seedling stage is very vulnerable, sexual reproduction is very energy intensive (because producing flowers, pollen, and fruit takes energy and most seeds and seedlings don't survive), and genetic recombination may separate adaptive traits.

   Asexual: Advantages include the hardiness of vegetative reproduction and the maintenance of genetically well-adapted plants in a given environment. Disadvantages relate to the advantages of sexual reproduction: There is no genetic variability from which to "choose" should conditions change, and there are no seeds for dormancy or dispersal.

3. *Benefits:* Reduce use of chemical pesticides; remove weeds with herbicides instead of erosion-producing tillage; increase nutritional value of crops; produce disease-resistant crops.

   Dangers: Harm nontarget species; create hybrid "superweeds" due to crop-to-weed transgene escape; produce other unanticipated harmful results that cannot be stopped once GMOs are released into the environment.

ANSWERS TO TEST YOUR KNOWLEDGE

Fill in the Blanks:

1. ovary
2. sporophyte
3. biofuel
4. embryo sac
5. radicle
6. epicotyl
7. coleoptile
8. scion
9. microspore
10. callus

Multiple Choice:

1. d	4. d	7. a	10. d	13. c	16. a
2. c	5. a	8. a	11. c	14. c	17. d
3. b	6. b	9. c	12. a	15. d	

Chapter 39: Plant Responses to Internal and External Signals

INTERACTIVE QUESTIONS

39.1 **a.** The signal for de-etiolation is light; the receptor is a phytochrome located in the cytosol.
b. Light-activated phytochrome activates at least two pathways, one producing the second messenger cGMP and one leading to an increase in cytosolic Ca^{2+} levels. Both activate specific kinases. Some kinase cascades may activate various transcription factors that change gene expression.
c. The plant response involves growth changes (slowing of stem elongation, expansion of leaves, elongation of roots). The de-etiolation process involves changes in levels of growth-regulating hormones and the activation or new production of enzymes involved in producing chlorophyll or in photosynthesis.

39.2 **a.** proton (H^+) pumps
b. lower
c. expansins
d. osmotic uptake of water
e. increased turgor pressure
f. loosened walls (increased expandability of walls)
g. cytoplasm
h. cell wall material

39.3 **a.** ethylene
b. cytokinins
c. abscisic acid
d. jasmonates
e. auxin
f. brassinosteroids
g. strigolactones
h. gibberellins

39.4 The relative amounts of red and far-red light are communicated to a plant by the ratio of the two forms of phytochrome. Canopy trees absorb red light, and a shaded tree will have a higher ratio of P_r to P_{fr}, which induces the tree to grow taller. Direct sunlight shifts the ratio toward the P_{fr} form, stimulating branching and inhibiting vertical growth.

39.5 Free-running periods are circadian rhythms that occur when an organism is kept in a constant environment without environmental cues. They usually vary from exactly 24 hours.

39.6 **a.** not flower
b. flower
c. flower
d. not flower
e. not flower
f. flower
g. flower
h. not flower
i. not flower
j. flower

In the next-to-last light regimen, a flash of far-red light cancels the effect of the red flash, and the red flash in the last one cancels the effect of the far-red flash.

39.7 **a.** A stem exhibits positive phototropism and negative gravitropism.
b. In response to contact, a tendril exhibits thigmotropism, in which mechanical stimulation causes unequal growth rates of cells on opposite sides of the tendril.

39.8 **a.** A plant may reduce transpirational water loss by stomatal closing and wilting, or produce heat-shock proteins to stabilize plant proteins.
b. A plant may respond to cold, wet weather by increasing the unsaturated fatty acids in its membranes; in addition, air tubes may develop in the root cortex.

39.9 **a.** Insects harm plants by eating plant tissues and reducing photosynthetic capacity and by causing wounds through which pathogens can enter a plant.
b. Insects benefit plants by pollinating plants and by eating other insects that are eating plant tissues.

SUGGESTED ANSWERS TO STRUCTURE YOUR KNOWLEDGE

1. Some agricultural uses of plant hormones are cytokinin spray used to keep flowers fresh; ethylene used to ripen stored fruit; synthetic auxins used to induce seedless fruit set; gibberellins sprayed on Thompson seedless grapes; and 2,4-D (auxin) used as herbicide.

2.

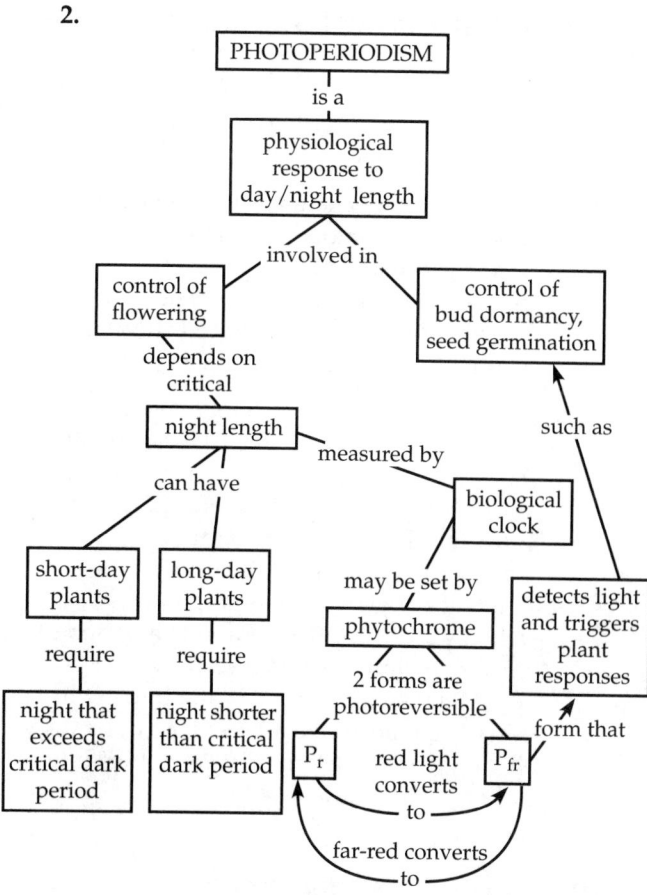

PHOTOPERIODISM

is a

physiological response to day/night length

involved in

control of flowering

control of bud dormancy, seed germination

depends on critical

night length — measured by — biological clock

such as

can have

short-day plants

long-day plants

may be set by

detects light and triggers plant responses

phytochrome

2 forms are photoreversible

require

require

form that

night that exceeds critical dark period

night shorter than critical dark period

P_r   red light converts — to   P_fr

far-red converts — to

molecules, sealing off of infected areas, and destruction of infected cells. The alarm signal methylsalicylic acid travels to distant cells, which convert it to salicylic acid, activating a systemic acquired resistance. Molecules are produced that protect the cells against a diversity of pathogens for several days.

ANSWERS TO TEST YOUR KNOWLEDGE

True or False:

1. False—change *abscisic acid* to *gibberellin*; or use *The removal of abscisic acid*.
2. True
3. False—change *thigmomorphogenesis* to *triple response*; or use *Thigmomorphogenesis is a change in growth form in response to mechanical stress*.
4. True
5. False—change *circadian rhythm* to *photoperiodism*; or change *day length* to *on a 24-hour cycle*.
6. True
7. False—change *PAMP-triggered immunity* to *effector-triggered immunity* and change *Toll-like receptor* to *R protein*; or change the end of the sentence (*specific effector proteins and the triggering of a hypersensitive response and systemic acquired resistance*) to *pathogen-associated molecular patterns (common to certain pathogens) and the production of phytoalexins*.

3. Specific resistance begins with the recognition of a pathogen molecule (effector) by a specific plant resistance (R) protein, which triggers a signal transduction pathway that initiates a hypersensitive response and production of an alarm signal. The hypersensitive response includes production of antimicrobial

Multiple Choice:

1. b	**6.** a	**10.** c	**14.** a
2. c	**7.** a	**11.** b	**15.** a
3. d	**8.** b	**12.** b	**16.** c
4. d	**9.** d	**13.** d	**17.** b
5. b			

Chapter 40: Basic Principles of Animal Form and Function

INTERACTIVE QUESTIONS

40.1 Complexity allows for more specialization. Specialized organs, which can be protected within the body, can meet the animal's needs and control the composition of the interstitial fluid. A complex body is especially advantageous in highly variable terrestrial environments.

40.2 a. Stratified squamous; lining of mouth; thick layer is protective, and new cells produced at the basal surface replace those abraded off the apical surface.
b. Simple columnar; lining of intestines; cells with large cytoplasmic volumes are specialized for secretion or absorption.

40.3 a. osteon
b. central canal
c. bone
d. red blood cells
e. white blood cells
f. plasma
g. blood

Bone has a hard, mineral matrix; blood has a liquid matrix.

40.4 Hormones are released and travel more slowly throughout the body in the bloodstream. Nerve impulses travel rapidly through neurons directly to target cells. Hormone signals last longer and usually mediate responses of longer duration. The endocrine system coordinates functions such as growth, reproduction, and metabolism. The nervous system regulates rapid responses such as locomotion and behavior.

40.5 Positive feedback is a mechanism in which a change in a variable serves to amplify rather than reverse the stimulus. An example is the increasing uterine contractions during childbirth. Homeostatic mechanisms tend to return a fluctuating variable to the normal range. Thus, positive feedback does not typically contribute to homeostasis.

40.6 Because the environmental temperature determines its body temperature, an ectotherm that lives where temperatures fluctuate very little will be homeothermic. If environmental temperatures are very high, an ectotherm may have a body temperature higher than that of an endotherm, which may have mechanisms for cooling the body.

40.7 a. Cooling mechanisms include vasodilation of surface blood vessels and activation of sweat glands.
b. Warming mechanisms include vasoconstriction of surface vessels and shivering by contraction of skeletal muscles.

40.8 a. In hibernation, the body temperature is maintained at a much lower level, greatly conserving energy and allowing the animal to withstand long periods of cold temperature and decreased food supply.
b. In estivation, animals survive long stretches of elevated temperature and diminished water supply by entering a period of inactivity and lowered metabolism.
c. Many small mammals and birds with very high metabolic rates enter a daily torpor during periods when they are not feeding or when temperatures are low.

SUGGESTED ANSWERS TO STRUCTURE YOUR KNOWLEDGE

1. All organs are covered with epithelial tissue, and internal lumina, ducts, and blood vessels are lined with epithelia. Most organs contain some types of connective tissue. Muscle tissue would be present if the organ must contract. Nervous tissue innervates most organs.

2. Endotherms, which warm their bodies with metabolic heat, have thermoregulatory mechanisms that enable them to maintain a constant internal temperature across a wider range of environmental temperatures. Even though ectotherms may use behavioral mechanisms to reduce temperature fluctuations, they may still be exposed to greater internal fluctuations as environmental temperatures change.

3. Due to the higher energy cost of maintaining a warm body temperature, the BMR of an endotherm would always be higher than the SMR of a similarly sized ectotherm. Energy for reproduction would vary by sex, time of year, size and number of offspring produced, and amount of parental care. The cost of thermoregulation for an endotherm varies depending

on the environment and the thermoregulatory adaptations of the animal. The growth and activity expenditures vary depending on whether the animal is growing and the amount of activity required for obtaining food, building shelter, or escaping predators. Even as the proportions (pieces of the pie) devoted to each activity change depending on the organism and the environment, the overall energy expenditure per unit mass is almost always higher for smaller animals.

ANSWERS TO TEST YOUR KNOWLEDGE

Multiple Choice:

1. b	5. d	9. c	13. d	17. b
2. a	6. d	10. a	14. d	18. a
3. a	7. b	11. c	15. b	19. c
4. c	8. d	12. c	16. a	

Chapter 41: Animal Nutrition

INTERACTIVE QUESTIONS

41.1 Because most plant proteins are deficient in one or more essential amino acids, vegetarians must eat a combination of plant foods that have complementary amino acids, such as corn and beans.

41.2 **a.** The most common type of malnutrition is protein deficiency, usually as a result of relying on a single food source such as rice or corn, which are incomplete proteins.
b. Calcium and phosphorus are required by vertebrates for bone and tooth construction.

41.3 The carbohydrates, proteins, nucleic acids, and fats that make up food are too large to pass through cell membranes, and the molecules in food are not identical to those an animal assembles for its own functions and tissues.

41.4 An alimentary canal can have specialized compartments for the sequential processing of nutrients. Because food moves in one direction, an animal can ingest more food while still digesting a previous meal.

41.5 **a.** Polysaccharides (starch and glycogen) are broken down in the mouth by salivary amylase to smaller polysaccharides and maltose; hydrolysis continues until salivary amylase is inactivated by the low pH in the stomach.
b. Proteins are hydrolyzed by pepsin to smaller polypeptides in the stomach.

41.6 **a.** The digestion of starch and glycogen into disaccharides is continued by pancreatic amylase. Disaccharidases, which are enzymes of the intestinal epithelium, split disaccharides into monosaccharides.

b. Protein digestion is completed in the small intestine by pancreatic trypsin and chymotrypsin, enzymes specific for peptide bonds adjacent to certain amino acids, and by pancreatic carboxypeptidase, as well as enzymes of the epithelium: dipeptidases, carboxypeptidase, and aminopeptidase, which split off one amino acid at a time.
c. Pancreatic nucleases hydrolyze DNA and RNA into nucleotide monomers. Nucleotidases, nucleosidases, and phosphatases dismantle nucleotides.
d. The digestion of fats is aided by bile salts, which coat or emulsify tiny fat droplets so they do not coalesce, leaving a greater surface area for pancreatic lipase to hydrolyze the fat molecules into glycerol, fatty acids, and monoglycerides.

41.7 Carnivore dentition is characterized by sharp incisors, fanglike canines, and jagged premolars and molars. Herbivores have broad premolars and molars that grind plant material and incisors and canines modified for biting off vegetation. Carnivorous vertebrates often have large, expandable stomachs. The longer intestine of an herbivore facilitates digestion of plant material, and its cecum may contain cellulose-digesting microorganisms.

41.8 **a.** insulin
b. glucagon
c. uptake of glucose
d. store glucose as glycogen
e. pancreatic islets
f. hydrolyze glycogen and release glucose
g. blood glucose level

SUGGESTED ANSWERS TO STRUCTURE YOUR KNOWLEDGE

1. **a.** salivary glands—produce saliva
    b. oral cavity—teeth and tongue mix food with saliva
    c. pharynx—throat, opening to esophagus and trachea
    d. esophagus—peristalsis moves bolus to stomach
    e. stomach—churns food with gastric juice
    f. small intestine—digestion and absorption of nutrients and water
    g. large intestine (colon)—houses bacteria, compacts feces
    h. anus—exit of alimentary canal
    i. rectum—stores feces until expelled
    j. pancreas—produces enzymes, bicarbonate, and hormones
    k. gallbladder—stores bile from liver
    l. liver—produces bile; processes nutrients; detoxifies

ANSWERS TO TEST YOUR KNOWLEDGE

Matching:

1.	J	**3.**	C	**5.**	H	**7.**	I
2.	E	**4.**	D	**6.**	B	**8.**	F

Multiple Choice:

1.	a	**5.**	d	**9.**	d	**13.**	b	**17.**	b
2.	c	**6.**	c	**10.**	a	**14.**	a	**18.**	c
3.	c	**7.**	b	**11.**	c	**15.**	c		
4.	c	**8.**	a	**12.**	b	**16.**	b		

Chapter 42: Circulation and Gas Exchange

INTERACTIVE QUESTIONS

42.1 **a.** Hemolymph is basically the interstitial fluid that bathes body tissues and is pumped through open-ended vessels into body sinuses. Blood refers to the circulatory fluid in a closed circulatory system, which remains within vessels and exchanges materials with body cells via the interstitial fluid.
b. Open circulatory systems take less energy to operate. Higher blood pressures enable more efficient delivery of nutrients and O_2 in closed circulatory systems, which also allow for greater regulation of blood distribution.

42.2 **a.** In the single circulation typical of fish, blood is pumped only once. It flows first through the gill capillaries, through a vessel to the systemic capillaries, and then back to the heart. In double circulation, blood is pumped twice as it circulates through the body: once to the gas exchange organs, from which it returns to the heart, and again to the systemic circuit.
b. The amphibian heart has three chambers: two atria and a ventricle. A ridge in the ventricle helps to direct oxygen-rich blood from the left atrium into the systemic circuit, and oxygen-poor blood from the right atrium into the pulmocutaneous circuit. In the four-chambered heart of mammals, the right side of the heart receives and pumps oxygen-poor blood to the lungs, and the left side of the heart receives and pumps oxygen-rich blood to the systemic capillaries.

42.3 **a.** capillaries of head and forelimbs
    b. left pulmonary artery
    c. aorta
    d. capillaries of left lung
    e. left pulmonary vein
    f. left atrium
    g. left ventricle
    h. aorta
    i. capillaries of abdominal organs and hind limbs
    j. inferior vena cava
    k. right ventricle
    l. right atrium
    m. right pulmonary vein
    n. capillaries of right lung
    o. right pulmonary artery
    p. superior vena cava

The pulmonary veins, left side of the heart, and aorta and arteries leading to the systemic capillary beds carry oxygen-rich blood. See text Figure 42.5 for numbers. Note that the aorta splits and oxygen-rich blood travels either to the head region or lower body, returning to the heart through the superior or inferior vena cava, respectively. Thus, numbers 7–10 in Concept 42.2 do not represent the sequential flow of blood.

42.4 a. atrioventricular valve
b. semilunar valve
c. SA (sinoatrial) node
d. AV (atrioventricular) node

42.5 a. elastic artery walls
b. diastole
c. systole
d. diastolic pressure
e. systolic pressure
f. arterioles
g. vasoconstriction
h. vasodilation
i. nerves, hormones (NO, endothelin)

42.6 A protein deficiency could result in a decrease in the production of blood plasma proteins. Blood pressure forces fluid out of capillaries by bulk flow. The osmotic pressure due to the proteins that remain within the capillary tends to counter this fluid movement out of the capillary. A reduction in plasma proteins would reduce this counterbalancing force and result in an increased loss of fluid from capillaries and fluid accumulation in body tissues.

42.7 a. plasma
b. blood electrolytes
c. osmotic balance, buffering, muscle and nerve functioning
d. plasma proteins
e. buffers, osmotic factors, lipid escorts, antibodies, clotting factors
f. nutrients, wastes, respiratory gases, hormones
g. erythrocytes (red blood cells)
h. transport O_2; help transport CO_2
i. leukocytes (white blood cells)
j. phagocytes
k. immunity
l. platelets
m. blood clotting

42.8 Smoking, not exercising, and a diet rich in animal fats or *trans fats*

42.9 An absence of surfactants, which are not produced by fetal lungs until about 33 weeks of development

42.10 a. Fish: water; gill; gill cover (operculum) and mouth pump water into mouth, across gills, and out of body, setting up counter-current exchange
b. Grasshopper: air; tracheoles of tracheal system; body movements compress and expand air tubes
c. Frog: air (and water); lungs and moist skin; positive pressure breathing: fill oral cavity, close nostrils, raise floor of oral cavity, and blow up lungs
d. Human: air; alveoli in lungs; negative pressure breathing: lower diaphragm and raise ribs, thereby increasing volume and decreasing pressure of lungs; air flows in

42.11 This graph shows the Bohr shift. The normal pH of blood is 7.4. The dissociation curve for hemoglobin is shifted to the right at a lower pH, meaning that at any given partial pressure of O_2, hemoglobin is less saturated with O_2. A rapidly metabolizing tissue produces more CO_2, which lowers blood pH, and hemoglobin will unload more of its O_2 to that tissue.

SUGGESTED ANSWERS TO STRUCTURE YOUR KNOWLEDGE

1. a. aorta
    b. left pulmonary artery
    c. left pulmonary veins
    d. left atrium
    e. left ventricle
    f. right ventricle
    g. inferior vena cava
    h. right atrioventricular valve
    i. semilunar valve at base of pulmonary artery
    j. right atrium
    k. superior vena cava
    Blood flow from venae cavae → right atrium → right ventricle → pulmonary artery → lung → pulmonary vein → left atrium → left ventricle → aorta. Check your answer with Figure 42.6 in the text.

2. Nasal cavity → pharynx → through glottis to larynx → trachea → bronchus → bronchiole → alveoli → interstitial fluid → alveolar capillary → hemoglobin molecule → venule → pulmonary vein → left atrium → left ventricle → aorta → renal artery → arteriole → capillary in kidney

ANSWERS TO TEST YOUR KNOWLEDGE

Multiple Choice:

1. c	6. c	11. d	16. d	21. d	26. a
2. d	7. b	12. d	17. d	22. d	
3. c	8. d	13. b	18. c	23. d	
4. c	9. b	14. c	19. a	24. a	
5. d	10. d	15. d	20. c	25. b	

Chapter 43: The Immune System

INTERACTIVE QUESTIONS

43.1 **a.** Double-stranded viral RNA is recognized as foreign by the Dicer-2 enzyme, which cuts the RNA into short fragments. The Argo protein complex binds to an RNA fragment and then binds to a viral mRNA with a complementary sequence. The Argo complex then cuts the viral mRNA, preventing translation.
b. Double-stranded RNA is not normally found in eukaryotic cells.

43.2 **a.** phagocytic white blood cells that circulate in the blood
b. large phagocytes that engulf microbes
c. white blood cells that attack parasitic worms with enzymes
d. white blood cells located in skin that move to lymph nodes and stimulate adaptive immunity
e. white blood cells that attack the body's infected or cancerous cells
f. release histamine to initiate inflammatory response
g. causes vasodilation and increased permeability of blood vessels
h. enzyme that attacks bacterial cell walls; in mammals, found in tears, saliva, and mucus
i. proteins released by virus-infected cells that stimulate neighboring cells to produce substances that inhibit viral replication
j. set of blood proteins that cause lysis of microbes and are involved with innate and adaptive defenses
k. signaling proteins that enhance an immune response and promote blood flow to injury site

43.3 B cell antigen receptors recognize epitopes of intact antigens that are either molecules on the surfaces of infectious agents or molecules free in the body. T cell antigen receptors recognize pieces of antigens that have complexed with an MHC molecule inside a cell and are then presented on the cell surface.

43.4 **a.** During development of B cells and T cells, the *V* and *J* gene segments of one light-chain allele and one heavy-chain allele are randomly recombined. The joining of these diverse light and heavy chains produces a huge number of different antigen-binding specificities among B and T cells.

b. As lymphocytes mature in the thymus or bone marrow, they are tested for self-reactivity, and those that react against self-components are inactivated or destroyed.
c. Once an antigen receptor binds its antigen, the selected cell divides into a clone of effector cells and memory cells specific for that antigen.
d. Immunological memory is an enhanced response to a previously encountered foreign molecule that arises from the presence of memory B and T cells, which rapidly form clones of effector cells upon reexposure to that specific antigen.

43.5 **a.** antigen-presenting cell
b. pathogen
c. antigen fragment
d. class II MHC molecule
e. T cell antigen receptor
f. accessory protein (CD4)
g. helper T cell
h. cytokines stimulate helper T cells, cytotoxic T cells, and B cells
i. cytotoxic T cell
j. B cell
k. cell-mediated immunity (attack on infected cells)
l. humoral immunity (secretion of antibodies by plasma cells)

43.6 **a.** neutralization—neutralizes the ability of a virus to infect a host cell by blocking binding sites; binds to and neutralizes toxins in body fluids
b. opsonization—the binding of multiple antibodies to bacterium makes it more recognizable to macrophages or neutrophils; the cross-linking of bacteria, virus particles, or antigens into aggregates facilitates phagocytosis
c. complement activation and formation of membrane attack complex—antigen–antibody complexes on microbes activate the complement system. Complement proteins form a membrane attack complex, which produces a pore in the membrane of a foreign cell and causes it to lyse.

43.7 **a.** Cytokines released from helper T cells stimulate cytotoxic T cells, B cells, and themselves.
b. Cytotoxic T cells release perforin molecules that form pores in the target cell and granzymes that initiate apoptosis of the target cell.

43.8 Because the red blood cells of individuals with type O blood do not carry A or B antigens on their surfaces, they are not recognized by anti-A or anti-B antibodies. Type O red blood cells can be transfused into individuals of any other blood type. (Note that there are other blood group systems in addition to the ABO system, and these other groups may result in incompatibility with type O red blood cells.)

43.9 **a.** HIV destroys helper T cells, thereby crippling the humoral and cell-mediated immune systems and leaving the body unable to fight opportunistic diseases such as Kaposi's sarcoma and *Pneumocystis* pneumonia.
b. HIV is readily passed through unprotected sex and needle sharing. The frequent mutational changes during replication generate drug-resistant strains of HIV. And frequent mutational changes in surface antigens have made development of an effective vaccine difficult.

SUGGESTED ANSWERS TO STRUCTURE YOUR KNOWLEDGE

1.

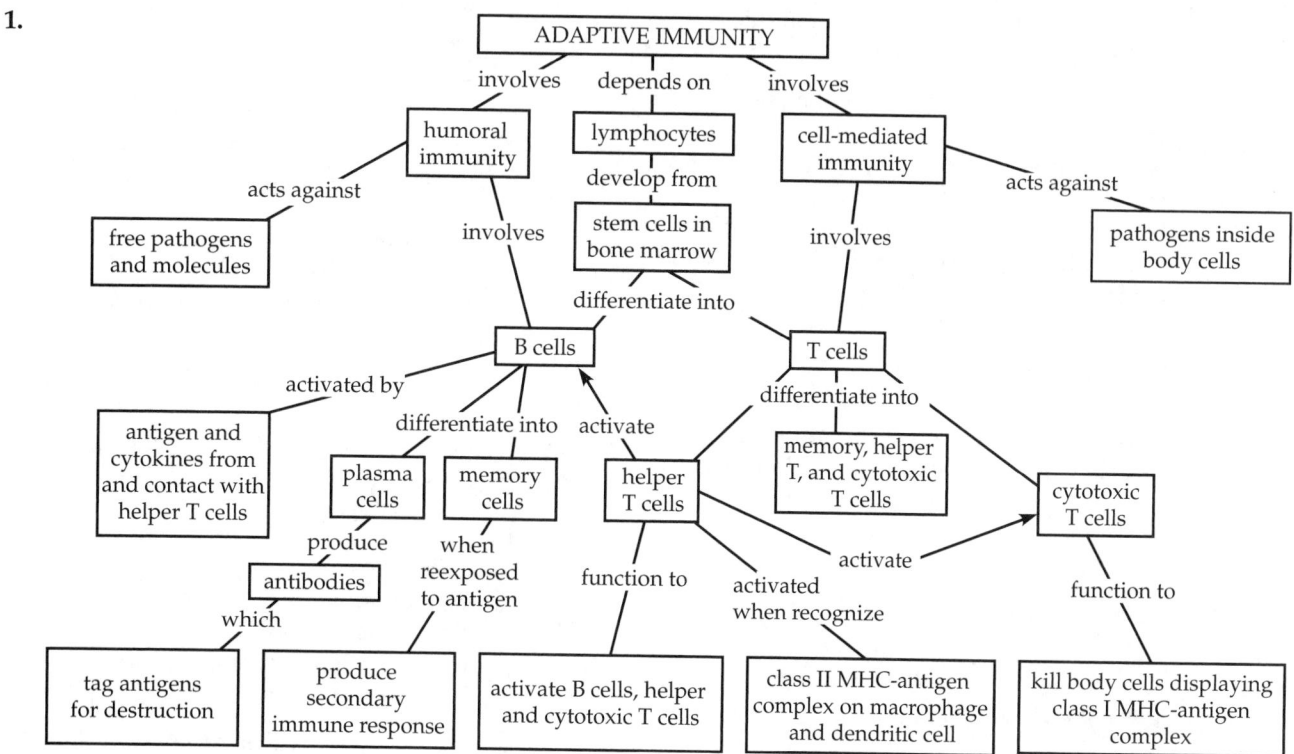

2. An antibody is a Y-shaped protein that consists of two identical light polypeptide chains and two identical heavy chains, held together by disulfide bonds. The amino acid sequences in the variable sections of the light and heavy chains in the arms of the Y account for the specificity in binding between antibodies and epitopes of an antigen. The two arms of an antibody allow it to cross-link antigens, which enhances phagocytosis. The constant region of the antibody tail determines its effector function; there are five classes of antibodies.

3. Macrophages and dendritic cells recognize and engulf pathogens in response to broad categories of molecules (innate immunity), but they also present peptide antigens in their class II MHC molecules to helper T cells. This interaction, along with synthesis of cytokines by these phagocytes, helps to activate helper T cells. The complement system functions in both innate and adaptive immunity. Some activated complement proteins promote inflammation or stimulate phagocytosis. The complement system may lyse microbes that are bearing antigen–antibody complexes.

ANSWERS TO TEST YOUR KNOWLEDGE

Multiple Choice:

1. d	**5.** d	**9.** c	**13.** c	**17.** b	**21.** a
2. d	**6.** a	**10.** b	**14.** b	**18.** b	**22.** b
3. c	**7.** c	**11.** d	**15.** c	**19.** d	
4. b	**8.** b	**12.** a	**16.** d	**20.** d	

Chapter 44: Osmoregulation and Excretion

INTERACTIVE QUESTIONS

44.1 **a.** Marine invertebrate: isosmotic; conformer but may regulate specific solutes
b. Shark: slightly hyperosmotic; maintain urea and TMAO at high concentrations; salt removed by kidneys and in feces
c. Marine bony fish: hypoosmotic; drink water to compensate for loss to seawater; gills pump out salt; little urine excreted
d. Freshwater fish: hyperosmotic: copious dilute urine to compensate for osmotic gain; may pump salts in through gills

44.2 **a.** An endotherm—it must eat more food to maintain its higher metabolic rate.
A carnivore—it eats more protein and thus produces more nitrogenous wastes.
b. *Uric acid* can be stored safely within the egg while the embryo develops. *Urea* is soluble and less toxic than ammonia and can be removed by the mother's blood. Urea is less energetically expensive to produce than uric acid.

44.3 Filtration is least selective because it allows a variety of small molecules and ions to pass into Bowman's capsule. Reabsorption and secretion are much more selective, involving specific molecules and ions.

44.4 **a.** water, salts, nitrogenous wastes, glucose, vitamins, and other small molecules
b. blood cells and large molecules such as plasma proteins

44.5 **a.** Bowman's capsule
b. glomerulus
c. proximal tubule
d. descending limb, loop of Henle
e. ascending limb, loop of Henle
f. distal tubule
g. collecting duct

1. HCO_3^-, NaCl*, H_2O, nutrients*, and K^+ are reabsorbed from the proximal tubule. H^{+*} and NH_3 are secreted into the proximal tubule.
2. Water is reabsorbed from the descending limb of the loop of Henle.
3. NaCl diffuses from the thin segment of the ascending limb and is pumped out* of the thick segment.

4. NaCl*, H_2O, and HCO_3^{-*} are reabsorbed from the distal tubule. K^{+*} and H^{+*} are secreted into the tubule.
5. NaCl* is reabsorbed; urea and H_2O diffuse out of the collecting tubule.
* indicates active transport.

44.6 **a.** Juxtamedullary nephrons; the longer the loop of Henle, the more hyperosmotic the urine can become.
b. Mammals and birds; loops of Henle are especially long in those taxa inhabiting dry habitats.

44.7 **a.** high blood osmolarity
b. ADH, antidiuretic hormone
c. distal tubules and collecting ducts
d. JGA, juxtaglomerular apparatus
e. renin
f. adrenal glands
g. aldosterone
h. Na^+ and water reabsorption
i. increase in blood pressure or volume
j. ANP, atrial natriuretic peptide

SUGGESTED ANSWERS TO STRUCTURE YOUR KNOWLEDGE

1. **a.** water, salts, glucose, amino acids, vitamins, urea, other small molecules
b. ammonia (to combine with secreted H^+), drugs and poisons, H^+, K^+
c. water, glucose, amino acids, vitamins, K^+, NaCl, bicarbonate

2. Two solutes, NaCl and urea, contribute to the high osmolarity of the medulla. As the filtrate moves up the ascending limb, NaCl first diffuses and is then pumped out, contributing to the gradient of NaCl in the medulla. As the filtrate flows through the collecting duct, some urea diffuses out and also contributes to the osmotic gradient. As the filtrate in the collecting duct moves through the high osmolarity of the medulla, water flows out by osmosis, creating a more concentrated urine.

ANSWERS TO TEST YOUR KNOWLEDGE

Multiple Choice:

1. a	**4.** b	**7.** d	**10.** a	**13.** d	**16.** c
2. c	**5.** a	**8.** d	**11.** b	**14.** d	**17.** b
3. d	**6.** b	**9.** c	**12.** c	**15.** a	**18.** d

Chapter 45: Hormones and the Endocrine System

INTERACTIVE QUESTIONS

45.1
a. water-soluble hormone
b. receptor protein
c. signal transduction produces cytoplasmic response
d. signal transduction leads to gene regulation
e. lipid-soluble hormone
f. transport protein
g. receptor protein
h. hormone–receptor complex (transcription factor)
i. gene regulation
j. cytoplasmic response

45.2 In negative feedback, the response reduces the initial stimulus, thereby turning off the response. For example, the rise in pH in the duodenum in response to bicarbonate released from the pancreas removes the stimulus and stops the release of secretin. The release of insulin and glucagon in the regulation of blood glucose levels are both controlled by negative feedback. In positive feedback, the stimulus is reinforced, leading to an even greater response. Oxytocin's stimulation of milk secretion in response to a baby's sucking or of uterine muscle contraction during childbirth are examples of positive feedback regulation.

45.3 A tropic hormone (*tropin*) is one that acts on an endocrine gland or endocrine cells. Many of the anterior pituitary hormones are tropic hormones, conveying signals from the hypothalamus to endocrine glands. Examples are thyroid-stimulating hormone (TSH), adrenocorticotropic hormone (ACTH), and the gonadotropins FSH and LH.

45.4 Insufficient iodine results in reduced production of T_3 and T_4. Thus, there is no negative feedback to turn off production of TRH by the hypothalamus and of TSH by the anterior pituitary. TSH continues to stimulate the thyroid, which enlarges.

45.5
a. Nervous stimulation from the hypothalamus to the adrenal medulla causes the secretion of epinephrine and norepinephrine, which increase blood pressure, rate and volume of heartbeat, breathing rate, and metabolic rate; stimulate glycogen hydrolysis and fatty acid release; and change blood flow patterns.

b. A releasing hormone from the hypothalamus stimulates ACTH release from the pituitary, which stimulates the adrenal cortex to release corticosteroids. Glucocorticoids increase blood glucose through conversion of proteins and fats. Mineralocorticoids increase blood volume and pressure by stimulating the kidney to reabsorb sodium ions and water.

SUGGESTED ANSWERS TO STRUCTURE YOUR KNOWLEDGE

1. Water-soluble signaling molecules that bind to plasma membrane receptors initiate signal transduction pathways that may activate cellular enzymes or affect gene expression. Steroid hormones and other lipid-soluble molecules bind with protein receptors inside a cell, and the hormone–receptor complex acts as a transcription factor to turn on (or off) specific genes.

2. A hormone cascade pathway involves a hormone sequence in which the hypothalamus secretes releasing hormones that stimulate the anterior pituitary to secrete a tropic hormone, which then stimulates target cells to secrete hormones that produce some physiological or developmental effect. For example, in response to a decrease in thyroid hormone level, the hypothalamus secretes TRH, which stimulates the anterior pituitary to secrete TSH, which acts on the thyroid gland to stimulate release of thyroid hormone, which increases metabolic rate. Thyroid hormone then acts by negative feedback to block the further release of TSH and TRH.

ANSWERS TO TEST YOUR KNOWLEDGE

Matching:

1. I g	**5.** A c	**9.** C d, released from h
2. F a	**6.** E b	**10.** J d, released from h
3. H e	**7.** L k	
4. B i	**8.** K f	

Multiple Choice:

1. d	**5.** a	**9.** c	**13.** a	**17.** d
2. c	**6.** d	**10.** b	**14.** d	**18.** c
3. b	**7.** a	**11.** c	**15.** c	**19.** b
4. d	**8.** a	**12.** b	**16.** a	**20.** e

Chapter 46: Animal Reproduction

INTERACTIVE QUESTIONS

46.1 **a.** Perpetuation of successful genotypes in stable habitats; elimination of need to locate mate; production of a larger number of offspring over time because all offspring can reproduce.

b. Varying genotypes and phenotypes of offspring may enhance reproductive success and adaptation of a population to fluctuating environments. Harmful alleles may be removed from a population more readily as genes are shuffled during sexual reproduction.

c. Sexual reproduction because it involves union of egg and sperm, and offspring will be genetically varied due to genetic recombination during meiosis and fertilization.

46.2 **a.** courtship behaviors that trigger the release of gametes and may provide a means for mate selection

b. pheromones that function as mate attractants or other chemical signals

c. environmental cues (temperature, day length)

46.3 **a.** vas deferens
b. erectile tissue of penis
c. urethra
d. glans
e. prepuce (foreskin)
f. scrotum
g. testis
h. epididymis
i. bulbourethral gland
j. prostate gland
k. seminal vesicle
l. oviduct
m. ovary
n. uterus
o. clitoris
p. opening of vagina
q. vagina
r. cervix

46.4 **a.** Spermatogenesis occurs continuously as stem cells and spermatogonia continue to divide from puberty onward. A female's supply of primary oocytes is established before birth.

b. Each meiotic division produces four sperm but only one egg (along with polar bodies that disintegrate).

c. Spermatogenesis is an uninterrupted process. Oogenesis occurs in stages: prophase I before birth; after puberty, meiosis I and II up to metaphase II in a maturing follicle just before ovulation; and completion of meiosis II (in humans) when a sperm cell penetrates the oocyte.

46.5 **a.** GnRH (gonadotropin-releasing hormone)
b. FSH (follicle-stimulating hormone)
c. LH (luteinizing hormone)
d. androgens (primarily testosterone)
e. negative feedback loops involving testosterone and inhibin

46.6 **a.** LH
b. FSH
c. follicular phase
d. ovulation
e. luteal phase
f. estradiol
g. progesterone
h. menstrual flow phase
i. proliferative phase
j. secretory phase

46.7 **a.** HCG, secreted by the embryo, maintains the corpus luteum (and thus progesterone) in the first trimester.

b. Progesterone, secreted by the corpus luteum and later by the placenta, regulates the growth and maintenance of placenta, formation of the mucus plug in the cervix, cessation of ovulation, and growth of uterus.

c. Oxytocin, secreted by the posterior pituitary, stimulates uterine contractions during birth and release of milk during nursing.

46.8 **a.** prevent release of gametes—sterilization, hormonal contraception of birth control pills (or injection)

b. prevent fertilization—abstinence, natural family planning, condom, diaphragm, spermicides, progestin injection or minipill, IUD, *coitus interruptus*

c. prevent implantation of embryo—IUD The most effective methods are abstinence, sterilization, and hormonal contraception. Least effective are natural family planning and *coitus interruptus*.

SUGGESTED ANSWERS TO STRUCTURE YOUR KNOWLEDGE

1. Path of sperm from formation to fertilization:

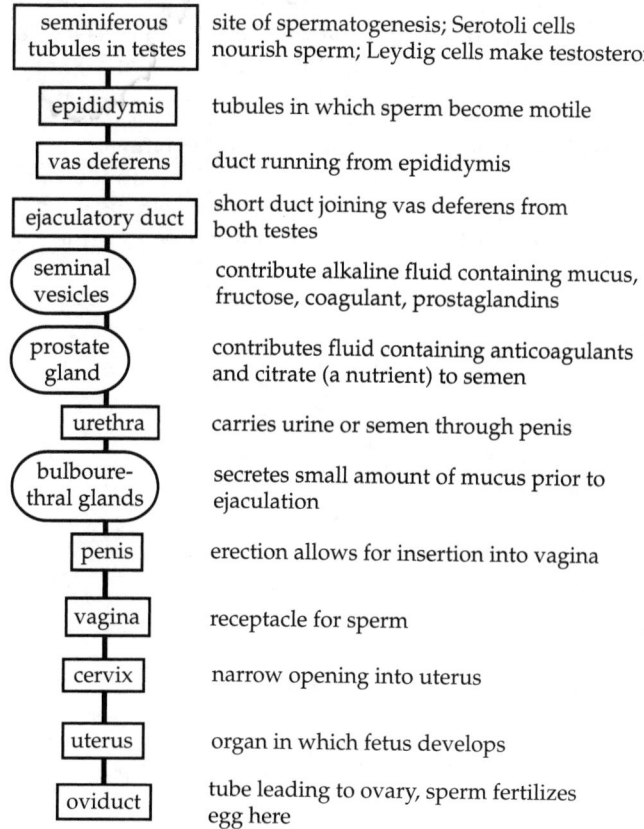

seminiferous tubules in testes	site of spermatogenesis; Serotoli cells nourish sperm; Leydig cells make testosterone
epididymis	tubules in which sperm become motile
vas deferens	duct running from epididymis
ejaculatory duct	short duct joining vas deferens from both testes
seminal vesicles	contribute alkaline fluid containing mucus, fructose, coagulant, prostaglandins
prostate gland	contributes fluid containing anticoagulants and citrate (a nutrient) to semen
urethra	carries urine or semen through penis
bulbourethral glands	secretes small amount of mucus prior to ejaculation
penis	erection allows for insertion into vagina
vagina	receptacle for sperm
cervix	narrow opening into uterus
uterus	organ in which fetus develops
oviduct	tube leading to ovary, sperm fertilizes egg here

2. **a.** GnRH is a releasing hormone of the hypothalamus that stimulates the anterior pituitary to secrete FSH and LH.
b. FSH stimulates growth of follicles.
c. The LH surge is caused by an increase in GnRH production that resulted from increasing levels of estradiol produced by the developing follicle.
d. The LH surge induces the maturation of the follicle and ovulation, and it transforms the ruptured follicle into the corpus luteum.
e. LH maintains the corpus luteum, which secretes estradiol and progesterone.
f. High levels of estradiol and progesterone act on the hypothalamus and pituitary to inhibit the secretion of LH and FSH.
g. The lack of LH causes the corpus luteum to disintegrate. Thus, the production of estradiol and progesterone ceases, which allows LH and FSH secretion to begin again. The drop in estradiol and progesterone also causes arteries in the endometrium to constrict and the uterine lining to disintegrate, leading to menstruation.

3. Birth control pills, which contain a combination of synthetic estrogen and progestin, act by negative feedback to prevent the release of GnRH by the hypothalamus and FSH and LH by the pituitary, thereby preventing the development of follicles and ovulation. RU486 blocks progesterone receptors in the uterus, thereby preventing progesterone from maintaining the endometrium.

ANSWERS TO TEST YOUR KNOWLEDGE

Fill in the Blanks:

1. budding
2. pheromones
3. parthenogenesis
4. hermaphrodite
5. cloaca
6. estrous cycle
7. *in vitro* fertilization
8. urethra
9. vasocongestion
10. menopause

Multiple Choice:

1. c
2. d
3. a
4. c
5. a
6. b
7. d
8. c
9. b
10. c
11. a
12. e
13. c
14. b
15. b
16. a
17. b

Chapter 47: Animal Development

INTERACTIVE QUESTIONS

47.1 a. Specific binding between receptors on the egg plasma membrane and surface proteins on the acrosomal process assure fertilization by sperm of the correct species.
b. Polyspermy is prevented by the fast block to polyspermy caused by membrane depolarization following the acrosomal reaction and by the slow block to polyspermy caused by formation of the fertilization envelope during the cortical reaction.

47.2 a. 8-cell stage
b. blastula
c. animal pole
d. gray crescent
e. vegetal pole
f. blastocoel

47.3 a. ectoderm
b. mesenchyme cells
c. endoderm
d. archenteron
e. blastopore (becomes anus)

f. ectoderm
g. mesoderm
h. endoderm
i. archenteron
j. yolk plug in blastopore

47.4 a. expanding region of trophoblast—forms fetal portion of placenta
b. epiblast—forms three primary germ layers
c. hypoblast
d. trophoblast

47.5 a. neural tube
b. neural crest cells
c. somite

d. archenteron
e. notochord

47.6 Localized cytoplasmic determinants in the sea urchin egg must have an unequal polar distribution. The developmental fates of cells in the animal and vegetal halves of the embryo are determined by the first horizontal division.

47.7 The developing limb would be a mirror image, with posterior digits facing both forward and backward. This is the same effect as transplanting a donor ZPA to that location.

SUGGESTED ANSWERS TO STRUCTURE YOUR KNOWLEDGE

1.

Event	Function
Acrosomal Reaction — Sperm contacts jelly coat, hydrolytic enzymes released from acrosome digest through jelly coat	Allows acrosomal process to penetrate jelly coat
Protein on acrosomal process attaches to receptor on egg cell membrane	Assures that egg is fertilized by sperm of same species
Sperm and egg plasma membranes fuse	Membrane fusion causes ion channels to open
Ion channels open, Na$^+$ flows into egg, and membrane depolarizes	Depolarization creates **fast block to polyspermy**
Cortical Reaction — Calcium released in egg cytoplasm; sperm nucleus enters egg cytoplasm	Triggered by depolarization, calcium release causes cortical reaction and egg activation; two haploid nuclei in the same cell
Exocytosis by cortical granules; fertilization envelope forms	Enzymes snip off sperm-binding receptors and cause fertilization envelope to form, creates **slow block to polyspermy**
Activation of Egg — Rates of cellular respiration and protein synthesis increase	Activation of egg, synthesis of proteins from mRNA stockpiled in egg
Sperm and egg nucleus fuse; DNA replication begins	Prepares for first cell division, ready for development

2.

Animal	Cleavage	Blastula	Gastrulation
Sea urchin	Holoblastic; equal-sized blastomeres	Hollow ball of cells, one cell thick, surrounding blastocoel	Invagination through blastopore to form endoderm of archenteron; migrating mesenchyme forms mesoderm
Frog	Holoblastic, but yolk impedes divisions; unequal-sized blastomeres	Blastocoel in animal hemisphere; walls several cells thick	Involution of cells at dorsal lip of blastopore; migrating cells form mesoderm and endoderm around archenteron
Bird	Meroblastic, only in cytoplasmic disk on top of yolk	Blastocoel is cavity between epiblast and hypoblast	Migration of epiblast cells through primitive streak forms mesoderm and endoderm; lateral folds join to form archenteron
Mammal	Holoblastic, blastomeres equal	Blastocyst with inner cell mass (layers are epiblast and hypoblast); trophoblast surrounds embryo	Migration of epiblast cells through primitive streak forms mesoderm and endoderm

ANSWERS TO TEST YOUR KNOWLEDGE

Multiple Choice:

1. b	**3.** b	**5.** d	**7.** d	**9.** d	**11.** a	**13.** a	**15.** b	**17.** c	**19.** a
2. c	**4.** b	**6.** a	**8.** c	**10.** d	**12.** d	**14.** a	**16.** d	**18.** a	**20.** c

Chapter 48: Neurons, Synapses, and Signaling

INTERACTIVE QUESTIONS

48.1
  a. dendrites
  b. cell body
  c. axon hillock
  d. axon
  e. synaptic terminal. The impulse moves from the axon hillock along the axon to the synaptic terminals. Neurotransmitter is released into the synapse with another cell at the synaptic terminal.

48.2
  a. K^+ inside the cell; Na^+ outside the cell
  b. The inside of the cell is more negative compared to outside the cell.
  c. The membrane potential became more negative and closer to E_K; more potassium channels may have opened, and/or sodium channels may have closed.

48.3
  a. sodium channel
  b. potassium channel
  c. K^+
  d. Na^+
  e. sodium inactivation loop
  f. membrane potential (mV)
  g. time

  h. resting potential
  i. threshold
  j. action potential

1. Resting state: Voltage-gated Na^+ and K^+ channels are closed.
2. Depolarization: Some Na^+ channels open; depolarization opens more Na^+ channels. If threshold is reached, action potential is triggered.
3. Rising phase: Most Na^+ channels open and Na^+ influx makes inside of cell positive.
4. Falling phase: Na^+ channels become inactivated, K^+ channels open, K^+ leaves cell, and inside of cell becomes negative.
5. Undershoot: Na^+ channels closed (many are still inactivated), K^+ channels still open, and membrane becomes hyperpolarized. As these channels close, membrane returns to resting potential.

    See textbook Figure 48.11 for location of the numbered phases on the graph.

48.4 Refer to textbook Figure 48.13 to check your drawing.

48.5
  a. presynaptic membrane
  b. Ca^{2+} flowing in through voltage-gated calcium channel

c. synaptic cleft
d. synaptic vesicle containing neuro-transmitter
e. postsynaptic membrane
f. receptor with bound neurotransmitter
g. closed ligand-gated ion channel

48.6 a. Temporal
b. Spatial

48.7 The type of postsynaptic receptor and its mode of action determine neurotransmitter function. Binding of acetylcholine to metabotropic receptors in heart muscle activates a signal transduction pathway that makes it more difficult to generate an action potential, thereby reducing the strength and rate of contraction. The acetylcholine ionotropic receptor in skeletal muscle cells opens ion channels and depolarizes the muscle cell membrane.

SUGGESTED ANSWERS TO STRUCTURE YOUR KNOWLEDGE

1.

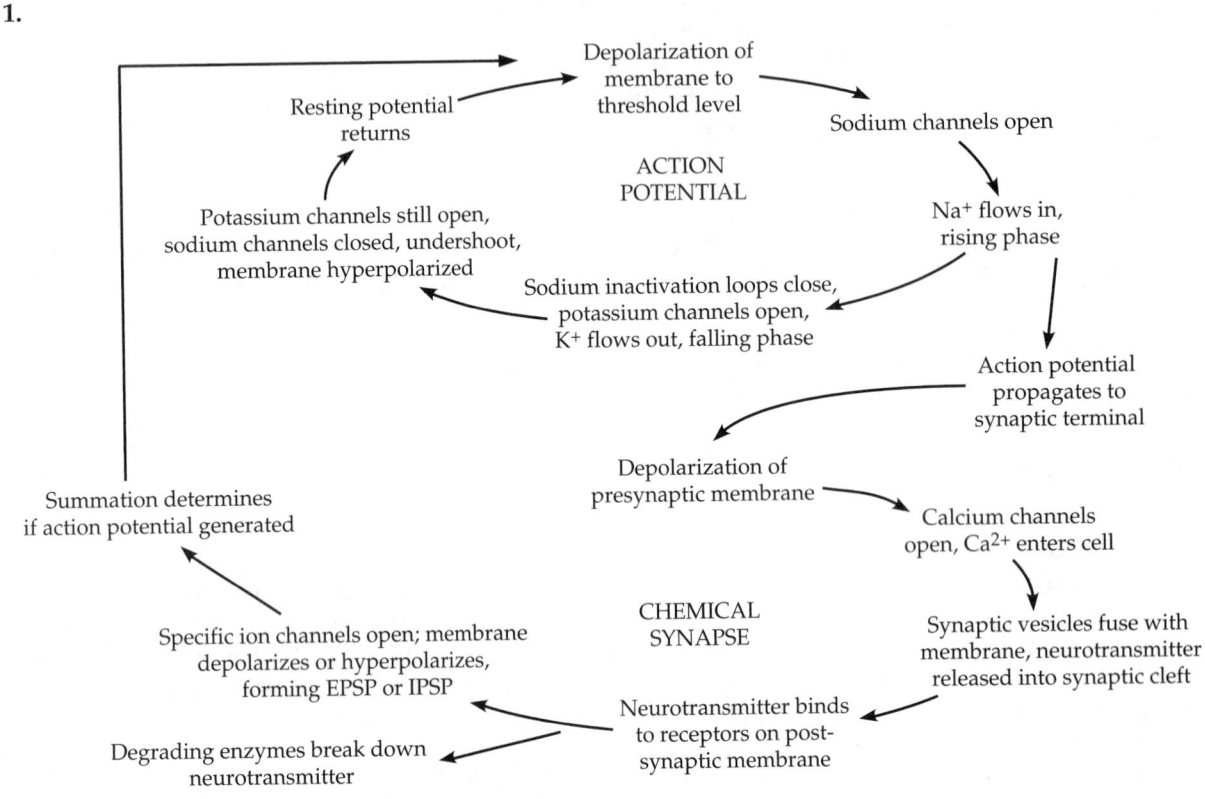

ANSWERS TO TEST YOUR KNOWLEDGE

Multiple Choice:

1. d	**3.** b	**5.** a	**7.** b	**9.** a	**11.** c	**13.** c	**15.** d	**17.** b
2. a	**4.** d	**6.** c	**8.** d	**10.** c	**12.** b	**14.** b	**16.** d	**18.** a

Chapter 49: Nervous Systems

INTERACTIVE QUESTIONS

49.1 Sessile molluscs such as clams have little cephalization and only simple sense organs. Cephalopod molluscs such as squid have large brains and image-forming eyes, which correlate with their active predatory lifestyle and their ability to learn.

49.2 a. afferent neurons
b. sensory receptors

c. efferent neurons
d. autonomic nervous system
e. motor system
f. enteric
g. parasympathetic
h. sympathetic
i. slows heart, conserves energy
j. skeletal muscles

49.3 **a.** A tuna is an active swimmer and has a larger cerebellum than the more stationary lamprey.
b. In birds and mammals, the relative size of the forebrain is larger and the overall size of the brain relative to body size is larger than in other groups.

49.4 **a.** cerebrum 5
b. thalamus 3
c. cerebellum 1
d. medulla 4
e. pons 2
f. midbrain 7
g. hypothalamus 6

49.5 The primary somatosensory cortex receives information from somatosensory receptors in the body. The motor cortex generates motor commands that travel to skeletal muscles. The neurons are distributed in an orderly fashion, from those associated with the legs and feet toward the midline of the parietal or frontal lobe, to those associated with the head located along the sides. The proportion of the primary somatosensory cortex and primary motor cortex devoted to each part of the body is correlated with the importance of that area.

49.6 Glutamate released by the presynaptic neuron opens NMDA receptors, but they remain blocked by magnesium ions until the postsynaptic membrane is depolarized by activity at nearby synapses. The flow of Ca^{2+} through the now-open NMDA receptor channels triggers insertion of AMPA glutamate receptors in the membrane. With both types of receptors in the membrane, glutamate release by the presynaptic neuron is now sufficient to initiate action potentials without input from other synapses.

SUGGESTED ANSWERS TO STRUCTURE YOUR KNOWLEDGE

1. **a.** diverse network of neurons extending through the brainstem; filters sensory information going to the cortex
b. in the hypothalamus; functions as biological clock for circadian rhythms
c. clusters of neurons deep in white matter of cerebrum; centers for motor coordination and planning and learning movements
d. functional center in parts of thalamus, hypothalamus, and inner cerebral cortex (olfactory bulb, hippocampus, and amygdala); centers of emotions and memories in humans
e. two structures in inner cerebral cortex; involved in limbic system and memory; amygdala is involved in forming emotional memories; hippocampus functions in short-term memory and transfer to long-term memory

ANSWERS TO TEST YOUR KNOWLEDGE

Multiple Choice:

1. b	**4.** d	**7.** a	**10.** a	**13.** c
2. a	**5.** c	**8.** b	**11.** c	**14.** d
3. d	**6.** b	**9.** d	**12.** b	

Chapter 50: Sensory and Motor Mechanisms

INTERACTIVE QUESTIONS

50.1 **a.** sensory reception: the detection of the energy of a stimulus by sensory cells
b. sensory transduction: the conversion of stimulus energy to a change in membrane potential of a sensory receptor cell
c. transmission: the passage of an action potential along sensory neurons to the brain
d. perception: the interpretation of sensory input by the brain

50.2 These drugs inhibit prostaglandin synthesis.

50.3 **a.** auditory canal
b. malleus (hammer)
c. incus (anvil)
d. stapes (stirrup)
e. semicircular canals

f.   auditory nerve
g.   cochlea
h.   round window
i.   Eustachian tube
j.   oval window
k.   tympanic membrane (ear drum)
1.   auditory canal and middle ear, which houses malleus, incus, and stapes
2.   cochlea, and the organs that sense equilibrium and movement: semicircular canals, utricle, and saccule

50.4 a.   sclera
b.   choroid
c.   retina
d.   fovea (center of visual field)
e.   optic nerve
f.   optic disk (blind spot)
g.   vitreous humor
h.   lens
i.   aqueous humor
j.   pupil
k.   iris
l.   cornea
m.   suspensory ligament

50.5 a.   small
b.   fovea

50.6 Animals use their senses of taste and smell to find mates, recognize territory, navigate, communicate, and locate and taste food.

50.7 a.   sarcomere
b.   Z line
c.   M line
d.   thin filaments (actin)
e.   thick filament (myosin)
The Z lines will be closer together as the overlap between thick and thin filaments increases. This sliding of filaments occurs by the following repeated sequence: Binding of ATP releases a myosin head from the actin filament; hydrolysis of ATP converts the head to a high-energy form that binds to actin, forming a cross-bridge; myosin head returns to low-energy configuration, pulling the thin filament toward the center of the sarcomere.

50.8 a.   fast-twitch glycolytic, fast-twitch oxidative, slow-twitch oxidative
b.   fast-twitch glycolytic
c.   slow-twitch oxidative

50.9 a.   Hydrostatic skeleton: no extra materials needed, good shock absorber; not much protection or support for lifting animal off the ground; cnidarians, flatworms, nematodes, and annelids

b.   Exoskeleton: quite protective; in arthropods, must be molted in order for the animal to grow, restricts size; most molluscs and arthropods
c.   Endoskeleton: various types of joints allow for flexible movement in vertebrates, good structural support; may not be very protective; sponges, echinoderms, and vertebrates

50.10 a.   Swimming is the most energy efficient. Overcoming gravity is not a problem, but friction is. A fusiform body shape reduces the drag exerted by the high density of water.
b.   Running is the most energy expensive. Air offers little resistance, but the animal must support itself against gravity. Strong skeletons and muscles are adaptions for movement on land, and energy-storing tendons facilitate the next step.

SUGGESTED ANSWERS TO STRUCTURE YOUR KNOWLEDGE

1.   Light enters the eye through the pupil and is focused by the lens onto the retina. When rhodopsin (visual pigment in rods) absorbs light energy, it sets off a signal transduction pathway that decreases the receptor cell's permeability to sodium and hyperpolarizes the membrane. The reduction in the release of neurotransmitter by a rod cell releases some connected bipolar cells from inhibition, which in turn generate action potentials in ganglion cells—the sensory neurons that form the optic nerve. Horizontal and amacrine cells provide lateral integration of visual information.

2.   Sound waves are collected by the *pinna* and travel through the *auditory canal* to the *tympanic membrane*, where they are transmitted by the *malleus, incus,* and *stapes*. Vibration of the stapes against the *oval window* sets up pressure waves in the perilymph in the *vestibular canal* within the *cochlea*, from which they are transmitted to the *tympanic canal* and then dissipated when they strike the *round window*. The pressure waves vibrate the *basilar membrane*, on which the *organ of Corti* is located within the *cochlear duct*. Tips of the hair cells arising from the organ of Corti are embedded in the *tectorial membrane*. When vibrated, they bend, triggering a depolarization, release of neurotransmitter, and initiation of action potentials in the sensory neurons of the *auditory nerve*, which leads to the *cerebral cortex*.

3. A motor neuron releases acetylcholine into the synapse with a muscle fiber, initiating an action potential that spreads into the center of the muscle cell along the transverse tubules. The action potential changes the permeability of the sarcoplasmic reticulum membrane, which releases Ca^{2+} into the cytosol. The calcium binds with troponin, moving tropomyosin and exposing the myosin-binding sites of the actin molecules. Heads of myosin molecules in their high-energy form bind to these sites, forming cross-bridges. The myosin heads bend, pulling the thin filament toward the center of the sarcomere. When ATP binds to myosin, the cross-bridge breaks. Hydrolysis of the ATP returns the myosin to its high-energy form, and it binds farther along the actin molecule. This sequence continues as long as there is ATP and until calcium is pumped back into the sarcoplasmic reticulum, when the tropomyosin–troponin complex again blocks the actin sites.

ANSWERS TO TEST YOUR KNOWLEDGE

Multiple Choice:

1. c	5. c	9. d	13. d	17. b	21. c
2. a	6. a	10. d	14. a	18. d	22. b
3. b	7. c	11. b	15. a	19. b	23. a
4. d	8. c	12. b	16. a	20. a	24. d

Chapter 51: Animal Behavior

INTERACTIVE QUESTIONS

51.1 Mammals are mostly nocturnal; birds are diurnal.

51.2
  a. imprinting
  b. associative learning (operant conditioning)
  c. cognition
  d. spatial learning
  e. social learning

51.3 Researchers measured the average number of drops required to break whelk shells at various heights and calculated the average total effort required by computing the *total flight height* (average number of drops × height per drop). They measured the average flight height for crows in their whelk-breaking behavior (5.23 m) and found it was very close to the 5-meter height predicted based on an optimal compromise between energy gained versus energy expended.

51.4 If a male's parental care resulted in his greater reproductive success, then the genes for his behavior would increase in frequency in a population. If he were parenting offspring that included those of other males, then he would not pass his "parental care" genes on to the next generation in a higher proportion because he was also helping to perpetuate the "non-parental-care" genes of other males.

51.5 Orange throats cannot successfully defend all their females from sneaky yellow throats, so yellow throats mate more often, and their frequency increases in the population. But blue throats restrict the access of yellow throats to their small number of females, and blue throat numbers will increase. The aggressive orange throats can take over blue throat territories, and once again, orange throat numbers will increase.

51.6 Raising animals in the lab enables researchers to search for evidence of genetic bases of behavior by controlling for environmental influences.

51.7 Altruistic behavior is more likely toward a parent or a sibling, because the individual shares more genes in common with them. The coefficient of relatedness is 0.5, whereas it is only 0.125 for a cousin. Thus, with Hamilton's rule $rB > C$, the benefit to the recipient is discounted less by a larger value of r.

SUGGESTED ANSWERS TO STRUCTURE YOUR KNOWLEDGE

1. The scientific study of animal behavior seeks to identify those behaviors that are innate and genetically programmed as well as those that can be modified through experience and learning. Fixed-action patterns are clearly developmentally fixed. In other cases, genetics may set the parameters for an organism's behavior; however, experience can modify behavior, and learning is clearly evident.

2. According to the concept of fitness, an animal's behavior should help to increase its

chance of survival and production of viable offspring. Survival behaviors that are genetically programmed would be most likely to be passed on, and natural selection would refine innate behaviors for foraging, migrating, mating, and care of offspring. The evolution of cognitive ability would help individuals deal with novel situations. Social behaviors and communication may increase an individual's fitness. Parental investment and certainty of paternity may result in differences in mating systems and parental care that influence an individual's reproductive success. Altruistic

behavior may be explained on the basis of kin selection; the inclusive fitness of an animal increases if its altruistic behavior benefits related animals who share many genes.

ANSWERS TO TEST YOUR KNOWLEDGE

Multiple Choice:

1. c	5. a	9. a	13. d
2. b	6. c	10. d	14. c
3. a	7. b	11. a	15. d
4. b	8. d	12. c	16. c

Chapter 52: An Introduction to Ecology and the Biosphere

INTERACTIVE QUESTIONS

52.1 **a.** South-facing slopes in the Northern Hemisphere receive more sunlight and are warmer and drier than north-facing slopes.
b. Air temperature drops with an increase in elevation, and high-altitude communities may be similar to communities at higher latitudes.
c. The windward side of a mountain range receives much more rainfall than the leeward side. The warm, moist air rising over the mountain releases moisture, and the drier, cooler air absorbs moisture as it descends the other side.

52.2 **a.** desert
b. temperate grassland
c. tropical forest
d. temperate broadleaf forest
e. northern coniferous forest
f. arctic and alpine tundra

52.3 **a.** + − −
b. − + +
c. + − −
d. − + −
e. − + +
f. + + +

52.4 7. abyssal   2. neritic
6. aphotic   3. oceanic
8. benthic   5. pelagic
1. intertidal   4. photic

52.5 **a.** dispersal: an area may be beyond the dispersal ability of a species

b. habitat selection behavior: insect larvae may be able to feed on more plants, but females lay their eggs on a single type of plant
c. biotic factors: the presence of predators, herbivores, parasites, pathogens, or competitors may restrict a species' range
d. abiotic factors: sunlight, water, oxygen, temperature, salinity, and soil characteristics may determine whether a species can inhabit an area

SUGGESTED ANSWERS TO STRUCTURE YOUR KNOWLEDGE

1. **a.** Ecology is the study of how organisms interact with their abiotic and biotic environments.
   b. The interactions of organisms with their environments can result in changes in the gene pool of a population, or evolution. Interactions occurring within an ecological time frame translate into adaptations to the environment that are evident on the scale of evolutionary time.

2. **a.** Biomes are characteristic ecosystem types—usually identified by the predominant vegetation for terrestrial biomes or the physical environment for aquatic biomes—which range over broad geographic areas.
   b. Convergent evolution, common adaptations of organisms of different evolutionary lineages to similar environments, accounts for similarities in life forms within geographically separated biomes.

ANSWERS TO TEST YOUR KNOWLEDGE

Matching:

1.	H	**3.**	D	**5.**	A	**7.**	B
2.	C	**4.**	E	**6.**	G	**8.**	F

Multiple Choice:

1.	b	**4.**	c	**7.**	c	**10.**	d	**13.**	c
2.	c	**5.**	a	**8.**	d	**11.**	d	**14.**	b
3.	b	**6.**	b	**9.**	d	**12.**	b	**15.**	a

Chapter 53: Population Ecology

INTERACTIVE QUESTIONS

53.1 If 10 of the 30 voles in the second trapping were marked, she would estimate that ⅓ of the vole population is marked. Because 25 voles were initially marked, the total population size is estimated at 75 (25/0.333). Using the textbook formula, $N = sn/x$, or [(number marked and released × total second catch)/number of marked recaptures], also yields 75 [(25 × 30)/10].

53.2 **a.** Type I is typical of species that produce relatively few offspring and provide parental care, such as humans and many large mammals.
b. Type II has a constant death rate over the organism's life span, such as Belding's ground squirrels and some other rodents, some annual plants, various invertebrates, and some lizards.
c. Type III is typical of species that produce many offspring, most of which die off rapidly, such as many fishes and marine invertebrates, and long-lived plants.

53.3 The per capita rate of increase—when r is greater than 0 and growth is constant, population growth will be exponential.

53.4 **a.** exponential growth; $dN/dt = rN$
b. logistic growth; $dN/dt = rN(K-N)/K$; K is 1,500

53.5 **a.** An organism has limited resources to divide between growth, survival, and reproduction.
b. r-selected; K-selected

53.6 **a.** competition for resources, availability of territories, disease, accumulation of toxic wastes, predation, intrinsic limiting factors
b. extremes in weather, natural disasters, fires

53.7 A large proportion of individuals of reproductive age or younger, as seen in the bottom-heavy Afghanistan pyramid, may result in more rapid population growth now and in the near future. The small base of the Italy pyramid predicts a population decrease in Italy.

SUGGESTED ANSWERS TO STRUCTURE YOUR KNOWLEDGE

1.

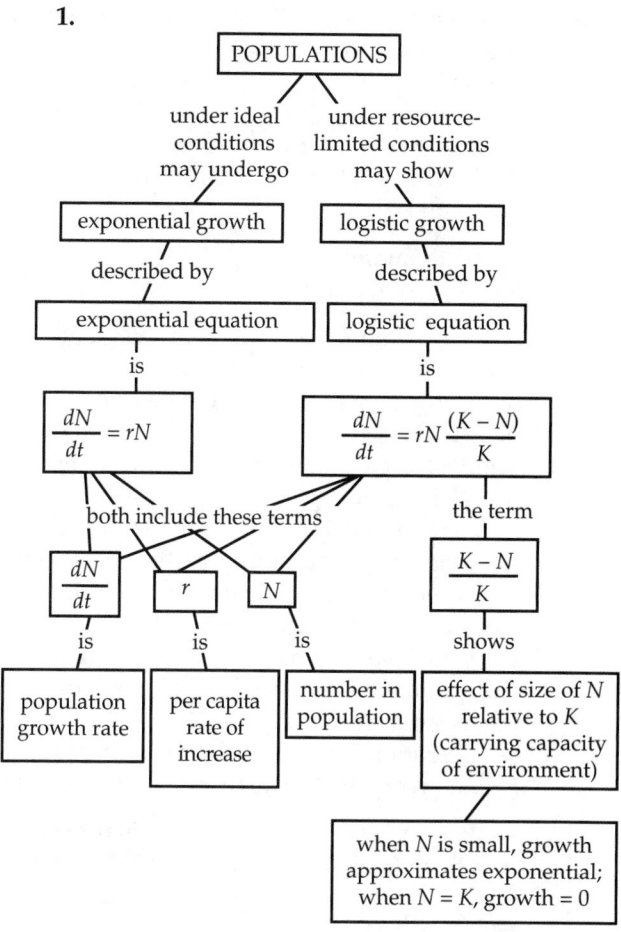

2. Reproductive success is measured in the number of offspring that survive and live to reproduce. Many "choices" are available in life history traits: big-bang (semelparity) versus repeated (iteroparity) reproduction, age at first reproduction, number of reproductive episodes,

number of offspring, parental investment in size of offspring or care. There are always trade-offs between reproduction and survival due to limited energy budgets. In addition, different environments (with varying degrees of stability and diverse biotic and abiotic factors) and different population densities (both influencing K- or r-selection) create different selection pressures on a population. Thus, there is no one "best" reproductive strategy.

ANSWERS TO TEST YOUR KNOWLEDGE

Multiple Choice:

1.	a	**6.**	c	**11.**	d	**16.**	a	**20.**	d
2.	b	**7.**	c	**12.**	c	**17.**	b	**21.**	c
3.	a	**8.**	d	**13.**	b	**18.**	c		
4.	d	**9.**	a	**14.**	b	**19.**	d		
5.	b	**10.**	a	**15.**	a				

Chapter 54: Community Ecology

INTERACTIVE QUESTIONS

54.1 Where these two similar spiny mouse species coexist, behavioral changes result in a temporal partitioning of resources. The niche of the naturally nocturnal *A. russatus* (as shown by laboratory studies as well as this removal study) changes to diurnal when in competition with *A. cahirinus*.

54.2 **a.** Batesian mimicry
 b. Müllerian mimicry

54.3 **a.** competition: weeds and garden plants compete for nutrients and water
 b. mutualism: mycorrhizae, flowering plants and pollinators, ants on acacia trees, cellulose-digesting microorganisms in termites and ruminants
 c. commensalism: cattle egrets and cattle that flush insects (although cattle may benefit when ectoparasites are eaten or when the birds warn of predators)
 d. parasitism: a type of symbiosis in which endo- and ectoparasites feed in or on host; tapeworms, ticks
 e. predation: animal predators killing prey
 f. herbivory: herbivores eating parts of plants

54.4 **a.** Pool 1: $H = -(0.4 \ln 0.4 + 0.3 \ln 0.3 + 0.3 \ln 0.3) = -(-0.37 + -0.36 + -0.36) = 1.09$
Pool 2: $H = -(-0.25 + -0.23 + -0.23 + -0.23) = 0.94$
 b. Pool 2 has more species and thus a greater species richness. Pool 1 has the higher diversity index.

54.5 The $+/-$ cascade that would be needed to end with a decrease in algae would require an increase in zooplankton, which would require a decrease in primary predators as a result of an increase in top predators. More top predators could be added to the lake, or primary predators could be removed.

54.6 Soil nitrogen levels begin quite low but rise during the alder stage due to the symbiotic nitrogen-fixing bacteria associated with alder.

54.7 More interspecific interactions would have had time to develop, and this longer span of evolutionary time would allow for more speciation events to have occurred.

54.8 **a.** As the number of species increases, the rate of immigration of new species would be predicted to decrease and the rate of extinction (due to competitive exclusion) to increase.
 b. The island with the higher number of species would probably be larger and closer to the mainland.

54.9 They are testing for the presence of the H5N1 strain of avian flu virus in migrating waterfowl to monitor the disease's possible entry into North America.

SUGGESTED ANSWERS TO STRUCTURE YOUR KNOWLEDGE

1. **a.** competition $(-/-)$
 b. resource partitioning/character displacement
 c. different niches
 d. mechanical or chemical defenses
 e. herbivory
 f. mutualism $(+/+)$
 g. food chain or food web

2. **a.** No two species with the same niche can coexist permanently in a habitat; the more competitive species will cause the local elimination of the other. The evidence of interspecific competition can be seen in resource partitioning and character displacement.

b. Food chains are limited to a few links because of the inefficiency of energy transfer (only about 10%) from one trophic level to the next.

c. Each trophic level controls the next higher level; adding nutrients will increase the biomass of all other trophic levels.

d. Community organization is controlled from the top by predation (or herbivory), with a cascade of $+/-$ effects down the trophic levels with changes in predator numbers.

e. Most communities do not reach a stable climax community but are constantly changing in composition due to the effects of disturbances.

f. The greatest species diversity will be found in communities subject to moderate levels of disturbance. Low levels may allow competitively dominant species to exclude other species, and high levels may exclude slow-growing or slow-colonizing species from the community.

g. The rates of immigration and extinction on "islands" of habitat are affected by the size of the island, the closeness to the "mainland," and the number of species currently on the island. When these rates are equal, an equilibrium number of species is reached.

ANSWERS TO TEST YOUR KNOWLEDGE

Multiple Choice:

1. d	**5.** b	**9.** b	**13.** c	**17.** d
2. a	**6.** d	**10.** d	**14.** d	**18.** a
3. d	**7.** c	**11.** d	**15.** c	**19.** b
4. d	**8.** b	**12.** c	**16.** a	**20.** c

Chapter 55: Ecosystems and Restoration Ecology

INTERACTIVE QUESTIONS

55.1 The mineral nutrient may limit production in that ecosystem.

55.2 Whereas energy makes a one-way trip through ecosystems, chemical elements move through the trophic levels and detritivores and are recycled back to primary producers.

55.3 **a.** tropical rain forest, coral reef, swamp and marsh, estuary

b. desert, tundra, open ocean

c. The open ocean covers 65% of Earth's surface area.

d. Upwellings in these cold seas bring nitrogen and phosphorus to the surface. The lack of these nutrients limits production in many tropical waters.

55.4 **a.** Much of a bird's or a mammal's assimilated energy is used for respiration to maintain a warm body temperature and is not available for net secondary production (growth and reproduction).

b. $\frac{1}{1000}$ (10% of 10% of 10%) or 0.1%

55.5 Water is essential to all organisms, and its availability in terrestrial ecosystems influences primary production and decomposition. Carbon forms the backbone for all organic molecules, which are essential to all organisms. Nitrogen is a component of amino acids and nucleic acids and is often a limiting plant nutrient. Phosphorus is found in nucleic acids, phospholipids, and ATP, and as a mineral in bones and teeth.

55.6 **a.** Nutrients cycle fastest in tropical rain forests because decomposition proceeds rapidly in the warm, wet climate, and nutrients are rapidly assimilated by new growth.

b. Nutrients cycle slowly in lakes and oceans; lack of oxygen slows decomposition and, unless there are upwellings, nutrients in sediments are not available to producers. Decomposition is also very slow in cold, wet peat lands.

c. Nutrients are not recycled and may leave the ecosystem through runoff.

55.7 Bioremediation uses organisms to detoxify or remove pollutants from ecosystems. Biological augmentation uses organisms to add essential materials to degraded ecosystems. Prokaryotes that can metabolize uranium are being used to treat contaminated groundwater. Plants that can fix nitrogen are often used in biological augmentation to enrich nutrient-poor soils and facilitate the recolonization of native species.

SUGGESTED ANSWERS TO STRUCTURE YOUR KNOWLEDGE

1. Primary production is the energy assimilated by the producers in an ecosystem. It can be expressed as $J/m^2 \cdot yr$ or $g/m^2 \cdot yr$. Primary production is limited in aquatic ecosystems by the penetration of light and the availability of nutrients, especially nitrogen and phosphorus. In terrestrial ecosystems, primary production is influenced by temperature and precipitation (reflected as evapotranspiration), as well as by nutrients such as nitrogen and phosphorus.

2. Secondary production is the production of new biomass from the energy contained in a consumer's food. Production efficiency is reduced by the energy used for cellular respiration and is lowest for endotherms. Trophic efficiency is the percentage of the energy stored in one tropic level that is transferred to the next trophic level. It is usually around 10%.

3. Photosynthetic and chemosynthetic organisms assimilate inorganic elements and compounds and incorporate them into organic compounds. Animals obtain their organic compounds by eating producers or other consumers. Elements are returned to inorganic form by the processes of respiration, excretion, and decomposition.

ANSWERS TO TEST YOUR KNOWLEDGE

Multiple Choice:

1.	b	4.	d	7.	b	10.	b	13.	c	16.	b
2.	c	5.	d	8.	a	11.	b	14.	a	17.	c
3.	c	6.	c	9.	c	12.	a	15.	c		

Chapter 56: Conservation Biology and Global Change

INTERACTIVE QUESTIONS

56.1   Many examples are provided in the text.

   a. About 70% of coral reefs have been damaged by human activities; 40–50% of the reefs could be lost in the next few decades. About a third of all marine fish species utilize these reefs.
   b. The introduction of the brown tree snake to Guam resulted in the extinction of multiple species of birds and lizards.
   c. Commercial harvest and/or illegal hunting have reduced populations of whales, the African elephant, and many fishes.
   d. It may take decades for aquatic ecosystems to recover from the damage done by acid precipitation, and damage continues to forests in Central and Eastern Europe.

56.2   Smaller, because usually not all individuals in a population successfully breed. Conservation programs attempt to sustain the effective population size above MVP in order to retain genetic diversity in a population.

56.3   The loss of genetic variation within a small population due to inbreeding and genetic drift can force it into an extinction vortex in which the population grows smaller and smaller until it becomes extinct. Promoting migration between small populations or introducing individuals from other populations to increase genetic variation is proposed as a conservation strategy.

56.4   This proactive approach relies on early detection of population decline. It then follows these logical steps: determination of the species' habitat needs; identification of all possible causes of the decline and the predictions that arise from each of these hypotheses; testing to determine which factors are contributing to the decline; and applying the results to the management of the threatened species.

56.5   Corridors promote dispersal between populations and may be essential to species that migrate between different habitats. However, they may also contribute to the spread of diseases.

56.6   Large reserves are required for large, farranging animals that require extensive habitats. They also have proportionally less border area and thus have fewer edge effects.

An advantage of smaller reserves that collectively have the same area as a large one is the slower spread of disease within a population.

56.7 Global warming may be slowed by agreements on an international strategy to reduce CO_2 emissions through increased energy efficiency, use of renewable solar and wind power, use of nuclear power, changes in industrial processes and personal lifestyles, and reductions in deforestation.

SUGGESTED ANSWERS TO STRUCTURE YOUR KNOWLEDGE

1. The destruction of habitat is probably the greatest threat to biodiversity. As humans alter and disrupt natural ecosystems, many species no longer have the physical or community resources necessary for survival. Introduced species can disrupt their new community by preying on or outcompeting local organisms. Because these organisms do not have the normal biological controls of predators, pathogens, or parasites, their populations may rapidly increase. The harvesting of wild species beyond their ability to reproduce may drive populations to extinction. Global changes to Earth's ecosystems in the form of acid precipitation, ozone depletion, and climate change are increasing threats to biodiversity.

2. Biodiversity is a natural resource from which we obtain medicines, crops, fibers, and other products; many potentially valuable species will become extinct before they are known to scientists. A loss of biodiversity may disrupt ecosystem processes in harmful ways. Humans have evolved within the context of living communities and may be affected in unknown ways by changes in our ecosystem.

3. Fragmentation of habitats produces more interfaces or *edges* between different ecosystems (e.g., between forests and cleared areas, between deserts and housing developments). The species that inhabit edges are able to use both types of ecosystems. As edges proliferate, edge-adapted species may become more dominant than the species in the adjoining habitats. Strips of quality habitat that connect fragmented habitat patches may serve as *movement corridors* that promote dispersal between isolated populations and help maintain genetic variation.

ANSWERS TO TEST YOUR KNOWLEDGE

Multiple Choice:

1. d	**4.** b	**7.** c	**10.** d	**13.** a
2. b	**5.** d	**8.** a	**11.** b	**14.** b
3. c	**6.** a	**9.** c	**12.** b	**15.** d